ROBERT E. LANDON

ANNALS OF THE NEW YORK ACADEMY OF SCIENCES

Volume 341

The New York Academy of Sciences
2 East 63rd Street
New York, New York 10021

ANION AND PROTON TRANSPORT

ANNALS OF THE NEW YORK ACADEMY OF SCIENCES

VOLUME 341

ANION AND PROTON TRANSPORT

Edited by William A. Brodsky

The New York Academy of Sciences
New York, New York
1980

Library of Congress Cataloging in Publication Data

Main entry under title:

Anion and proton transport.

(Annals of the New York Academy of Sciences; v. 341)
Papers presented at a conference held
June 18–21, 1979 by the New York Academy of Sciences.
1. Biological transport—Congresses. 2. Anions—Congresses. 3. Protons—Congresses. I. Brodsky, William A., 1918– II. Series: New York Academy of Sciences. Annals; v. 341.
Q11.N5 vol. 341 [QH509] 500s [591.87′6]
80–15917

PCP
Printed in the United States of America
ISBN 0–89766–070–6 (cloth)
ISBN 0–89766–071–4 (paper)

ANNALS OF THE NEW YORK ACADEMY OF SCIENCES

VOLUME 341

May 31, 1980

ANION AND PROTON TRANSPORT *

Editor and Conference Chairman

WILLIAM A. BRODSKY

Advisory Committee

ASER ROTHSTEIN, JUHA KOKKO, EFRAIM RACKER,
H. RONALD KABACH AND GEORGE SACHS

CONTENTS

* This series of papers is the result of a conference entitled Anion and Proton Transport, held 18–21 June, 1979 by The New York Academy of Sciences.

Part III. Regulatory Mechanisms for Anion and Proton Transport in Intestinal, Salivary, Pancreatic, Renal and Corneal Epithelia

Part IV. Transport Mechanism for Proton Secretion in Gastric Epithelium

Part V. Biochemical Aspects of Anion and Proton Transport Mechanisms in the Erythrocyte

Financial assistance was received from:

- BURROUGHS WELLCOME COMPANY
- HOFFMANN-LAROCHE INC.
- ICI AMERICAS INC.
- MERCK SHARP & DOHME POSTGRADUATE PROGRAM
- MERCK SHARP & DOHME RESEARCH LABORATORIES
- NATIONAL INSTITUTE OF ARTHRITIS
- NATIONAL INSTITUTES OF HEALTH-METABOLISM AND DIGESTIVE DISEASES
- NATIONAL SCIENCE FOUNDATION
- OFFICE OF NAVAL RESEARCH
- ORTHO PHARMACEUTICAL CORPORATION
- SCHERING CORPORATION
- SMITH KLINE & FRENCH LABORATORIES
- UNITED STATES ARMY RESEARCH OFFICE

OPENING REMARKS

William A. Brodsky

Mount Sinai School of Medicine
New York, New York 10029

The history of research and development in the field of anion and proton transport in biological systems dates back to work at the turn of the century on plant cells and animal erythrocytes. Thus Lundegardh's development of the redox mechanism of active anion transport in root cells and Hamburger's characterization of the chloride:bicarbonate exchange in the erythrocyte have stimulated much of the subsequent interest and experimentation in the field.

The more recent contributions of the participants at this Conference have incorporated some of the basic concepts of physical chemistry, biochemistry, and physiology (e.g., those of Gibbs and Guggenheim, Onsager, Kedem and Katchalsky on classical and irreversible thermodynamics) into their own experimental work. Sophisticated techniques and new approaches have been developed for obtaining a more precise characterization of biomembranes in general and of anion and proton transport processes in particular.

In what follows, we shall see that some of the new advances are related to the presence of anion-selective and proton-selective channels, carriers, and pump mechanisms in the membranes of intact transporting systems and in isolated bilayer membranes in vesicular or planar form; and that the operations of these components, together with those of certain membrane-bound enzymes are sufficient to account for most of the currently documented transmembrane flows of anions and protons in subcellular and cellular systems.

Much of the success of this conference has been due to the tireless efforts and enthusiasm of Drs. Aser Rothstein, Juha Kokko, Efraim Racker, Ronald Kaback and George Sachs, who comprised the Program Advisory Committee. For any shortcomings, I alone am responsible.

Part I. General Aspects of Ion-Transport; Theory, Approaches, and Emphasis on Anion and Proton Transport

MODELS FOR ELECTROGENIC PROTON TRANSPORT MECHANISMS *

Warren S. Rehm

Department of Medicine, Division of Nephrology
University of Louisville
Louisville, Kentucky 40232

I originally planned to present a paper on anion transport in gastric mucosa; but at the request of the conference committee, I agreed to discuss electrogenic proton pumps instead. I will present models for an electrogenic oxidation-reduction proton pump and ATP-driven proton pumps, but I will not attempt to determine the validity of these models for given biological systems.

Oxidation-Reduction Proton Pump

In Figure 1, the upper left hand model depicts an electrogenic oxidation-reduction model.[1] AH_2/A, XH_2/X and B/BH_2 represent oxidation-reduction systems and no attempt is made to identify them with actual components of cells. AH_2 donates two hydrogen atoms to a carrier X in the membrane forming XH_2. The hydrogen atoms are translocated to the other side of the membrane where the reaction $XH_2 \rightarrow X + 2H^+ + 2e^-$ occurs; the H^+ move out into the lumen and electrons are transported through the membrane and accepted by B. For H^+ to be pumped to the outside, the electron-donating tendency of XH_2 must be greater than that of the B system. For an efficient pump,[1] the magnitude of the oxidation-reduction potential of the A system should approximate that of the X system, so that the oxidation-reduction span between XH_2/X and B/BH_2 is approximately equal to that between A/AH_2 and B/BH_2.

In the upper right hand diagram, a simple electrochemical model for this proton pump is presented.[1] A sheet of platinum separates two aqueous compartments: The upper one contains the X system and the lower one the B system. The chemical reaction is also given in Figure 1. Further, we assume that there is no completed circuit, i.e., the parallel paths have infinite resistances, hence the system is at equilibrium and no pumping occurs. There is a potential difference with the outside positive to the inside. For the reaction to proceed, a completed circuit is needed (the parallel limbs no longer have an infinite resistance) and then pumping will occur. The decrease in the Gibbs free energy ($-\Delta G$) will be equal to the electrical energy output (nFE), which is given by the second equation in Figure 1. The emf of the pump E_{OR} is given in the third equation. The terms other than the H^+ are collected in E^*_{OR}, which is defined by the last equation, where K is the equilibrium constant for the reaction when one Faraday is transferred per gram mole of H^+. For tissues that produce an H^+ concentration of about 10^{-4} mM, the value of E^*_{OR} would be about 200 mV (inside is maintained at pH 7.0). This model would be a viable

* Supported in part by National Science Foundation Grant PCM-7828018 and by National Institutes of Health Grant 7 RO1 EY 03161–01.

0077–8923/80/0341–001 $01.75/0 © 1980, NYAS

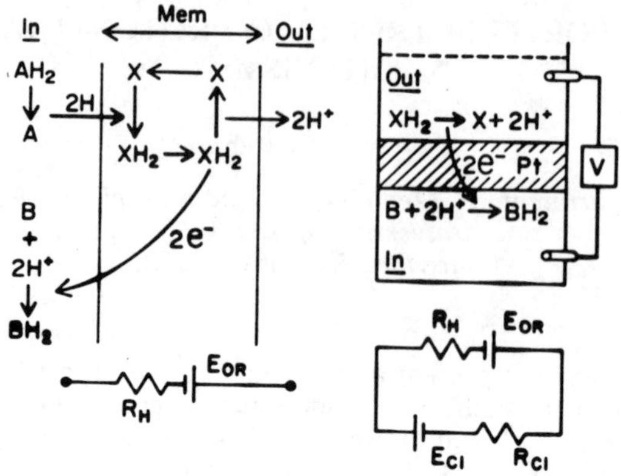

$$XH_2 + B + 2H^+_{in} \rightleftharpoons X + BH_2 + 2H^+_{out}$$

$$-\Delta G = nFE = RT \ln K + RT \ln \frac{react}{prod}$$

$$E_{OR} = E^*_{OR} - \frac{RT}{F} \ln \frac{[H^+]_{out}}{[H^+]_{in}}$$

$$E^*_{OR} = \frac{RT}{F} \ln \left[\frac{[XH_2][B]}{[X][BH_2]} \right]^{1/2} K$$

FIGURE 1. In upper left, note a diagram of an oxidation-reduction proton transport mechanism (proton pump). With parallel pathways having an infinite resistance, the equivalent circuit is shown immediately below where E_H and R_H are the emf and resistance of this pump. *In* represents cytoplasm for the pump in plasma membrane and mitochondrial matrix for pump in inner mitochondrial membrane. *Out* represents lumen or fluid outside the mitochondria. Upper right represents conventional electrochemical model for the pump. Circuit below is equivalent circuit for electrogenic theory of HCl secretion, the Cl^- limb completes the circuit and E_{Cl} and R_{Cl} represent the emf and resistance of the active Cl^- mechanism. Brackets in equations represent activities. See text for details.

one for tissues such as the kidney tubule [2] or the turtle bladder.[3] However, for the gastric mucosa when the H^+ ratio is about 10^7, E^*_{OR} would have to be about 400 mV[1]. With the pH inside (pH = 7) maintained constant, increasing the H^+ concentration on the outside would decrease the magnitude of E_{OR}, the emf of the pump (by 60 mV for each 10-fold increase in the H^+). When the ratio of H^+ concentrations reaches approximately 10^7, then the emf of the pump would be approximately zero. The actual magnitude of E^*_{OR} needed would be a function of the emfs of the return circuit (see equivalent circuit in FIGURE 1). In the dog stomach with an intact blood supply, the blood side is about 60 mV positive to the lumen side so the magnitude of E^*_{OR} could be less than in the absence of a potential difference across the stomach.[4] However, the sum of the emf of the proton pump and the emfs in the return circuit must be large enough to drive the current around the loop.

So we see that the oxidation-reduction model of a proton pump could be applicable to various biological tissues, with the magnitude of the oxidation-reduction potential difference between the A and B systems being a function of the magnitude of the H^+ concentration of the secretion.

ATPase Proton Pump Driven by ATP Hydrolysis

A model [1] showing such a pump is shown in FIGURE 2. In this model OH^- can enter the membrane from the outside and no other solutes can enter or exit from this side. On the other side of the ATPase system, H^+ and ATP enter and ADP and Pi exit—the net transport of charge is one proton entering the ATPase system from the inside. Again, as in the development of the equation for the oxidation-reduction pump, we assume that initially the resistance of parallel limbs is infinite, so there will be no net pumping. The system will be in equilibrium.

I chose for presentation of the development of the equation for the emf E_{ATP} what I believe is the simplest one. E_{ATP} of the system would equal the sum of E_1 and E_2, which are both given by the Nernst equation, as shown in FIGURE 2. Using the equation for water dissociation (third equation), and after substitution and rearrangement, the equation (fourth line) has the same form as that for the emf of the oxidation-reduction system. The second term is the same as that in the oxidation-reduction system and E^*_{ATP} is given by the fourth line. For a $-\Delta G$ of about 9,200 calories and with one Faraday per mole of ATP hydrolysis, the emf of this pump would be about 400 mV. This model would provide sufficient energy for the production of H^+ by the gastric mucosa. However, it would be inefficient for the kidney or the turtle bladder, each of which would need only about 200 mV. Again, the system without a completed

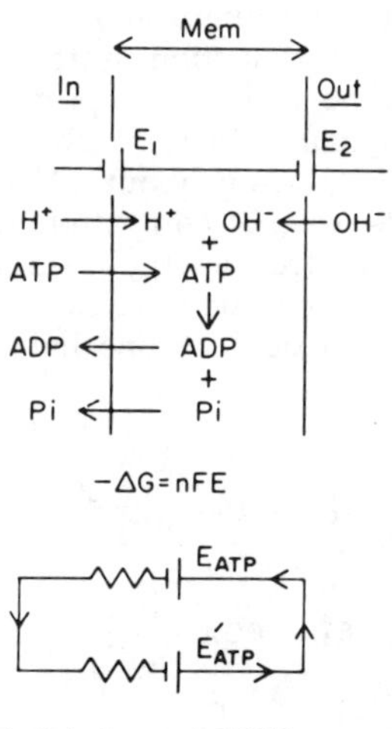

$$E_1 = 60 \log \frac{[H^+]_{in}}{[H^+]_M}$$

$$E_2 = 60 \log \frac{[OH^-]_{out}}{[OH^-]_M}$$

$$[H^+][OH^-] = K[H_2O]$$

$$E_{ATP} = E^*_{ATP} - 60 \log \frac{[H^+]_{out}}{[H^+]_{in}}$$

$$E^*_{ATP} = 60 \log \frac{[H_2O]_{aq}}{[H_2O]_M}$$

$$E^*_{ATP} \simeq 400\ \text{MV}\ (-\Delta G = 9200)$$

FIGURE 2. Model for an ATPase proton pump. The emf of the pump $E_{ATP} = E_1 + E_2$, and E_1 and E_2 are given by the Nernst equation. Subscript M refers to membrane, $-\Delta G$ is decrease in Gibbs free energy. All of $-\Delta G$ is converted to electrical energy when pump is active, i.e., with a completed circuit. In equivalent circuit E'_{ATP} is assumed to be greater than E_{ATP}. Subscripts M refer to membrane ATPase and aq to aqueous solutions and brackets denote activities. See text for details.

circuit would be at equilibrium and for it to pump H^+, a completed circuit would be needed.

Now the following is a rather important point. If one is dissatisfied with the foregoing development and prefers another one, and if the emfs of the two developments are not identical, a dilemma arises. The boundary conditions would be the same in any development of this model; that is, only OH^- could get in or out on one side and, on the other side, there is a net transport of H^+ and ATP in one direction, and ADP and Pi in the other direction. If we assume the emfs due to the two developments were not identical and the two models were placed in parallel as in the lower left circuit, then the system with the higher emf would be a proton pump, in which ATP would be hydrolyzed; while in the adjacent system, with the lower emf, the system would run backwards and ATP would be synthesized. There would be no change in the composition of the inside or the outside solutions since the movements through the two pumps would be in opposite directions. Obviously, we could obtain useful work if the emfs were not the same, which would clearly violate the laws of thermodynamics. Therefore, the equations for this proton pump given in the fourth and fifth lines of FIGURE 2 are valid.

Proton Pumps and the Chemiosmotic Theory

If these two proton pumps are placed in parallel as in FIGURE 3, then we have the essential concept of the chemiosmotic theory of oxidative phosphorylation.[5] It is assumed in this theory that the inner membrane of mitochondria contain oxidation-reduction pumps and an ATPase pump is parallel. The equations for the two pumps are given in FIGURE 3. When E_{OR}, the emf of the oxidation-reduction pump (or pumps) is greater than the emf of the ATP pump, current is driven back through the ATP pump and ATP is synthesized and the electrons move down the oxidation-reduction system in the normal direction. The E^* in this nomenclature is the same as the protomotive force of Mitchell and Moyle.[6, 7] In Mitchell's original model, which is essentially the same as the ATPase model in FIGURE 2, E^*_{ATP} was about 400 mV. It became apparent that a value of E^*_{ATP} of 400 mV was not compatible with the experimental findings. However, the decrease in free energy ($-\Delta G$) equals nFE and in the model of FIGURE 2, $n = 1$, i.e., 1 Faraday per mole ATP. Now, if n equals 2 then for the same $-\Delta G$, E^*_{ATP} would be one half of 400 mV or

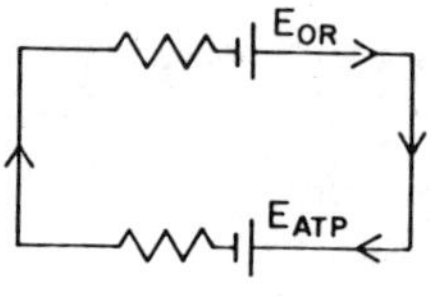

$$E_{OR} = E^*_{OR} - 60 \log \frac{[H^+]_{out}}{[H^+]_{in}}$$

$$E_{ATP} = E^*_{ATP} - 60 \log \frac{[H^+]_{out}}{[H^+]_{in}}$$

$$-\Delta G = nF\,E^*_{ATP} \qquad E^*_{ATP} = \frac{-\Delta G}{nF}$$

FIGURE 3. Equivalent circuit representing the essential concept of the chemiosmotic theory. When $E^*_{OR} > E^*_{ATP}$ then $E_{OR} > E_{ATP}$ and current will flow in indicated direction; ATP will be synthesized and electron flow will be in normal direction via the electron transport system. See text.

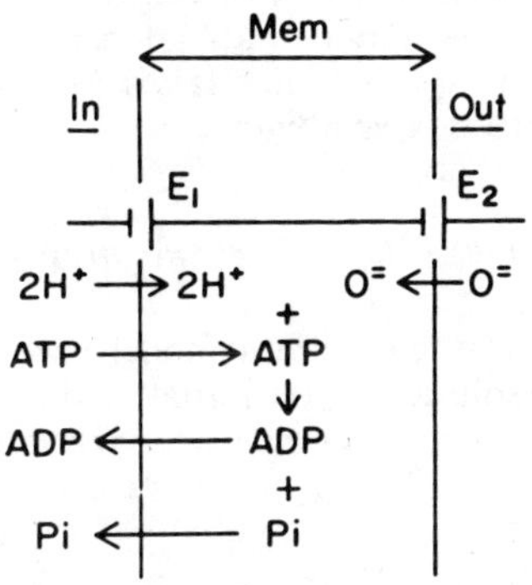

$$E_1 = \frac{RT}{2F} \ln \frac{[H^+]_{in}^2}{[H^+]_M^2}$$

$$E_2 = \frac{RT}{2F} \ln \frac{[O^=]_{out}}{[O^=]_M}$$

$$[O^=][H^+]^2 = K'[H_2O]$$

$$E_{ATP} = E^*_{ATP} - \frac{RT}{F} \ln \frac{[H^+]_{out}}{[H^+]_{in}}$$

$$E^*_{ATP} = 30 \log \frac{[H_2O]_{aq}}{[H_2O]_M}$$

FIGURE 4. Model for ATPase proton pump with two Faradays per mole of ATP hydrolyzed. $E_{ATP} = E_1 + E_2$. E^*_{ATP} (bottom line) is one half that for the ATPase pump of FIGURE 2. See text.

200 mV. A value of 200 mV is much more compatible with the experimental findings.

In FIGURE 4, a model is given for an ATP proton pump in which there are two Faradays per mole of ATP hydrolyzed. In the right hand side of the ATPase, 0^{2-} enters and on the other side 2 H^+ and 1 ATP enter, and 1 ADP and 1 Pi exit. Again, everything is reversed when current is driven through the ATPase pump from the right to the left. The development of the equation for the emf is along the same lines as for the previous ATPase. The K′ for water is given in the third line of FIGURE 4 and by a similar analysis to that of FIGURE 2, the emf of the proton pump is given in the fourth line where E^*_{ATP} is now about 200 mV for a $-\Delta G$ of about 9,000 calories. One may object to the concept of 0^{2-} entering or exiting the membrane, but in order to satisfy the stoichiometry there must be a net movement of an 0^{2-}. Both ATPase models satisfy the stoichiometry. In some papers in the biochemical literature diagrams show H^+ conducting the current across both sides of the ATPase. This, of course, is wrong because the stoichiometry is not satisfied. In the hydrolysis of ATP, there must be a source of oxygen and vice versa for the synthesis.

The ATP proton pump in FIGURE 4 with E^*_{ATP} equal to about 200 mV would be a reasonable model for those acid-producing biological systems that can establish a H^+ ratio no greater than about 10^{3}.[5] The pump in FIGURE 4 is obviously not a viable model for gastric secretion. Parenthetically, the K^+-ATPase of the gastric mucosa appears to be a proton pump with a value of E^*_{ATP} of only about 200 mV. This problem will undoubtedly be discussed in two papers in this annal, one by John Forte and the other by George Sachs.

As I earlier pointed out, I was only going to present models of electrogenic proton pumps and not discuss in detail the merits of the application of these models to given biological systems. However, there is one point I will touch upon that I don't think has been handled adequately in the literature. The chemiosmotic theory has been challenged on the basis of studies on the ratio of the transport of H^+ to that of other ions, and also on the basis of the number of H^+ ions transported per oxygen consumed. The point I want to make is concerned with the number of ions transported in violation of electroneutrality when the potential difference across the inner mitochondrial membrane is changed. If this is a significant amount and it is not taken into account, then the interpretation of these ion ratios must be reevaluated.

Ratio of H^+ to Other Ions Transported When Electroneutrality Is Violated

In the upper left hand corner (FIGURE 5), note a model with a membrane separating 100 mM and 10 mM K^+ solutions. If, initially, the membrane is impermeable to everything and then K^+ conductance channels are established

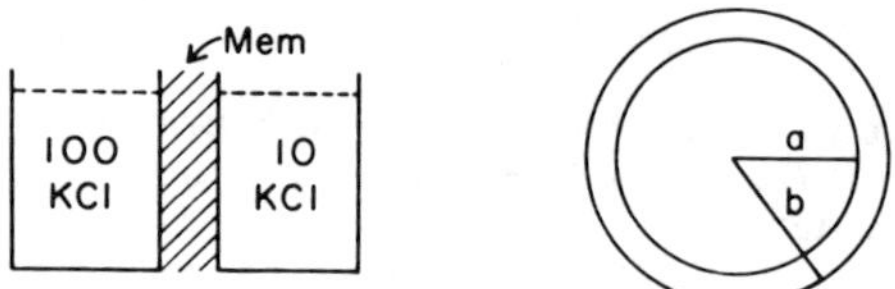

$$C = \frac{4\pi \epsilon a b}{b - a}$$

Diameter			10X	100X
μm	0.1	1.0	1.0	1.0
pH	4.7	5.7	4.7	3.7

FIGURE 5. Upper left shows membrane between two compartments containing 100 mM and 10 mM KCl. Equation is capacitance C for spherical capacitor (upper right). ϵ is dielectric constant$=k\epsilon_0$ where k is dimensionless dielectric coefficient (assumed to equal 10 for calculations) and ϵ_0 is permitivity of free space and in mks units$=8.85\times10^{-14}$ Farads/*cm*. $b-a=50\times10^{-8}$ *cm* for calculations. C is calculated for vesicles 0.1 and 1.0 μm diameters and for vesicles of 1.0 μm with $10\times$ and $100\times$ increase in area due to folding (e.g., the presence of christae in mitochondria). Quantity of charge Q (in Equivalents) moving out of vesicles would be given by: $Q=C\Delta\psi\ A/F$ where $\Delta\psi$ is potential difference across membrane, F the Faraday constant and A ratio of actual area to that of a smooth sphere. Change in outside concentration$=Q$/outside volume. The pH in the figure is final outside pH for $\Delta\psi=180$ mV. See text.

in the membrane, K^+ will move from the high concentration to the low concentration until a potential difference of about 60 mV is established. The 60 mV acts to drive the K^+ uphill with the same force as a ratio of 10 attempts to drive it downhill. The number of K^+ moving would be so small that it would be impossible to detect chemically the change in K^+ concentration in the two vessels.

However, we made similar calculations with respect to vesicles.[8, 9] The ratio of surface to volume is enormous for small vesicles. Calculations revealed that with small vesicles a chemical change could be easily detected.

In the upper right hand corner we have a spherical vesicle, the outer radius being b and the inner radius a. Assume the vesicle contains a high H^+ concentration and initially the membrane is impermeable to everything, and the external volume is 100 times the volume of the vesicles and the external solution is 0.1 N NaCl and has a pH of 7.0. Proton conductance channels are then established in the vesicle membrane. H^+ will exit from the vesicles to the external solution—the membrane is impermeable to all other ions—and hence no anions move out nor cations in. Calculations reveal, for the establishment of a potential difference (*PD*) across the vesicle membrane, the outside solution pH would be reduced below 7. For a *PD* change of 180 mV (outside becoming more positive), and then for vesicles with the diameter of 0.1 μm, the pH outside would drop from 7.0 to 4.7 (FIGURE 5). With the vesicles of a diameter of 1 μm (comparable to the diameter of mitochondria) the external pH would then be reduced from 7.0 to 5.7. However, if there were considerable folding of the membrane as in the inner membrane of the mitochondria, and if this folding increased the area tenfold, then the pH would drop from 7.0 to 4.7; with an increased area of 100-fold the pH would drop from 7.0 to 3.7. Simple inspection of the EMs of mitochondria indicate to me that the area of the inner membrane due to the presence of cristae is substantially more than 10 times an area of a comparable smooth sphere and probably closer to 100 times or even more.

Probably the workers dealing with the problem, within the framework of the Mitchell theory, of the number of protons transported to the number of other ions transported, have taken into consideration the concepts presented in this figure and have adequately disposed of it. However, from what I have read, I am not convinced that they have done this and I would like to be taken by the hand and be led, step by step, through an analysis, which would demonstrate that the charging or discharging of the capacitances of the membrane due to changes in potential differences is of only trivial magnitude in determining the ratio of protons to other ions transported.

REFERENCES

1. REHM, W. S. 1972. Proton transport. *In* Metabolic Pathways. L. Hokin, Ed.: 187–241. Academic Press. New York, N.Y.
2. RECTOR, F. C., JR. 1973. Acidification of the urine. *In* Handbook of Renal Physiology. R. W. Berliner & J. Orloff, Eds. Sec. **8:**431–454 Am. Physiol. Soc., Washington, D.C.
3. BRODSKY, W. A. & T. P. SCHILB. 1967. Fed. Proc., Fed. Amer. Soc. Exp. Biol. **26:**1314–1321.
4. REHM, W. S. 1945. Am. J. Physiol. **144:**115–125.
5. MITCHELL, P. 1961. Nature (London) **191:**144–148.
6. MITCHELL, P. 1967. Fed. Proc., Fed. Amer. Soc. Exp. Biol. **26:**1370–1379.
7. MITCHELL, P. & J. MOYLE. 1969. Eur. J. Biochem. **7:**471–484.
8. REHM, W. S., S. G. SPANGLER, G. M. SCHOEPFLE & G. SACHS. 1976. Possible fallacy in use of electroneutrality principle in studies on vesicles. Biophysical J. **10:**139a.
9. SCHOEPFLE, G. M., W. S. REHM & J. H. YOUNG. 1978. Bull. Math. Biol. **40:** 411–424.

DISCUSSION

W. A. BRODSKY: What is the justification for the assumption of two Faradays per proton in the Mitchell II hypotheses?

W. S. REHM: The ATPase I of the Mitchell theory is very similar to the first ATP proton pump which I described. It has one faraday per ATP and a protomotive force or E* of about 400 mV. An ATPase proton pump for mitochondria with an E* of about 400 mV was not compatible with subsequent experimental data. The data were more compatible with an ATPase proton pump in which E* is about 200 mV and in which there are two Faradays per ATP.

UNIDENTIFIED QUESTIONER: I wonder if Dr. Rehm would kindly go over an apparent violation of electroneutrality when he had a striking difference of size between the inside and the outside particles?

REHM: With a flat membrane separating two solutions, the surface-to-volume ratio is small and it is essentially impossible to detect a change in chemical concentration due to violation of electroneutrality. However, with small vesicles the ratio of surface to volume becomes enormous. Calculations reveal that with small vesicles, a violation of electroneutrality can be detected easily for certain conditions. For example, consider a system of vesicles in which initially the vesicular membranes were impermeable to everything, and with a high hydrogen ion concentration inside, and a bathing solution outside at pH of 7 with no buffer. Now assume that in some way proton conductance channels were established in the membranes. Then hydrogen ions would move out until there was electrochemical equilibrium. Calculations show that the pH outside would be significantly reduced. We assume that the membranes have a capacitance of about one microfarad per square centimeter of surface area. One might assume that all of the hydrogen ions that move from the inside to the outside would be concentrated at the outside surface. However, if the external bathing solution was an isotonic sodium chloride solution, then the ratio of sodium at the surface to sodium in the bulk solution would be the same as the corresponding ratio for hydrogen ions. These ratios are independent of the mobility of the ions. Therefore most of the hydrogen ions would be in the outside bulk solution.

P. D. MITCHELL (*Glynn Research Institute, Bodmin, Cornwall, England*): What does the term, E*, mean? It looked to me to be the same as what the chemiosmoticists call the electric membrane potential. I thought I heard you say that the E*s were at least in one case, the proton motive force. I think we should be careful not to get confused about that. The proton motive force is not strictly analogous to an electromotive force. The electric membrane potential is a kind of fiction, meaning the electric effect of force on an electrically-charged particle. It doesn't include the other potential that would be due to the concentration difference of that particle; so if you have an electromotive force, you would include the electronic concentration difference as well as the electric potential difference of the electron.

The proton motive force is defined by both the electric energy difference of the proton and the concentration difference (or activity difference, given by the PH difference). So I reiterate the question because the E*s looked to me like the purely electromotive force that would be produced either by the redox system or by the ATP hydrolysis if there were no concentration difference term.

A second question is concerned with the apparent violation of the electroneutrality principle in dealing with small vesicles. Were you making the point that your measurements of concentration differences are or are not meaningful in such systems?

REHM: What I did was develop a formula for the actual emf of these proton pumps, an oxidation-reduction pump and ATPase pumps. The actual electromotive force of the pump is equal to two terms, i.e., RT log of the ratio of the hydrogen ion activities and another term in which I have collected all the other terms under E*. When the ratio of the hydrogen ion activities is unity, then E* is equal to the actual emf of the pump. When the hydrogen ratio is increased, E* is constant but the emf of the pump then changes. As it pumps protons against a higher and higher concentration ratio, there will be a decrease (60 mV per 10 × change in the ratio) of the pump. On the basis of your papers, I feel sure that E* is the same as your proton motive force.

MITCHELL: Your equations indicated that the total redox energy available was E* plus the concentration term and not the other way round. It didn't look to me that E* was the whole lot.

REHM: The actual emf of the oxidation-reduction pump is the same as E* when the ratio of hydrogen activities is unity. Now if the inside hydrogen ion concentrations stays fixed and the outside hydrogen ion concentration increases then the emf of the pump decreases 60 mV for every tenfold change and if it increases enough then the actual emf would be reversed.

MITCHELL: I think it's just a case of nomenclature if the chemiosmotocists were using what was called the ordinary membrane potential, $\Delta\psi$, as E*.

Again, are the concentration differences in very small vesicles falsified by the lack of electrical neutrality? Is this what you were suggesting?

REHM: In one of your papers you pointed out that if the system is perturbed and the perturbation results in a change in the potential difference across the membrane, then some of the ions that were involved in charging the capacitors would come out or in and one would have to be careful in interpreting the meaning of the net fluxes of ions. It was surprising to me when the calculations indicated that with small vesicles one could demonstrate quite clearly that a chemical change in concentrations could be detected. At times I have been asked whether I would perform the appropriate experiments. I then point out that I have not made myself clear because the experiments have been done by Coulomb, Helmholtz, Faraday and many others. I am not going to test the laws of electricity on a biological system. If the laws of electricity apply, then there is no way out. Under certain conditions one can detect chemically the change in hydrogen ion concentration when electroneutrality is violated in small vesicles.

MITCHELL: But surely this is okay. We have in classical biochemistry perfectly good examples of very tiny vesicles. Protein molecules that have a charged surface; and the double layer about the ionic cloud of that surface has quite a different composition than the major part of the surrounding medium. Normally one overcomes this kind of difficulty by making appropriate measurements, such as looking at the amount of chloride within such a cloud because you find in a concentrated solution of such a protein that there are different amounts of the ions of a dilute medium than those in the neighborhood of the protein molecule. I can't see that there is any different problem from that which we have in ordinary solution chemistry when we're dealing with large molecules.

REHM: This is not a relevant system. If protons leave a protein molecule they are replaced by other cations. For the protein, plus the Gouy-Chapman layers surrounding the protein, there would be no violation of electroneutrality. What I have been attempting to get across is that with small vesicles an ion could go out of the vesicles and not be accompanied by an anion or associated with cations going in the other direction and this violation of electroneutrality could be detected chemically when the surface-to-volume ratio of the system is enormous. The pH in the model I gave would decrease on the outside from something like 7 by several pH units. If this has not been taken into consideration in interpreting the ratios of potassium to hydrogen ions in organelles like mitochondria, then the usual interpretations that have been used by some opponents of your theory would not necessarily be valid. This may be a trivial matter but I would like the experimenters in this field to provide us with enough data to show that it is trivial.

A. R. CROFTS (*University of Illinois, Urbana-Champaign, Ill.*): As will be pointed out in my talk, we have been able to measure the uncompensated proton transfers for some years. We can also measure the movement of electrical charge across the membrane from the electrochromatic change, and one can calculate that it fits very nicely with the capacitance expected from the member of charges that can be moved across the membrane before saturating the electric capacity of that membrane.

REHM: You have made similar extrapolations and can detect this chemically?

CROFTS: Yes, but it's a much smaller pH change than you indicated, largely because of the buffering capacity of endogenous groups on the chromaticols.

REHM: Yes, I said the outside solution would be a pure sodium chloride solution of pH 7.0 and without buffers. If there were buffers present, there would be much less change in pH. Is that the point?

CROFTS: Yes, it's a fraction of 0.1 pH unit.

REHM: In that case, yes. But we may not be talking about the same phenomenon. In my model the outside solution is a nice tidy sodium chloride solution at pH 7 with no buffers. With buffers present there would obviously be less of a change in pH. From what I understand, you have done this experimentally under conditions in which there is a violation of electroneutrality. You have ruled out the possibility of concurrent movement of anions in the same direction, or of cations in the opposite direction. The results of my calculations have been criticized on the basis that if hydrogen ions move out of vesicles, they would saturate the capacitor and not be present in the bulk solution, and hence would not be observed. However, with a substantial amount of cations other than hydrogen ions in the bulk solution, most of the hydrogen ions moving out from the vesicles would be in the bulk solution. The ratio of sodium at the membrane to the sodium in the bulk solution would be the same as that for hydrogen ions. The mobility of the ions would not be a factor in determining their distribution.

MITCHELL: The old indicator error for proteins was due to a cloud of hydrogen ions around the surface of a protein molecule. This caused an effective change in the degree of titration of the surface groups. If a change of the surface charge were suddenly imposed on such a molecule, there would be a change of local pH and a titration effect. Such a phenomenon might happen on a membrane surface in a mitochondrium or chloroplast. In the chloroplast, a flash of light could induce a charging-up effect that could conceivably give rise to a local delay of titration, or a speeding up of titration of certain groups. Again, by studying the kinetics of the process and knowing what charge changes are there make it possible to analyze this indicator error.

REHM: Well, I am not quite sure that I understood the question. If one does something to create an emf, like shine light on a chloroplast, then with the appropriate circuit there would be a flow of current and there will be continuity of the current through every part of the system, even through the dielectric capacitors in the membrane (Maxwell's Displacement Current). The transport of current up to and away from the membrane would involve the movement of ions. And, hence, there would be an accumulation of cations on one side and an accumulation of anions on the other side of the membrane. And if hydrogen ions move from the inside to the outside of vesicles and were not accompanied by anions moving in the same direction, nor cations in the opposite direction, this would be the situation in which there would be violation of electroneutrality and it could be (in the models that I have presented) detected chemically. I do not believe that the situation you described about a protein molecule with the displacement of one ion by another would be an example of what I am talking about.

THE CYCLIC ELECTRON TRANSPORT CHAIN OF PHOTOSYNTHETIC BACTERIA AND ITS ROLE AS A PROTON PUMP *

A. R. Crofts and J. R. Bowyer

Department of Physiology and Biophysics
University of Illinois at Urbana-Champaign
Urbana, Illinois 61801

Introduction

It is now generally accepted that electron transport chains function as proton pumps,[1–5] but the precise molecular mechanism, and even in some cases the stoichiometry, remain controversial. These problems are most readily studied in photosynthetic systems in which electron transport can be started by a flash of light. This has two immediate experimental advantages. Firstly, no addition of electron donor or acceptor is involved, so that problems of changed pH, buffering capacity, and mixing are avoided. Secondly, with a sufficiently short, intense flash, all the photochemical reaction centers can be made to turnover once, and only once, so that a single reducing electron and oxidizing electron hole can be introduced virtually instantaneously into each electron transport chain. In addition, in many photosynthetic systems, some of the photosynthetic pigments themselves undergo an electrochromic color change in response to the electrical fields generated across,[6–12] and within,[13–15] the membrane; and this change can be used to follow electrogenic events associated with electron transport. In this paper we wish to summarize our current understanding of the electron transport chains of *Rhodopseudomonas capsulata* and *Rps. sphaeroides,* and to discuss two problems of current interest. These are 1.) the role of ubiquinone, and the involvement of four distinct populations of ubiquinone in the electron transport chain, and 2.) the role of the Rieske-type,[16, 17] Fe-S center, and its identification as the reductant for cytochrome c_2.

Components of the Chain

The kinetically identified components of the cyclic electron transfer chain of *Rps. capsulata* are shown against a scale of redox midpoint potential at pH 7 (Em7) in FIGURE 1. The components of *Rps. sphaeroides* are similar, but with minor differences in the value of Em7 for some components. The values in parentheses beside each component are stoichiometries with respect to the reaction center. Where two values are shown, the first is a kinetically determined value, and the second that measured by redox potentiometry. The values shown against arrows connecting components are measured half times for the reaction indicated.

Arrangement of the Chain in the Membrane

We have discussed at length elsewhere the evidence that suggests the cyclic electron transport chain is arranged in the membrane as two proton pumps,[2, 3, 19]

* The work reported was supported by USPHS Grant GM-2630501 to A.R.C., and by a grant from the Research Board of the University of Illinois to J.R.B.

0077–8923/80/0341–0012 $01.75/0 © 1980, NYAS

operating in series with respect to electron flow, and in parallel across the membrane. The scheme of FIGURE 1 shows the chain connected in a linear sequence of reactions. In FIGURE 2, the same linear pathway is shown with components arranged anisotropically so as to form two proton pumps. The evidence for the arrangement of the components shown close to the reaction center is unambiguous, and is based on evidence from a variety of approaches.[18-20] The distribution of the components of the "dark" chain connect-

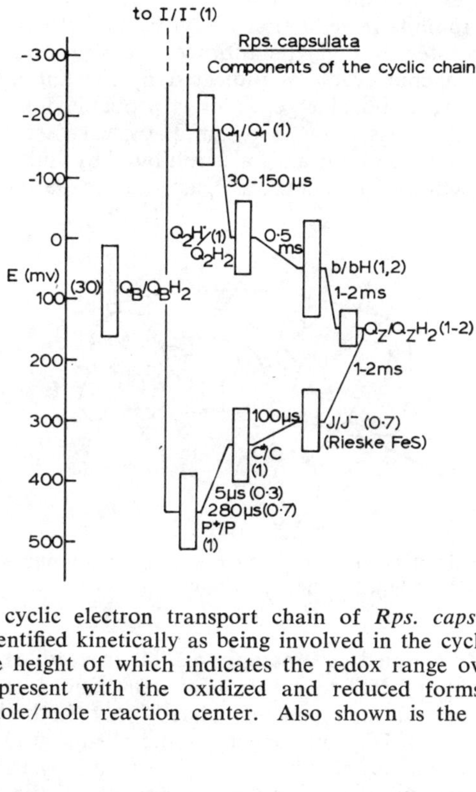

FIGURE 1. The cyclic electron transport chain of *Rps. capsulata.* Components shown are those identified kinetically as being involved in the cycle. They are identified by a block the height of which indicates the redox range over which the component would be present with the oxidized and reduced forms at concentrations greater than 0.1 mole/mole reaction center. Also shown is the bulk pool of ubiquinone.

ing the two sides of the reaction center is less sure, and the arrangement shown is somewhat speculative. The evidence is summarized briefly below.

1. The electron transport chain is similar in many respects to that part of the mitochondrial pathway involved in the oxidation of succinate and reduction of cytochrome oxidase.
2. Components of appropriate redox potential include *b*- and *c*-type cytochromes, a Rieske-type iron center, a lower potential iron center and ubiquinone$_{10}$. Of these components, only three have previously been shown unambiguously to be involved kinetically—ubiquinone, a *b*-cytochrome (cyt b_{50}, $E_{m7} \sim 50$mV, α peak at 560 nm in the reduced-oxidized difference

spectrum), and a *c*-cytochrome (cyt c_2, $E_{m7} \sim$ 340mV in *Rps. capsulata*, α peak at 551 nm for difference spectrum).

3. At appropriate redox potentials, flash illumination causes rapid oxidation of reduced cyt c_2, followed by its re-reduction, and a transient reduction of cyt b_{50} followed by its reoxidation. Electron transfer through the chain leads to the uptake of two protons/one electron, and to fast and slow phases of the carotenoid electrochromic change. The fast phase is associated with the electrogenic event accompanying the photochemical reactions. The slow phase is associated with electron transport through the antimycin sensitive site.
4. Antimycin inhibits re-reduction of cyt c_2, and the oxidation of cyt b_{50}.
5. Electron transfer through the antimycin-sensitive site is associated with a second electrogenic event, as indicated by the antimycin-sensitive slow phase of the carotenoid change. This is probably linked to oxidation of cyt b_{50}, since cyt c_2 is reduced by an H-carrier (see below).
6. Uptake of the second proton is also inhibited by antimycin.
7. The re-reduction of cytochrome c_2 appears to be the rate-limiting step

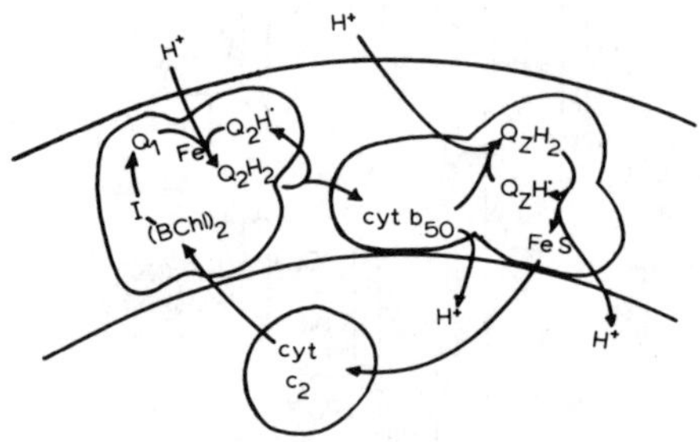

FIGURE 2. The cyclic electron transport chain as a proton pump showing a possible arrangement of the components in the membrane.

in the cycle. In coupled chromatophores under steady-state conditions, the rate of electron transfer to cyt c_2 is much reduced compared with that following a single flash, but is stimulated when both the electrical ($\Delta\psi$) and chemical ($-z\Delta pH$) components of the proton gradient are dissipated by appropriate uncoupling conditions. Electron flow in the steady-state is not markedly stimulated if either component of the gradient alone is dissipated.[4, 21]

Role of Ubiquinone

Our kinetic studies of *Rps. capsulata* and those of Dutton and colleagues on the closely related *Rps. sphaeroides* have shown some interesting properties of the electron transfer between cytochromes b_{50} and c_2. Over a range of ambient redox potentials around 150mV (at pH7) the rate of electron transport speeds up dramatically as the potential is lowered.[21–23] This effect has been attributed to reduction of a component Q_Z, previously assumed to be the physiological reductant for cyt c_2. The component Q_Z has not been identified spectroscopically but many of its properties have been assayed by observing effects of redox potential, pH, transient reactant concentration, etc., on the antimycin sensitive reduction of cyt c_2. In both *Rps. sphaeroides* and *Rps. capsulata,* over a wide range of values of pH, the component has the redox properties of an H-carrier of the quinone type (half-cell reaction $QH_2 \rightarrow Q + 2H^+ + 2e^-$).[24] The antimycin sensitive reduction of cyt c_2 involves a second-order process, and we have suggested that Q_ZH_2 might be present in several fold excess over ferricytochrome c_2. Dutton *et al.*[25] found that the concentration of Q_Z/Q_ZH_2 was constant and close to 0.7 Q_Z/reaction center; its concentration with respect to photooxidized ferricytochrome c_2 would therefore depend on the ratio of cyt c_2 to reaction center in any given preparation.

Many of the properties of the component Q_Z suggested that it may be a special ubiquinone in an environment separate from that of the bulk quinone. In collaboration with the Melandris and following their earlier work,[26, 27] we have recently studied the effect of extraction of ubiquinone on the reaction between cyt b_{50} and c_2 in chromatophores from *Rps. capsulata.* The results were somewhat clouded because the lyophilization, which was a necessary part of the extraction procedure, had a partial inhibitory effect on the rates of electron transport. However, we were able to show a clear inhibition of the reaction by which cyt b_{50} was oxidized, which was reversed when ubiquinone was added back. This observation would be consistent with the identity of Q_Z with ubiquinone.

Our results also showed that despite extensive extraction, electron transport to cyt b_{50} was not eliminated, and the re-reduction of cyt c_2 was only partially inhibited. We concluded that if the intermediate between the primary acceptor and cyt b_{50}, and the component Q_Z, were ubiquinone, then they must both be in an environment distinct from that of the bulk quinone. We had earlier shown that in the presence of antimycin, electrons from the photochemical reaction equilibrate rapidly with a small pool (~ 3 equivalents), and only slowly with the large pool, presumably of bulk ubiquinone.[28] Takamiya and Dutton have recently made more extensive studies of the effects of ubiquinone extraction on electron transport in *Rps. sphaeroides.*[29] They showed that 25–30 moles of ubiquinone were present per mole of reaction center. Of these, ~90% could be readily extracted without markedly affecting the pattern of electron transport, electrochromic changes, or H^+-uptake observed after a single flash. Redox titration of this readily extractable UQ showed that most of it behaved homogeneously, with an $E_{m7} \sim 90$mV, varying by -60mV/pH unit between pH5 and 9. On further extraction, removal of another mole of UQ was sufficient to completely inhibit those reactions in which Q_Z is implicated (reduction of cyt c_2 and the slow phase of the carotenoid change), and these could be restored by adding back a small quantity of ubiquinone. These results leave little doubt that Q_Z is a molecule of ubiquinone. It seems likely that it corresponds to one of the quinone-proteins (QP-C) identified in mitochondria by King *et al.*,[30, 31] and also with one of the EPR visible quinones of the b-c_1 complex observed by Ohnishi *et al.* (personal communication).

The reduction of ferricytochrome c_2 is a single electron process, so the overall reaction discussed above would be of the type:

$$Q_ZH_2 + \text{cyt } c^{3+} \longrightarrow Q_ZH^{\cdot} + \text{cyt } c^{2+} + H^+. \quad \textbf{(1)}$$

Our own results show that reduction of cyt c_2 by electron flow through the antimycin block (presumably by Q_ZH_2) is closely integrated with the oxidation of cyt b_{50}, since under normal circumstances these two processes, and the slow phase of the carotenoid change, appear to be kinetically matched.[21] We have suggested that this may reflect the fact that the semiquinone species, $Q_ZH^{\cdot}$, produced in reaction [1] above, acts as the oxidant for reduced cyt b_{50},

$$\text{cyt } b^{2+} + Q_ZH^{\cdot} + H^+ \longrightarrow Q_ZH_2 + \text{cyt } b^{3+} \quad \textbf{(2)}$$

and that this reaction is electrogenic.

Dutton and colleagues have suggested in contrast that the semiquinone acts as the reductant for oxidized cyt b_{50},[4, 5, 32]

$$\text{cyt } b^{3+} + Q_ZH^{\cdot} \longrightarrow \text{cyt } b^{2+} + Q_Z + H^+. \quad \textbf{(3)}$$

This reaction, together with that of Equation **(1)** are proposed by Dutton and Prince,[4, 5] to form part of a "Q-cycle" as suggested by Mitchell, but this appears to be incompatible with our observations on the kinetics of cyt b_{50} (see below), and we rejected such a mechanism some years ago on these grounds.

We have recently made a detailed study of the kinetics and spectra of the absorption changes following flash activation in *Rps. capsulata* and *Rps. sphaeroides* and have been able to obtain detailed spectra of the components contributing to the changes in the region between 400 and 620 nm. By appropriate choice of wavelength, and by using the pure spectra to provide a set of normalizing factors, we have been able to measure the kinetics and time resolved spectra of cyt b_{50} and cyt c_2 without any contamination from other components. Our results show unambiguously that, under a wide variety of conditions of redox poise and in the presence or absence of antimycin, cyt b_{50} goes reduced before it goes oxidized, that the rate of reduction is more rapid than the apparent rate of re-reduction of cyt c_2 by electron flow through the antimycin block, and that antimycin does not inhibit the rate of reduction of cyt b_{50}. The difference between the changes in the presence and absence of antimycin may be taken to reflect the rate of electron flow through the antimycin site from cyt b_{50} to cyt c_2. These results seem to preclude a mechanism in which $Q_ZH^{\cdot}$ acts as a reductant for cyt b_{50}, and to favor a simple linear scheme of the type shown in FIGURE 2. However, some recent experiments using a new inhibitor throw a new light on the nature of the reductant for cyt c_2.

Inhibition of Electron Flow by a Ubiquinone Analogue

The evidence summarized above made use of the inhibitor antimycin, assumed to block electron transfer between cytochromes b_{50} and c_2. In addition to antimycin the effects of several other inhibitors have been particularly informative. We show in FIGURE 3a the effects of a ubiquione analogue synthesized by Dr. K. Folkers (5-n-undecyl 6-hydroxy,4,7-dioxobenzothiazole, UHDBT), on the kinetics of cytochromes b_{50} and c_2, and on the reaction center changes. In the presence of the inhibitor, the reduction and oxidation of cyt b_{50} were completely eliminated, and the re-reduction of cyt c_2 was also inhibited (FIGURE 3a). When UHDBT was added to chromatophores previously inhibited by antimycin, reduction of cyt b_{50} was inhibited, and an increase in the extent of flash induced cyt c_2 oxidation was observed (FIGURES 3b, 3a). This latter effect was unexpected, since it had previously been thought that antimycin blocked reduction of cyt c_2 by its immediate donor, Q_ZH_2.[33, 34] The simplest explanation for the extra oxidation was that this assumption was wrong, and that an addi-

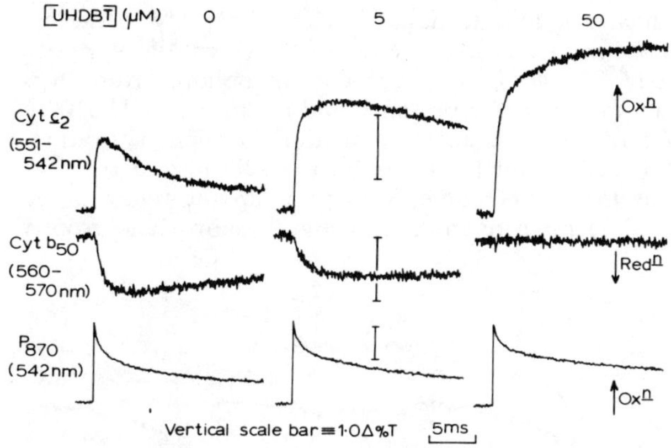

FIGURE 3A. Effect of UHDBT on the single turnover kinetics of cyt c_2, cyt b_{50} and P_{870} in the absence of antimycin A. Chromatophores (15 μM BChl) were suspended in 50 mM MOPS, 100 mM KCl at pH 7.0 in the presence of 2 μM valinomycin in a stirred anaerobic cuvette. UHDBT (in ethanol solution) was added to the concentrations shown. Traces were an average of 4. The instrument response time was 20 μs.

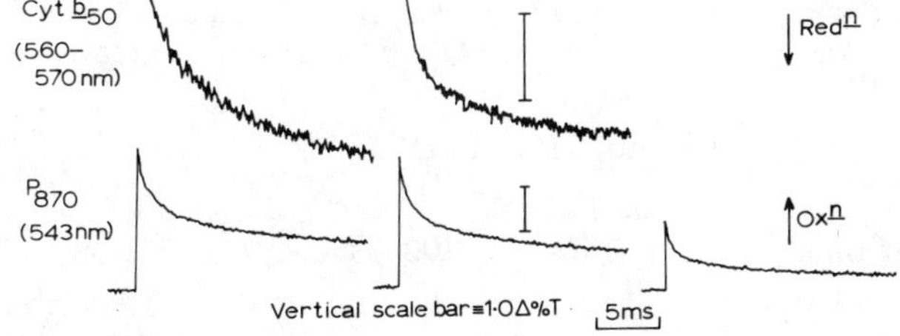

FIGURE 3B. Effect of redox potential on the single turnover kinetics of cyt c_2, cyt b_{50} and P_{870} in the presence of antimycin A. Chromatophores (15 μM BChl) were suspended in 50 mM MOPS, 100 mM KCl, pH 7.0 in the presence of 2 μM valinomycin and 5 μM antimycin A in a stirred anaerobic cuvette. Redox mediators were: DAD, 1.0 μM; PMS, 0.5 μM; 1,4-naphthoquinone, 10 μM; 1,2-naphthoquinone, 10 μM; and 2-OH-1,4-naphthoquinone, 10 μM. Redox potential was adjusted by addition of ferricyanide or dithionite solution. Traces were an average of 4 and the instrument response time was 20 μs.

tional component, J, existed between the site of action of antimycin and cytochrome c_2, which was able to donate electrons rapidly to the cytochrome.[35-37] In FIGURE 4a, we show a redox titration of the extent of cyt c_2 oxidation for the first and second of two flashes, for chromatophores from *Rps. sphaeroides* inhibited by antimycin in the presence and absence of UHDBT. The points in FIGURE 4b are from a computer simulation of such a titration in which we assumed that J, cyt c_2 and P870 came to equilibrium at the time of measurement, and that their stoichiometries and midpoint redox potentials were as stated. It would seem from the reasonable fit shown and from the pattern of

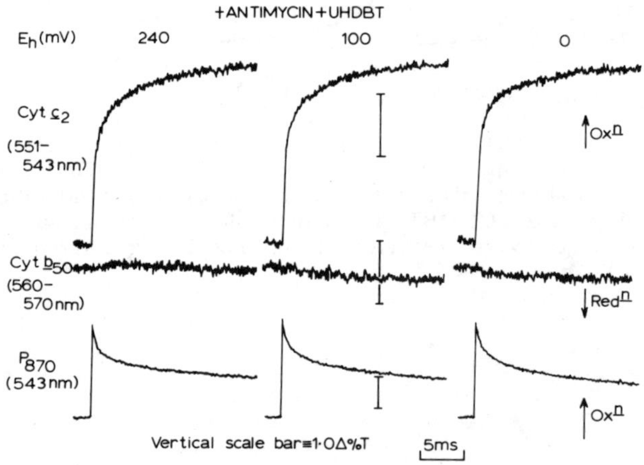

FIGURE 3C. Effect of redox potential on the single turnover kinetics of cyt c_2, cyt b_{50} and P_{870} in the presence of antimycin A and UHDBT. The conditions were as for FIGURE 3B except that UHDBT was added to 25 μM.

inhibition that we can account rather well for the results in terms of a simple linear electron transport chain between Q_Z and P as shown below:

	antimycin ↓↓	UHDBT ↓↓		20 μs (0.3)
half-time	1–2 ms	200 μs		200 μs (0.7)
	Q_Z ———→	J ———→	cyt *c*	———→ P870
relative molar ratio	0.7–2	0.7	1	0.9

In view of the similarity between the component J and the Rieske type Fe-S center,[16, 17] we have previously suggested that the two components might be identical, and that if this were so we had demonstrated for the first time the role of the Fe-S center in electron transport.[36, 37] Recently, in collaboration

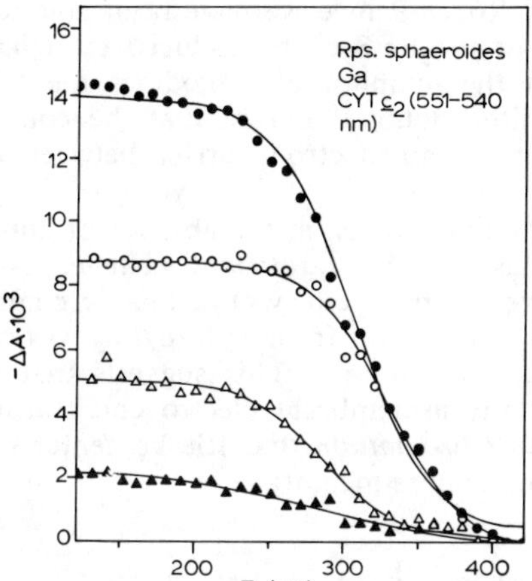

FIGURE 4A. (a) Redox titrations of extent of cytochrome c_2 photo-oxidation caused by 2 flashes 50 ms apart. With antimycin A: *Rps. sphaeroides* Ga chromatophores were suspended to ~1.1 μM reaction centers in 50 mM MOPS, 100 mM KCl (pH 7.0) containing 2 μM antimycin A and 2 μM valinomycin. For traces E_h 300 mV, redox mediators were ≥200 μM potassium ferricyanide and ferrocyanide and 7 μM TMPD. Below E_h 300 mV, redox mediators were 7 μM DAD, 7 μM 1,4-naphthoquinone, 7 μM 1,2-naphthoquinone and 7 μM 2-hydroxy,1,4-naphthoquinone. Signals were not averaged. The instrument response time was 100 μs. (○) Extent of oxidation induced by flash 1, recorded ~40 ms after the flash. (△) Extent of oxidation induced by flash 2, recorded ~40 ms after the flash. With antimycin A and UHDBT: As above except that 40 μM UHDBT was also present, and 5 μM TMPD was used above E_h 300 mV instead of 7 μM. (●) Flash 1-induced cyt c_2 oxidation. (▲) Flash 2-induced cyt c_2 oxidation.

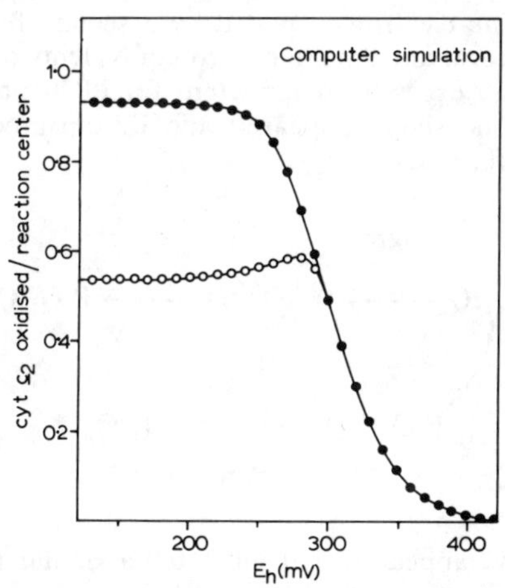

FIGURE 4B. Computer simulation of the titration shown in FIGURE 4A. Data for the first flash only are shown. (○) With antimycin. (●) With antimycin and UHDBT. The stoichiometry of J:cyt c_2: P870 was set at 0.7: 1·1: 1.0, and the midpoint redox potentials were set at 285 mV, 295 mV and 440 mV respectively.

with P. L. Dutton and Roger Prince we have been able to show that UHDBT inhibits the oxidation of the Fe-S center induced by a flash or by continuous illumination, and that the inhibitor also modifies the EPR spectrum of the center. There seems little doubt therefore that the component J is the Fe-S center, and that it acts as an electron carrier between Q_ZH_2 and cyt c_2 as shown above.

In carotenoid containing strains, in the absence of antimycin, UHDBT also inhibited the slow phase of the carotenoid change. No phase sensitive to UHDBT but insensitive to antimycin, with a kinetic component similar to the expected rate of electron donation from J^- to cyt c_2 was observed in the carotenoid change under any conditions. This suggests that the electron transfer from J to cyt c_2 is not a transmembrane electrogenic transfer and is compatible with the observation in *Chromatium* that Rieske center is located towards the inner side of the chromatophore membrane.[52]

Accumulation of Reducing Equivalents in the Reactions of the Secondary Acceptor

The photochemical reaction centers of a number of photosynthetic bacteria have been isolated as pure proteins. The reaction center from the carotenoid deficient R26 mutant of *Rps. sphaeroides* has been studied in great detail; this work has been reviewed at length elsewhere.[38-40] We will be concerned here with the reactions by which the first stable acceptor, the ubiquinone-iron complex (Q1·Fe) of the reaction center, usually referred to as the primary acceptor, donates its electrons to loosely bound ubiquinone (Q2) present as a secondary acceptor. Wraight and Vermeglio [41, 42] showed independently that electrons were transferred to bulk ubiquinone through a mechanism in which the electron on the quinone of the primary acceptor (Q1) following the first of a series of saturating flashes reduced a second equivalent of ubiquinone (Q2) to the semiquinone anion. The electron provided by a second flash then reduced the semiquinone with concomitant uptake of two equivalents of H^+. On subsequent flashes, provided that an excess of a reductant for $P870^+$, and an acceptor were present, the semiquinone signal appeared and disappeared, and protons were taken up, on alternate flashes.

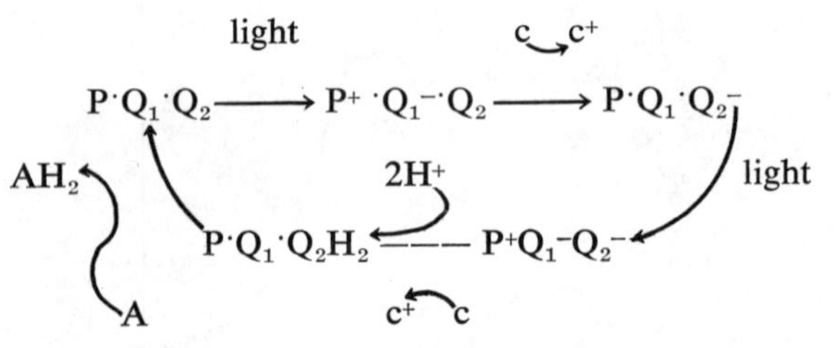

Several reports have appeared that indicated a similar two-stage mechanism may operate *in vivo*. Clayton and colleagues,[43] have shown a periodicity of two in the flash-induced uptake of protons on illumination from the dark state, but could observe this only with the carotenoid-free mutant R26 of *Rps. sphaeroides*. In contrast, de Grooth *et al.*[44] observed a periodicity of two in flash-induced changes ascribed to the semiquinone anion, and in carotenoid changes, in chromatophores from a carotenoid-containing strain. Fowler[45] reported briefly that although no periodicity of two was apparent in the uptake of protons by chromatophores (confirming previous observations by several other groups), a marked periodicity could be observed in the release of protons to an uncoupler accessible phase (presumed to be the internal aqueous phase).

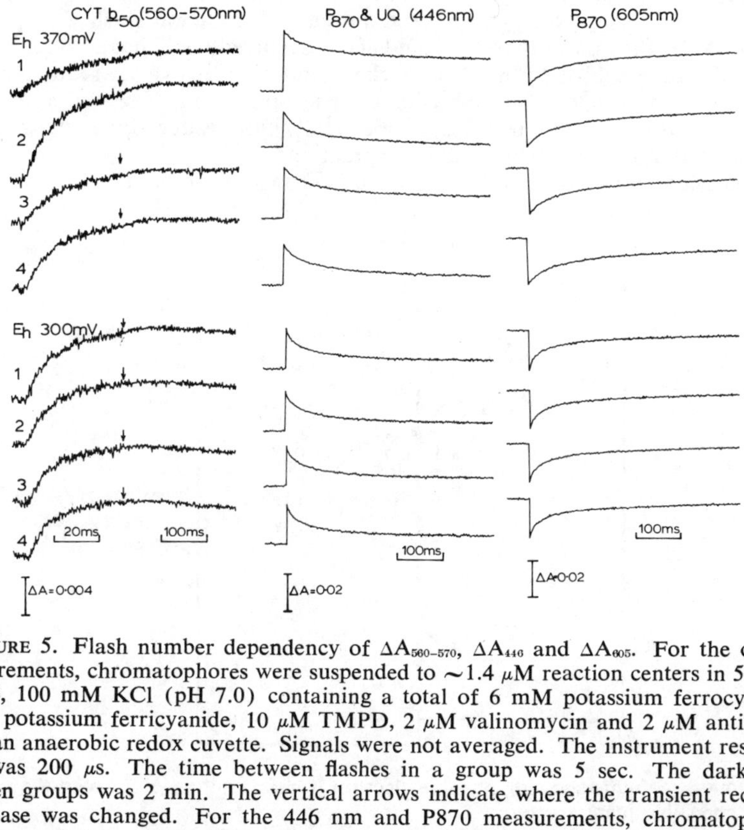

FIGURE 5. Flash number dependency of $\Delta A_{560-570}$, ΔA_{446} and ΔA_{605}. For the cyt b_{50} measurements, chromatophores were suspended to ~1.4 μM reaction centers in 50 mM MOPS, 100 mM KCl (pH 7.0) containing a total of 6 mM potassium ferrocyanide, 6 mM potassium ferricyanide, 10 μM TMPD, 2 μM valinomycin and 2 μM antimycin A, in an anaerobic redox cuvette. Signals were not averaged. The instrument response time was 200 μs. The time between flashes in a group was 5 sec. The dark time between groups was 2 min. The vertical arrows indicate where the transient recorder time base was changed. For the 446 nm and P870 measurements, chromatophores were suspended to ~1.4 μM reaction centers in 50 mM MOPS, 100 mM KCl (pH 7.0) containing a total of 2 mM potassium ferrocyanide and ferricyanide, 2 μM valinomycin and 2 μM antimycin A in an anaerobic redox cuvette. Signals were not averaged. The instrument response time was 500 μs. The time between flashes in a group was 10 sec. The dark time between groups was 2 min.

We have recently investigated these effects further, and have shown that over a limited range of redox potential, cyt b_{50} appears to act as an acceptor for electrons from the binary gating mechanism.[46] The flash-induced reduction of cyt b_{50} showed oscillations of period two as a function of flash number which were out of phase with the signal ascribed to the semiquinone anion (FIGURE 5).

The gating effect, as observed in cyt b_{50} and the semiquinone anion signal, was obvious only over the redox potential range between 300 and 450 mV. At the upper end of this range the photochemical reaction center was largely oxidized, but we do not understand yet why the effect disappeared at the lower end of the range. We have eliminated a number of the possible explanations we had previously suggested.[46] These centered around the possible role of cytochrome c_2, which becomes chemically reduced over this range, either as an indirect effector, or as a source of electrons in a double-hit process. We have now shown that the gating effects show essentially similar behavior when a laser (500 ns flash duration) is used instead of a xenon flash (10 μs flash duration), so that no double-hit could have occurred. We have also observed that in the presence of DBMIB, a marked inhibition of cyt b_{50} reduction after the first flash occurs, but full reduction on a second flash is observed at an E_h of 220 mV;[52] at this potential, cyt c_2 is chemically fully reduced, yet the behavior suggests that a binary gate may still be operating.

A possible explanation for these effects is shown in FIGURE 6.

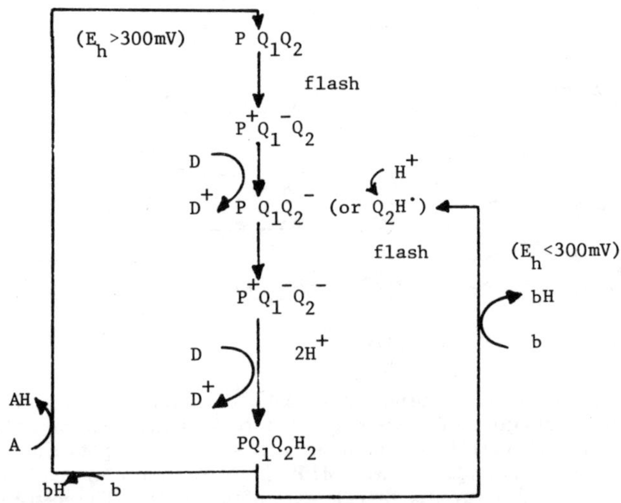

FIGURE 6. Scheme showing possible electron transfer pathways between the reaction center and subsequent acceptors.

In this scheme (FIGURE 6), the pathway shown on the left is essentially that of Wraight[47] but with cyt b_{50} as an acceptor for one of the two equivalents of electrons from Q_2H_2. We envisage this reaction pathway as occurring at high E_h (>300 mV), and being responsible for the binary oscillations. The reaction pathway on the right represents a one-electron cycle, and shows cyt b_{50} as the acceptor for one equivalent of electrons from Q_2H_2 as in the other pathway. However, in this case, we assume that the semiquinone product ($Q_2H^{\cdot}$) is stable, and can react as an acceptor of electrons from Q_1^-, produced in a subsequent flash, to form Q_2H_2.

If a reaction sequence such as this is to be put forward as the explanation for the disappearance of the oscillations, then some implications for the redox chemistry of the Q_2 couples must also follow. We would have to assume that the couple $Q_2/Q_2H^{\cdot}$ (or Q_2/Q_2^-) has a redox midpotential of about 300 mV, i.e., that the disproportionation equilibrium constant strongly favors the semiquinone. If this were the case, the couple $Q_2H^{\cdot}/Q_2H_2$ (or Q_2^-/Q_2H_2) would have a low redox midpotential (~0mV, assuming $E_mQ_2/Q_2H_2 \sim 150$mV), and the acceptor of electrons from Q_1^- under physiological conditions would be $Q_2H^{\cdot}$ (or Q_2^-), rather than the fully oxidized quinone (Q_2) as usually represented. Interestingly, this is the situation implied in Mitchell's earlier suggestions for a dual role of quinone either in a linear chain or a Q-cycle loop for the similar portion of the mitochondrial electron transfer chain.[48]

Conclusions

The scheme of Figure 2 summarizes our model for the general arrangement of the components of the cyclic electron transport chain and the pathway by which electron flow is coupled to proton pumping. The mechanism is, in principle, very similar to that originally suggested by Mitchell for mitochondria.[49, 50] However, some quite fundamental questions remain unanswered. It is clear that at least four different populations of ubiquinone are involved in the chain. Three of these are probably present at stoichiometries close to 1/RC: —Q_1, the first stable acceptor; Q_2, the second ubiquinone of the reaction center; and Q_Z, the reductant for cytochrome c_2 and J. The bulk quinone (Q_B) probably reacts relatively slowly with the main pathway, and functions mainly as a redox buffer. It seems likely that the functional quinones are bound in a protein environment which specifically determines their redox properties. They are prosthetic groups on enzymes and they form a class of redox proteins just as do the cytochromes and flavins. However, we know very little about the properties of this newly recognized class of redox protein, either in terms of protein chemistry, redox potential or reaction pathway.

The inhibitor UHDBT has enabled us to identify the role of the Rieske Fe-S center as the immediate electron donor to cyt c_2 in the electron transport chain. Taken together with the recent studies of Trumpower and Edwards,[51] who have isolated a protein containing the center that is required for reduction of cyt *c* by the mitochondrial succinate, and ubiquinol-cyt *c* reductase complexes; and with the EPR studies from a number of laboratories, especially that of Ohnishi and colleagues on this complex, these results seem to offer a real promise of understanding the immediate pathway of cyt *c* reduction in some detail.

The role of cyt b_{50} (and of the *b* cytochromes in mitochondria) is still rather puzzling.[2-5, 22] Under most circumstances, the cyt b_{50} in the bacteria behaves as would be expected from a component in a linear chain, but there are anomalies still to be resolved.[32] In particular, we do not understand the effect of redox potential on the two-electron gate between the reaction center and cyt b_{50}. Our early observation that the rate of cyt b_{50} reduction accelerates markedly over the redox potential range when Q_Z becomes reduced before the flash[22] is still unexplained, and we do not yet know the site of inhibition by UHDBT of cyt b_{50} reduction. These problems all seem to be connected with the involvement of cyt b_{50} as an acceptor of electrons from the reaction center, and presumably reflect hidden complications in the mechanism of the reaction between Q_1Q_2 and cyt b_{50} that have yet to be understood. What has become clear is that these problems can be recognized and studied in the photosynthetic bacteria, and that such study offers the possibility of experimental resolution. Presumably, similar problems remain to be identified in the mitochondrial chain, an area in which these problems will be less amenable to experimental study.

Summary

The photosynthetic electron transport chains of *Rhodopseudomonas capsulata* and *Rps. sphaeroides* are similar: the "dark" chain connecting the two sides of the photochemical reaction center (RC) is also similar to the mitochondrial chain between the dehydrogenases and cytochrome (cyt) oxidase. The photosynthetic chain is more amenable to kinetic study; in addition, electrogenic events can be followed by the carotenoid shift, and H^+ changes by use of indicator dyes. These studies, in conjunction with redox potentiometry and poising, and use of specific inhibitors, show that the chain is arranged as two proton pumps in series, with the RC and the oxidation of cyt b contributing two electrogenic steps. The H-carrying steps are associated with three distinct pools of ubiquinone that are extracted less readily than the bulk ubiquinone, and are probably protein-bound. Two of these (Q_1 and Q_2) are associated with a two electron-gating mechanism by which electrons are delivered from the RC to cyt *b*. The third (Q_Z) is an electron donor to cyt c_2. The Rieske FeS center acts as a carrier between Q_Z and cyt c_2 and is the site of inhibition by a ubiquinone analogue, UHDBT.

Acknowledgment

We wish to thank Dr. Colin Wraight for useful discussions.

References

1. Boyer, P. D., B. Chance, L. Ernster, P. Mitchell, E. Racker & E. C. Slater. 1977. Annu. Rev. Biochem. **46:**955–1026.
2. Crofts, A. R. & P. M. Wood. 1978. Curr. Top. in Bioenerg. **7:**175–244.
3. Crofts, A. R. & S. Saphon. 1978. *In* Frontiers in Physicochemical Biology. B. Pullman, Ed.: 459–483. Academic Press. New York, N.Y.
4. Dutton, P. L. & R. C. Prince. 1978. *In* The Photosynthetic Bacteria. R. K. Clayton & W. R. Sistrom, Eds.: 525–570. Plenum Press. New York, N.Y.
5. Dutton, P. L. & R. C. Prince. 1978. *In* The Bacteria. L. N. Ornston & J. R. Sokatch, Eds. **6:**523–584. Academic Press. New York, N.Y.
6. Symons, M., C. Swysen & C. Sybesma. 1977. Biochim. Biophys. Acta. **462:** 706–717.
7. deGrooth, B. G. & J. Amesz. 1977. Biochim. Biophys. Acta **462:**237–246.
8. deGrooth, B. G. & J. Amesz. 1977. Biochim. Biophys. Acta **462:**247–258.
9. Holmes, N. G. & A. R. Crofts. 1977. Biochim. Biophys. Acta **459:**492–505.
10. Holmes, N. G. & A. R. Crofts. 1977. Biochim. Biophys. Acta **461:**141–150.
11. Junge, W. & H. T. Witt. 1968. Z. Naturforsch. Teil B **23:**244–254.
12. Sewe, K.-U. & R. Reich. 1977. Z. Naturforsch. **32c:** 161–171.
13. Heathcote, P., A. Vermeglio & R. K. Clayton. 1977. Biochim. Biophys. Acta **461:**358–364.
14. Cogdell, R. J., S. Celis, H. Celis & A. R. Crofts. 1977. FEBS Lett. **80:** 190–194.

15. SYMONS, M. & A. R. CROFTS. Unpublished data.
16. PRINCE, R. C., J. G. LINDSAY & P. L. DUTTON. 1975. FEBS Lett. **51:**108–111.
17. PRINCE, R. C. & P. L. DUTTON. 1976. FEBS Lett. **65:**117–119.
18. WRAIGHT, C. A., R. J. COGDELL & B. CHANCE. 1978. *In* The Photosynthetic Bacteria. R. K. Clayton & W. R. Sistrom, Eds.: 471–511. Plenum Press. New York, N.Y.
19. CROFTS, A. R. 1974. *In* Perspectives in Membrane Biology. S. Estrado-O & & C. Gitler, Eds.: 373–412. Academic Press. New York, N.Y.
20. PARSON, W. W. & R. J. COGDELL. 1975. Biochim. Biophys. Acta **416:**105–149.
21. CROFTS, A. R., D. CROWTHER & G. V. TIERNEY. 1975. *In* Electron Transfer Chains and Oxidative Phosphorylation. E. Quagliariello, S. Papa, E. Palmieri, E. C. Slater & N. Siliprandi, Eds.: 233–241. North-Holland. Amsterdam, the Netherlands.
22. EVANS, E. H. & A. R. CROFTS. 1974. Biochim. Biophys. Acta **357:**89–102.
23. PRINCE, R. C., W. H. VAN DEN BERG, K.-I. TAKAMIYA, C. L. BASHFORD & P. L. DUTTON. 1978. *In* Frontiers of Biological Energetics. P. L. Dutton, J. S. Leigh & A. Scarpa, Eds. **1:**201–209. Academic Press. New York, N.Y.
24. PRINCE, R. C. & P. L. DUTTON. 1977. Biochim. Biophys. Acta **462:**731–747.
25. PRINCE, R. C., C. L. BASHFORD, K.-I. TAKAMIYA, W. H. VAN DEN BERG & P. L. DUTTON. 1978. J. Biol. Chem. **253:**4137–4142.
26. BACCARINI-MELANDRI, A. & B. A. MELANDRI. 1977. FEBS Lett. **80:**459–464.
27. BOWYER, J. R., A. BACCARINI-MELANDRI, B. A. MELANDRI & A. R. CROFTS. 1978. Z. Naturforsch. **33c:**704–711.
28. CROFTS, A. R. & J. BOWYER. 1978. *In* The Proton and Calcium Pumps. G. F. Azzone, *et al.*, Eds.: 55–64. Elsevier/North-Holland Biomedical Press. Amsterdam, the Netherlands.
29. TAKAMIYA, K.-I. & P. L. DUTTON. 1979. Biochim. Biophys. Acta **546:**17–42.
30. YU, C. A., S. NAGAOKA, L. YU & T. E. KING. 1978. Biochem. Biophys. Res. Commun. **82:**1070–1078.
31. KING, T. E., L. YU, S. NAGAOKA, W. R. WIDGER & C. A. YU. 1978. *In* Frontiers of Biological Energetics. P. L. Dutton, J. S. Leigh & A. Scarpa, Eds.: 174–182. Academic Press. New York, N.Y.
32. VAN DER BERG, W. H., R. C. PRINCE, C. L. BASHFORD, K.-I. TAKAMIYA, W. D. BONNER & P. L. DUTTON. 1979. J. Biol. Chem. **254:**8594–8604.
33. DUTTON, P. L. & J. B. JACKSON. 1972. Eur. J. Biochem. **30:**364–374.
34. PRINCE, R. C. & P. L. DUTTON. 1975. Biochim. Biophys. Acta **387:**609–613.
35. BOWYER, J. R. & A. R. CROFTS. 1978. *In* Frontiers of Biological Energetics. P. L. Dutton, J. S. Leigh & A. Scarpa, Eds. **1:**326–333. Academic Press. New York, N.Y.
36. BOWYER, J. R., G. V. TIERNEY & A. R. CROFTS. 1979. FEBS Lett. **101:**207–212.
37. CROFTS, A. R. 1979. *In* Light-induced Charge Separation in Biology and Chemistry. H. Gerischer & J. J. Katz, Eds. :389–410. Dahlem Konferenzen. Berlin, West Germany.
38. BLANKENSHIP, R. E. & W. W. PARSON. 1978. Ann. Rev. Biochem. **47:**635–653.
39. DUTTON, P. L., R. C. PRINCE, D. M. TIEDE, K. M. PETTY, K. J. KAUFMANN, T. L. NETZEL & P. M. RENTZEPIS. 1976. Brookhaven Symp. Biol. **28:**213–237.
40. DUTTON, P. L., R. C. PRINCE & D. M. TIEDE. 1978. Photochem. Photobiol. **28:** 939–949.
41. WRAIGHT, C. A. 1977. Biochim. Biophys. Acta **459:**525–531.
42. VERMEGLIO, A. 1977. Biochim. Biophys. Acta **459:**516–524.
43. BAROUCH, Y. & R. K. CLAYTON. 1977. Biochim. Biophys. Acta **462:**785–788.
44. DEGROOTH, B. G., R. VANGRONDELLE, J. C. ROMIJK & M. P. J. PULLES. 1978. Biochim. Biophys. Acta **503:**480–490.
45. FOWLER, C. F. 1976. *In* Int. Conf. Primary Electron Transport and Energy Transduction in Photosynthetic Bacteria. Abstr. WB9. C. Sybesma, conference organizer. Brussels, Belgium.
46. BOWYER, J. R., G. V. TIERNEY & A. R. CROFTS. 1979. FEBS Lett. **101:**201–206.

47. Wraight, C. A. 1979. Biochim. Biophys. Acta. Submitted.
48. Mitchell, P. 1975. FEBS Lett. **56:**1–6.
49. Mitchell, P. 1966. Chemiosmotic Coupling in Oxidative Phosphorylation. Glynn Research. Bodmin, England.
50. Mitchell, P. 1968. Chemiosmotic Coupling in Oxidative Phosphorylation. Glynn Research. Bodmin, England.
51. Trumpower, B. L. & C. A. Edwards. 1979. FEBS Lett. **100:**13–16.
52. Bowyer, J. R. 1980. Dissertation for Ph.D. degree, University of Bristol, U.K.

Discussion

E. Racker (*Cornell University, Ithaca, N.Y.*): Would you care to speculate as to why the analog of ubiquinone seems to inhibit the reaction between the nonheme iron center, which, I believe is the same as what has been found in mitochondria, the so-called oxidation factor, and cytochrome *c*?

A. R. Crofts: I suggest that the protein of the iron-sulfur center must have a reaction site on it for the donor quinone; presumably, the quinone analog binds to that site.

Racker: You suggest that the quinone actually interacts with and sticks to the nonheme iron protein?

Crofts: Speculatively, yes.

STUDIES ON ANION TRANSPORTERS *

E. Racker, J. A. Belt, W. W. Carley and J. H. Johnson

Section of Biochemistry, Molecular and Cell Biology
Cornell University
Ithaca, New York 14853

TABLE 1 contains a list of the anion transport systems that we have studied during the past few years. Since the work on the first three systems has been published,[1–4] we shall discuss the last three systems still currently under investigation.[5–8]

Lactate Transport

Cells that produce lactate and either do not metabolize it or do not use it as fast as they make it, have to excrete it. Erythrocytes represent the first group, many tumor cells the second. We have shown [5] that erythrocytes have a specific carboxylic acid anion transporter that excretes a proton together with each lactate. This transport is distinct from the common ion transporter of erythrocytes, which is sensitive to stilbene derivatives, such as 4-acetamido-4′-isothiocyano-2,2′-disulfonic stilbene (SITS).[1] SITS actually also affects lactate transport, but indirectly, by inhibiting the proton gradient generated by the movement of chloride ions in exchange for bicarbonate ions via the common anion transporter. When chloride ions move into erythrocytes and bicarbonate ions move out, the proton gradient thus generated stimulates lactate excretion. When chloride ions move out and the cells become more alkaline inside, lactate excretion is inhibited. Consequently, SITS inhibits lactate excretion when chloride moves in and stimulates it when chloride moves out.

In Ehrlich ascites tumor cells, the same lactate/H^+ symport mechanism is operative,[9] but without the influence of a SITS-sensitive anion channel. All other properties we have examined are similar in these two cell types, including pH dependence, activation energies of lactate efflux, and uptake and sensitivity to bioflavonoid inhibitors.

We are particularly interested in lactate transport and hope to design a compound that will specifically inhibit it. It should result in the acidification of the cytoplasm of cells and therefore should be more toxic to cells that have a high aerobic glycolysis. Such a suicide compound should be more toxic for tumor cells than for erythrocytes or other normal cells, which generate lactate much more slowly. Erythrocytes have a great buffering capacity because of their high content in hemoglobin; moreover, they have the chloride anion channel which serves as an alternative mechanism to excrete protons.

* This work was supported by Grant No. BC-156 from the American Cancer Society; Grant No. CA-08964, Grant No. CA-14454, Fellowship No. F32 CA-05484 (JAB), Fellowship No. F32 CA-06336 (JHJ), awarded by the National Cancer Institute, DHEW.

0077–8923/80/0341–0027 $01.75/0 © 1980, NYAS

TABLE 1

ANION TRANSPORT IN CELLS, PLASMA MEMBRANE VESICLES AND RECONSTITUTED SYSTEMS

Anions	Location	Material
Adenine nucleotides	Inner membrane of mitochondria	Submitochondrial particles and reconstituted liposomes [1]
Phosphate	Inner membrane of mitochandria	Submitochandrial particles and reconstituted liposomes [2, 3]
Chloride	Plasma membrane of stomach cells	Plasma membrane vesicles [4]
Lactate	Plasma membrane of erythrocytes and Ehrlich ascites tumor cells	Intact cells [5, 6]
Phosphate	Plasma membrane of Ehrlich ascites tumor cells	Vesicles [7]
Phosphate	Sarcoplasmic reticulum	Vesicles and reconstituted liposomes [8]

Search for an Ideal Suicide Inhibitor

In the course of our studies with various inhibitors of lactate transport, we have become increasingly aware of what we need. Obviously, the inhibitor should be specific; it should not inhibit other metabolic pathways, particularly those utilizing monocarboxylic acids. The compound should be effective in the presence of serum and should preferably act from the outside of the cell; it should form a covalent bond with the transporter and the inhibitor should be irreversible. We have not found such an ideal compound, but we have come fairly close. We shall discuss two types of lactate transport inhibitors, the bioflavonoids and lactate anhydrides.

Bioflavonoids

We discovered the effect of bioflavonoids on lactate transport when we encountered two representatives of methoxyflavonoids [10] that inhibited glycolysis without affecting Na^+K^+ ATPase activity, which is responsible for the delivery of ADP and P_i, the rate-limiting step in glycolysis. Tests with dextran sulfate-treated cells, which are permeable to small ions and glycolyze only in the presence of externally added P_i and ADP, revealed no inhibition of glycolysis by the methoxybioflavonoids, ruling out a direct inhibition of glycolytic enzymes and suggesting involvement of a membrane-associated process. Lactate transport [9] looked like a good candidate. TABLE 2 contains a partial list of bioflavonoid inhibitors of lactate transport.

Lactate Anhydrides

We have synthesized an anhydride of lactic acid that served as an effective inhibitor of lactate transport, both in erythrocytes and Ehrlich ascites tumor cells.[11] In contrast to bioflavonoids, it appeared to be specific and its inhibitory effect was not reversed by washing the cells with serum albumin. The compound inhibited lactate transport 50% at 0.5 μg per mg of protein and had no effect on phosphate transport in the same cells. The effect of the anhydride on glycolysis resembled that observed with bioflavonoids. While the appearance of lactate in suspension fluid was greatly depressed, the intracellular lactate concentration rose precipitously.

We suspected that the effect of the anhydride may be caused by an interaction with an SH group in the transporter, resulting in the formation of a thiolester. In line with this possibility was the observation that exposure of cells to hydroxylamine partly reversed the inhibition of transport (TABLE 3).

TABLE 2

EFFECT OF BIOFLAVONOIDS ON LACTATE UPTAKE *

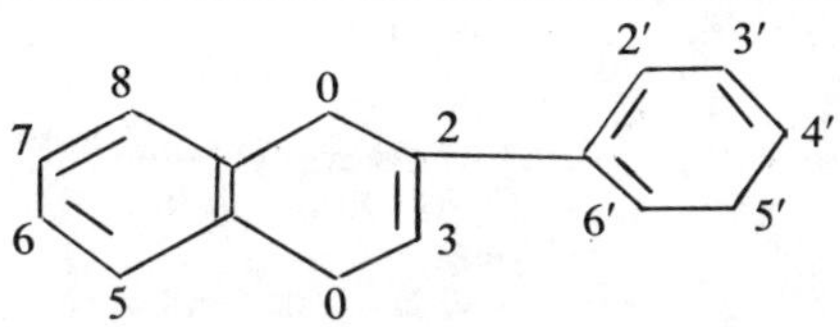

Flavone

Common Name	Hydroxyl pattern and substituents	Lactate uptake nmol/mg protein min	(%) Inhib.
None		25.8	
------†	3,3′,4′–OH	27.6	(+7)
Apigenin ‡	5,7,4′–OH	17.8	30
Fisetin ‡	3,7,3′,4′–OH	12.6	52
Morin §	3,5,7,2′,4′–OH	11.4	56
Quercetin ¶	3,5,7,3′,4′–OH	8.5	67
Kaempferid ‖	3,5,7–OH, 4′–OCH_3	31.3	(+21)
K_3 ‖	5,7,4′–OH, 3,6–OCH_3	13.4	48
------‖	5,7,3′–OH, 3,6,4′–OCH_3	21.6	16
Quercetrin †	5,7,3′,4′–OH, 3-rhamnose	25.0	3
Rutin §	5,7,3′,4′–OH, 3-rhamnose-glucose	30.8	(+19)
Dihydroquercetin #	2,3 dihydro, 3,5,7,3′,4′–OH	27.2	(+6)

* $[^{14}C]$Lactate uptake was determined as described previously.[6] The protein concentration in the assay was 2 mg/ml, lactate was 10 mM and bioflavonoids were 6 μM (equivalent to 0.9 μg quercetin/mg protein). A zero-time value of 9.1 nmol/mg protein has been substracted from all values.

† Supplied by Dr. T. A. Geissman.

‡ Supplied by Aldrich Chemical Co., Milwaukee, Wis.

§ Supplied by J. T. Baker Chemical Co., Phillipsburg, N.J.

¶ Supplied by Sigma Chemical Co., St. Louis, Mo.

‖ Supplied by Dr. E. Wollenweber.

Supplied by Dr. J. Chopin.

Inhibition by hydroxylamine in control cells was kept at a minimum by using a low concentration. A more effective reversal was achieved by incubation with dithiothreitol, presumably via chemical acyl transfer.

Since we have impressed our biochemical colleagues with our successes in the reconstitution of membranous transport processes, it is time we report a failure. We have thus far not succeeded in the reconstitution of the lactate transporter from Ehrlich ascites tumor cells. Or perhaps we should say we have succeeded too well. Crude membrane fractions incorporated into liposomes greatly increased the rate of lactate entry. However, the transport was not sensitive to inhibition by either bioflavonoids or lactate anhydride. We concluded that we were dealing with nonspecific leakage and that we cannot use the biological transport assay during purification of the lactate transporter. The lactate anhydride—if we can make it highly radioactive—represents a promising alternative; provided such an affinity-label assay will not mislead us, but guide us in the purification of a reconstitutively active lactate transporter.

TABLE 3

REVERSAL OF ANHYDRIDE-INDUCED INHIBITION OF LACTATE UPTAKE IN EHRLICH ASCITES TUMOR CELLS *

Inhibitor	Restoration	nmoles lactate × min^{-1} × mg $protein^{-1}$
None	None	20.0
None	0.3 M Hydroxylamine	11.0
None	0.05 M Dithiothreitol	18.0
5 μg Anhydride	None	0
5 μg Anhydride	0.3 M Hydroxylamine	6.8
5 μg Anhydride	0.05 M Dithiothreitol	11.0

* Ascites cells were harvested, washed, and resuspended at 10 mg protein per ml as described previously.[6] Anhydride was added as an ethanotic solution (5 μl/ml cell suspension) and the cells were incubated for 1 hr at 0°–5° C. For restoration experiments, the cells were washed twice in buffer containing 1 mg defatted BSA per ml; resuspended in 0.3 M hydroxylamine pH 7.4 or 0.05 M dithiothreitol, 105 mM NaCl, 5 mM KCl, 1 mM $MgCl_2$, 4 mM NaP_i, pH 7.4; incubated for 3 hrs at 0°–5° C; and washed twice in buffer containing 1 mg defatted BSA per ml. Measurements of ^{14}C-lactate uptake (inhibited by 2.5 mM α-cyano,4-hydroxycinnamate) were performed as described.[6]

Phosphate Uptake into Plasma Membrane Vesicles Isolated from Ehrlich Ascites Tumor Cells

Plasma membranes prepared by conventional procedures[12] exhibit a low uptake of P_i (TABLE 4). Following a dilution with 12 volumes of 10 mM Tris-HEPES (pH 7.4), vesicles are sedimented at 180,000 × g and suspended in 150 mM NaCl — 50 mM Tris-HEPES. These vesicles now catalyze an uptake of P_i that is considerably more rapid than in the original vesicles. Since the total activity of phosphate transport also recovered increases, it seems likely

TABLE 4

PHOSPHATE TRANSPORT IN VESICLES OF PLASMA MEMBRANES *

	$^{32}P_i$ Uptake	
	Specific Act. nmoles $\times$ min^{-1} $\times$ mg protein^{-1}	Total Act. nmoles $\times$ min^{-1}
Plasma membranes	0.09	0.34
Plasma membrane vesicles	0.46	0.92

* Plasma membranes were prepared from Ehrlich ascites tumor cells by the method of Brunette and Till [12] and stored at $-70°$ C in 0.3 M sucrose and 1 mM EDTA pH 7.4. Plasma membrane vesicles were prepared by diluting the thawed plasma membranes (8 mg/ml) with 12 volumes of 10 mM Tris-HEPES pH 7.4 and centrifuging for 30 min at 180,000$\times$g. The pellet was resuspended in 10 mM Tris-HEPES pH 7.4 at 2 mg protein per ml and then diluted with an equal volume of 300 mM NaCl and 100 mM Tris-HEPES pH 7.4. After standing for 1 hr on ice, the plasma membrane vesicles were centrifuged at 200$\times$g for 2 min to remove precipitated protein. Assays: 50 μl of plasma membranes or plasma membrane vesicles were added to 50 μl of 150 mM NaCl, 50 mM Tris-HEPES pH 7.4 and 20 mM Tris-$^{32}P_i$. After 5 min at 20° C, 75 μl of the suspension was centrifuged through a precentrifuged 1 ml Sephadex G-50 fine column [13] and the eluate was counted in Liquiscent.

that the newly-formed vesicles were either less leaky or that an inhibitor of phosphate transport has been removed during dilution and sedimentation. The uptake of P_i was biphasic (FIGURE 1). As can be seen from TABLE 5, P_i transport in these vesicles was sensitive to phloretin, particularly at low external P_i concentrations, but insensitive to cyclohexylmaleimide, which is known to inhibit P_i transport in mitochondria or submitochondrial particles.[14] Thus, we have ruled out P_i transport by contamination with mitochondrial membranes.

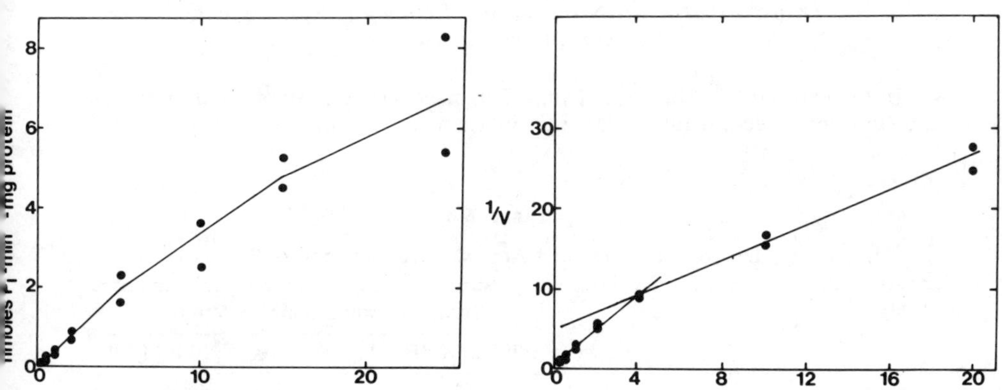

FIGURE 1. Phosphate titration of uptake in plasma membrane vesicles. Plasma membrane vesicles were prepared and assayed for P_i uptake as described in the legend of TABLE 4. Assays were performed for 1 min at 20° C with the indicated concentration of Tris-$^{32}P_i$.

TABLE 5

EFFECT OF INHIBITORS ON P_i UPTAKE IN PLASMA MEMBRANE VESICLES *

	nmoles × min^{-1} × mg protein^{-1}	
	0.1 mM $^{32}P_i$	10 mM $^{32}P_i$
Exp. 1		
Control vesicles	0.090	0.68
+0.5 mM CHM †	0.068	0.78
Exp. 2		
Control vesicles	0.047	0.92
+1 mM phloretin	0.017	0.64

* Plasma membrane vesicles were prepared as described in TABLE 4. After a 20 min incubation at room temperature with or without the inhibitors, phosphate uptake was determined with 0.1 mM Tris-^{32}P (700 cpm/pmol) for 1 min or 10 mM Tris-^{32}P (20 cpm/pmol) for 5 min.

† CHM-cyclehexylmaleimide.

Digitonin at 0.05% released most of the P_i that had been transported into the vesicles at either low or high P_i concentrations (TABLE 6). Accumulated counts were also released by exposure of the vesicles to 3% cholate.

Finally, as shown in FIGURE 2, P_i uptake was stimulated by Na^+. The rate of entry was also more rapid at pH 6.8 than at 7.4. These observations of P_i transport into vesicles from Ehrlich ascites tumor cells, are similar to previous reports on P_i transport of vesicles isolated from other cells.[15, 16] They also confirm the early observations [17] made with intact Ehrlich ascites tumor cells that indicated operation of a Na-P_i symport mechanism.

Transport of P_i in Sarcoplasmic Reticulum Vesicles and Reconstituted Liposomes

In the experiments shown in TABLE 7, frozen sarcoplasmic reticulum vesicles and liposomes reconstituted with sarcoplasmic reticulum were assayed for the

TABLE 6

DIGITONIN RELEASE OF ACCUMULATED PHOSPHATE *

	nmoles × mg protein^{-1} × 5 min^{-1}	
	0.1 mM $^{32}P_i$	10 mM $^{32}P_i$
Control vesicles	0.134	3.6
Vesicles + 0.05% digitonin at 4.5 min	0.040	0.8

* Plasma membrane vesicles were prepared as described in TABLE 4. Assays were for 5 min at room temperature with the indicated concentration of Tris-$^{32}P_i$.

ability to transport P_i. In both systems, P_i transport was dependent on the presence of Mg^{2+} and ATP. Trans-1:2-diaminocyclohexantetra-acetic acid, (CDTA), a hydrophobic chelator for divalent cations, inhibited P_i transport 90% in the presence of ATP and Mg^{2+}. *N*-ethylmaleimide (NEM), inhibited P_i transport in both systems, but the inhibition was much more striking in the reconstituted system. Ethyleneglycol-bis-N,N′-tetraacetic acid, (EGTA) at 500 μM inhibited P_i transport in the reconstituted system over 80% (TABLE 8). A-23187, on the other hand, had virtually no effect on the reconstituted system, while it distinctly diminished the rate of P_i transport in the sarcoplasmic reticulum vesicles. A-23187 did not inhibit P_i transport in sarcoplasmic reticulum vesicles when the Ca^{2+} concentration was raised to 800 μM. The ineffectiveness

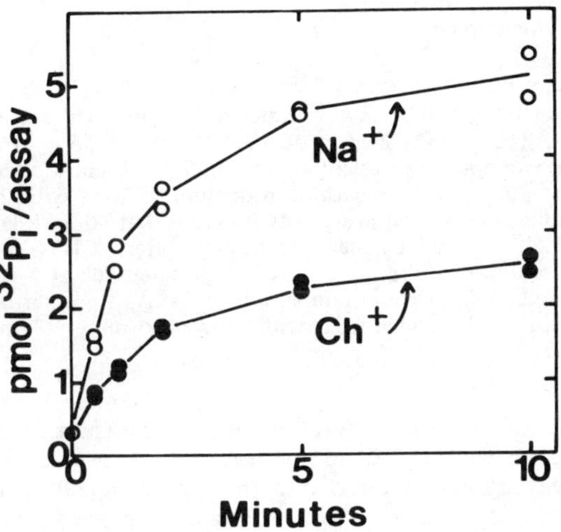

FIGURE 2. Sodium dependence of phosphate uptake in plasma membrane vesicles. Plasma membrane vesicles were prepared as described in the legend of TABLE 4 except that choline chloride (ChCl) was substituted for NaCl. Uptake assays were initiated by adding 50 μl of vesicles (ca 40 μg protein) to 50 μl of 50 mM Tris-HEPES (pH 7.4), 0.2 mM Tris-$^{32}P_i$ and 150 mM NaCl or ChCl.

of A-23187 is a significant observation because it indicates that P_i transport does not require a Ca^{2+} gradient. It, therefore, seems highly unlikely that the role of ATP-Mg is to provide a Ca^{2+} trap for the accumulation of P_i and we must search for a more direct effect on P_i transport.

From these experiments we draw the tentative conclusion that P_i transport requires Mg^{2+}, and Ca^{2+}, as well as ATP. Attempts to demonstrate a P_i-dependent ATPase activity have, thus far, not been successful, but these failures may have been caused by the overriding presence of the Ca^{2+}-dependent ATPase activity of the Ca^{2+} pump. Attempts to demonstrate activation of P_i transport by ATP, e.g., by phosphorylation of the phosphate transporter, have also been unsuccessful.

TABLE 7

EFFECT OF ATP-MG ON P_i TRANSPORT IN SARCOPLASMIC RETICULUM AND RECONSTITUTED LIPOSOMES *

Assay system	nmoles × min⁻¹ × mg protein⁻¹ Sarcoplasmic Reticulum Vesicles	Reconstituted Liposomes
Complete	309	233
Mg^{2+} deleted	50	121
ATP deleted	12	78
Both deleted	9	35
Complete + CDTA (20 mM)	10	28
Complete + A-23187 (5 μg/ml)	39	214
Complete + *N*-ethylmaleimide	14	61
Complete (zero time)	12	7

* Assays were performed in a final volume of 0.25 ml in the presence of 100 μM $CaCl_2$, 10 mM NaATP, 10 mM $MgCl_2$, 50 mM K $^{32}P_i$ (1550 cpm/nmole), 700 mM sorbitol and 30 μg sarcoplasmic reticulum at pH 7.5. Reconstituted liposomes were prepared by mixing 30 μg of sarcoplasmic reticulum vesicles with 2 mg of sonicated soybean phospholipid as described previously,[19] except that 40 mM Tris-maleate, 2 mM DTT and 0.2 M KCl were used instead of the P_i buffer. CDTA and A-23187 were added prior to the addition of the vesicles. *N*-ethylmaleimide at 5 mM was incubated with the vesicles for 10 min at room temperature. Assays were performed for 1.5 min at 36° C. Uptake of $^{32}P_i$ was measured with Dowex-formate columns as described.[18]

What conclusions can we draw about anion transport from these studies? There is clearly one common denominator: These processes are not cheap. Thus far we have not encountered one that we can call facilitated diffusion comparable to the duty-free transport of glucose into red cells. In each case something has to be paid. In the case of the nucleotide transporter of mitochondria, we are dealing with an electrogenic process. To keep the nucleotides moving, ADP in and ATP out, we have to collapse the membrane potential that they generate. This is in line with the dogma. But when we examine the

TABLE 8

EFFECT OF EGTA ON ATP-DEPENDENT P_i TRANSPORT IN RECONSTITUTED LIPOSOMES *

Assay conditions	P_i transport μmoles × min⁻¹ × mg protein⁻¹
Complete system	247
Complete system plus 10 μM EGTA	192
Complete system plus 100 μM EGTA	168
Complete system plus 500 μM EGTA	39
Complete system plus 2000 μM EGTA	8

* Experimental conditions were as described in TABLE 7. EGTA was added to the reconstituted liposomes 30 sec prior to assay.

P_i transporter of mitochondria, we no longer move in the straight path. All experts agree that P_i transport in mitochondria is electrically neutral, one hydroxyl being exchanged for each P_i (or one proton moving with P_i). Our reconstituted vesicles are stimulated by valinomycin-K^+, suggestive of an electrical imbalance. In any case, it is agreed that P_i transport is driven by a pH gradient. Similarly, lactate export is dependent on a pH gradient. Chloride transport in the stomach vesicles is stimulated by ATP, so is P_i transport in sarcoplasmic reticulum. Although it is by no means clear that these are pump processes, i.e., that stoichiometric amounts of ATP are cleaved for each P_i or chloride that is translocated, there is some form of energy expenditure.

Those of us concerned with ion transport, be it anion or cation transport, should start counting our pennies (or our ATP). It is remarkable how little we know about what portion of the total energy budget is spent on transportation. It looks from our studies, particularly with tumor cells, that ion transport consumes a major portion of the cellular energy budget. Maintenance of the "unnatural" distribution of K^+ and Na^+ requires continuous work, so does the movement of anions. As the great philosophers in Vienna used to say: Nothing in life is free, only death, and that costs you your life.

References

1. SHERTZER, H. & E. RACKER. 1976. Reconstitution and characterization of the adenine nucleotide transporter derived from bovine heart mitochondria. J. Biol. Chem. **251:**2446–2452.
2. BANERJEE, R. K., H. G. SHERTZER, B. I. KANNER & E. RACKER. 1977. Purification and reconstitution of the phosphate transporter from bovine heart mitochondria. Biochem. Biophys. Res. Commun. **75:**772–778.
3. BANERJEE, R. K. & E. RACKER. 1979. Partial purification and characterization of the phosphate transporter from bovine heart mitochondria. Membrane Biochem. **2:**203–225.
4. SOUMARMON, A. & E. RACKER. 1978. Chloride transport in gastric cells and microsomes. *In* Frontiers of Biological Energetics **1:**555–562. Academic Press.
5. DUBINSKY, W. P. & E. RACKER. 1978. The mechanism of lactate transport in human erythrocytes. J. Membrane Biol. **44:**25–36.
6. BELT, J. A., J. A. THOMAS, R. N. BUCHSBAUM & E. RACKER. 1979. Inhibition of lactate transport and glycolysis in Ehrlich ascites tumor cells by bioflavonoids. Biochemistry. **18:**3506–3511.
7. BELT, J. A. 1979. Phosphate uptake in plasma membrane vesicles from Ehrlich ascites tumor cells. Fed. Proc. **38:**246.
8. CARLEY, W. W. & E. RACKER. 1979. Phosphate transport in rabbit sarcoplasmic reticulum. XIth International Congress of Biochemistry. Toronto, Canada. In press.
9. SPENCER, T. L. & A. L. LEHNINGER. 1976. L-lactate transport in Ehrlich ascites tumor cells. Biochem. J. **154:**405–414.
10. RACKER, E. 1976. A New Look at Mechanisms in Bioenergetics. Academic Press, New York.
11. JOHNSON, J. H. 1979. Inhibition of lactic acid translocation in Ehrlich ascites tumor cells. Fed. Proc. **38:**246.
12. BRUNETTE, D. M. & J. E. TILL. 1971. A rapid method for the isolation of L-cell surface membranes using an aqueous two-phase polymer system. J. Membrane Biol. **5:**215–224.
13. PENEFSKY, H. 1977. Reversible binding of P_i by beef heart mitochondrial adenosine triphosphatase. J. Biol. Chem. **252:**2891–2899.

14. RHODIN, T. R. & E. RACKER. 1974. Phosphate transport in submitochondrial particles. Biochem. Biophys. Res. Commun. **61:**1207–1212.
15. LEVER, J. E. 1978. Active phosphate ion transport in plasma membrane vesicles isolated from mouse fibroblasts. J. Biol. Chem. **252:**2081–2084.
16. HAMILTON, R. T. & M. NILSEN-HAMILTON. 1978. Transport of phosphate in membrane vesicles from mouse fibroblasts transformed by simian virus 40. J. Biol. Chem. **253:**8247–8256.
17. SCHOLNICK, P., D. LANG & E. RACKER. 1973. Regulatory mechanisms in carbohydrate metabolism. IX. Stimulation of aerobic glycolysis by energy-linked ion transport and inhibition by dextran sulfate. J. Biol. Chem. **248:**5175–5182.
18. ZIMNIAK, P. & E. RACKER. 1978. Electrogenicity of Ca^{2+} transport catalyzed by the Ca^{2+}-ATPase from sarcoplasmic reticulum. J. Biol. Chem. **253:**4631–4637.
19. GASKO, O. D., A. F. KNOWLES, H. G. SHERTZER, E-M. SUOLINNA & E. RACKER. 1976. The use of ion-exchange resins for studying ion transport in biological systems. Analytical Biochem. **72:**57–65.

DISCUSSION

J. M. SALHANY (*University of Nebraska Medical Center, Omaha, Neb.*): Have you done the lactate transport experiments in resealed ghosts in which the system was well buffered? From your paper, one could assume that there would be no lactate transport. Is that correct?

E. RACKER: The lactate transport is measured in a highly buffered system.

SALHANY: Comparing your data on red cells with those of Steck's on resealed ghosts, one notes that he was buffering on both sides, and that there was no proton gradient. I don't see how you could drive lactate transport in a resealed ghost system that is well buffered when there is no proton gradient; in red cells there is a proton gradient, but is that coincidental?

RACKER: I'm not sure I understand your question; under physiological conditions the cell is operating with a gradient because it produces lactic acid and a proton transfer down the gradient is utilized to push the lactate out. It is possible that in Steck's experiments with resealed ghosts without gradient some other mechanism is operative, but our experiments suggest that in intact erythrocytes the H^+/lactate symport mechanism is the major pathway.

J. D. ROBINSON (*Upstate Medical Center, Syracuse, N.Y.*): Could you comment on the relationship between phosphate and calcium transport as far as the stoichiometry and the electrogenicity is concerned?

RACKER: In what system?

ROBINSON: In the sarcoplasmic reticulum.

RACKER: What we formulate is purely speculative; there is no direct physiological evidence, as far as I know, that the phosphate does go into the sarcoplasmic reticulum together with the calcium. However, we get about two to three hundred nanomoles of phosphate transported per miligram of protein into a vesicle and the rate of calcium transport is about the same. We know now that the entry of Ca^{2+} is electrogenic. We cannot make a statement about phosphate entry.

PHOSPHATE ION TRANSPORT IN FIBROBLAST PLASMA MEMBRANE VESICLES *

Julia E. Lever

Department of Biochemistry and Molecular Biology
The University of Texas Medical School
P.O. Box 20708
Houston, Texas 77025

Vesicles formed from biological membranes and isolated from the remainder of cellular constituents have provided a simplified system to study mechanisms of nutrient and ion transport,[1-3] compared with the complexities encountered in the use of intact cells. Transport data obtained with plasma membrane vesicles isolated from eukaryotic cells can be interpreted without ambiguities arising from transport, binding or compartmentalization of substrates and ions by intracellular membrane systems, metabolism by intracellular enzymatic systems, or effects of cell shape and attachment. Furthermore, by this technique a medium of known chemical and osmotic composition can be imposed on both the inner and outer surfaces of the membrane, facilitating the quantitative interpretation of transport data.

Clonal fibroblast cell cultures provide a more uniform source of membrane material than that available from mixed cell types in a tissue or tumor cell population, with additional advantages derived from experimental control of cell proliferative state.

Membrane vesicles isolated from the surface membrane of nontransformed and Simian virus 40 (SV40)-transformed mouse fibroblast cultures catalyze transport of several nutrients and ions, including nucleosides,[4] amino acids,[5-7] glucose,[8] calcium ion,[9] and phosphate ion.[10, 11]

This discussion will focus on the elucidation of mechanisms of active phosphate transport across the plasma membrane of animal cells using the membrane vesicle approach. Models of the phosphate ion transport cycle derived from these studies extend concepts previously proposed [12, 13] for active transport of organic nutrients via Na^+ symport mechanisms to explain active transport of an inorganic charged solute.

Isolation of Plasma Membranes

The most widely-used method for isolation of transport-competent plasma membrane vesicles from fibroblast cultures is based on a nitrogen cavitation procedure. A batch of 9×10^9 SV3T3 cells processed by this method would produce enough purified plasma membrane material (17 mg protein) for a maximum of 200 transport assays.[6] Cells are detached from the culture vessel by scraping with a rubber blade rather than using digestion with trypsin, in order to avoid degradation of surface membrane components, then washed and

* This work was supported by National Cancer Institute Grant CA05174 and U.S. Public Health Service Grants GM 25006 and GM 27055.

0077–8923/80/0341–0037 $01.75/0 © 1980, NYAS

suspended in an isotonic sucrose solution. Homogenization is achieved using a nitrogen pressurization apparatus (Artisan, Model B-plus). Cells are equilibrated with 600 to 800 lb/in^2 nitrogen gas, and then shifted back to normal atmospheric pressure after the suspension is extruded through a small orifice. As a result, closed unilaminar vesicles (0.1–0.2 μ diameter) are produced by pinching-off right-side out [14] from the plasma membrane, whereas inverted vesicles encompassing a similar size range are formed from the endoplasmic reticulum. This method offers several advantages, such as enhanced preservation of intact nuclei, mitochondria and lysozomes, the use of inert atmosphere, and prevention of local heating.

Two categories of plasma membrane preparations based on purity assessed by marker enzymes were used—purified mixed vesicles and purified plasma membrane vesicles, containing 20% and 3%, respectively, of an endoplasmic reticulum marker, NADH oxidase (EC 1.6.99.3), and less than 0.5% mitochondrial contamination evaluated by succinate-cytochrome c reductase (EC 1.3.99.1) activity. The former type of preparation can be prepared in 2–3 hr and has the advantage of 50% yield of plasma membrane material evaluated using 5′-nucleotidase (EC 3.1.3.5) and Na^+, K^+ ATPase (EC 3.6.1.3) as markers of plasma membrane material.

Phosphate Transport in Animal Cells: Coupling to Intracellular Metabolism

By contrast with phosphate carriers of mitochondrial membranes [15–19] or the anion transport system of erythrocyte membranes,[20, 21] the mechanism of active phosphate ion accumulation across the plasma membrane has been poorly understood.

Early studies of phosphate uptake in Ehrlich ascites cells suggested that a specific carrier for phosphate ion occurred in the plasma membrane.[22] Subsequent studies using intact cells suggested a relationship between Na^+ concentration and phosphate uptake. Addition of Na^+ to the cell suspension medium enhanced uptake of phosphate ion by Ehrlich cells.[23] This observation was later extended to phosphate uptake in fibroblast cultures.[24] The sigmoid dependence of phosphate uptake on Na^+ concentration suggested that more than one Na^+ participated in each transport cycle.[24] A partial contribution of Na^+, K^+, ATPase activity was implicated by observations that ouabain caused a partial inhibition of phosphate uptake in mouse fibroblast cultures under conditions in which $^{86}Rb^+$ uptake via the Na^+, K^+ ATPase system was completely inhibited.[25]

Regulation of rates of phosphate uptake plays an important role in cellular economy. Rates of phosphate uptake in resting fibroblast cultures show rapid increases when DNA synthesis is stimulated by serum or certain hormones.[25–30] Greenberg *et al.*[30] showed that phosphate uptake stimulation after mitogen addition to resting fibroblasts could be experimentally dissociated from subsequent initiation of DNA synthesis, suggesting these two events did not share a causal relationship. A similar rapid incorporation enhancement in response to mitogenic stimulation was observed accompanying blast transformation of lymphocytes.[31]

Pools of intracellular phosphate serve an important role in regulating glycolysis and oxidative phosphorylation.[32–34] It has been suggested that stimulation of phosphate uptake by serum may be accounted for by enhanced phosphate utilization in phosphorylation of small compounds rather than enhancement of the transport step.[35] Levinson and Gordon[36] had observed that 30% of the total rate of phosphate uptake in Ehrlich ascites cells was dependent on metabolic phosphorylation and sensitive to the action of metabolic inhibitors.

Phosphate Ion Transport in Plasma Membrane Vesicles

The first direct evidence for the mechanism of energy coupling of phosphate ion transport across the plasma membrane of animal cells was provided by studies using membrane vesicles from SV40-transformed mouse fibroblast cultures.[10]

As a first step in constructing these lines of evidence, it was necessary to prove that phosphate ion enters vesicles via a specific membrane component or carrier. When vesicles were loaded with arsenate, a competitive inhibitor of phosphate ion transport, stimulation of $^{32}P_i$ uptake by exchange diffusion was observed,[10] as predicted for carrier-mediated uptake. Retention of phosphate ion was sensitive to differences in osmotic pressure between the external suspension medium and the intravesicular space. This observation implied that phosphate ion was not bound to membranes, but was retained inside vesicles that shrink and swell in response to osmotic differences.

Phosphate ion was not metabolized in membrane vesicles. Over 99.9% of $^{32}P_i$ radioactivity transported into vesicles remained unchanged as inorganic phosphate and no detectable ^{32}P radioactivity was converted to nucleotides or sugar phosphates.[10] Furthermore, no covalent incorporation of ^{32}P into membranes occurred under these incubation conditions.[10] Therefore transport of phosphate ion in this cell-free system could be evaluated dissociated from complications resulting from subsequent metabolism.

Contribution of Na^+ Gradients

Active transport of phosphate ion in vesicles required a gradient of Na^+ across the membrane. It could be calculated from estimates of average intravesicular volume that $^{32}P_i$ accumulated eightfold in vesicles above its external concentration when a 50 mM NaCl gradient (external > internal) was artificially imposed across the membrane (FIGURE 1A). When NaCl was replaced by chloride, uptake was significantly decreased. Stimulation was specific for Na^+ under these conditions; K^+, Rb^+ or Li^+ were ineffective.

The property of transport stimulation by Na^+ provided an approach to prove that P_i transport in these membrane preparations represented a transport function specific to the plasma membrane. Whereas all of the subcellular membrane fractions exhibited phosphate transport activity, the subcellular distribution of the property of stimulation by Na^+ of phosphate transport activity closely paralleled the distribution of 5′-nucleotidase activity, a plasma membrane marker enzyme (TABLE 1). The property of Na^+ stimulation differentiated plasma membrane phosphate ion transport from both types of phosphate carriers of the mitochondrial inner membrane, the P_i/OH^- exchange system and the P_i/dicarboxylate exchange system, neither of which show stimulation by Na^+.[15–19]

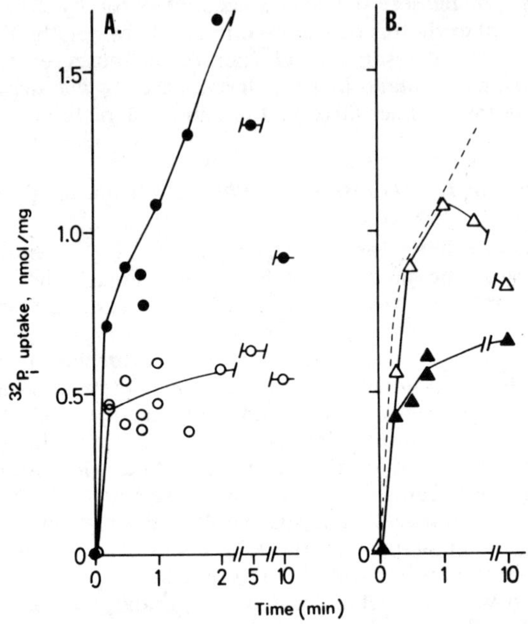

FIGURE 1. Stimulation by a Na^+ gradient of phosphate transport in membrane vesicles. *A*. Vesicles were diluted into mixtures containing 10 mM Tris-MES, pH 6.5, 0.1 mM Tris-[^{32}P-] phosphate and either 50 mM NaCl (●) or 50 mM choline chloride (○). *B*. As described for *A*., except vesicles were first incubated 15 min with either 50 mM NaCl (▲) or 50 mM KCl (△) before assay. Dashed line indicates control uptake without salt preincubation. (From Lever.[10] By permission of *Journal of Biological Chemistry*.)

TABLE 1

Na^+ STIMULATION OF PHOSPHATE TRANSPORT IN ISOLATED MEMBRANES ACCOMPANIES A PLASMA MEMBRANE MARKER DURING SUBCELLULAR FRACTIONATION OF SV 3T3 FIBROBLASTS

Fraction *	Phosphate Transport Activity NaCl	Phosphate Transport Activity Choline Cl	5′-Nucleotidase Activity
	pmol/min. mg protein		nmol/min. mg protein
Homogenate	180	92	2.8
Crude nuclear	83	47	1.3
Crude mitochondrial	180	74	6.0
Crude mixed vesicles	550	180	3.4
Purified plasma membranes	860	330	20.9
Crude endoplasmic reticulum	230	59	4.9

* Fractions were prepared and assayed as described previously.[10]

Stimulation of phosphate transport in vesicles by Na^+ was greatly decreased under conditions expected to dissipate a Na^+ gradient across the membrane. This effect was achieved either by loading vesicles with Na^+ by preincubation before the transport assay (FIGURE 1B) or by addition of an ionophore such as monensin which has affinity for Na^+ (TABLE 2). Thus, a difference in Na^+ concentration across the membrane provides a driving force for active phosphate ion transport.

Additional lines of evidence reinforced the assumption that mitochondrial phosphate transport systems did not contribute to Na^+-stimulated phosphate transport activity. Addition of succinate did not stimulate phosphate transport in plasma membrane fractions but greatly enhanced P_i transport by crude mitochondrial fractions. Thus, measurable amounts of mitochondrial inner membrane dicarboxylate/P_i exchange activity were not found in these plasma membrane vesicle preparations. Reynafarje and Lehninger[37] recently described an additional mitochondrial transport exchange pathway for phosphate and adenine nucleotides. Addition of either Mg-ATP (1 mM) or Mg-ADP (1 mM) to plasma membrane vesicles did not affect P_i transport activity. Furthermore, the inhibitors antimycin A (50 μg/ml) and oligomycin (50 μg/ml) did not alter Na^+-stimulated phosphate transport activity in vesicles. As a further difference between the mitochondrial and plasma membrane systems, the residual Na^+-independent phosphate transport activity of the plasma membrane fraction did not respond appreciably to variation in pH from 6 to 8, whereas mitochondrial transport exhibits a marked pH dependence in this pH range in the absence of Na^+.

Ouabain, an inhibitor of Na^+, K^+, ATPase activity that inhibits P_i transport in intact cells, was without effect on phosphate transport in vesicles, indicating P_i carrier activity can be dissociated from Na^+ pump activity.[10]

Although clearly a chemical gradient of Na^+ could provide a major driving force for active phosphate ion transport, the possibility exists that multiple driving forces can be utilized. The range of experimental manipulation possible using the cell-free system provided by membrane vesicles permitted the design of experiments to test this question.

Contribution of pH

Transport of phosphate ion in vesicles in the presence of Na^+ exhibited a marked pH dependence, in contrast with the relative pH-independence of Na^+-independent phosphate transport (FIGURE 2).

The pH profile for Na^+-stimulated transport shown in FIGURE 2 closely parallels the expected changes in concentration of $H_2PO_4^-$ as a function of pH. At pH values above 7, phosphate transport showed much less stimulation by Na^+ compared with replacement by choline chloride. These findings are consistent with a hypothesis that monovalent phosphate ion ($H_2PO_4^-$) is the species of orthophosphate transported across the plasma membrane in the presence of a Na^+ gradient.

An initial pH gradient across the membrane can be artificially created by suspending vesicles loaded with buffer at pH 7.5 into a solution buffered at a different pH. The question of whether such a proton gradient can contribute energetically to phosphate ion transport was investigated using ionophores. TABLE 2 shows the results of adding either FCCP, nigericin or monensin, all

TABLE 2

DIFFERENTIAL RESPONSE OF PHOSPHATE TRANSPORT TO IONOPHORES IN THE PRESENCE AND ABSENCE OF Na^+

Addition	Phosphate Transport Activity NaCl	Choline Cl
	pmol P_i/mg protein	
Control	1100 (6.1) †	220 (1.2)
Nigericin	880 (4.9)	190 (1.1)
FCCP	910 (5.1)	200 (1.1)
Monensin	720 (4.0)	210 (1.2)
Monensin+FCCP	520 (2.9)	180 (1.0)

† Numbers in parentheses refer to the ratio of the apparent intravesicular P concentration to its external concentration. Total transport in the presence of NaCl is shown without subtraction of Na^+-independent uptake. Assays were performed as described previously.[10]

ionophores capable of dissipating proton gradients, to membranes. Monensin, an ionophore that preferentially exchanges Na^+ for H^+ across the membrane, caused the greatest reduction in Na^+-stimulated phosphate ion transport. By contrast, nigericin, which greatly accelerates K^+/H^+ exchange, and FCCP, a proton-conductor, each had a lesser effect. None of the ionophores completely prevented Na^+-stimulated phosphate accumulation. Phosphate transport in the absence of Na^+ was not appreciably inhibited by these ionophores, another point of difference from the mitochondrial P_i/OH^- carrier.

Based on these observations that phosphate transport activity is reduced by conditions expected to produce Na^+ gradient dissipation—such as addition of

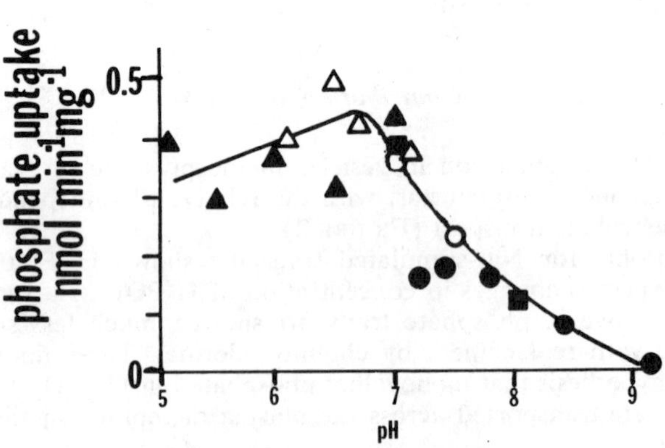

FIGURE 2. pH dependence of Na^+-stimulated phosphate ion transport in membrane vesicles. Vesicles were incubated 30 s after dilution into 50 mM NaCl, 0.1 mM Tris/[^{32}P] phosphate, and 50 mM of either Tris-HCl (●), sodium succinate (▲), glycine (■), 4-(2-hydroxyethyl)-1-piperazine-ethanesulfonic acid (○), 2-[bis(2-hydroxyethyl)-amino]-ethanesulfonic acid (△) or Tris-MES (□) at the indicated pH.

monensin or preincubation of vesicles with Na^+—compared with the small effect of proton ionophores, it can be concluded that Na^+ gradients provide the major driving force for active phosphate transport across the plasma membrane. Furthermore, dissipation of proton gradients does not prevent the transport stimulation by Na^+, although coupling of transport to Na^+ is pH-dependent, as evidenced by the pH profile shown in FIGURE 2.

Proton transport can accompany Na^+/P_i cotransport, since monovalent phosphate ion accumulated in vesicles can dissociate if the vesicle interior is more alkaline. The slight inhibition of Na^+-stimulated phosphate transport by proton-conducting ionophores may reflect such a response to artificially-imposed pH gradients. It remains to be investigated whether regulation of cytoplasmic pH can represent a physiologically significant means for modulating plasma membrane Na^+/P_i cotransport.

Contribution of Membrane Potential

A cellular mechanism for active phosphate transport must overcome the problem of accumulating a negatively-charged molecule against its own concentration gradient and opposing large magnitudes of interior-negative cellular membrane potentials. Similarly, in the cell-free system, under certain of the conditions employed in these studies using membrane vesicles, such as the imposition of chloride ion gradients across the membrane, measurable values of interior-negative membrane potentials were produced.[38]

There are several possibilities to surmount this energy barrier:

1. Transport could take place by an electrogenic mechanism mediated by binding of phosphate ion and Na^+ to the carrier resulting in formation of a complex with a net positive charge. This type of mechanism would permit the transport system to utilize the driving force derived from interior-negative membrane potentials in addition to the driving force derived from the Na^+ gradient as I have discussed previously for neutral amino acid transport.[38] Thus, in this model, Na^+-stimulated phosphate transport would be further stimulated by an interior-negative membrane potential. Based on the properties described thus far for phosphate transport, this type of model predicts that, in simplest form, two molecules of Na^+ would be required for each molecule of $H_2PO_4^-$ transported, assuming the uncomplexed carrier is electroneutral. The resulting quaternary complex would bear a single positive charge in this electrogenic mechanism.

2. Alternately, transport could be mediated by an electroneutral mechanism. In this case, Na^+/P_i transport would be unaffected by either positive or negative values of membrane potential. Cotransport of 1 Na^+ per molecule of $H_2PO_4^-$ via formation of an electroneutral ternary complex with uncharged carrier could fit such a model. Alternately, an electroneutral transport cycle can be achieved by exchange, coupled to exit of other charged species.

These possibilities were tested in the experiment shown in FIGURE 3. Known magnitudes of interior-negative membrane potential, measured by accumulation of the lipophilic cation triphenylmethylphosphonium ion shown in panel A, were imposed across the membrane by means of a valinomycin-induced K^+ diffusion potential. In control experiments, this potential was dissipated by maintaining equal K^+ concentrations on each side of the membrane.

Na^+-stimulated phosphate transport in vesicles was not affected by values of interior-negative membrane potential up to —41 mV (FIGURE 3). This value for the electrical potential difference across the vesicle membrane was calculated from the fivefold ^{3}H-triphenylmethylphosphonium ion ($TPMP^+$) accumulation shown in FIGURE 3. Using the Nernst equation, membrane potential, $\Delta\psi$, was estimated from permeant cation distribution as follows:

$$\Delta\psi = -58.8 \log (TPMP^+_{in}/TPMP^+_{out})$$

This value of $\Delta\psi$ achieved artificially in vesicles is higher than the range of values estimated from microelectrode measurements in living cells. Lassen *et al.*[39] reported a value of —24 mV for Ehrlich ascites tumor cells. Similarly, phosphate transport was not affected by interior-positive membrane potentials, achieved by reversing the direction of K^+ diffusion potentials in an experiment analogous to that shown in FIGURE 3.

The properties of phosphate transport demonstrated in these studies can be summarized in terms of the working hypothesis shown in FIGURE 4, in which active phosphate transport across the plasma membrane is mediated by an electroneutral $Na^+/H_2PO_4^-$ cotransport. This model predicts a stoichiometry of 1 Na^+ per molecule of phosphate transported.

This mechanism is analogous to the Na^+ cotransport mechanism proposed to explain active transport of *A* system neutral amino acids [12, 13] and verified in studies using membrane vesicles.[6, 38] However, these mechanisms differ in at least one respect, which is that amino acid transport is electrogenic, stimulated

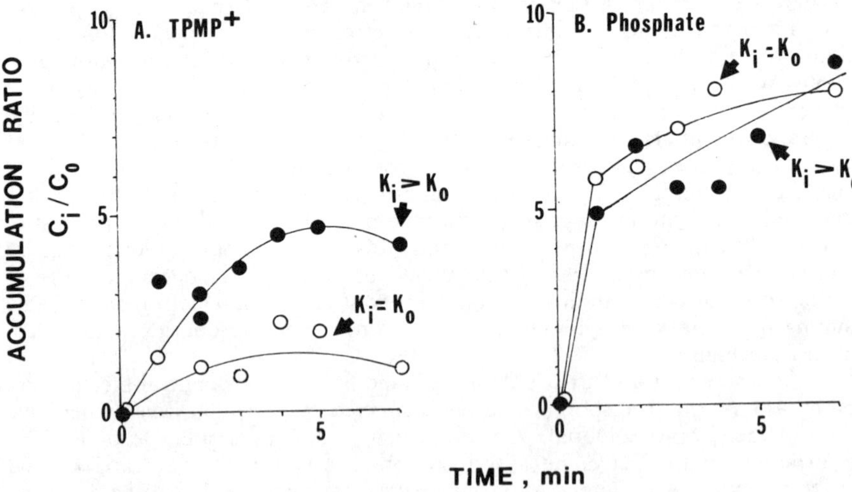

FIGURE 3. Effect of artificial generation of an interior-negative membrane potential on Na^+-stimulated phosphate ion accumulation in membrane vesicles from SV3T3. Vesicles were first incubated with 50 mM KCl, 50 mM choline Cl and valinomycin (100 μg/ml). Then, vesicles were diluted tenfold into a solution containing 50 mM NaCl, 10 mM Tris-MES, pH 6.5, 5 mM $MgCl_2$ and either 5 mM KCl plus 50 mM choline Cl ($K_i^+ > K_o^+$) (●) or 50 mM KCl plus 5 mM choline Cl ($K_i^+ = K_o^+$) (○). In panel *A*, uptake of 23 μM ^{3}H-TPMP$^+$ is shown. In panel *B*, uptake of 0.1 mM [^{32}P]-Tris-phosphate is shown.

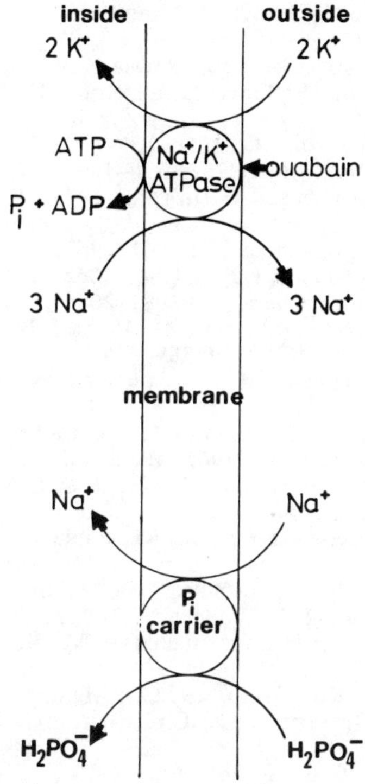

FIGURE 4. A $Na^+/H_2PO_4^-$ symport model for active phosphate ion transport across the plasma membrane.

by interior-negative membrane potentials, whereas phosphate transport, unaffected by membrane potential, must occur via an electroneutral mechanism at this pH. However, since both phosphate and *A* system neutral amino acid transport systems are energized by Na^+ gradients, it can be predicted transport of *A* system amino acids can dissipate cellular capacity for Na^+-stimulated phosphate transport. This type of competition between different transport systems mediated by a common mode of coupling to a Na^+ gradient has been demonstrated for intestinal transport systems catalyzing Na^+-stimulated transport of glucose and amino acids.[40] Evidence for Na^+-stimulated phosphate transport has also been obtained using intestinal membranes.[41]

Since $H_2PO_4^-$ is a proton-donating anion, in contrast to chloride, phosphate transport and distribution across the plasma membrane may also play a role in regulating cytoplasmic pH in response to Na^+ gradients generated by Na^+, K^+, ATPase activity.

REFERENCES

1. KABACK, H. R. 1974. Science **186:**882–892.
2. LEVER, J. E. 1980. *In* CRC Critical Reviews in Biochemistry. G. Fasman, Ed. CRC Press. Palm Beach, Fla. **7:**187–246.

3. Hopfer, U., K. Sigrist-Nelson & H. Murer. 1975. Ann. N.Y. Acad. Sci. **264:** 414–427.
4. Hochstadt, J., D. C. Quinlan, R. L. Rader, C.-C. Li & J. D. Dowd. 1975. *In* Methods in Membrane Biology. E. D. Korn, Ed. **5:**117–162 Plenum Press. New York, N.Y.
5. Lever, J. E. 1976. Proc. Natl. Acad. Sci. USA **73:**2614–2618.
6. Lever, J. E. 1977. J. Biol. Chem. **252:**1990–1997.
7. Hamilton, R. T. & M. Nilsen-Hamilton. 1976. Proc. Nat. Acad. Sci. USA **73:**1907–1911.
8. Lever, J. E. 1979. J. Biol. Chem. **254:**2961–2967.
9. Moore, L. & I. Pastan. 1977. J. Biol. Chem. **252:**6304–6309.
10. Lever, J. E. 1978. J. Biol. Chem. **253:**2081–2084.
11. Hamilton, R. T. & M. Nilsen-Hamilton. 1978. J. Biol. Chem. **253:**8247–8256.
12. Crane, R. K. 1960. Physiol. Rev. **40:**789–825.
13. Riggs, T. R., L. M. Walker & H. N. Christensen. 1958. J. Biol. Chem. **233:** 1479–1484.
14. Olden K., M. Willingham & I. Pastan. 1976. Cell **8:**383–
15. Chappell, J. B. & A. R. Crofts. 1965. Biochem. J. **95:**393–402.
16. Hoek, J. B., N. E. Lofrumento, A. J. Mayer & J. M. Tager. 1971. Biochim. Biophys. Acta **226:**297–308.
17. Palmieri, F., E. Quagliarello & M. Klingenberg. 1970. Eur. J. Biochem. **17:**230–238.
18. Chappell, J. B. & K. N. Haarhoff. 1967. *In* Biochemistry of Mitochondria. E. C. Slater, Z. Kaniuga & L. Wojtczak, Eds.: 75. Academic Press. London.
19. Palmieri, G. Prezioso, E. Quagliarello & M. Klingenberg. 1971. Eur. J. Biochem. **22:**66–74.
20. Cabantchik, Z. I. & A. Rothstein. 1972. J. Membr. Biol. **10:**311–330.
21. Wolosin, J. M., H. Ginsberg & Z. I. Cabantchik. 1977. J. Biol. Chem. **252:** 2419–2427.
22. Levinson, C. 1972. J. Cell. Physiol. **79:**73–78.
23. Scholnick, P., D. Lang & E. Racker. 1973. J. Biol. Chem. **248:**5175–5182.
24. Brown, K. D. & J. F. Lamb. 1975. J. Physiol. (London) **251:**58P–59P.
25. Lever, J. E., D. Clingan & L. Jimenez de Asua. 1976. Biochem. Biophys. Res. Commun. **71:**136–143.
26. Weber, M. J. & G. Edlin. 1971. J. Biol. Chem. **246:**1828–1833.
27. Cunningham, D. D. & A. B. Pardee. 1969. Proc. Natl. Acad. Sci. USA **64:** 1049–1056.
28. Hilborn, D. A. 1975. J. Cell. Physiol. **87:**111–122.
29. Jimenez de Asua, L., E. Rozengurt & R. Dulbecco. 1974. Proc. Nat. Acad. Sci. USA **71:**96–98.
30. Greenberg, D. B., G. S. Barsh, T.-S. Ho & D. D. Cunningham. 1976. J. Cell. Physiol. **90:**193–210.
31. Fisher, D. B. & G. C. Mueller. 1968. Proc. Nat. Acad. Sci. USA **60:**1396.
32. Lehninger, A. L. 1970. Biochem. J. **119:**129–138.
33. Lehninger, A. L. 1974. Proc. Nat. Acad. Sci. USA **71:**1520–1524.
34. Koobs, D. H. 1972. Science **178:**127–133.
35. Jullien, M. & L. Harel. 1976. Exp. Cell. Res. **97:**23–30.
36. Levinson, C. & E. D. Gordon. 1971. J. Cell. Physiol. **78:**257–264.
37. Reynafarje, B. & A. L. Lehninger. 1978. Proc. Nat. Acad. Sci. USA **75:** 4788–4792.
38. Lever, J. E. 1977. Biochemistry **16:**4328–4334.
39. Lassen, U. V., A.-M. T. Nielsen, L. Pape & L. O. Simonsen. 1971. J. Membrane Biol. **6:**269–288.
40. Murer, H., K. Sigrist-Nelson & U. Hopfer. 1975. J. Biol. Chem. **250:**7392–7396.
41. Berner, W., R. Kinne & H. Murer. 1976. Biochem. J. **160:**467–474.

Discussion

M. Colombini (*Albert Einstein College of Medicine, Bronx, N.Y.*): Do you attribute the pH effect on phosphate transport to a dissociation constant of the phosphate molecule itself?

J. Lever: Yes, this is just a working model at the moment because we have not rigorously excluded other possibilities.

Colombini: Would you then explain its effect on the accumulation in the steady state.

Lever: We are not really looking at a steady state. If the system were at true steady state, we would have no accumulation since we are starting with an artificially-imposed gradient that is dissipating as a function of time. So at true equilibrium, the gradient would be completely dissipated and we would have no phosphate accumulation. What I refer to as quasi-steady state (after 15 min incubation) probably has an appreciable contribution derived from rates of uptake.

W. S. Rehm: In what form do you think the phosphate goes in?

Lever: In the present working model, phosphate enters in the $H_2PO_4^-$ form—the monovalent anion.

THE SEPARATION OF APICAL FROM BASAL-LATERAL PLASMA MEMBRANES OF EPITHELIAL CELLS: A TOOL TO IDENTIFY TRANSPORT SYSTEMS *

Eva Kinne-Saffran and Rolf Kinne

Max-Planck-Institut für Biophysik
Heinrich-Hoffmann-Strasse 7
D-6000 Frankfurt am Main 71, West Germany

For the last two years the relation between membrane bound ATPase and ion transport in the kidney have been of prime interest to us. Since studies on isolated whole plasma membrane envelopes of epithelial cells provide more general information on membrane transport processes and do not reveal the cellular locus of the process, we attempted a separation of luminal and contraluminal membranes of the proximal tubule of renal cortex. In the case of this tissue, a separation by density gradient centrifugation yielded in brush border membrane fractions with an appreciable degree of $Na^+ + K^+$-ATPase activity. It was the use of the free flow electrophoresis that made it possible to separate the membranes of the two cell poles from each other and from other cell organelles, such as mitochondria and endoplasmic reticulum.[1]

The advantage of a separation like this allows one to examine two criteria that indicate whether an enzyme may participate in the ion transfer across the plasma membrane; namely, its localization in the plasma membrane itself and its distribution between the basal and the apical part of the cell.

In FIGURE 1, the distribution of $Na^+ + K^+$-ATPase is shown as a marker enzyme for basal-lateral membranes and the distribution of alkaline phosphatase is shown as a marker enzyme for brush border membranes after separation of a crude plasma membrane fraction by free flow electrophoresis. In the lower part of this figure, the distribution of a Ca^{++}-ATPase is shown, which follows the $Na^+ + K^+$-ATPase,[2] and of a Mg^{++}-ATPase stimulated by anions with a same distribution pattern as alkaline phosphatase.[3] Thus, a Ca^{++}-ATPase is located exactly where an active step is required in the transepithelial transport of calcium, and a Mg^{++}-ATPase stimulated by anions is present in the luminal membrane where active proton transport is thought to take place. The differences observed in fractions 30 to 37 between $Na^+ + K^+$-ATPase and Ca^{++}-ATPase, on one hand, and between alkaline phosphatase and the Mg^{++}-ATPase are most probably due to the contribution of mitochondria to both enzyme activities, since mitochondria accumulate in these fractions as shown in FIGURE 2. The activity of the succinic dehydrogenase is already reduced in the crude plasma membrane fraction to at least 10%, compared with the activity found in the homogenate before the plasma membranes are injected into the electrical field. Furthermore, for all studies with basal-lateral or brush border membranes only the fractions starting with the maximal activity and then collected to the left or to the right have been used, so that the contamination with mitochondria is really reduced to a negligible amount.

* This paper has been supported by the Deutsche Forschungsgemeinschaft, Grant Fa 48/21.

0077–8923/80/0341–0048 $01.75/0

In the following, we want to describe our attempts to measure a direct coupling between ATP hydrolysis and ion transport. In micropuncture studies, it has been demonstrated that in the rat kidney proximal tubule active Ca^{++} reabsorption occurs.[4] Due to the low intracellular concentration of free calcium, it is obvious that the exit of calcium from the cell is the step that requires energy. With the isolated basal-lateral plasma membranes at hand, and with

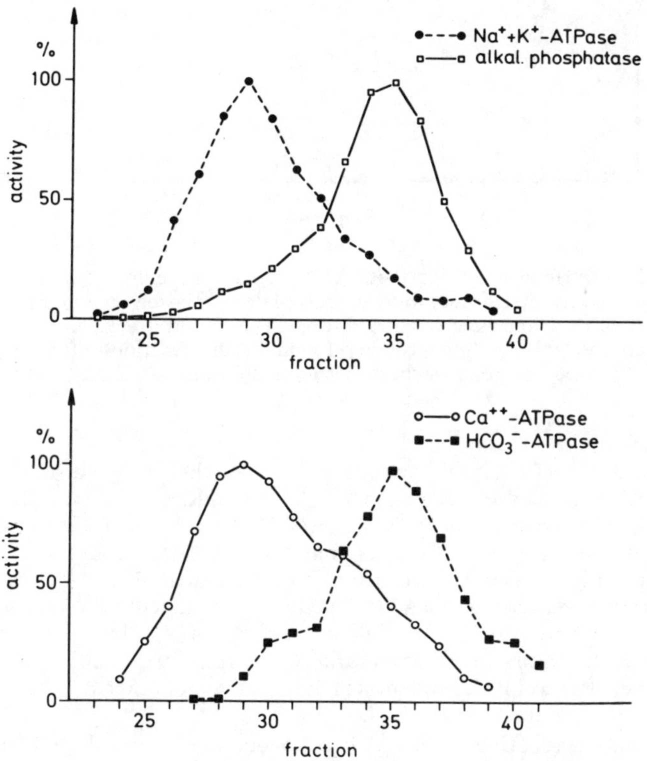

FIGURE 1. Distribution of $Na^{+}+K^{+}$-ATPase, Ca^{++}-ATPase, alkaline phosphatase and HCO_3^--ATPase activity after separation of basal-lateral plasma membranes from brush border microvilli by free flow electrophoresis. The values are given as percent of the maximal activity (mU/ml) found in the fractions after electrophoresis separation of purified rat renal cortical plasma membranes.[1] (From Kinne-Saffran and Kinne.[2, 3])

the awareness of a contraluminal localization of a Ca^{++}-ATPase, we studied the role of ATP in calcium transport of the membranes.

Basal-lateral membranes vesiculate spontaneously into right-side out and into inside-out oriented vesicles that make it possible to study an ATP driven Ca^{++} transport into membrane vesicles by addition of external ATP because the ATP hydrolyzing site of the Ca^{++}-ATPase is present on the outer surface of the reversed or inside-out vesicles.

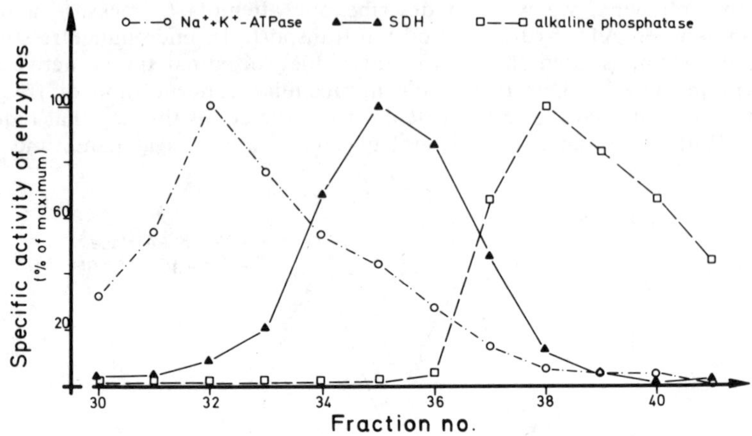

FIGURE 2. Distribution of $Na^{+}+K^{+}$-ATPase, succinic dehydrogenase and alkaline phosphatase activity after separation of basal-lateral plasma membranes from brush border microvilli by free flow electrophoresis. The values are given as percent of the maximal activity (mU/ml/mg protein) found in the fractions after electrophoresis separation of purified rat renal cortical plasma membranes.

FIGURE 3 shows the rapid uptake of Ca^{++} in the presence of ATP into the vesicles, which is unobserved when no ATP is added, or when the ionophore A 23187 is added simultaneously with ATP. The lack of Ca^{++} uptake in the presence of the ionophore suggests that the ATP-driven uptake occurs most probably as an intravesicular accumulation, which is abolished by the ionophore. The distribution pattern of an ATP-dependent Ca^{++} accumulation and of Na^{+} + K^{+}-ATPase activity during free flow electrophoresis are very similar and, furthermore, inhibitors of mitochondrial Ca^{++} transport, such as oligomycin do not influence the ATP-dependent calcium movement across the basal-lateral membranes significantly.

These data and other data not mentioned here [5] provide evidence for the existence of an ATP-driven transport system for calcium in the basal-lateral membranes of rat proximal tubular epithelial cells. This system is probably of minor importance for the transepithelial transport of calcium that is sodium-dependent, but justifies our conclusions about the presence of an active calcium transport system in the basal-lateral membranes from the distribution of a Ca^{++}-ATPase.

Now we would like to turn to the question of whether H^{+} secretion in the proximal tubule of the kidney is exclusively driven by Na^{+} ions entering from the lumen into the cell, or if there is an additional action of an ATPase, possibly the Mg^{++}-ATPase stimulated by anions that are located in the brush border membrane.

Recently, physiological data derived from micropuncture and electrophysiological studies in our institute provided evidence for the existence of two different mechanisms involved in the process of acidification in the kidney.[6, 7] This encouraged us again to investigate the involvement of the enzyme in H^{+} transport over the luminal membrane in an analogy to the ATPase present in mitochondria that are capable of performing ATP-driven proton transport. On the other hand, it has been just the existence of a mitochondrial Mg^{++}-ATPase that has caused controversy about the exact cellular localization or about the existence of an intrinsic brush border membrane-bound anion stimulated Mg^{++}-ATPase in general.[8]

In addition to the evidence for the existence of a Mg^{++}-ATPase stimulated by anions in the brush border membrane, such as parallel distribution of the enzyme and alkaline phosphatase; and parallel enrichment during the free flow electrophoresis, the use of an inhibitor would be an excellent tool to elucidate the role of the enzyme in transepithelial transport, as in the case of ouabain, $Na^+ + K^+$-ATPase and sodium transport by erythrocytes.

Inspired by the work of De Kruijff[9] from the Netherlands, we tested the effect of filipin, a polyene antibiotic, on the anion-stimulated Mg^{++}-ATPase activity of a mitochondrial fraction and on the Mg^{++}-ATPase of brush border membranes. Filipin forms complexes with cholesterol and cholesterolesters, and it should therefore be expected to act exclusively on the cholesterol-rich brush

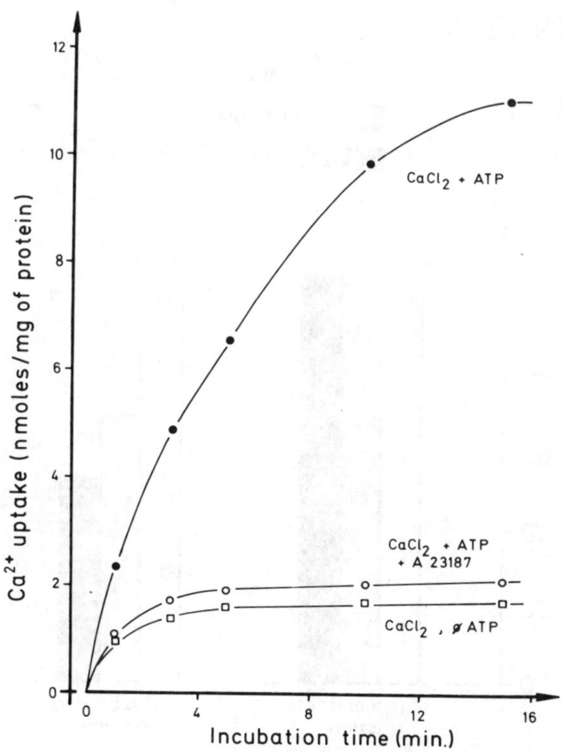

FIGURE 3. Influx of Ca^{++} into basal-lateral membrane vesicles: with (●) or without (□) 5 mMATP (di-TRIS salt). ○, influx in the presence of ATP and ionophore A 23187 (1 μg/mg protein). (From Gmaj, Murer & Kinne.[5])

border membrane, and not on the cholesterol-poor inner mitochondrial membrane. As can be seen in FIGURE 4, this was found to be the case. Filipin inhibits only the Mg^{++}-ATPase in the luminal membrane fraction, and has no influence on the enzyme activity of mitochondria.[10] Thus, filipin can be regarded as a *membrane*-specific inhibitor in contrast to *enzyme*-specific inhibitors such as ouabain.

Brush border microvilli disrupted from the cell seem to vesiculate at a very early step and they vesiculate to a very high degree into right-side out vesicles. To investigate a possible involvement of the ATPase in ion transport, the vesicles had to be loaded with ATP in order to have it at the active site of the enzyme; namely, on the cytoplasmic face of the membrane. The loading of the brush border vesicles with ATP was achieved by adding ATP, magnesium, phosphoenolpyruvate, pyruvic kinase, potassium and oligomycin to the medium in which the rat kidney cortex is homogenized initially. As determined by the luciferin-luciferase method, the isolated membrane fraction contains approximately 2 nmoles ATP per mg membrane protein. Evidence that we deal indeed with ATP, which is present in the intravesicular space, has been derived from two different experiments. First, even repeated washing of the membrane suspension did not alter the amount of ATP found in the membrane fraction, and second, the addition of hexokinase and glucose to the outside of the

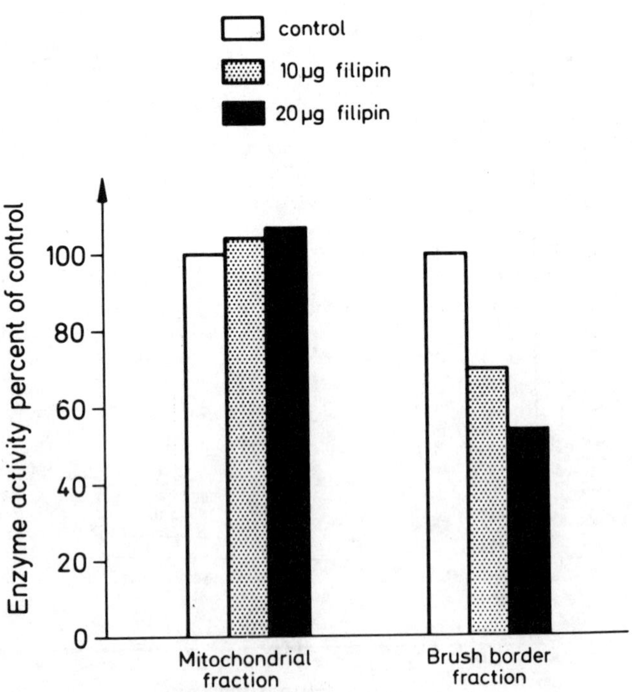

FIGURE 4. Effect of filipin on the Mg^{++}-ATPase activity of mitochondria and of brush border membranes isolated from rat kidney cortex. The enzyme activity in the absence of filipin is plotted as 100%.

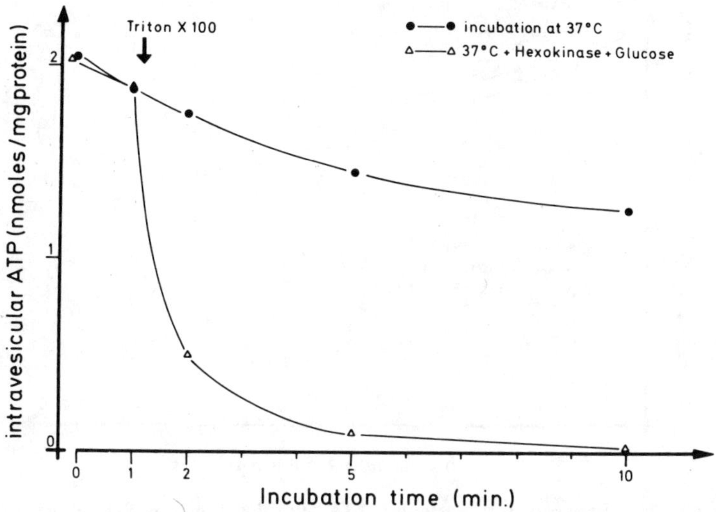

FIGURE 5. Hydrolysis of intravesicular ATP (nmoles/mg protein) during incubation at 37° C in the absence (●) of hexokinase and glucose and in the presence (△) of hexokinase (50 μg/mg protein) and 20 mM glucose. At ↓ Triton X 100 (0.5%) is added to the membrane suspension.

vesicles, which should have abolished any ATP present at the outside, did not alter the ATP content of the membrane fraction.

In FIGURE 5, the ATP content of the vesicles is compared in the absence of and in the presence of hexokinase and glucose at the outside of the vesicles. Within the first minute of incubation at 37° C there is no significant difference to the control. If, however, the vesicles are opened up by Triton X 100, ATP is released and enters the reaction with glucose and hexokinase, as indicated by the sharp decrease in ATP one or two minutes after the addition of Triton. These conditions allowed us now to study the ATPase reaction in the vesicles by measuring the intravesicular ATP level under different experimental situations.

FIGURE 6 shows the effect of incubations at different temperatures. When the vesicles are incubated at 0° C, the ATP content remains essentially constant, whereas incubation at 37° C leads to a decrease in ATP. This decrease can be inhibited by filipin, indicating that the decrease in ATP is due to the action of an ATPase located in a cholesterol-rich membrane, the brush border membrane. The rate of the ATPase reaction as judged by the decrease in vesicular ATP is not constant. It is faster in the first minutes and slows down in the later period of the experiment. The same has been observed with vesicles of the inner mitochondrial membrane and has been explained in the following way: an electrogenic proton ATPase generates an electrical potential difference and a pH gradient across the membrane that subsequently inhibits the ATPase reaction. The inhibition can be overcome by the addition of a protonophor, which facilitates the dissipation of the electrochemical potential difference for protons across the membrane and therefore stimulates the ATPase activity.[11] A similar phenomenon can be observed with brush border membrane vesicles (FIGURE 7).

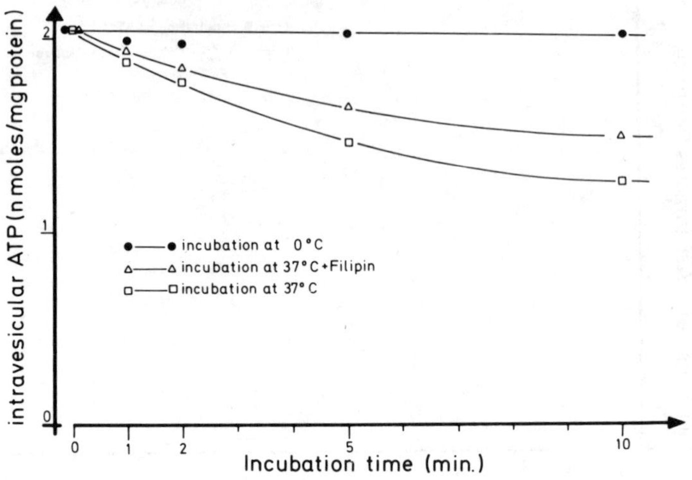

FIGURE 6. Hydrolysis of intravesicular ATP (nmoles/mg protein): effect of incubation at 0° C (●), 37° C (□). △, effect of filipin (100 μg/mg protein) during incubation at 37° C.

When carbonyl cyanide p-trifluoromethoxyphenylhydrazone (CFCCP) is added to the vesicles, the rate of the ATP hydrolysis increases strikingly; again, this ATPase reaction is inhibited by filipin. Since all experiments are performed in the presence of oligomycin—to inhibit any residual mitochondrial ATPase activity—and in the presence of ouabain—to inhibit any $Na^+ + K^+$-ATPase

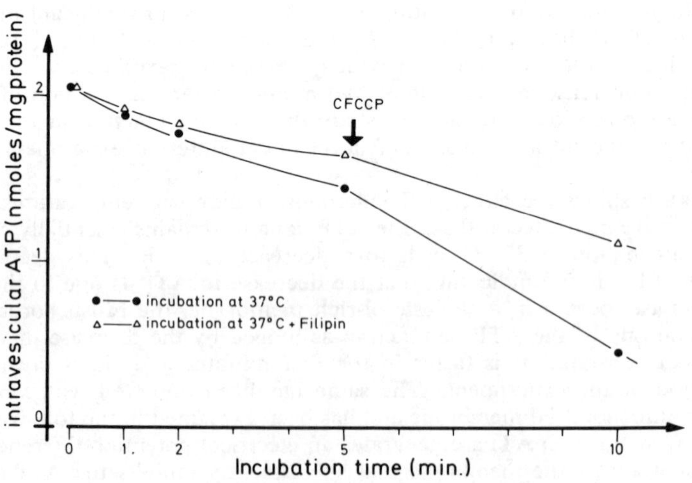

FIGURE 7. Hydrolysis of intravesicular ATP (nmoles/mg protein) during incubation at 37° C in the absence (●) of filipin and in the presence (△) of filipin (100 μg/mg protein). At ↓ CFCCP (5 μg/mg protein) was added.

activity—we would like to interpret the results as evidence for the presence of an electrogenic ATP-driven proton pump in the brush border membranes.

Conclusion

The distribution of ATPases stimulated by ions has led in the past to assumptions about an identical distribution of ATP driven ion transport systems. These assumptions now can be supported by the use of membrane vesicles derived from the luminal and the contraluminal membrane. Thus, the separation of the membranes from the two cell poles has led to the cellular localization and characterization not only of sodium gradient dependent transport systems,[12] but also of ATP dependent transport systems.

Acknowledgments

The authors thank Mrs. B. Moewes for her excellent artwork and Professor Klaus Ring for his continuous support in the ATP measurements.

References

1. Heidrich, H.-G., R. Kinne, E. Kinne-Saffran & K. Hannig. 1972. J. Cell Biol. **54:**232–245.
2. Kinne-Saffran, E. & R. Kinne. 1973. J. Membr. Biol. **17:**263–274.
3. Kinne-Saffran, E. & R. Kinne. 1974. Proc. Soc. Exper. Biol. Med. **146:**751–753.
4. Ullrich, K. J., G. Rumrich & S. Klöss. 1976. Pflügers Arch. **364:**223–228.
5. Gmaj, P., H. Murer & R. Kinne. 1978. Biochem. J. **178:**549–557.
6. Frömter, E. 1974. Fortschr. Zool. **23:**248–257.
7. Ullrich, K. J., G. Capasso, G. Rumrich, F. Papavassiliou & S. Klöss. 1977. Pflügers Arch. **368:**245–252.
8. van Amelsvoort, J. M. M., J. J. H. H. M. de Pont, A. L. H. Stols & S. L. Bonting. 1977. Biochim. Biophys. Acta **471:**78–91.
9. De Kruijff, B. & R. A. Demel. 1974. Biochim. Biophys. Acta **339:**57–70.
10. Kinne-Saffran, E. & R. Kinne. 1979. J. Membr. Biol. **49:**235–251.
11. Kozlov, J. A. & V. P. Skulachev. 1977. Biochim. Biophys. Acta **463:**29–89.
12. Kinne, R. & I. L. Schwartz. 1978. Kidney Int. **14:**547–556.

Discussion

E. Racker: Are you proposing a plasma membrane calcium pump that is responsible for the pumping out of the calcium?

E. Kinne-Saffran: Yes.

Racker: Did you try calmodulin on your pump? This is a calcium binding protein that is required for the calcium pump in red blood cells.

Kinne-Saffran: No.

Racker: Is there bicarbonate stimulation?

Kinne-Saffran: It is a magnesium ATPase that has one of the properties stimulated by anions, such as bicarbonate, but I don't think that we can postulate that this enzyme, because it is stimulated by the bicarbonate, is involved in the transport of bicarbonate.

RACKER: But have you ruled out that what you observed is due to a mitochondrial system, that there's no cytochrome oxidase in your particle?

KINNE-SAFFRAN: Yes.

RACKER: So you've ruled out mitochondrial contamination.

KINNE-SAFFRAN: Yes.

P. R. STEINMETZ (*University of Iowa, Iowa City, Iowa*): It was interesting that the presumed proton translocating ATPase was oligomycin-insensitive and thereby different, from mitochondrial ATPases.

KINNE-SAFFRAN: With high concentration of oligomycin, there will be about 10% inhibition.

STEINMETZ: This goes along with some observations in turtle bladder in which there is anaerobic proton transport that can be inhibited by DCCD, but is not inhibited by oligomycin. Is your transport system inhibited by DCCD?

KINNE-SAFFRAN: I haven't done it yet.

W. S. REHM: Have you tried azide?

KINNE-SAFFRAN: Azide is ineffective.

COUPLING OF ION FLOWS IN CELL SUSPENSION SYSTEMS *

P. Geck

Gustav Embden Z.B.C.
J. W. Goethe University
6000 Frankfurt 70, West Germany

E. Heinz

Department of Physiology
Cornell University Medical College
1300 York Avenue
New York, New York, 10021

A reliable test for cotransport is the "degree of coupling," q, between the solute flows concerned, as defined in terms of Thermodynamics of Irreversible Processes. E.g., for the coupling between the flows of solute i and j, the phenomenological flow equations are:

$$J_i = L_{ii}X_i + L_{ij}X_j \qquad \textbf{(1a)}$$

$$J_j = L_{ij}X_i + L_{jj}X_j \qquad \textbf{(1b)}$$

and the degree of coupling is

$$q = \frac{L_{ij}}{\sqrt{L_{ii} \cdot L_{jj}}} \qquad \textbf{(2a)}$$

For tight stoichiometric coupling the intrinsic q, i.e., that of the truly coupled flows, should be unity. The experimental q may, however, be much lower, depending on the leakage flows. This can be visualized by the "quasichemical" notation, which tries to separate between coupled and leakage flows:

$$q = \frac{\nu_i \nu_j L_r}{\sqrt{(\nu_i^2 L_r + L_i^u)\ (\nu_j^2 L_r + L_j^u)}} \qquad \textbf{(2b)}$$

in which L_r refers to the truly coupled flow, L_i^u and L_j^u to the leakage flows of i and j, respectively. ν_i and ν_j are stoichiometric coefficients.

The degree of coupling, q, is conveniently assessed by the "pulse-response" procedure according to the following equation:

$$q^2 = \left(\frac{\partial J_i}{\partial J_j}\right)_{X_i} \cdot \left(\frac{\partial J_i}{\partial J_j}\right)_{X_j} \qquad \textbf{(3)}$$

that is derived from the equations given above. The procedure is based on the response of each flow J_i to a pulse imposed on the other, under the strict condition that the electrochemical potential difference of the responding solute species, X_i or X_j, respectively, is kept constant. This condition may be met with suffi-

* Supported by Deutsche Forschungsgemeinschaft Grant No. HE 102/14 and by USPHS, NIH Grant No. RO1 GM 26554–01.

0077–8923/80/0341–0057 $01.75/0

cient accuracy if at least one of the two solutes is electroneutral, but poses great difficulties when both solutes are free ions. This is because the pulse of an ion flow is usually produced by a sudden increase of concentration gradient, which is bound to change the electrical potential, and hence the electrochemical potential difference of the responding species, unless this is prevented by external potentiostat device, as it is not possible with a cell supension. Ignoring these pd. changes, one obtains false q-values, which would indicate electrostatic interactions between the ion flows concerned rather than true coupling.

The above difficulty does, however, not apply to the extent that the coupled ion flows are nonrheogenic, i.e., if they are forcibly and electroneutrally linked to other ion flows, via symport and antiport. A nonrheogenic ion flow does not respond to an electrical pd. change, nor should the pulse of a nonrheogenic ion flow induce such a change. The true degree of coupling can therefore be obtained for the coupling between such ion flows by the pulse-response method, while only the chemical pd., not the electrical pd., need be maintained constant; provided however, that rheogenic leakage flows of the ions concerned can be eliminated or minimized, e.g., by specific inhibitors of the coupled ion flows. If such an inhibitor is specific enough not to affect the leakage permeability of the membrane, leakage flows can be "filtered out" by subtracting the residual flow during inhibition from the control flow. The difference flows thus obtained can then be tested for coupling by the pulse-respond method without regard of electrical pd. changes.

The above principles could be verified for the furosemide-inhibitable portions of the flow of Na^+, K^+ and Cl^- in Ehrlich cells.[3] These portions are easily obtained from the difference between flows in the presence and absence of maximally inhibiting furosemide concentrations, preferably measured in the presence of ouabain in order to exclude interference by the Na, K pump.

The furosemide-inhibitable flows of all three ions, measured individually at constant concentration differences were found to be insensitive towards changes in electrical potential, whether such changes are induced by the gradual depolarization of the plasma membrane by valinomycin, or by gradual inhibition of the electrogenic pump by ouabain (TABLE 1).[4] Hence these flows are nonrheogenic, and rheogenic leakage components must be negligible. This tells us only that each of the three flows concerned is coupled to an electrically-equivalent flow of other ion species, but not whether and to what extent they are coupled to each other. To answer this question, the degree of coupling, q has been determined by the pulse-response procedure for each combination of two among the three flows as follows: In each of three sets of experiments two ion species were kept at constant concentrations, whereas the third one was "pulsed" by subjecting it to a sudden rise of its concentration difference. The results are shown in FIGURE 1. It is seen that there is a linear relationship between each response flow and the corresponding pulse flow for all three possible combinations. The q values obtained are listed in TABLE 2. None of them is significantly different from unity. These results show that all three furosemide-sensitive flows tested are tightly coupled to each other, and that neither of them is significantly coupled to any other flow and that leakage is negligible for each of them. The signs of the slopes further indicate that the coupled flows are all in the same direction. Since all furosemide-sensitive flows are electrically silent, it follows that

$$\nu_1 + \nu_2 - \nu_3 = 0 \tag{4}$$

TABLE 1

INFLUENCE OF CHANGES IN ELECTRICAL POTENTIAL DIFFERENCE ON FUROSEMIDE-SENSITIVE TRANSPORT *

	a	b
$(\partial \ln J_{Na}/\partial \ln R_{TPP})_{[Na^+]}$	−0.03±0.32	0.10±0.13
$(\partial \ln J_K/\partial \ln R_{TPP})_{[K^+]}$	−0.03±0.02	0.02±0.01
$(\partial \ln J_{Cl}/\partial \ln R_{TPP})_{[Cl^-]}$	−0.09±0.01	0.12±0.05
$(\partial \ln_{H_2O}/\partial \ln R_{TPP})$	−0.11±0.06	0.12±0.21

* The ascites were harvested from mice six days after transplantation. The cells after three washings in K^+-free Krebs-Ringer-phosphate buffer pH 7.4 with 1% albumin are preincubated for 20 min in the cold and thereafter for 20 min at 37° C in this K^+-free buffer. After a further 5 min preincubation in new medium at 37° C, the incubation was started by addition of K^+ (30 mM). After 5 min, the incubation was terminated by cooling the suspension to −4° C. The electrical potential difference (pd) was varied in two different ways: By altering the rate of the Na^+, K^+-pump using graded concentrations of ouabain, and by varying the extracellular glycine concentration to gradually shunt the pump potential via the rheogenic Na^+-glycine-cotransport. The potential dependences are not significantly different from zero: (A symmetrical Eyring's mechanism for transport of monovalent ions would lead to a slope of 0.5).

ν_1, ν_2, ν_3 being the stoichiometric coefficients of Na, K, and Cl, respectively, within the coupled flow. As there are infinite stoichiometric ratios that satisfy this equation, the determination of the stochiometry of the coupling requires additional information, which is already available from the slopes of the pulse-response curves. Having ascertained that all three flows are tightly coupled, we may base the further analysis on the quasichemical notation of Nonequilibrium Thermodynamics.[2] Accordingly we treat the furosemide-sensitive flows of the three ions as belonging to the same coupled process:

$$\nu_1\ \mathrm{Na}' + \nu_2\ \mathrm{K}' + \nu_3\ \mathrm{Cl}' \rightarrow \nu_1\ \mathrm{Na}'' + \nu_2\ \mathrm{K}'' + \nu_3\ \mathrm{Cl}'' \qquad \textbf{(5)}$$

with the rate

$$J_r = L_r\ (\nu_1 X_{\mathrm{Na}} + \nu_2 X_{\mathrm{K}} + \nu_3 X_{\mathrm{Cl}}) \qquad \textbf{(6)}$$

Hence the individual flows are

$$J_{\mathrm{Na}} = \nu_1 J_r + J_{\mathrm{Na}}{}^{\mathrm{u}} \qquad \textbf{(7a)}$$

$$J_{\mathrm{K}} = \nu_2 J_r + J_{\mathrm{K}}{}^{\mathrm{u}} \qquad \textbf{(7b)}$$

$$J_{\mathrm{Cl}} = \nu_3 J_r + J_{\mathrm{Cl}}{}^{\mathrm{u}} \qquad \textbf{(7c)}$$

As $J_{Na}{}^{u}$, $J_K{}^{u}$, $J_{Cl}{}^{u}$, the (uncoupled) leakage flows, can be neglected as far as the furosemide-sensitive flows are concerned, we obtain

$$J_{\mathrm{Na}} = \nu_1{}^2 L_r X_{\mathrm{Na}} + \nu_1\nu_2 L_r X_{\mathrm{K}} + \nu_1\nu_3 L_r X_{\mathrm{Cl}} \qquad \textbf{(8a)}$$

$$J_{\mathrm{K}} = \nu_1\nu_2 L_r X_{\mathrm{Na}} + \nu_2{}^2 L_r X_{\mathrm{K}} + \nu_2\nu_3 L_r X_{\mathrm{Cl}} \qquad \textbf{(8b)}$$

$$J_{\mathrm{Cl}} = \nu_1\nu_3 L_r X_{\mathrm{Na}} + \nu_2\nu_3 L_r X_{\mathrm{K}} + \nu_3{}^2 L_r X_{\mathrm{Cl}} \qquad \textbf{(8c)}$$

Further information concerning the stoichiometry is obtained from the slopes of FIGURE 1, the same ones that have been used to derive the *q* values. These slopes are listed in TABLE 3.

Each of the ratios $\frac{\nu_2}{\nu_1}$, $\frac{\nu_3}{\nu_1}$, and $\frac{\nu_3}{\nu_2}$ can be obtained from two independent

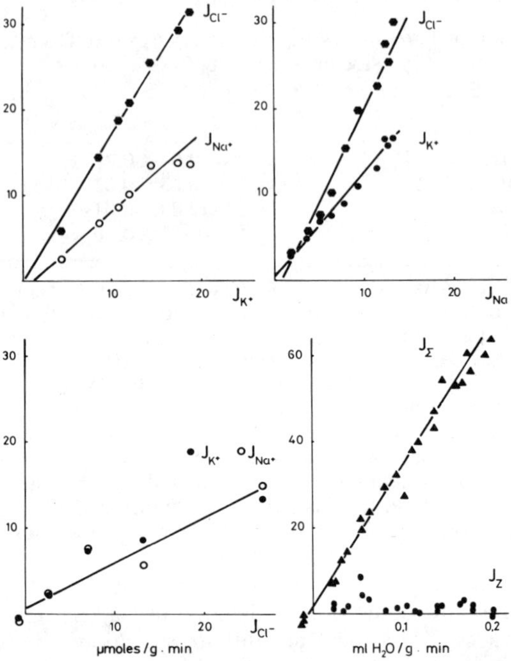

FIGURE 1. In pulse-response experiments with Na^+, K^+, and Cl^- Na^+-rich, K^+-depleted cells were used as described in the footnote to TABLE 1. The furosemide-sensitive transport (μmoles/g·min) of the responding (constant) ion is plotted versus that of the pulsed one. The furosemide sensitive ion movements are defined as the uptake with 1 mM ouabain minus that with 1 mM ouabain and 2 mM furosemide. To vary the concentration of the corresponding pulsed ion species, Na^+ and K^+ were partially substituted by tetraethylammonium and Cl by gluconate. For the different types of experiments the following ion concentrations were kept constant: Upper left: 84 mM Na^+, 135 mM Cl^-; upper right: 28 mM K^+, 135 mM Cl^-; and lower left: 116 mM Na^+, 29 mM K^+. Further, the slopes of the regression lines are given in TABLE 2, while, on the lower right, the results of all three types of experiments are pooled. The triangles show the sum of the three furosemide-sensitive ion fluxes ($J_{Na}+J_K+J_{Cl}$) (μmoles/g·min) plotted against the associated water movements (ml/g·min). The slope of the line indicating the osmolarity of the transported mixture in these three ion species is 321±8 mM. The points give the flux of electrical charge ($J_{Na}+J_K+J_{Cl}$) (μmoles/g·min). The slope of the regression line is not significantly different from zero (1±5 mM).

TABLE 2

DEGREE OF COUPLING IN THE NA^+, K^+, CL^--COTRANSPORT

$q_{1,}*_{2}=0.98\pm0.06$
$q_{1,\ 3}=1.10\pm0.16$
$q_{2,\ 3}=0.93\pm0.05$

* The subscripts 1, 2, 3 refer to the furosemide-sensitive net flows of Na^+, K^+ and Cl^-, respectively.

TABLE 3

THE SLOPES OF THE PULSE-RESPONSE CURVES BETWEEN Na^+ (1), K^+ (2), AND Cl^- (3) ‡

$(\partial J_1/\partial J_2)X_1^c\ X_3^c = \nu_1/\nu_2 = 0.80 \pm 0.09$
$(\partial J_2/\partial J_1)X_1^c\ X_3^c = \nu_2/\nu_1 = 1.20 \pm 0.07$
$(\partial J_1/\partial J_3)X_1^c\ X_2^c = \nu_1/\nu_3 = 0.51 \pm 0.13$
$(\partial J_3/\partial J_1)X_2^c\ X_3^c = \nu_3/\nu_1 = 2.36 \pm 0.14$
$(\partial J_2/\partial J_3)X_1^c\ X_2^c = \nu_2/\nu_3 = 0.49 \pm 0.05$
$(\partial J_3/\partial J_2)X_1^c\ X_3^c = \nu_3/\nu_2 = 1.75 \pm 0.05$

‡ The slopes of the pulse-response curves between Na^+ (1), K^+ (2) and Cl^- (3) are obtained from the experimental pulse response curves depicted in FIGURE 1, according to Equation 3. The subscripts X_1^c, X_2^c, X_3^c refer to the chemical potential differences of Na^+, K^+ and Cl^-, respectively, kept constant during the pulse response measurements.

slopes, a straight one and a reciprocal one. The geometric means of these ratios are listed on TABLE 4. (The arithmetic means are essentially the same). It follows that $\nu_1:\nu_2:\nu_3 = 1:1.14:2$, as indicated, in line with Equation 4 that the stoichiometry of the transport is:

$$Na:K:Cl = 1:1:2$$

Several kinds of ion cotransport systems have been described for various cells, such as NaK cotransport and NaCl cotransport and a KCl cotransport. It is likely, though not certain, that all these different systems of cotransport refer to the same system, similar to that described above for Ehrlich cells.

The physiological significance of this kind of electrically-silent cotransport is likely to concern the regulation of cellular volume. By introducing osmolarity into the cell it tends to cause swelling, thereby counteracting the shrinking effect resulting from the Na,K pump. This cotransport is indeed activated by a shrinkage of the cells and inactivated by swelling. Its osmotic effect seems to be very strong as each transport cycle leads to the translocation of four osmolar units from the medium into the cell.

TABLE 4

THE STOICHIOMETRY OF THE Na^+, K^+, Cl^--COTRANSPORT

$\nu_2/\nu_1 = 1.22 \pm 0.08$
$\nu_3/\nu_1 = 2.15 \pm 0.28$
$\nu_3/\nu_2 = 1.89 \pm 0.10$
i.e. $Na^+:K^+:Cl^- = 1:1:2$ §

§ The ratios are the geometric means of the pairs of the corresponding ratios listed on TABLE 3.

SUMMARY

A valid test for cotransport between solutes is a demonstration that the degree of coupling between all coupled solute flows concerned, q defined in terms of Irreversible Thermodynamics, is sufficiently close to unity. The usual method to determine q kinetically by pulses and responses of flows can not simply be applied to rheogenic ion flows, as electrical potential difference changes due to the pulses can hardly be avoided. If, however, the ion flows are all electrically silent, changes in electrical potential difference (*PD*.) should not interfere with the determination of q. This holds for the furosemide-sensitive fluxes of Na^+, K^+, and Cl^- in Ehrlich cells, each of which could be shown to be unaffected by a change in electrical *PD* and vice versa. Hence the q values could be determined for any pair of the three ion flows concerned and none differed significantly from unity. These results appear to indicate a furosemide-sensitive, electrically silent ternary symport mechanism for Na^+, K^+, Cl^- with the stoichiometry 1:1:2, which is active but does not utilize ATP. It is assumed to function as a very efficient regulator of cellular volume and may be identical with other previously described binary symport systems.

References

1. KEDEM, O & R. CAPLAN. 1965. Trans. Faraday Soc. **61:**1897–1911.
2. HEINZ, E. 1974. *In* Current Topics in Membranes and Transport. A. Kleinzeller and F. Bronner, Eds.: **5:**137–159. Academic Press. New York, N.Y.
3. GECK, P., E. HEINZ, C. PIETRZYK & B. PFEIFFER. 1978. *In* Cell Membrane Receptors for Drugs and Hormones. R. W. Straub and L. Bolis, Eds.: 301–307. Raven Press. New York, N.Y.
4. GECK, P., B. C. JACOBI, C. PIETRZYK, B. PLEIFFER & E. HEINZ. 1980. Biochim. Biophys. Acta. In press.

Discussion

T. H. WILSON (*Harvard Medical School, Boston, Mass.*): How strong is the evidence that this is a single transport system and not a two-transport system, one for sodium chloride and one for potassium chloride?

E. HEINZ: I think that the evidence is rather strong, because the q-values could not be unity in all three combinations. If there were separate systems, then there should be at least one q-value that is very low. Let's say if there were separate flows of sodium chloride and potassium chloride, then between the sodium and potassium flows there could not be a significant degree of coupling.

J. FORTE (*University of California, Berkeley, Calif.*): Do you rule out the possibility that there are chloride:hydroxyl and hydrogen:sodium exchanges going on that could further complicate it by adding two more ions? But you restrict your analysis to three ions, and, therefore, you get the coupling coefficients for those three, but I believe this doesn't necessarily rule out the possibility that there are other couplings. . . .

HEINZ: I think the answer is very simple: the present system does not rule out the chloride-hydroxyl exchange, it rather seems that they have nothing to do with each other. This can be shown by the fact that the chloride-hydroxyl exchange mechanism cannot distinguish between chloride and nitrate, but the present system can. Chloride can only be exchanged to some extent by bromide, but not by any other anion we tried, including NO_3^-.

F. H. EPSTEIN (*Harvard Medical School, Boston, Mass.*): Do you have any information as to whether the transport system you describe depends on or utilizes energy? That's the first question. And secondly, why should the cell need a mechanism to swell it when cells naturally tend to swell if nothing is done and it seems like a waste when all they would need to do would be to shut off the mechanism used to shrink them?

HEINZ: The first question is, if I understand correctly, whether this mechanism is active.

EPSTEIN: Yes.

HEINZ: I think the answer is yes and no. The mechanism is active insofar as it still transports if the overall force, i.e., the sum of all individual driving forces is zero or in the opposite direction. In other words, plotting the transport versus the sum of the driving forces, the line does not go through the origin. So it would appear that additional energy is necessary as that derived merely from the electrochemical potential differences of the three species concerned is not sufficient.

Furthermore, metabolic inhibitors like antimycin A, and others, completely inhibit this mechanism. On the other hand, the transport does not appear to utilize energy. In my former laboratory, Dr. Pietrzyk and Dr. Geck very carefully studied the ATP utilization, but did not find any effect of furosemide. Hence, unless the stoichiometry between ATP and this transport mechanism is excessively high, it looks that at least ATP is not the source of energy. Whether the ATP comes from respiration or glycolysis doesn't make any difference. Why ATP is not being used up by this mechanism is an open question. If there is an energy source involved, it must be other than ATP.

EPSTEIN: What about cold? Does it go on in the cold?

HEINZ: We haven't tried cold but I would expect that it does inhibit since metabolic inhibitors do so. Your second question was: What is the mechanism good for? I just said before that the sodium potassium pump naturally tends to shrink the cell because there are three Na^+ plus one Cl^- pump out for two K^+ pumped in. So with each cycle, one sodium chloride comes out of the cell and nothing gets in for it. The present mechanism would counteract that and apparently it does so only when the cell is shrinking, and that it gets very weak, if the cell has reached its normal volume. But if you ask me how this regulation works, I'm not able to answer that.

E. RACKER: I'm not quite sure how you have ruled out the participation of ATP. Is that done in the presence of antimycin because you say antimycin inhibits? My first question is how would you measure the intracellular ATP steady state concentration that is continuously generated by oxidative phosphorylation? My second question is the same question my predecessor asked, what is the sense of doing those things? You say such an approach will counteract the sodium pump, but the sodium pump is a very good physiological action, and, if it is counteracted, we end up with nothing.

HEINZ: May I answer the last question first? The cotransport does not counteract the sodium pump. It only counteracts the shrinking effect of the sodium pump, not the sodium pump itself.

RACKER: How do you do one without the other?

HEINZ: I think that the Na pump is very important. As we need it also in order to explain our views on amino acid transport, I would not dream of ruling out the sodium pump. But it's unavoidable that the sodium pump moves osmolarity out of the cell, and there must be a mechanism to prevent the cell from shrinking. And I think this might be the explanation.

RACKER: Could you give us some idea about the relative rates of the sodium pump versus that of this mechanism.

HEINZ: The relative rates are of similar order of magnitude. In earlier times, we thought that there was only a potassium chloride cotransport, the sodium is sort of a newcomer in this mechanism. The sodium-potassium pump is activated by outside potassium, but beyond a level of potassium between 10 and 20 milimolars the sodium-potassium pump cannot be stimulated any further. The present mechanism, however, is within these ranges not saturable. So if you increase the potassium concentration outside, then this mechanism keeps going up, and if you have reached a level of, let's say, 30 milimolar potassium outside, there's already more potassium pumped in by this mechanism than by the potassium sodium pump. So I would expect that the rates are of somewhat comparable magnitude.

As to your other question concerning the ATP consumption, I said already this is something that is not completely solved. I can only tell you what has been observed so far. If you, for instance, follow the ATP level inside the cell, measured by the firefly method, after further generation of ATP has been stopped, you get a course of decay of the ATP level. If one then turns on the sodium potassium pump by giving either potassium or taking away the ouabain, one quite distinctly sees that the slope of decay gets steeper because more ATP is being used up per time unit. But if you do the same thing in the presence and absence of furosemide, the slopes remain the same within experimental error.

A. A. EDDY (*University of Manchester, Manchester, England*): I think at one time both our labs found that there is a relatively fast uptake of KCl in Ehrlich cells if you soak them in a potassium Ringer's medium in the absence of sodium. Were you, at one time, suggesting that this was the mechanism whereby chloride uptake could be relatively fast but nonelectrogenic?

HEINZ: Yes, actually its true that previously everything looked like a symport between potassium and chloride and that, of course, would account for your observation. But it should be added that those experiments were going for ten minutes and during ten minutes some rearrangement is likely to occur, for instance Cl^- exchange for OH^-. That's the only explanation I can give at the present moment, but I agree with you we are still somewhat puzzled why in previous times this sodium effect did not show. I hope we find out sometime.

RACKER: Is it possible that potassium can substitute for the sodium? That you have two potassiums against two chlorides or do you know that it is absolutely specific for sodium and potassium?

HEINZ: This would indeed be an explanation, but the specificity studies are not yet complete for the sodium. We found out that the potassium site appears to be rather specific for potassium, which can only be replaced by rubidium but not by sodium. But we have not tested yet whether potassium can replace sodium at the sodium site. That would be a very interesting experiment.

REHM: What would happen if you applied this analysis to Donnan equilibrium in which there's sodium, potassium and chloride?

HEINZ: In the Donnan equilibrium all driving forces are zero, and, therefore, you cannot apply irreversible thermodynamics.

REHM: Suppose you would perturb the system to throw it off equilibrium?

HEINZ: If determined correctly, there should not be any coupling other than by electrostatic forces. If, however, you are able to filter out these coupled electrostatic flows, you shouldn't get any q-value whatsoever.

RACKER: I'd like to return to the ATP issue because of some contradiction there. Why should antimycin inhibit? You said you probably have other energy that is probably not going to work in the cold and still you don't think that it is ATP directed. How do you know whether when you have antimycin present glucose will substitute? Have you done those experiments?

HEINZ: Yes, glycolysis doesn't substitute either, at least as far as the ATP is concerned. So I think the energy question is still a great puzzle and we would be very grateful for some suggestion of where we could look for it.

E. FRÖMTER (*Max Planck Institut für Biophysik, Frankfurt, West Germany*): I haven't been able to completely understand your arguments regarding the electroneutrality. With any set of ion fluxes the electroneutrality condition must hold. Do you have direct arguments or data regarding the potential changes under operation or nonoperation of this mechanism?

E. HEINZ: Since neither the presence nor the absence of furosemide has any effect on the electrical potential, it was concluded that the furosemide-sensitive part of these flows (which is what we're talking about) has no effect on the electric potential.

C. LEVINSON (*University of Texas Health Science Center, San Antonio, Texas*): At the furosemide concentrations used in these experiments, was there an effect upon all fluxes equally?

HEINZ: No, only part of these fluxes are inhibited by furosemide.

LEVINSON: What percent of the total unidirectional flux is affected by furosemide? In other words, what fraction of the total flux can be accounted for in terms of the mechanism?

HEINZ: With the potassium, I think it changes about 30% to 40%; then another 30% or 40% of the potassium unidirectional flow is due to the Na^+-K^+ pump and the leakage flow of potassium was found to be very low, maybe about 10% at the most.

LEVINSON: Does the mechanism than apply to only 30% to 40% of the total flux?

HEINZ: Approximately.

J. A. ZADUNAISKY (*New York University, New York, N.Y.*): Is this system sensitive to SITS or DIDS?

HEINZ: No. These drugs do not work very strongly in Ehrlich cells; and they inhibit a chloride-OH exchange, which we think is distinctly different from the present system.

M. C. GRUSTIN (*Yale University, New Haven, Conn.*): Do you have any evidence from tissues in which sodium potassium ATPase is particularly high, e.g., from shark rectal gland? Would you expect to find a high furosemide sensitive ion flux as well as a high ouabain-sensitive flux?

HEINZ: Unfortunately, I cannot answer the question because we studied only Ehrlich cells. It ought to be maybe extended to other cells and we wouldn't be surprised if something similar happened there.

A. M. WEINSTEIN (*National Institutes of Health, Chevy Chase, Md.*): I'm not sure I understood the reply to Dr. Frömter's question. Would you reexplain how you assured through the course of these experiments that you maintained a constant electrical potential across the membrane?

HEINZ: The electrical potential, of course, cannot be maintained constant, but as I said we "filtered out" all flows other than the furosemide-sensitive ones and only these are claimed to be electrically silent. Of course, if you apply a pulse to the total flows, you will effect the potential. Does that answer your question?

UNIDENTIFIED QUESTIONER: In your system potassium, chloride and sodium are tightly coupled. What happens if you substitute thiocyanate for chloride?

HEINZ: It doesn't work. Even bromide can hardly be substituted for chloride. There may be other ions that can substitute for Cl^- but those that we tried, including thiocyanate, did not.

UNIDENTIFIED QUESTIONER: When you use thiocyanate, is the sodium exchange blocked?

HEINZ: Not entirely. Only if chloride is present, of course, you get the furosemide effect and the thiocyanate transport is not inhibited by furosemide.

S. ALPER (*Yale University Medical School, New Haven, Conn.*): You mentioned that there's a different K^+-dependence of the ouabain-sensitive and the furosemide-sensitive fluxes in these cells. Did you find saturation with external K of the furosemide sensitive flux?

HEINZ: No, at least not in the range of K^+ concentrations we tried, which were pretty high.

ALPER: With sodium or with chloride?

HEINZ: We did not see any saturation within the concentration ranges used for these ions, though the concentrations were fairly high.

RACKER: Have you found a compound other than furosemide that inhibits the sodium potassium ATPase in Ehrlich ascites cells?

HEINZ: We did not see much of such an inhibition.

RACKER: No? We have published that it does inhibit.

HEINZ: As to other inhibitors we tried several compounds that are related to furosemide in their diuretic effect, like piretanide, bumetanide and others. They do all the same as does furosemide.

M. HASS (*Duke University, Durham, N.C.*): How high did you actually go with the external sodium potassium and chloride concentrations?

HEINZ: Between 50 and 100 mM.

F. H. EPSTEIN: Dr. Heinz, I must have missed how you measured the electrical potential, would you mention the method?

HEINZ: We derived it from the tetraphenylphosphonium distribution. We also did it with fluorescent quenching, but that works only if the mitochondria are paralyzed.

PART II. ANION AND PROTON TRANSPORT MECHANISMS IN KIDNEY TUBULES, TURTLE BLADDER AND RELATED EPITHELIA

A NEW MICRO METHOD FOR DETERMINATION OF CO_2 TENSION AND ITS APPLICATION IN THE STUDY OF LUMINAL FLUID ACIDIFICATION IN VARIOUS NEPHRON SEGMENTS *

Thomas D. DuBose, Jr. and Juha P. Kokko

Department of Internal Medicine
The University of Texas Health Science Center at Dallas
5323 Harry Hines Boulevard
Dallas, Texas 75235

INTRODUCTION

An important feature in the analysis of the mechanism mediating renal acidification and the reabsorption of bicarbonate is the direct measurement of tubular fluid pCO_2 and an evaluation of its determinants. Most previous micropuncture studies supporting hydrogen ion secretion as the mechanism of this process have relied on the demonstration of a negative or acid disequilibrium pH.[1, 2] The disequilibrium pH represents the difference between the *in situ* pH of tubule fluid and the pH measured after the tubule fluid is removed and allowed to reach equilibrium:

$$pH_{Diseq} = pH_{Eq} - pH_{in\ situ}$$

Previously, in measuring equilibrium pH (pH_{Eq}), the sample has been equilibrated at a CO_2 tension similar to the animal's systemic arterial pCO_2.[1, 2] Thus, one of the necessary assumptions inherent in this calculation, as previously applied, has been that the pCO_2 of tubule fluid and systemic arterial blood are equal. Brodsky and Schilb[3] noted, however, that if one assumes a pH of 6.71 units, a bicarbonate concentration of 8 mM (findings consistent with numerous studies utilizing several different techniques), and reaction equilibrium in the proximal tubule, the calculated pCO_2 for these values would be 65 mm Hg. These investigators emphasized, therefore, that the pCO_2 of proximal tubule fluid might exceed that of systemic arterial blood, and that the determination of the actual disequilibrium pH requires knowledge of the *in situ* pCO_2.

In the few studies in which the pCO_2 of tubule fluid has been estimated indirectly,[4, 5] or determined by microelectrode techniques,[6, 7] the calculated[4, 5] or measured[6, 7] CO_2 tension in proximal tubule fluid significantly exceeded that of systemic arterial blood. In the studies of Karlmark and Danielson[4] and Sohtell,[5] the pCO_2 of tubule fluid was not determined directly, but was calculated by indirect techniques (Astrup technique) from measurements of *in situ* pH. Nevertheless, the studies suggested an elevated pCO_2 in the early portion of the proximal tubule and that a declining profile for CO_2 tension existed along the length of this segment. Although the calculation of an elevated pCO_2 in

* These studies were supported in part by NIAMDD Grant 1 RO1 AM14677, 1 RO1 AM25730, and by Biomedical Research Support Grant 5 SO7 RR05426.

0077-8923/80/0341-0067 $01.75/0

proximal tubule fluid was later supported by direct determination,[6, 7] a similar measurement of pCO_2 and its profile along the length of the proximal tubule was not reported. Sohtell has recently reported that the pCO_2 of efferent arteriolar blood was significantly less than the pCO_2 in both proximal tubule fluid and systemic arterial blood.[7] Thus, this [7] and other studies [4, 8] have supported the concept that the proximal tubular epithelium might serve as a "diffusion barrier" for CO_2. This hypothesis has been challenged, however, by the determination of a high CO_2 permeability coefficient for rabbit proximal straight tubules perfused *in vitro*.[9]

We have employed a new pCO_2 microelectrode, modified, as recently reported,[10, 11] from that originally described by Caflisch and Carter.[12] The specific purposes of these studies were first, to compare proximal tubular pCO_2 with that of other portions of the nephron as well as the stellate vessel. Second, to evaluate the possibility of a declining profile for CO_2 along the length of the proximal tubule in order to determine if this epithelium could serve as a diffusion barrier for CO_2. Third, to evaluate the relationship between the magnitude of bicarbonate reabsorption and renal cortical CO_2 generation; and fourth, to determine both *in situ* pH and pCO_2 in segments of the proximal tubule before and after carbonic anhydrase inhibition. In addition, in an attempt to validate the magnitude of the pCO_2 determined by the pCO_2 microelectrode, we have compared the bicarbonate concentration calculated from pH and pCO_2 with the concentration measured by the microcalorimetric technique of Vurek *et al.*[13]

pCO_2 *Microelectrode Characteristics*

The *in situ* pCO_2 of proximal and distal tubular fluid in the control Sprague-Dawley rat was determined with pCO_2 microelectrodes of 4–6 μm tip diameter. The electrode utilized in these studies has been reported in detail.[10, 11] This newly-developed pCO_2 microelectrode represents a significant improvement of the two types of microelectrodes previously reported by Sohtell and Karlmark [6] and Caflisch and Carter.[12] A schematic of the pCO_2 microelectrode assembled in our laboratory and utilized in these studies is represented in FIGURE 1. Note that there are five component parts:

1. An outer shell, or body of the microelectrode that is pulled from a 2 mm (o.d.) borosilicate capillary and beveled to the desired tip dimeter (4–6 μm);
2. A silicon rubber, gas-permeable, liquid-impermeable membrane (GE-RTV 102, 1 part plus 7 parts toluene) of 10–50 μm length from the electrode tip which is allowed to cure in a dry environment for 20 minutes to assure a liquid tight seal;
3. The filling solution (1 mM $NaHCO_3$ plus 100 mM NaCl);
4. A glass membrane pH microelectrode [14] characterized by near theoretic slope (59 mV/pH U at 22° C) which is advanced under microscopic control to a point no more than 10 μm behind the CO_2 membrane; and
5. A silver:silver chloride reference wire.

When constructed in this manner the pH of the bicarbonate solution will be determined by, and directly proportional to, the partial pressure of the CO_2 gas in solution. That is to say, as the pCO_2 of the solution is varied, the pH of the bicarbonate solution will vary in the manner predicted by the Henderson-Hasselbalch equation. This electrode obtains near theoretic slope (57–61 mV/$\log_{10} pCO_2$), has a response time of 1–3 minutes, and a drift of less than 3 mV

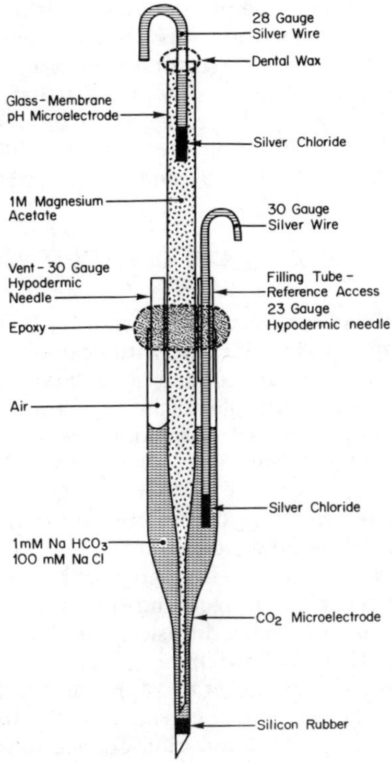

FIGURE 1. Schematic representation of pCO_2 microelectrode. Component parts include an outer shell into which a weak bicarbonate solution is placed behind a gas-permeable, liquid-impermeable silicon rubber membrane. Changes in pH as a result of increases or decreases in the concentration of dissolved CO_2 in this solution are recorded by a glass-membrane pH electrode. The circuit is completed by a silver:silver chloride reference wire. (From DuBose *et al.*[10] By permission of *Journal of Clinical Investigation.*)

per 30 minutes. For comparison, the first microelectrode described[12] was characterized by a sensitivity of 35–45 mV/$\log_{10}$ pCO_2. The pCO_2 microelectrode described by Sohtell and Karlmark[6] employed an antimony pH sensor and was characterized by a sensitivity of 37–47 mV/$\log_{10}$ pCO_2. Although the inaccuracies of antimony as a pH sensor of biological solutions has been well documented,[15] when employed as a component of a pCO_2 microelectrode (because the solution measured remains ionically constant) the anion error is not of major concern. However, the two factors that have the most deleterious effects on the antimony potential, temperature and gases (O_2 and N_2) in solution, are potential sources of major error.[15, 16]

The actual assembly, calibration and *in vivo* utilization of this pCO_2 microelectrode in rat renal micropuncture experiments has been described in detail.[10] Close agreement between this microelectrode and a standard macroelectrode have been demonstrated for blood and other biological solutions.[10] Electrodes with a sensitivity of less than 57 mV/$\log_{10}$ pCO_2 were rejected. Calibration in 10 mM $NaHCO_3$ solutions equilibrated with gases of analyzed content (2, 5, 10, and 20% CO_2, balance O_2) were performed at 37° C immediately before and after obtaining 3–5 data points *in vivo*. During both the pre- and post-puncture calibrations, the electrode was always immersed in pH 4.0 buffer to assure that liquid pathways in the silicone rubber membrane were not present. For example, if liquid pathways exist, pH 4.0 buffer would enter the weak $NaHCO_3$ solution and rapidly decrease pH. Conversely, if the membrane is gas permeable, the CO_2 within the electrode would rapidly diffuse into the pH 4.0 solution (pCO_2 of pH 4.0 = 0) and increase pH.

Determination of pH, pCO_2 *and* [HCO_3^-] *in Control Rats*

The direct microelectrode determination of pH and pCO_2 has been performed in 40 adult Sprague-Dawley rats with normal systemic acid-base parameters. These results are summarized and schematically displayed in FIGURE 2. First, note that values for *in situ* pH in the segments of the superficial proximal and distal tubules closely agree with previously reported values.[1, 2, 17] However, the pCO_2 in the proximal tubule significantly exceeded the pCO_2 in systemic arterial and renal arterial blood by 25.9 ± 0.92 mm Hg ($p < 0.001$). Furthermore, the values for pCO_2 along the length of the proximal tubule, the distal tubule, and stellate vessel blood were similarly elevated.[10]

Equal values for pCO_2 between the early proximal tubule and stellate vessel, and the failure to demonstrate a declining profile for pCO_2 along the nephron indicate that CO_2 gas is in or near diffusion equilibrium throughout the cortex. Thus, the hypothesis that the proximal tubular epithelium could serve as a "diffusion barrier" for CO_2 [4, 5] cannot be supported by this data.

Since the demonstration of an elevated pCO_2 throughout the cortex by microelectrodes raises important questions concerning methodology, we have attempted to validate these values by comparison of directly determined values of pH and pCO_2 and thus the calculated total CO_2 concentration with the total CO_2 concentration measured by the microcalorimetric method (picapnotherm). Such a comparison is displayed in FIGURE 3. This comparison demonstrates

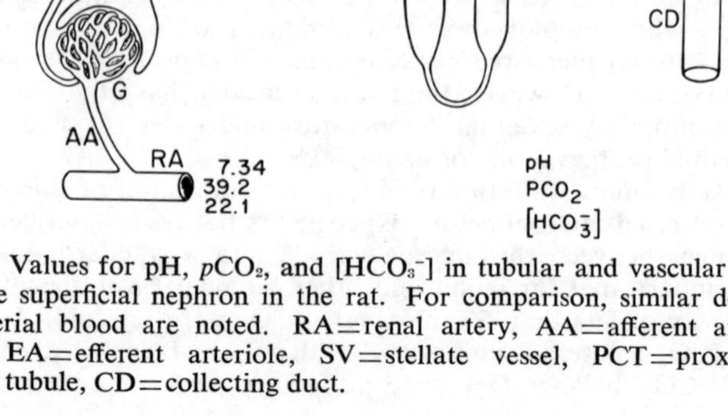

FIGURE 2. Values for pH, pCO_2, and [HCO_3^-] in tubular and vascular structures of the accessible superficial nephron in the rat. For comparison, similar determinations in renal arterial blood are noted. RA=renal artery, AA=afferent arteriole, G=glomerulus, EA=efferent arteriole, SV=stellate vessel, PCT=proximal tubule, DCT=distal tubule, CD=collecting duct.

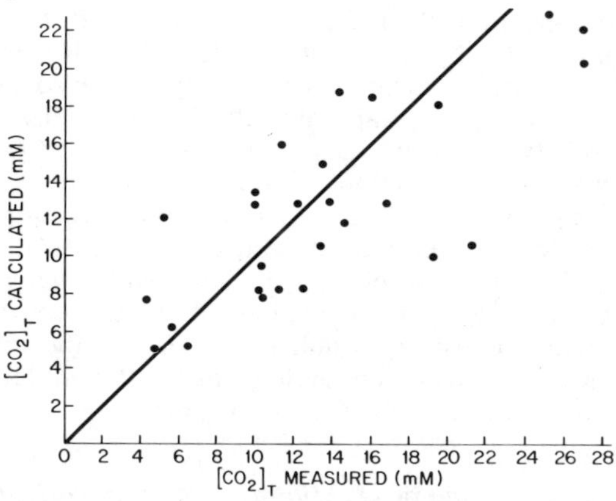

FIGURE 3. Comparison of calculated and measured total CO_2 concentration in the PCT. Abscissa represents values calculated from measured pH and pCO_2 in the PCT. Ordinate represents $[CO_2]_T$ measured by microcalorimetric method. The ratio of calculated to measured total $[CO_2]$ does not differ from 1.0 ($p > 0.05$). (From DuBose *et al.*[10] By permission of *Journal of Clinical Investigation.*)

that the two techniques are not significantly different and thus strongly supports the values for pCO_2 obtained in this nephron segment. A similar comparison in the stellate vessel also demonstrated close agreement.

Relationship between Bicarbonate Reabsorption and Renal Cortical CO_2 Generation

Although the explanation of elevated values for CO_2 gas tension in, or near, equilibrium throughout the cortex is not immediately apparent, we have attempted to indirectly assess the determinants of CO_2 generation by determination of pCO_2 in these same structures during conditions designed to either increase or decrease bicarbonate reabsorption by the proximal tubule. In seven rats, bicarbonate reabsorption was increased by a combination of acute hypercapnia, volume depletion, and hypokalemia, while in seven additional rats the converse conditions were imposed in an attempt to decrease bicarbonate reabsorption. The results of these experiments are displayed in FIGURE 4. These studies demonstrate that variations in the magnitude of bicarbonate reabsorption result in parallel changes in the pCO_2 gradient between arterial blood and the superficial cortex (ΔpCO_2). In addition to the studies depicted in FIGURE 4, bicarbonate reabsorption was also decreased by the systemic administration of benzolamide (2.0 mg/kg) and again ΔpCO_2 fell significantly.[10] The fact that two different methods, each depressing bicarbonate reabsorption by separate mechanisms, resulted in a reduction of ΔpCO_2 suggests that this reduction was not a function of the specific manner by which bicarbonate transport was depressed, but rather of the extent of residual bicarbonate reabsorption.

In this regard, it is of interest to note that the magnitude of reduction of ΔpCO_2 during combined hypocapnia and volume expansion greatly exceeded the reduction after benzolamide. Such differences might occur if H^+ secretion into bicarbonate-containing tubular fluid could proceed at a greater rate during benzolamide administration than during hypocapnia if this latter condition reduced the supply of H^+ by increasing intracellular pH. The demonstration of continued H^+ secretion during luminal application of benzolamide has recently been reported by our laboratory.[18] These studies, then, support a strong relationship between bicarbonate reabsorption and pCO_2 generation; and suggest that luminal CO_2 production, as a consequence of H_2CO_3 dehydration formed from secreted H^+ and filtered HCO_3^-, is the major source of renal cortical CO_2 generation. Parenthetically, variations in ΔpCO_2 of the magnitude of that depicted in FIGURE 4, would not be expected if metabolic CO_2 production was the major source of this elevation. Metabolic sources for CO_2 production of this magnitude also seem unlikely in view of the studies of Schwartz *et al.*,[19, 20] in which a contribution of less than 2.0 mm Hg pCO_2 could be attributed to this source in the turtle urinary bladder.

Use of pH and pCO_2 *Microelectrodes as a Means of Assessing the Mechanism of Renal Acidification*

The demonstration of elevated values for CO_2 tension in both proximal and distal tubules indicate that previous calculations of the disequilibrium pH, assuming equilibrium between systemic blood and the renal cortex, were not entirely correct. However, it should be emphasized that the magnitude of the pCO_2 observed in the distal tubule could not account for the disequilibrium pH of —0.85 units previously reported.[1]

Recently Sohtell has calculated the pCO_2 along the proximal tubule from the measured *in situ* pH before and after acetazolamide administration.[5] Based on these indirect studies, which require the assumption of reaction equilibrium

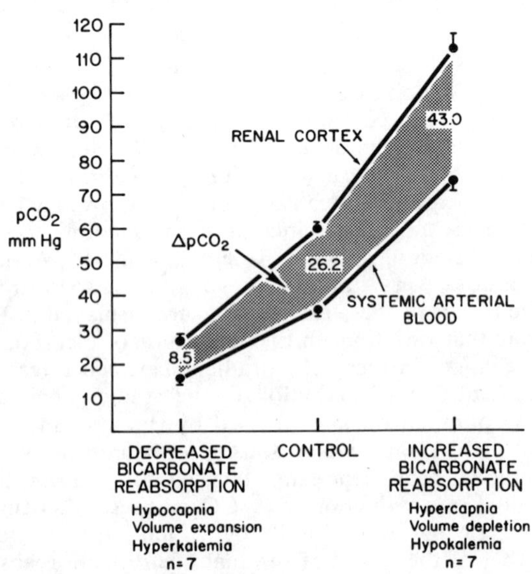

FIGURE 4. Relationship between magnitude of renal cortical pCO_2 and variations in proximal tubular bicarbonate reabsorption. Upper line represents renal cortical pCO_2 obtained from PCT, DCT, and SV; lower line represents systemic arterial pCO_2, and stippled area represents difference (ΔpCO_2).

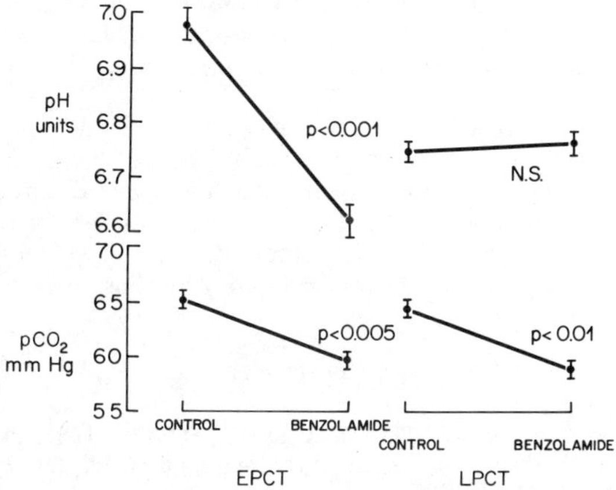

FIGURE 5. Effect of benzolamide on *in situ* pH and pCO_2 in segments of the proximal tubule. pH (upper portion of abscissa) and pCO_2 (lower portion) are displayed during control and benzolamide periods in the early (EPCT) (left panel) and late (LPCT) (right panel). (From DuBose *et al.*[21] By permission of *Kidney International.*)

in proximal tubule fluid after carbonic anhydrase inhibition, it was proposed that the disequilibrium pH observed in this segment after carbonic anhydrase inhibition was the result of accumulation of CO_2 rather than H_2CO_3.[5] Since such an assumption is contrary to previous studies in which the disequilibrium pH has been measured during this condition,[1, 2, 17] and since direct measurement of both pH and pCO_2 would be required to assess the means by which a disequilibrium pH is generated, we have recently reported simultaneous determinations of these parameters in both the early and late segments of the proximal convoluted tubule.[21] The results of these studies are displayed in FIGURE 5. In both the early and late proximal tubule, pCO_2 (bottom panel) decreased significantly after benzolamide administration. This finding demonstrates that accumulation of CO_2 in proximal tubule fluid could not contribute to the disequilibrium pH observed in this segment during inhibition of carbonic anhydrase. Furthermore, in additional studies, we have also reported that benzolamide treatment decreased pCO_2 in the distal tubule and stellate vessel as well.[10] Values for *in situ* pH before and after benzolamide administration are also displayed in FIGURE 5. In the early proximal tubule, pH decreased significantly from 6.98 ± 0.03 to 6.62 ± 0.03 units ($p < 0.001$). Thus a significant decrease in both pH and pCO_2 indicates clearly that benzolamide administration and inhibition of carbonic anhydrase results in accumulation of carbonic acid in excess of the concentration predicted at equilibrium. That is, the disequilibrium pH in the proximal tubule occurs as a result of accumulation of carbonic acid, not CO_2.

It should be cautioned that the true magnitude of the disequilibrium pH cannot be determined from these studies since equilibrium pH was not measured. However, based on the observed fall in pH in the EPCT and the concentration of bicarbonate in this segment, during this condition, a minimum value of −0.40 units can be estimated. Since the first 1.0 mm of the proximal tubule is not accessible in this species, it is possible that values exceeding −0.40 units are attainable in this early portion of the nephron. An equilibrium electrode that allows aspiration of tubule fluid and equilibration at the prevailing pCO_2 *in vivo* is currently being utilized in our laboratory. Determination of *in situ* pH and equilibrium pH would then allow calculation of the "true" disequilibrium pH.

It can be stated with certainty, however, that the bulk of evidence to date supports a major role for H^+ secretion in the process of bicarbonate reabsorption by the proximal tubule. Since it appears that the "true" disequilibrium pH may not reach values as high as −1.0 pH units, a value necessary to account for all of bicarbonate reabsorption by H^+ secretion,[22] a component of bicarbonate reabsorption per se cannot be entirely ruled out by these experiments.

References

1. Rector, F. C., Jr., N. W. Carter & D. W. Seldin. 1965. The mechanism of bicarbonate reabsorption in the proximal and distal tubules of the kidney. J. Clin. Invest. **44**:278–290.
2. Vieira, F. A. & G. Malnic. 1968. Hydrogen ion secretion by rat renal cortical tubules as studied by an antimony electrode. Am. J. Physiol. **214**:710–718.
3. Brodsky, W. A. & T. P. Schilb. 1974. The means of distinguishing between hydrogen secretion and bicarbonate reabsorption: theory and applications to the reptilian bladder and mammalian kidney. Curr. Top. Membr. Transp. **5**: 161–224.
4. Karlmark, B. & B. G. Danielson. 1974. Titratable acid, pCO_2, bicarbonate, and ammonium ions along the rat proximal tubule. Acta Physiol. Scand. **91**:243–258.
5. Sohtell, M. 1979. CO_2 along the proximal tubules in rat kidney. Acta Physiol. Scand. **105**:146–155.
6. Sohtell, M. & B. Karlmark. 1976. *In vivo* micropuncture pCO_2 measurements. Pflügers Arch. Gesamte Physiol. **363**:179–180.
7. Sohtell, M. 1979. pCO_2 of the proximal tubule fluid and the efferent arteriolar blood in the rat kidney. Acta. Physiol. Scand. **105**:137–145.
8. Karlmark, B. 1971. Renal tubular pCO_2. Proc. Int. Union Physiol. Sci. **12**:206. (Abstr).
9. Warnock, D. G. & F. C. Rector, Jr. 1977. CO_2 permeability of the rabbit proximal straight tubule. Abstr. Am. Soc. Neph. **10**:124A (Abstr.)
10. DuBose, T. D., L. R. Pucacco, D. W. Seldin, N. W. Carter & J. P. Kokko. 1978. Direct determination of pCO_2 in the rat renal cortex. J. Clin. Invest. **62**:338–348.
11. Pucacco, L. R. & N. W. Carter. 1978. An improved pCO_2 microelectrode. Ann. Biochem. **90**:427–434.
12. Caflisch, C. R. & N. W. Carter. 1974. A micro pCO_2 electrode. Ann. Biochem. **60**:252–257.
13. Vurek, G. C., D. G. Warnock & R. Corsey. 1975. Measurement of picomole amounts of carbon dioxide by calorimetry. Ann. Chem. **47**:765–767.
14. Pucacco, L. R. & N. W. Carter. 1976. A glass-membrane pH microelectrode. Ann. Biochem. **73**:501–512.
15. Caflisch, C., L. R. Pucacco & N. W. Carter. 1978. The manufacture and utilization of antimony pH electrodes. Kidney Int. **14**:12–27.
16. Perley, G. A. 1939. Characteristics of the antimony electrode. Industrial and Engineering Chemistry (Analytical edit.). **11**:319–322.
17. Malnic, G., M. Mello Aires & G. Giebisch. 1972. Micropuncture study of renal tubular hydrogen ion transport in the rat. Am. J. Physiol. **222**:147–158.
18. Lucci, M. S., T. D. DuBose, L. R. Pucacco, J. P. Kokko & N. W. Carter. 1979. Effect of carbonic anhydrase inhibitors of varying permeabilities on proximal tubular pH and total CO_2 reabsorptive rate. Clin. Res. **27**:423A (Abstr).

19. SCHWARTZ, J. H., J. T. FINN, G. VAUGHN & P. R. STEINMETZ. 1974. Distribution of metabolic CO_2 and the transported ion species in acidification by turtle bladder. Am. J. Physiol. **226:**283–289.
20. SCHWARTZ, J. H. & P. R. STEINMETZ. 1971. CO_2 requirements for H^+ secretion by the isolated turtle bladder. Am. J. Physiol. **220:**2051–2057.
21. DUBOSE, T. D., L. R. PUCACCO, D. W. SELDIN, N. W. CARTER & J. P. KOKKO. 1979. Microelectrode determination of pH and pCO_2 in rat proximal tubule after benzolamide: Evidence for hydrogen ion secretion. Kidney Int. **15:** 624–629.
22. WALSER, M., G. H. MUDGE. 1960. Renal excretory mechanisms. *In* Mineral Metabolism. C. L. Comar & F. Bronner, Eds. : 287–336. Academic Press. New York, N.Y.
23. FORSTER, R. E. & E. D. CRANDALL. 1975. J. Applied Physiol. **38:**710.

DISCUSSION

M. BURG (*National Institutes of Health, Bethesda, Maryland*): Are you proposing that the high pCO_2 in the cortex might be the result of hydrogen ion secretion and subsequent reabsorption of CO_2? It is generally thought that the source of secreted hydrogen ions is the CO_2 in cells via the hydration-dehydration and there isn't any CO_2 produced from the series of reactions you have proposed because the CO_2 produced in the lumen would be consumed in the cell. Do your findings indicate that there is a non-CO_2 source of secreted hydrogen ions?

T. DUBOSE: Our data cannot be explained entirely by the classical view of hydrogen secretion, as you have indicated. It has also been shown that hydrogen ions can be supplied by the splitting of water at the luminal membrane. Hydrogen secretion into bicarbonate containing fluid from this source would be expected to generate CO_2 in the tubule lumen; and if this CO_2 is then added immediately to the peritubular capillary, and is either separated from or consumed slowly by hydroxyl or bicarbonate then our observation of an elevated pCO_2 and acid pH in stellate vessel blood could be explained. Recent studies by Forster and Crandall[23] indicate that the half-time for equilibration of H^+ when CO_2 is added to a red cell suspension is quite prolonged and thus compatible with our hypothesis.

W. BRODSKY: Do the calculations of Crandall and Forster provide an answer to the following problem: Assuming that a known increment of CO_2 is added to the peritubular blood and stellate vessel, how long a time is required for that increment to be hydrated and reach equilibrium with respect to the H_2CO_3, HCO_3^-, and H^+ in the renal venous blood? Does this time coincide with the transit time for a blood cell to travel from the stellate capillary to the renal vein? What is the contribution of the intra-erythrocyte carbonic anhydrase to the CO_2 hydration rate in the blood (flowing from the stellate vessel to the renal vein)?

DUBOSE: I think that these are very important questions. Let me just expand upon the Crandall-Forster data. They measured the rate of change of pH when the red cells are hemolyzed and carbonic anhydrase present. This rate is very rapid when compared to that found when carbonic anhydrase is not present (as in the case in plasma). I can't say what might occur in the blood of the stellate vessel for sure. Recently, however, we have found that the pH of the blood in this structure increases after carbonic anhydrase has been infused into the renal artery.

E. FRÖMTER: There is quite a distance between the glomerular capillary and the stellate vessel in which your measurements were made. Is it not possible that there is a build up of CO_2 pressure, as a result of a tubular-vascular counter-current system? This would indicate that in some portion of the kidney other than the cortex the CO_2 tension should be low and similar to the pCO_2 in venous blood.

DUBOSE: We have recently measured the pCO_2 of the fluids in the structures of the exposed papilla of the Munich-Wistar rat, and, in fact, do find a lower pCO_2, as you have anticipated. The pCO_2 was 35 mm Hg in the papillary collecting duct and in the loop of Henle.

FUNCTIONAL ORGANIZATION OF PROTON AND BICARBONATE TRANSPORT IN TURTLE URINARY BLADDER *

Philip R. Steinmetz,† Loren H. Cohen,
Russell F. Husted ‡ and Allan Mueller

Laboratory of Epithelial Transport and Kidney Physiology
Department of Internal Medicine
University of Iowa College of Medicine
Iowa City, Iowa 52242

Introduction

The bladder of the fresh water turtle, *Pseudemys scripta*, is a relatively "tight" epithelium that resembles amphibian urinary bladders and frogskin in its capacity to absorb sodium by active transport. The major advantage of the turtle bladder preparation is it contains a potent transport system for urinary acidification [1-3] that lends itself to close examination under simplified and controlled conditions *in vitro*. Substantial evidence is now available to suggest that the energy-dependent step in acidification is located at the luminal cell membrane and involves the translocation of protons from the cell into the urinary compartment.[2-5] The nature of the acidification pump remains to be defined precisely, but recent evidence [6, 7] points toward a proton-translocating ATPase that shares some of the characteristics of H^+-ATPases in mitochondria, chloroplasts and bacterial membranes. Since net transport across epithelia requires translocation of ions across two cell membranes in series, the organization of the acidification pathway is more complex than in single membrane systems. As shown in Figure 1, the extrusion of H^+ ions across the luminal membrane results in an accumulation of OH^- ions within the cell. In the presence of CO_2 and carbonic anhydrase, HCO_3^- will be generated that moves "downhill" across the basolateral cell membrane into the serosal solution.[2-4] This acidification pathway was first defined in a system free of exogenous CO_2 and HCO_3^- in which the production and distribution of metabolic CO_2 and HCO_3^- were measured by means of a conductometric technique.[5, 8] Studies by Steinmetz, Omachi and Frazier [9] and Schwartz and Steinmetz [8] demonstrated that in this system H^+ transport is electrogenic and not coupled directly to the transport of sodium. Accordingly in Figure 1, the two transport systems are shown separately. When net sodium absorption is abolished by addition of ouabain or removal of sodium the electrical potential difference across the tissue reverses so that the lumen becomes positive with respect to the serosa and the reversed short-circuit current becomes equal to the rate of H^+ secretion as measured by pH stat titration.[8, 9, 10]

Aside from the sodium and H^+ pumps, Figure 1 depicts an exchange trans-

* Supported by NIH Grants AM 17568 and HL 14388.

† Josiah Macy Jr. Foundation Faculty Awardee and Visiting Professor, Department of Physiology, Cornell University Medical College, New York, New York 10021, during part of this work.

‡ Recipient of a post doctoral fellowship under NIH Grant AM 07140.

0077–8923/80/0341–0077 $01.75/0 © 1980, NYAS

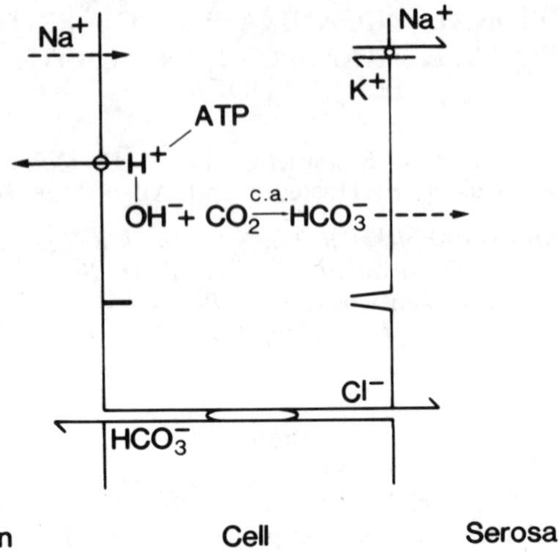

FIGURE 1. Transport systems of turtle urinary bladder. The pumps for H^+ and Na^+ transport are indicated by circles at the luminal and serosal membranes respectively. Since the site of coupling to metabolic energy for the transcellular Cl-HCO_3 exchange flows remains to be determined, it is placed temporarily between the two cell membranes. The hydroxylation of CO_2 is catalyzed by carbonic anhydrase (c.a.).

port pathway in which the absorption of Cl^- is coupled to HCO_3^- secretion. This transport system was discovered when exogenous HCO_3^- was added to the serosal solution.[11, 12] Net Cl^- absorption, previously described by Brodsky and Schilb [13] and Gonzalez *et al.*,[14] was observed to occur only if HCO_3^- was present on the serosal or both sides of the epithelium.[11, 15] Conversely, HCO_3^- transport from serosa to mucosa was dependent on the presence of Cl^- in the mucosal solution.[11] Several lines of evidence indicate that in the steady state the overall operation of this transport system is electroneutral. Thus, Leslie *et al.*[11] explored the effects of adding HCO_3^- to the serosal solutions under conditions of zero net Na^+ transport and zero net H^+ secretion. Na^+ transport was abolished by ouabain and H^+ secretion by lowering the luminal pH to the point at which the H^+ pump stops. Under these conditions of zero current, serosal addition of 20 mM HCO_3^- caused alkali entry into the luminal solution, which was easily measurable by pH stat titration. Almost all the HCO_3^- was transported by an electroneutral mechanism as judged from the simultaneously measured short-circuit current. Measurement of Cl^- fluxes by Leslie *et al.*[11] and Husted *et al.*[15] demonstrated that the increment in the absorptive Cl^- flux brought about by HCO_3^- addition is approximately equal to the oppositely directed HCO_3^- flux and, thereby, support the interpretation that the two transepithelial flows are coupled about one to one. Many aspects of this transport system remain to be clarified. Neither the precise site of coupling nor the way metabolic energy is supplied to this exchange has been well defined. Perhaps the most important issue is the relationship between the Cl-HCO_3 exchange pathway and the acidification pathway. How are these two acid-base transport systems organized within the epithelium? The acidification pathway includes the disposition of OH^- ions and the efflux of HCO_3^- across the serosal membrane. The Cl-HCO_3 exchange involves movement of HCO_3^- in the opposite direction.

In this paper we focus first on the H^+ pump and the disposition of OH^- ions generated in series with the pump. How is the transport rate controlled from the cytoplasmic side of the pump? Then, we provide evidence that the acidification pathway is at least in part separate from the Cl-HCO_3 exchange pathway.

Proton Pump at the Luminal Cell Membrane and the Disposition of OH^- and HCO_3^- in Series with the Pump

Previous studies by Steinmetz and Lawson,[16] Al-Awqati, Mueller and Steinmetz,[17] and Beauwens and Al-Awqati [18] have shown how the transport rate of protons is a linear function of the applied pH gradient across the epithelium. Since the passive H^+ permeability of the luminal cell membrane and the epithelium as a whole is very low,[16, 18] the applied pH gradient primarily affects the active transport rate. If, as we have suggested, the transport of H^+ occurs by an electrogenic mechanism, one would predict that the pump rate would be altered by imposed electrical potential gradients, $\Delta\psi$.

FIGURE 2 shows that the imposition of a voltage opposing H^+ secretion (J_H) indeed inhibits the active transport rate in a manner almost identical to an equivalent ΔpH.[17] Similarly, a favorable $\Delta\psi$ such as the spontaneous PD generated by Na^+ transport accelerates J_H.[2] The equivalence of the effects of electrical and chemical potential gradients on J_H may be compared with similar results obtained in mitochondria [19] and bacterial membrane vesicles.[20] In these preparations, the energy available from the total electrochemical potential for protons, $\Delta\tilde{\mu}_H$, is utilized for ATP synthesis or substrate transport. The close control of active proton transport in turtle bladder by $\Delta\tilde{\mu}_H$ is consistent with recent evidence suggesting that the pump is a proton-translocating ATPase.

Dixon and Al-Awqati [6] have presented an interesting study in which they were able to drive protons in the reverse direction through the pump by means of an imposed $\Delta\tilde{\mu}_H$, and increase the ATP content of the epithelial cells. Another experimental approach has taken advantage of the capacity of the turtle

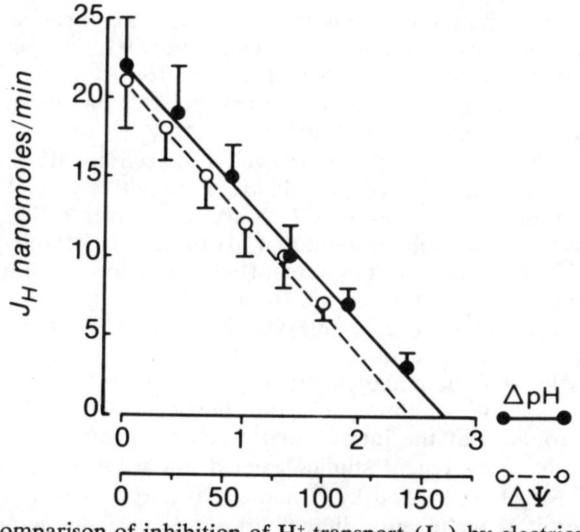

FIGURE 2. Comparison of inhibition of H^+ transport (J_H) by electrical and chemical potential gradients. The slopes and intercepts obtained by the $\Delta\psi$ and ΔpH methods were not different by paired analysis. (From Al-Awqati, Mueller and Steinmetz.[17])

bladder to transport protons anaerobically. During anaerobiosis, J_{II} is no longer inhibitable by sodium cyanide and becomes independent of mitochondrial ATPase function. Steinmetz, Mueller and Beauwens [7] examined the effects of certain known inhibitors of H^+-ATPases in mitochondria and chloroplasts in the absence of mitochondrial ATPase function. Dicyclohexylcarbodiimide (DCCD), and Dio-9, two inhibitors of proton flow in the proteolipid or "channel" part of the enzyme caused dramatic inhibition of anaerobic J_{II}. In contrast, oligomycin failed to inhibit the transport rate. These studies suggest that the luminal plasma membrane may contain a H^+-ATPase that shares some, but not all, of the characteristics of mitochondrial H^+-ATPase. In yeast, the plasma membrane H^+-ATPase has a lower molecular weight and a simpler subunit structure than mitochondrial ATPase.[21] Both ATPases are inhibited by DCCD and Dio-9, but only the mitochondrial ATPase is oligomycin sensitive.[22] Similarly, the ATPase of luminal membrane vesicles of gastric mucosa is sensitive to DCCD and not to oligomycin.[23] The present state of knowledge appears to indicate ATPases that translocate H^+ or OH^- do exist in the plasma membranes of epithelial cells equipped for acid-base transport. In turtle bladder, such a H^+-ATPase would be clearly different from the $(Na^+ + K^+)$-ATPase of the basolateral cell membrane since H^+ secretion is not inhibited by ouabain. Preliminary studies [37] of a crude vesicle preparation of the luminal membrane of turtle bladder reveal an abundance of Mg^{++}-dependent ATPase with almost no $(Na^+ + K^+)$-stimulated or ouabain-inhibitable ATPase. About a third of the ATPase activity is oligomycin-sensitive, the remaining oligomycin insensitive ATPase may represent H^+-ATPase of the cell membrane.

These studies pointing to a proton-translocating ATPase as the acidification pump in turtle bladder are consistent with previous studies by Green *et al.*[24] and Schwartz *et al.*[5] on the transported ion in urinary acidification. Since there is as yet no unanimity of view about the ion that is transported across the luminal cell membrane of turtle bladder, we also refer the reader to studies by Schilb and Brodsky [1, 25-27] that suggest HCO_3^- is the transported ion. For the purposes of this paper, we have focussed on the control of the transport rate of acidification from the cell side of the pump. How is the pump rate controlled by the supply of protons to it? Or, put in another way, how are the OH^- ions generated behind the pump disposed? Since OH^- reacts with CO_2 to form HCO_3^-, the disposition of HCO_3^- and its transport across the basolateral cell membrane become important factors in the control of the pump rate.

Studies with the weak acid 5,5-dimethyl-2,4-oxazolidinedione (DMO) have shown that the epithelial cells of turtle bladder are relatively alkaline even when the luminal solution is acid.[4] FIGURE 3 shows the concentration profile of H^+ and OH^- across the epithelium at a luminal pH of 5.27 ± 0.04 and a serosal pH of 7.40. At this luminal pH, the rate of H^+ secretion is inhibited but not abolished. Hence, H^+ secretion across the luminal cell membrane takes place against a steep concentration gradient. On the other hand, the efflux of alkali across the serosal membrane can be accounted for by a concentration driving force. Little is known about the electrical profile along the acidification pathway under these specific conditions. In the short-circuited bladder, Hirschhorn and Frazier [28] found that the intracellular electrical potential was negative by about 10 mV. In more recent studies carried out using microelectrodes with small tips by Nagel and Brodsky,[38] somewhat larger negative values were observed. Since it is technically difficult to obtain micropuncture data in the turtle bladder and since acidification may take place in a population of specialized epithelial cells, the information on the electrical profile remains of

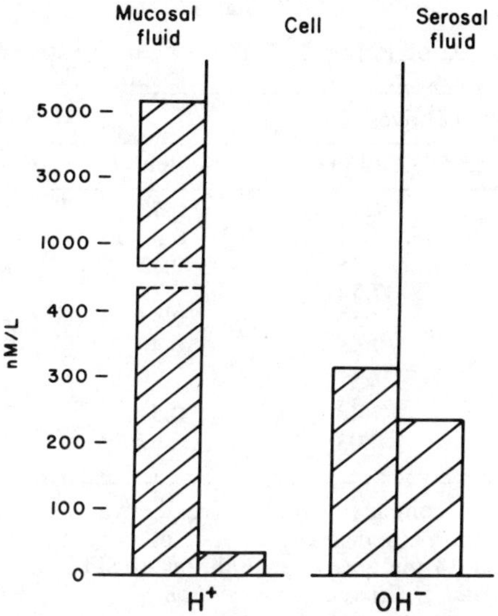

FIGURE 3. Concentration profile of H^+ and OH^- across the mucosal and serosal borders of the epithelium. (From Steinmetz.[4] By permission of the *Journal of Clinical Investigation.*)

necessity limited. From what is known in turtle bladder and other epithelia, it seems reasonable to estimate that the intracellular electrical potential would be negative. The electrical profile, therefore, increases the magnitude of the uphill step for H^+ transport into the mucosal solution and provides an additional passive driving force for the efflux of HCO_3^- across the serosal membrane. The serosal cell membrane is thought to be highly permeable to HCO_3^- and is thought to contain transport sites, which facilitate HCO_3^- transfer and which can be blocked by serosal addition of the disulfonic stilbene, SITS (4-acetamido-4′-isothiocyano-2,2′ disulfonic stilbene). It is of interest that SITS inhibits H^+ secretion (J_H) only when added to the serosal, but not the mucosal side of the epithelium.[29, 30] Since SITS is an inhibitor of anion transport and probably exerts its effect on a transport protein from the outside of the cell membrane, the observed inhibition of J_H was attributed to inhibition of the exit step for HCO_3^- rather than of the H^+ pump directly.[30] If this interpretation is correct, one would predict that the cell would become more alkaline as a result of HCO_3^- accumulation. Cohen *et al.*[30] measured the apparent cell pH by means of the DMO method in seven experiments in which bladder halves exposed to 0.2 mM SITS were compared to control halves. The results are given in TABLE 1. In each of the experiments, there was an increase in the overall cell pH. Although the pH increment was highly significant, it was small: only 0.13 ± 0.02 pH units, a ΔpH that would have little effect on J_H if it were imposed at the luminal side of the H^+ pump. These results raise several interesting issues. First, the DMO method measures the mean OH^- concentration of the cell water of the entire epithelium. If the region of alkalinity in series with the H^+ pump is small, the overall pH would underestimate the pH increment. For example, if H^+ secretion were limited to a population of cells with a volume of 10% of the cell water and if only this population were affected by SITS, it would take a 0.7 pH unit increment in this cell population to give the

TABLE 1

EFFECT OF SEROSAL ADDITION OF SITS (2×10^{-4} M) ON J_H AND APPARENT CELL pH *

	Control Halves		SITS Halves		
	J_H μmol/hr	Cell pH′	J_H μmol/hr	Cell pH′	ΔpH
1	1.81	7.48	0.55	7.60	0.12
2	1.04	7.59	0.00	7.72	0.13
3	0.70	7.35	0.32	7.60	0.25
4	1.42	7.54	0.77	7.60	0.06
5	1.36	7.50	0.54	7.62	0.12
6	0.66	7.46	0.40	7.60	0.14
7	1.20	7.44	0.66	7.51	0.07
Mean	1.36	7.48	0.54	7.61	0.13†
± SE	0.14	0.03	0.07	0.02	0.02

* Control halves are compared to their respective SITS-treated halves. J_H was measured by the pH stat method in the absence of exogenous CO_2. Apparent cell pH (pH′) was determined 30 minutes after addition of SITS. (From Cohen *et al.*[30] By permission of the *Journal of Clinical Investigation.*)

† $p < 0.005$.

overall result obtained. The second issue deals with the control of the pump rate by the electrochemical potential gradient, $\Delta\tilde{\mu}_H$. When the pH is lowered at the luminal side of the pump from 7.4 to about 4.4 while the serosal side is kept at pH 7.4, J_H is reduced to zero. Under these circumstances, cell pH remains relatively constant.[4] What is the effect of cell alkalinity on J_H? Does it require an equivalent ΔpH or $\Delta\psi$ on the cell side of the pump to stop transport? This question was examined by increasing the HCO_3^- concentration and the pCO_2 on the serosal or nutrient side of the bladder.[31] Under these conditions § the cellular acid-base milieu is imposed from the serosal solution. CO_2 is highly diffusible and the basolateral (but not the luminal) cell membrane is quite permeable to HCO_3^-.[3] Cohen and Steinmetz[31] found that J_H measured as the short-circuit current in ouabain-treated bladders decreased as serosal [HCO_3^-] was increased from 20 mM to 60 mM at a constant pCO_2 of 5% in the gas phase. The decrease in J_H was linear with either increasing serosal [HCO_3^-] or mean cell [OH^-] measured by the DMO method. If these results are examined in terms of the overall cell pH, it becomes evident that J_H is much more sensitive to pH changes on the cell side of the pump than on the luminal side. FIGURE 4 shows this asymmetry in the control of J_H by ΔpH. The ΔpH values represent the changes from pH 7.4, the control condition of level secretion. In all instances the transepithelial PD was maintained nullified by an external voltage clamp. On the right of the ordinate the linear decrease in J_H brought about by lowering luminal pH is shown schematically from previous studies.[16–18] On the left, the effects of the imposition of an unfavorable pH gradient on the cell side are shown. Cell pH was increased by increasing serosal [HCO_3^-] at constant pCO_2.[31] A small ΔpH on the cell side had a large effect on J_H. A similar response of J_H was observed when cell pH was altered in 0.3 pH unit steps by

§ In the absence of exogenous CO_2 and HCO_3^- cell pH is governed more by the production of metabolic CO_2 and HCO_3^- and by other cellular acid-base events than by the pH of the serosal bulk solution.[3]

acute changes in the CO_2 tension between 5% and 20% CO_2 at constant serosal [HCO_3^-]. The results of these maneuvers to alter cell pH suggest that the pump rate is not symmetrically controlled by ΔpH. The possibility remains, however, that if we would be able to follow the changes in the total electrochemical proton potential, $\Delta\tilde{\mu}_H$, at the critical sites across the energy barriers within the pump pathway, the control of J_H would prove to be symmetrical. In terms of the organization of the transport system, the sidedness of the pH control must reflect the fact that the input of metabolic energy as well as the supply of protons take place from the cytoplasm; these supply reactions, which are kinetic factors, dominate the observed inhibition of J_H.

In summary, the transport rate of protons by this "tight" urinary epithelium is regulated by negative feedback from the electrochemical potential of protons in the urinary compartment; an increase in $\tilde{\mu}_H$ of about 180 mV will stop the pump. Within the cell, small pH changes have large effects on the transport rate.

Organization of Transport Pathways for Acidification and for Cl-HCO$_3$ Exchange

Bicarbonate moves in two directions across the basolateral surface of the epithelium. The HCO_3^- formed in the acidification pathway flows downhill from the cell into the serosal compartment whereas serosal HCO_3^- is transported from the serosal (S) to the mucosal (M) compartment in exchange for absorbed Cl^-. Do these HCO_3^- flows use common or separate transport sites in the basolateral membrane? If the sites on this cell membrane are separate, are the pathways across the entire epithelium separate or is there a common HCO_3^- pool for the two transport systems? Not all the answers to these questions are

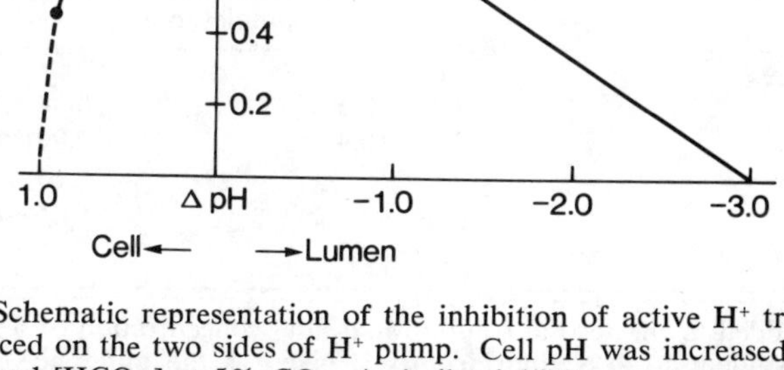

FIGURE 4. Schematic representation of the inhibition of active H^+ transport by pH changes produced on the two sides of H^+ pump. Cell pH was increased (left side) by increasing serosal [HCO_3^-] at 5% CO_2. A similar inhibitory response was observed in another series of experiments in which the CO_2 concentration was lowered in steps from 20% to 5% at constant serosal [HCO_3^-]. The linear inhibition of J_H with decreasing luminal pH (right side) is drawn from previous studies.[16, 17] J_H was normalized to 1.0 at level secretion in the different experiments, see text.

in, but several lines of evidence indicate that at least parts of the two transport paths are separate.

Husted, Cohen and Steinmetz [15] examined the effects of SITS on the Cl-HCO_3 exchange transport system under conditions of zero net H^+ secretion. Bicarbonate transport from S to M was measured by pH stat titration following the addition of HCO_3^- to S in a final concentration of 20 mM. Unidirectional M to S Cl^- fluxes were measured simultaneously. TABLE 2 shows that under control conditions J_{HCO_3} was about equal to the increment in the M to S Cl^- flux caused by HCO_3^- addition, in accord with previous results reported by Leslie *et al.*[11] Serosal addition of 5x10^{-4}M SITS had no significant effect on either the HCO_3^- or the Cl^- flux. The effects of serosal addition of SITS were also explored in bladders bathed with symmetrical HCO_3^- solutions in which net Na^+ transport was abolished by 5×10^{-4} M ouabain. Net Cl^- fluxes were obtained from the difference between the two unidirectional Cl^- fluxes. Under these conditions H^+ secretion (J_H) can be measured as the short-circuit current. As shown in FIGURE 5, serosal SITS addition caused a marked inhibition of J_H but had no effect on the net Cl^- flux. When 5×10^{-4} M SITS was added to the mucosal solution the unidirectional and net Cl^- fluxes were again unaffected.[15] These studies showing that SITS inhibits the exit of HCO_3^- in the acidification pathway, but not the oppositely directed HCO_3^- flow of the anion exchange system, strongly suggest that the two HCO_3^- flows take place at separate transport sites in the serosal membrane.

Another line of evidence suggesting the Cl-HCO_3 exchange transport system is independent of the acidification pump is provided by Cohen and Karp [32] in a study of water turtles that were bicarbonate loaded *in vivo*. The bladders of bicarbonate-loaded turtles exhibited increased HCO_3^- secretory rates *in vitro*, but had unaffected acidification rates. The two transport systems appear to have separate adaptive responses to loading.

The final observation deals with the effects of serosal addition of SITS on the characteristics of the luminal surface of the bladder. In control segments

TABLE 2

EFFECT OF SEROSAL SITS ON S → M BICARBONATE ($J_{HCO_3^-}$) AND BICARBONATE-DEPENDENT M → S CHLORIDE FLUX (J_{Cl^-}) *

	Control	Experimental
	μmoles/hour	
Control (n=8)		
$J_{HCO_3^-}$	1.16±0.22	0.93±0.18
J_{Cl^-}	1.07±0.47	0.90±0.32
SITS (n=7)		
$J_{HCO_3^-}$	0.99±0.10	1.05±0.13
J_{Cl^-}	0.96±0.23	0.95±0.24

* SITS was added to the serosal solution in a final concentration of 5×10^{-4} M. The bicarbonate-dependent M → S chloride flux is given as the difference in the M → S chloride flux with and without 20 mM HCO_3^- in the serosal solution. $J_{HCO_3^-}$ was measured by pH stat titration in the mucosal solution at $(M\ pH)_{J_H=0}$. (From Husted *et al.*[15])

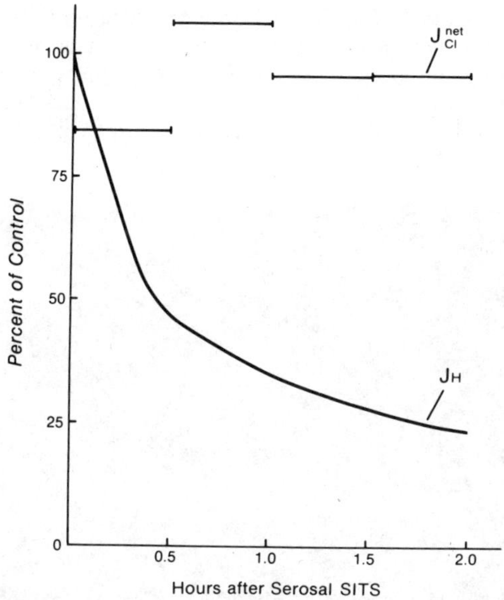

FIGURE 5. Selective inhibition of H^+ secretion (J_H) by 5×10^{-4} M SITS. Net chloride flux (J_{Cl}^{net}) measured simultaneously was unaffected. (From Husted *et al.*[15] By permission of the *Journal of Membrane Biology*.)

of turtle bladder, scanning electron microscopy reveals that about 12 percent of the surface epithelial cells have a luminal surface that is characterized by a prominent network of microridges or microplicae. Two of these cells with microplicae are clearly shown in the upper panel of FIGURE 6. They appear round compared to the hexagonal granular cells. On transmission electron microscopy they are often flask-shaped and rich in mitochondria. Mueller *et al.*[33] compared the surface appearance of control segments of bladder with that of segments in which H^+ secretion was inhibited by exposure to 5×10^{-4}M SITS on the serosal side for two hours. One example of such a comparison is shown in FIGURE 6. In the SITS-exposed segments, cells with microplicae were no longer detectable; the microplicae had disappeared from the surface of the mitochondria-rich cells. The quantitation of these changes and the description of the morphologic fate of the cells with microplicae after SITS exposure are the subject of another study. These morphologic observations are consistent with the interpretation that cells with microplicae specialize in urinary acidification. One explanation of the insensitivity of the Cl-HCO_3 exchange flows to SITS would be that these anion flows occur in cells with different surface characteristics. Schwartz *et al.*[34] and Rosen *et al.*[35] have suggested that the carbonic anhydrase-rich cells of the turtle bladder and Colombian toad bladder might be responsible for urinary acidification on the basis of a correlation between carbonic anhydrase histochemistry and acidifying capacity. Since the carbonic anhydrase-rich cells correspond to the mitochondria-rich cells and may be characterized by either prominent microplicae or microvilli, it is not possible to conclude that SITS affects all carbonic anhydrase-rich cells. It is of interest that, in contrast to SITS, the carbonic anhydrase inhibitor acetazolamide inhibits H^+ secretion as well as Cl-HCO_3 exchange transport.[2, 11, 34] The greater selectivity of the physiologic effects of SITS may permit further correlation between the morphology of the epithelial cells and their transport functions.

The Cl-HCO_3 exchange flows are not only relatively independent of the

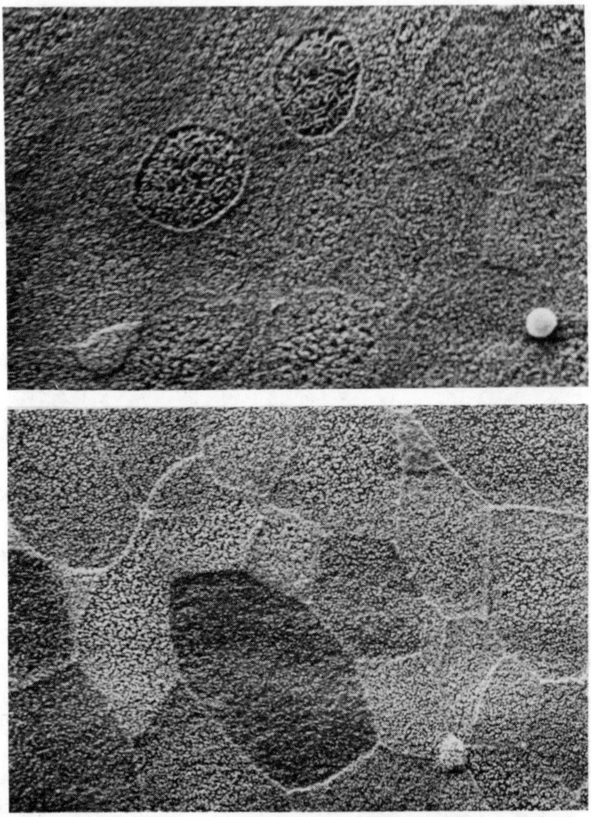

FIGURE 6. Scanning electron micrographs of the luminal surface of two segments on a turtle bladder fixed after mounting in lucite chambers ($\times 1200$). The micrograph of the control segment (upper panel) shows two cells with a rounded or oval border with characteristic microplicae. Two hours after exposure of bladder segments to 5×10^{-4} M SITS on the serosal side H^+ secretion is markedly reduced and cells with microplicae are no longer detectable as such on the micrograph (lower panel), see text.

process of acidification but also of sodium transport. In contrast to the electrogenic Cl^- transport systems of many epithelia that are inhibited by ouabain and represent secondary active transport,[36] the electroneutral Cl-HCO_3 exchange system of the turtle bladder is not inhibited by ouabain [11, 15] and not dependent on ambient sodium.[39]

In summary, we have presented several lines of evidence suggesting that the acidification pathway in the turtle bladder is at least in part separate from the electroneutral transport system for bicarbonate secretion.

References

1. Schilb, T. P. & W. A. Brodsky. 1966. Acidification of mucosal fluid by transport of bicarbonate ion in turtle bladders. Am. J. Physiol. **210:**997–1008.
2. Steinmetz, P. R. 1967. Characteristics of hydrogen ion transport in urinary bladder of water turtle. J. Clin. Invest. **46:**1531–1540.
3. Steinmetz, P. R. 1974. Cellular mechanisms of urinary acidification. Physiol. Review **54:**890–956.
4. Steinmetz, P. R. 1969. Acid-base relations in epithelium of turtle bladder: site of active step in acidification and role of metabolic CO_2. J. Clin. Invest. **48:** 1258–1265.
5. Schwartz, J. H., J. T. Finn, G. Vaughn & P. R. Steinmetz. 1974. Distribution of metabolic CO_2 and the transported ion species in acidification by turtle bladder. Am. J. Physiol. **226:**283–289.
6. Dixon, T. E. & Q. Al-Awqati. 1979. Urinary acidification in turtle bladder is due to a reversible proton-translocating ATPase. Proc. Natl. Acad. Sci. USA **76:**3135–3138.
7. Steinmetz, P. R., A. Mueller & R. Beauwens. 1979. Effects of ATPase inhibitors on anaerobic H^+ transport by turtle urinary bladder. Fed. Proc. **38:**1121.
8. Schwartz, J. H. & P. R. Steinmetz. 1971. CO_2 requirements for H^+ secretion by the isolated turtle bladder. Am. J. Physiol. **220:**2051–2057.
9. Steinmetz, P. R., R. S. Omachi & H. S. Frazier. 1967. Independence of hydrogen ion secretion and transport of other electrolytes in turtle bladder. J. Clin. Invest. **46:**1541–1548.
10. Schwartz, J. H. 1976. H^+ current response to CO_2 and carbonic anhydrase inhibition in turtle bladder. Am. J. Physiol. **231:**565–572.
11. Leslie, B. R., J. H. Schwartz & P. R. Steinmetz. 1973. Coupling between Cl^- absorption and HCO_3^- secretion in turtle urinary bladder. Am. J. Physiol. **225:**610–617.
12. Oliver, J. A., S. Himmelstein & P. R. Steinmetz. 1975. Energy dependence of urinary bicarbonate secretion in turtle bladder. J. Clin. Invest. **55:**1003–1008.
13. Brodsky, W. A. & T. P. Schilb. 1966. Ionic mechanisms for sodium and chloride transport across turtle bladders. Am. J. Physiol. **210:**987–996.
14. Gonzalez, C. F., Y. E. Shamoo & W. A. Brodsky. 1967. Electrical nature of active chloride transport across short-circuited turtle bladders. Am. J. Physiol. **212:**641–650.
15. Husted, R. F., L. H. Cohen & P. R. Steinmetz. 1979. Pathways for bicarbonate transfer across the serosal membrane of turtle urinary bladder: Studies with a disulfonic stilbene. J. Membrane Biol. **47:**27–37.
16. Steinmetz, P. R. & L. R. Lawson. 1971. Effect of luminal pH on ion permeability and flows of Na^+ and H^+ in turtle bladder. Am. J. Physiol. **220:**1573–1580.
17. Al-Awqati, Q., A. Mueller & P. R. Steinmetz. 1977. The transport of H^+ against electrochemical gradients in turtle urinary bladder. Am. J. Physiol. Renal Fluid Electrol. Physiol. **2:**F502–F508.
18. Beauwens, R. & Q. Al-Awqati. 1976. Active H^+ transport in the turtle urinary bladder: coupling of transport to glucose oxidation. J. Gen. Physiol. **68:**421–439.

19. RACKER, E. 1976. A new look at mechanisms in bioenergetics. Academic Press. New York, N.Y.
20. KABACK, H. R. 1976. Molecular biology and energetics of membrane transport. J. Cell. Physiol. **89:**585–594.
21. DUFOUR, J. P. & A. GOFFEAU. 1978. Solubilization by lysolecithin and purification of the plasma membrane ATPase of the yeast *Schizosaccharomyces pombe*. J. Biol. Chem. **253:**7026–7032.
22. DELHEZ, J., J. P. DUFOUR, D. THINES & A. GOFFEAU. 1977. Comparison of the properties of plasma membrane-bound and mitochondria-bound ATPases in the yeast *Schizosaccharomyces pombe*. Eur. J. Biochem. **79:**319–328.
23. SACHS, G., H. H. CHANG, E. RABON, R. SCHACKMAN, M. LEWIN & G. SACCOMANI. 1976. A nonelectrogenic H^+ pump in plasma membranes of hog stomach. J. Biol. Chem. **251:**7690–7698.
24. GREEN, H. H., P. R. STEINMETZ & H. S. FRAZIER. 1970. Evidence for proton transport by turtle bladder in the presence of ambient bicarbonate. Am. J. Physiol. **218:**845–850.
25. SCHILB, T. P. & W. A. BRODSKY. 1972. CO_2 gradients and acidification by transport of HCO_3^- in turtle bladders. Am. J. Physiol. **222:**272–281.
26. SCHILB, T. P. 1980. A new macromethod for estimating CO_2 tension and its use in identifying the acidification mechanism in turtle bladder or other epithelia. This volume.
27. BRODSKY, W. A. 1980. Active secretion and reabsorption of HCO_3 in turtle bladder. This volume.
28. HIRSCHHORN, N. & H. S. FRAZIER. 1971. Intracellular electrical potential of the epithelium of turtle bladder. Am. J. Physiol. **220:**1158–1161.
29. EHRENSPECK, G. & W. A. BRODSKY. 1976. Effects of 4-acetamido-4′-isothiocyano-2,2′-disulfonic stilbene on ion transport in turtle bladders. Biochim. Biophys. Acta. **419:**555–558.
30. COHEN, L. H., A. MUELLER & P. R. STEINMETZ. 1978. Inhibition of bicarbonate exit step in urinary acidification by a disulfonic stilbene. J. Clin. Invest. **61:** 981–986.
31. COHEN, L. H. & P. R. STEINMETZ. 1979. Cell [OH^-] as a determinant of H^+ secretion by turtle urinary bladder: asymmetry in the control of the H^+ pump. Clin. Res. **27:**520A.
32. COHEN, L. H. & P. H. KARP. 1978. Adaptive response to metabolic alkalosis in turtle bladder. Kidney Intern. **14:**754.
33. MUELLER, A. L., R. G. KESSEL, R. F. HUSTED, L. H. COHEN & P. R. STEINMETZ. 1978. Effects of a disulfonic stilbene (SITS) on surface morphology and urinary acidification in turtle bladder. Fed. Proc. **37:**511.
34. SCHWARTZ, J. H., S. ROSEN & P. R. STEINMETZ. 1972. Carbonic anhydrase function and the epithelial organization of H^+ secretion in turtle urinary bladder. J. Clin. Invest. **51:**2653–2662.
35. ROSEN, S., J. A. OLIVER & P. R. STEINMETZ. 1974. Urinary acidification and carbonic anhydrase distribution in bladders of Dominican and Colombian toads. J. Membrane Biol. **15:**193–205.
36. FRIZZEL, R. A., M. FIELD & S. G. SCHULTZ. 1979. Sodium-coupled chloride transport by epithelial tissues. Am. J. Physiol. **236:**F1–F8.
37. STEINMETZ, P. R. & H. R. KABACK. Unpublished observations.
38. NAGEL, W. & W. A. BRODSKY. Personal communication.
39. HUSTED, R. F. & E. D. EYMAN. Unpublished observations.

DISCUSSION

R. B. GUNN (*University of Chicago, Chicago, Ill.*): The asymmetry of inhibition of transport by proton gradients on the inside of the cell and on the luminal side may be due to negative surface charges on the inside of the cell. Have you looked for that same kind of asymmetry in the other transport system—the one that secretes HCO_3^-? Do the two transport systems share the same HCO_3^- pool?

P. R. STEINMETZ: The rate of HCO_3^- secretion must be measured by pH stat titration and this is possible only when H^+ secretion is first brought to zero. It has been difficult, therefore, to examine the dependence of HCO_3^- secretion on concentration gradients. Leslie *et al.*[11] have shown that the rate of HCO_3^- secretion is not voltage-dependent. In our hands this is an electroneutral system.

In response to your second question, the acidification pathway is at least in part separate from the pathway for Cl-HCO_3 exchange. There is probably little interaction between the HCO_3^- pools of the two systems.

M. BURG: Do you have information on the specificity of the chloride-bicarbonate exchange operation, e.g. can any other anions be substituted for chloride?

STEINMETZ: Dr. Loren Cohen in our laboratory has substituted bromide for chloride in the luminal solution. Bromide seems to work less well, causing a clear reduction in bicarbonate secretion. When sulfate is substituted for chloride (and chloride concentrations are reduced to about 0.5 mEq/L), there is a marked decrease in bicarbonate secretion. These effects are reversible. Fluoride and iodide have not been studied because of their toxic actions. I should mention that the Cl-HCO_3 exchange is unaffected by ouabain or sodium removal. In that respect, it differs from the electroneutral NaCl cotransport described in other epithelia.

T. P. SCHILB (*Louisiana State University, New Orleans, La.*): When you inhibit acidification by placing bicarbonate in the serosal solution and removing HCO_3 from the luminal solution, is it not possible that this maneuver might inhibit bicarbonate reabsorption just as effectively as proton secretion? Can you exclude this possibility?

STEINMETZ: The issue of the transported ion in acidification was not dealt with in these experiments (see Ref. 5).

G. EHRENSPECK (*Ohio University College of Osteopathic Medicine*): Did you do experiments with bicarbonate on both sides?

STEINMETZ: Yes. In one group of experiments with symmetrical bicarbonate solutions net chloride transport was unaffected by SITS. H^+ secretion measured in the same bladders was inhibited by some 80% (see FIGURE 5).

A NEW MACRO METHOD FOR ESTIMATING CO_2 TENSION AND ITS USE IN IDENTIFYING THE ACIDIFICATION MECHANISM IN TURTLE BLADDER OR OTHER EPITHELIA

Theodore P. Schilb

Department of Physiology
Louisiana State University
New Orleans, Louisiana 70119

INTRODUCTION

The method to be described has the following features. The bladder may be filled with, and incubated in, solutions that have been pre-equilibrated with the chosen gas mixture. Once the gas-equilibrated, lumen fluid enters the apparatus, it is never in contact with any substance other than glass, mercury, and the mucosa of the bladder. Changes in the composition of the lumen fluid are solely due to the action of the bladder wall. When the bladder is replaced with Laytex,® no changes in composition are detected. The only dead space in the system is a small residual volume in the bladder itself that cannot be perfectly evacuated. Serial measurements of the composition of lumen fluid can be made at will.

pCO_2s are determined directly with a mercury manometer. Sample size is determined from the weight of mercury displaced by the volume of the sample. When measurements are made at constant temperature, the relative pCO_2s of serial samples can be estimated directly from the raw readings of the manometer and the beam balance by a formula having the following form: $(P_1 - P_2)$ mm Hg/g Hg displaced. No constants of any kind are required.

FIGURES 1 and 2 show schematic diagrams of the apparatus. The configuration of the stopcocks in FIGURE 1 is appropriate for drawing a sample of lumen fluid into the reaction chamber for analysis. By performing the appropriate maneuvers with the height of the reservoir and the stopcocks it is also possible to: 1. Transfer gas-equilibrated fluid from the cup to the lumen of the bladder; 2. Transfer aliquots of lumen fluid to the cup for pH measurements; and 3. Equilibrate samples of other fluids, such as 0.1 N H_2SO_4, in the cup with the same gas as that perfusing the incubation chamber and to analyze them while the bladder is incubating. (During the incubation of lumen fluid in the bladder, the side arm is filled with mercury so that there is no dead space.)

The Analytical Procedure

In essence, the procedure is as follows: A sample is drawn into the reaction chamber by opening the syphon. The displaced mercury is weighed. An appropriate volume of degassed distilled water is added to the sample. The sample is degassed by drawing a vacuum. The gas phase containing free CO_2, O_2 and water vapor is compressed to a fixed volume and P_1 is read on the manometer. If the sample contains bicarbonate, one ml of 0.6 N H_2SO_4 is then added and the procedure repeated to obtain P_2, which represents the total

0077-8923/80/0341-0090 $01.75/0

CO_2, O_2 and water vapor. Then 1 ml of 4 normal NaOH is added, and the procedure is repeated to obtain P_3, which represents the O_2 and water vapor. The free CO_2 is estimated from the difference in P_1 and P_3. The bicarbonate is estimated from the difference in P_2 and P_1.

In order to determine if this method for determining free CO_2 is valid, and whether there might be substances in the lumen fluid that interfere, the following experiment was done. Twelve aliquots of lumen fluid that had been in contact with a bladder for 5 hours were paired with aliquots of 0.1 N H_2SO_4

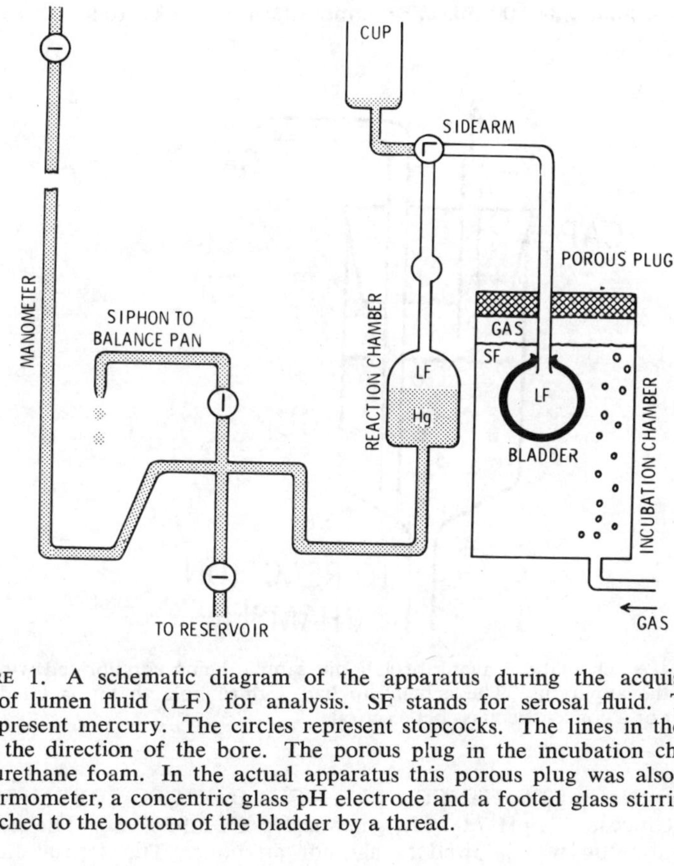

FIGURE 1. A schematic diagram of the apparatus during the acquisition of a sample of lumen fluid (LF) for analysis. SF stands for serosal fluid. The shaded areas represent mercury. The circles represent stopcocks. The lines in the stopcocks indicate the direction of the bore. The porous plug in the incubation chamber was of polyurethane foam. In the actual apparatus this porous plug was also penetrated by a thermometer, a concentric glass pH electrode and a footed glass stirring rod that was attached to the bottom of the bladder by a thread.

and coequilibrated with the same gas. Each pair of samples was analyzed on the same day and at the same temperature, but various pairs of samples were analyzed on different days at different temperatures and the mucosal fluid samples had been in contact with different bladders. Thus, the conditions for the controls were identical to those of the real experiments. The mean difference between the concentration of dissolved free CO_2 determined in the sulfuric acid and in the mucosal fluid was 0.016 ± 0.020 mM and the standard deviation of the individually estimated difference was 0.067 mM. These values represent the lower limit for the detection of a temporal or transmural change in the concentration of dissolved free CO_2 and demonstrated the extent to which the method is valid.

Experimental Protocol

Bladders were excised, rinsed, filled with and immersed in gas-equilibrated bathing fluids. The luminal fluid volume was 6 ml. The composition of the lumen fluid expressed in mM was: $NaHCO_3$, 5; NaCl, 95, pH 6.5. The volume of the serosal fluid was 500 ml. The composition of the serosal fluid expressed

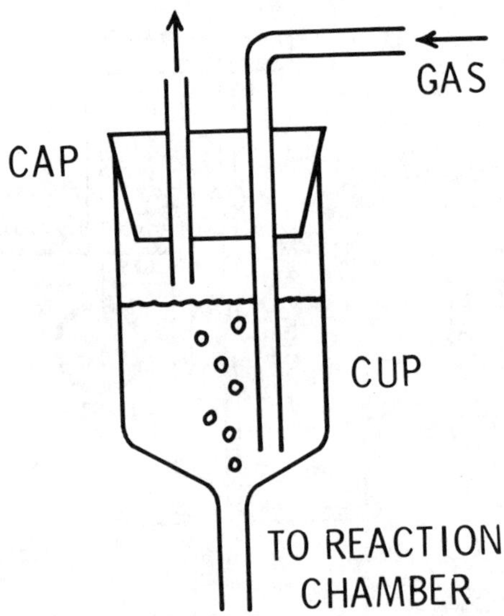

FIGURE 2. A schematic drawing of a liquid sample being equilibrated with gas in the cup of the apparatus. The actual cap had a third hole that was fitted with a thermometer or a concentric glass pH electrode.

in mM was: NaCl, 83.5; $NaHCO_3$, 17.7; KCl, 4.0; KH_2PO_4, 0.8; $MgSO_4$, 0.8; CaCl, 2.0; Glucose, 11, pH 7.0. The gas was 7% CO_2, 93% O_2. The assembled experimental setup was allowed to age for an hour. The serosal fluid was gassed continuously. At the end of the hour, the lumen fluid was completely replaced by fresh gas equilibrated lumen fluid. After five minutes an "initial sample" of lumen fluid was drawn and analyzed. The system was allowed to age for four to five hours. Three serial 0.7 ml "final" lumen fluid samples were withdrawn and analyzed.

As the initial luminal and serosal pCO_2s were equal, and the serosal pCO_2 stayed constant, the magnitude and direction of temporal changes in the lumen pCO_2 and of differences in the transmural pCO_2 at the end of the incubation should be in the same direction and have about the same magnitude. This was found to be the case. In order to ascertain this, the mean value of the final mucosal fluid pCO_2 was compared to: the pCO_2 found in the initial sample of lumen fluid; the pCO_2 found in 0.1 N H_2SO_4 that had been coequilibrated with the same gas as was perfusing the serosal fluid; and, the pCO_2 of the serosal fluid, as estimated from the pH and total CO_2 by the Henderson-Heselbalk equation. In general, the value of pCO_2 found in the initial lumen sample, the coequilibrated acid and the serosal fluid were the same, within experimental error. If there was a significant difference between these three measurements, the value obtained from the coequilibrated sulfuric acid was assumed to be correct.

Principle

According to Ficks law, if a lumen fluid is surrounded by a CO_2 producing tissue that is submerged in a solution having a fixed pCO_2, the pCO_2 of the lumen fluid will come into concentration equilibrium with the pCO_2 of the cells adjacent to the lumen fluid. When this occurs, all of the metabolic CO_2 will leave the tissue by diffusing into the serosal fluid. This means that the pCO_2 at every point in the tissue is higher than that of the serosal fluid and that the pCO_2 of the innermost cells adjacent to the lumen fluid is highest of all. Thus, in the absence of an acidifying mechanism in such a system, the pCO_2 of the cells adjacent to the lumen is the base line. If the lumen fluid contains HCO_3, a hydrogen secretory mechanism will generate CO_2 in the lumen fluid and luminal pCO_2 will rise above that of the cells. A bicarbonate absorption mechanism will cause hydration of CO_2 in the lumen fluid and luminal pCO_2 will fall below that of the cells. Unfortunately, we have no way of measuring the pCO_2 in the cells. This means that if we find the luminal pCO_2 to become and remain greater than that of the serosal fluid we will never know whether the result is due to H^+ secretion or the diffusion of metabolic CO_2 into the lumen fluid.

On the other hand, if we find that the luminal pCO_2 becomes and remains less than that of the serosal fluid, we are guaranteed by Ficks law that the pCO_2 of the lumen fluid is less than that of the cells, whatever that value may be. Thus, when the luminal pCO_2, pH and $[HCO_3]$ all go down together we have proof of HCO_3 absorption. The CO_2 could not have left the lumen by diffusion and if it had left as H_2CO_3, the fluid would become alkaline instead of acid.[2]

When the pCO_2 of the lumen fluid becomes less than that of the serosal fluid or even less than that of the cell fluid, we are looking at the balance of two dynamic processes: the diffusion of metabolic CO_2 into the lumen fluid and the hydration of CO_2 in the lumen fluid, subsequent to the removal of HCO_3. It is now recognized that there are several factors that can influence this balance in such a way that the luminal pCO_2 will remain equal to, or greater than, that of the serosal fluid. Some of these are: the ratio of the lumen fluid volume to the mass of surrounding tissue,[6] the buffer poise,[2] and the metabolic state of the turtle from which the bladder is obtained.[7] In addition, there may be other important factors not yet considered. For these reasons so long as the tissue pCO_2 is not directly measured, experiments that result in a lumen pCO_2 greater than that of the serosal fluid fail to provide evidence of bicarbonate absorption, but do not militate against the bicarbonate transport postulate and do not contribute evidence of H^+ secretion.

FIGURE 3 presents data that clearly demonstrate the presence of a bicarbonate absorption mechanism and the effect of the metabolic state of the *in vivo* turtle on the balance between the rate of diffusion of metabolic CO_2 into the lumen fluid, and the rate of hydration of CO_2 in the lumen fluid subsequent to the absorption of HCO_3 during the *in vitro* incubation.

The data show the effect of the housing temperature of the turtles on the final transmural gradient developed during the *in vitro* incubation. In order to obtain these results, it was necessary to adapt the turtles to the chosen temperature for at least five days. The turtles were fed as much whole raw fish as they would eat. For each 6° increase in temperature the overall metabolic rate of the turtle doubled, the bladder urine became more acid, and the *in vitro* rate of

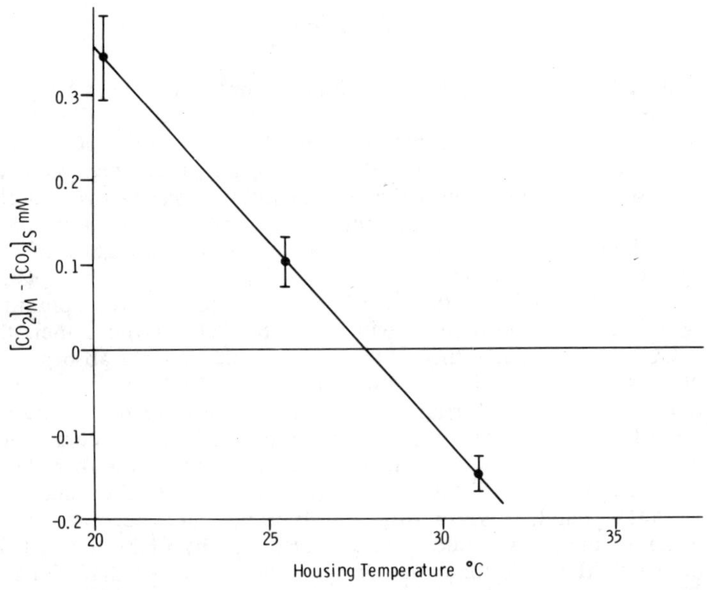

FIGURE 3. A plot of the final millimolar transmural differences in the concentration of free CO_2 ($[CO_2]_{LF}$ minus $]CO_2]_{SF}$) that developed during the *in vitro* incubation of bladders, versus the housing temperature to which the donor turtles had been adapted. The vertical bars represent the standard error of the mean. At 20° C, N=6; 26° C, N=11; 32° C, N=13.

acidification increased. In each case, the bladder was allowed to acidify the lumen fluid to a pH of about 6.0 and consume about 60% of the HCO_3. On the average, this required 5 hours for bladders from 20° turtles and 4.5 and 4.0 hours for bladders taken from 26° and 32° turtles, respectively. The number of turtles at 20°, 26°, and 32° were 6, 11, and 13, respectively. Varying the temperature of the *in vitro* incubation rather than that of the turtles had no apparent effect on the final transmural CO_2 gradient.

Data on the bladders taken from 32° turtles show that the luminal pCO_2 is significantly less than that of the serosal fluid and clearly demonstrate that the mechanism of acidification is one of HCO_3 transport from lumen to serosa. These findings confirm and extend previous work by Schilb and Brodsky.[1, 3–5]

References

1. Brodsky, W. A. & T. P. Schilb. 1967. Mechanism of acidification in turtle bladder. Fed. Proc. Symposium **26:**1314–1321.
2. Brodsky, W. A. & T. P. Schilb. 1973. The means of distinguishing between hydrogen secretion and bicarbonate reabsorption: Theory and applications to reptilian bladder and mammalian kidney. *In* Current Topics in Membranes and Transport. A. Kleinzeller & F. Bronner, Eds. Academic Press, New York, N.Y.
3. Gonzales, C. F. & T. P. Schilb. 1969. Acetazolamide-sensitive short-circuiting versus mucosal HCO_3 concentration in turtle bladders. Biochim. Biophys. Acta **193:**419–429.
4. Schilb, T. P. & W. A. Brodsky. 1963. The acidification mechanism of isolated turtle bladder. Proc. 2nd Intern. Cong. of Nephrol., Excerpta Medica Found. Series, **78:**103–105.
5. Schilb, T. P. & W. A. Brodsky. 1966. Acidification of mucosal fluid by transport of bicarbonate ion in turtle bladders. Am. J. Physiol. **210:**997–1008.
6. Schilb, T. P. & W. A. Brodsky. 1972. CO_2 gradients and acidification by transport of HCO_3 in turtle bladders. Am. J. Physiol. **22:**272–281.
7. Schilb, T. P. 1978. Bicarbonate ion transport: A mechanism for the acidification of urine in the turtle. Science **200**(4338):208.

Discussion

P. R. Steinmetz: At room temperature, some of the results with your new technique are now similar to our previously reported results with pCO_2 electrodes, while others are not similar. Do you mean to imply that the pump changes when turtles are first maintained in a cool environment and then in a hot environment? I disagree to some extent with your statement on metabolic CO_2. In our studies, the observed increase in mucosal pCO_2 was much larger than can be accounted for by metabolic CO_2, which was (Schwartz *et al.* 1974. *Am. J. Physiol.* **226:**283) estimated from that produced by bladder sacs without acidification and that were maintained in solutions free of exogenous CO_2.

T. P. Schilb: Despite differences and similarities between the pCO_2 electrode and ours with the manometric method, I am confident that if you use the pCO_2 electrode, after controlling the temperature and feeding the turtle as I have described, you will find that the luminal pCO_2 will often become less than the serosal pCO_2 during the *in vitro* acidification of lumen fluid. This lowering of the luminal pCO_2 can only be explained by bicarbonate absorption. If neither of us finds a lowering of luminal pCO_2 under a different set of conditions, this does not exclude the bicarbonate reabsorption hypothesis.

Steinmetz: The way you test the question with your approach you can get only one result.

Schilb: That is correct in part: The low luminal pCO_2 under previously described conditions in acidified luminal fluids is unique evidence for the HCO_3 pump. But as you suspect, an elevated luminal pCO_2 is as consistent with the HCO_3 pump as it is with a proton secretory pump, as has been dealt with in detail previously (1974. Current Concepts in Membrane Transport, Vol. 5). However, the minimal luminal pH achieved by the turtle bladder is 4.0 for a tissue like gastric mucosa, which acidifies its secretion to pH 1.0, the bicarbonate reabsorption can be excluded on a priori grounds alone.

STEINMETZ: Have you done the stomach studies again with your improved technique to make sure it shows bicarbonate reabsorption?

SCHILB: No. There is no doubt that the stomach secretes H^+.

STEINMETZ: Nevertheless, the 1% CO_2 used in your frog stomach experiments in your original paper would not meet any of your requirements for establishing the existence of HCO_3 reabsorption by the stomach. I would be curious to see the results if you used 5% CO_2, instead of 1%, in the stomach experiments. However, we still think that our evidence for H^+ secretion by the turtle bladder is internally consistent.

EFFECT OF INHIBITORS AND THE MECHANISMS OF ANION TRANSPORT IN THE PROXIMAL RENAL TUBULE OF RATS

E. Frömter and K. J. Ullrich

Max-Planck-Institut für Biophysik
Kennedyallee 70
6000 Frankfurt am Main 70, West Germany

As we have learnt from the outstanding work of Ussing and collaborators, all characterization of ion transport mechanisms starts with the distinction of active and passive fluxes. Working with frog skin, Ussing achieved this by measuring the unidirectional tracer fluxes in a electrically short-circuited preparation.[1] Since this technique is not applicable in mammalian kidney tubules, we had to find an alternative method to determine active and passive fluxes. Our approach was based on irreversible thermodynamics. As described by Sauer,[2] under certain restrictions the net flow equation can be written as

$$J_i = (1 - \sigma_1)\bar{c}_i J_v + P_i\,(\Delta c_i + \bar{c}_i \frac{z_i F}{RT}\Delta\phi) + J_i^{\mathrm{act}} \qquad \textbf{(1)}$$

where J_i and J_i^{act} are the net flow and active transport of a substance i; J_v is the volume flow; Δc_i is the concentration difference of i across the membrane; c_i is the mean concentration of i averaged over both outer compartments; σ_i and P_i are reflexion coefficient and permeability coefficient respectively; $\Delta\phi$ is the electrical potential difference across the membrane measured with KCl-bridges, and z_iF and RT have their usual meaning. If we add a poorly permeable nonelectrolyte, such as raffinose to the isotonic salt solution in the tubular lumen, water will flow into the lumen until osmotic equilibrium is reached. The water, which is, so to speak, trapped with the nonelectrolyte will effectively dilute the salt concentrations of the lumen fluid, and salt will leak in from the peritubular surface along its concentration gradient. As active salt efflux from the lumen continues, a situation will finally develop in which the active salt outflux equals the passive salt backflux, until all net fluxes cease and also the volume flux approaches zero (zero net flux situation).

In this situation, equation 1 reduces to

$$J_i^{\mathrm{act}} = -P_i\,(\Delta c_i + \bar{c}_i \frac{z_i F}{RT}\Delta\phi) \qquad \textbf{(2)}$$

which means that we can calculate the rate of active solute transport if we measure the transepithelial concentration difference of i and the electrical potential difference at zero net flux, and if we know the permeability of the solute in question. We have applied this approach to the principal ions Na^+, Cl^- and HCO_3^-[3] as well as to Ca^{2+}.[4] Zero net flux transepithelial concentration differences were also determined for other substances, including most anions of physiological interest. Before presenting the data we want to point out, however, that recently a number of alternative approaches have been used by others[5,6] to arrive at a distinction between active and passive fluxes. These approaches are essentially based on a comparison of the net volume absorption under

0077–8923/80/0341–0097 $01.75/0 © 1980, NYAS

various conditions of poisoning. When trying to interpret such data, one runs the risk of not knowing whether the poison is indeed as specific as expected, whether the applied dose inhibits the transport mechanism completely or not, and whether or not the drug has side effects, inhibiting not only the active component but possibly also some passive component that might be coupled to it.

TABLE 1 shows the results of our attempts to determine the active transport rates of proximal tubular epithelium and compares these active fluxes with net fluxes, which can be observed under conditions simulating free flow in proximal tubular loops. Besides the anions, HCO_3^- and Cl^-, we have also listed the data for Na^+ and Ca^{2+}. It can be seen that the rate of active bicarbonate transport is greater than net bicarbonate absorption; and that active Cl^- transport, if present at all, is insignificant compared to passive Cl^- transport. This means, in other words, that Cl^- transport is mainly, if not exclusively, passive. These results agree with the observations from free-flow studies, which show that bicarbonate is preferentially absorbed over Cl^- in the proximal tubule, resulting in the buildup of tubular fluid chloride concentrations in excess of interstitial fluid chloride concentrations.[7] The cations, Na^+ and Ca^{2+}, undergo both active and passive transport. For a detailed discussion of these observations, see references 3, 4 and 8.

FIGURE 1 shows steady state zero net flux concentrations in the tubular fluid for various anions as observed in further experiments.[9-13] It can be seen that the concentrations of phosphate, sulfate, lactate, and aspartate decrease in the tubule fluid below those in the peritubular perfusate, while that of *p*-amino-hippurate (PAH) increases above the peritubular level. Although we are not able to apply equation 2 in these cases to derive the rate of active transport of these anions quantitatively (because we do not know the respective permeability coefficients), we conclude from these data that all these anions are actively transported across the tubular epithelium; phosphate, sulfate, lactate and aspartate (as well as glutamate, as can be inferred from further observations) are actively absorbed while PAH is actively secreted.

To obtain some insight into the mechanism of active anion transport, experiments were performed to test whether we are dealing with primary active transport, or with secondary active transport: the latter being defined as arising from local passive flux coupling of the anion with another actively transported substance such as, for example, Na^+.[14] These experiments comprise measurements of zero net flux concentrations under ouabain (to block the active Na^+

TABLE 1

ACTIVE AND PASSIVE COMPONENTS OF Na^+, Cl^-, HCO_3^- AND Ca^{2+} ABSORPTION FROM SHRINKING DROPLETS IN RAT KIDNEY PROXIMAL TUBULE *

	J_i^{act}	J_i^{pass}	J_i^{net}
Na^+	1.5	3.7	5.2
Cl^-	0.3	4.0	4.3
HCO_3^-	2.1	−1.2	0.9
Ca^{2+}	0.03	0.03	0.06

* All fluxes are in 10^{-11} mol/s cm (length of tubule). The data are taken from references 3 and 4.

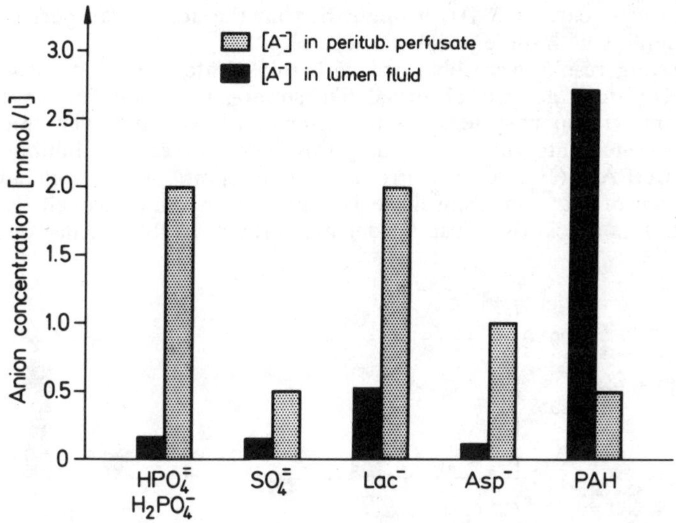

FIGURE 1. Steady state zero net flux concentrations for different anions in rat proximal tubule. The right stippled columns represent the anion concentrations (in mmol/l) in the peritubular perfusate, the left dark columns represent the anion concentrations (in mmol/l) in the lumen fluid, which develop under conditions of $J_l=0$ and $J_v=0$ and remain stable. Note that the luminal concentrations are smaller than the peritubular ones in case of phosphate, sulfate, lactate and aspartate indicating active absorption, but larger in case of hippuric acid (PAH) indicating active secretion. The data are taken from references 9–13.

pump), acetazolamide (to reduce HCO_3^- transport) or anion transport inhibitors such as 4-acetamido-4-isothiocyano-stilbene-2,2′ disulfonic acid (SITS) as well as experiments during substitution of all Na^+ by choline, or of all HCO_3^- by Cl^-. The results of experiments with Na^+-free solutions and with ouabain are depicted in FIGURE 2. It can be seen that omission of Na^+ reduced or abolished the active transport (as deduced from the steady state zero net flux concentrations) of all anions tested, except PAH, while ouabain (as tested in the more sensitive golden hamster) influenced only phosphate and sulfate, but left PAH and glycodiazine (which can be considered as a substitute for HCO_3^- transport) largely unaffected. Lactate and aspartate were not tested with ouabain so far. These data suggest that the transport of phosphate, sulfate, and aspartate is coupled to active Na^+ transport and may be driven, as initially suggested by Crane for intestinal sugar absorption in the brush border membrane, by local flux coupling between the respective solute and the actively transported Na^+ ions. With respect to HCO_3^- transport, however, the situation is less clear. Although the Na^+ dependence was documented, it is possible that HCO_3^- does not simply and exclusively undergo secondary active transport in coupling with Na^+ flux, but that a bicarbonate-stimulated ATPase is also involved that utilizes metabolic energy directly. Supporting evidence against HCO_3^- transport being exclusively driven by Na^+ transport was also obtained in electrical measurements.[15] In the case of PAH, it might be that the active transport is achieved by flux coupling with some metabolic substrate.

Interesting results were obtained with thiosulfate. In the presence of Na^+ ions, thiosulfate is actively absorbed like sulfate, with both anions competing for the same transport system. In the absence of Na^+, however, thiosulfate is actively secreted into the tubule, and this secretion can be inhibited by the presence of PAH (FIGURE 3). From these observations, we conclude that in the presence of Na^+ ions thiosulfate is driven from lumen to cell by coupling to the Na^+ flux across the brush border membrane. In the absence of Na^+ ions,

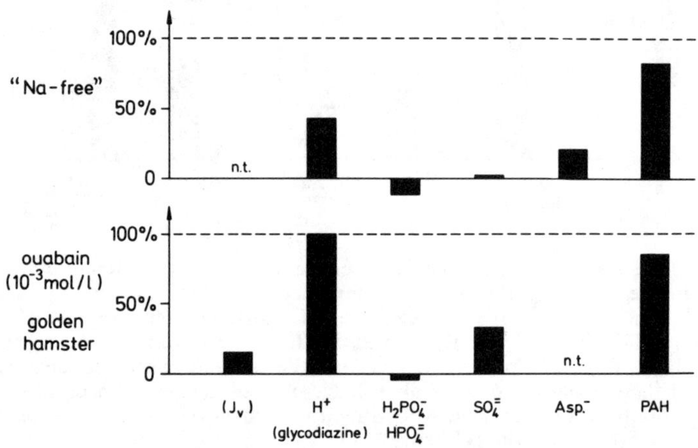

FIGURE 2. Effect of "Na^+-free" solutions and ouabain on active anion transport. Active anion transport is expressed in percent of control, assuming that the passive permeability of the anions (see equation 2) remains constant. The first column gives the effect on volume absorption (J_v) n.t.: not tested. The experiments in the upper half of the figure were performed with Na^+-free solutions, but during peritubular perfusion there was an influx of Na^+ into the perfusion fluids resulting in Na^+ concentrations of the order of 5 to 8 mmol/l. The ouabain experiments in the lower half of the figure were performed on kidneys of golden hamster, because this animal is more sensitive to ouabain than the rat. Note that Na^+-free solutions drastically reduce all active anion transport with the exception of PAH, while ouabain does neither affect PAH transport nor HCO_3^- transport. The data are taken from references 9, 10, 12, 13, and 16.

this transport ceases and thiosulfate is taken up by the acid transport system (which also transports PAH) at the peritubular cell membrane into the cell where it enters the tubule lumen via an unspecific Na^+-independent pathway to yield overall net secretion.

As shown in FIGURE 4, acetazolamide reduced volume absorption most probably as a result of its inhibition of HCO_3^- transport. The latter effect was more directly demonstrated in experiments in which an all HCO_3^- buffer was replaced by a glycodiazine buffer. In addition, acetazolamide slightly reduced phosphate transport but had no influence on sulfate transport. SITS reduced volume flow and glycodiazine transport, but only if applied from the peritubular

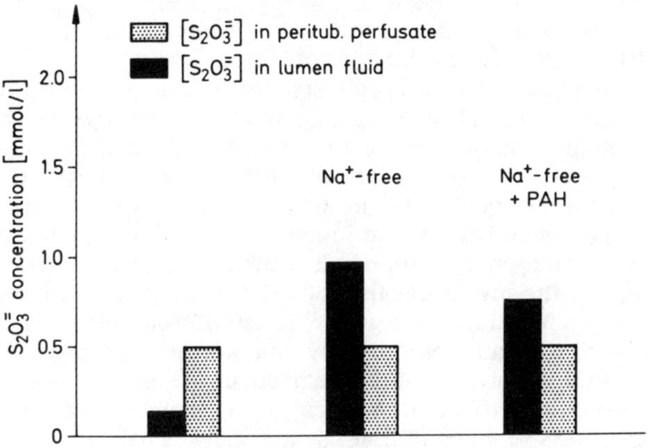

FIGURE 3. Zero net flux concentrations of thiosulfate in the presence and absence of Na^+ and in the presence of PAH. Details as in FIGURE 1. Note that thiosulfate is actively absorbed in Na^+-Ringer's (first pair of columns) and actively secreted in Na^+-free (choline)-Ringer's solution (second pair of columns) and that the latter transport is inhibited by the presence of PAH (third pair of columns), suggesting competition for a common carrier system in the peritubular cell membrane.

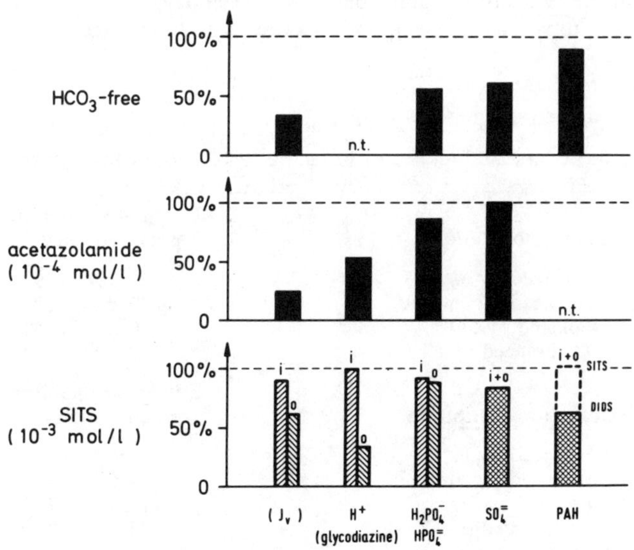

FIGURE 4. Effect of HCO_3^--free solutions, acetazolamide and SITS on active anion transport. Details as in FIGURE 2. The side of drug application is denoted by i-lumen or o-peritubular perfusate. Note that acetazolamide and SITS yield approximately the same inhibition pattern. The data are taken from reference 16.

surface. These observations suggest that SITS acts preferentially on the mechanism of H^+ secretion (or HCO_3^-, respectively glycodiazine buffer absorption). Its congener DIDS also slightly inhibited PAH transport.

To further elucidate the transport mechanism and to define the individual transport steps in the luminal and peritubular cell membrane, we have carried out electrophysiological studies on rat proximal tubular cells, looking at potential responses to ion substitutions in the luminal or the peritubular fluid. Theoretically, such experiments should allow us to determine ion permeability ratios or transference numbers of the individual cell membranes for individual ions. In practice, however, this approach meets some constraints that concern the quantification of the results because of the presence of a paracellular shunt in proximal tubular epithelium, whose ion selectivity properties are not exactly known. Moreover, we can observe only ion permeation processes that are associated with current flow. Coupled electroneutral ion transport mechanisms cannot be observed directly, but their existence may be inferred by proper reasoning under some special conditions, as we shall see below.

Before discussing the anion data, it seems appropriate to briefly mention the findings with cations (TABLE 2). Like in muscle cell membrane, we find that both the luminal and peritubular cell membrane of proximal tubular cells exhibit a significant electrically measurable K^+ permeability, but we find only a small independent Na^+ permeability (i.e., substitution of Na^+ by choline$^+$ does not produce large depolarizations in kidney cells, provided the solutions are free of organic solutes such as glucose, or amino acids). This contradistinguishes the luminal membrane of proximal tubule cells from the luminal membrane of urinary bladder cells or from the outer membrane of frog skin, and raises the question of how the Na^+ ions enter the cell before they are actively expelled from the cell into the peritubular fluid (see below).

With regard to the anions, the findings are as follows: Replacing all Cl^- by

TABLE 2

CELL MEMBRANE POTENTIAL DURING LUMINAL OR PERITUBULAR ION SUBSTITUTIONS †

A. Luminal Substitution	Luminal membrane potential (mV)
Na^+ replaced by K^+	-45 ± 7 (19)
Na^+ replaced by choline	-92 ± 3 (9)
Cl^- replaced by $SO_4^=$	-60 ± 1.5 (14)
HCO_3^- replaced by Cl^-	-72 ± 1 (5)
B. Peritubular substitution	Peritubular membrane potential (mV)
Na^+ replaced by K^+	-10 ± 3 (10)
Na^+ replaced by choline$^+$	-64 ± 3 (7)
Cl^- replaced by $SO_4^=$	-73 ± 1 (4)
HCO_3^- replaced by Cl^-	-41 ± 9 (7)

† The data are calculated from the observed potential changes referring always to the same control cell membrane potential of −74 mV. The numbers indicate mean values ± SD with the numbers of observations in parenthesis. The data are taken from reference 17.

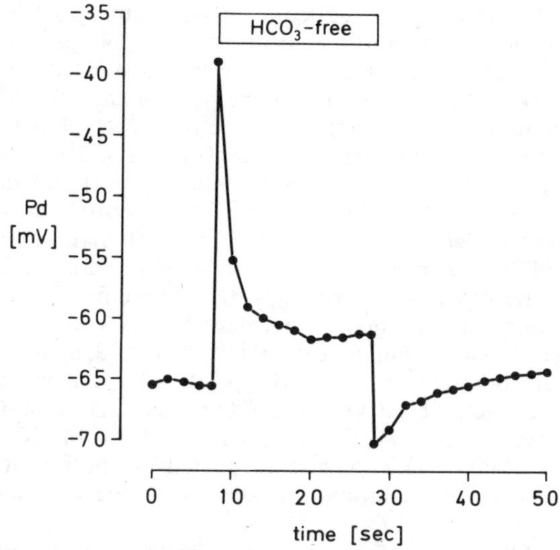

FIGURE 5. Effect of peritubular capillary perfusion with HCO_3^--free solution on proximal tubular cell membrane potential. Abscissa: time in sec, ordinate: cell membrane potential in mV. Note rapid depolarization on removal of peritubular HCO_3^- and hyperpolarization on readmission of bicarbonate. From reference 21.

SO_4^{2-}, either in the tubular lumen or in the peritubular fluid, does not produce cell potential changes in excess of what can be attributed to effects from the paracellular shunt. This indicates either that all chloride transport is electroneutral (in the sense of cotransport with some cation or countertransport with other anions) or that both the luminal and the peritubular cell membrane are only poorly permeable to chloride. Although the first possibility holds for intestine cells, where electroneutral NaCl cotransport has been documented by various methods,[18-20] it does not seem to hold for kidney cells. The strongest argument against it comes from a comparison of pH measurements in response to Na^+ salt gradients in preparations of brush border membrane vesicles. While in intestinal preparations,[19] the Na^+-induced pH shifts collapse rapidly with the relaxation time constant depending on the presence of permeable anions, no comparable phenomenon is observed in kidney brush border preparations.[31] The other alternative, however, is appealing because it fits well with the above reported finding of Cl^- transport being largely, if not entirely, passive. Passive chloride absorption should proceed via the paracellular shunt (which has a large chloride conductance) so that the passive Cl^- permeabilities of the cell membranes may be low.

The data with bicarbonate are even more interesting. Replacing all HCO_3^- by Cl^- had no effect in the tubular lumen, but it produced a typical time-dependent cell depolarization if the change was effected in the peritubular perfusate[21] (compare FIGURE 5). The potential transient resembled the transients observed by Hodgkin and Horowicz[22] on replacing the permeable Cl^- by the poorly penetrating sulfate in frog muscle cell preparations. The initial potential change varied linearly with the logarithm of the peritubular HCO_3^- concentration, yielding a slope of 27 mV per tenfold concentration change. We interpret these observations as indicating that the peritubular cell membrane is highly permeable to a bicarbonate buffer. Further experiments showed somewhat surprisingly that the bicarbonate permeability of the peritubular cell membrane can be inhibited by carbonic anhydrase inhibitors, such as acetazolamide and by SITS. As shown in FIGURE 6 acetazolamide in concentrations of 10^{-4} mol/l reduced the potential response to bicarbonate removal, yielding a slope of only 12 mV per tenfold concentration change. Furthermore, the peritubular cell membrane resistance increased under acetazolamide.[23] The fact that similar observations were made with SITS, which has a molecular weight of 522 and does not seem to penetrate into the cells, argues strongly for a specific direct effect of these drugs on the peritubular bicarbonate permeability as compared to a possible indirect action (e.g., via changing cell pH). It would appear therefore that passive bicarbonate permeation across the peritubular cell membrane involves the dehydration reaction of HCO_3^- into CO_2^- and OH^- that is catalyzed by carbonic anhydrase, and that HCO_3^- passes in the form of CO_2^- and OH^- (or H^+ in opposite direction). This view is supported by the recent demonstration of membrane-bound carbonic anhydrase activity in the peritubular cell membrane.[24]

With regard to the entry of bicarbonate from the lumen into the cell, the electrical measurements are not immediately useful; however, the absence of an electrical response to changing luminal HCO_3^- concentrations together with the irresponsiveness to changing luminal Na^+ concentrations suggests that both species migrate across the luminal membrane in a coupled electroneutral fashion, in agreement with what has been concluded from experiments with brush-border membrane vesicles.[25] As pointed out already above, it is likely that besides a simple Na^+/H^+ exchange mechanism, which could be responsible for driving the coupled $NaHCO_3^-$ inflow, there exists a H^+ pump that is linked to metabolic energy directly and that is predominantly active in early proximal tubular loops.[26]

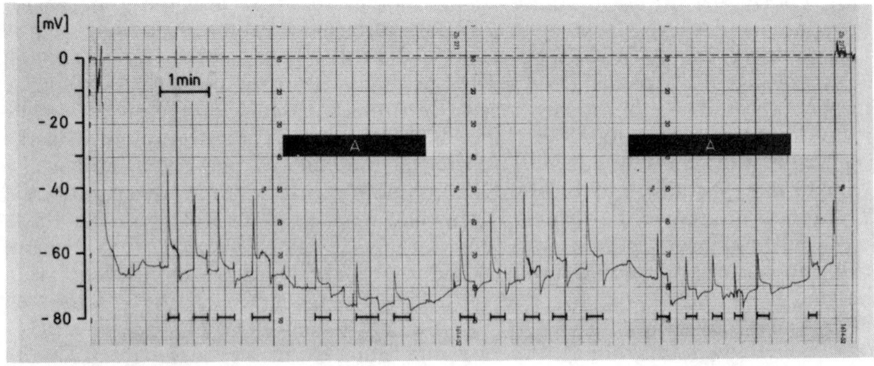

FIGURE 6. Effect of acetazolamide on cell potential response to peritubular HCO_3^- removal. Abscissa and ordinate as in FIGURE 5. During the light bars (⊢⊣) the peritubular capillaries were intermittently perfused with HCO_3^--free NaCl-Ringer's solution. During the heavy bar (A) the lumen was perfused with Ringer's solution containing 10^{-4} mol/l acetazolamide. Note reversible reduction of cell potential response to HCO_3^- removal. Trace record taken from reference 21.

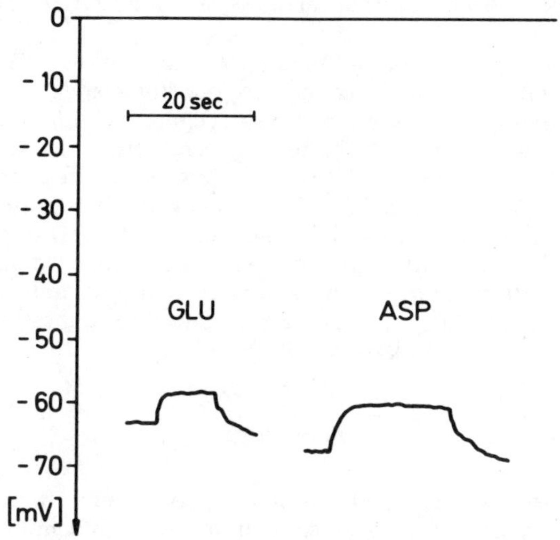

FIGURE 7. Effect of luminal application of glutamate or aspartate on proximal tubular cell membrane potential. Details as in FIGURE 5. The depolarization suggests that these amino acids induce a net positive ion current from lumen to cell indicating that they traverse the brushborder together with excess positive charge (possibly two Na^+ ions per anion). Similar effects on the cell potential are observed with glucose or neutral amino acids [23] as well as with primary phosphate.

The transport mechanisms of the remaining anions, phosphate, lactate, sulfate, aspartate (and glutamate) were also investigated with electrical methods.[32] Phosphate was applied in concentrations of 2 mmol/l in the lumen perfusate. It depolarized the tubular cell by 3.7 S.D. $\pm$ 1.3 (mV) ($n = 61$) with a time course similar to that observed during luminal perfusion with glucose or amino acid containing Ringer's solution. However, the depolarization in response to phosphate occurred only if the lumen perfusate was acid (pH 6.5). In alkaline solutions (pH 8.5), no significant depolarization could be detected. These observations suggest that phosphate is transported across the brush border membrane by a Na^+-dependent, flux-coupling mechanism that carries two Na^+ ions together with one phosphate ion. If the phosphate bears one negative charge (pH 6.5), the transport is associated with current flow; if it bears two negative charges (pH 8.5), the transport is electroneutral.

Similar cell depolarizations were also seen with aspartate, glutamate (FIGURE 7) and lactate; but not with sulfate, suggesting again that the transport mechanisms of these anions transfer two positive charges together with one anion—so that the overall flux is electrogenic: if the anion bears a single net negative charge, or electroneutral: if it bears two negative charges. It is interesting to note that a discrepancy exists between our findings with glutamate and aspartate, and respective observations on brush border membrane vesicles from rat and rabbit kidney.[27, 28] With the latter technique all investigators have reported that glutamate (and aspartate) transport is electroneutral, whereas we find an electrogenic component. It is possible that this discrepancy reflects different functional states of the carrier system in the *in vitro* and *in vivo* system.

As far as the transport of phosphate, lactate, sulfate and aspartate across the peritubular cell membrane is concerned, our knowledge is far less advanced. The only additional information that we can report is the finding that glutamate and aspartate do also depolarize the tubular cell membranes if they are applied from the peritubular surface and that this depolarization depends also on the presence of Na^+. We conclude from this observation that these two amino acids can be taken up into the cell by sodium-dependent transport mechanisms both from the luminal and peritubular surface. This peculiar distribution of Na^+-dependent transport mechanisms may explain the enormous accumulations of these amino acids reaching cell-to-plasma concentration ratios of up to 200, which have been observed in kidney cells.[29, 30]

Conclusion

We have briefly summarized our findings on anion transport mechanisms in the rat kidney proximal tubular epithelium. Our data indicate that chloride transport is mainly, if not exclusively, passive and proceeds along the paracellular transport route between the cells. The absorption of bicarbonate, phosphate, sulfate, thiosulfate, lactate, aspartate and glutamate is active and proceeds across the cell membranes. With the exception of bicarbonate, which seems to undergo both primary active and secondary active transport (in coupling with Na ions), and of PAH, which is actively secreted into the tubule, all remaining anion transports must be classified as secondary active via local passive flux coupling with primary active Na^+ transport. The flux coupling mechanisms, which are situated in the brush border membrane, are electroneutral in case of bicarbonate, secondary phosphate and sulfate; but rheogenic in case of primary phosphate, lactate, aspartate and glutamate. Anion transport mechanisms across the peritubular cell membrane have been identified so far only for bicarbonate and the dicarboxylic amino acids. Bicarbonate flow across the peritubular cell membrane is not coupled to other ion flows, and can be inhibited by carbonic anhydrase inhibitors and by the disulfonic stilbene derivative SITS. Aspartate and glutamate are also taken up into the cell from the peritubular surface by a Na-dependent transport mechanism.

References

1. Ussing, H. H. & K. Zehran. 1951. Active transport of sodium as the source of electric current in the short-circuited isolated frog skin epithelium. Acta Physiol. Scand. **23:**110–127.
2. Sauer, F. 1973. Non-equilibrium thermodynamics of kidney tubule transport. *In* Handbook of Physiology, Sec. 8, Renal Physiology. R. W. Berliner & J. Orloff, Eds. 399–414. Amer. Physiol. Soc. Washington, D.C.
3. Frömter, E., G. Rumrich & K. J. Ullrich. 1973. Phenomenologic description of Na^+, Cl^- and HCO_3^- absorption from proximal tubules of the rat kidney. Pflügers Arch. Gesamte Physiol. **343:**189–220.
4. Ullrich, K. J., G. Rumrich & S. Klöss. 1976. Active Ca^{2+} reabsorption in the proximal tubule of the rat kidney. Dependence on sodium and buffer transport. Pflügers Arch. Gesamte Physiol. **365:**223–228.

5. Green, R., J. H. V. Bishop & G. Giebisch. 1979. Ion requirements of proximal tubular sodium transport. III. Selective luminal anion substitution. Am. J. Physiol. **236:**F268–F277.
6. Lucci, M. S. & D. G. Warnock. 1979. Effects of anion transport inhibitors on NaCl reabsorption in the rat superficial proximal convoluted tubule. J. Clin. Invest. In press.
7. Litchfield, J. G. & P. A. Bott. 1962. Micropuncture study of renal excretion of water, K, Na and Cl in the rat. Amer. J. Physiol. **203:**667–670.
8. Frömter, E. 1974. Electrophysiology and isotonic fluid absorption of proximal tubules of mammalian kidney. *In* Kidney and Urinary Tract Physiology, MTP Int. Rev. of Science, Physiol. Series I, K. Thurau, Ed. **6:**1–38. Butterworths, London. University Park Press, Baltimore, Md.
9. Baumann, K., C. DeRouffignac, N. Roinel, G. Rumrich & K. J. Ullrich. 1975. Renal phosphate transport: Inhomogeneity of local proximal transport rates and sodium dependence. Pflügers Arch. Gesamte Physiol. **356:**287–297.
10. Ullrich, K. J., G. Rumrich & S. Klöss. 1979. Sulphate reabsorption in the proximal convolution of the rat kidney. Pflügers Arch. Gesamte Physiol. **379:**R17.
11. Höhmann, B., P. P. Frohnert, R. Kinne & K. Baumann. 1974. Proximal tubular lactate transport in rat kidney: A micropuncture study. Kidney Int. **5:**261–270.
12. Ullrich, K. J., G. Rumrich & S. Klöss. 1974. Sodium dependence of the aminoacid transport in the proximal convolution of the rat kidney. Pflügers Arch. Gesamte Physiol. **351:**49–60.
13. Ullrich, K. J. 1974. Recent progress in renal physiology: Transport of hexoses, aminoacid, phosphate and paraaminohippurate in the proximal tubule. *In* Atti del Congresso Riva del Garda. Soc. Ital. Fisiol. : 1–26.
14. Frömter, E. 1978. Primary and secondary active mechanisms in rat renal proximal tubule. New Aspects of Renal Function. H. G. Vogel & K. J. Ullrich, Eds. Excerpta Med. Amsterdam, the Netherlands.
15. Frömter, E. & K. Gessner. 1975. Effect of inhibitors and diuretics on electrical potential differences in rat kidney proximal tubule. Pflügers Arch. Gesamte. Physiol. **357:**209–224.
16. Ullrich, K. J., G. Capasso, G. Rumrich, F. Papavassiliou & S. Klöss. 1977. Coupling between tubular transport processes. Studies with ouabain, SITS and HCO_3^- free solutions. Pflügers Arch. Gesamte Physiol. **368:**245–252.
17. Frömter, E., K. Sato & K. Gessner. 1976. Electrical studies on the mechanism of H^+/HCO_3^- transport across rat kidney proximal tubule. VI. Int. Cong. Nephrol. S. Giovanetti, V. Bonommi & G. D'Amico, Eds. 108–112. S. Karger. Basel, Switzerland.
18. Nellans, H. N., R. A. Frizzell & S. G. Schultz. 1974. Brushborder processes and transepithelial Na and Cl transport by rabbit ileum. Amer. J. Physiol. **226:** 1131–1141.
19. Liedtke, C. M. & U. Hopfer. 1977. Anion transport in brushborder membranes isolated from rat small intestine. Biochem. Biophys. Research Comm. **76:** 579–585.
20. Armstrong, W. McD., W. R. Bixenman, K. F. Frey, J. F. Garcia-Diaz, M. G. O'Regan & J. L. Owens. 1979. Energetics of coupled Na^+ and Cl^- entry into epithelial cells of bullfrog small intestine. Biochim. Biophys. Acta **551:**207–219.
21. Frömter, E. & K. Sato. 1976. Electrical events in active H^+/HCO_3^- transport across rat kidney proximal tubular epithelium. *In* Gastric Hydrogen Ion Secretion. D. K. Kasbekar, G. Sachs & W. S. Rehm, Eds. 382–403. Dekker. New York, N.Y.
22. Hodgkin, A. L. & P. Horowicz. 1959. The influence of potassium and chloride ions on the membrane potential of single muscle fibers. J. Physiol. **148:**127–160.

23. FRÖMTER, E. Magnitude and significance of the paracellular shunt path in rat kidney proximal tubule. *In* Intestinal Permeation: 166–178. M. Kramer & F. Lauterbach, Eds. Excerpta Med. Amsterdam, the Netherlands.
24. WISTRAND, P. J. & R. KINNE. 1977. Carbonic anhydrase activity of isolated brush border and basal-lateral membranes of renal tubular cells. Pflügers Arch. Gesamte Physiol. **370**:121–126.
25. MURER, H., U. HOPFER & R. KINNE. 1976. Sodium/proton antiport in brush border membrane vesicles isolated from rat small intestine and rat kidney. Biochem. J. **154**:597–604.
26. FRÖMTER, E. & K. GESSNER. 1974. Active transport potentials, membrane diffusion potentials and streaming potential across rat kidney proximal tubular epithelium. Pflügers Arch. Gesamte Physiol. **351**:85–98.
27. SCHNEIDER, E. G., M. R. HAMMERMAN & B. SACKTOR. 1978. The Na^+-dependent transport of L-glutamic acid by renal brush border membranes: Evidence for an electroneutral mechanism. Fed. Proc. **37**:A1295.
28. MURER, H. & G. BURKHARDT. Personal communication.
29. SHALHOUB, R., W. WEBBER, S. GLABMAN, M. CANESSA-FISCHER, J. KLEIN, J. DEHAAS & R. F. PITTS. 1963. Extraction of amino acids from and their addition to renal blood plasma. Am. J. Physiol. **204**(2):181–186.
30. CHAN, A. W. K., H. B. BURCH, T. R. ALVEY & O. H. LOWRY. 1975. A quantitative histochemical approach to renal transport. I. Aspartate and glutamate. Am. J. Physiol. **229**:1034–1044.
31. BARAC-NIETO, M., H. MURER & R. KINNE. Unpublished observations.
32. SAMARŽIJA, I. & E. FRÖMTER. Unpublished observations.

DISCUSSION

E. RACKER: Did you imply from the last few sentences that in your system the lactate goes in by a symport mechanism with sodium ions?

E. FRÖMTER: Yes. This was demonstrated also more directly by Drs. Barac-Nieto, Murer and Kinne in experiments with brush border membrane vesicles from rat kidney. However, the transport across the cell membrane facing the blood side is not sodium-dependent.

T. MAREN (*University of Florida, Gainesville, Fla.*): These experiments show that Diamox appears to alter the HCO_3 permeability of the basolateral membrane of the proximal tubule. As will be shown later in this conference, these membranes contain a Diamox-sensitive carbonic anhydrase. At the concentrations used by Frömter (down to 2×10^{-5} M), I know of no action of Diamox other than that on the enzyme. Perhaps this effect is a secondary one, due to inhibition of HCO_3 accumulation at the basolateral membrane. The role of the membrane-bound enzyme might be to build up high HCO_3 concentrations at the basolateral membrane, and that your results might be explained in these terms.

E. FRÖMTER: Generally speaking this could be true, insofar as the partial conductance or respectively the transference number of an ion depends on its mean concentrations in the ambient solutions. Thus, if intracellular HCO_3^-—inspite of the increasing pH—decreases after diamox, because the CO_2 rehydration reaction does no longer reach its equilibrium, the transference number of HCO_3^- could indeed fall, as was observed. However a number of additional experiments suggests that diamox rather inhibits a permeation step *within* the peritubular cell membrane. In my view therefore it seems more likely that HCO_3^- permeates in form of CO_2 and OH^- and that diamox interferes with the permeation by blocking the generation of CO_2 and OH^- in the cell membrane.

MAREN: This is, of course, the role of the enzyme, and if HCO_3^- exit from the cell proceeds through its dissimilation to $CO_2 + OH^-$, your are correct, and the experiment can be explained in the usual pattern of carbonic anhydrase inhibition. However, I had thought that your experiment had to be interpreted in terms of movement of HCO_3^- ion as such, particularly since, if I recall correctly, you obtained similar results with other anions.

E. FRÖMTER: The data which you mention do not necessarily favor HCO_3^- ion permeation as such. In these experiments we have found that practically all observations with HCO_3^- could be duplicated thus far with a series of chemically different buffers, which have in common with the HCO_3^- buffer that their non-ionic form is highly lipidsoluble. These experiments, therefore, seem to support rather the notion of separate permeation mechanism for the hydroxyl ion and the non-ionic buffer species, and the action of diamox in the presence of glycodiazine could be possibly explained, if the hydroxyl permeation pathway in the peritubular cell membrane was an integral part of the molecule which exhibits the carbonic anhydrase activity.

A. M. SPENCER (*Department of Biochemistry, Columbia University, New York City*): I was puzzled by your finding that ouabain did not inhibit PAH transport in the hamster since it has been shown to inhibit PAH transport in chicken, rabbits and snake. I wondered if you had any explanation for this?

FRÖMTER: I would like to hand this question over to Professor Ullrich, because it is his data you are asking about.

K. J. ULLRICH (*Max Planck Institut für Biophysik, Frankfurt, West Germany*): In our experiments on the proximal tubules of golden hamster, 1 mmol/l ouabain inhibited the transport of Na^+, glucose, amino acids, phosphate and calcium almost completely, but not that of H^+/glycodiazine buffer and very little that of PAH (14%). Furthermore, when Na^+ was omitted from the perfusates, all tested transport systems were inhibited by more than 60%; the PAH system, however, only by 20%. Furthermore, Murer and Kinne (Fed. Proc. 1979) could demonstrate Na^+-substrate cotransport in membrane vesicles for all mentioned substances, but not for organic acids (PAH). Thus, we have little evidence that the PAH secretion is driven by a sodium gradient. As confirmation of that, we found recently that thiosulfate is reabsorbed via the Na^+ sulfate cotransport mechanism when the solutions contained Na^+ but it is secreted via the PAH transport system when the perfusates were Na^+-free. The discrepancy between our results and those of other investigators who found a highly diminished PAH transport under ouabain or when Na^+ was omitted from the perfusates, seems to be caused by the duration of exposure to these solutions. Our tubules are exposed only for up to three minutes to ouabain or the sodium-free solutions, while isolated tubules, kidney slices of the perfused organ, are exposed for a much longer time. Since ouabain and Na^+-free solutions affect all Na^+-coupled transport processes, also that of lactate and other metabolites, this must finally also affect the PAH transport via changed cell metabolism. At the moment, therefore, we favor the hypothesis that the transport for PAH might be a tertiary active transport, i.e., a countertransport against a substance X, possibly a metabolite, which by itself or as a precursor is taken up into the cell by a Na^+ coupled secondary active transport system.

RACKER: If you talk about ouabain sensitivity using rodents like mice and rats, do you have to use high concentrations before you get a clear effect on the Na^+/K^+ pump?

ULLRICH: That is correct. Therefore we applied the rather high ouabain concentration of 1 mmol/l. Furthermore, we used in our ouabain experiments, golden hamsters, which are somewhat more sensitive to ouabain than are rats.

CO_2 PERMEABILITY OF THE NEPHRON IN RELATION TO THE NATURE OF LUMINAL FLUID ACIDIFICATION

Bertil Karlmark

Departments of Clinical Physiology, Physiology and Medical Biophysics
Uppsala University
S–751 23 Uppsala, Sweden

If we take a nephron luminal fluid sample out of its natural site and measure the pH *in vitro,* we will probably get a value that is not identical to that measured *in vivo.* There are at least two explanations for this "*in vivo–in vitro* pH difference": 1. The bicarbonate buffer system is not in chemical equilibrium *in vivo,* but has the time to become so when the *in vitro* measurement is performed. This condition of a chemical nonequilibrium is sometimes also named a chemical "disequilibrium." 2. The *in vitro* measurement is made at the wrong pCO_2, usually due to the fact that local pCO_2 values at cellular values are not easily obtained.

In the rat kidney tubule it is now recognized that the intraluminal pCO_2 is not identical to that in systemic arterial blood.[1,2] It is, however, not clear if the "*in vivo–in vitro* pH difference" noticed in the tubule during different physiological[3,4] and pharmacological conditions[5,6] can fully be explained by a high pCO_2 or if, besides, a chemical nonequilibrium condition exists. Provided the *in vivo* bicarbonate activity is known and the constants in the Henderson-Hasselbalch equation are known for the particular fluid under study, a chemical nonequilibrium condition theoretically can be quantitatively translated into a ΔpCO_2 condition. Of course, both circumstances can occur simultaneously, a topic unfortunately discussed only infrequently in the literature.

To distinguish between the two conditions in the kidney tubule, the measurement of local pCO_2 *in vivo* is a useful approach. The first measurements performed using microelectrodes[2] showed a 17 mm Hg higher pCO_2 in the proximal convoluted tubule of the rat as compared to that in systemic arterial blood. An extension of this study[1] showed a somewhat higher figure, 26 mm Hg, under comparable conditions. In addition, in the distal tubule as well as in the star vessel, nearly identical values of 27 mm Hg and 25 mm Hg, respectively—higher values as compared to systemic arterial blood—were obtained. The interpretation in the later study[1] was a diffusion equilibrium for CO_2 all over the cortex, similar to what is proposed for NH_3. Since, however, a net transport of CO_2 takes place from the lumen to the blood, a pCO_2 difference across the cell must exist. Accordingly, a 29 mm Hg lumen to blood pCO_2 difference was found in the rat proximal cortical tubule,[8] in sharp contrast to the aforementioned finding of 1 mm Hg. A calculation was made for this difference on the basis of findings from the isolated rabbit proximal convoluted tubule and an estimation of this difference of, at best, 8.7 mm Hg was obtained.[7]

0077–8923/80/0341–0111 $01.75/0 © 1980, NYAS

Carbonic Anhydrase Inhibitors

With benzolamide, the cortical ΔpCO_2 dropped some 10 mm Hg as compared to the control state.[1] Thus, an increase in luminal pH would be expected, parallel to the effect of carbonic anhydrase inhibition at the level of final urine. It is, however, convincingly shown[4, 6, 9–11] that proximal pH drops about 0.4 pH units with adequate doses of carbonic anhydrase inhibitors. In the late proximal tubule, the pH drop is considerably smaller due probably to the bicarbonate accumulation in this segment.

These facts raise one main question: Is the proximal pH drop after carbonic anhydrase inhibition primarily due to the accumulation of carbonic acid (a chemical nonequilibrium) or an accumulation of CO_2 with a maintained intraluminal chemical equilibrium in the bicarbonate buffer system? The former alternative requires that the brush border carbonic anhydrase catalyzes the luminal fluid under control conditions, to prevent the accumulation of carbonic acid. The latter alternative requires that the drug blocks the transport rate of CO_2 from lumen to blood.

To further elucidate this question, we made two carbonic anhydrase inhibitors, which do not enter the tubular cell, but stay strictly intraluminally if given in a tubular perfusion.[12] One inhibitor was coupled with a dextran molecule and the other inhibitor was sulfanilamide coupled with a quaternary ammonium compound. These inhibitors were tested *in vitro* against the rat brush border enzyme and were found to have the adequate potency. *In vivo,* however, they had no effect at all on the acidification rate, which makes the occurrence of a luminally active carbonic anhydrase unlikely. When acetazolamide was given in the same way, however, a significantly reduced acidification rate was obtained. This study favors the view that the effects of carbonic anhydrase inhibition in the rat kidney should be explained in terms of blocking the intracellular carbonic anhydrase activity and not a luminal enzyme.

In a comparison among different routes for the administration of carbonic anhydrase inhibitors, the intravenous, the local application on the kidney surface, and the local intraluminal perfusion,[10] it was revealed that the by-far quickest way to reach the enzyme is via an intravenous injection. This finding also favors the view that the effect on the cellular enzyme and not a luminal enzyme is responsible for the luminal pH drop.

Carbonic Anhydrase Localization

Recent histochemical electron microscopic studies[13] have demonstrated the localization of carbonic anhydrase throughout the rat kidney tubule. In the proximal convoluted tubule, high enzyme activity was noted in the brush border, and in the basal-lateral membrane. In the cytoplasm, however, the activity was lower when compared to the membrane structures. So there is good evidence that the brush border enzyme and the cytoplasmic enzyme are two different types[14, 15] with different sensitivity to inhibitors. The function of the brush border enzyme is unknown but it might participate in the CO_2 diffusion from the lumen as well as in the hydroxylation of CO_2, indirectly promoting the splitting of water and also hydrogen ion secretion. The brush border enzyme is less sensitive to sulfanilamide[14] and seems to require very high concentrations of acetazolamide to be inhibited,[16] a condition that gave rise to an intracellular acidification in the rabbit isolated proximal convoluted tubule. Lower concentrations of acetazolamide (less than $2 \cdot 10^{-3}$ M) gave rise to an alkalinization of the cell pH in the same study, favoring the view that the basal-lateral membrane enzyme promotes the bicarbonate transfer from the cell to the peritubular blood.[17] According to a rising body of evidence,[18] one of the roles of carbonic anhydrase might be to facilitate the diffusion of CO_2 from lumen to blood, thus giving credit to the three localizations of the enzyme in the proximal tubule. The recent finding of carbonic anhydrase in the capillary wall in the lung seems to support a similar view.[19]

In view of these aforementioned new findings of the localization of carbonic anhydrase in the brush border, what about the direction of the active site of the enzyme: is it towards the cell or towards the lumen? In the latter case, the concept of a chemical nonequilibrium during carbonic anhydrase inhibition was supported, but was also contradicted.[10, 12] If the face of the enzyme is in the membrane or towards the cytoplasm, a reduced "facilitated diffusion" takes place during the enzyme inhibition, leading one to believe that a raised local pCO_2 in the proximal lumen was responsible for the significant drop in pH. This view is, however, not supported by the finding of a decreased intraluminal pCO_2 with benzolamide.[1]

CO_2 Diffusivity

If the production rate of CO_2 in the lumen, due to the titration of bicarbonate, is higher than the diffusion rate (with or without carbonic anhydrase inhibition), an accumulation of CO_2 in the lumen must occur. On basis of equilibrium studies it was proposed[20] that this CO_2 could stay elevated in the urine until it reached the end of the collecting duct. If this is the case, we have also an alternative explanation for the high pCO_2 found in the distal tubule.[1] This interpretation of data requires a diffusion "barrier" for CO_2 in the nephron. Mainly for technical reasons, the permeability for CO_2 in the rat cortical nephron was not measured. In the isolated rabbit cortical tubule, the permeability coefficient was estimated and found to be very close to that in free water.[21] Indirect measurements of CO_2 transfer across the rat cortical tubule, however, showed a 20–30 times slower disappearance rate for CO_2 as compared to the production rate; the latter measured as the uncatalyzed dehydration rate of carbonic acid.[22] If this estimate is correct, by necessity it means a luminal CO_2 accumulation in the absence of a carbonic anhydrase inhibitor. It has been known for decades that considerable pCO_2 differences (100 mm Hg or more) could be maintained by the collecting-duct epithelia as well as the ureter.[23] That the rate-limiting step in the gastric acid secretion might be the CO_2 diffusion is also shown.[24] Recent works on erythrocytes[25] show a considerable diffusion barrier for CO_2 in red blood cells, too. The validity of estimates of CO_2 permeability in membranes other than those in the rat cortical nephron is questioned for use in this presentation; however, other cell membranes at least show the capacity to separate considerable pCO_2 differences.

Unifying Ideas

If the major part of the bicarbonate is reabsorbed via the hydrogen ion secretion, which at least seems to be true in the rat cortical nephron, we should expect this titration to occur in the brush border of the luminal membrane. A radial concentration gradient is then the obligatory result, with the highest concentration in the brush border or the cell membrane itself. A set of model experiments [26] favors this asymmetry of bicarbonate titration and also indirectly favors the view of a locally increased pCO_2 in the proximal cell membrane. With an electrode or pipette located in the center of the lumen we will not, in such a case, get a measurement or a sample that is representative for what, in general, is going on in the lumen.

The reasons for the pH decrease in the rat proximal tubule at carbonic anhydrase inhibition remains to be elucidated. Clarifying attempts must consider that the enzyme inhibition strikes the enzymes in the brush border, the cytoplasm as well as in the basal-lateral membrane. One such attempt is to postulate that the enzyme inhibition impairs the lumen to blood diffusion of CO_2.

References

1. DuBose, T. D., L. R. Pucacco, D. W. Seldin, N. W. Carter & J. P. Kokko. 1978. J. Clin. Invest. **62:**338–348.
2. Sohtell, M. & B. Karlmark. 1976. Pflügers Arch. **363:**179–180.
3. Karlmark, B. & B. G. Danielson. 1974. Acta Physiol. Scand. **91:**243–258.
4. Sohtell, M. 1979. Acta Physiol. Scand. **105:**146–155.
5. Rector, F. C., N. W. Carter & D. W. Seldin. 1965. J. Clin. Invest. **44**(2): 278–290.
6. Vieira, F. L. & G. Malnic. 1968. Amer. J. Physiol. **214:**710–718.
7. Warnock, D. G. & F. C. Rector, Jr. 1979. Ann. Rev. Physiol. **41:**197–210.
8. Sohtell, M. 1979. Acta Physiol. Scand. **105:**137–145.
9. Quehenberger, P. 1977. Pflügers Arch. Deutsche Physiol. Ges. (Abstr.).
10. Lang, F., P. Quehenberger, R. Greger & H. Oberleitner. 1978. Pflügers Arch. **375:**39–43.
11. DuBose, T. & J. Kokko. 1979. This volume.
12. Karlmark, B., B. Ågerup & P. Wistrand. 1979. Acta Physiol. Scand. In press.
13. Lönnerholm, G. & Y. Ridderstråle. 1979. Uppsala J. Med. Sci. Suppl. 26. 3rd European Colloquium on Renal Physiol. (Abstr. 49).
14. Wistrand, P. J. & R. Kinne. 1977. Pflügers Arch. **370:**121.
15. Maren, T. H. 1979. This volume.
16. Paillard, M., M. Bichara, J. P. Gardin, F. Leviel & M. Fouchard. 1979. Uppsala J. Med. Sci. Suppl. 26. 3rd European Colloquium on Renal Physiol. (Abstr. 59).
17. Frömter, E. & K. J. Ullrich. 1979. This volume.
18. Suchdeo, S. R. & J. S. Schultz. 1974. Biochem. Biophys. Acta (Amst.) **352:** 412–440.
19. Crandall, E. D. & J. E. O'Brasky. 1978. J. Clin. Invest. **62:**618–622.
20. Uhlich, E., C. A. Baldamus & K. J. Ullrich. 1968. Pflügers Arch. **303:**31–48.
21. Warnock, D. G. & F. E. Rector, Jr. 1977. Abstr. Am. Soc. Neph. **10:**124 A.
22. Malnic, G. & M. DeMello-Aires. 1971. Am. J. Physiol. **220**(6)**:**1759–1767.
23. Reid, E. L. & A. G. Hills. 1969. Clin. Sci. **37:**381–393.
24. Kidder, G. W. & C. W. Montgomery. 1974. Am. J. Physiol. **227**(2):300–304.
25. Silverman, D. N., T. Chingkuang & G. C. Wynns. 1976. J. Biol. Chem. **251** (14):4428–4435.
26. Sohtell, M. 1979. Acta Physiol. Scand. **105:**129–136.

DISCUSSION

T. DUBOSE: As you've noted, there are many similarities in our findings, particularly in those on the pCO_2 gradient between proximal tubule and systemic arterial blood. I should like to comment about the values that I have measured in the stellate vessel. First, we measured pCO_2 and pH, then calculated the bicarbonate concentration. This was then compared with the measured bicarbonate concentration in the stellate vessel and there was close agreement in the results from the two techniques. (In addition, we have measured the chloride concentration in the stellate vessel, as have Afadan and Weinstein, and we find that those two ions are reciprocally related.)

I think a major question is raised about your thesis that the proximal tubule is a considerable diffusion barrier. Do you accept the permeability coefficient for CO_2, which has been measured to, say, 10^{-3} across the proximal tubular epithelia? Can the proximal tubule, in fact, sustain a gradient for CO_2 tension of the magnitude that you measure to approximately 30 mm Hg?

I should like to say that we are also very interested in the functional differences of carbonic anhydrase and the types of carbonic anhydrase present in the kidney. Using an *in vivo* microperfusion technique, we have recently shown that the luminal application of both benzolamide and acetazolamide, as well as methazolamide, almost completely inhibits the flux of $HCO_3 + CO_2$ (or J_{TCO_2}), but in the presence of benzolamide there is a more acid pH in the proximal tubule than in its absence (controls). This contrasts with results obtained when acetazolamide is applied in the lumen. This pH does not differ from control conditions; we suppose that the permeability of the tubule to benzolamide differs from that of methazolamide or acetazolamide. When benzolamide is applied in the lumen, only the luminal carbonic anhydrase is inhibited so that the hydrogen ion secretion continues being supplied, of course, from the cellular carbonic anhydrase reaction. In acetazolamide application, however, no change in pH would indicate that both cellular and luminal forms are inhibited. So I would also take issue with your hypothesis that this equilibrium-pH in the proximal tubule is in fact due to an accumulation of CO_2. Finally, how can you explain the disequilibrium pH as resulting from an accumulation of CO_2 when we show that it falls?

B. KARLMARK: The data we have from the inhibition with nonreabsorbed inhibitors do not show any evidence of a luminal carbonic anhydrase, which was supposed to be inhibited in your experiments as an inhibition of the dehydration of carbonic acid. We have no evidence of this enzyme at all. Of course, the $pCO_2$is too high to account for the disequilibrium measured, which is around 0.3–0.5 pH units. The value of 0.85 has never been repeated as far as I know. I do not think that the luminal pCO_2 values measured are necessarily representative of what is going on in the tubule. The pCO_2 in the center of the lumen might well differ from that in the periphery at or near the brush border.

To comment on your first question, I expect that we have a pCO_2 difference between the lumen and the blood. We know that the net effect in the proximal tubule is a bicarbonate reabsorption or CO_2 transport. So we expect, in fact, a difference. Dr. Warnock has calculated that this difference could be 7–9 mm Hg pCO_2 difference which, I think is beyond the resolution of the pCO_2 electrodes.

DUBOSE: I never stated, nor have I implied, that there would *not* be a gradient between lumen and peritubular capillary. I have simply stated that we cannot measure one, and again the question is can the proximal tubule sustain a gradient as high as you have measured if you accept the permeability coefficient that Dr. Warnock and Dr. Rector have reported? I would agree with your response, namely that there should be a gradient and that perhaps it is beyond the resolution of the two microelectrodes.

KARLMARK: As long as we cannot give a permeability coefficient for CO_2 in the rat proximal tubule, we cannot answer the question.

T. MAREN: Do we mean by luminal that it is facing the lumen, or do we mean that it is on the luminal side. That is a very, very important difference. Dr. Karlmark's experiments are rather clear cut and unfortunately he did not have time to elaborate on them and Dr. Karlmark very properly asked what was going on. In one case, they coupled dextran to the most powerful carbonic anhydrase inhibitor known, a close analogue of benzolamide. When they put it on the luminal side, they got absolutely no effect at all. In a similar experiment, they coupled a quaternary ammonium compound, again being reasonably assured that there was no permeation. This is another result that militates against a carbonic anhydrase facing the luminal side. Dr. Lucci's experiments are a little less precise, because benzolamide permeates the luminal membrane to some degree. All we know is that there are some differences among the drugs that have to be pinpointed.

M. BURG: Dr. Warnock had previously made some measurements in San Francisco and Dr. Schwartz, in our laboratory in Bethesda, has made an extensive series of measurements of CO_2 permeability on the proximal convoluted tubule of the rabbit. From these measurements with the imposed transtubular gradients the result was the same. The CO_2 permeability was large enough to indicate there should not be this large transtubular concentration difference, even if the CO_2 had to get entirely across the epithelium to be neutralized, rather than from just the lumen, into the cells. Thus the direct measurements of CO_2 permeability don't agree with your theory that it could be an important barrier for CO_2 permeation.

KARLMARK: I am aware of your measurements and I am sorry I did not cite them. As long as we do not have the proper measurements from rat preparations, I think we cannot draw extensive conclusions from figures obtained from the isolated convoluted rabbit tubule.

BURG: There is nothing apparently wrong with the rabbit tubule.

USE OF STIMULANTS AND INHIBITORS FOR STUDYING THE MECHANISMS OF Cl TRANSPORT *

Oscar A. Candia

Departments of Ophthalmology, and Physiology and Biophysics
Mount Sinai School of Medicine
City University of New York
New York, New York 10029

The isolated frog cornea develops a potential difference (PD) of about 25 mV (the tear side negative with respect to the aqueous humor side), and a short-circuit current (SCC) of about 13 $\mu a/cm^2$. This electrical activity is due to a net transport of Cl by the cornea's outer layer, the corneal epithelium, from the stroma to the tear side.[1]

It has been recently shown that Cl enters the epithelial cell on the basolateral side against its electrochemical gradient and leaves the cell down the electrochemical gradient.[2, 3] It has been suggested that the entry step mechanism is a NaCl cotransport system.[4] A Na:K pump is also located in the basolateral membrane of the corneal epithelium[5, 6] and the following model, shown in FIGURE 1, has been proposed to explain ion transport in this epithelium. Na is pumped out from, and K is pumped into the cellular compartment. The apical side is rather impermeable to Na and K, so K diffuses back towards the stromal side. Na gets back into the cell compartment utilizing the cotransport system and driving Cl with it. Finally Cl diffuses out across the apical side. The overall result is a recirculation of Na and K across the basolateral membrane and a transepithelial movement of Cl.

Even though a "primary Cl pump" is not present, the corneal epithelium transports Cl against an electrochemical gradient and produces a net flow in the absence of it. In spite of a lack of an intrinsic Cl pump, an equivalent driving force, E_{Cl}, could be calculated from the unidirectional fluxes J_1 and J_2 using the flux ratio equation:

$$E_{Cl} = (RT/zF)\ ln\ J_1/J_2$$

where R, T, z and F are the gas constant, absolute temperature, the charge of the ion and the Faraday constant, respectively.

From the ratio $E_{Cl}/SCC = R_a$, the resistance of the active pathway can be calculated. It can be considered that R_a consists of two elements in series: R_{a1}, the entry step representing the cotransport system and R_{a2}, the exit step across the apical border (FIGURE 2).

In calculating E_{Cl}, only transepithelial fluxes across the active pathway should be considered. However, the experimentally measured unidirectional Cl fluxes represent the addition of the flux across the active pathway plus the flux across the paracellular pathway.

One possible way of separating the cellular and paracellular components of the flux is by using inhibitors of transcellular Cl transport. It can be hypothesized that when a pharmacological agent has reduced the SCC to near zero and

* Supported by Grants EY-00160 and EY-01867 from the National Eye Institute.

0077–8923/80/0341–0117 $01.75/0 © 1980, NYAS

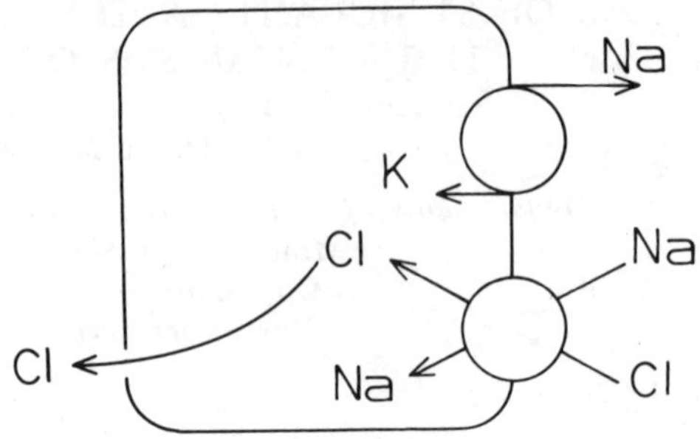

FIGURE 1. Model of NaCl cotransport to explain transepithelial Cl transport in the corneal epithelium.

when the two opposite unidirectional fluxes are nearly equal, these fluxes represent the paracellular pathway. This hypothesis requires that the drug has no effect on the paracellular pathway, a fact difficult to prove. Nevertheless, we have tested the effects of three inhibitors of Cl transport in the cornea; bumetanide, ouabain, and tryptamine, on their effects on unidirectional Cl fluxes. The results are shown in TABLE 1. As can be seen, the three agents reduce the SCC and the forward Cl flux, but their effects on the backward Cl fluxes are different. Thus, it is difficult to decide which of the three drugs, if any, is having an effect on the active pathway only.

We would like to present a different approach to the problem. It will be shown how E_{Cl} can be calculated from measurement of unidirectional fluxes and electrical resistance. In the electrical model shown in FIGURE 2, J_{a1} and J_{a2} are the forward and backward Cl fluxes across the active pathway. J_p repre-

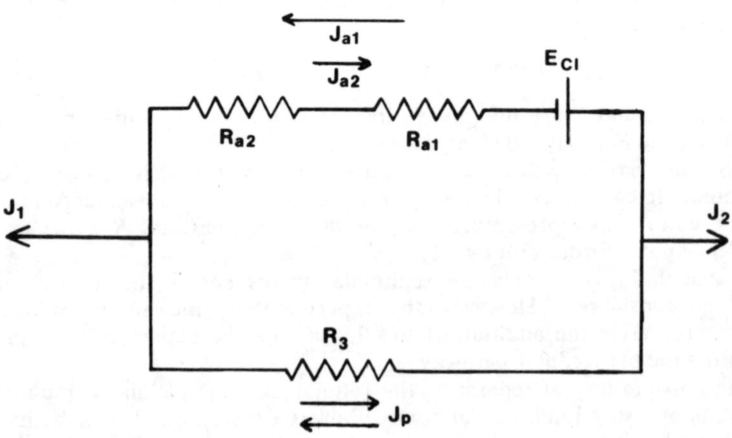

FIGURE 2. Electrical equivalent for transcellular and paracellular ionic movement across corneal epithelium.

sents all passive fluxes across R_s, the paracellular pathway. These fluxes are mainly those of Na and Cl. Also,

$$J_1 = J_{a1} + J_p \text{ and} \tag{1}$$

$$J_2 = J_{a2} + J_p \tag{2}$$

Under short-circuit conditions with identical solutions on both sides of the membrane,

$$J_p = \frac{RT/zF}{R_s} \tag{3}$$

$$J_{a1} = \frac{RT/zF \cdot f}{R_a} \quad \text{(4); where } f = \frac{x}{1-e^{-x}}$$

$$\text{and } x = \frac{E_{Cl}\ zF}{RT}$$

$$J_{a2} = \frac{RT/zF \cdot b}{R_a} \quad \text{(5); where } b = \frac{f}{e^x}$$

$$J_1 = \frac{(RT/zF)f}{R_a} + \frac{RT/zF}{R_s} \tag{6}$$

$$J_2 = \frac{(RT/zF)b}{R_a} + \frac{RT/zF}{R_s} \tag{7}$$

$$\frac{1}{R_t} = \frac{1}{R_a} + \frac{1}{R_s} \tag{8}$$

Thus, combining (6) and (8)

$$\frac{1}{R_t} - \frac{J_1}{RT/zF} = \frac{1-f}{R_a} \tag{9}$$

and similarly

$$\frac{1}{R_t} - \frac{J_2}{RT/zF} = \frac{1-b}{R_a} \tag{10}$$

Dividing (9) by (10) we obtain

$$\frac{1-f}{1-b} = \frac{(RT/zF) - (R_t \cdot J_1)}{(RT/zF) - (R_t \cdot J_2)} = N \tag{11}$$

Since $f = b \cdot e^x$, there is only one value of x for each value of N.

Knowing x, E_{Cl} can be calculated.

In FIGURE 4 of the appendix, values of N for E_{Cl} between zero and 378 mV have been tabulated.

Effect of Isoproterenol

Isoproterenol is a β-agonist that stimulates Cl transport in the cornea.[7] The increase is associated with a reduction of the electrical resistance. Values for

TABLE 1

EFFECTS OF BUMETANIDE, OUABAIN, AND TRYPTAMINE ON CL FLUXES AND SHORT-CIRCUIT CURRENT ACROSS THE ISOLATED FROG CORNEAL EPITHELIUM

	SCC *		Forward Cl Flux *		Backward Cl Flux *	
	C †	E ‡	C	E	C	E
10^{-5} M Bumetanide	13.5	1.1	19.0	5.6	6.0	6.0
10^{-3} M Ouabain	18.0	1.0	25.0	15.4	7.0	14.6
10^{-3} M Tryptamine	14.0	2.1	17.2	6.5	5.1	3.8

* Values are in μa/cm^2.
† C: Control.
‡ E: Experimental.

the various fluxes and electrical parameters are shown in TABLE 2. Using the control values from the table, a ratio N of $-22.94/6.54 = -3.50$ is obtained. This N value corresponds to an E_{Cl} of 107 mV. After isoproterenol, the ratio is $-29.10/8.30 = -3.50$, also corresponding to an E_{Cl} of 107 mV. It can be calculated that the main effect of isoproterenol was to reduce R_a from 9.72 $K\Omega \cdot cm^2$ to 4.86 $K\Omega \cdot cm^2$. The paracellular resistance also decreased from 3.70 $K\Omega \cdot cm^2$ to 2.61 $K\Omega \cdot cm^2$. There was no effect on Na permeability as shown by the Na fluxes. Since E_{Cl} represents the Na gradient that drives Cl across the entry step, it seems that isoproterenol did not change the properties of the entry step and only the resistance of the exit step across the apical barrier.

Effect of the Calcium Ionophore A23187

A23187 produces a large stimulation of Cl transport and a marked reduction of the electrical resistance.[8] The maximum response is obtained about 10 minutes after addition of A23187 and the SCC returns to its control value in

TABLE 2

EFFECT OF 2×10^{-6} M ISOPROTERENOL ON UNIDIRECTIONAL IONIC FLUXES, SCC AND ELECTRICAL RESISTANCE IN THE ISOLATED FROG CORNEAL EPITHELIUM

	Control	Isoproterenol
Unidirectional Na fluxes *	5.0	5.0
Forward Cl flux *	13.0	27.0
Backward Cl flux *	2.0	5.0
SCC *	11.0	22.0
Resistance (K$\Omega\cdot$cm^2)	2.68	1.70
N	−3.50	−3.50
E_{Cl} (mV)	107	107
R_a (K$\Omega\cdot$cm^2)	9.72	4.68

* Values are in μa/cm^2.

about 70 minutes. However, at this point, even though the net Cl flux is equal to the control, the unidirectional Cl fluxes are larger, and the electrical resistance is smaller than the control.

Values for the various parameters under control conditions, 10 minutes and 70 minutes after A23187 are shown in TABLE 3. N for the control is equal to $-20.90/5.35 = -3.90$ corresponding to $E_{Cl} = 119$ mv. $R_a = 9.52$ $K\Omega \cdot cm^2$ and $R_s = 2.69$ $K\Omega \cdot cm^2$. At the peak of the effect of A23187, the N ratio is -1.91 corresponding to an E_{Cl} of 51 mv. R_a is 2.17 $K\Omega \cdot cm^2$. The calcium ionophore does not change Na transport in the cornea. Thus, the reductions of R_a and E_{Cl} can be interpreted as an effect on the exit step (as isoproterenol) plus an effect on the NaCl cotransport system that increases cellular Na concentration reducing the equivalent driving force for Cl.

After 70 min, the SCC and the N ratio $(-16.70/4.30)$ had returned to their

TABLE 3

EFFECT OF 10^{-6} M A23187 ON UNIDIRECTIONAL IONIC FLUXES, SCC AND ELECTRICAL RESISTANCE IN THE ISOLATED FROG CORNEAL EPITHELIUM

	Control	A23187 10 min	A23187 70 min
Unidirectional Na fluxes *	5.0	5.0	5.0
Forward Cl flux *	17.0	31.0	20.0
Backward Cl flux *	4.5	7.5	7.5
SCC *	12.0	23.5	12.5
Resistance ($K\Omega \cdot cm^2$)	2.10	1.23	1.68
N	−3.90	−1.91	−3.88
E_{Cl} (mV)	119	51	118
R_a ($K\Omega \cdot cm^2$)	9.52	2.17	9.52

* Values are in $\mu a/cm^2$.

control values. However the unidirectional Cl fluxes are larger, and the electrical resistance smaller, indicating an additional effect on the paracellular Cl permeability (from 2.69 to 2.04 $K\Omega \cdot cm^2$).

Inhibitors of Cl Transport

Equation 11 can be used to find the value of E_{Cl} as long as J_1 is larger than J_2. However, when using inhibitors of Cl transport such as bumetanide, ouabain, and tryptamine, the SCC and net Cl transport are reduced to near zero and opposite unidirectional fluxes are almost identical. In this situation, the absolute values of the numerator and denaminator of equation 11 approach equality and the limit of N is equal to -1; thus E_{Cl} is indeterminate (see appendix).

Nevertheless, by examining the value of the numerator (or denominator), it is still possible to infer if all ionic pathways are conductive; that is to say, if all ionic movements contribute to the electrical resistance. When $J_1 = J_2$, the product $J_1 \times R_t$ should be equal to RT/zF, if all pathways are conductive.

$J_1 \times R_t$ larger than RT/zF is an indication that a fraction of the total unidirectional flux moves across a nonconductive pathway. Examples of this are exchange diffusion, and facilitated diffusion, of which cotransport is a special case.

The diuretic bumetanide is a potent inhibitor of Cl transport.[9] As shown in TABLE 4, in bumetanide-treated corneas $RT/zF \approx J_1 \cdot R_t$. This suggests that $R_t = R_s$, and that the active pathway R_a has been completely blocked. Additional evidence examined elsewhere [4] indicates that the bumetanide effect is on the exit step R_{a2}.

It has been shown that ouabain [10] and tryptamine [11] inhibit the Na:K pump in the corneal epithelium. As a result of this inhibition, cellular Na increases and the gradient that drives the Cl transport is dissipated. However, the active pathway R_a remains open, and due to the presence of the cotransport system, J_1 is larger than predicted by the conductance equation making $RT/zF < J_1 \cdot R_t$. As seen in TABLE 4, tryptamine increases R_t more than ouabain and the discrepancy between RT/zF and $J_1 \cdot R_t$ is less with tryptamine. This suggests that tryptamine partially increases the resistance of the active pathway.

We have also examined the effect of orthovanadate on active Cl transport in the corneal epithelium. Vanadate is an inhibitor of the (Na-K)ATPase from several tissues.[12, 13] Since Cl transport in the cornea is dependent upon the

TABLE 4

COMPARISON OF THE $J_1 \times R_t$ PRODUCT WITH RT/zF AFTER SCC HAS BEEN INHIBITED TO NEAR ZERO VALUES BY BUMETANIDE, OUABAIN, OR TRYPTAMINE

	$J_1{}^* \times R_t{}^\dagger$
Bumetanide	$11.2 \times 2.30 \approx RT/zF$ ‡
Ouabain	$21.4 \times 1.95 \gg RT/zF$
Tryptamine	$11.2 \times 2.91 > RT/zF$

* J_1 is in μa/cm^2.
† R_t is in K$\Omega \cdot$cm^2.
‡ RT/zF is in mV.

Na gradient maintained by the Na:K pump, vanadate should depress Cl transport if it indeed inhibited the (Na-K)ATPase of the corneal epithelium. Vanadate, 10^{-4} M, inhibited the Cl-SCC by 45 percent and in 10^{-3} M concentration inhibited the Cl-SCC by 81 percent.

It has been reported that vanadate penetrates the red cell via the ion exchange system,[13] and dinitrostilbene disulphonate, an inhibitor of ion exchange, prevents the initial uptake of ^{48}V-vanadate in red cells and slows down the onset of inhibition of (Na^+, K^+)ATPase by vanadate. We tested the disulfonic-stilbene, DIDS, for its capacity to prevent the inhibitory effect of vanadate on Cl transport in the cornea in a total of five experiments. DIDS, 5×10^{-4} M, produces only a small and temporary stimulation of the SCC (FIGURE 3). Vanadate, 10^{-4} M, added 20 to 60 minutes after DIDS failed to inhibit the SCC. 10^{-3} M vanadate (in the presence of 5×10^{-4} M DIDS) had no effect, or a small inhibition of Cl transport. DIDS does not inhibit active Cl transport in the cornea, but effectively prevents the vanadate inhibition. This supports the notion that, as in the red cell, vanadate must penetrate the cell membrane by the anion exchange system to exert its action from the cytoplasmic side. As shown in FIGURE 3, if vanadate is allowed to inhibit transport first, DIDS is unable to restore the control Cl transport.

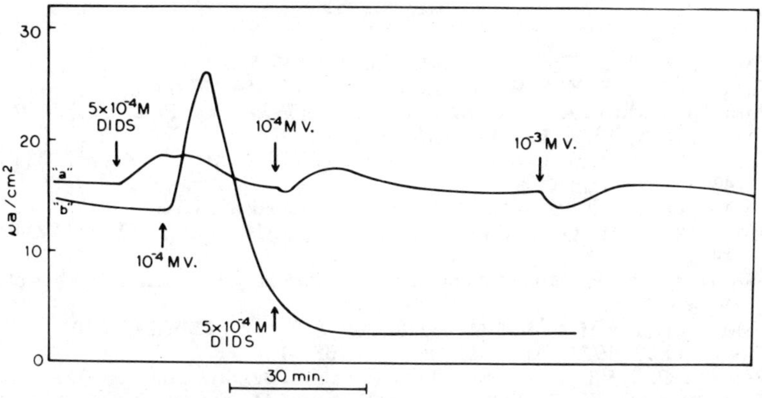

FIGURE 3. Short-circuit current of corneas "a" and "b" indicating active chloride transport. "a": DIDS prevents the inhibitory effect of 10^{-4} M and 10^{-3} M vanadate (V). "b": DIDS does not restore Cl transport that has been inhibited by 10^{-4} M vanadate (V).

Acknowledgments

I wish to thank my collaborators Drs. Peter Reinach and Howard Schoen for their contributions to this work.

Appendix

Values of E_{Cl} have been plotted as a function of N for values of N between -1 and -14 in FIGURE 4, below.

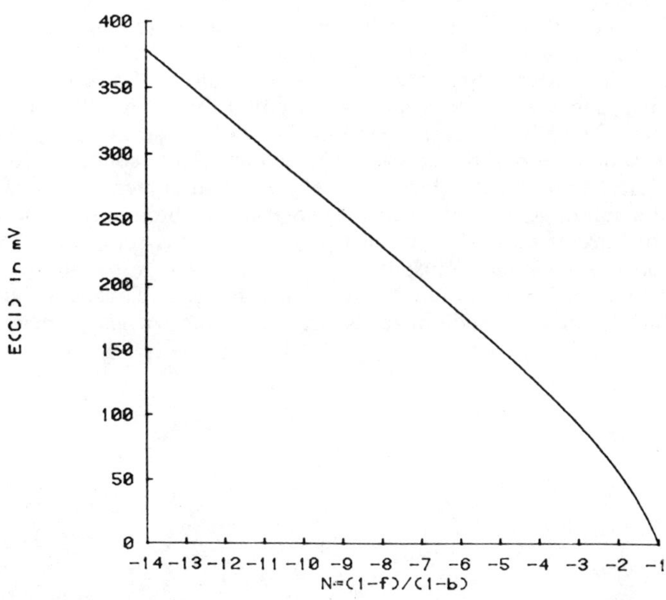

References

1. Zadunaisky, J. A. 1966. Am. J. Physiol. **211:**506–512.
2. Klyce, S. D. & R. K. S. Wong. 1977. J. Physiol. **266:**777–799.
3. Zadunaisky, J. A., K. R. Spring & T. Shindo. 1979. Fed. Proc. **38**(3)**:**1059.
4. Candia, O. A. 1979. Fed. Proc. In press.
5. Candia, O. A., P. J. Bentley & P. I. Cook. 1974. Am. J. Physiol. **226:**1438–1444.
6. Candia, O. A. & P. Reinach. 1978. The Physiologist. **21:**16.
7. Montoreano, R., O. A. Candia & P. Cook. 1976. Am. J. Physiol. **230:**1487–1493.
8. Candia, O. A., R. Montoreano & S. M. Podos. 1977. Am. J. Physiol. **233:** F94–F101.
9. Candia, O. A. & H. F. Schoen. 1978. Am. J. Physiol. **234:**F297–F301.
10. Candia, O. A. 1972. Am. J. Physiol. **223:**1053–1057.
11. Reinach, P. S. & O. A. Candia. 1978. Biochim. Biophys. Acta **510:**327–338.
12. Cantley, L. C., Jr., L. Josephson, R. Warner, M. Yanagisawa, C. Lechene & G. Guidotti. 1977. J. Biol. Chem. **252:**7421–7423.
13. Cantley, L. C., Jr., M. D. Resh & G. Guidotti. 1978. Nature. **272:**552–554.

Discussion

E. Racker: I would like to emphasize that vanadate is a very nonspecific inhibitor. It doesn't just inhibit the sodium potassium pump. We are now submitting a paper showing it inhibits a calcium pump. It inhibits several other ATPases as some of the very nonspecific inhibitors.

What is your proposition in your system on how vanadate acts?

O. A. Candia: I have some additional experiments. We see a 30% inhibition of Na transport by 10^{-5} M vanadate.

E. Frömter: If I understand it correctly, vanadate does not behave in your system like ouabain. Right?

Candia: No, in our system, their effects are similar. If you remember from the first figure, there is a sodium potassium pump on the basolateral side. The permeability of the apical side is very low for sodium, so normally there is no transport of sodium across the cornea. This is the normal condition for chloride transport. After removal of chloride, the short-circuit current decreases to zero; and after the subsequent addition of amphotericin B, both the Na permeability and the short-circuit current increase again, but in this state, the short-circuit current represents sodium transport. If we then add vanadate (10^{-5} M to 10^{-3} M) in this condition, sodium transport is inhibited. Thus vanadate has the same inhibitory effect on sodium transport across the cornea as does ouabain.

THE ROLE OF ANIONS IN THE REGULATION OF PROXIMAL TUBULAR SODIUM AND FLUID TRANSPORT

R. Green and S. L. Greenwood

Department of Physiology
University of Manchester
Manchester M13 9PT, England

G. Giebisch

Department of Physiology
Yale University
New Haven, Connecticut 06510

Reabsorption of fluid from the renal proximal tubule is a purely passive phenomenon depending on net ion reabsorption. For many years, it was assumed that the primary ionic movement was active reabsorption of sodium, but more recently the role of anions in this net reabsorption has been more thoroughly investigated and it has become apparent that anions can alter fluid transport in many different ways.

In a wide variety of epithelial tissues, the concentration of bicarbonate in the fluid bathing the basal membrane of the cell has a profound effect on reabsorption,[1-6] and a similar effect has been described in renal proximal convoluted tubules by several groups of workers.[7-10] At the peritubular membrane, anion inhibitors have also been shown to have a dramatic effect.[11, 12] There is less agreement about the effects of anions at the luminal membrane of the tubular cells, however.

Because of preferential bicarbonate reabsorption in the early part of the proximal tubule, there is a chloride concentration gradient in later parts of the proximal tubule with the chloride concentration in the luminal fluid being 20–25 mmole litre^{-1} greater than that in the plasma. This constitutes a driving force for anion reabsorption, and sodium and fluid are passively reabsorbed. While there is no dispute about the mechanism, the magnitude of its contribution to total fluid reabsorption is not clear. Estimates range from 20%–30% [10, 13] to at least 60 percent.[14]

A direct effect of anions on the reabsorptive process has been more difficult to demonstrate. Substitution of chloride by bromide or nitrate in isolated perfused rabbit proximal tubules, *in vitro,* did not significantly alter reabsorption of fluid; [15] but partial substitution of chloride by larger anions, such as cyclamate or methylsulphate, dramatically reduced the net flux of sodium and water.[13, 14] There was apparently a preferential anionic series for supporting sodium and fluid reabsorption such that bicarbonate > chloride > acetate > cyclamate when they were partially substituted for chloride in the luminal fluid.[13]

Although it is becoming apparent that intraluminal anions can have a marked effect on sodium and water reabsorption, the mechanisms underlying this process are not understood. Indeed, in spite of increased understanding of anionic transport in other tissues, in the mammalian renal proximal tubule it is still not clear whether chloride, the most abundant anion in tubular fluid, is reabsorbed passively, actively, in exchange for another anion, or in concert

0077–8923/80/0341–0125 $01.75/0 © 1980, NYAS

with a cation, such as sodium. In all probability, some combination of these mechanisms occurs.

To study chloride movement and its effects on net sodium and water reabsorption in renal proximal tubule, we have used a specific blocker, SITS (4-acetamido-4′ isothiocyano-stilbene-2,2′ disulphonic acid), an amino reactive agent that has been shown to bind to a variety of cell membranes. It is a known inhibitor of anion fluxes in red cells,[16, 17] of chloride efflux and bicarbonate uptake in squid axon,[18] and snail neurones,[19] and of the anion-dependent part of the short circuit current across turtle bladder,[20] and is thought to be specific for its action on anions. It has been demonstrated not to cross cell membranes although experimental evidence is confined to the red cell.[16] Its action on the kidney is not, however, as well documented.

All our experiments have been performed on male Sprague-Dawley rats weighing 150–200 g each and prepared for micropuncture using conventional techniques.[21] The initial experiments, which were designed to test the effects of a number of inhibitors, measured net fluid flux by calculating the rate of shrinkage of droplets of Ringer's solution isolated within the tubular lumen by castor oil blocks.[21] The composition of the Ringer's solution used was (in mmole $litre^{-1}$) NaCl, 135; KCl, 5; $NaHCO_3$, 5; $CaCl_2$, 1.8; raffinose, 10; and to this were added the inhibitors SITS (10^{-3} M), acetazolamide (10^{-3} M) or SITS and acetazolamide (both 10^{-3} M). The rate of shrinkage of the droplet was measured using a computer program on a Magiscan system (Joyce Loebl Ltd., Team Valley, Gateshead, England) [22] and the net reabsorptive fluid flux was calculated using the correction described by Györy.[23]

The results of this experiment are shown in FIGURE 1, in which the data are presented as means ± s.e.m. The normal reabsorptive rate with no inhibitors present was 1.77 ± 0.12 nl min^{-1} mm^{-1}, which is a little lower than what has been described for other strains of rats from different laboratories. However, it is consistent with data obtained in several other ways from this strain of rat in our own laboratory. Regardless of these lower control values, however, the net reabsorptive flux was significantly diminished by intraluminal

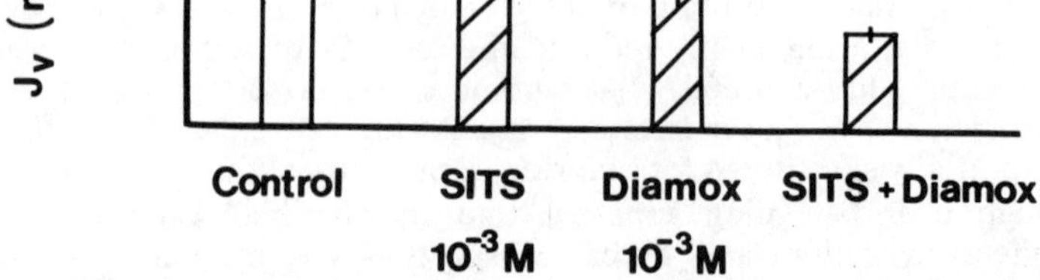

FIGURE 1. Rate of fluid reabsorption from split droplets containing Ringer's solution (see text). Histograms represent mean±s.e.m. Inhibitors were added to the intraluminal fluid. For controls, $n=66$; for SITS, $n=32$; for diamox, $n=30$, and for combined diamox and SITS, $n=17$.

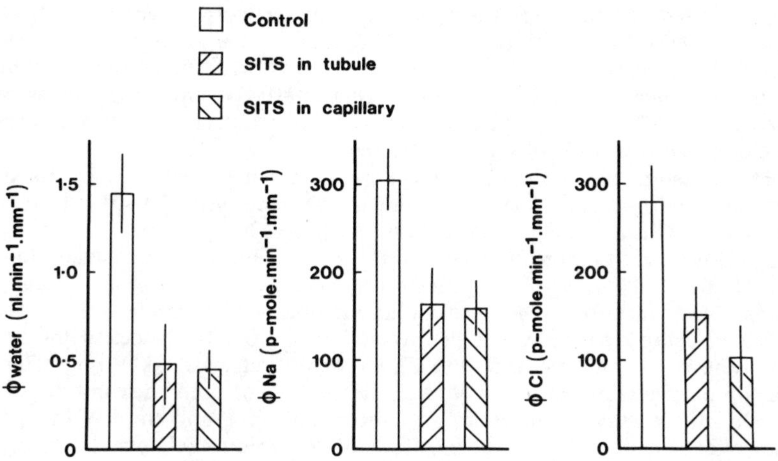

FIGURE 2. Net reabsorptive fluxes of water, sodium and chloride in proximal convoluted tubules artificially perfused with chloride Ringer's solution. Peritubular capillaries were perfused with bicarbonate Ringer's solution. SITS was added either to the tubule lumen or to the capillary perfusate. Histograms represent mean ± s.e.m. For controls, $n=18$; for SITS in tubules, $n=14$, and for SITS in capillaries, $n=16$.

SITS (by 33%, $p < 0.001$). Acetazolamide, a known inhibitor of bicarbonate reabsorption and net fluid flux, reduced the net fluid flux rather more than did SITS (by 54%) and together they reduced the net fluid reabsorption by 67%. This is significantly less ($p < 0.05$) than either inhibitor acting alone.

It has been shown previously [24, 25] that acetazolamide maximally inhibits fluid reabsorption at a dose of 10^{-3} M so it seems likely that 10^{-3} M used here would give maximal inhibition. Since the two inhibitors are additive in this situation we suggest this is tentative evidence that they are acting, at least in part, on different mechanisms contributing to the transport process.

Using this technique, however, it was not possible to measure net fluxes of ions directly and we investigated net ionic fluxes in a further series of experiments on randomly selected proximal tubules, which were artificially perfused *in vivo* at 30 nl min^{-1} using a Hampel pump; at the same time the capillaries were artificially perfused,[21] thus enabling us to apply SITS independently to either side of the cell. The tubular perfusate was the same as described above, while the capillary perfusate contained (in mmole $litre^{-1}$), NaCl 115; KCl, 5; $NaHCO_3$, 25; $CaCl_2$, 1.8; raffinose, 10. SITS was added to either peritubular or tubular perfusate at a concentration of 10^{-3} M. Because SITS was added to the peritubular perfusate, this solution did not have added albumin as is usual in the peritubular capillaries [21] since SITS would bind to the amino groups of the albumin and any effect on membranes would be unpredictable.

As a consequence of the omission of albumin, the net reabsorptive fluid flux (FIGURE 2) in control experiments is lower than previously reported reabsorptive fluxes,[10] but is consistent with the flux obtained when low albumin concentrations were used in the peritubular capillary perfusate.[21] The effects of SITS in these experiments confirmed the previous findings since again intraluminal

SITS inhibited net fluid reabsorption, and this reflects a reduced net sodium flux and a reduced net chloride flux. The nonchloride anionic flux (calculated as the difference between the net sodium and chloride fluxes) was small and was not significantly altered by SITS. Quantitatively similar results were obtained when SITS was applied to the peritubular capillaries, as net fluid, sodium and chloride fluxes were all significantly reduced.

These results are similar in some respects to those obtained by Lucci and Warnock[26] with intraluminal SITS; but at variance with the split drop results reported by Ullrich and his coworkers,[27] who, although they found an effect with peritubular SITS on fluid transport were unable to demonstrate an intraluminal effect. The peritubular effects have been explained as an effect on the bicarbonate transport at the peritubular membrane,[27] a result which has been confirmed using electrophysiological techniques,[12] but the effect at the luminal membrane is less well understood. Theoretically, at least, SITS might have an effect on paracellular pathways, increasing the leak of fluid back into the tubule; secondly, there may be an effect on cation transport, and thirdly, there might be an effect on anionic transport, either on an anion exchange mechanism or some carrier for anion transport.

To investigate the first possibility, viz., a general effect on paracellular pathways, we measured the rate of disappearance of sucrose from droplets of Ringer isolated between oil blocks. Randomly selected proximal convoluted tubules were punctured with a double barrelled pipette and the oil injected from

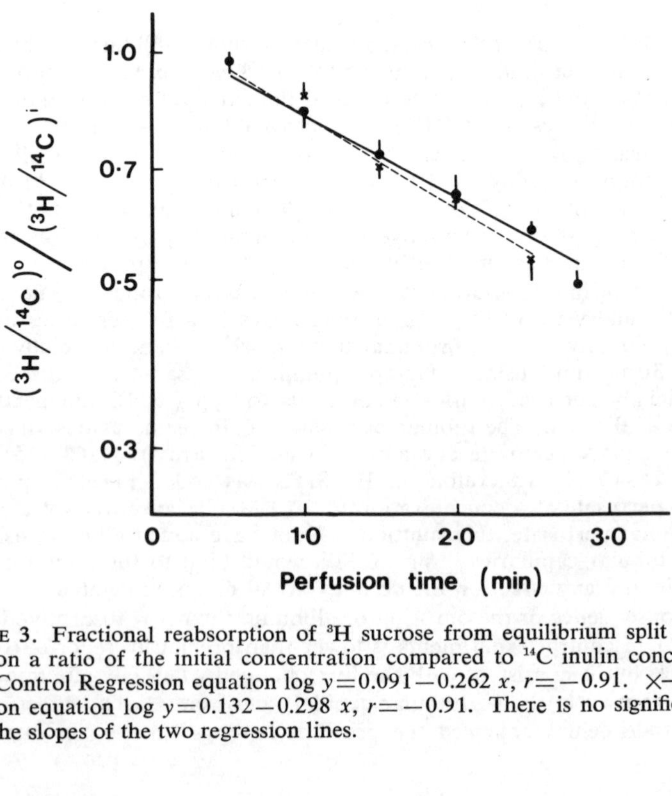

FIGURE 3. Fractional reabsorption of ^{3}H sucrose from equilibrium split drops (expressed on a ratio of the initial concentration compared to ^{14}C inulin concentration). ●—● Control Regression equation log $y=0.091-0.262\ x$, $r=-0.91$. ×– –× SITS Regression equation log $y=0.132-0.298\ x$, $r=-0.91$. There is no significant difference in the slopes of the two regression lines.

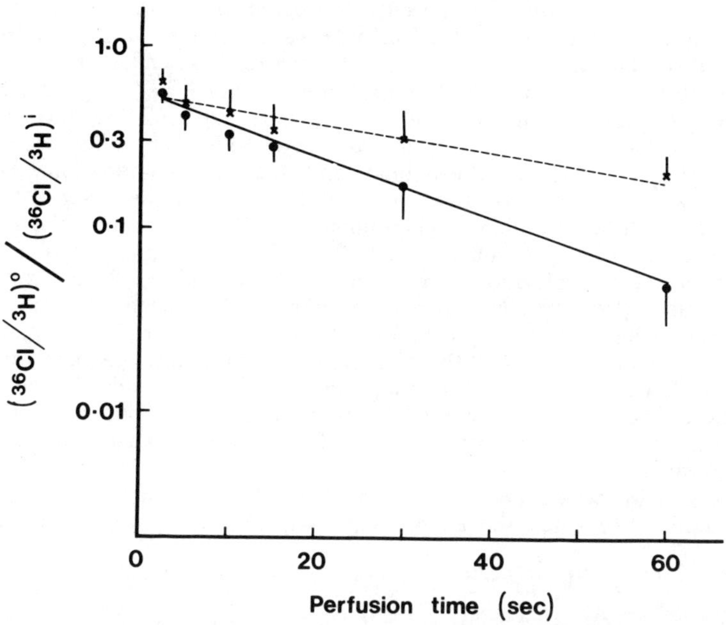

FIGURE 4. Fractional reabsorption of ^{36}Cl from equilibrium split drops (expressed as a ratio of the initial concentration when compared to ^3H inulin concentration). ●——● Control Regression equation $\log y = -0.622 - 0.039\ x$, $r = -0.82$. ×— —× SITS Regression equation $\log y = -0.636 - 0.018\ x$, $r = -0.54$. The slopes of the two regression lines are significantly different. $p < 0.001$.

one barrel was split by fluid injected from the second barrel. To reduce net fluid flow, the injected solution was a so-called equilibrium solution containing (in mmole litre^{-1}) NaCl, 110; $NaHCO_3$, 5; KCl, 5; $CaCl_2$, 1.8; raffinose 66 and as a marker ^3H sucrose (0.5 mCi ml^{-1}). To correct for any changes in sucrose concentration which might be due to residual net fluid movement, ^{14}C inulin (0.125 mCi ml^{-1}) was added and all ^3H sucrose measurements were expressed as a ratio of ^3H sucrose: ^{14}C inulin. After specified time intervals, ranging from 0.5 to 3.0 minutes, the fluid was reaspirated with a second pipette; three recollected drops were pooled to give sufficient fluid for analysis and they were analyzed for ^3H and ^{14}C using a liquid scintillation spectrometer. The ratio of ^3H:^{14}C when compared to the initial ratio gives the rate of disappearance of sucrose independent of any effects of water movement, and, since sucrose is thought to move through paracellular pathways only,[29] this gives a measure of general non-ionic paracellular permeability.

The results (FIGURE 3) show that SITS has no effect on paracellular permeability, at least as measured in this way; so the altered net reabsorption of fluid is unlikely to be due to an increased backflux of salt and water and indicates that cellular mechanisms may be primarily involved. There still remains a possibility, however, that some of the effects may be mediated through an alteration of the anion selectivity of the paracellular pathway,[28] and that this would secondarily alter transport of sodium and fluid.

The second possibility, viz., an effect on cation transport, seems inherently unlikely because of the lack of effect of SITS on cationic transport in other tissues but there is no direct evidence pertaining to these effects in kidney.

The third possibility for the luminal action of SITS, i.e., a direct effect on anion reabsorption, and in the case of the present experiments, chloride reabsorption, we have attempted to assess directly. Ringer's solution was placed between oil blocks in the renal proximal convoluted tubule and recollected after intervals varying from 2.5–60 seconds, using the technique described above. The Ringer solution contained (in mmole litre^{-1}) NaCl, 110; $NaHCO_3$, 5; KCl, 5; $CaCl_2$, 1.8, and raffinose, 66. ^{36}Cl (0.37 mCi ml^{-1}) was added to monitor the rate of chloride disappearance and ^3H inulin (0.8 mCi ml^{-1}) was used to correct for any changes in fluid volume. The fractional reabsorption of chloride (the ^{36}Cl:^3H ratio compared to the initial ratio) is plotted in FIGURE 4, which shows that there is an exponential disappearance of ^{36}Cl. The chloride permeability is a function of the slope of the line and with SITS the slope of the line is significantly different ($p < 0.001$) from control values, indicating that SITS has reduced the rate of ^{36}Cl disappearance and the permeability of the tubule to chloride.

To determine whether this effect was dependent on simultaneous transport of sodium, we repeated the experiments when sodium was initially absent.

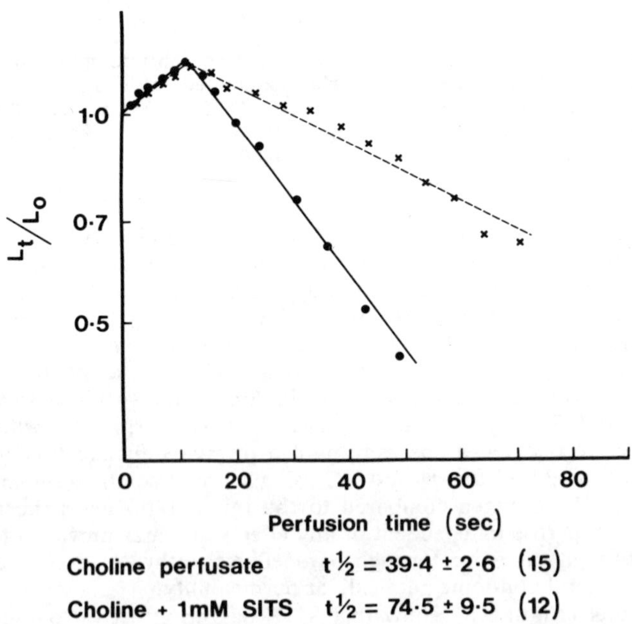

FIGURE 5. Rate of change of length of equilibrium split drops when sodium is substituted by choline in two representative experiments. Note both with SITS (×— —×) or without (●——●), there is initial swelling of the drops and subsequent shrinkage. The half time of the shrinkage phase for the series is given by the $t_{½}$ values. They are significantly different between control and SITS ($p < 0.001$). The figures in parenthesis represent the numbers of experiments.

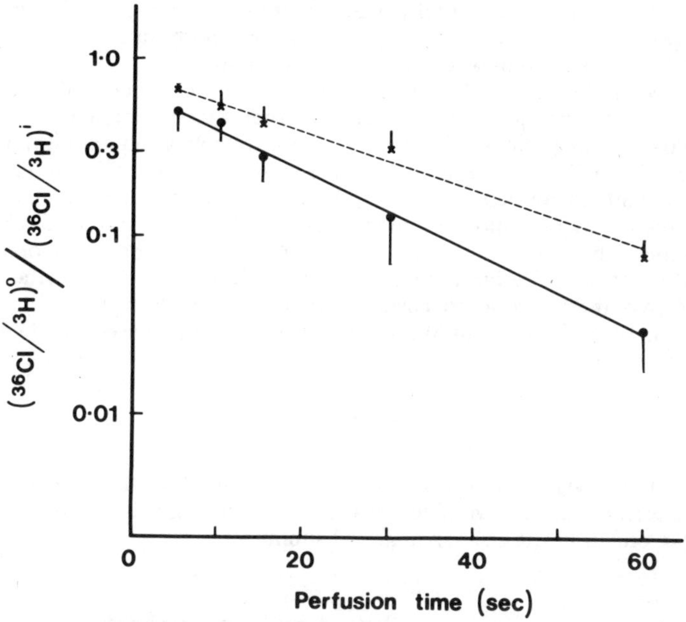

FIGURE 6. Fractional reabsorption of ^{36}Cl from equilibrium split drops where sodium is substituted by choline with SITS (×– –×) or without (●——●). All ratios are expressed with reference to ^{3}H inulin concentrations to avoid difficulties with net movement of fluid (FIGURE 5). The regression equations for: Control, $\log y = -0.439 - 0.052\,x$ $r = -0.86$; SITS, $\log y = -0.213 - 0.036\,x$ $r = -0.85$. The slopes of the line are significantly different. $p < 0.001$.

Experiments were performed as above except that the sodium in the Ringer's solution in the droplet was completely replaced by choline. Because of the influx of sodium into these droplets and subsequent reabsorption of choline, the droplets first swell and then shrink, as is shown in FIGURE 5, in which the shrinkage is plotted (as length at time *t* compared with length at time *0*) against time. Both with and without SITS, the rate of swelling was similar: being maximal at 13.5 seconds and the length increasing by a factor of 1.34. SITS slowed the subsequent shrinkage of the droplet (as it had slowed sodium chloride droplet shrinkage in the first series presented) demonstrating again an effect on fluid reabsorption ($p < 0.001$). When chloride permeability was measured under identical conditions, the rate of ^{36}Cl disappearance was reduced when SITS was present ($p < 0.001$), demonstrating an effect on chloride permeability (FIGURE 6). It is impossible to say that this is in the complete absence of sodium, since when droplets swell sodium enters; but further analysis showed that for both control and SITS droplets, the slope of the regression line over the initial 10 seconds (when the drop was swelling) was not significantly different from the slope during the later shrinkage phase, indicating that chloride permeability was not markedly influenced by the presence or absence of sodium. Although this conclusion requires much more rigorous proof in experiments in which sodium is completely excluded from the system, it does give an indication that sodium may not be essential for all chloride movement in the mammalian renal proximal convoluted tubule, even though Kimura and Spring [30] have postulated that this is so in *Necturus* proximal tubule.

Thus, we have shown that anion inhibitors reduce net reabsorptive fluid fluxes from the renal proximal tubule and that this is not a consequence of altered paracellular permeability, but probably follows a reduction in anion permeability. Furthermore, we have tentative evidence that the inhibition of chloride permeability is not dependent on sodium. We suspect that there are multiple mechanisms for reabsorption of chloride and that SITS is able to inhibit one or more of these; however, inhibition of fluid reabsorption by acetazolamide and SITS is not entirely through a common mechanism. This data further substantiates the hypothesis proposed earlier [13] that anion reabsorption can modulate reabsorption of sodium and fluid from the proximal convoluted tubule of the rat.

Acknowledgments

The authors wish to thank Professor S. Thomas for stimulating discussion, A. Bakes, K. Fletcher and S. Vockings for technical assistance, and the Medical Research Council (M.R.C.) for financial support.

References

1. Diamond, J. M. 1965. Transport of salt and water in rabbit and guinea pig gall bladder. J. Gen. Physiol. **48:**1–14.
2. Gonzales, C. F., Y. E. Shamoo & W. A. Brodsky. 1969. The accelerating effect of serosal bicarbonate on sodium transport in short circuited turtle bladders. Biochim. Biophys. Acta. **193:**403–418.
3. Hokin, M. R. 1969. Respiration and ATP and ADP levels during Na^+ transport in salt gland slices. Life Sci. **5:**1829–1837.
4. Knauf, H. 1972. The minimum requirements for the maintenance of active sodium transport across the isolated salivary duct epithelium of the rabbit. Pflügers Arch. Gesamte Physiol. **333:**326–336.
5. Schultz, I. 1971. Influence of bicarbonate–CO_2 and glycodiazine buffer on the secretion of the isolated cats pancreas. Pflügers Arch. Gesamte Physiol. **329:** 283–306.
6. Wright, E. M. 1972. Mechanisms of ion transport across the choroid plexus. J. Physiol. **226:**545–571.
7. Ullrich, K. J., H. W. Radtke & G. Rumrich. 1971. The role of bicarbonate and other buffers on isotonic fluid absorption in the proximal convolution of the rat kidney. Pflügers Arch. Gesamte Physiol. **330:**149–161.
8. Kokko, J. P., F. C. Rector & D. W. Seldin. 1970. Mechanism of salt and water reabsorption in proximal convoluted tubule. Abst. Ann. Meeting Soc. Nephrol. 4th **:** 42.
9. Maude, D. 1974. The role of bicarbonate in proximal tubular sodium transport. Kidney Int. **5:**253–260.
10. Green, R. & G. Giebisch. 1975. Ionic requirements of proximal tubular sodium transport I. Bicarbonate and chloride. Am. J. Physiol. **229:**1205–1215.
11. Frömter, E. 1977. Mechanism of H/HCO_3^- transport in rat kidney proximal tubule and mode of action of carbonic anhydrase inhibitors. Proc. Int. Union Physiol. Sci. **12:**204. Paris, France.
12. Edelman, A., J. Teulon & T. Anagnostopoulos. 1978. The effect of a di-

sulphonic acid stilbene on proximal cell membrane potential in *Necturus* kidney. Biochim. Biophys. Acta. **514:**137–144.
13. GREEN, R., J. H. V. BISHOP & G. GIEBISCH. 1979. Ionic requirements of proximal tubular sodium transport III. Selective luminal anion substitution. Am. J. Physiol. **236:**F268–F277.
14. NEUMANN, K. H., & F. C. RECTOR, JR. 1976. Mechanisms of NaCl and water reabsorption in the proximal convoluted tubule of rat kidney. J. Clin. Invest. **58:**1110–1118.
15. BURG, M. B. & N. GREEN. 1976. Role of monovalent ions in the reabsorption of fluid by isolated perfused proximal renal tubules of the rabbit. Kidney Int. **10:**221–228.
16. MADDY, A. H. 1964. A fluorescent label for the outer components of the plasma membrane. Biochim. Biophys. Acta. **88:**390–399.
17. CABANTCHIK, Z. I. & A. ROTHSTEIN. 1972. The nature of the membrane sites controlling anion permeability of human red blood cells as determined by studies with disulfonic stilbene derivatives. J. Memb. Biol. **10:**311–330.
18. RUSSELL, J. M. & W. F. BORON. 1976. Role of chloride transport in regulation of intracellular pH. Nature **264:**73–74.
19. THOMAS, R. C. 1977. The role of bicarbonate, chloride and sodium ions in the regulation of intracellular pH in snail neurones. J. Physiol. **273:**317–338.
20. EHRENSPRECK, G. & W. A. BRODSKY. 1976. Effects of 4-acetamido-4′-isothiocyano 2-2′ disulphonic stilbene on ion transport in turtle bladders. Biochim. Biophys. Acta. **419:**535–558.
21. GREEN, R., E. E. WINDHAGER & G. GIEBISCH. 1974. Protein oncotic pressure effects on proximal tubular fluid movement in the rat. Am. J. Physiol. **226:** 265–276.
22. GARLAND, H. O., J. M. BRUNT, C. TAYLOR & R. GREEN. 1979. Computerized analysis of split drop micropuncture data. Pflügers Arch. Gesamte Physiol. **381:**11–14.
23. GYÖRY, A. Z. 1971. Reexamination of the split oil drop method as applied to kidney tubules. Pflügers Arch. Gesamte Physiol. **324:**328–343.
24. BURG, M. B. & N. GREEN. 1977. Bicarbonate transport by isolated perfused rabbit proximal convoluted tubule. Am. J. Physiol. **233:**F307–F314.
25. LUCCI, M. S., D. G. WARNOCK & F. C. RECTOR, JR. 1979. Carbonic anhydrase dependent-bicarbonate reabsorption in the rat proximal tubule. Am. J. Physiol. **236:**F58–F65.
26. LUCCI, M. S. & D. G. WARNOCK. 1979. Effects of anion transport inhibitors on NaCl reabsorption in rat superficial proximal convoluted tubule. J. Clin. Invest. **66.**
27. ULLRICH, K. J., G. CAPASSO, G. RUMRICH, F. PAPAVASSILIOU & S. KLÖSS. 1977. Coupling between proximal tubular transport processes. Studies with ouabain, SITS and HCO_3^--free solutions. Pflügers Arch. Gesamte Physiol. **368:**245–252.
28. BERRY, C. A., D. G. WARNOCK & F. C. RECTOR, JR. 1978. Ion selectivity and proximal salt reabsorption. Am. J. Physiol. **235:**F234–F245.
29. BERRY, C. A. & E. L. BOULPAEP. 1975. Non-electrolyte permeability of the paracellular pathways in *Necturus* proximal tubule. Am. J. Physiol. **228:**581–595.
30. KIMURA, G. & K. R. SPRING. 1978. Transcellular and paracellular tracer chloride fluxes in *Necturus* proximal tubule. Am. J. Physiol. **235:**F617–F625.

SPECIFIC INHIBITION BY "LOOP" DIURETICS OF AN ANION-DEPENDENT $NA^{+}+K^{+}$ COTRANSPORT SYSTEM IN AVIAN ERYTHROCYTES *

H. C. Palfrey and P. Greengard

Department of Pharmacology
Yale University School of Medicine
New Haven, Connecticut 06510

P. W. Feit

Leo Pharmaceutical Products
Ballerup, Denmark

"Loop" diuretics such as furosemide are known to inhibit a number of anion and cation transport processes in both epithelial and nonepithelial tissues. For example, "active" chloride transport in the shark rectal gland [1] and frog cornea,[2, 3] NaCl transport processes in Ehrlich ascites tumor cells [4] and squid giant axon,[5] and chloride self-exchange in human erythrocytes [6, 7] are all blocked by millimolar concentrations of furosemide. The primary mode of action of these agents in the kidney, as revealed by perfused isolated tubule studies, is thought to be inhibition of active chloride pumping from lumen to peritubular fluid in the thick ascending limb of the loop of Henle.[8, 9] Despite the diversity of tissues and transport phenomena affected by "loop" diuretics, little attempt has been made to investigate structure-activity relationships of this class of inhibitors in any system. Recently, the synthesis of a series of chemically related compounds, which are 3,4-disubstituted derivatives of 5-sulfamoylbenzoic acid, has been described, and several of these compounds shown to be more potent as diuretics than furosemide in the dog *in vivo*.[10] The availability of this group of compounds potentially allows a detailed analysis of structure-activity profiles in different tissues sensitive to this class of diuretics.

As a convenient model system to investigate such effects, we have used the avian erythrocyte, which possesses a Na^{+} plus K^{+} cotransport mechanism activated by catecholamines or cyclic AMP,[11–13] and which is inhibitable by furosemide.[13] Upon stimulation, this system transports Na^{+} and K^{+} bidirectionally across the membrane in an obligatorily coupled manner; and has been shown to be independent of the Na^{+}, K^{+} exchange pump, being totally insensitive to ouabain.[11–13] An important but poorly understood feature of this transport system, which was first reported in duck erythrocytes,[13] is the dependence of hormonally stimulated net cation movements on extracellular chloride. This requirement has been confirmed by us [14] using measurements of unidirectional cation influx in turkey and other species of avian erythrocytes.

We have estimated the activity of the Na^{+} plus K^{+} cotransport system in turkey erythrocytes in the presence of different concentrations of the chemically

* This study was supported by Public Health Service Grants DA-01627, MH-17387 and NS-08440 (to P. Greengard) and grants from Hoffmann-La Roche and the McKnight Foundation (to P. Greengard). H. C. Palfrey was the recipient of a Muscular Dystrophy Association of America fellowship.

0077–8923/80/0341–0134 $01.75/0 © 1980, NYAS

related "loop" diuretics described above. Routinely, unidirectional ^{86}Rb (as a tracer for K^+) influx was assessed, following a preincubation period of 15 min in the presence of 1 mM 8-bromo-cyclic AMP (8BrcAMP) plus various concentrations (10^{-10} — 10^{-3} M) of each diuretic. IC_{50} values derived from dose-response curves for six such compounds are shown in TABLE 1. Concomitantly, the natriuretic potency of these agents *in vivo* in dogs was tested by measurement of Na^+ excreted in a 3-hr time period following a single intravenous administration of 0.25 mg/kg. As shown in TABLE 1, three of these compounds were effective inhibitors of cation cotransport in the turkey erythrocyte and also highly active as diuretics in dogs *in vivo*. Relatively small modifications of the 3′-substituent (R_1), such as the addition of a methoxy group on the benzylamine ring, resulted in a compound with severely reduced diuretic potency and cotransport inhibitory capacity (compare A with E in TABLE 1). Furosemide was intermediate in potency in both systems. These results suggest a high degree of structural similarity between the binding site for these diuretics on the avian erythrocyte and the putative binding site in the lumen of the loop of Henle. Investigation of a more extensive series of compounds has confirmed that a good correlation exists between diuretic potency and inhibition of avian cation cotransport.[15] Structurally dissimilar diuretics, such as amiloride and acetazolamide, as well as the related thiazide group of compounds, were ineffective in the avian erythrocyte system. Interestingly, one of the compounds in TABLE 1 (Compound C: bumetanide) has also been shown to be a more effective inhibitor of active chloride transport than furosemide in the frog cornea,[3] the differential potency (about 100-fold) being close to that found in the turkey erythrocyte.

We have shown elsewhere that some of the properties of the catecholamine- and cyclic AMP-stimulated Na^+ plus K^+ cotransport mechanism in erythrocytes of goose, pigeon and duck are similar to those of the turkey red cell.[14] These include interdependence of Na^+ and K^+ influx, insensitivity to ouabain, and loss of stimulation upon replacement of extracellular chloride with impermeant anions, such as gluconate. Consequently, we investigated the sensitivity of erythrocytes from these different species to various members of the diuretic series described above. As shown in TABLES 1 and 2, IC_{50} values for the four sulfamoylbenzoic acid derivatives tested were comparable for the different species. This constitutes further evidence that the cotransport mechanism in these various species is, in fact, similar.

If the characteristics of the diuretic "receptor" in the nephron and avian erythrocyte are similar, it might be anticipated that the transport processes inhibited would themselves be related. At present, insufficient information is available to draw any firm conclusion on this point. A Na^+ requirement for active chloride transport in rabbit cortical diluting segments has been demonstrated,[9] and a suggestion has been made that coupled NaCl transport occurs in this kidney segment based on similarities with other NaCl transporting epithelia,[16] but the issue remains unresolved. Analogously, as mentioned above, an extracellular chloride requirement has been shown for operation of cation cotransport in the avian erythrocyte. However, it is not clear whether chloride is itself cotransported or affects some other cellular parameter such as membrane potential that secondarily influences the activity of Na^+ plus K^+ cotransport, as has been postulated for duck red cells.[13] What can be ruled out is participation of the anion self-exchange system of the red cell membrane, since an inhibitor of this mechanism, SITS (0.5 mM), has very little effect on

TABLE 1

EFFECTS OF STRUCTURAL MODIFICATION OF DIURETICS ON 8BRCAMP-STIMULATED ^{86}RB UPTAKE IN TURKEY ERYTHROCYTES AND ON NATRIURETIC POTENCY IN DOGS

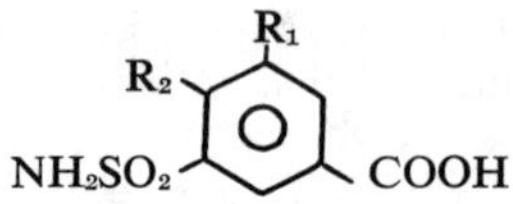

Compound	R_1	R_2	Turkey Erythrocyte IC_{50} (M) *	Natriuretic Potency †
A	$-NHCH_2-$ 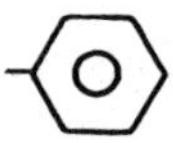	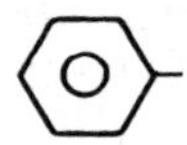 O-	6.1×10^{-8}	4.1
B	$-NHCH_2-$ 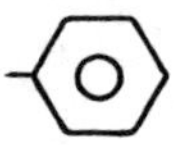	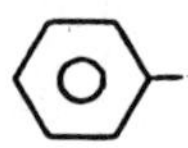 S-	4.6×10^{-8}	3.6
C	$-NH(CH_2)_3CH_3$ (bumetanide)	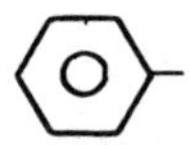 O-	2.5×10^{-7}	4.0
D	$-NH_2$	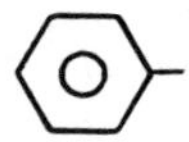 O-	$> 10^{-3}$	0 ‡
E	$-NHCH_2-$ 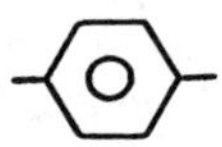 $-OCH_3$	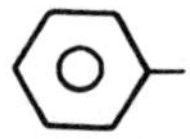 O-	3×10^{-4}	0 ‡
Furosemide	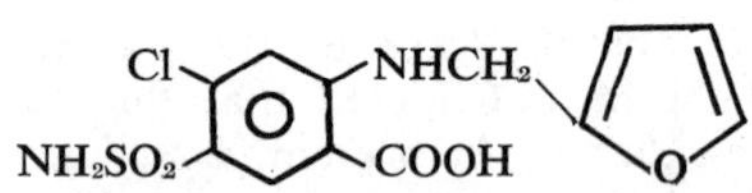		2.3×10^{-5}	1.7

* Cells (5% hematocrit) were preincubated for 15 min at 37° with 1 mM 8BrcAMP and various concentrations of the respective diuretics in a medium containing 157.5 mM NaCl, 2.5 mM KCl, 10 mM HEPES (pH 7.3) and 0.2% glucose. K^+ uptake was then assayed for 7 min by the addition of tracer ^{86}Rb in the presence of 1 mM ouabain. Control 8BrcAMP-stimulated influx was 9.85 mM/liter cells/hr.

† As mEquiv. Na^+ excreted/3 hr following i.v. administration of 0.25 mg/kg to dogs.

‡ Not significantly different from control.

TABLE 2

IC$_{50}$ VALUES FOR VARIOUS DIURETICS ON 8BrcAMP-STIMULATED ^{86}Rb INFLUX * INTO RED CELLS FROM DIFFERENT AVIAN SPECIES

Diuretic	Duck	Goose	Pigeon
		IC$_{50}$ (M)	
Compound B †	1.4×10^{-8}	8.5×10^{-9}	2.3×10^{-8}
Bumetanide	3.7×10^{-7}	1.7×10^{-7}	2.1×10^{-7}
Furosemide	2.3×10^{-5}	1.1×10^{-5}	1.2×10^{-5}
Compound E †	2.7×10^{-4}	1.7×10^{-4}	3.5×10^{-4}

* Incubation medium and conditions were as in TABLE 1 except that the hematocrits and influx times used were: duck (1%, 5 min); goose (2%, 5 min); pigeon (5%, 10 min). Control values for 8BrcAMP-stimulated K^+ influx, calculated from ^{86}Rb tracer influx, in the presence of 1 mM ouabain were (mM/liter cells/hr): duck, 76.1; goose, 35.8; pigeon, 2.8.

† See TABLE 1.

8BrcAMP-stimulated ^{86}Rb influx in the turkey erythrocyte (TABLE 3). Conversely, the SITS-sensitive anion-exchange mechanism (studied as $^{35}SO_4$ efflux in turkey red cells) is totally unaffected by the new "loop" diuretics studied here at concentrations that block cation cotransport (TABLE 3). Furosemide, however, has effects on both systems when used at 1 mM. This agent, frequently employed at this concentration in epithelial transport studies, is thus relatively nonspecific and may not be able to inhibit selectively either anion self-exchange or cation-anion cotransport processes in these preparations. The higher potency and specificity of the active loop diuretics (*e.g.*, bumetanide) could be useful in discriminating between the different modes of anion transport found in various tissues. These agents did not affect Na^+ pump, hormone-sensitive adenylate cyclase, or cyclic AMP-dependent protein phosphorylation activities in the intact turkey erythrocyte at concentrations that effectively

TABLE 3

DIFFERENTIAL EFFECTS OF DIURETICS AND SITS ON ANION SELF-EXCHANGE ($^{35}SO_4$ EFFLUX) AND $Na^+ + K^+$ COTRANSPORT (8BrcAMP-STIMULATED ^{86}Rb INFLUX) IN THE TURKEY ERYTHROCYTE

Agent	Anion self-exchange *	$Na^+ + K^+$ cotransport †
	(%)	(%)
Control	100	100
SITS (5×10^{-4} M)	0.5	85
Bumetanide (10^{-5} M)	100	1
Furosemide (10^{-3} M)	54	0

* Cells (10% hematocrit) were loaded for 3 hr at 37° C with $^{35}SO_4$ (20 μCi/ml) in a standard medium (TABLE 1) modified by substituting 6.25 mM Na_2SO_4 for 12.5 mM NaCl. They were then washed in the same medium at 0° C and efflux performed at 20° C, at a hematocrit of 0.5%. Control rate constant $k = 0.025$ min.$^{-1}$

† Assayed as in legend to TABLE 1.

blocked cation cotransport.[15] The similarity of pharmacological potency in the avian erythrocyte and mammalian kidney leads us to propose that the site of diuretic action is probably at the level of the respective transport systems themselves. Our results suggest that the avian erythrocyte may serve as a useful system for physiological and pharmacological analysis of "loop"-diuretic sensitive transport systems in general.

References

1. Epstein, F. H., P. Silva & J. Stoff. 1978. Curr. Prob. Clin. Biochem. **8:**107–122.
2. Candia, O. 1973. Biochim. Biophys. Acta **298:**1011–1014.
3. McGahan, M. C., T. Yorio & P. J. Bentley. 1977. J. Pharm. Exp. Ther. **203:** 97–102.
4. Heinz, E. & P. Geck. 1980. This volume.
5. Russell, J. M. 1979. J. Gen. Physiol. **73:**801–818.
6. Deuticke, B. 1970. Naturwiss. **57:**172–179.
7. Brazy, P. & R. B. Gunn. 1976. J. Gen. Physiol. **68:**583–599.
8. Rocha, A. & J. P. Kokko. 1973. J. Clin. Invest. **52:**612–623.
9. Burg, M. & J. E. Bourdeau. 1978. *In* New Aspects of Renal Function. H. G. Vogel & K. J. Ullrich, Eds. Hoechst Workshop Conferences **6:**91–102. Excerpta Medica. Amsterdam, the Netherlands.
10. Nielsen, O. B. T. & P. Feit. 1978. *In* Diuretic Agents. E. J. Cragoe, Jr., Ed. ACS Symp. Ser. **83:**12–23.
11. Gardner, J. D., G. D. Aurbach, A. M. Spiegel & E. M. Brown. 1976. Rec. prog. Horm. Res. **32:**567–595.
12. Kregenow, F. M. 1977. *In* Transport Processes in Red Cells. J. C. Ellory & V. Lew, Eds. : 383–426. Academic Press. New York, N.Y.
13. McManus, T. J. & W. F. Schmidt, III. 1978. *In* Membrane Transport Processes. J. F. Hoffman, Ed. **1:**79–106. Raven Press. New York, N.Y.
14. Palfrey, H. C. & P. Greengard. 1980. In preparation.
15. Palfrey, H. C., P. Feit & P. Greengard. 1980. Am. J. Physiol. In press.
16. Frizzell, R. A., M. Field & S. G. Schultz. 1979. Am. J. Physiol. **236:**F1–F8.

Part III. Regulatory Mechanisms for Anion and Proton Transport in Intestinal, Salivary, Pancreatic, Renal and Corneal Epithelia

THE ROLE OF BICARBONATE IONS AND OF ADENOSINE 3′,5′-MONOPHOSPHATE (cAMP) IN CHLORIDE TRANSPORT BY EPITHELIAL CELLS OF BULLFROG SMALL INTESTINE *

W. McD. Armstrong and S. J. Youmans

Department of Physiology
Indiana University School of Medicine
Indianapolis, Indiana 46223

The mechanism of Cl^- transport by the small intestine and other epithelia has been intensively investigated in recent years.[1] *In vitro* studies,[2] using the classical short-circuit technique of Ussing and Zerahn,[3] have shown that net mucosal to serosal ($m \rightarrow s$) transepithelial Cl^- transport in the small intestine of the bullfrog (*Rana catesbeiana*) follows a similar pattern to that seen in a variety of epithelial systems, e.g., mammalian[4] and teleost[5] small intestine, mammalian gallbladder,[6,7] and the proximal tubules of amphibian kidney.[8] All these epithelia are characterized by the capacity to absorb large amounts of Na^+ and Cl^- accompanied by an isotonic equivalent of water. In addition they belong to a class of epithelial systems that is usually described as "leaky" because the systems possess relatively low-resistance paracellular shunt pathways for water and ions.[9] In bullfrog small intestine,[2] as in the other epithelial systems described above,[1] a major component of transcellular Cl^- absorption is an electroneutral (1/1) coupled Na^+ and Cl^- entry process across the apical membranes of the absorptive cells. This process results in the maintenance, within these cells, of Cl^- activities significantly higher than that predicted for "passive" Cl^- distribution across this membrane.[10,11] Thus, Cl^- entry into the absorptive cells is an energy-requiring process and a considerable amount of evidence now points to the transapical Na^+ electrochemical difference as the probable source of this energy.[11,12] Conversely, direct measurements of intracellular Cl^- activity (a_{Cl}^i) in bullfrog small intestine[10,11] indicate that net Cl^- exit across the basolateral cell membrane occurs down a gradient of electrochemical potential but the mechanism responsible remains unclear. This report describes some experiments performed with the object of gaining further insights into the factors controlling the exit of Cl^- across the baso-lateral membrane of the absorptive cells in bullfrog small intestine. Experiments on the effect of adenosine 3′,5′-monophosphate (cAMP), a known inhibitor of coupled transapical Na/Cl entry in a number of leaky epithelia,[1,13] on Na^+ and Cl^- transport by bullfrog small intestine *in vitro* are also described.

Intestinal Cl^- Transport—Effects of HCO_3^-

Although widely investigated, the role of HCO_3^- in the transcellular transport of ions by the small intestine remains somewhat ill-defined. In particular, there are apparent inconsistencies between intestinal transport of HCO_3^- *in vivo*

* This work was supported by USPHS Grants AM12715 and HL23332.

0077-8923/80/0341-0139 $01.75/0 © 1980, NYAS

and *in vitro,* even in the same tissue, that have not yet been satisfactorily explained. The formidable technical problems associated with the direct measurement of HCO_3^- fluxes in living tissues have contributed, in no small measure, to current uncertainties concerning intestinal transport of this ion. A detailed discussion of intestinal HCO_3^- transport *in vivo* and *in vitro* lies outside the scope of this report and can be found elsewhere.[13] Of more immediate interest for our purposes are a number of reports that HCO_3^- is implicated in the intestinal absorption of Cl^-. Turnberg *et al.*[14] showed that in normal human ileum, perfused *in situ,* Na^+ and Cl^- are absorbed and HCO_3 is secreted. Coupling between Cl^- absorption and HCO_3^- secretion was suggested by the observation that increasing the luminal HCO_3^- concentration to 60 mM or more resulted in an absorption of HCO_3^- and a secretion of Cl^-. Similarly, when luminal Cl^- was replaced by SO_4^-, absorption of HCO_3^- was observed. To account for these findings, Turnberg *et al.*[14] proposed an electroneutral Cl^-/HCO_3^- exchange that could function reversibly, depending on the relative proportions of Cl^- and HCO_3^- in the luminal perfusate. The technical constraints inherent in *in vivo* experiments of this kind precluded any conclusions concerning the exact location, in the absorptive cells, of this postulated exchange process. A dependence of bicarbonate secretion on luminal Cl^- has also been reported in the rat ileum *in vivo.*[15]

Field *et al.*,[5] working with isolated segments of flounder (*Pseudopleuronectes americanus*) intestine, stripped of external muscle layers and maintained under short-circuit conditions—between identical Ringer's solutions in an Ussing chamber—found that when the HCO_3^- concentration of the medium was reduced from 20 mM to 5 mM there was an approximately two-fold reduction in the rate of active Cl^- absorption. This reduction was entirely attributable to a decline in the net mucosal to serosal unidirectional Cl^- flux (J_{ms}^{Cl}) and suggests a strong dependence of Cl^- absorption, in this tissue, on external HCO_3^- even though, in these experiments, no evidence was found for significant net transepithelial HCO_3^- transport. With the same technique and using a similar stripped preparation of *amphiuma* small intestine, Gunter-Smith and White[16] observed that replacement of either Cl^- or HCO_3^- in the bathing medium reversed the sign of the normally serosal-negative transepithelial potential difference (*PD*) and short-circuit current. They suggested that, in this tissue, Cl^- and HCO_3^- transport are linked, possibly through a Cl^-/HCO_3^- exchange mechanism. Finally, the observation[17] that replacement of 25 mM Cl^- by HCO_3^- in the Ringer's solution bathing isolated segments of stripped bullfrog small intestine induced a significantly serosal negative I_{sc}, and a large net absorption of Cl^- under short-circuit conditions, prompted us to postulate the existence, as an important component of active Cl^- transport by this tissue, of an electrogenic Cl^-/HCO_3^- exchange across the basolateral membrane of the absorptive epithelial cells. The experiments that led to this conclusion, and further related experiments, are described in the following section.

Effect of HCO_3^- on Transepithelial Electrical Parameters and Na^+ and Cl^- Fluxes in Isolated Bullfrog Small Intestine

To study the response to HCO_3^- of the transepithelial electrical parameters and the unidirectional Na^+ and Cl^- fluxes across the isolated small intestine of the bullfrog, two sets of experiments were performed. In these experiments, segments of small intestine from frogs that were kept for 1–2 weeks in captivity without food were stripped, by blunt dissection, of their external muscle layers,[18] and mounted in Ussing chambers between identical Ringer's solutions at 23° C. Transepithelial *PD*, I_{sc}, and unidirectional mucosal to serosal and serosal to mucosal Na^+ and Cl^- fluxes (J_{ms}^{Na}, J_{ms}^{Cl}, J_{sm}^{Na} and J_{sm}^{Cl}) were measured by standard electrophysiological and radio-tracer techniques.[2] Two sets of experiments were performed. In one of these, the bathing medium was a phosphate buffered (pH 7.2) HCO_3^--free solution[2] that contained 105 mM Na^+, 105 mM Cl^-, and 5.4 mM K^+. During experiments, this medium was gassed continuously with 100% O_2. The medium employed in the second set of experiments differed from this with 25 mM NaCl replaced by 25 mM $NaHCO_3$. This medium was gassed with a 95/5% O_2/CO_2 mixture. The pH was 7.4.

The first two rows in TABLE 1 summarize the values obtained for *PD* and I_{sc} in the presence and in the absence of HCO_3^-. The *PD* and I_{sc} measurements

TABLE 1

TRANSEPITHELIAL *PD* AND I_{sc} IN BULLFROG SMALL INTESTINE †

Immersion Medium ($Na/Cl/HCO_3$, mM)	Number of Experiments	*PD*	I_{sc}
A. 105/105/0	24	0.3±0.2 *	0.1±0.1 *
B. 105/80/25	23	−0.8±0.3	−1.0±0.3
C. 105/80/25+ cAMP (6 mM)	23	1.3±0.4	0.7±0.2

† Average *PD* (mV) and I_{sc} ($\mu Eq/cm^2 \cdot hr$) values (with respect to the mucosal medium) ± S.E.M. are given for the number of experiments indicated; B and C are paired experiments.

* Denotes values that did not differ significantly from zero ($p > 0.05$).

in this table include values from the paired flux experiments described below (TABLE 2).

In HCO_3^--free media, both *PD* and I_{sc} had low, serosal-positive, average values that did not, in the present experiments, differ significantly from zero. Similar low values for these parameters were observed previously in experiments with unstripped intestinal segments.[2] By contrast, in media containing 25 mM HCO_3^-, both the *PD* and I_{sc} were serosal-negative and had mean values that differed significantly from zero. A similar reversal of the I_{sc}, from a serosal-negative to a smaller serosal-positive value was observed in *amphiuma* small intestine when Cl^-, HCO_3^- or both these ions were omitted from the bathing medium.[16]

The serosal-negative I_{sc} elicited by HCO_3^- in bullfrog small intestine indicates a significant net absorption of anions or secretion of cations under these conditions. To determine the ionic basis of this current, unidirectional transepithelial Na^+ and Cl^- fluxes were measured under short-circuit conditions. In these experiments paired stripped segments were prepared from contiguous regions of the small intestine and mounted in Ussing chambers as described above. ^{22}Na and ^{36}Cl were used to determine J_{ms} and J_{sm} for both these ions. These isotopes were added to the fluid bathing the mucosal side of one member of each pair of intestinal segments and to the fluid bathing the serosal side of its companion. Following mounting in the chamber and the attainment of an electrical steady-state, the ratio I_{sc}/PD was compared in each pair. Addition of the radio-labels and sampling for flux determinations were performed only for those pairs whose I_{sc}/PD ratios did not differ by more than 35%. Furthermore, if, following the addition of radio-isotopes, the I_{sc}/PD ratios of the paired tissues differed by more than 35% at any time, the experiment was abandoned.

TABLE 2 illustrates the results obtained in two sets of paired flux experiments, one performed in the absence and the other in the presence of external HCO_3^-. In the absence of HCO_3^- the average J_{net}^{Na} suggests that Na^+ was being absorbed at a low rate. However such absorption, if it occurred, was not statistically detectable in the experiments shown in TABLE 2. Similarly, there was no detectable net movement of Cl^- under these conditions. As in the more extensive series of electrical measurements summarized in TABLE 1, the very low serosal positive I_{sc} observed in the experiments under discussion did not differ significantly from zero. Thus, under these conditions, there was no detectable net flux of any other ion or ions; that is, the residual net flux, J_R, was essentially zero (TABLE 2).

These results indicate that in the absence of external substrates, Na^+ and Cl^- absorption take place at a minimal rate, if at all, in stripped *in vitro* intestinal segments from fasting animals. This is at variance with earlier data reported from our laboratory for Na^+ and Cl^- transport across isolated segments of unstripped bullfrog under identical conditions.[2] In unstripped segments, although the transepithelial *PD* and I_{sc} in an HCO_3^--free medium had very low serosal-positive values, and J_{ms} and J_{sm} were smaller than they were in the stripped preparation, there was a significant net m → s flux of Na^+ and Cl^- under these conditions. The net Na^+ flux (1.2 μEq/cm^2 hr) exceeded that of Cl^- (0.9 μEq/cm^2 hr), and the difference between them accounted quantitatively for the observed I_{sc}.

The origin of this discrepancy is not clear. A possible explanation is, however, suggested by the following considerations. There seems to be little doubt that, *in vitro,* unstripped preparations of mammalian small intestine are seriously hypoxic[19] and it seems reasonable to suppose that, despite the less exacting demands for O_2 that are characteristic of amphibian tissues compared to those of mammals, unstripped preparations of bullfrog intestine may also be somewhat hypoxic under *in vitro* conditions. A link between hypoxia and Na^+ and Cl^- absorption in the absence of external HCO_3^- is suggested by unpublished experiments by W. McD. Armstrong and S. J. Cohen in this laboratory. In these experiments, steady-state unidirectional Na^+ and Cl^- fluxes were compared in segments of unstripped bullfrog small intestine when the bathing medium (identical to the HCO_3^--free medium used in the present study) was gassed alternately with 100% O_2 and 100% N_2. Under N_2, the net fluxes of both Na^+ and Cl^- were increased in short-circuited preparations.

If, indeed, hypoxic or anoxic conditions can induce a net absorption of Na^+ and Cl^- in bullfrog small intestine, the mechanism underlying such absorption remains to be uncovered. It is of some interest, however, to note that intestinal Na^+ and Cl^- absorption in another amphibian species, *amphiuma,* is apparently unimpaired by anoxic conditions.[20]

In the presence of 25 mM HCO_3^-, the pattern of ionic fluxes across the

TABLE 2

TRANSEPITHELIAL Na^+ AND Cl^- FLUXES, I_{sc}, AND NET RESIDUAL FLUXES (J_R), IN BULLFROG SMALL INTESTINE †

Immersion Medium ($Na/Cl/HCO_3$, mM)	Number of Experiments	Na^+ J_{ms}	Na^+ J_{sm}	Na^+ J_{net}	Cl^- J_{ms}	Cl^- J_{sm}	Cl^- J_{net}	I_{sc}	J_R
A. 105/105/0	8	5.2 ± 1.1	4.7 ± 1.1	0.5 * ± 0.9	3.1 ± 0.8	3.3 ± 0.8	−0.2 * ± 1.0	0.2 * ± 0.1	--
B. 105/80/25	5	9.0 ± 0.6	9.0 ± 0.8	--	8.8 ± 0.7	4.6 ± 0.8	4.2 ± 0.8	−1.4 ± 0.5	−2.8 ± 0.8
C. 105/80/25 + cAMP (6 mM)	5	6.8 ± 0.4	7.5 ± 0.8	−0.8 * ± 0.6	5.8 ± 0.4	4.1 ± 0.5	1.8 ± 0.4	0.2 * ± 0.1	−2.8 ± 0.9
Δ (C-B)		−2.2 ± 0.8	−1.5 * ± 0.6	0.8 * ± 0.6	−3.0 ± 0.5	−0.5 * ± 0.5	−2.5 ± 0.7	1.6 ± 0.4	--

† All values shown are in $\mu Eq/cm^2 \cdot hr \pm$ S.E.M. Paired unidirectional fluxes (J_{ms} and J_{sm}) are given for the number of experiments indicated. Net flux (J_{net}) is calculated from individual values of ($J_{ms} - J_{sm}$). J_R is calculated from the difference between I_{sc} and ($J^{Na}_{net} + J^{Cl}_{net}$) in each experiment. In addition, sets B and C represent measurements made in the same tissue samples, and Δ values are the averages of the changes observed in individual experiments. I_{sc} values are with respect to the mucosal solution; negative signs before net fluxes indicate secretion, and negative Δ values represent decreases in magnitude.

* Indicates values that did not differ significantly from zero ($p < 0.05$).

stripped, oxygenated small intestine of the bullfrog is dramatically changed from that seen in the complete absence of external HCO_3^- (TABLE 2). Both J_{ms} and J_{sm} for Na^+ were much larger in the presence of HCO_3^- than in the absence of HCO_3^-. They were, however, of equal magnitude so that net Na^+ absorption was again zero. Similarly, in the presence of external HCO_3^-, J_{ms}^{Cl} was substantially larger than in the absence of this ion. However, in contrast to the situation described above for Na^+, there was a large and highly significant net Cl^- absorption under these conditions. As seen in TABLE 2, this net Cl^- flux was significantly larger than the corresponding serosal negative I_{sc}; that is to say, under these conditions there is a large and highly significant residual net flux, J_R, that is consistent with a secretion of anions into the mucosal solution. Since this flux is absolutely dependent on external HCO_3^-, we conclude that it reflects a net s → m flux of HCO_3^-.

A net secretion of HCO_3^-, similar to that observed in the present experiments, is observed in the short-circuited rabbit ileum when the external HCO_3^- concentration is relatively high (25 mM),[18] but is, apparently, absent in media containing relatively low HCO_3^- concentrations.[21] Despite its apparent dependence on external HCO_3^-, it has been suggested [13] that this secretion is largely, if not exclusively, mediated by endogenous HCO_3^- derived from oxidative metabolism. A different interpretation for the origin of J_R in our experiments is suggested by the fact that when 2 mM 4-acetamido-4′-isothiocyano-2,2′-stilbene disulfonate (SITS), a compound that is known to inhibit Cl^-/HCO_3^- exchange in a variety of cell types,[22, 23] is added to the serosal bathing medium, the negative I_{sc} observed in the presence of HCO_3^- is abolished. In seven experiments, in which SITS was added to the serosal medium, following the establishment of an electrical steady state, I_{sc} decreased from -2.4 ± 0.6 to -0.4 ± 0.3 $\mu Eq/cm^2$ hr. The average steady state values in the presence of SITS did not differ significantly from zero. When added to the mucosal medium, SITS had little effect on I_{sc} in media containing HCO_3^-. These experiments prompted us [24] to propose the existence of a Cl^-/HCO_3^- exchange in the basolateral membrane of the absorptive cells as an important regulatory mechanism in transepithelial Cl^- and HCO_3^- transport by bullfrog small intestine.

In summary, our view of net Na^+ and Cl^- transport by epithelial cells of bullfrog small intestine under short-circuit conditions, and of the role of HCO_3^- in transcellular ion transport by this tissue can be represented by the conceptual model shown in FIGURE 1. In this model, two routes for Na^+ entry across the brush border membrane are shown. One of these is the electroneutral Na/Cl coupled entry pathway already discussed. The other is a Cl^--independent pathway [2] that functions as a component of net Na^+ transport when Cl^- is replaced, for example, by $SO_4^=$. Although net m → s Cl^- transfer in the absence of external Na has been observed in this tissue,[25] the coupled Na/Cl entry process is the only pathway indicated in FIGURE 1 for Cl^- entry into the cell. This is because transcellular Cl^- transport in the absence of external Na^+, though finite, is extremely small under aerobic conditions and in the absence of exogenous metabolites. It only becomes of major importance when a metabolically utilizable substrate, e.g., glucose, is present in the external medium under anaerobic conditions or in the presence of inhibitors of oxidative metabolism.[25] Similarly, the coupled mucosal entry route for Na^+ that comes into play in the presence of actively transported solutes, such as sugars and amino acids, is not shown. Finally, since the experiments discussed herein are concerned with transcellular pathways for ion transport, the paracellular shunt pathway, despite its importance in the overall electrophysiological and ionic transfer characteristics of bullfrog small intestine [26] is not explicitly considered.

The model shows two exit processes across the baso-lateral cell membrane. One of these is the ubiquitous ATP-ase driven Na^+/K^+ pump for which evidence has been found in bullfrog intestine,[27] as in many other tissues. The other is the coupled Cl^-/HCO_3^- exchange, already referred to, that becomes apparent in the presence of external HCO_3^-. An exit route for HCO_3^-, across the brush border membrane, as shown in the model, is a necessary requirement for HCO_3^- secretion. However, our results to date do not permit a detailed characterization of HCO_3^- transport across this membrane. Finally, since there is no evidence for net transcellular K^+ transport, under the conditions to which the model is supposed to apply,[2] K^+ transport, except as a component of the baso-lateral Na^+/K^+ pump, is ignored.

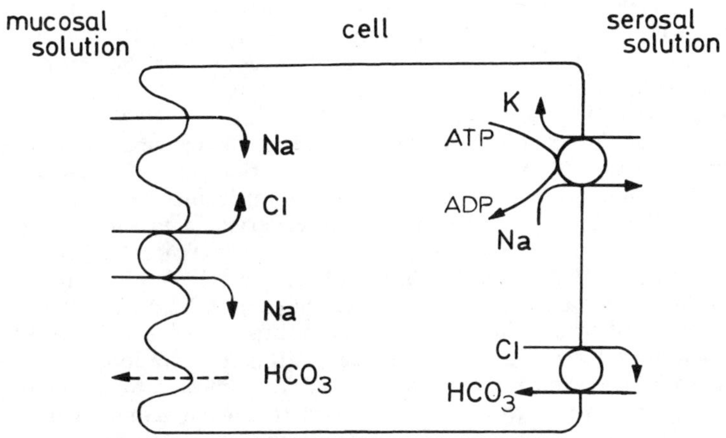

FIGURE 1. A model for *in vitro* ion transport by absorptive cells of short-circuited bullfrog small intestine. Where this is known, the curved portions of the arrows indicate whether net ionic transfer is energetically uphill or downhill with respect to the appropriate transmembrane electrochemical potential gradient.

In terms of the model, the data shown in TABLE 2 for Na^+ and Cl^- transport, in the presence and absence of HCO_3^-, can be interpreted as follows: In the absence of HCO_3^-, the lack of significant Na^+ or Cl^- absorption implies that under steady-state conditions, coupled entry of these ions across the apical cell membrane must be proceeding at a relatively low rate and that the greater part of the unidirectional $m \rightarrow s$ fluxes of Na^+ and Cl^- measured under these conditions occurs by way of the paracellular route. We suggest that an important limiting step in the transcellular transport of Cl^-, and possibly of Na^+, under these conditions may be the net rate of Cl^- efflux across the baso-lateral cell membrane. Little is known about the mechanism of Cl^- transport across this membrane. Direct measurements, with Cl^- selective microelectrodes, of intracellular Cl^- activity (a_{Cl}^i) in absorptive cells of bullfrog small intestine [10, 11] indicate that net Cl^- exit across the baso-lateral cell membrane is energetically downhill, since $a_{Cl}{}^{i}$ exceeds the level required for electrochemical equilibrium across this membrane. However, data obtained in gallbladder, an epithelial system that is similar in many respects to the small intestine, indicate that the Cl^- conductance of the baso-lateral membrane is very low [28] so that, unless the efflux of Cl^- across this membrane is coupled to the movement of another ion, net efflux of Cl^- from the absorptive cells could be slow.

In terms of the model shown in FIGURE 1, a major effect of external HCO_3^- on Cl^- transport is activation of the Cl^-/HCO_3^- exchange mechanism in the baso-lateral cell membrane, thus permitting a more rapid movement of Cl^- out of the cells. This is consistent with the large net absorption of Cl^- observed in the presence of HCO_3^- (TABLE 2). As a result of this, one would predict that coupled Na^+ and Cl^- entry across the mucosal membrane should increase and, indeed, as shown in TABLE 2, $J_{ms}{}^{Na}$ and $J_{ms}{}^{Cl}$ are greatly increased when HCO_3^- is present in the external medium.

If, as the model suggests, HCO_3^- secretion is mediated by baso-lateral Cl^-/HCO_3^- exchange, the magnitudes of $J_{net}{}^{Cl}$ and J_R (TABLE 2) suggest that the Cl^-/HCO_3^- exchange ratio is greater than unity, i.e., Cl^-/HCO_3^- exchange must be an electrogenic process. This could account for the serosal negative PD and I_{sc} observed in the presence of HCO_3^- (TABLE 1).

The marked increase in $J_{sm}{}^{Na}$ induced by external HCO_3^- (TABLE 2) is difficult to explain. Since it is likely that in bullfrog small intestine, as in mammalian small intestine [29] $J_{sm}{}^{Na}$ takes place principally, if not solely, by the paracellular route, an increase in J_{sm} could reflect in increase in the ionic conductance of the shunt pathway. From the electrical data shown in TABLE 1, one can calculate that the average total tissue conductance (the ratio I_{sc}/PD) in the presence of HCO_3^- (26.5 ± 1.7 m.ohm^{-1} cm^2) is significantly larger than the corresponding value in the absence of this ion (19.2 ± 1.7 m.ohm^{-1} cm^2). This is consistent with a decrease in shunt resistance, since shunt resistance, in effect, determines total tissue resistance in this preparation.[30] However, a generalized increase in shunt resistance should be reflected by increases in both $J_{sm}{}^{Na}$ and $J_{sm}{}^{Cl}$. Although $J_{sm}{}^{Cl}$ was in fact somewhat larger in the presence than in the absence of HCO_3^-, the increase in this parameter was much smaller than the corresponding increase in $J_{sm}{}^{Na}$ and fell far short of statistical significance ($p > 0.4$).

Effect of cAMP on Transepithelial Na⁺ and Cl⁻ Transport

Much evidence now points to a central role for adenosine 3′,5′-monophosphate (cAMP) as a mediator of salt and water secretion by the small intestine *in vivo* and *in vitro*.[13] Interest in the effects of this agent and related secretagogues, e.g., theophylline and prostaglandins (both of which are known to increase intracellular cAMP levels in the intestine) on intestinal ion transport has been stimulated by a number of similarities between their effects and those elicited by cholera toxin and other bacterial enterotoxins. For example, cholera toxin mimics the effects of cAMP on intestinal transport *in vitro* and increases intestinal cAMP levels. Furthermore, infusion of theophylline of prostaglandins into the mesenteric artery induces a cholera-like secretion of water and electrolytes by canine jejunum *in vivo*.[13] These and other findings suggest that cAMP is a key factor in the secretory responses induced by cholera toxin and other bacterial enterotoxins in the *in vivo* intestine. Since these responses underlie the massive losses, from the gastrointestinal tract, of water and electrolytes that characterize a number of enterobacterial infestations, cAMP has been considered to play a central role in the pathophysiology of a number of gastrointestinal disorders.

In vitro experiments with short-circuited preparations of stripped small intestine of several animal species have uncovered interesting variations in their response to cAMP and other secretagogues. Inhibition of coupled mucosal Na/Cl influx appears to be a rather general effect.[13] In the rabbit intestine *in vitro,* cAMP and theophylline induce a net secretion of Cl^- but have no effect on HCO_3^- secretion.[31, 32] In flounder intestine, there is no secretion of Cl^- or HCO_3^- in response to these agents.[33] Since flounder intestine lacks functional crypts of Lieberkühn, it has been suggested that, in the rabbit, Cl^- secretion may be mediated by crypt cells rather than by the absorptive cells.[33] On the other hand, a secretory response to theophylline has been demonstrated in the isolated small intestine of *amphiuma* in which the presence of functional crypts has not been definitely established.[16] Furthermore, HCO_3^- rather than Cl^- is secreted in this instance, an intriguing finding in view of the fact that in mammalian small intestine *in vivo,* HCO_3^-, not Cl^-, is involved in the secretory response to cAMP and related secretagogues.[13]

We have examined the electrical and ionic transfer responses induced by cAMP in isolated stripped segments of bullfrog small intestine in the presence of external HCO_3^-. Preliminary experiments established that cAMP altered the transepithelial electrical parameters of this tissue only when it was added to the serosal medium; and, the effects of cAMP were maximal at a concentration of 3 mM. Increasing the concentration to 10 mM had no further effect. Subsequent experiments were carried out at a serosal cAMP concentration of 6 mM and were performed on the same tissues that were used to establish the effects of HCO_3^- already discussed. Therefore, the results of our experiments with cAMP are shown in TABLES 1 and 2, and the changes elicited by this agent are also listed in these tables.

Considering first the electrical changes induced by cAMP (TABLE 1), it is evident that this agent reverses the sign of the (serosal-negative) *PD* and I_{sc} seen in media containing HCO_3^-. The final steady-state values of *PD* and I_{sc}, when cAMP was added to the serosal medium were serosal-positive and highly significant statistically.

TABLE 2 shows the effects of cAMP on Na^+ and Cl^- fluxes in a medium containing HCO_3^-. It will be seen from this table that cAMP reduces both J_{ms}^{Na} and J_{ms}^{Cl} by approximately equivalent amounts. That is, the decrease in J_{ms}^{Na} did not differ significantly from the decrease in J_{ms}^{Cl}. By contrast, cAMP had no significant effect on J_{sm}^{Cl}. J_{sm}^{Na} was somewhat reduced by cAMP, but less so than J_{ms}^{Na}. The change in J_{sm}^{Na} fell just short of statistical significance ($0.1 > p > 0.05$) in our experiments, but this may reflect the relatively small number of observations involved.

Considering the effect of cAMP on net fluxes, it is apparent from TABLE 2 that net Cl^- absorption (J_{net}^{Cl}) is greatly reduced, though not completely abolished by this agent. A small cAMP-induced net secretion of Na^+ is suggested by the data in TABLE 2, but this was not statistically significant. If one takes account of this in calculating J_R, it appears that the latter parameter is not affected by cAMP (TABLE 2), and even if J_{net}^{Na} is assumed to be zero, the change in J_R following the addition of cAMP (-0.9 ± 0.7 $\mu Eq/cm^2 \cdot hr$) is not significant ($p > 0.1$).

In terms of the model shown in FIGURE 1, the effects of cAMP on transepithelial transport may be interpreted as follows: The statistically identical decreases in J_{ms}^{Na} and J_{ms}^{Cl} suggest that a primary effect of cAMP is to reduce coupled mucosal Na^+ and Cl^- entry into the absorptive cells. As already mentioned, a similar effect of cAMP has been observed in the small intestine of other animal species. The identity of ΔJ_{ms}^{Na} and ΔJ_{ms}^{Cl} in our experiments also indicates that this coupled entry process is highly selective for Cl^-, i.e., under the conditions of these experiments, mucosal HCO_3^- does not appear to compete for entry by this route. This is in accord with other studies performed in our laboratory. For example, in extensive *in vitro* studies in which $SO_4^=$ replaced Cl^- as the major external anion [26] no evidence for coupled $SO_4^=$ transport by this route was uncovered. Similarly, it was found [35] that a number of other anions, e.g., Br^-, NO_3^-, and SCN^- resembled $SO_4^=$ rather than Cl^- in their effects on *PD* and I_{sc} in this tissue.

Since coupled Na^+ and Cl^- entry is electrically silent, a reduction in this process could not *per se* account for the reversal of the *PD* and I_{sc} observed in the presence of cAMP (TABLES 1 and 2). Similarly, the small secretion of Na^+ suggested by the data of TABLE 2, if real, could only serve to increase the serosal negativity of these parameters. We therefore suggested [24] that since, in the presence of HCO_3^-, Cl^- exits readily from the cell *via* the Cl^-/HCO_3^- exchange pathway shown in FIGURE 1, a secondary consequence of the inhibition by cAMP of coupled mucosal Na/Cl entry is a reduction in intracellular Cl^- that, in turn, slows the exit of Cl^- in exchange for external HCO_3^-. The reduced net Cl^- flux observed in the presence of HCO_3^- is consistent with this supposition. Since, however, J_R, which we tentatively identify as net HCO_3^- secretion, is apparently unchanged by cAMP (TABLE 2), this assumption requires that not only the rate of HCO_3^- coupled Cl^- efflux but also the stoichiometry of the coupled exchange process should depend on the intracellular Cl^- concentration. A decreased rate of Cl^- transport across the baso-lateral cell membrane, together with an unaltered net $s \rightarrow m$ movement of HCO_3^- could account, to a large extent, for the electrical effects of cAMP summarized in TABLE 1.

Effect of cAMP on Ionic Content of Bullfrog Small Intestine

In these experiments, segments of stripped intestinal tissue were immersed for 3 hrs at 23° C, with and without 6 mM serosal cAMP, in the HCO_3^- medium used in the electrical and tracer flux measurements described above. During immersion, the medium was gassed continuously with a 95/5% O_2/CO_2 mixture. Following immersion, the segments were gently blotted, weighed while wet, dried to constant weight at 105° C and extracted for 72 hr in 0.1 N HNO_3.[11] Tissue Na^+ and K^+ were determined in aliquots of the HNO_3 extract by atomic absorption spectrophotometry (Perkin-Elmer model 303). Tissue Cl^- was determined in separate aliquots of this extract by conductometric titration (Aminco-Cotlove chloride titrator). Duplicate segments of intestine from the same animal were used in each experiment. One was immersed in the medium containing cAMP, the other served as a control. A second series of experiments was performed in the HCO_3^--free medium described above.

To monitor directly the effect of cAMP on cell Cl^-, intracellular Cl^- activity (a_{Cl}^i) was measured with Cl^--selective microelectrodes. Because of their insensitivity to HCO_3^-, the solid-state silver microelectrodes described by Armstrong *et al.*[10] were used in this study. a_{Cl}^i was calculated from the relationship $a_{Cl}/a_{Cl}^i = 10^{\Delta E/S}$, where a_{Cl} is the calculated Cl^- activity of the external medium, ΔE is the change (corrected for the mucosal membrane potential) in the potential recorded when the tip of the Cl^--selective microelectrode is moved from the mucosal bathing solution to the interior of an absorptive cell, and S is the slope (in mV) of the microelectrode potential as a function of log a_{Cl}. S was determined by calibrating the Cl^- selective microelectrodes in solutions containing different known Cl^- activities.[10]

To measure a_{Cl}^i, segments of stripped intestine were mounted in the divided chamber described by White and Armstrong.[30] Both the mucosal and serosal aspects of the tissue were perfused continuously with the external medium. The chamber used is designed to permit rapid changes in the perfusate, so that the serosal perfusate could be changed from a cAMP-free medium to one containing cAMP or vice versa during an experiment. Mucosal membrane potentials

TABLE 3

Na^+, K^+ AND Cl^- CONTENT OF STRIPPED SEGMENTS OF BULLFROG SMALL INTESTINE *In Vitro* †

Immersion Medium ($Na/Cl/HCO_3$, mM)	Na^+	K^+	Cl^-
A. 105/105/0	60.7±2.7 (12)	66.5±2.2 (12)	58.6±0.8 (12)
B. 105/80/25	69.3±2.3 (16)	62.4±1.7 (16)	43.6±1.7 (12)
C. 105/80/25+ 6 mM cAMP	63.4±2.3 (16)	68.5±2.2 (16)	38.8±1.5 (12)
Δ (C-B)	−5.9±1.6 *	6.1±2.7 *	−4.6±1.1 *

† Average concentrations (±S.E.M.) are given in μEq/g wet weight for the numbers of experiments shown in parentheses; experiments B and C are paired.
* Indicates changes that differed significantly from zero ($p < 0.05$).

were measured with conventional open-tip microelectrodes. A parallel series of experiments (but without cAMP) was performed in the HCO_3^--free medium described above. Details of the experimental protocols for measuring a_{Cl}^i, the electrical circuitry employed, the methods used to calibrate Cl^--selective microelectrodes, and the criteria applied to evaluate the acceptability of impalements with open-tip and Cl^--selective microelectrodes are given elsewhere.[10, 11, 30]

The results of these experiments are summarized in TABLES 3 and 4. Turning first to TABLE 3, which shows the tissue Na^+, K^+, and Cl^- contents found in these experiments, a number of findings emerge. First the unpaired sets of experiments *A* and *B* show a significant increase ($p < 0.05$) in tissue Na^+ in the presence of HCO_3^-. The concomitant decrease in K content was, however, not significant ($p > 0.1$). This may reflect the increased mucosal Na^+ entry in the presence of HCO_3^- suggested by the flux measurements shown in TABLE 2. Unfortunately, because of the difference in external Cl^- concentrations, a meaningful comparison between the tissue Cl^- content in sets *A* and *B* of TABLE 3 is not possible.

TABLE 4

CHLORIDE ACTIVITIES (a^i_{Cl}, mM) AND THEIR RATIO (R) TO THE EQUILIBRIUM CHLORIDE ACTIVITY IN EPITHELIAL CELLS OF BULLFROG SMALL INTESTINE—EFFECT OF cAMP *

Immersion Medium (Na/Cl/HCO_3, mM)	a_{Cl} (medium)	a^i_{Cl}	R
105/105/0	81	31±2 (4:15)	1.5±0.1
105/80/25	61.5	34±5 (5:13)	2.1±0.3
105/80/25+ 6 mM cAMP	61.5	29±4 (5:17)	1.8±0.2

* Mean activities and R values ± S.E.M. are given for the number of animals and total number of impalements shown in parentheses.

A comparison of the paired sets of data designated B and C in TABLE 3 reveals highly significant decreases in tissue Na^+ and Cl^- when cAMP is added to the medium containing HCO_3^-. Furthermore, these decreases are statistically identical. This is consistent with an inhibition by cAMP of the coupled mucosal entry of Na^+ and Cl^-. At the same time the K^+ content of tissue increases to a highly significant extent.

TABLE 4 explores a_{Cl}^i under the various conditions employed in these experiments. At the outset, it should be noted that there was no significant difference in the mucosal membrane potentials (average value, −35 mV) recorded under the three sets of conditions represented in this table. Turning next to the values obtained for a_{Cl}^i, one may note first that the average value of this parameter found in the absence of HCO_3^- (31 ± 2 mM) is in excellent agreement with an earlier estimate (33 ± 2 mM) under identical conditions.[11] Second, although the replacement of 25 m.mol/l of Cl^- by HCO_3^- in the external medium lowered a_{Cl} from 81 to 62 mM, a_{Cl}^i in the presence of HCO_3^- did not differ significantly from the value obtained in the absence of this ion. Since the mucosal membrane potential did not change, this means that the ratio of a_{Cl}^i to its equilibrium concentration, calculated from the Nernst equation, is significantly higher in the presence than in the absence of external HCO_3^- (TABLE 4). In terms of the mechanism proposed for increased Cl^- entry into the cells in the presence of HCO_3^-; that is, increased entry could be solely due to an increased baso-lateral exit of Cl by way of the postulated Cl^-/HCO_3^- exchange pathway shown in FIGURE 1, this is an unexpected finding. It suggests a direct stimulatory effect of HCO_3^- on coupled mucosal Na/Cl entry in addition to the effect of this ion on Cl^- transport across the baso-lateral cell membrane.

As one would predict for a substance that inhibits mucosal entry of Na^+ and Cl^-, and in agreement with the data shown in TABLE 3, cAMP caused a decline in a_{Cl}^i. Unlike the cAMP induced decrease in tissue Cl^- (TABLE 3), the decrease in a_{Cl}^i found in the experiments shown in TABLE 4 was not statistically significant ($p > 0.1$). However, it is entirely possible that, in these experiments, our failure to find a statistically significant decrease in a_{Cl}^i is due to the degree of scatter observed in a relatively small number of unpaired measurements.

Does cAMP Stimulate the Basolateral Na^+-K^+ Pump in Bullfrog Small Intestine?

The observation (TABLE 3) that cAMP caused a significant increase in tissue K^+ raised the possibility that, in addition to its inhibitory effect on mucosal Na^+ and Cl^- entry, this compound may stimulate the baso-lateral Na^+-K^+ pump in bullfrog small intestine. This possibility is, at first glance, somewhat surprising since there is evidence, on one hand, that cAMP may exert its effects on intestinal transport through an increase in cytosolic Ca^{++},[13] and, on the other hand, that elevated intracellular Ca^{++} levels inhibit net Na^+ transport in a variety of epithelia.[34] Nevertheless, it appears to derive further support from a series of experiments in which the ionic dependence of the *PD* and I_{sc} response to cAMP was explored. The rationale underlying these experiments was as follows: If the effect of cAMP on *PD* and I_{sc} is determined solely by alterations in the stoichiometry of an electrogenic Cl^-/HCO_3^- exchange in the basolateral cell membrane; then, in the complete absence of these ions from the external medium, these responses should be completely absent or negligible. To test this hypothesis, we measured the effect on cAMP-induced changes in *PD* and I_{sc} of removing HCO_3^-, Cl^- or both these ions from the bathing medium. The results for I_{sc} are shown in TABLE 5. It is evident from this table that removal of HCO_3^- from the bathing medium attenuates the I_{sc} response to cAMP to about 35% of its value in a medium containing 25 m.mol/l of this ion. Removal of both HCO_3^- and Cl^-, at constant Na^+ concentration, further attenuates this response by a factor of 2 and removal of Na^+ and HCO_3 completely abolishes it.

A salient point in these results is that, in the complete absence of external Cl^- and HCO_3^- from the bathing medium, but with Na^+ present, the I_{sc} response to cAMP, though small, is finite, and that this response is completely abolished when Na^+ is removed from the medium. This argues rather strongly in favor of a direct, if quantitatively small, effect of cAMP on Na^+ transport.

To examine this response further, steady-state Na^+ fluxes were measured

TABLE 5

IONIC DEPENDENCE OF I_{sc} RESPONSE TO cAMP †

Immersion Medium ($Na/Cl/HCO_3$, mM)	Number of Experi-ments	I_{sc} Control	I_{sc} cAMP	ΔI_{sc}	ΔI_{sc} (%)
105/80/25	23	−1.0±0.3 (1)	0.7 ±0.2 (1)	1.7 ±0.2	100
105/105/0	24	0.1±0.1 (1)	0.7 ±0.1	0.6 ±0.1	35
105/0/0 (2)	14	0.1±0.07	0.4 ±0.1	0.3 ±0.1	17.5
0/105/0 (3)	5	0.1±0.1	0.14±0.1	0.05±0.1 *	3

† Mean I_{sc} values ($\mu Eq/cm^2$ hr) ±S.E.M. are given for numbers of experiments indicated: (1) data from TABLE 1; (2) 52.5 mM Na_2SO_4 substituted for NaCl, mannitol added to maintain constant osmolality; (3) 105 mM Tris (tris hydroxymethyl aminomethane) Cl substituted for NaCl.

* Not significant ($p > 0.05$).

under short-circuit conditions in the presence and absence of 6 mM serosal cAMP. As before, stripped intestinal segments were used in these experiments. They were immersed in a medium containing 105 mM Na^+ and 52.5 mM $SO_4^=$ (TABLE 5) but no Cl^- or HCO_3^-. The medium was gassed continuously with O_2. The results obtained are summarized in TABLE 6. It is apparent from this table that cAMP induces a highly significant net absorption of Na^+ under these conditions and that this cAMP-induced absorption is due entirely to an increase in J_{ms}^{Na}. This is consistent with a stimulatory effect of cAMP on the baso-lateral Na^+-K^+ pump. In contrast to the results observed in media containing Cl^- and HCO_3^- (TABLE 2), J_{sm}^{Na} in a Na_2SO_4 medium was not affected by cAMP.

Finally, the results shown in TABLE 6 indicate that, under these conditions, J_{net}^{Na}, in the presence of cAMP, greatly exceeds I_{sc}. That is, there is a highly significant J_R that is consistent with a net absorptive movement of anions or a net secretion of cations. It may be that J_R in these experiments represents a net movement, across the baso-lateral cell membrane, of HCO_3^- generated by oxidative metabolism. Alternatively, it could reflect a secretion of H^+ across the

TABLE 6

Na^+ FLUXES IN Na_2SO_4 RINGER-EFFECT OF cAMP (N=7) †

Experimental Conditions	J_{ms}	J_{sm}	J_{net}	I_{sc}	J_R
Control	7.5±0.7	7.3±0.8	0.2±0.7	0.1±0.1	−0.1±0.6
cAMP (6 mM)	8.3±0.8	7.0±0.6	1.3±0.5 *	0.4±0.1	−0.9±0.4
Δ	0.8±0.3 *	−0.3±0.5	1.0±0.7	0.3±0.1	0.8±0.6

† Mean values ($\mu Eq/cm^2 \cdot hr$) ±S.E.M. given for all parameters. For composition of medium, see TABLE 5.

* Significantly different from zero ($p < 0.05$).

luminal membrane of the absorptive cells. In either event, the identity of J_R under these conditions remains to be established.

SUMMARY

In an HCO_3^--free medium, isolated segments of bullfrog small intestine, stripped of their external muscle layers, displayed a small, serosal positive *PD* that did not, on the average, differ significantly from zero. Similarly, in this medium, the mean values of I_{sc} and of net Na^+ and Cl^- absorption under short-circuit conditions did not differ significantly from zero. External HCO_3^- (25 mM) induced a highly significant serosal negative *PD* and I_{sc} and a large net absorption of Cl^-. Net Cl^- absorption exceeded I_{sc}, i.e., there was a significant net flux, J_R, which was consistent with a net secretion of HCO_3^-. The ratio of the internal Cl^- activity of the absorptive cells (a_{Cl}^i) to its equilibrium value was larger in the presence than in the absence of HCO_3^-. In the presence of HCO_3^-, cAMP, added to the serosal medium, reversed the serosal negative *PD* and I_{sc}, and inhibited, though it did not completely abolish, net Cl^- absorption.

J_R was unchanged; tissue Cl^- and $a_{Cl}{}^i$ were reduced, and tissue Na^+ decreased and tissue K^+ increased. When $HCO_3{}^-$ and Cl^- were removed from the bathing medium, the electrical response of the tissue to cAMP, though greatly attenuated, was not completely abolished. Under these conditions, cAMP induced a significant net Na^+ absorption. A model for ion transport in the absorptive cells of the small intestine is proposed that is consistent with these findings.

Acknowledgments

We thank W. R. Bixenman for his excellent assistance in the Cl^- activity measurements. The tracer flux experiments described herein form part of a dissertation submitted by S. J. Youmans to Indiana University in partial fulfillment of the requirements for the Ph.D. degree.

References

1. Schultz, S. G., M. Field & R. A. Frizzell. 1979. Am. J. Physiol. **236**(1): F1–F8.
2. Quay, J. F. & W. McD. Armstrong. 1969. Am. J. Physiol. **217:**694–702.
3. Ussing, H. H. & K. Zerahn. 1951. Acta Physiol. Scand. **63:**141–155.
4. Nellans, H. N., R. A. Frizzell & S. G. Schultz. 1974. Am. J. Physiol. **226:** 1131–1141.
5. Field, M., K. J. Karnaky, Jr., P. L. Smith, J. E. Bolton & W. B. Kinter. 1978. J. Membrane Biol. **41:**265–293.
6. Cremaschi, D. & S. Henin. 1975. Pflügers Arch. **361:**33–41.
7. Frizzell, R. A., M. Dugas & S. G. Schultz. 1975. J. Gen. Physiol. **65:**769–795.
8. Spring, K. A. & G. Kimura. 1978. J. Membrane Biol. **38:**233–254.
9. Frömter, E. & J. M. Diamond. 1972. Nature New Biol. **235:**9–13.
10. Armstrong, W. McD., W. Wojtkowski & W. R. Bixenman. 1977. Biochim. Biophys. Acta **465:**165–170.
11. Armstrong, W. McD., W. R. Bixenman, K. F. Frey, J. F. Garcia-Diaz, M. G. O'Regan & J. L. Owens. 1979. Biochim. Biophys. Acta **551:**207–219.
12. Duffey, M. E., K. Turnheim, R. A. Frizzell & S. G. Schultz. 1978. J. Membrane Biol. **42:**229–245.
13. Nellans, H. N., R. A. Frizzell & S. G. Schultz. 1974. Ann. Rev. Physiol. **36:**51–91.
14. Turnberg, L. A., F. A. Bieberdorf, S. G. Morawski & J. S. Fordtran. 1970. J. Clin. Invest. **49:**557–567.
15. Hubel, K. A. 1969. Am. J. Physiol. **217:**40–45.
16. Gunter-Smith, P. J. & J. F. White. 1979. Am. J. Physiol. **236**(6):E775–E783.
17. Youmans, S. J. & W. McD. Armstrong. 1977. Physiologist **20:**104.
18. Field, M., D. Fromm & I. McColl. 1971. Am. J. Physiol. **220:**1388–1396.
19. Frizzell, R. A., L. Markscheid-Kaspi & S. G. Schultz. 1974. Am. J. Physiol. **226:**1142–1148.
20. White, J. F. 1977. Biochim. Biophys. Acta **467:**91–102.
21. Schultz, S. G. & R. Zalusky. 1964. J. Gen. Physiol. **47:**567–584.
22. Ehrenspeck, G. & W. A. Brodsky. 1976. Biochim. Biophys. Acta **419:**555–558.
23. Cabantchik, Z. I., P. A. Knauf & A. Rothstein. 1978. Biochim. Biophys. Acta **515:**239–302.
24. Youmans, S. J. & W. McD. Armstrong. 1978. Physiologist **21:**131.
25. Armstrong, W. McD., T. K. Suh & G. A. Gerencser. 1972. Biochim. Biophys. Acta **255:**647-662.

26. ARMSTRONG, W. McD. 1976. Bioelectric parameters and sodium transport in bullfrog small intestine. *In* Intestinal Ion Transport. J. W. L. Robinson, Ed. : 19–40. MTP Press Ltd., Lancaster, England.
27. GERENCSER, G. A. & W. McD. ARMSTRONG. 1972. Biochim. Biophys. Acta **255:** 663–674.
28. REUSS, L. 1979. Federation Proc. **38:**962.
29. FRIZZELL, R. A. & S. G. SCHULTZ. 1972. J. Gen. Physiol. **59:**318–346.
30. WHITE, J. F. & W. McD. ARMSTRONG. 1971. Am. J. Physiol. **221:**194–201.
31. NELLANS, H. N., R. A. FRIZZELL & S. G. SCHULTZ. 1973. Am. J. Physiol. **225:** 467–475.
32. FIELD, M. 1971. Am. J. Physiol. **221:**992–997.
33. FIELD, M. & P. L. SMITH. 1975. Bull. M. Desert Isl. Biol. Lab. **15:**14.
34. TAYLOR, A. & E. E. WINDHAGER. 1979. Am. J. Physiol. **236**(6):F505–F512.
35. ARMSTRONG, W. McD. & S. J. COHEN. Unpublished observations.

DISCUSSION

E. RACKER: Do you imply that the entry of the sodium is down the electrical gradient and that chloride follows? If so, how do you visualize a nonspecific process like this to be influenced by cyclic AMP?

W. McD. ARMSTRONG: I don't know why you describe the coupled entry process as nonspecific. I think the coupling between sodium and chloride is relatively specific. If sulphate, glutamate or other anions are substituted for chloride, one does not find evidence for any sodium-coupled transport of these non-chloride anions. Energetically, the coupled process works very simply. The transapical electrochemical potential gradient for sodium, in some unspecified way, provides the free energy needed to drive chloride uphill.

RACKER: My point is that these processes are, in all probability, *not* specific. If we have either carriers or channels for a variety of things, then, I can visualize how cyclic AMP connects with this process. I have the impression from the physiologists that sodium goes down the gradient and chloride follows, as if this were an uncatalyzed process; and I don't believe that.

ARMSTRONG: Well, I don't know. The evidence available to us and others is that the electrochemical potential difference for sodium across the apical membrane provides the driving force for chloride accumulation. Exactly how this is accomplished we don't know.

RACKER: I don't deny this, but something else must be involved in this process. What is the effect of ouabain on your system?

ARMSTRONG: We did not use ouabain in these experiments. We do know from our own work and that of others who have studied these kinds of epithelia that ouabain does nothing if it is added to the mucosal side. Ouabain does inhibit Na entry when added to the basolateral side by its action on the basolateral sodium-potassium pump. Therefore, any inhibitory effect of ouabain on coupled sodium and chloride entry would be due to an indirect feedback secondary to its direct action on the basolateral pump.

M. BERRIDGE (*University of Cambridge, Cambridge, England*): It has been suggested that in intestines from other animals, cyclic AMP may act by mobilizing calcium. Do you know, in your system, whether there is any involvement of calcium in these effects?

ARMSTRONG: I might be able to answer that question next year. As of now, we are aware of the involvement of calcium in the effects of cyclic AMP on mammalian intestine, but we have not yet looked at this in the bullfrog intestine.

J. A. YOUNG (*University of Sydney, Sydney, Australia*): It is interesting that your effects of cyclic AMP are only produced when you supply the substance from the serosal surface. We have found that cyclic AMP produces effects on salivary secretion that resemble those you find in the bullfrog intestine; but these effects are exclusively obtained from the luminal surface of the salivary duct and are evoked with concentrations of 1×10^{-5} molar. In our hands, cyclic AMP doesn't appear to enter the cell from the basal side. Since this cyclic AMP effect mimics those of adrenergic agonists, we may have "tuned in" on the pathway of some β-adrenergic stimulation. Why do you think the drug works only on the serosal surface in your system?

ARMSTRONG: Well, it is a fact it does; I don't know why and I am not as educated about the salivary function as I am about the intestine. Other epithelia, such as the cornea, work differently. Zadunaisky's data lead me to picture the cornea as an inside-out intestine or gallbladder. My own question is what specific entry processes exist on the luminal side of salivary gland cells, as compared to those in the intestine?

M. BERRIDGE: Your model for absorption in the intestine would do quite well for that in the salivary duct cell!

ARMSTRONG: Except that cyclic AMP works only from the luminal surface in the salivary duct. Did you use cyclic AMP itself or dibutyryl cyclic AMP? Cyclic AMP may well work at an internal site, and dibutyryl cyclic AMP may well penetrate cells much more rapidly than the parent compound, and therefore, be more effective at lower external concentrations. In our preparation, the cyclic AMP site might be accessible from the serosal side only and there are probably permeability differences between your system and ours. Because of these factors, cyclic AMP enters the cells from the luminal side in your case, but I don't know the real answer as yet.

F. H. EPSTEIN: In the perfused rectal gland of the dogfish shark, we also find that cyclic AMP increases the concentration of intracellular potassium, decreases that of intracellular sodium, and turns on ouabain-inhibitable oxygen consumption. All of this suggests that it turns on sodium potassium ATP-ase at the same time as it induces secretion of chloride. However I don't know whether or not chloride is secreted in the bullfrog intestine. You were going to tell us that.

ARMSTRONG: We did not find such a chloride secretion in the bullfrog. We found an inhibition of chloride reabsorption from mucosal to cell fluid.

I am aware of your data from the shark rectal gland; and there are some interesting similarities between the effects of cyclic AMP on the isolated intestine and the effects of this compound on the perfused shark rectal gland.

THE ROLE OF CYCLIC NUCLEOTIDES AND CALCIUM IN THE REGULATION OF CHLORIDE TRANSPORT

Michael J. Berridge

Department of Zoology
A.R.C. Unit of Invertebrate Chemistry and Physiology
University of Cambridge
Cambridge, CB2 3EJ, England

Introduction

The transport of chloride across epithelia is an important component of many secretory and absorptive processes. Chloride movement, which may be either active or passive, is often under hormonal or nervous control. Both cyclic AMP and calcium have been implicated as second messengers in the control of chloride movement across the insect salivary gland,[1, 2] the mammalian intestine,[3, 4] and the cornea.[5] In many of these examples, however, there is still considerable uncertainty concerning the precise nature of the intracellular events responsible for regulating chloride movement. The absence of detailed information on the actual mechanism of chloride movement across both surfaces of the epithelium is partly responsible for this uncertainty, but there are also gaps in our knowledge concerning which of these second messengers functions in regulating ion transport.

An intriguing aspect of the regulation of ion transport in many epithelia is the apparent integrated control by both cyclic AMP and calcium. There are numerous examples in which the action of a hormone or neurotransmitter can be mimicked by either adding cyclic AMP or by adding the divalent ionophore A23187 to artificially elevate the intracellular level of calcium. It is not yet clear whether such observations indicate a duality of control, so transport can be regulated either through cyclic AMP or calcium, or whether these two second messengers are organized as part of a sequence. The role of these two second messengers in regulating ion transport has been studied extensively in the salivary glands of the adult blowfly in an attempt to define more precisely their functional involvement in the case of a secretory cell.

Salivary glands of the adult blowfly consist of two long tubes which extend down the length of the body.[6] These glands are responsible for producing a watery saliva during feeding. When removed from the fly, the glands secrete an isosmotic fluid composed mainly of potassium and chloride. The rate of fluid secretion *in vitro* is regulated by 5-hydroxytryptamine (5-HT),[7] which induces a large increase in the parallel flow of both potassium and chloride. These glands thus provide an ideal opportunity for studying how second messengers function to regulate both the mechanism and the control of anion transport.

The Role of Chloride in Fluid Secretion

In order to understand how chloride transport is regulated, it is necessary to first establish how chloride moves across the epithelium. Electrophysiological

0077–8923/80/0341–0156 $01.75/0 © 1980, NYAS

techniques were used to characterize the ionic mechanisms responsible for the secretion of both potassium and chloride.[1, 2, 7–9] Under resting conditions, the potential across the gland is 15 mV to 30 mV with the lumen being positive; but it depolarizes rapidly to 0 mV soon after the application of 1×10^{-8} M 5-HT (FIGURE 1). The traces in FIGURE 1 also include resistance measurements obtained by passing constant current pulses across the epithelium. The voltage deflections caused by these current pulses are a measure of resistance and they indicate that there is a marked decrease in resistance

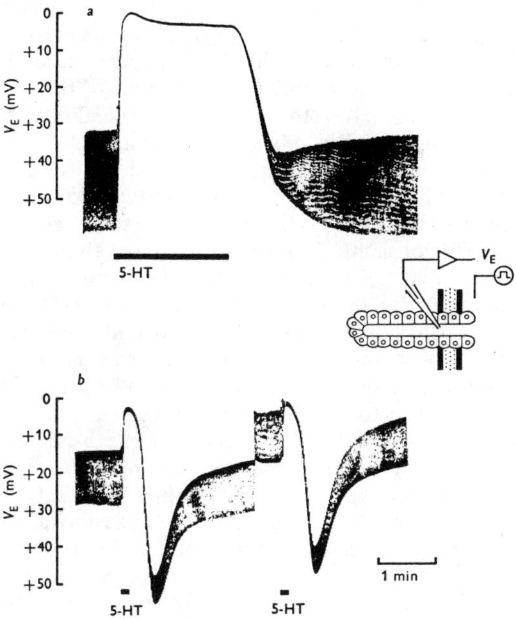

FIGURE 1. Changes in transepithelial potential across a *Calliphora* salivary gland during stimulation with 1×10^{-8} M 5-HT (solid bars) for 120 secs (*a*) or 10 secs (*b*). The potential responses (V_e) were recorded by means of a microelectrode inserted in the lumen (see inset). The deflections were produced by 0.11 μA current pulses used to measure resistance. In *b,* the direction of the current pulses was reversed between the two periods of stimulation.[2]

during the action of 5-HT. During a prolonged 5-HT treatment (FIGURE 1a) the resistance remains low and the potential lies close to 0 mV. Both parameters return to resting conditions with a slight undershoot soon after 5-HT is withdrawn. The undershoot is much more pronounced after brief 5-HT treatments (FIGURE 1b) and the reason for this biphasic response will be described later during the discussion of the role of cyclic AMP and calcium in these 5-HT effects.

The biphasic electrical response to short pulses of 5-HT has proved particularly useful in analyzing the changes which occur when secretion is being stimulated. The initial rapid depolarization seems to result from an increase in chloride permeability whereas the positive undershoot can be attributed to the activation of an electrogenic potassium pump. During brief treatments with 5-HT the very marked temporal difference in the activation of these two affects give rise to the pronounced biphasic response. This temporal separation can probably be traced back to a difference in the time course of the action of the second messengers responsible for regulating secretion. The calcium signal, which is thought to be responsible for the increase in chloride permeability, switches on quickly ($t_{1/2} = 2$ secs) to give the rapid depolarization and decrease in resistance (FIGURE 1). 5-HT also acts to raise the level of cyclic AMP, which seems to be associated in some way with the activation of the potassium pump responsible for the positive undershoot. During a short treatment of 5-HT the effect on calcium develops and fades before the cyclic AMP signal; so that, towards the end of the response, the latter continues to activate potassium transport after the chloride permeability has switched off, thus resulting in the pronounced positive undershoot.

Current-clamp experiments [2] have substantiated the idea that this biphasic response is due to an initial change in permeability followed by the activation of a potassium pump (FIGURE 2). By passing current between the two outer chambers of the perfusion bath it was possible to alter the position of the resting potential that was recorded by means of a microelectrode positioned in the lumen (FIGURE 2). In the unclamped condition, 5-HT induced a typical biphasic response. If the resting potential was made less positive, the early phase of the response was reduced in size, but was greatly increased when the resting potential was made more positive. When the lumen was made negative, the early response to 5-HT actually reversed (FIGURE 2). However, despite the fact that the negative going phase of the 5-HT response could be reversed, there was no reversal of the secondary positive phase. These observations are thus consistent with the model that the early depolarization is caused by a change in chloride permeability whereas the positive phase is the result of an increase in the activity of the potassium pump.

Another way of demonstrating that the decrease in resistance is caused by an increase in chloride permeability is to study the effect of replacing chloride with an impermeant anion such as isethionate.[8] When glands are stimulated in such a chloride-free medium, there is no change in resistance and instead of a negative going response there is a large increase in positivity. In the absence of the short-circuiting effect of chloride, the ability of 5-HT to activate the potassium pump is unveiled and this accounts for the large increase in lumenal positivity. The time course of this positive going response ($t_{1/2} = 30$ secs) is very much slower than the negative going potential, due to the increase in chloride permeability. As described earlier, it is this temporal difference in the action of the two second messengers that may account for the biphasic response to short pulses of 5-HT.

Ion-selective microelectrodes have provided further details of how the intracellular levels of potassium and chloride vary during the action of 5-HT.[10, 11] When glands were perfused in a saline containing 10 mM potassium, the intracellular level was 137 mM and increased to 144 mM during stimulation with 1×10^{-8} M 5-HT.[10] This increase in the internal level of potassium occurs despite the fact that the cells are rapidly extruding potassium across the apical membrane in order to initiate a flow of secretion. At equilibrium, the entry of potassium across the basal membrane must match the extrusion of this ion across the opposite surface. In most cells, potassium entry is coupled to sodium extrusion through the classical Na^+-K^+-activated ATPase pump mechanism. However, this pump seems to play no role in the net secretion of potassium in the insect salivary gland because fluid secretion is completely insensitive to ouabain.[10] The ability of 5-HT to increase the intracellular level of potassium was equally unaffected by ouabain. Fluid secretion can also occur in the complete absence of sodium.[12] Similarly, fluid secretion can also occur in a potassium-free medium where sodium is the sole monovalent cation. Electron microprobe analysis of glands incubated in potassium-free media reveal that almost all of the intracelluar potassium has been replaced by sodium which accounts for the observation that the glands begin to secrete saliva rich in sodium rather than potassium.[11] The fact that

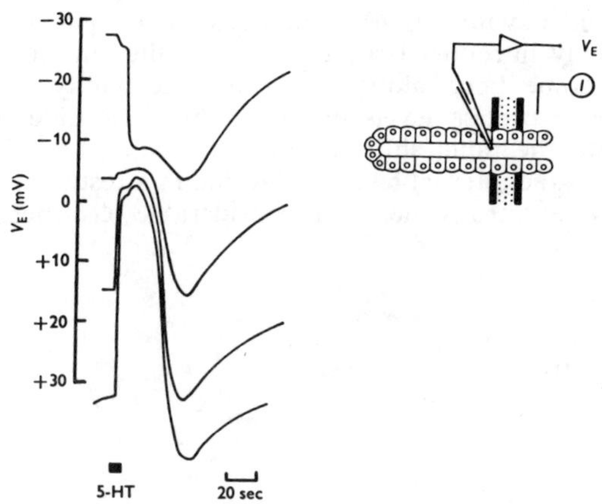

FIGURE 2. The effect of varying the resting potential of a *Calliphora* salivary gland (using a current-clamp) on its response to 8 sec pulses of 1×10^{-8} M 5-HT. The second trace from the bottom is the normal response in the unclamped condition.[2]

these glands can secrete fluid when provided with a single monovalent cation [12] seems to finally rule out any role for a coupled pump on the basal membrane during the action of 5-HT. In a resting gland, there is some evidence that a ouabain-sensitive sodium pump acts in a custodial manner to maintain high potassium levels in the absence of stimulation. However, such a pump cannot supply the large amounts of potassium that are required when 5-HT induces high rates of secretion. The problem, therefore, is to determine how potassium enters the cell across the basal plasma membrane during a normal secretory response.

A comparison of membrane potential and the potassium equilibrium potential during the course of a secretory response suggests that potassium may enter the cell passively. During the action of 5-HT, the potential across the basal plasma membrane hyperpolarizes [9] and this increase in membrane potential may be responsible for pulling potassium into the cell.[10] Since the potential was found to hyperpolarize in a normal way in a potassium-free medium,[12] it is reasonable to argue that the increase in membrane potential during the action of 5-HT is not caused by the increase in potassium concentration. The next question to consider, therefore, is why the membrane hyperpolarizes during the action of 5-HT. Since the membrane becomes permeable to chloride, it is conceivable that 5-HT may hyperpolarize the membrane by lowering the intracellular level of this anion.

There are some remarkable shifts in the intracellular level of chloride when 5-HT stimulates secretion. All measurements were done with a double-barrelled chloride selective microelectrode prepared in much the same way as the potassium selective electrodes used previously.[10] The larger barrel was back-filled with a short column of chloride-sensitive resin (Corning No. 477315). The remaining part of the tip and shank was filled with 0.5 M KCl. The reference barrel which was filled with 1.0 M sodium sulphate, had a tip potential of 2-5 mV that probably introduced some error into the chloride measurements. Although there is some uncertainty concerning the ability of these chloride electrodes to monitor precise levels of intracellular chloride, they provide a valuable insight into the rapid shifts in chloride induced by 5-HT. The first point to notice is that the internal level of chloride in a resting gland is relatively low (FIGURE 3), which means there is a considerable electrochemical gradient

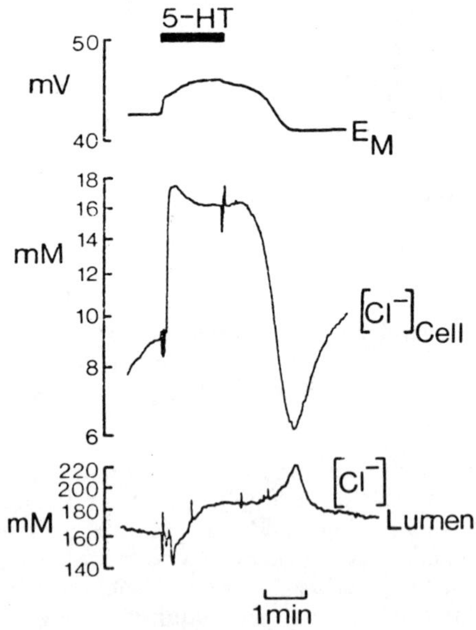

FIGURE 3. Changes in cellular and lumenal chloride concentration during stimulation of *Calliphora* salivary glands with 2×10^{-8}M 5-HT. The membrane potential (E_m) hyperpolarizes in two stages, an initial rapid change followed by a slower phase that corresponds with a gradual decline in the intracellular level of chloride. Note that during recovery the intracellular chloride level undershoots to 6 mM which corresponds to an overshoot in the lumenal concentration. The potassium concentration in the saline was 20 mM.

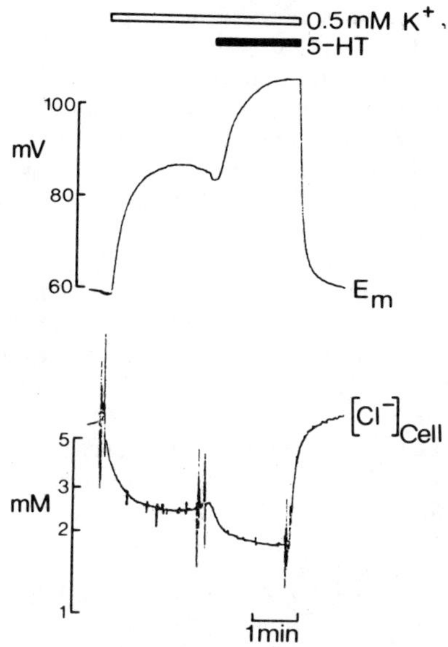

FIGURE 4. Changes in the intracellular level of chloride during stimulation of *Calliphora* salivary glands with (2×10^{-8}M). 5-HT in a low potassium medium. On changing from the normal potassium level of 10 mM to 0.5 mM the membrane potential (E_m) hyperpolarized with a corresponding fall in the level of chloride. Stimulation with 5-HT caused a further lowering of the chloride concentration together with a large membrane hyperpolarization.

for chloride entry into the cell. This gradient is responsible for the very rapid increase in the internal chloride level that occurs during the action of 5-HT (FIGURE 3). An interesting feature of this initial influx of chloride is that most of it seems to come from the lumen as electrodes positioned there recorded a marked fall in chloride concentration coincidential with a sudden rise in the intracellular compartment.

During the early stages of secretion, therefore, there is a flux of chloride from the lumen into the cell that subsequently reverses when the cation pump begins to transport potassium into the lumen. This flow of potassium into the lumen pulls chloride out of the cell causing the intracellular level to fall to a lower value. It is this decline in the internal level of chloride that may be responsible for the hyperpolarization that develops with a time course similar to that for the change in chloride level. When 5-HT is withdrawn, the chloride level falls back down to its low resting value, often displaying a marked undershoot (FIGURE 3).

The relationship between membrane potential and intracellular chloride concentration can also be demonstrated by varying the external concentration of potassium (FIGURE 4). When the membrane is hyperpolarized in a low potassium medium (0.5 mM), there is corresponding fall in the intracellular level of chloride. Subsequent stimulation of glands in such low potassium media leads to a further decline in the chloride level with a concomitant hyper-polarization considerably larger than that seen in normal saline (10 mM potassium) (FIGURE 4). It is this membrane hyperpolarization that seems to be important for driving the entry of potassium into the cell.[10]

In addition to providing the counter ion for electrolyte secretion, chloride may play a crucial role in facilitating cation transport by coupling active potassium extrusion across the apical membrane to passive entry of potassium across the basal membrane (FIGURE 5). The transport of potassium into the lumen is followed by chloride, which lowers the internal level of chloride, resulting in the hyperpolarization necessary to draw potassium into the cell. As yet there is no evidence for the transport of chloride being active because the flux across both membranes appears to be passive. The nature of the channels responsible for chloride conductance have not been investigated. A freeze-fracture study has revealed the presence of a very high density of intramembranous particles (approx. $4500/u^2$) in both the apical and basal membranes.[13] Such a high density is very reminiscent of that found in erythrocytes,[14] in which they are thought to function in anion movement. Some of the membrane proteins that constitute the particles in the insect salivary gland may thus be the sites for chloride movement.

Control of Chloride Permeability

Epithelia present special control problems because transport rates have to be regulated at both surfaces of the cell, even though the hormones or neurotransmitters arrive at only one surface. It is not too surprising, therefore, to find that second messengers, such as the cyclic nucleotides and calcium, feature significantly in conveying information between the two surfaces in a number of transporting epithelia, such as the intestine,[3,4] mammalian salivary glands,[15-18] cornea,[5] and the insect salivary gland.[1,7,19] However, the precise way in which the second messengers act within many of these epithelia is unclear because the nature of the secretory mechanisms, particularly those concerned with ion transport, have not been described. In the case of the insect salivary gland, the mechanism of secretion has been described in sufficient detail to provide a firm basis for analyzing the mode of action of these second messengers. As described earlier, both cyclic AMP and calcium have been implicated in the control of electrolyte transport.

A role for cyclic AMP in the action of 5-HT is suggested by several pieces of evidence. Firstly, there is a significant increase in the intracellular level of cyclic AMP when salivary glands are stimulated with 5-HT.[20] Secondly, the phosphodiesterase inhibitor theophylline can markedly potentiate the stimulatory effect of 5-HT and is also capable of inducing a modest stimulatory effect in the absence of hormones.[21] Finally, cyclic AMP by itself can stimulate a flow of saliva almost equal to that produced by 5-HT. The evidence for a second messenger role for calcium also depends on several lines of evidence. These is now substantial proof that 5-HT acts on the membrane to stimulate the entry of calcium. Preliminary experiments showed 5-HT was capable of stimulating an influx of ^{45}Ca.[20] These observations have been extended through the development of a technique for continuously monitoring the permeability of the cell to calcium.[22] There is a dose-dependent increase in calcium flux (FIGURE 6) with maximal rates of fluid secretion being achieved at relatively low rates of calcium entry. As the dose of 5-HT increases above 0.01 μM, there is little further change in fluid secretion whereas calcium entry goes on

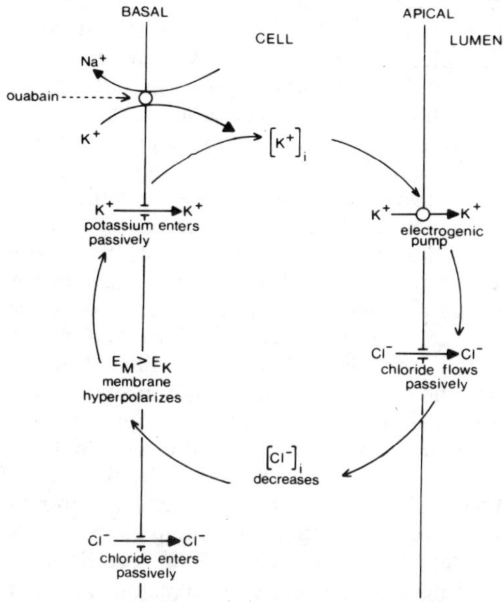

FIGURE 5. A model of the ionic mechanisms responsible for fluid secretion by the salivary glands of *Calliphora*. An electrogenic pump on the apical membrane transports potassium into the lumen and chloride follows passively. This flow of chloride into the lumen depletes the cell of chloride that causes the membrane to hyperpolarize sufficiently for potassium to enter the cell passively. See text for further details.

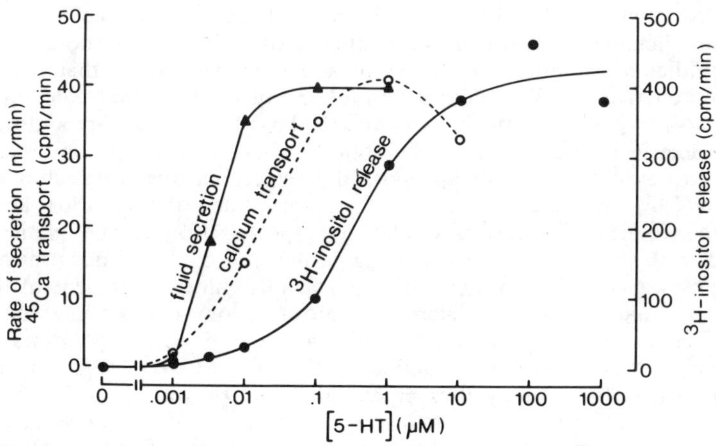

FIGURE 6. The relationship between 5-HT concentration and the rate of fluid secretion, calcium transport and the release of inositol. The latter is an indirect measure of the rate of phosphatidylinositol hydrolysis.[23]

increasing, which may reflect the existence of spare receptors. A maximal rate of fluid secretion can thus be achieved when relatively few of the available receptors have been occupied. Further increases in agonist concentration result in the formation of more second messengers (i.e. the gating of calcium) without any change in fluid secretion. The ability of 5-HT to gate calcium seems to depend on the hydrolysis of phosphatidylinositol (PI).[23, 24] This hydrolysis of PI in the insect salivary gland has been monitored by measuring the efflux of ^{3}H-inositol from prelabeled glands in which most of the label was incorporated into PI. The dose-response relationship for ^{3}H-inositol release shown in FIGURE 6 also illustrates the existence of spare receptors. Further studies have confirmed that this phospholipid may play an important role in coupling the 5-HT receptor to the calcium gate. A more difficult question to answer concerns the level of intracellular calcium and whether or not this changes during the action of 5-HT. The small size of most secretary cells has precluded the use of the calcium-sensitive photoprotein aequorin to monitor the intracellular level of calcium. However, some preliminary measurements using a calcium-sensitive double-barrelled microelectrode have indicated that there are significant increases in the intracellular level of calcium during the action of 5-HT.[25] The ability of the divalent cation ionophore A23187 to stimulate secretion in a calcium-dependent manner provides further proof for an involvement of calcium in the control of secretion.

An understanding of how cyclic AMP and calcium act within the cell has developed in conjunction with an understanding of the ways in which potassium and chloride are moved across the epithelium. As outlined earlier, chloride movement across both surfaces of the cell appears to be passive and under the control of calcium. The evidence for such an action of calcium is somewhat indirect and is mainly derived from observations carried out in calcium-free conditions. When calcium is removed from the bathing medium, there is a marked change in the normal electrical response. Instead of depolarizing, the potential across the epithelium increases due to a marked hyperpolarization of the apical membrane.[1] The potential increases because the potassium pump increases its activity in the absence of the normal calcium-induced increase in chloride conductance. When glands are stimulated with the ionophore A23187 the transepithelial potential depolarizes in a manner similar to that observed during the action of 5HT. Another way of trying to show that calcium acts to alter chloride permeability, is to compare the effects of different doses of 5-HT on electrical potential under conditions where chloride is limiting. As described earlier, the transepithelial potential hyperpolarizes when chloride is replaced by the impermeant anion isethionate. When the chloride level is reduced to 40 mM, there is a similar hyperpolarization if the glands are stimulated with 5×10^{-8} M 5-HT. Increasing the 5-HT concentration to 10^{-6} M causes a considerable depolarization in this chloride-deficient medium, which is not maintained if calcium is removed from the bathing medium (FIGURE 7). In the absence of calcium, the potential hyperpolarizes even further than that seen during the action of 5×10^{-8} M 5-HT. The depolarization seen with 10^{-6} M may thus be explained by the fact that higher concentration of 5-HT cause a much larger entry of calcium which can drive more chloride movement under these conditions where diffusible anions are severely limiting. This ability of calcium to regulate chloride permeability is not unique to salivary glands because a similar effect has been described in the cornea [5] and in the intestine.[3, 4] Calcium also has a marked effect on potassium conductance in a large number of cells,[26] including secretory cells such as the parotid gland.[27, 28] The way in which calcium might regulate the permeability of membranes to chloride or potassium is unknown. Since many of the actions of calcium in cells seem to depend on the intervention of a specific receptor protein (the calcium dependent regulator or calmodulin), it is reasonable to propose that this protein might mediate the stimulatory effect of calcium on chloride permeability. Schulman and Greengard [29] have already noted that the ability of calcium to phosphorylate member proteins is mediated by calmodulin. Ilundain and Naftalin [30] have suggested that calmodulin is responsible for mediating the ability of calcium to increase chloride permeability in the intestine. Another possible mechanism for altering ion permeability is to alter the state of membrane phospholipids. In a number of cells, calcium seems to be responsible for activating the hydrolysis of tri-

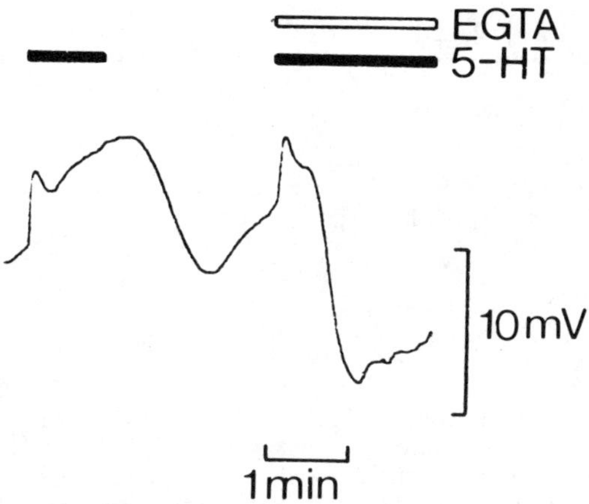

FIGURE 7. Effect of excess 5-HT (1×10^{-6} M, solid bars) on the transepithelial potential of a gland previously stimulated with 1×10^{-8} M 5-HT in a low chloride medium (40 mM). The second high 5-HT treatment was carried out in a calcium-free medium containing 5 mM EGTA.

phosphoinositides and diphosphoinositides,[31-34] and it has been suggested that the hydrolysis of these polyphosphoinositides might mediate some of the effects of calcium on ion permeability.[31] However, in red blood cells, Allan and Michell [34] have presented evidence that the ability of calcium to trigger potassium efflux is apparently not related to the hydrolysis of these phospholipids. The way in which calcium regulates the ionic permeabilities of cell membranes remains as a challenging problem for the future.

While there appears to be a clear role for calcium in regulating chloride conductance, the precise function of cyclic AMP has proved difficult to uncover. The problem is that cyclic AMP seems to be capable to raising the intracellular level of calcium by releasing this ion from internal reservoirs that makes it difficult to separate the effects of these two second messengers. Such a mobilization of calcium from internal reservoirs by cyclic AMP has been proposed for a number of transporting epithelia including the insect salivary gland,[20] the parotid gland,[17] the frog cornea,[5] the rabbit colon[4] and the liver.[35] The observations on liver are particularly interesting in the light of recent studies showing that calcium is as effective as cyclic AMP in promoting the hydrolysis of glycogen. Both second messengers can activate phosphorylase kinase, cyclic AMP by phosphorylating the enzyme while calcium acts directly on the enzyme through calmodulin. This duality of control through either cyclic AMP or calcium might also be applicable to other systems, such as the salivary gland. The fact that fluid secretion can be stimulated by cyclic AMP or by the ionophore A23187, which raises calcium levels in the gland without altering the cyclic AMP level, lends some support to the idea that these two second messengers might exert similar effects. Electrophysiological measurements have shown that cyclic AMP acts very differently to 5-HT or the inophore in that it causes the transepithelial potential to hyperpolarize in contrast

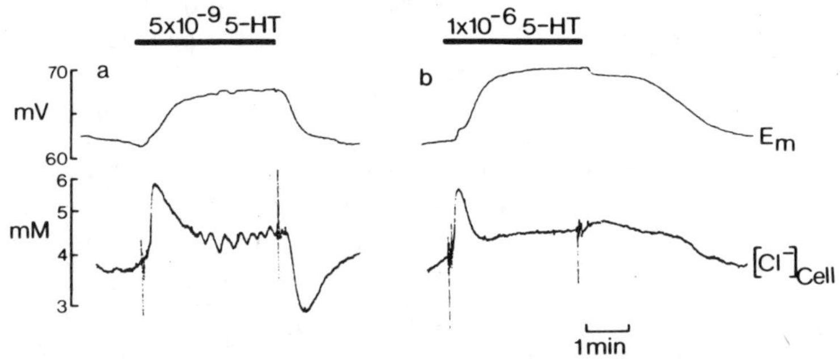

FIGURE 8. (a) Oscillations in the intracellular level of chloride (lower trace) during stimulation of *Calliphora* salivary gland with 5×10^{-9}M 5-HT (solid bar). The overall response is similar to that described in FIGURE 3. (b) No such oscillations were recorded when the same gland was stimulated with 1×10^{-6}M 5-HT. Note that the change in membrane potential (upper traces) took much longer to recover after stimulation with the higher dose of 5-HT. (The potassium concentration of the bathing medium was 10 mM).

to the depolarization caused by the other two agents. Cyclic AMP also has no effect on resistance so it does not alter chloride permeability and its ability to increase the potential is attributable to an increase in the potassium pump. Another expression of this ability of cyclic AMP to activate the potassium pump is evident in calcium-free experiments in which 5-HT causes the potential to hyperpolarize.

Since the removal of calcium does not alter the ability of 5-HT to generate cyclic AMP, the large hyperpolarization is once again explicable in terms of cyclic AMP activating the potassium pump in the absence of the short-circuiting action of calcium that normally causes the membrane to depolarize. As described earlier, however, it is not clear whether cyclic AMP acts directly on the potassium pump or whether it acts indirectly by releasing calcium from some internal reservoir. The amount of calcium released by cyclic AMP must be rather small because the intracellular concentration does not reach the threshold level necessary to switch on chloride permeability. If the pump does turn out to be sensitive to calcium, it is necessary to postulate that there may be a hierarchy of calcium-sensitive processes in that potassium transport seems to require low levels of calcium whereas higher concentrations seem to be necessary to increase chloride permeability. A similar hierarchy may exist in the mammalian parotid gland in which the calcium concentration necessary for amylase release seems to be less than that for potassium release.

A possible consequences of the dynamic interactions which exist between the second messengers is that their levels may oscillate under certain conditions.[36] The transepithelial potential in these insect salivary glands often displays regular oscillations especially at submaximal 5-HT concentrations.[8, 36, 37] The depolarizing phase of the oscillation is associated with a large decrease in resistance suggesting that membrane instability is driven by variations in chloride permeability caused by an underlying oscillation in the intracellular level of calcium. These regular fluctuations in chloride permeability result in oscillations in the intracellular level of chloride (FIGURE 8). When glands are stimulated with a low 5-HT concentrations, which routinely gives rise to potential oscillations, the intracellular level of chloride oscillates but remains constant when stimulated with a much higher dose of 5-HT. These regular fluctuations in the intracellular level of chloride illustrate the dynamic nature of the intracellular control mechanisms responsible for regulating ion transport across epithelia.

REFERENCES

1. PRINCE, W. T. & M. J. BERRIDGE. 1973. The role of calcium in the action of 5-hydroxytryptamine and cyclic AMP on salivary glands. J. Exp. Biol. **58:** 367–384.
2. BERRIDGE, M. J., B. D. LINDLEY & W. T. PRINCE. 1975. Membrane permeability changes during stimulation of isolated salivary glands by 5-hydroxytryptamine. J. Physiol. **244:** 549–567.
3. BOLTON, J. E. & M. FIELD. 1977. Ca ionophore-stimulated ion secretion in rabbit ileal mucosa: relation to actions of cyclic 3′,5′-AMP and carbamylcholine. J. Membrane Biol. **35:** 159–173.
4. FRIZZELL, R. A. 1977. Active chloride secretion by rabbit colon: calcium-dependent stimulation by ionophore A23187. J. Membrane Biol. **35:** 175–187.
5. CANDIA, O. A., R. MONTOREANO & S. M. PODOS. 1977. Effect of the ionophore A23187 on chloride transport across isolated frog cornea. Am. J. Physiol. **233:** F94–F101.
6. OSCHMAN, J. L. & M. J. BERRIDGE. 1970. Structural and functional aspects of salivary fluid secretion in *Calliphora.* Tissue & Cell **2:** 281–310.
7. BERRIDGE, M. J. & W. T. PRINCE. 1972. The role of cyclic AMP and calcium in hormone action. Adv. Insect Physiol. **9:** 1–49.
8. BERRIDGE, M. J. & W. T. PRINCE. 1972. Transepithelial potential changes during stimulation of isolated salivary glands with 5-hydroxytryptamine and cyclic AMP. J. Exp. Biol. **56:** 139–153.
9. PRINCE, W. T. & M. J. BERRIDGE. 1972. The effects of 5-hydroxytryptamine and cyclic AMP on the potential profile across isolated salivary glands. J. Exp. Biol. **56:** 323–333.
10. BERRIDGE, M. J. & W. R. SCHLUE. 1978. Ion-selective electrode studies on the effects of 5-hydroxytryptamine on the intracellular level of potassium in an insect salivary gland. J. Exp. Biol. **72:** 203–216.
11. GUPTA, B. L., M. J. BERRIDGE, T. A. HALL & R. B. MORETON. 1978. Electron

microprobe and ion-selective microelectrode studies of fluid secretion in the salivary glands of *Calliphora*. J. Exp. Biol. **72:** 261–284.
12. BERRIDGE, M. J., B. D. LINDLEY, & W. T. PRINCE. 1976. Studies on the mechanisms of fluid secretion by isolated salivary glands of *Calliphora*. J. Exp. Biol. **64:** 311–322.
13. SKAER, H. LEB., M. J. BERRIDGE & W. M. LEE. 1975. A freeze-fracture study of adult *Calliphora* salivary glands. Tissue & Cell **7:** 677–688.
14. PINTO, DA SILVA P. & D. BRANTON. 1970. Membrane splitting in freeze-etchings. Covalently bound ferritin as a membrane marker. J. Cell. Biol. **45:** 598–605.
15. BUTCHER, F. R. 1975. The role of calcium and cyclic nucleotides in amylase release from slices of rat parotid: studies with the divalent cation ionophore A23187. Metabolism **24:** 409–418.
16. SCHRAMM, M. & Z. SELINGER. 1975. The functions of cyclic AMP and calcium as alternative second messengers in parotid gland and pancreas J. Cyclic Nucleotide Res. **1:** 181–192.
17. PUTNEY, J. W., S. J. WEISS, B. A. LESLIE & S. H. MARRIER. 1977. Is calcium the final mediator of exocytosis in the rat parotid gland? J. Pharm. Exp. Ther. **203:** 144–155.
18. MARTIN, C. J., E. FRÖMTER, B. GEBLER, H. KNAVE & J. A. YOUNG. 1973. The effects of carbachol on water and electrolyte fluxes and transepithelial electrical potential differences of the rabbit submaxillary main duct perfused *in vitro*. Pflügers Arch. **341:** 131–142.
19. BERRIDGE, M. J. 1975. The interaction of cyclic nucleotides and calcium in the control of cellular activity. Adv. Cyclic Nucleotide Res. **6:** 1–98.
20. PRINCE, W. T., M. J. BERRIDGE & H. RASMUSSEN. 1972. Role of calcium and adenosine -3′:5′-cyclic monophosphate in controlling fly salivary gland secretion. Proc. Natl. Acad. Sci. U.S.A. **69:** 553–557.
21. BERRIDGE, M. J. 1970. The role of 5-hydroxytryptamine and cyclic AMP in the control of fluid secretion by isolated salivary glands. J. Exp. Biol. **53:** 171–186.
22. BERRIDGE, M. J., & H. LIPKE. 1979. Changes in calcium transport across *Calliphora* salivary glands induced by 5-hydroxytryptamine and cyclic AMP. J. Exp. Biol. **78:**137–148.
23. FAIN, J. N. & M. J. BERRIDGE. 1979. Relationship between hormonal activation of phosphatidylinositol hydrolysis, fluid secretion and calcium flux in the blowfly salivary gland. Bioch. J. **176:** 45–58.
24. BERRIDGE, M. J. & J. N. FAIN. 1979. Inhibition of phosphatidylinositol synthesis and the inactivation of calcium entry after prolonged exposure of the blowfly salivary gland to 5-hydroxytryptamine Biochem. J. **176:** 56–69.
25. BERRIDGE, M. J. 1979. Relationship between calcium and the cyclic nucleotides in ion secretion. Kroc Foundation Symposium. Mechanism of Intestinal Secretion. H. Binder, Ed.: 65–81. Alan R. Liss. New York, N.Y.
26. MEECH, R. W. 1978. Calcium-dependent potassium activation in nervous tissues. Ann. Rev. Biophys. Bioeng. **7:** 1–18.
27. SELINGER, Z., S. EIMERL, & M. SCHRAMM. 1974. A calcium ionophore stimulating the action of epinephrine on the adrenergic receptor. Proc. Natl. Acad. Sci U.S.A. **71:** 128–131.
28. PUTNEY, J. W. 1978. Role of calcium in the actions of agents affecting membrane permeability. *In* Calcium in Drug Action. G. B. Weiss, Ed.: 173–194. Plenum Press.
29. SCHULMAN, H. & GREENGARD, P. 1978. Ca^{2+}- dependent protein phosphorylation system in membranes from various tissues, and its activation by calcium dependent regulator Proc. Natl. Acad. Sci. U.S.A. **75:** 5432–5436.
30. ILUNDAIN, A. & R. J. NAFTALIN, 1979. Role of Ca^{2+}-dependent regulator protein in intestinal secretion. Nature. **279:** 446–448.
31. ABDEL-LATIF, A. A., R. A. AKHTAR & J. N. HAWTHORNE. 1977. Acetylcholine

increases the breakdown of triphosphoinositide of rabbit iris muscle prelabelled with (^{32}P) phosphate. Bioch. J. **162:** 61–73.
32. HAWTHORNE, J. N. & M. R. PICKARD. 1979. Phospholipids in synaptic function. J. Neurochem. **32:** 5–14.
33. GRIFFIN, H. D. & J. N. HAWTHORNE. 1978. Calcium-activated hydrolysis of phosphatidyl-*myo*-inositol 4-phosphate and phosphatidyl-*myo*-inositol 4,5-bisphosphate in guinea pig synaptosomes. Biochem. J. **176:** 541–552.
34. ALLAN, D. & R. H. MICHELL. 1978. A calcium-activated polyphosphoinositide phosphodiesterase in the plasma membrane of human and rabbit erythrocytes. Biochem. Biophys. Acta **508:** 277–286.
35. FRIEDMANN, N. 1972. Effects of glucagon and cyclic AMP on ion fluxes in the perfused liver. Biochem. Biophys. Acta **274:** 214–225.
36. RAPP, P. E. & BERRIDGE, M. J. 1977. Oscillations in calcium-cyclic AMP control loops form the basis of pacemaker activity and other high frequency biological rhythms. J. Theor. Biol. **66:** 497–525.
37. BERRIDGE, M. J. & P. E. RAPP. 1979. A comparative survey of the function, mechanism and control of cellular oscillators. J. Exp. Biol. **81:**217–279.

DISCUSSION

W. MC D. ARMSTRONG: What is the response time of these Simon-type calcium microelectrodes?

M. J. BERRIDGE: Relatively fast, approximately 1 second in the micromolar range.

ARMSTRONG: What's the resistance?

BERRIDGE: I haven't measured their resistance.

ARMSTRONG: If this resistance were very high, that would be quite an astonishingly fast time.

BERRIDGE: It really is very fast; and furthermore, the measurement of intracellular calcium with these electrodes provides a good indicator of the quality of electrode penetrations. With a successful penetration, the calcium electrode rapidly responds to the low cellular calcium level and leads to a stable membrane potential. Sometimes, the membrane potential does not remain stable, but begins to fall after a poor penetration. This is reflected in a sudden change of the calcium level, which shoots up towards the external level. Apparently, external calcium rushes into the cell through spaces between the electrode and the membrane. If the position of the electrode is adjusted slightly, the membrane usually reseals itself around the electrode and the potential returns to a stable value while the intracellular calcium concentration rapidly returns to a typical level of approximately 10^{-7} M.

ARMSTRONG: Have you looked at calcium ionophores, such as A 23187, and do they mimic the effects of 5-HT?

Lastly, with respect to the chloride response, You used the expression "sucking chloride out of the cell." I wonder if what you are seeing is not initially a rise in chloride activity above the point of electrochemical equilibrium and then a decrease back toward that equilibrium after the cyclic AMP comes into play.

With those membrane potentials, I suspect that your equilibrium activity of chloride is probably around 8–10 mM. I haven't done the calculations, but rather than a "sucking out," I think that it is a "falling down" to the equilibrium level rather than a "sucking out."

BERRIDGE: First, we have found in several experiments that A23187 almost exactly mimics the natural hormone, i.e., it can induce fluid secretion. It can also alter chloride membrane permeability leading to membrane potential changes identical to those seen with 5-HT. Therefore, it almost mimics the action of the hormone, but not completely because it does not evoke full rates of fluid secretion. This is apparently a characteristic feature of the calcium ionophore that has been described in other systems in which it often does not exactly mimic the natural hormone.

With regard to the decrease in cell chloride, I have tried to calculate the equilibrium potential, but this is rather difficult because one cannot put too much faith in the absolute value of these chloride measurements.

Therefore, it is not particularly meaningful to discuss chloride movement in terms of the equilibrium potential. When considering chloride movement in these glands, one must always bear in mind that ionic fluxes are dominated by the powerful electrogenic potassium pump system located on the apical membrane. When this is activated by 5-HT there is a rapid flux of potassium from the cell into the lumen with chloride following passively which has thus led to the notion that chloride is sucked out of the cell.

UNIDENTIFIED QUESTIONER: Is the chloride conductance voltage- or pH-dependent, and do you know of any pharmacological agents that inhibit chloride conductance on the outside of the cells?

BERRIDGE: The rate of fluid secretion is practically independent of pH levels between 6.0 and 9.0. Chloride conductance also seems to be independent of voltage. All the evidence suggests that this conductance is calcium-dependent. The obvious question is, does calcium mediate changes in chloride conductance through the ubiquitous calcium-binding protein, calmodulin? With respect to inhibitors, I have tried some of the classical anion blockers on this system, but I might get some additional ideas about them at this conference.

E. H. EPSTEIN: Can you tell us whether in insects, in general, and this insect, in particular, sodium-potassium ATP-ase inhibited by ouabain plays any role at all in their economy? Do they have it?

BERRIDGE: Curiously, many transporting epithelia in insects are resistant to ouabain, which may or may not reflect the existence of the sodium-potassium ATPase system. Such an enzyme might be present in the intestinal tract and in the nervous system.

However, in the salivary glands it is possible to demonstrate the existence of a sodium potassium ATP-ase, which plays a custodial role, but apparently is not involved in secretion during 5-HT stimulation. If you put these glands in a potassium-free medium, within about five minutes all the potassium in the gland is completely removed and replaced with sodium and, in fact, the gland just switches over and starts to secrete sodium chloride. If you now return to normal saline to determine whether the gland can reconstitute its potassium in the presence of ouabain, you find that it is not able to do so. It is only able to reaccumulate a small amount. If however, you stimulate with hormone, you will find a massive membrane hyperpolarization with potassium being sucked into the cell despite the presence of ouabain. If you don't add the hormone, the gland is not able to sequester potassium. So, I think that under resting conditions there is a custodial sodium-potassium ATP-ase performing its normal function; but, when you want a massive entry of potassium, the gland must switch to this potential-dependent process of getting potassium into the cell; and this process seems to be very efficient.

O. CANDIA: Have you tried some of the serotonin analogs on your system? Tryptamine, for example, inhibits sodium and chloride transport in the cornea.

BERRIDGE: Yes. I have tried tryptamine and it has classical pharmacological effects in that you simply get a shift to the right of the dose-response curve. Otherwise, it acts in much the same ways as 5-HT.

CANDIA: Have you measured the chloride fluxes in the opposite direction? I understand you only measured in one direction.

BERRIDGE: No. All the studies so far have concentrated on chloride movement into the lumen.

TRANSPORT OF BICARBONATE AND OTHER ANIONS IN SALIVARY SECRETION *

J. A. Young, R. M. Case, A. D. Conigrave, and I. Novak

Department of Physiology
University of Sydney
New South Wales 2006, Australia

INTRODUCTION

In general, compound exocrine glands are composed of two distinct types of epithelia, one always secretory and the other either secretory or reabsorptive, and it is the interaction of the transport systems of the two epithelia that determines the composition of the final secretion. All major salivary glands conform to this pattern, producing an isotonic primary secretion in the endpiece-intercalated duct region of the gland, and then modifying it by processes of electrolyte secretion and reabsorption (but with minimal fluid movement) during the passage of the primary fluid along the gland excurrent duct system.[54, 61] The interaction of these two epithelial transport systems makes interpretation of the underlying secretory mechanisms more difficult than with simpler epithelial organs, such as amphibian skin, particularly since certain important investigative techniques, such as intracellular potential recording and measurement of short-circuit current, are not equally adaptable to each type of epithelium. Thus, in the case of the primary (secretory) epithelium of the gland endpieces, a considerable body of information, based on intracellular recording *in vivo* and *in vitro,* is available,[42] whereas such information is almost completely lacking in the case of the ductal (reabsorptive) epithelium.[61] Again, largely for technical reasons, most workers have concentrated on cation transport across salivary epithelia—so that a systematic investigation of anion and proton transport has not been undertaken. The purpose of the present review is to summarize what is known of salivary anion and proton transport in the hope of stimulating more interest in the topic.

The Primary Fluid: Composition and Formation

Composition

Thaysen,[54, 55] by analyzing electrolyte excretion curves from the intact salivary gland, was the first to postulate that primary saliva would resemble a plasma ultrafiltrate in its electrolyte composition, although he was fully aware that the fluid was the result of a secretory, not a filtration process. Subsequently Martinez[41] introduced the technique of gland micropuncture and by analyzing the electrolyte content of primary saliva from the rat mandibular gland was able to confirm Thaysen's hypothesis directly.

* The authors' experimental work was supported by the National Health and Medical Research Council of Australia and by the Cystic Fibrosis Association of New South Wales.

0077-8923/80/0341-0172 $01.75/0 © 1980, NYAS

Table 1 shows the available data concening primary fluid anion composition for the five species studied so far. Unfortunately, direct measurement of pH *in situ*, or gasometric determination of HCO_3 in aspirated samples (volume $\sim 10^{-10}$ l), has not yet proven technically possible; indeed, only one attempt at measuring the pH of aspirated samples has been made.[5] In all three major salivary glands of the rat, the primary chloride concentrations were similar

TABLE 1

ELECTROLYTE COMPOSITION OF PRIMARY SALIVA OBTAINED BY GLAND MICROPUNCTURE IN VARIOUS SPECIES *

Gland		[Na+K]	[Cl]	[HPO_4/ H_2PO_4]	[RA]	Osmolality	Source
			Rat				
Mandibular	U	145	105	—	40	308	62–63
	S	143	100	—	43	299	62–63
Parotid	U	147	113	—	34	303	32–33
	S	146	—	—	—	298	32–33
Sublingual	U	138	99	—	39	—	37
	S	141	96	—	45	—	37
			Mouse				
Mandibular	U	157	125	—	32	297	34
	S	162	123	—	39	301	34
Parotid	U	147	113	—	34	283	34
	S	149	114	—	35	287	34
			Rabbit				
Mandibular	U and S	165	128	—	37	308	35
Parotid	U and S	163	132	—	31	311	35
			Cat				
Mandibular	U	169	127	—	42	—	17
	S	174	121	—	53	—	17
Sublingual	U	145	128	—	17	—	16
	S	155	122	—	33	—	16
			Sheep				
Parotid							
Na-replete	U	132	53	26	53	312	5
	S	142	50	25	67	296	5
Na-depleted	U	95	59	22	14	308	5
	S	94	54	23	17	288	5

* Concentrations are given in mequiv l^{-1}, osmolality in mosm kg^{-1}. RA indicates the difference between the summed Na and K concentrations and the Cl (plus PO_4 when available) concentration. U=unstimulated and S=stimulated (cholinergic).

to those of plasma, in fact, lying somewhere between the concentration in plasma and the concentration found in a plasma ultrafiltrate. There appears to be a good linear correlation between the Cl concentration in primary fluid and that in a plasma ultrafiltrate (FIGURE 1). Similar, but slightly higher, values have been reported for the mouse and rabbit (mandibular and parotid glands) and for the cat (mandibular and sublingual glands). Stimulation,

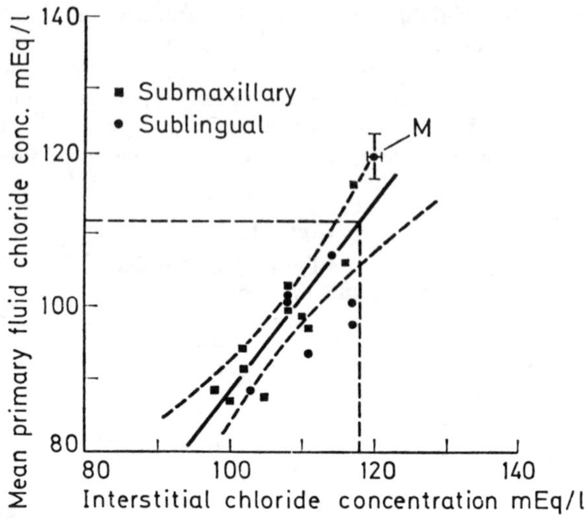

FIGURE 1. Relation between Cl concentration in interstitium and in primary fluid of rat salivary glands. Least-square regression and 95 percent confidence limits shown (From Young and Martin [63]; point labelled "M" from Martinez [41]).

either with parasympathomimetic agents (TABLE 1), or with sympathomimetic agents,[61] has no effect on these concentrations. Since only one study of the transepithelial potential difference of a salivary endpiece is available (—2 mV, cat sublingual gland [31]), some caution is needed in interpreting these results, but it would appear likely that, in each of these glands, chloride is passively distributed between interstitium and endpiece lumen. In one study, however, the chloride concentration of the primary fluid has been reported as considerably lower than in plasma, viz., in the sheep parotid (50–60 mmol l^{-1}). In this case, the corresponding monovalent cation concentrations were also low, so that the residual anion concentration was not greatly increased, and it would thus not appear that Cl was low because of the secretion of another anion in exchange for it. Further, in this case, one cannot make any assumption about the likely transepithelial potential difference (*PD*) so it is not possible to state whether the Cl concentration has been lowered passively or actively.

In most salivary glands, the inorganic phosphate of the final saliva is three times that in plasma, but in sheep parotid saliva it is 12 times greater. This encouraged Compton *et al.*[5] to attempt to measure phosphate concentrations in primary saliva: they obtained values of 12–14 mmol l^{-1} at pHs of 7.5–8.0, which suggests a charge concentration of 22–26 mequiv l^{-1}. Cholinergic nerve stimulation did not change the concentration. Although, as for Cl, our interpretation is hampered by the lack of *PD* measurements, it seems likely that phosphate must have been actively secreted in view of the steep concentration between lumen and plasma, which would otherwise require an equilibrium *PD* of around 66 mV, lumen positive.

As mentioned above, no direct measurements on primary fluid HCO_3 content (or on pH *in situ*) are available but, nevertheless, it is possible to set plausible limits on what the HCO_3 concentration might be by calculating the residual anion concentration (the summed concentrations of measured cations, less measured anions). These concentrations are shown in TABLE 1. It is evident that the concentrations normally range between 30–40 mequiv. l^{-1} although values outside these limits are found for the sheep parotid. Stimulation appears not to alter the concentration appreciably. Apart from Na and K, the only other cations that could be expected to be present in primary saliva in more than trace concentrations would be Ca and Mg, but, since neither of these is present in final saliva in concentrations greater than in plasma,[59] the primary concentrations seem unlikely to be very high.† Hence, the residual anion concentration should represent the summed concentration of HCO_3, PO_4 and various organic anions: even assuming that the phosphate concentration of primary saliva is always about as high as in final saliva (see above), it seems likely that HCO_3 would account for more than half of the residual anion concentration, i.e., more than 20 mmol l^{-1}.

Another approach to assessing the possible HCO_3 content of primary fluid comes from studies on the isolated perfused salivary gland. FIGURE 2 depicts the results of experiments in which the isolated rabbit mandibular gland was perfused *in vitro* with solutions lacking HCO_3 or containing it in concentrations of 12.5, 25, or 50 mmol l^{-1}. It can be seen that the final salivary HCO_3 concentration was approximately the same as in the perfusate, at least up to the concentrations normally seen in plasma, although at the higher perfusate concentration (50 mmol l^{-1}), the salivary concentration was only 36 mmol l^{-1}. In the rabbit mandibular gland, the HCO_3 transport activity of the ducts is relatively low, so that, at high secretory rates, the primary and final salivary HCO_3 concentrations can be expected to approximate one another. Hence, the data suggest that, over the physiological range of plasma HCO_3 concentrations,

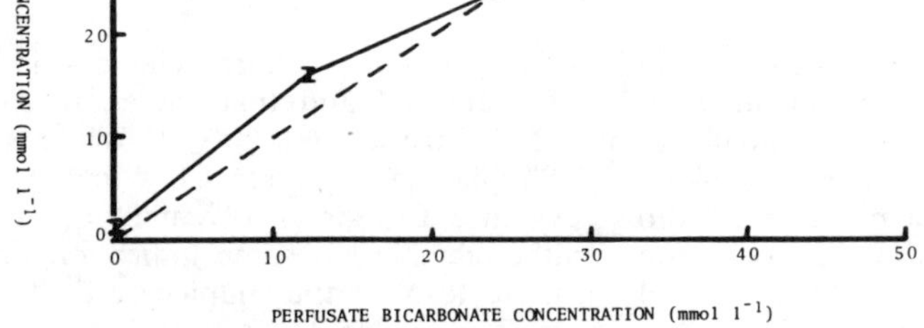

FIGURE 2. Relation between HCO_3 concentration in perfusate and in final saliva of the rabbit mandibular gland perfused *in vitro*. Each point is the mean ± S.E.M. of 4–8 observations; the line of identity is also shown (From Conigrave *et al.*[66]).

† In one recent micropuncture study on the rat parotid,[36] a primary Ca concentration of 2.94 mequiv l^{-1} was reported.

the primary fluid level is also likely to reflect the perfusate concentration closely.

Formation

Although a considerable body of information is available to indicate how the primary fluid is secreted,[42, 61] very little work has been carried out to determine what role active transport of anions may have in the process. Originally Lundberg,[31] who investigated the effect of stimulation on the basal membrane potential of salivary endpiece cells, postulated that the secretion was the result of the active uptake of Cl from the interstitium into the cytosol. Subsequently, however, many studies have shown that this theory is untenable,

TABLE 2

ONE-HOUR SECRETORY RESPONSE OF THE ISOLATED MANDIBULAR GLAND OF THE RABBIT PERFUSED *in vitro* *

	1-hr response (ml)	S.E.M.	n
Bicarbonate-containing			
50 mmol l^{-1}	2.23	0.18	8
25 mmol l^{-1}	2.71	0.16	16
12.5 mmol l^{-1}	2.36	0.13	8
Bicarbonate-free			
Phosphate buffer	3.05	0.29	9
HEPES buffer	2.23	0.12	6
Acetate buffer	2.94	0.32	4

* With physiological salt solutions containing HCO_3 in various concentrations, or solutions containing no HCO_3 but buffered to pH 7.4 with phosphate (1.6 mmol l^{-1}), HEPES (10 mmol l^{-1}) or acetate (25 mmol l^{-1}); the pHs of the HCO_3-containing solutions were held at 7.4 by adjustment of the CO_2 tension. Secretion was evoked by continuous intra-arterial infusion of acetylcholine (8×10^{-5} mol l^{-1}); n=number of glands studied.

and it is now generally held that stimulation effects changes in the basal membrane conductance to Na, Ca and K, and that these are followed by activation of a luminally located Na extrusion pump.[42, 61] While it is known that secretion can be blocked by replacing perfusate Cl with SO_4 or NO_3,[42] there appears to be no direct evidence for the involvement of an active Cl pump in the secretory process although there is some indication that the ion may be concentrated actively in the cytosol of the endpiece cells.[16, 43, 51, 61]

Equally little is known about the role of HCO_3 or proton transport in secretion. On the one hand, secretion appears not to be influenced by inhibition of carbonic anhydrase[65] and, on the other, the gland appears able to secrete in the absence of HCO_3 in the vascular perfusate. TABLE 2 shows the secretory response of the isolated rabbit mandibular gland perfused *in vitro* with solutions containing HCO_3 in concentrations of 50, 25 or 12.5 mmol l^{-1} as well as solutions containing no HCO_3, buffered at pH 7.4 with phosphate

(1.6 mmol l^{-1}), acetate (25 mmol l^{-1}) or HEPES (10 mmol l^{-1}). It is evident that the gland's capacity to secrete fluid is in no way impaired by alterations in the HCO_3 concentration of the perfusate provided the pH is held at 7.4. One is drawn to the conclusion that the catalytic hydration/dehydration reaction of CO_2/H_2O is not a rate-limiting step in the fluid secretory process and that the secretion rate is independent of the transport of HCO_3 or protons. However, recent studies of ours on the isolated perfused mandibular gland of the rabbit indicate that methazolamide and acetazolamide can reduce total output of HCO_3 by more than 30% even though fluid output is unchanged. Similarly, in the sheep parotid gland, acetazolamide has been found to cause a reduction in HCO_3 output.[68]

Ductal Modification of the Saliva

Apart from the intercalated ducts, which perforce must be studied in conjunction with the endpieces,[60] salivary glands have two distinct types of excurrent ducts, striated (largely intralobular) and excretory (largely extralobular) ducts. The anatomy of the striated ducts is such that these structures are only rarely accessible to a micropuncture pipette and in consequence have not been studied by microperfusion.[60, 61] The excretory ducts, particularly the main ducts, being extralobular, are accessible and can be punctured so that microperfusion, both *in vivo* and *in vitro* is possible. There are good grounds for believing that the electrolyte transport properties of striated and excretory ducts may be similar,[61] but direct proof of this contention is lacking and it would seem surprising if some difference of importance between the properties of the two ducts were not eventually to be uncovered. From a technical point of view, it is relatively simply to study the transport of any of the common salivary ions across the duct epithelium; but, in contrast to endpieces, it has not yet been proven possible to make electrical recordings from duct cells, so that the analysis of ductal transport properties has been seriously hampered. What follows is an attempt at summarizing what is known about proton and anion transport across salivary excretory ducts, but, of necessity, attention must also be paid to Na and K transport since transport of these ions seems to be linked to proton transport.

Ductal Permeability Properties

In all salivary ducts so far studied, the transepithelial *PD* has exhibited a dependency on luminal Na concentration arising because the luminal cell membrane is selective for Na relative to anions, both univalent, such as Cl and HCO_3, as well as divalent, such as SO_4.[11, 19, 22, 23, 45, 64] In the rat mandibular duct, the relative selectivities for Na and Cl are such that, during perfusion with isotonic NaCl, the plot of *PD* against log luminal Na concentration has a slope of 35 mV per decade, but in the rabbit mandibular duct the Cl conductance is higher and the electrode slope, correspondingly lower, about 5 mV per decade.[19, 64] In both species, replacement of Cl with SO_4 results in an electrode slope close to the ideal 61 mV per decade. Very few studies are available on the Na selectivity of the interstitial surface of the duct epithelium, but it is generally agreed that the duct is relatively impermeable to Na, e.g., replacing Na in the extracellular medium with choline does not much alter the *PD*. The overall permeability coefficient of the epithelium to passive sodium movement (TABLE 3) has been estimated as 5.3×10^{-6} cm s^{-1}. With such a low value, it is not surprising that the ducts can develop steep transepithelial concentration gradients for Na.[11]

The permeability properties of salivary ducts for K are the reverse of those for Na: the luminal cell membrane of the rat mandibular duct is rather impermeable to the ion while the basal cell membrane is highly permeable.[19, 24, 25, 46, 48, 64] Nevertheless, the luminal membrane permeability to K is greater than to choline.[25, 28, 64] Essentially similar results have been reported for the rabbit mandibular duct[24, 61] and for the main ducts of the human parotid and submandibular glands.[22] The overall permeability coefficient of the epithelium to passive K movement (TABLE 3) has been estimated[61] to be $\sim 30 \times 10^{-6}$ cm s^{-1}.

Information is available on the duct permeability to two salivary anions, Cl and HCO_3. The main duct of the rat mandibular gland is a classic example of the "tight" class of epithelium, transporting salt hypertonically, having a low hydraulic conductance and maintaining high electrical potential differences and ionic electrochemical gradients across the epithelium.[12] For this reason, it is to be expected that it would have a relatively low Cl conductance and that substitution of SO_4 for Cl would make little difference to the transepithelial *PD*.[45, 58] The transepithelial Cl conductance, estimated during duct perfusion with a variety of electrolyte solutions,[45, 58, 61] ranges from $13\text{–}65 \times 10^{-4}$ Ω^{-1} cm^{-2} and the passive permeability coefficient (see TABLE 3) is about 15×10^{-6} cm s^{-1}.

The rabbit mandibular duct, in contrast to that of the rat, is quite atypical as an epithelium,[12, 19, 39] having many features characteristic of a tight epithelium (low hydraulic conductance, and steep electrochemical gradients for actively transported ion species[61]) but having a lower transepithelial potential difference and a very high Cl conductance.[2, 3, 13, 19, 24, 52] Thus, while perfusion of the duct with Na_2SO_4 solutions reveals that the luminal cell membrane has near-ideal Na electrode characteristics, perfusion with NaCl solutions indicates that Cl can shunt out the Na diffusion potential and reduce the electrode slope from 61 to less than 5 mV per decade.[13, 19, 24, 52] Perfusion with isotonic solutions containing mixtures of NaCl and Na_2SO_4 (FIGURE 3),[14] or with various concentrations of NaCl,[13] reveals that the luminal surfaces of the duct can behave like an ideal Cl electrode.

Early attempts at measuring the electrical conductance of the duct epithelium were not entirely successful since the duct polarizes markedly in response to square-wave current pulses.[19] Frömter[13] overcame this problem and estimated the Cl conductance from the zero-current membrane resistance, measured during NaCl or Na_2SO_4 perfusion: the value he obtained (85×10^{-3} Ω^{-1} cm^{-2}) is more than an order of magnitude greater than for the rat duct (see above). Similar values have also been reported by Augustus *et al.*[3] The estimated Cl permeability coefficient for the duct is also rather high, viz., 55×10^{-6} cm s^{-1} (see TABLE 3).[52] The cause of this high Cl conductance has been investigated carefully in several studies,[2, 3, 13, 52] and it now seems well established that it is not due to a low resistance shunt across the epithelial tight junction, but due to an abnormally low Cl resistance across the epithelial cell membranes. It is thus possible to conclude that the rabbit mandibular duct does belong to the "tight" class of epithelia but is atypical in having cell membranes with low Cl resistances.

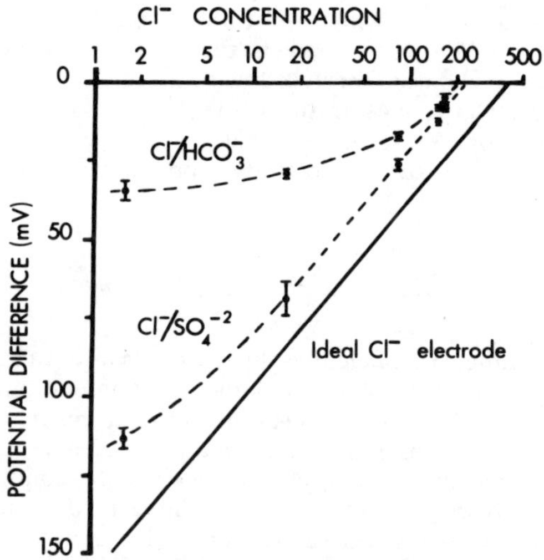

FIGURE 3. Cl-selectivity characteristics of the luminal face of rabbit mandibular duct epithelium perfused with isotonic mixtures of NaCl and either Na_2SO_4 or $NaHCO_3$. Each point is mean±S.D. (From Grant *et al.*[14]).

TABLE 3

TRANSPORT PARAMETERS OF THE MAIN DUCTS OF THE MANDIBULAR GLANDS OF RAT AND RABBIT PERFUSED *in vivo* OR *in vitro* *

	Rat [11, 38, 58, 61, 64–65]		Rabbit [2–3, 13–14, 19, 39, 44, 51, 61]	
	in vivo	*in vitro*	*in vivo*	*in vitro*
Steady-state *PD* and electrolyte concentrations				
Na (mmol l^{-1})	2	2	2	2
K "	135	140	88	63
Cl "	78	–	12	17
HCO_3 "	55	–	80	48
PD (mV)	−11	–	−33	−47
PD and net electrolyte fluxes during perfusion with bicarbonate saline				
Na (mol cm^{-2} s^{-1} 10^{-9})	+ 3.2	+ 2.8	+ 9.4	+11.4
K "	− 2.5	− 1.9	− 0.5	− 0.5
Cl "	− 0.7	+ 1.7	+ 8.7	+10.5
HCO_3 "	− 0.3	–	+ 0.12	+ 0.50
PD (mV)	−71	−64	−14	−13
Permeability properties				
L_p (ml cm^{-2} s^{-1} atm^{-1} 10^{-6})	1.6	4.2	2.9	14.0
P_{Na} (cm s^{-1} 10^{-6})	–	5.3	–	–
P_K "	29.1	–	–	33.0
P_{Cl} "	15.0	–	–	55.0
Duct dimensions				
area (cm^2)	0.130	0.032	0.140	0.051
radius (cm)	0.010	0.010	0.015	0.015

* With isotonic bicarbonate-saline, or under steady-state conditions of zero net flux of ions and water. The symbols L_p and P indicate, respectively, the hydraulic conductivity and the permeability coefficient (From Young & van Lennep[61]).

The only other anion for which information on duct permeability is available is bicarbonate.[14] FIGURE 3 depicts the results of experiments in which rabbit mandibular ducts were perfused with isotonic mixtures of NaCl and $NaHCO_3$. At low HCO_3 concentrations, the duct is so impermeable to HCO_3 that it gives a *PD* slope close to the ideal for a Cl electrode but, as luminal concentration increases, the Cl electrode slope deviates markedly from the ideal.

Ductal Electrolyte Transport

When contact time is sufficient, salivary ducts are capable of developing steep concentration gradients between lumen and interstitium for Na, K, Cl, HCO_3 and Ca.[36, 45, 61, 64] TABLE 3 lists the luminal concentrations of Na, K, Cl and HCO_3 and the transepithelial potential differences (*PD*) that develop when the duct is permitted to establish steady-state conditions of zero net flux. By using the Nernst equation, it is possible to demonstrate that Na, K and HCO_3 are all subject to active transport (Na reabsorbed, K and HCO_3 secreted) but that Cl is distributed across the epithelium according to the prevailing electrochemical potential gradient.

The Na transport rates observed when the salivary ducts are perfused with bicarbonate-saline (TABLE 3) are extremely high,[61] among the highest reported for any mammalian epithelium, and, when steady-state conditions develop, the pumping mechanism can maintain a Na gradient of 6.5–10 kJ mol^{-1}. The transport process is dependent on the activity of a basally located K-activated, ouabain-sensitive ATPase that appears to pump electrogenically;[1, 24, 27] the pump activity seems also to be inhibited by reduction of the periductal HCO_3 concentration[20, 26] and its activity is stimulated by adrenal corticosteroids.[15, 21, 26, 29] At the apical cell membrane, the high Na selectivity mentioned above appears to depend on an unusally complex membrane conductance channel that operates, in part, to exchange Na for K and H and, in part, as a NaCl carrier. This mechanism can be blocked by luminal application of amiloride and triamterene[25, 28, 40, 47] or by replacement of luminal Cl with SO_4 or HCO_3;[30] it operates to exchange Na for K and can be blocked by raising luminal K.[30] The mechanism exhibits saturation kinetics with respect to luminal Na concentration.[11, 30] There is no evidence to indicate that the apical carrier is capable of doing direct transport work—it seems most likely that it is driven by the movement of Na down its electrochemical gradient from the lumen into the cytosol.[61]

As was observed in the case of Na, the duct appears capable of establishing a steep electrochemical gradient for K (TABLE 3), about 7.5 kJ mol^{-1} for the rat, and 2.5 kJ mol^{-1} for the rabbit.[61] It is thought that the basal membrane ATPase referred to above is responsible for concentrating K in the duct cell and that it enters the lumen across the apical membrane passively via two distinct mechanisms: a Na-K exchange carrier (referred to above)[21, 25, 28, 47, 64] and a K-H transfer mechanism that effects coupled secretion of K and HCO_3.[25, 26, 28, 48] This latter mechanism appears to be inhibited by luminally-applied cytochalasin-B. It is clear from the data in TABLE 3 that the rabbit duct secretes K at very much lower rates than the rat duct, but, in both the rat and the rabbit the secretion process appears not to be greatly influenced by the prevailing electrical gradient but to depend only on electroneutral exchange mechanisms.[6, 30]

The equilibrium HCO_3 concentrations developed by both rat and rabbit ducts (TABLE 3) indicate unmistakably that these ducts secrete HCO_3 actively. This fact is especially clear in the case of the rabbit duct studied *in vitro*[44] where we calculate that the steady-state electrochemical gradient for HCO_3 (6.1 kJ mol^{-1}) is almost as large as for Na. In contrast, however, the ducts appear to have secreted little or no HCO_3 when perfused under nonsteady-state conditions with an isotonic bicarbonate-saline solution (TABLE 3); indeed, in the rabbit duct perfused *in vivo* or *in vitro*, it is common to observe a slight but significant net absorption of HCO_3 (FIGURE 4).[39] In the rat

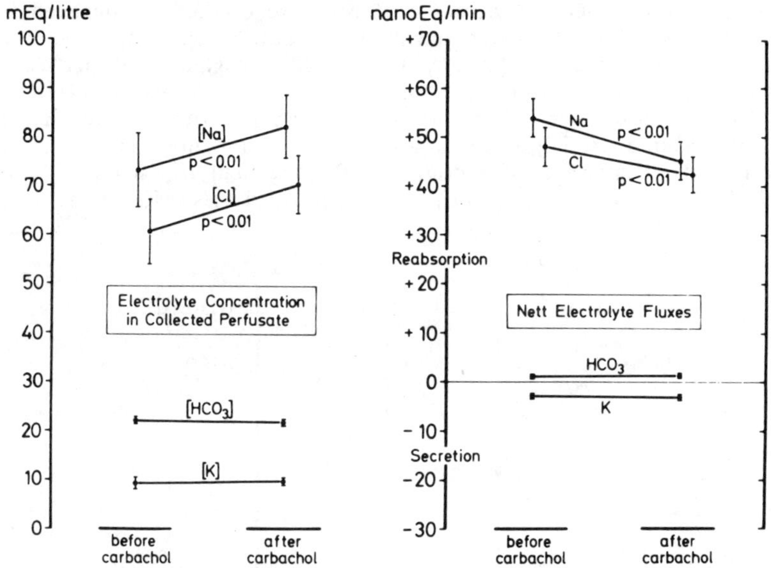

FIGURE 4. Effect of carbachol on luminal electrolyte concentrations and on transepithelial net fluxes in rabbit mandibular duct perfused *in vitro* with bicarbonate-saline. Note that there is significant net absorption of HCO_3 (From Martin and Young[38]).

duct undergoing perfusion with isotonic or hypotonic (25 per cent strength) bicarbonate-saline, a small but significant degree of net HCO_3 secretion has been seen[38, 65] but, during perfusion with isotonic mannitol or KCl,[10] the net HCO_3 secretory rates were much greater. It appears that ductal HCO_3 secretion can be increased by lowering the luminal Na concentration, and ductal HCO_3 secretion can be decreased, or even reversed, when luminal Na is high.

Knauf[21, 25, 28, 30, 56] has focussed attention on the luminal K-H exchange mechanism by studying ductal secretion of K and HCO_3 in the rat under conditions when Na absorption had been blocked, either by removing the ion from the luminal perfusate or by administration of drugs such as amiloride or triamterene or the steroid antagonist, spironolactone. He has provided very convincing evidence that these two ions are secreted by a tightly coupled K-H transport process but that additional K is normally secreted in exchange for absorbed Na. The K-H exchange mechanism may be associated with a membrane-bound HCO_3-activated ATPase.[56] FIGURE 5 depicts Knauf's current model for salivary duct electrolyte transport, incorporating the various features just described. Initially, as the NaCl-rich primary fluid passes along the ducts, there will be net absorption of NaCl and some net secretion of K, but HCO_3 flux will be negligible. As the luminal Na concentration falls, the activity of the K-H exchange pump will become relatively more conspicuous and net secretion of K and HCO_3 will be observed. The electrical behavior of the rat and rabbit ducts as they effect these transformations of the luminal electrolyte concentrations is shown in FIGURE 6.

In contrast to the electrolyte transport processes operating across the end-piece cells, those operating across duct cells are markedly influenced by the anion composition of the interstitial fluid. FIGURE 7 depicts the results of experiments in which the isolated rabbit mandibular gland was perfused with fluids containing no HCO_3 (see also TABLE 2).[66] Replacement of HCO_3 with Cl (and adjustment of the pH to 7.4 with phosphate buffer, 1.6 mmol l^{-1}) resulted in a striking decrease in ductal Na reabsorption and a smaller but quite definite decrease in K secretion.‡ Replacement of HCO_3 with acetate restored ductal Na and K transport to control levels whereas replacement

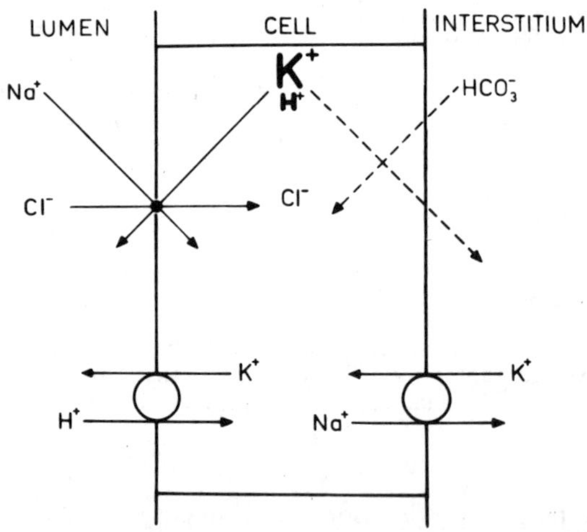

FIGURE 5. Hypothetical model to account for electrolyte transport by the main duct of the rat mandibular gland. The luminal Na conductance site appears also to transport Cl, K and H, and it is sensitive to amiloride. The luminal K-H exchange pump is associated with an HCO_3-ATPase and is sensitive to cytochalasin-B. The basal Na-K pump is electrogenic, and is ouabain-sensitive and K-activated (From H. Knauf).

‡ Micropuncture data are not yet available, but the magnitude of the changes in concentration is too great to be explained solely in terms of changes in the primary Na and K concentrations.

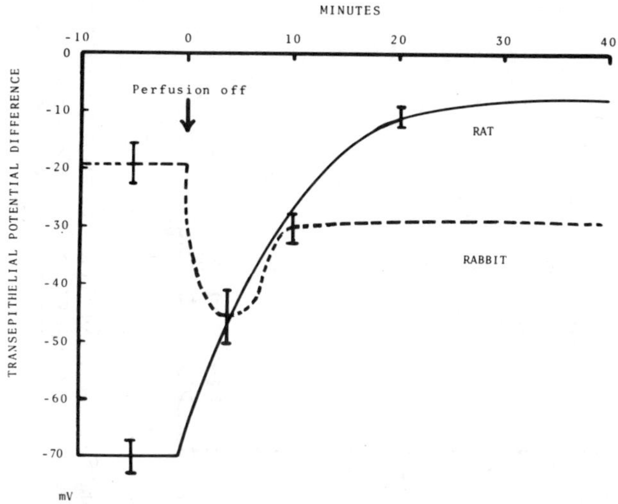

FIGURE 6. Transepithelial *PD* of rat and rabbit mandibular duct during perfusion with bicarbonate-saline and following cessation of perfusion. Each point is mean ± S.D. In the rabbit duct, the *PD* during perfusion is low due to the Cl shunt, and, as NaCl is reabsorbed following cessation of perfusion, the *PD* becomes more negative; later, as luminal HCO_3 increases, the *PD* falls again to a steady-state value. In the rat, the *PD* is high during perfusion and falls monophasically following cessation of perfusion to a steady-state value, since NaCl absorption and $KHCO_3$ secretion take place at comparable rates.

with the extracellular anion, HEPES, failed to restore Na absorption. From analysis of the Na, K, Cl and HCO_3 concentration in saliva, it appears that HEPES does not enter the saliva whereas acetate is present in concentrations comparable to those seen for HCO_3 in control studies. Whether it enters at the primary or the ductal level is unknown, but it is clear that removal of extracellular HCO_3 reduces Na absorption unless it is replaced with an intracellular buffer anion, even though the extracellular pH is kept at 7.4. It is interesting to note that earlier *in vivo* studies in the rat show that metabolic acidosis, in which, of course, both pH and extracellular HCO_3 are low, causes a reduction in ductal Na reabsorption and in secretion of K and HCO_3.[21, 26]

Autonomic Control of Ductal Transport

Since nerve endings need not penetrate the epithelial basement lamina in order to establish fully functional neuroeffector junctions with glandular epithelial cells, it is difficult to decide on purely morphological ground whether or not salivary excretory ducts are under autonomic control, but the combined morphological, pharmacological and physiological evidence provides very strong support for such a proposition.[60, 61] Cholinergic agonists, in both the rat and rabbit, inhibit ductal Na reabsorption,[8, 38, 39] probably by acting on the luminal Na-K(H) exchange site (FIGURE 5). Simultaneously, at least in

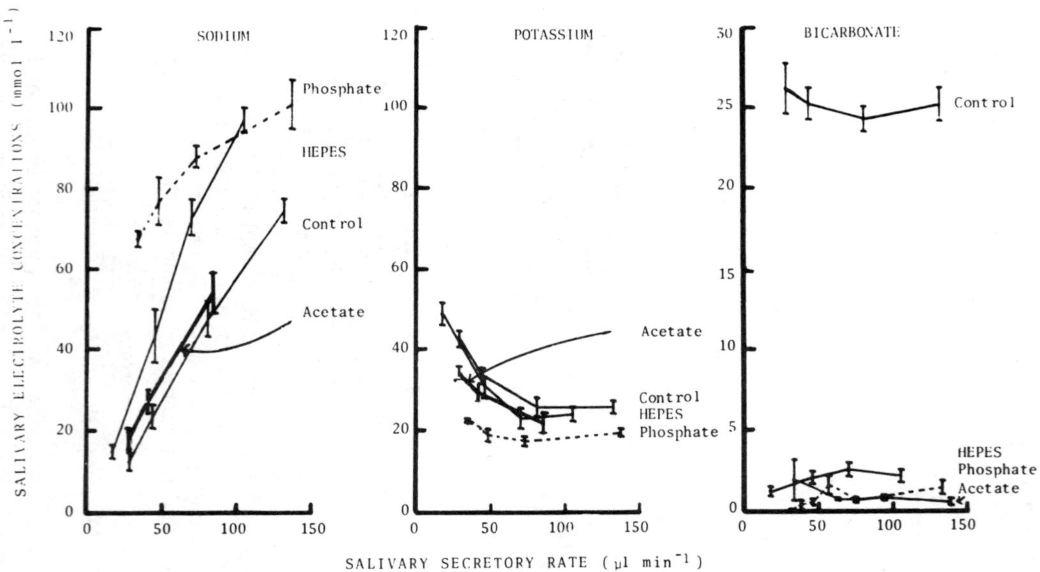

FIGURE 7. Concentrations of Na, K and HCO_3 in the final saliva of the rabbit mandibular gland isolated and perfused with solutions, buffered to pH 7.4, containing acetylcholine, 8×10^{-5} mol l^{-1}. The label "Control" indicates perfusion with a physiological salt solution containing HCO_3 (25 mmol l^{-1}); "Phosphate," "HEPES" and "Acetate" refer to perfusion fluids containing Cl but no HCO_3 and buffered, respectively, with phosphate (1.6 mmol l^{-1}), HEPES (10 mmol l^{-1}) and acetate (25 mmol l^{-1}). Each point is the mean ± S.E.M. of 9–113 samples (From Conigrave *et al.*[66]).

the rat, they seem to stimulate the luminal K-H exchange mechanism[38,61] so that the net effect is to inhibit Na absorption, to stimulate HCO_3 secretion and to have a variable effect on K secretion (according to the prevailing luminal Na concentration). Adrenergic agonists appear to have similar actions,[8, 38] although they have been less extensively studied. Direct secretomotor nerve stimulation, both parasympathetic and sympathetic, indicates that these effects are not simply of pharmacological interest.[49, 50] Studies on the isolated perfused mandibular gland of the rabbit show that K and HCO_3 excretion curves are very sensitive to changes in the perfusate concentration of acetylcholine.[67]

Secretion by the Intact Gland

Having examined what is known about the transport of anions and protons across the epithelia of salivary endpieces and ducts, let us now look at the excretion patterns of these ions for the intact gland in order to see to what extent they can be explained. The common ions of saliva are Na, K, Cl and HCO_3 which, together, make up more than 90% of the total osmotic activity of saliva, although inorganic phosphate also makes a significant contribution in the saliva of animals of the order Bovidae.[5, 61] Since all of these ions are present in the primary saliva in approximately constant concentrations (TABLE 1), their concentrations in final saliva (and their dependence on salivary secretion rate) will reflect the manner in which the ducts handle the ions.[59] Hence, if our analysis of endpiece and ductal function is reasonably comprehensive, it should be possible to predict the general shape of the various salivary electrolyte excretion curves.

Whereas the salivary excretion curves for Na and K (particularly K) are similar in the salivary glands of the 10 or so mammalian species studied, the excretion curves for HCO_3 exhibit a bewildering variety.[59] FIGURE 8 depicts in schematic form the HCO_3 excretion curves for the parotid and mandibular glands of eight different mammalian species and FIGURE 9 depicts the raw data on which one of these curves, that of the rabbit mandibular gland,[9] is based. It is apparent that salivary HCO_3 content may always exceed the plasma concentration in some glands (rat mandibular and sheep parotid) and be invariably lower than in plasma in others (mouse mandibular) but in most, the HCO_3 concentration is below plasma at low secretory rates and rises with increasing secretory rates to values greater than in plasma. In a few glands (mandibular of rat, rabbit and cat and the mouse parotid) the HCO_3 concentration is high at low flow rates and, as flow rate increases, it first falls towards or below plasma and then rises again as maximum flow rates are approached.

It is difficult to account for such variety with a single secretory model particularly when, as has been pointed out above, data on primary fluid pH and HCO_3 content are incomplete or lacking. If we may accept that the primary HCO_3 concentration is normally less than 40 mmol l^{-1}, perhaps close to the concentration in a plasma ultrafiltrate ($\sim$ 30 mmol l^{-1}), then it is possible to make some predictions concerning what the shape of the HCO_3 excretory curve might be. Most salivary glands do not secrete spontaneously but only in response to nerve stimulation and the secretory response to basal levels of stimulation is quite low.[61] It follows that the primary fluid secreted under resting conditions is in contact with the duct epithelium for relatively long time periods and the electrolyte concentrations will have to approach steady-state values (TABLE 3). Hence the HCO_3 concentration in "resting" saliva can be expected to be high whenever the ducts possess HCO_3 secretory mechanisms capable of establishing steep concentration gradients for the ion, as is evidently the case in the mandibular gland of the rat and rabbit (TABLE 3). Of course, the HCO_3 concentration would remain low if the resting flow rate were too rapid or whenever a ductal HCO_3 secretory mechanism was lacking, as appears to be the case for the cat sublingual gland.[16] Since duct cells can possess two opposing H transfer mechanisms (FIGURE 5), a secretory mechanism coupled to passive Na absorption (which can be inhibited by neurotransmitters) and an absorptive mechanism coupled to the active secretion of K (which can be stimulated by neurotransmitters), it is evident that salivary HCO_3 may drop below the primary level whenever Na reabsorption is conspicuous, as it must be whenever salivary flow is brisk and large volumes of Na-rich fluid are being delivered to the ducts. However, since stimulation of primary fluid formation is likely to be accompanied by simultaneous duct stimulation, the Na-H mechanism may be partially inhibited and the K-H mechanism stimulated, so that at high flow rates the HCO_3 concentration may rise to and perhaps surpass the primary level. A "classic" HCO_3 excretion curve is depicted in FIGURE 9: we note a high HCO_3 concentration in saliva collected at the lowest secretory rates (due to K-H exchange) that first falls rapidly as flow rate increases and then rises again as flow rate approaches its maximum. It can be viewed as the sum of two components, one attributable to each H exchange pump, the precise shape of each being influenced by

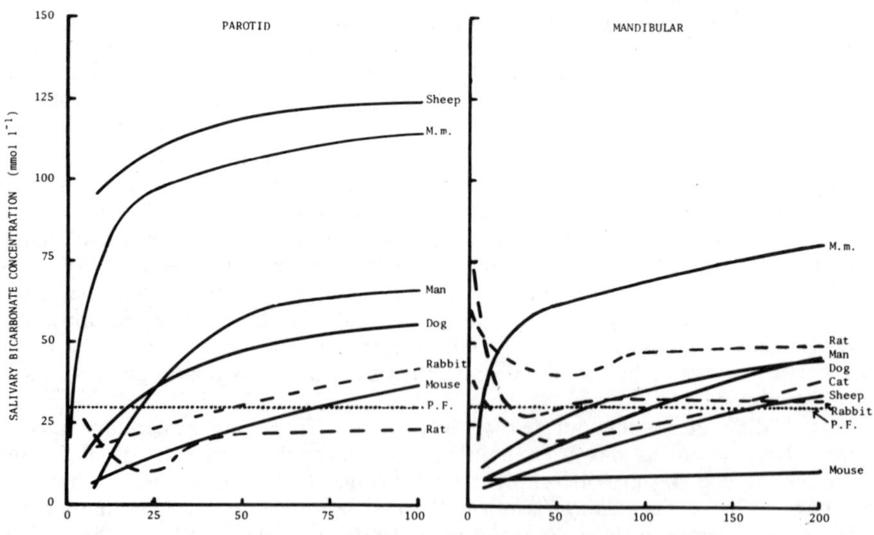

FIGURE 8. Schematized HCO_3 excretion curves for salivary glands of various species. Horizontal broken lines indicate the likely upper limit for primary fluid HCO_3 concentrations. The curves are based on original data for the sheep,[5,18] rhesus monkey (M.m.),[4] man,[7,54,55] dog,[57] cat,[17] rabbit,[9,35] mouse[34] and rat,[32,33,53,62] as summarized by Young.[59,61]

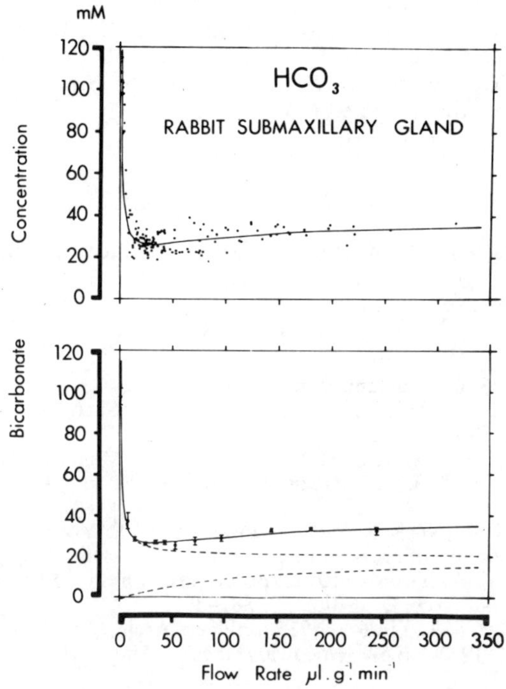

FIGURE 9. Relation between HCO_3 concentration and secretory rate of final saliva from rabbit mandibular glands stimulated *in vivo* with carbachol and pilocarpine. Raw data are shown above and binned data (means ± S.E.M.) below. The broken lines represent two rectangular hyperbolae ($y=20+110/x$ and $y=21-2700/(x+120)$) which, when summed, fit the data best according to a least-squares method. The first hyperbola may describe the HCO_3 secretory activity of the K-H exchange pump and the second hyperbola may describe the action of the Na-H exchange mechanism; progressive inhibition of the Na-H pump and stimulation of the K-H pump by the administered parasympathomimetic drugs would influence the precise shape of each curve and the value of the apparent asymptote of the summed curve (From Endre[9]).

cholinergic stimulation. The curve exhibited by the rat mandibular gland (FIGURE 8) [62] is similar except that stimulation of ductal K-H exchange is much more apparent.

Most of the curves depicted in FIGURE 8 can be explained in terms of the above mechanisms, the principal variables being the presence, and strength, of the K-H exchange pump, and the extent to which the Na carrier transports Cl or K rather than H. The curves for the sheep parotid, however, present some problems of interpretation. The salivary HCO_3 concentration (~ 100–120 mmol l^{-1}) is greater than the possible maximum in the primary fluid (TABLE 1), so one must invoke ductal HCO_3 secretion. However, in Na-replete animals, the gland is unusual in giving no evidence whatsoever of ductal Na reabsorptive or K secretory activity: in consequence one must invoke isotonic ductal secretion of $NaHCO_3$ (cf. pancreas) or HCl absorption, which would be equivalent to Cl-HCO_3 exchange. Even in Na-depleted sheep, where ductal Na-K exchange is conspicuous, a similar mechanism has to be invoked since the Na-K exchange seems to be closely coupled in a 1:1 ratio. Whatever the HCO_3 secretion mechanism in the sheep parotid, it seems to be rather different from that depicted in FIGURE 5.

ACKNOWLEDGMENT

We thank Mrs. P. Fabiano for her expert technical assistance.

REFERENCES

1. AUGUSTUS, J. 1976. Biochem. Biophys. Acta **419:** 63–75.
2. AUGUSTUS, J., J. BIJMAN, C. H. VAN OS & J. F. G. SLEGERS. 1977. Nature **268:** 657–658. London, England.
3. AUGUSTUS, J., J. BIJMAN & C. H. VAN OS. 1978. J. Membrane Biol. **43:** 203–226.
4. BRUCKMAN, J. E. & J. A. MANGOS. 1970. Proc. Soc. Exp. Biol. Med. **134:** 606–609.
5. COMPTON, J. S., J. NELSON, R. D. WRIGHT & J. A. YOUNG. 1980. J. Physiol. London, England. In press.
6. COOK, D. I. & J. A. YOUNG. 1977. Proc. Aust. Physiol. Pharmacol. Soc. **8:** 3P.
7. DAWES, C. 1974. Arch. Oral Biol. **19:** 887–895.
8. DENNISS, A. R. & J. A. YOUNG. 1975. Pflügers Arch. **357:** 77–89.
9. ENDRE, Z. H. 1972. B.Sc. (medical) thesis. University of Sydney. Sydney, Australia.
10. FIELD, M. J. 1972. B.Sc. thesis. University of Sydney. Sydney, Australia.
11. FIELD, M. J. & J. A. YOUNG. 1973. Pflügers Arch. **345:** 207–220.
12. FRÖMTER, E. & J. M. DIAMOND. 1972. Nature New Biol. **235:** 9–13.
13. FRÖMTER, E., B. GEBLER, K. SCHOPOW & H. POCKRANDT-HEMSTEDT. 1974. *In* Secretory Mechanisms of Exocrine Glands. N. A. Thorn & O. H. Petersen, Eds.: 496–513. Munksgaard. Copenhagen, Denmark.
14. GRANT, S. B. J., Z. H. ENDRE & J. A. YOUNG. 1974. Proc. Aust. Physiol. Pharmacol. Soc. **5:** 206–208P.
15. GRUBER, W. D., H. KNAUF & E. FRÖMTER. 1973. Pflügers Arch. **344:** 33–49.
16. KALADELFOS, G. & J. A. YOUNG. 1973. Pflügers Arch. **341:** 143–154.
17. KALADELFOS, G. & J. A. YOUNG. 1974. Aust. J. Exp. Biol. Med. Sci. **52:** 67–79.
18. KAY, R. N. B. 1960. J. Physiol. **150:** 515–537. London, England.
19. KNAUF, H. 1972. Pflügers Arch. **333:** 82–94.
20. KNAUF, H. 1972. Pflügers Arch. **333:** 326–336.
21. KNAUF, H. 1974. *In* Secretory Mechanisms of Exocrine Glands. N.A. Thorn & O. H. Petersen, Eds.: 535–548. Munksgaard. Copenhagen, Denmark.
22. KNAUF, H. & E. FRÖMTER. 1970. Pflügers Arch. **316:** 238–258.
23. KNAUF, H. & E. FRÖMTER. 1970. Pflügers Arch. **316:** 259–274.
24. KNAUF, H. & E. FRÖMTER. 1971. *In* Electrophysiology of Epithelial Cells. G. Giebisch, Ed.: 187–206. Schattauer. Stuttgart, West Germany.
25. KNAUF, H. & R. LÜBCKE. 1975. Pflügers Arch. **361:** 55–60.
26. KNAUF, H., P. RÖTTGER, U. WAIS & K. BAUMANN. 1975. Fortschr. Zool. **23:** 307–321.
27. KNAUF, H., B. SIMON & U. WAIS. 1976. N.-S. Arch. Exp. Path. Pharmakol. **292:** 189–192.
28. KNAUF, H., U. WAIS, R. LÜBCKE & G. A. ALBIETZ. 1976. Eur. J. Clin. Invest. **6:** 43–50.
29. KNAUF, H., K. PANDER & S. FRANCK. 1976. Eur. J. Clin. Invest. **6:** 17–20.

30. Knauf, H., R. Lübcke, W. Kreutz & G. Sachs. 1980. Submitted for publication.
31. Lundberg, A. 1958. Physiol. Rev. **38:** 21–40.
32. Mangos, J. A. & N. R. McSherry. 1969. Proc. Soc. Exp. Biol. Med. **132:** 797–801.
33. Mangos, J. A., G. Braun & K. F. Hamann. 1966. Pflügers Arch. **291:** 99–106.
34. Mangos, J. A., N. R. McSherry, S. Nousia-Arvanitakis & K. Irwin. 1973. Am. J. Physiol. **225:** 18–24.
35. Mangos, J. A., N. R. McSherry, K. Irwin & R. Hong. 1973. Am. J. Physiol. **225:** 450–455.
36. Mangos, J. A., M. T. Garrish, R. Wells, W. Farnham & C. Bouchlas. 1978. J. Dent. Res. **57:** 818–825.
37. Martin, C. J. & J. A. Young. 1971. Pflügers Arch. **324:** 344–360.
38. Martin, C. J. & J. A. Young. 1971. Pflügers Arch. **327:** 303–323.
39. Martin, C. J., E. Frömter, B. Gebler, H. Knauf & J. A. Young. 1973. Pflügers Arch. **341:** 131–142.
40. Martinez, J. R. & A. M. Martinez. 1974. Arch. Oral Biol. **19:** 57–60.
41. Martinez, J. R., H. Holzgreve & A. Frick. 1966: Pflügers Arch. **290:** 124–133.
42. Petersen, O. H. 1976. Physiol. Rev. **56:** 535–577.
43. Roberts, M. L., N. Iwatsuki & O. H. Petersen. 1978. Pflügers Arch. **376:** 159–167.
44. Samuels, P. J. A. 1977. B.Sc. thesis. University of Sydney. Sydney, Australia.
45. Schneyer, L. H. 1968. Am. J. Physiol. **215:** 664–670.
46. Schneyer, L. H. 1969. Am. J. Physiol. **217:** 1324–1329.
47. Schneyer, L. H. 1970. Am. J. Physiol. **219:** 1050–1055.
48. Schneyer, L. H. 1974. Am. J. Physiol. **227:** 606–612.
49. Schneyer, L. H. 1976. Am. J. Physiol. **230:** 341–345.
50. Schneyer, L. H. 1977. Am. J. Physiol. **233:** F22–F28.
51. Schneyer, L. H. & C. A. Schneyer. 1960. Am. J. Physiol. **199:** 649–652.
52. Slegers, J. F. G., W. M. Moons, P. P. Idzerda & A. M. Stadhouders. 1975. J. Membrane Biol. **25:** 213–236.
53. Sommer, H. M., D. Kaiser & E. Drack. 1975. Pflügers Arch. **355:** 353–360.
54. Thaysen, J. H. 1960. Handb. Exp. Pharmak.**13:** 424–463.
55. Thaysen, J. H., N. A. Thorn & I. L. Schwartz. 1954. Am. J. Physiol. **178:** 155–159.
56. Wais, U. & H. Knauf. 1975. Pflügers Arch. **361:** 61–64.
57. Yoshimura, H., H. Iwasaki, T. Nishikawa & S. Matsumoto. 1959. Jap. J. Physiol. **9:** 106–123.
58. Young, J. A. 1968. Pflügers Arch. **303:** 366–374.
59. Young, J. A. 1979. Internat. Rev. Physiol. **19:** 1–58.
60. Young, J. A. & E. W. van Lennep. 1978. The Morphology of Salivary Glands. Academic Press. London, England.
61. Young, J. A. & E. W. van Lennep. 1979. *In* Membrane Transport in Biology. Vol. 4B: 563–674. Springer. New York, N.Y.
62. Young, J. A. & C. J. Martin. 1971. Pflügers Arch. **327:** 285–302.
63. Young, J. A. & E. Schögel. 1966. Pflügers Arch. **291:** 85–98.
64. Young, J. A., E. Frömter, E. Schögel & K. F. Hamann. 1967. Pflügers Arch. **295:** 157–172.
65. Young, J. A., C. J. Martin, M. Asz & F. D. Weber. 1970. Pflügers Arch. **319:** 185–199.
66. Conigrave, A. D., I. Novak, R. M. Case & J. A. Young. 1978. Proc. Aust. Physiol. Pharmacol. Soc. **9:** 115P.
67. Case, R. M., A. D. Conigrave, I. Novak & J. A. Young. 1980. J. Physiol. London, England. In press.
68. Wright, R. D. Personal communication.

Discussion

W. McD Armstrong: Is anything known about the thermodynamic requirements for transport in your system?

J. A. Young: There is an enormous literature on intracellular electrical potentials in the endpieces, but, with one exception, no one has succeeded in getting an electrode into the lumen. So, we know a lot about the basal membrane properties of the endpiece cells, but we don't know anything about the transepithelial properties. In the ducts, the position is exactly the reverse. The ducts that can be isolated *in vitro* and perfused are surrounded by a dense coat of connective tissue. The nerve supply to the gland, both cholinergic and adrenergic, comes in along the ducts and it has been so far impossible to make intracellular recordings from the duct cells. However, transepithelial potentials are readily obtained. Dr. Frömter may have found something new but, so far, I have seen no intracellular recordings from duct cells.

BICARBONATE TRANSPORT IN THE EXOCRINE PANCREAS

I. Schulz

Max-Planck-Institut für Biophysik
Kennedyallee 70
6000 Frankfurt am Main 70, West Germany

Introduction

Electrolyte secretion in the pancreas occurs in both acinar and duct cells. Whereas acinar cells secrete a fluid high in NaCl and low in $NaHCO_3$ in response to secretagogues of enzyme secretion,[1] a fluid rich in $NaHCO_3$ is secreted in duct cells elicited by secretin and related compounds.[2-4] In response to secretin, the flow rate of pancreatic juice increases and its HCO_3^- concentration reaches values about fivefold greater than those found in blood plasma. The polarity of electrical potential difference (lumen-negative against interstitium) together with the HCO_3^- concentration difference established by the stimulated cells allow the assumption that HCO_3^- transport into the duct lumen is an active process.

The Na^+ concentration of ~ 160 meq/1 does not change during stimulation and the secreted fluid is isotonic to blood plasma at all flow rates.[2] The sensitivity of electrolyte and fluid secretions to ouabain, and to change of the Na^+ concentration in the perfusate of an isolated pancreas, suggest that Na^+ is also actively transported. Electrolyte and fluid secretions can be elicited by the dibutyrylic analog of cAMP to the same extent, and with the same electrolyte pattern, as with secretin.[3] Similarly, cholera toxin, a stimulator of the catalytic unit of adenylate cyclase, can also elicit bicarbonate and fluid secretions.[5, 6]

Pancreata, in which acinar cells are destroyed due to a copper-deficient D-penicillamine diet given to rats for 100 days,[7, 8] respond to secretin with HCO_3^- and fluid secretion, but not to acetylcholine or pancreozymin with enzyme secretion. In these organs also, cAMP levels rise after secretin administration. Furthermore, an adenylate cyclase stimulated by secretin, but not by secretagogues of enzyme secretion like pancreozymin, is present in isolated duct cells.[9] These data indicate that duct cells are the site of $NaHCO_3$ secretion and that cAMP is the second messenger of secretin. Secretin receptors also have been found on acinar cells [10]; however, the function of these receptors has been attributed to being involved with enzyme secretion.[11] Although different attempts have been made to clarify the mechanism of secretin-stimulated $NaHCO_3$ secretion in duct cells, some of the main questions about the nature of HCO_3^- transport and the site of the active transport step in the duct cell remain unanswered. However, from all data available at present, a model involving H^+/HCO_3^- transport energized by ATP at the luminal cell membrane seems to be likely. In addition, at the contraluminal cell site, a secondary active transport of H^+ ion out of the cell in exchange for Na^+, whose electrochemical gradient is kept by the Na^+-K^+ ATPase, must be considered.

0077-8923/80/0341-0191 $01.75/0 © 1980, NYAS

Patterns and Site of Electrolyte Secretion

The most striking changes in the ionic composition of the secreted pancreatic juice involve HCO_3^- and Cl^-. At low secretory rates, the HCO_3^- concentration is similar to that in plasma. After stimulation with secretin, it rises with increasing flow rates asymptotically to a maximum that is different for different animal species (FIGURE 1). In carnivores such as cat and dog, it reaches 130 to 140 meq/1 [2, 12]; whereas in herbivores, like rat and rabbit, a maximum of about 80 meq/1 is reached.[1, 13] The Cl^- concentration pattern is mirror-like to that of HCO_3^-, the sum of both anions remaining constant at all flow rates. Secretion is iso-osmotic with plasma and the concentration of Na^+ is independent of flow rate, remaining at about 160 meq/1. In rats, the K^+ concentration of the secreted juice increases from a plasma level to twice

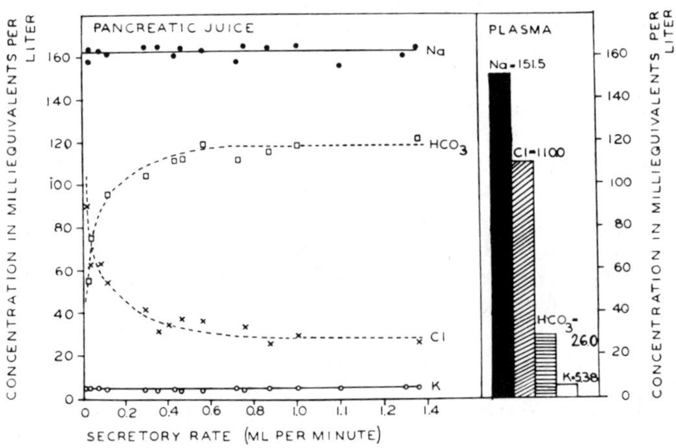

FIGURE 1. Relation between the rate of secretion and the concentration of sodium potassium, chloride and bicarbonate in the pancreatic juice of the dog. (From Bro-Rasmussen *et al.*[14])

that value when stimulated with secretin [1]; whereas in other animal species, like cat, dog and rabbit, it remains fairly constant at the plasma level.[14, 15] Species differences exist also between cat and rat concerning the response of electrolyte and fluid secretions to different secretagogues. While in cat, an HCO_3^--rich fluid secretion is evoked by secretin,[2] in rat only moderate fluid and HCO_3^- secretion can be stimulated by secretin.[1] In rat it is possible to stimulate fluid secretion with the enzyme secretagogues pancreozymin and caerulein even more than with secretin. But during pancreozymin stimulation, the HCO_3^- and Cl^- concentrations in the secreted fluid stay constant, similar to that in plasma (FIGURE 2).[1] These patterns of electrolyte secretions seem to reflect the site of their formation: pancreozymin- or acetylcholine-evoked secretion of plasma-like composition occurs in acinar cells, while secretin-evoked, bicarbonate-rich secretion is produced in duct cells. These conclusions have been confirmed by micropuncture experiments that demonstrate the

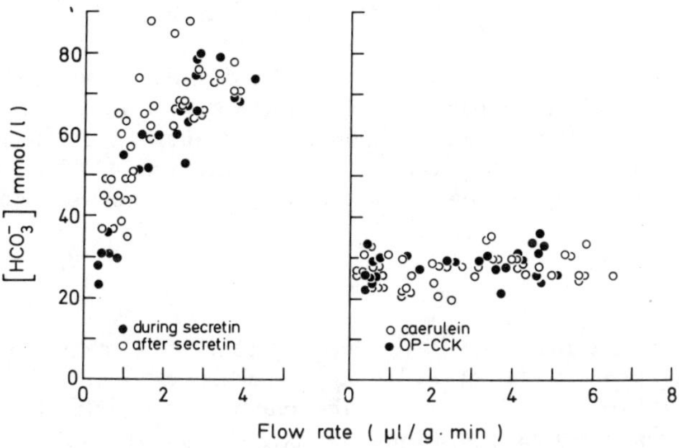

FIGURE 2. The electrolyte composition of pancreatic juice in the rat stimulated by, left, secretin and, right, caerulein, or OP-CCK, analogs of pancreozymin. (From Sewell and Young.[1])

primary secretion has a relatively high Cl^- concentration,[16, 17] which decreases along the duct system in response to secretin stimulation [16, 18] (FIGURE 3). The same conclusion has been derived by another approach involving selective destruction of acinar cells by feeding rats for about 2.5 months to 3 months with a copper-deficient diet supplemented with D-penicillamine.[7, 8] Such a pancreas, devoid of acinar cells, responds to secretin with an HCO_3^--rich fluid secretion; whereas secretion of plasma-like composition elicited by pancreozymin or acetylcholine, as well as enzyme secretion, cannot be evoked any more. These data also clearly demonstrate that $NaHCO_3$ together with fluid is secreted by duct cells.

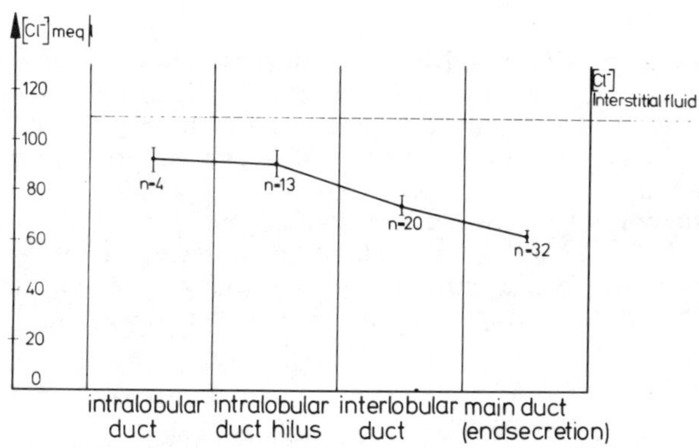

FIGURE 3. Free-flow chloride concentrations in micropuncture samples of pancreatic juice. The samples are obtained at different sites of the duct system after secretin stimulation. The chloride concentration of the interstitial fluid is marked by a horizontal dotted line. (From Schulz *et al.*[16]).

Ionic Requirements of Secretion

The ionic requirement for electrolyte and fluid secretions can best be studied in the artificially perfused cat pancreas[19] or in the rabbit pancreas kept *in vitro* and superfused with saline.[20] The question of the ionic requirement of secretion is closely related to the question of which of the main cations and anions appearing in pancreatic juice are actively transported.

Sodium Ions

If NaCl in the perfusate is replaced by sucrose, a transepithelial concentration difference for Na^+ of 110 meq/1 (perfusate 55 meq/1, secreted fluid 165 meq/1) can be maintained by the pancreas.[19] If Na^+ ions were to be transported passively, a transtubular electrical potential difference (*PD*) of up to 29 mV (lumen-negative) would be expected. The measured *PD* at free-flow conditions is only —5 to —9 mV[16, 18, 21] (TABLE 1), but it has not been measured at zero Na^+ and fluid net flux conditions. However, since the S/P ratio for Cl^- remains near 1.0,[19] it can be assumed that a rise of the transtubular electrical potential difference is not the reason for the accumulation of Na^+ ions in the duct system and active Na^+ transport is the more likely explanation.

TABLE 1

TRANSDUCTAL ELECTRICAL POTENTIAL DIFFERENCES (*PD*)

Site	At Rest	After Secretin	Researchers/Work Year
Intralob. duct	−5.0		Schulz *et al.*, 1969
(small)	−9.1	−9.1*	Swanson and Solomon, 1973
Interlob. duct	−2.4		Schulz *et al.*, 1969
(med. size)	−7.4	−7.4*	Swanson and Solomon, 1973
Main duct	−5.3	−6.0	Schulz *et al.*, 1969
(big)	+2.0	−4.9	Way and Diamond, 1970

* Uncorrected for diffusion potential.

Further support of active Na^+ transport comes from the observation that replacement of Na^+ by Li^+ causes inhibition of secretion by ~ 90 percent[15] and that the dose response curve of ouabain on pancreatic Na^+-K^+ ATPase coincides with that on resting fluid secretion of the isolated rabbit pancreas.[22]

Potassium Ions

Omission of potassium from the perfusate reduces secretion by 65 percent.[15] When the K^+ concentration in the perfusate is varied at the expense of Na^+, at 10 meq/1 K^+ in the perfusate, maximal Na^+ secretion is observed. At higher K^+ concentrations Na^+ secretion is diminished, paralleling the effect on secretory flow rate. The K^+ concentration in the juice is the same as in the perfusate over the whole concentration range. The effect of K^+ omission on one hand, and the stimulatory effect of 10 meq/1 of K^+ in the perfusate on secretion of Na^+ and fluid on the other hand, is to make the assumption that K^+ is also involved in active transport of Na^+. Since, however, the K^+ concentration in the secretion does not exceed that of the perfusate, one is rather inclined to attribute a secondary role to K^+; for instance, by an effect on the Na^+-K^+ ATPase, which is enriched in duct cells [40] and has been located at the contraluminal cell side.[23] If secondary active transport steps, as for instance a Na^+/H^+ exchange at the contraluminal cell side, are dependent on the electrochemical Na^+ gradient established by the Na^+-K^+ ATPase, anything that affects the Na^+-K^+ ATPase would also affect the secondary active Na^+/H^+ transport systems.

Chloride Ions

If Cl^- concentration in the perfusate is replaced by bromide or iodide, only a slight increase or decrease in the secretory rate was observed.[24] Even if Cl^- is replaced by sulfate or methylsulfate, secretion is maintained; although after some time there is a marked reduction in rate. In this situation, higher concentrations of bicarbonate appear at a given flow rate in the secretion. This behavior of tested anions indicates, too, that they play only a secondary role in electrolyte and fluid secretion.

Bicarbonate Ions

Bicarbonate secretion in the isolated cat pancreas is directly proportional to perfusate HCO_3^- concentration [26, 27] (FIGURE 4). Under normal conditions, almost all secreted bicarbonate is derived from blood and only a small extent (<6%) is contributed by CO_2 produced in pancreatic cells.[13, 18, 25–28] If HCO_3^- is omitted from the perfusate, secretion of electrolyte and fluid as stimulated with secretin stops.[26, 30] In contrast, NaCl and fluid secretion as stimulated by secretagogues of enzyme secretion in acinar cells is not dependent on HCO_3^- in the extracellular medium.[29] Since at highest secretory flow rates HCO_3^- concentration can reach values about 5 times the plasma concentrations, most interest has been focussed on the question about the mechanism of bicarbonate transport.

Measurements of transepithelial electrical potential differences have shown that the duct lumen is slightly electrically negative as compared to the interstitium (—5 to —9 mV) [16, 18] and a slight increase in electronegativity after secretin stimulation was observed. Way and Diamond [21] measured +2 mV at rest and —4.9 mV after secretin stimulation. Taken together, these data indicate that HCO_3^- transport occurs against an electrochemical gradient and is an active process.

Cellular Mechanism and Sidedness of Buffer Secretion

Concerning the mechanism and sidedness of buffer secretion, the following questions are of main interest:

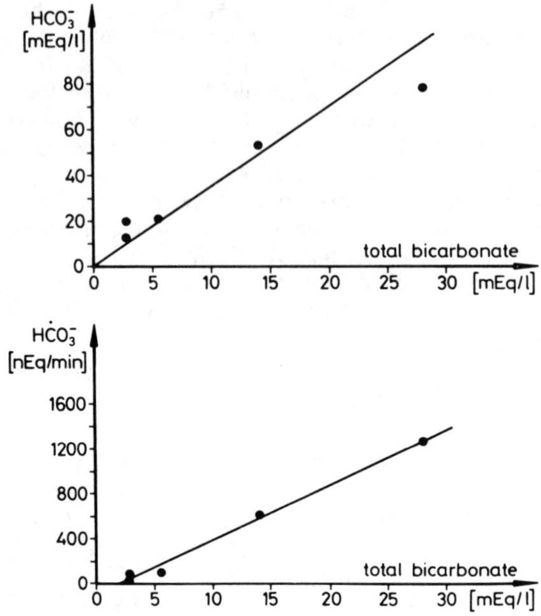

FIGURE 4. Top: Influence of bicarbonate concentration of the perfusate (x-axis) on the HCO_3^- concentration of the secretion (y-axis). Bottom: Influence of the bicarbonate concentration of the perfusate (x-axis) on the secretory rate of HCO_3^- (y-axis). (From Schulz.[30])

1. What buffer component is transported actively, i.e., either HCO_3^- secretion or H^+ reabsorption (OH^- secretion) respectively.

2. At which cell side does active transport take place?

3. a) Is it a primary active transport, i.e.: does it use ATP directly and by what ATPase? b) Is it a secondary active transport; e.g., Na^+/H^+ or HCO_3^-/Cl^- exchange: with the energy derived from an ion gradient that is maintained by a primary active transport system; e.g., the Na^+-K^+ ATPase? Thus far, these questions have been answered only indirectly.

In regard to the first question, evidence that H^+ or OH^- is transported primarily came from the finding that weak organic acids, such as sulfonamide glycodiazine[30] or the organic acid anions formate, propionate, butyrate and acetate[28, 31, 32] can substitute for bicarbonate in promoting pancreatic buffer and fluid secretions. Similarly, as observed with HCO_3^-,[26, 30] an increase in the buffer anion concentration (A^-) of glycodiazine[30] and of acetate[28] in the perfusate increased buffer secretion. But at a given bicarbonate, glycodiazine or acetate concentration, the buffer secretion increased if the perfusate was made more alkaline, i.e., if the concentration of the undissociated buffer was diminished (TABLE 2). Under special circumstances, however, an increase in the concentration of the undissociated buffer component in the perfusate also can augment buffer secretion. Thus, with sulfamerazine, a weak acid, of which only the undissociated buffer component is lipid soluble, secretion of the buffer was dependent on the concentration of its undissociated component in the perfusate.[27] Similarly an increase in pCO_2 of low HCO_3^- perfusate was also shown to stimulate HCO_3^- secretion.[30] These results were interpreted with a model, in which it was tentatively assumed that the active transport step is located at the luminal cell side, transferring H^+ from the lumen into the cell or OH^- ions from the cell into the lumen, respectively. In the case of lipid-soluble anion, such as glycodiazine, acetate or HCO_3^- itself, the contraluminal cell side could be penetrated and H^+ ions be buffered in the cell. Thus, the anion concentration of the perfusate is rate-limiting for buffer secretion. If only the undissociated component was lipid soluble; as in the case of sulfamerazine, this buffer component would determine the rate of buffer secretion, since it can penetrate both cell membranes and buffer OH^- ions produced at the luminal cell side.

Another attempt, also indirect, to localize the cell side of the active transport step, was based on the knowledge of transcellular electrical potential profile of small ducts and the subjective estimation of intracellular pH by means of the indicator, bromthymol blue. Schulz *et al.*[16] and Swanson and Solomon [18] reported values for the lumen and cell of small ducts of —5 and —9 mV (lumen-interstitium) and —50 and —40 mV (cell-interstitium), respectively. Knowing the intraluminal HCO_3^- concentration of maximal 100 meq/1 at a contraluminal HCO_3^- concentration of 28 meq/1, Schulz *et al.*[33] have argued that the active transport step must be located at the contraluminal cell side if the intracellular HCO_3^- concentration is higher than 18 meq/1, i.e., if the intracellular pH is higher than 7.27, but it must be located at the luminal cell side if the intracellular bicarbonate concentration is lower than 4.2 meq/1, i.e., if the intracellular pH is lower than 6.65 (FIGURE 5). From the subjective estimation of the cell color in experiments in which bromthymol blue has been added to the perfusate or was injected into the duct system, they estimated the intracellular pH of the small duct cells to be far lower than 7.0, and favored the hypothesis that the active transport step is located at the luminal cell side.

Since, obviously, the determination of the electrochemical potential difference for HCO_3^- or H^+/OH^- at each cell side of the duct cell is very difficult to achieve, and no direct determinations in duct cells have been made so far, we have used a biochemical approach, which involves:

1. Isolation and separation of single, dispersed duct cells.
2. Determination of transport enzymes in duct cells.

TABLE 2

EFFECT OF BUFFER COMPONENTS ON BUFFER SECRETORY RATE

Buffer	Rise in Perfusate A^- (AH=const.)	Rise in Perfusate pH (A^-=const.)	Researchers/Work Year
HCO_3^-/CO_2	↑ *	↑	Swanson and Solomon, 1975
Glycodiazine	↑	↑	Schulz, 1971
Acetate	↑		Swanson and Solomon, 1975
Acetate		↑	Hadi *et al.*, 1976

* ↑=Increase in secretion.

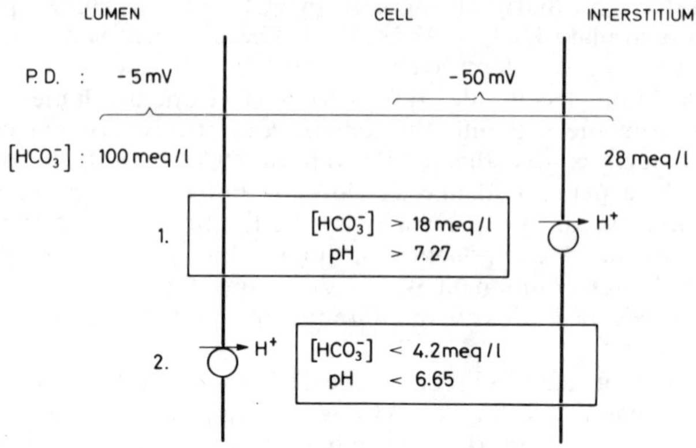

FIGURE 5. Theoretical possibilities for the location of the active transport step of H^+ ions in the duct cell. 1. If the active transport step is located at the contraluminal plasma membrane and the passive HCO_3^- distribution takes place over the luminal membrane, intracellular HCO_3^- concentration should be 18 meq/1 or higher (pH>7.27). 2. If active H^+ transport takes place at the luminal membrane and passive distribution at the contraluminal membrane, HCO_3^- concentration should be 4.2 meq/1 or lower (pH<6.65).

3. Estimation of the intracellular pH by measuring the distribution of nicotine bitartrate.

4. Evaluation of the effects of compounds of known action on intracellular pH and the activity of membrane-bound enzyme, and

5. Estimation of the cellular potential difference by measuring the distribution of triphenylphosphonium (TPP^+).

Preparation of Dispersed Duct Cells from the Pancreas and Enzymatic Characterization

Cells were isolated by standard methods using collagenase digestion with an interposed step of EDTA treatment to dissociate the cells. Heavier acinar cells were then separated from smaller cells by centrifugation through 4% albumin.[34] Cells which had not passed the albumin, including duct cells and smaller acinar cells, were further separated using a Beckman Elutriator rotor.

Purification of duct cells obtained by this method is shown in FIGURE 6. Electronmicroscopy shows well-preserved cells with morphological characteristics of acinar and duct cells (FIGURE 7 and 8, respectively). The most prominent features are the zymogen granules in acinar cells (FIGURE 7), the absence of zymogen granules and the presence of big nuclei with one or more indentations in duct cells (FIGURE 8).

Hormone-stimulated adenylate cyclase activity in membranes isolated from acinar and duct cells suggests the presence of well-preserved receptors in both types of cells (FIGURE 9). This enzyme showed very different characteristics in acinar and duct cells. The most striking difference is the absence of pancreozymin-stimulated adenylate cyclase activity in duct cells and the strong stimulation found in acinar cells. In contrast to pancreozymin, which stimulates only the acinar adenylate cyclase, secretin is a potent stimulant of both adenylate cyclases in acinar and duct cells. Both enzymes differ, however, in response to secretin. Half-maximal stimulatory secretin concentration is 10 times lower in duct than in acinar cells. Maximally stimulated enzyme activity, however, is about 30% lower in duct cells as compared with acinar cells.

The enzyme distribution pattern reveals that duct cells are enriched with Mg^{2+} ATPase, Na^+-K^+ ATPase, alkaline phosphatase and 5′-nucleotidase by factors of 26, 26, 7 and 6, respectively, as compared to activity/protein in unseparated cells (FIGURE 10). Since per number of cells, the protein content is ~10 times higher in acinar than in duct cells, the respective enrichment factors per an equal number of cells are 2.6, 2.6, 0.7, and 0.6. Most striking is the absence of anion-stimulated, Mg^{2+}-dependent ATPase in duct cells, usually called "HCO_3^- ATPase," which often had been attributed with involvement in HCO_3^- transport in the pancreas.[35, 36] This assumption was based on findings showing enrichment of both HCO_3^- ATPase and the plasma membrane marker enzyme Na^+-K^+ ATPase in a cell membrane fraction prepared from the total tissue homogenate.[36] Recently, however, Van Amelsvoort *et al.* presented evidence for exclusive location of the pancreatic HCO_3^- ATPase in mitochondria and not in plasma membranes.[37]

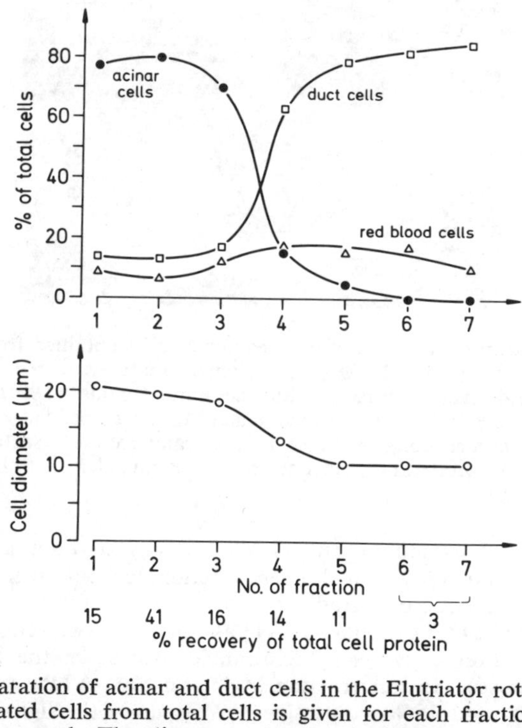

FIGURE 6. Separation of acinar and duct cells in the Elutriator rotor. Upper panel: Percent of separated cells from total cells is given for each fraction obtained from the rotor. Lower panel: The diameter of cells obtained and the recovery of cell protein in percent of total cell protein is given for each fraction.

TABLE 3 shows that a Mg^{2+} or Ca^{2+}-activated ATPase in the total homogenate of unfractionated pancreatic cells can be stimulated by oxyanion sulfite by 34% and 37%, respectively. Acinar cells, which comprise about 90% of pancreatic tissue, show similar degree of stimulation by sulfite as unseparated cells, whereas in duct cells stimulation by sulfite is absent. The Mg^{2+}-ATPase activity is rather slightly inhibited, whereas Ca^{2+} ATPase activity is slightly stimulated by SO_3^{2-} ions. However, since the specific activity of the Mg^{2+} and Ca^{2+} ATPase is higher in duct cells than in acinar cells, a small effect of SO_3^{2-} is difficult to assess; and, at present, we are not able to show a

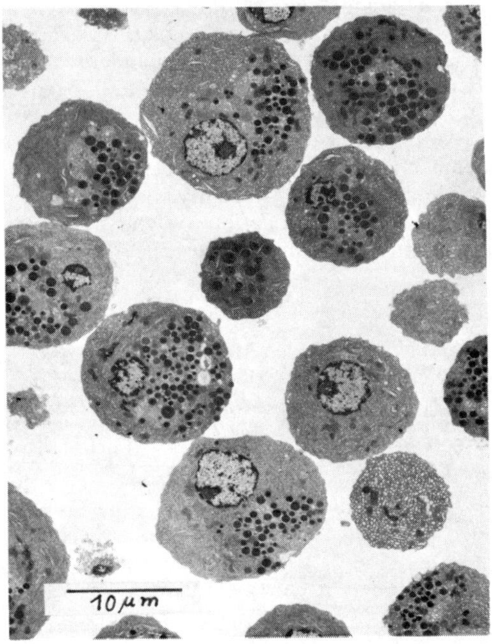

FIGURE 7. Electron micrograph of isolated acinar cells obtained from the Elutriator rotor. Cells were fixed with 2.5% glutaraldehyde buffered with 0.1 mol/1 sodium cacodylate (pH 7.4) and post-fixed with osmium tetroxide (1% OsO_4, 0.1 mol/1 cacodylate buffer, pH 7.4). Dehydrated material was embedded in a plastic resin (Spurr's medium) and sectioned with an LKB ultramicrotome. Sections were double stained with uranyl acetate and lead citrate and examined in a Philips 300 electron microscope, × 1,470.

significant effect of oxyanions on ATPase activity in duct cells. When cell homogenates of acinar cells had been fractionated following centrifugation through a three-step sucrose gradient (15%, 31%, 60% sucrose, w/w) for 3 hr, nearly total SO_3^{2-}-stimulated ATPase activity cofractionated with the mitochondrial marker enzyme, cytochrome *c* oxidase in the heavier fraction (31%/60% sucrose). In this fraction Mg^{2+} and Ca^{2+} ATPase were stimulated by SO_3^{2-} by 91% and 75%, respectively, whereas in the lighter fraction "cell membranes" only 7% and 6%, respectively, could be achieved. This suggests that in acinar cells SO_3^{2-} or (HCO_3^-)-stimulated ATPase is located in mitochondria.

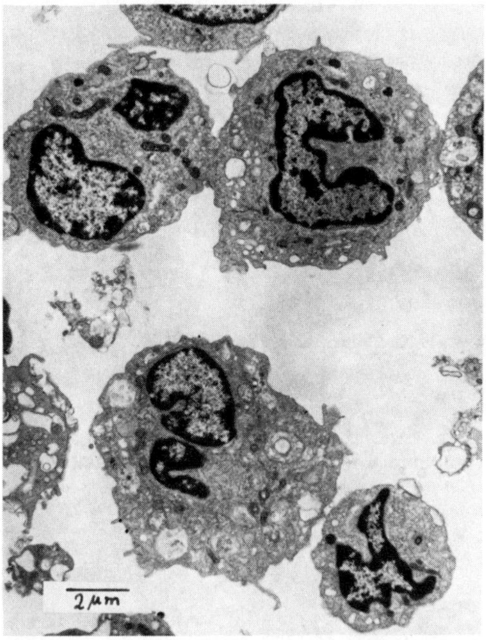

FIGURE 8. Electron micrograph of isolated duct cells from the Elutriator rotor. The preparation for electron microscopy was the same as that described in FIGURE 7, × 4,550.

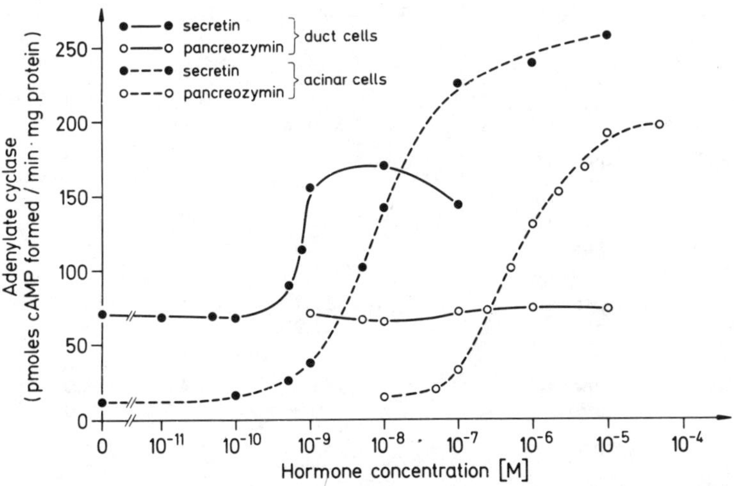

FIGURE 9. The effect of different concentrations of secretin and pancreozymin on adenylate cyclase activity in cell membranes of acinar and duct cells obtained from the Elutriator rotor. Cells were broken by hypotonic shock and mild homogenization and the membranes were purified by centrifugation on a 15/31% sucrose step gradient. Adenylate cyclase was determined as described.[36] Points represent the mean of four experiments and eight determinations.

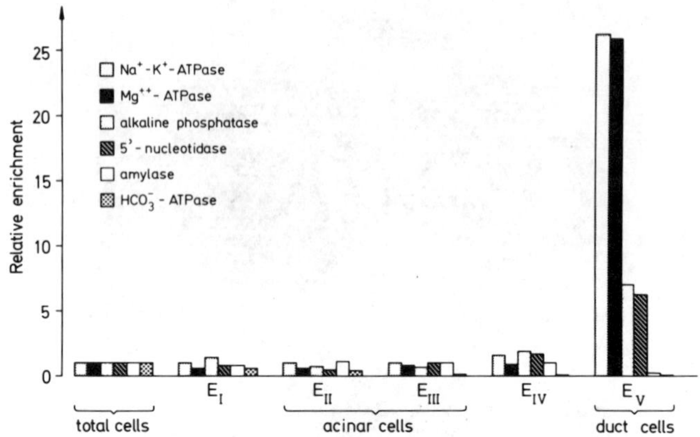

FIGURE 10. Relative enrichment of marker enzymes in cells separated on the Elutriator rotor as compared to unseparated total cells (activity/cell protein).

TABLE 3

DISTRIBUTION PATTERN OF ATPASES (μMOLES/MG PR. $\times$ H) AND CYTOCHROME *c* OXIDASE (U/MG PR.) IN PANCREATIC CELLS

		Mg^{2+} $(+Cl^-)$	Mg^{2+} $(+SO_3^{2-})$	Ca^{2+} $(+Cl^-)$	Ca^{2+} $(+SO_3^{2-})$	cyt. *c* oxidase
Pancreatic cells (unseparated)	spec. act.	5.9	7.9	4.4	6.0	0.32
	ratio		1.34		1.37	
Acinar cells	spec. act.	4.4	6.4	4.4	7.2	0.44
	ratio		1.44		1.64	
Acinar "cell membr." (15/31%)	spec. act.	20.2	21.7	20.0	21.2	0.02
	ratio		1.07		1.06	
Acinar "mitoch." (31/60%)	spec. act.	5.8	11.1	6.3	11.03	2.80
	ratio		1.91		1.75	
Duct cells	spec. act.	128.2	123.2	131.4	138.0	2.20
	ratio		0.96		1.05	
Duct "cell. membr." (15/31%)	spec. act.	302.3	284.0	278.3	285.2	1.29
	ratio		0.94		1.02	
Duct "mitoch." (31/60%)	spec. act.	145.2	145.2	167.8	167.8	7.00
	ratio		1.00		1.00	

In duct cells, further enrichment of Mg^{2+} and Ca^{2+} ATPase in the lighter membrane fraction can be achieved by fractionation. In the heavier fraction, enriched by cytochrome *c* oxidase, only a small increase in Mg^{2+} and Ca^{2+} ATPase activity was observed, suggesting its main location in cell membranes and not in mitochondria. SO_3^{2-} stimulation was not present in either fraction. Since HCO_3^- transport has been localized in duct cells,[16] the absence of an HCO_3^- ATPase in this type of cell supports the conclusion of Van Amelsvoort *et al.* that this enzyme is not involved in pancreatic bicarbonate secretion.[37]

Determination of Intracellular pH by the Distribution of ³H-Nicotine-Bitartrate

For determination of the intracellular pH (pH_i) nicotine was used, since in contrast to DMO (5,5′-dimethyl-2,4-oxazolidinedione) it accumulates in cells, whose content is acidic.[38]

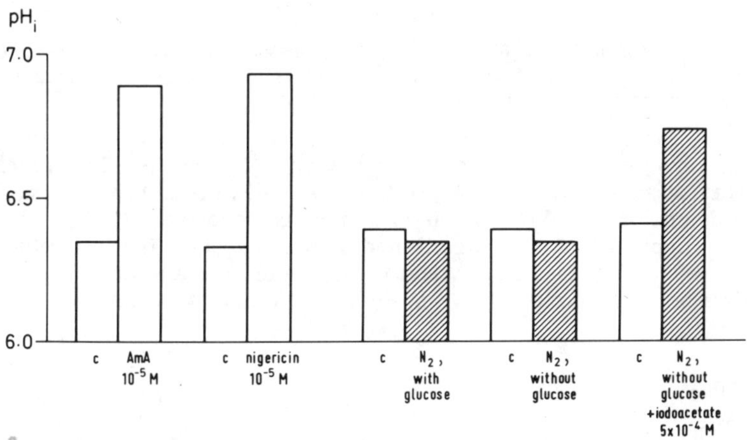

FIGURE 11. The effect of indicated substances and conditions on pH_i of duct cells. Each experiment is representative for three others, c=control, AmA=antimycin A.

Duct cells were incubated in Krebs-Ringer-bicarbonate solution (25 mmol/1 HCO_3^- 5% CO_2) in the presence of 10^{-5} mol/1 ³H-nicotine-bitartrate and ¹⁴C-inulin for 20 min. Then they were centrifuged in glass tubes of ~1 mm in diameter for 2 min at 10,000 rpm to obtain a tightly packed cell pellet. The tip of the glass tube was cut through the pellet and weighed to obtain the wet weight. After subtraction of the weight of the glass tip, the intracellular space was calculated after subtraction of the separately determined dry weight and correction for the inulin space.

The calculated value for intracellular pH (pH_i) is 6.4. In the presence of the redox inhibitor antimycin A and the ionophoric cation exchanges nigericin, alkalinization of the cells was observed (FIGURE 11). Inhibition of glycolysis by iodoacetate also caused alkalinization, whereas absence of oxygen in the

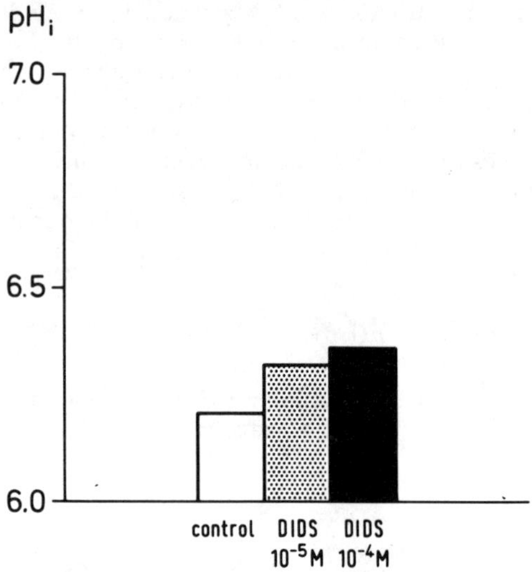

FIGURE 12. Intracellular pH of isolated pancreatic duct cells as calculated by nicotine distribution. The effect of DIDS on pH$_i$ of duct cells. This represents one experiment representative of three similar experiments.

presence or absence of glucose was not enough to increase the intracellular pH. The substance DIDS (4,4-diisothio cyano-dihydro stilbene-2,2′-disulfonic acid), an analog of SITS, which also inhibits pancreatic HCO_3^- and fluid secretion (unpublished observation) had a small effect on the intracellular pH of duct cells (FIGURE 12). Acetazolamide and ouabain had no effect on pH$_i$ (data not shown). As shown in FIGURES 13 and 14, secretin (10^{-6} mol/l)

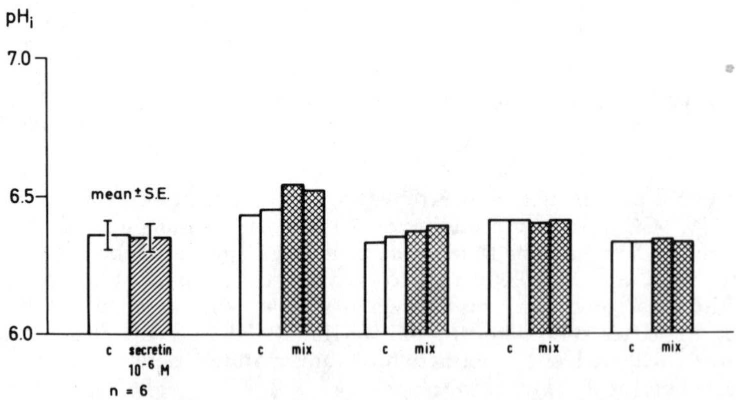

FIGURE 13. Intracellular pH (pH$_i$) in duct cells as evaluated according to the ^{14}C-nicotine distribution, the effect of secretin, and a mixture of secretin 10^{-6} mol/1+ dibutyryl cAMP 5×10^{-4} mol/1+theophylline 5×10^{-3} mol/1 (mix) on pH$_i$. For the effect of mix, four experiments, each made in duplicates, are shown.

or a mixture of secretin, dibutyrylic cAMP and theophylline had no effect on the intracellular pH, but it depolarized the cell by about 10 mV.

Model for HCO_3^- Secretion in the Pancreatic Duct Cell

Although insufficient knowledge about the single steps involved in HCO_3^- and fluid secretions in the duct system does not yet allow a definite description of the nature and site of the active transport step, the following model already proposed by Swanson and Solomon in 1975 [39] is consistent with the data obtained so far.

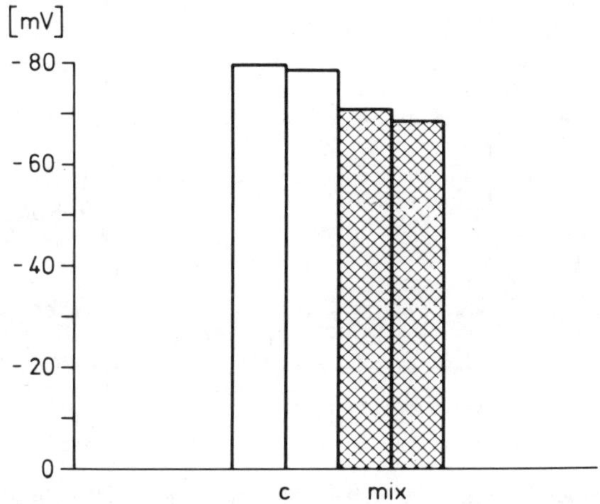

FIGURE 14. Cell potential in duct cells as evaluated according to TPP^+ distribution and the effect of mix on it. This represents one experiment representative of two similar experiments. c=control, mix=10^{-6}M secretin + 5×10^{-4}M dbc AMP + 5×10^{-3}M theophylline.

For buffer secretion in the duct cell, we have to postulate at least one active transport step through one cell side, combined with a passive step through the other. Considering enrichment of Na^+-K^+ ATPase in duct cells and the strong Na^+ dependence and ouabain sensitivity of buffer transport on the one hand, and the fact that active H^+ (or OH^-) transport rather than HCO_3^- transport takes place, the simplest model would consist of a combination of the Na^--K^+ ATPase and Na^+/H^+ exchange located at the contraluminal cell side with a passive exit step of HCO_3^- at the luminal cell side (FIGURE 15). Na^+ could move into the lumen paracellularly. If, however, the low intracellular pH of 6.4 measured in isolated duct cells is not solely due to metabolic acidification; but at least partly due to active H^+ transport into the cell, we would have to postulate, in addition to the above-mentioned model, an ATP-driven H^+ pump, possibly an electrogenic H^+ or HCO_3^- pump at the luminal cell side. Such a transport could be achieved by the Mg^{2+} ATPase, which we have found within duct cells in high activity. Inhibition of this ATPase by DIDS by about 30% (data not shown) as well as alkalinization of duct cells by DIDS would support such a model. Furthermore, stimulation of the pump by secretin could be responsible for the observed cell depolarization.

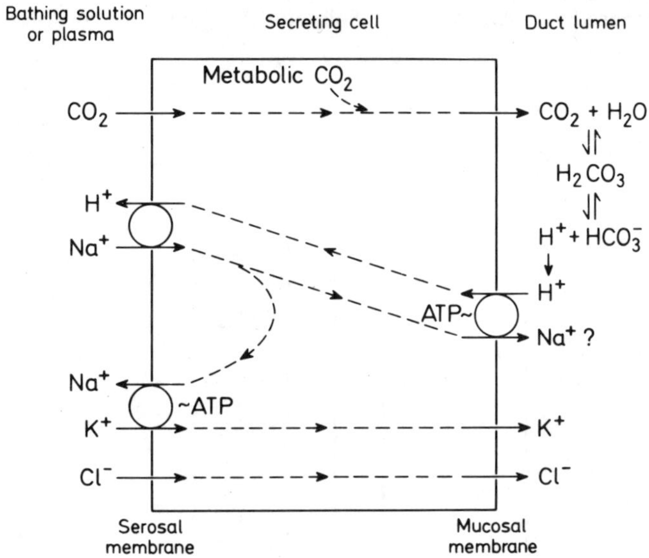

FIGURE 15. Model for the mechanism of pancreatic electrolyte and fluid secretion. A primary active electrogenic H^+ or OH^- transport is located at the luminal cell side. If it is coupled to Na^+, the stoichiometry of H^+/Na^+ should be higher than 1. The coupled Na^+/H^+ exchange at the contraluminal cell side is a secondary active transport process. This model is the same as that proposed by Swanson and Solomon.[32]

Acknowledgments

I want to thank Professor Karl Ullrich for valuable discussions and support. The technical assistance of Klaus Heil, Gerhard Rumrich, Sonja Klöss, Celia Hornung and Jutta Wiechmann is gratefully acknowledged. The electronmicroscopy was performed by Dr. Winfried Haase.

References

1. SEWELL, W. A. & J. A. YOUNG. 1975. Secretion of electrolytes by the pancreas of the anaesthetized rat. J. Physiol. **252:** 379–396.
2. CASE, R. M., A. A. HARPER & T. SCRATCHERD. 1969. The secretion of electrolytes and enzymes by the pancreas of the anaesthetized cat. J. Physiol. **201:** 335–348.

3. CASE, R. M. & T. SCRATCHERD. 1972. The actions of dibutyryl cyclic adenosine 3′,5′-monophosphate and methylxanthines on pancreatic exocrine secretion. J. Physiol. **223:** 649–667.
4. SCRATCHERD, T., R. M. CASE & P. A. SMITH. 1975. A sensitive method for the biological assay of secretin and substances with "secretin-like" activity in tissues and biological fluids. Scand. J. Gastroenterol. **10:** 821–828.
5. SMITH, P. A. & R. M. CASE. 1975. Effects of cholera toxin on cyclic adenosine 3′,5′-monophosphate concentration and secretory processes in the exocrine pancreas. Biochim. Biophys. Acta **399:** 277–290.
6. KEMPEN, H. J. M., J. J. H. H. M DE PONT & S. L. BONTING. 1975. Rat pancreas adenylate cyclase III. Its role in pancreatic secretion assessed by means of cholera toxin. Biochim. Biophys. Acta **392:** 276–287.
7. FÖLSCH, U. R. & W. CREUTZFELDT. 1975. Electrolyte secretion by a pancreatic duct model in the rat *in vivo* and accumulation of cyclic adenosine 3′,5′-monophosphate *in vitro* in response to gastrointestinal hormones. *In* Stimulus-Secretion Coupling in the Gastrointestinal Tract. R. M. Case & H. Goebell, Eds. pp. 381–384. MTP Press. Lancaster, England.
8. FÖLSCH, U. R. & W. CREUTZFELDT. 1977. Pancreatic duct cells in rats: secretory studies in response to secretin, cholecystokinin-pancreozymin and gastrin *in vivo*. Gastroenterology **73:** 1053–1059.
9. MILUTINOVIĆ, S., I. SCHULZ, G. ROSSELIN & H. FASOLD. 1977. The interaction of pancreatic secretagogues with pancreatic plasma membranes. *In* Hormonal Receptors in Digestive Tract Physiology. S. Bonfils, P. Fromageot & G. Rosselin, Eds. pp. 213–226. North-Holland. Amsterdam, the Netherlands.
10. CHRISTOPHE, J. P., T. P. CONLON & J. D. GARDNER. 1976. Interaction of porcine vasoactive intestinal peptide with dispersed pancreatic acinar cells from the guinea pig. J. Biol. Chem. **251:** 4629–4634.
11. GARDNER, J. D. & M. J. JACKSON. 1977. Regulation of amylase release from dispersed pancreatic acinar cells. J. Physiol. **270:** 439–454.
12. HART, W. M. & J. E. THOMAS. 1945. Bicarbonate and chloride of pancreatic juice secreted in response to various stimuli. Gastroenterology **4:** 409–420.
13. ROTHMAN, S. S. & F. P. BROOKS. 1965. Pancreatic secretion *in vitro* in "Cl^--free," "CO_2-free" and low Na^+ environment. Amer. J. Physiol. **209:** 790–796.
14. BRO-RASMUSSEN, F., S. A. KILLMANN & J. H. THAYSEN. 1956. The composition of pancreatic juice as compared to sweat, parotid saliva and tears. Acta Physiol. Scand. **37:** 97–113.
15. CASE, R. M. & T. SCRATCHERD. 1974. The secretion of alkali metal ions by the perfused cat pancreas as influenced by the composition and osmolality of the external environment and by inhibitors of metabolism and Na^+, K^+-ATPase activity. J. Physiol. **242:** 415–428.
16. SCHULZ, I., A. YAMAGATA & M. WESKE. 1969. Micropuncture studies on the pancreas of the rabbit. Pflügers Arch. **308:** 277–290.
17. MANGOS, J. A. & N. R. MCSHERRY. 1971. Micropuncture study of excretion of water and electrolytes by the pancreas. Amer. J. Physiol. **221:** 496–503.
18. SWANSON, C. H. & A. K. SOLOMON. 1973. A micropuncture investigation of the whole tissue mechanism of electrolyte secretion by the *in vitro* rabbit pancreas. J. Gen. Physiol. **62:** 407–429.
19. CASE, R. M., A. A. HARPER & T. SCRATCHERD. 1968. Water and electrolyte secretion by the perfused pancreas of the cat. J. Physiol. **196:** 133–149.
20. ROTHMAN, S. S. & F. P. BROOKS. 1965. Electrolyte secretion from the rabbit pancreas *in vitro*. Amer. J. Physiol. **208:** 1171–1176.
21. WAY, L. W. & J. M. DIAMOND. 1970. The effect of secretin on electrical potential differences in the pancreatic duct. Biochim. Biophys. Acta **203:** 298–307.
22. RIDDERSTAP, A. S. & S. L. BONTING. 1969. Na^+K^+ activated ATPase and exocrine pancreatic secretion *in vitro*. Amer. J. Physiol. **217:** 1721–1727.
23. POULSEN, J. H., M. BUNDGAARD & M. MØLLER. 1978. Localization of ($Na^+ + K^+$)

activated ATPase in the cat's pancreas. 11th Symp. European Pancreatic Club. Zurich. Aug. 24–26, 1978. Abstract.
24. CASE, R. M., A. A. HARPER & T. SCRATCHERD. 1968. Ionic requirements for pancreatic secretion. Proc. Int. Union Physiol. Sciences XXIV Washington. Abstr. 230.
25. BALL, E. C., H. F. TUCKER, A. K. SOLOMON & B. VENNESLAND. 1941. THE source of pancreatic juice bicarbonate. J. Biol. Chem. **140:** 119–129.
26. CASE, R. M., T. SCRATCHERD & R. D'A. WYNNE. 1970. The origin and secretion of pancreatic juice bicarbonate. J. Physiol. **210:** 1–15.
27. SCHULZ, I., F. STÖVER & K. J. ULLRICH. 1971. Lipid soluble weak organic acid buffer as "substrate" for pancreatic secretion. Pflügers Arch. **323:** 121–140.
28. HADI, N., J. HOTZ, T. SCRATCHERD & R. D. A. WYNNE. 1976. The secretion of organic anions by the isolated perfused pancreas of the cat. J. Physiol. **259:** 56P.
29. PETERSEN, O. H. & N. UEDA. 1977. Secretion of fluid and amylase in the perfused rat pancreas. J. Physiol. **264:** 819–835.
30. SCHULZ, I. 1971. Influence of bicarbonate CO_2^- and glycodiazine buffer on the secretion of the isolated cat's pancreas. Pflügers Arch. **329:** 283–306.
31. SWANSON, C. H. 1971. A micropuncture study of electrolyte secretion by the *in vitro* rabbit pancreas. Ph.D. thesis. Harvard University. Cambridge, Mass.
32. SWANSON, C. H. & A. K. SOLOMON. 1975. Micropuncture analysis of the cellular mechanisms of electrolyte secretion by the *in vitro* rabbit pancreas. J. Gen. Physiol. **65:** 22–45.
33. SCHULZ, I., R. PEDERSON, V. WIZEMANN & S. KONDO. 1974. Stimulatory processes of the exocrine pancreas and their inhibition by nonpenetrating SH reagent. *In* Secretory Mechanisms of Exocrine Glands. N.A. Thorn & O. H. Petersen, Eds. pp. 88–98. Academic Press. New York.
34. AMSTERDAM, A. & J. D. JAMIESON. 1972. Structural and functional characterization of isolated pancreatic exocrine cells. Proc. Nat. Acad. Sci. USA **69:** 3028–3032.
35. SIMON, B., R. KINNE & G. SACHS. 1972. The presence of a HCO_3^- ATPase in pancreatic tissue. Biochim. Biophys. Acta **282:** 293–300.
36. MILUTINOVIĆ, S., G. SACHS, W. HAASE & I. SCHULZ. 1977. Studies on isolated subcellular components of cat pancreas. I. Isolation and enzymatic characterization. J. Membrane Biol. **36:** 253–279.
37. VAN AMELSVOORT, J. M. M., J. W. C. M. JANSEN, J. J. H. H. M. DE PONT & S. L. BONTING. 1978. Is there a plasma membranes-located anion-sensitive ATPase? IV. Distribution of the enzyme in rat pancreas. Biochim. Biophys. Acta **512:** 296.
38. HANNAN, S. F. & P. M. WIGGINS. 1976. Intracellular pH of frog sartorius muscle. Biochim. Biophys. Acta **428:** 205–222.
39. SWANSON, C. H. & A. K. SOLOMON. 1975. Micropuncture analysis of the cellular mechanisms of electrolyte secretion by the *in vitro* rabbit pancreas. J. Gen. Physiol. **65:** 22–45.
40. SCHULZ, I. Unpublished.
41. SCHULZ, I. *et al.* Submitted for publication.

DISCUSSION

R. B. GUNN: Is there carbonic anhydrase in the luminal fluid or on the luminal surface?

I. SCHULZ: Yes, carbonic anhydrase is present in duct cells. Diamox inhibits bicarbonate and fluid secretion by 50%; however, Diamox had no detectable effect on the intracellular pH.

W. McD. Armstrong: In your final diagram you showed chloride going across by what looked like simple conductive pathways. Is there any evidence in this cell for coupling of chloride to other ions, specifically sodium?

Schulz: There is no evidence that chloride is transported by a primary or secondary active system. In principle, it might be possible that coupled sodium transport occurs across the basolateral cell side, and then the chloride bicarbonate exchange takes place, all driven by the sodium potassium ATPase, but that is a very complicated model. We have no evidence for an active chloride transport.

M. Burg: Did you determine whether removing sodium changes the pH in these isolated cells?

Schulz: We have not yet done this. Ouabain produced no effect, but we have not yet determined the effect of lowering the sodium.

BICARBONATE AND CHLORIDE TRANSPORT IN RELATION TO THE ACIDIFICATION OR ALKALINIZATION OF THE URINE

William A. Brodsky, John H. Durham, and Gerhard Ehrenspeck

Departments of Physiology and Biophysics
Mount Sinai School of Medicine
The City University of New York
New York, New York 10029

Background

Active mechanisms for the transport of Na, Cl, HCO_3 (and/or H) across the urinary bladder of *Pseudemys scripta* turtles are responsible for the dilution and acidification, and, under certain conditions, for the alkalinization of the mucosal fluid. When these bladders are isolated and incubated in Na-rich media under open-circuit conditions, the spontaneous transepithelial potential difference (*PD*) is oriented so that the serosal surface is electropositive to the mucosal surface while NaCl and $NaHCO_3$ are actively reabsorbed.[1–4] When incubated in Na-free (choline) media, the *PD* is oriented so that the serosa becomes electronegative to the mucosa, while choline Cl and choline HCO_3 are actively reabsorbed.[1–4]

Independence of Ion Flows

A mutual independence among three ion flows has been demonstrated in bladders under short-circuiting conditions during which individual ions (Na, Cl, and HCO_3), rather than ion pairs are actively reabsorbed. Each of these ion flows occurs singly or in concert with the others, depending upon the composition of the bathing media,[5, 6] or upon the exposure of the bladder to certain transport inhibitors.[7–12] The rate of Cl transport across the short-circuited bladder remains unchanged when choline is substituted for Na,[5] or when ouabain or amiloride is added to the media.[7, 12] Conversely, the rate of Na transport across the short-circuited bladder remains constant when SO_4 is substituted for HCO_3 in the mucosal fluid—providing that HCO_3 is not simultaneously removed from the serosal fluid.[6]

The turtle bladder also actively secretes HCO_3 from the serosal to the mucosal fluid, but a complete independence of this process from the concomitant reabsorption of Na and Cl has not been established. In fact, Na and Cl reabsorption are dependent on the presence of serosal HCO_3 (see below). What has been shown about HCO_3 secretion comes from two kinds of experiments. In one kind, a metabolically-dependent, but downhill flow of HCO_3 from the serosal to the mucosal fluid has been demonstrated in asosciation with a nearly equimolar reabsorption of Cl [13, 14]; and in another kind, a short-circuiting current and *PD* (oriented so that the mucosa is electronegative to the serosa) has been found in SITS-treated bladders bathed on both surfaces

0077–8923/80/0341–0210 $01.75/0

by identical HCO_3-rich media devoid of all exogenous Cl and supplemented with ouabain.[10]

Dependence of Ion Flows on Serosal HCO_3

The dependence of Na reabsorption on the presence of serosal HCO_3 has been demonstrated as follows: The addition of HCO_3 to the serosal fluid (but not to the mucosal fluid) of short-circuited bladders increases the rate of Na transport in the presence of serosal glucose under anerobic as well as aerobic conditions.[15] This effect, reproduced in the presence and in the absence of exogenous Cl, cannot be reproduced in the glucose-depleted, oxygenated bladder until glucose has been added to the serosal fluid (FIGURE 1). Conversely, the oxygenation of prenitrogenated, glucose-containing media produces no increase in Na reabsorption until HCO_3 has been added to the serosal fluid.[15]

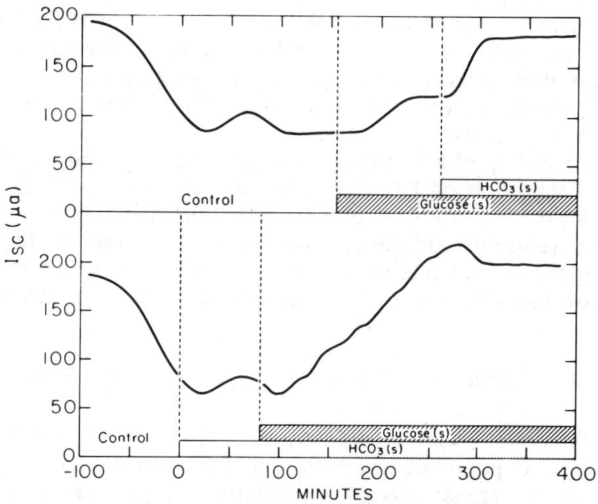

FIGURE 1. Glucose-dependent, HCO_3-induced acceleration of the I_{sc} across bladders bathed initially in HCO_3-free, Cl-free Na methylsulfate Ringer's solutions. The upper panel shows the HCO_3 effect in the presence of glucose; the lower panel, the lack of this effect until glucose has been added. (From Gonzalez *et al.*[15])

The reabsorption of Cl is also dependent on the presence of serosal HCO_3 in short-circuited bladders bathed by glucose-enriched, oxygenated media. The net Cl flux, maximal in short-circuited bladders bathed on both surfaces by identical (HCO_3 + Cl)-containing media, remains unchanged after the removal of mucosal HCO_3,[5] but decreases near-zero levels in the absence of exogenous HCO_3 from the serosal and mucosal media.[10, 13]

The aforementioned data on the independence and dependence of ion flows can be summarized and interpreted as follows:

1. With HCO_3 and glucose in the serosal fluid, the transepithelial flows

of Na, Cl, HCO_3 and/or H are demonstrably independent of each other in short-circuited turtle bladders.[1, 5, 6] This behavior, completely accounted for by the action or actions of discrete electrogenic ion pumps in parallel within the epithelial cell layer, neither requires nor excludes the concomitant action of electroneutral ion exchanges.

2. After the removal of HCO_3 from the serosal fluid alone or from both the serosal and mucosal fluids (by equivalent replacements with sulfate), the reabsorption of *both* Na and Cl along with the transepithelial electrical conductance are decreased; after HCO_3 is added back to the serosal fluid, all three parameters are increased to their original levels.

Whereas an electroneutral exchange of Cl reabsorption for HCO_3 secretion [13, 14] in the basal-lateral membrane [16] is sufficient to account for the effect of serosal HCO_3 on Cl transport, this mechanism, of itself, fails to account for the concomitantly-observed changes in electrical conductance in the experiments on Na transport [15] and in those on Cl transport.[10] Moreover, the electroneutral mechanism has no requirement for or relation to the effect of serosal HCO_3 on Na transport.[15] In short, the Cl:HCO_3 exchange hypothesis requires additional assumptions (lacking in experimental support) in order to be made consistent with the aforementioned findings on transepithelial conductance and Na transport.

An alternative hypothesis holds that the reabsorption of Na, Cl, and HCO_3, and the secretion of HCO_3 are consequences of four discrete electrogenic pumps in the membranes of this epithelial cell layer. Although this hypothesis also requires additional assumptions to account for the HCO_3-dependent changes in ion flux and conductance, these assumptions are supported, to a considerable extent, by experimental evidence, see below.

A General Model (for Two Special Hypotheses)

A "general" model has been constructed (FIGURE 2) in which electrical and/or nonelectrical pump components can be appropriately adjusted (some even eliminated) to represent two of the currently postulated systems of ion transport:

System 1 involves the actions of the four discrete electrogenic components with or without a contribution from any electroneutral anion exchanger.

System 2 involves the actions of only two of the four electrogenic components (the Na reabsorptive and the HCO_3 reabsorptive or proton secretory pumps) with a major contribution from an electroneutral anion exchanger.

System 1

If one assumes that Cl reabsorption and Na reabsorption are driven by discrete electrogenic pumps in the basal-lateral membrane, then the stimulatory effects of serosal HCO_3 and Cl and Na reabsorption [10, 13, 15] could be due in part to a secondary increase in the overall metabolic rate. This possibility was previously invoked because of the glucose dependence of the HCO_3-induced effect on Na transport.[15]

We now think that the observed stimulation of both the Na and the Cl reabsorption is a consequence of direct interactions of HCO_3 and glucose in the serosal fluid with HCO_3-selective and glucose-selective receptor sites of binding and/or carrier components on the outer surface of the basal-lateral membrane. The explicit assumptions, illustrated in part by the diagram in FIGURE 3, are the following:

1. The glucose and HCO_3-binding components are located between, and are in physical contact with, the Na pump and the Cl pump in the basal-lateral membrane. There is ample evidence for the basal-lateral location of the Na pump [7, 17] and presumptive evidence for that of the Cl pump.[13]

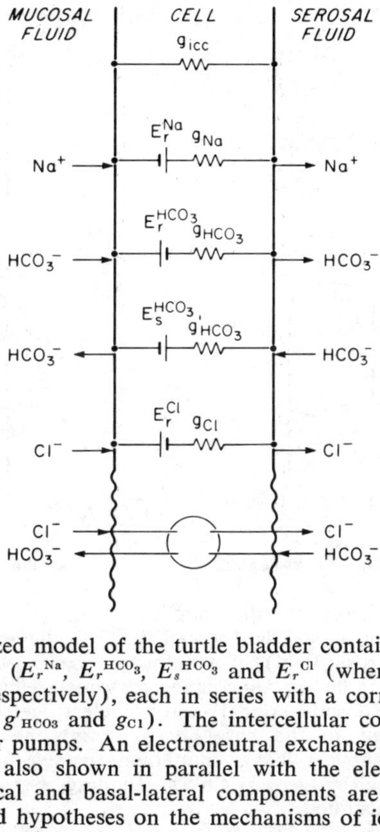

FIGURE 2. A generalized model of the turtle bladder containing a parallel array of four electrogenic pumps (E_r^{Na}, $E_r^{HCO_3}$, $E_s^{HCO_3}$ and E_r^{Cl} (where r and s denote reabsorption and secretion respectively), each in series with a corresponding ion-selective conductance (g_{Na}, g_{HCO_3}, g'_{HCO_3} and g_{Cl}). The intercellular conductance path, g_{icc}, is in parallel with the other pumps. An electroneutral exchange pump (Cl reabsorption for HCO_3 secretion) is also shown in parallel with the electrogenic pumps. This array, in which the apical and basal-lateral components are lumped, represents all of the currently-proposed hypotheses on the mechanisms of ion transport across this tissue. Moreover HCO_3 reabsorption and H secretion are electrically equivalent, so that either one can be used in this model.

2. After initial binding of serosal HCO_3 and glucose, the conformation of the binding component is changed so that its connections with each of the contiguous pump elements are modified. As a result, the Na-specific and Cl-specific conductances in series with the corresponding pump EMFs are increased in some unspecified way (FIGURE 3), which, in turn, produces the observed increases in Na and Cl transport as well as the increase in transepithelial conductance.

3. In addition to binding, it is assumed that these components passively cotransport glucose and HCO_3 from the serosal fluid into the cell fluid, where glucose is consumed by metabolic reactions to cause an increase in energy production and consequent increases in Na and/or Cl transport. The coentry of HCO_3 into the cell via this carrier operation is balanced by its ejection into the serosal fluid through a HCO_3-selective conductance path in the basal-lateral membrane. The HCO_3 ejection is driven by the electronegativity of the cellular interior, which in turn is generated and maintained by the Na pump

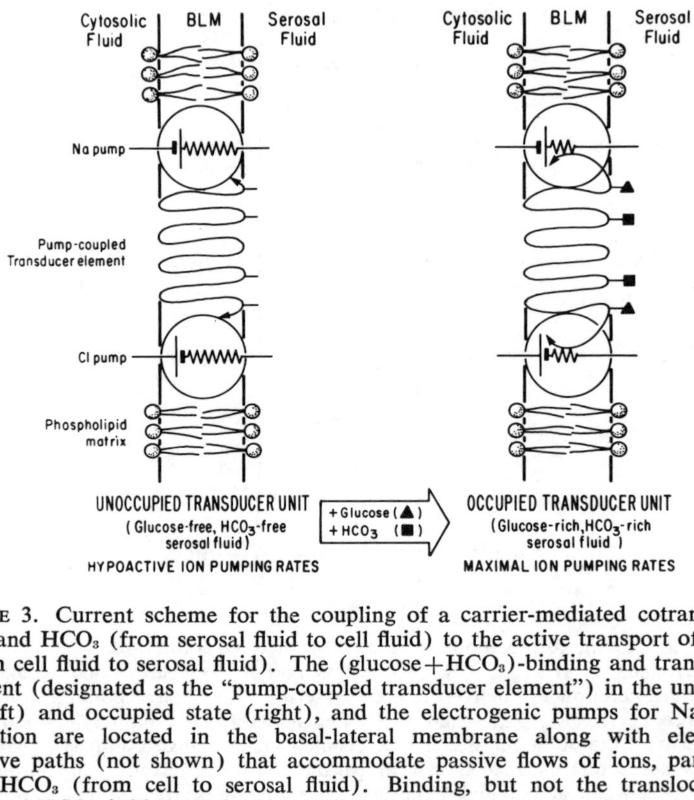

FIGURE 3. Current scheme for the coupling of a carrier-mediated cotransport of glucose and HCO_3 (from serosal fluid to cell fluid) to the active transport of Na and Cl (from cell fluid to serosal fluid). The (glucose+HCO_3)-binding and translocating constituent (designated as the "pump-coupled transducer element") in the unoccupied state (left) and occupied state (right), and the electrogenic pumps for Na and Cl reabsorption are located in the basal-lateral membrane along with electrically-conductive paths (not shown) that accommodate passive flows of ions, particularly that of HCO_3 (from cell to serosal fluid). Binding, but not the translocation of glucose and HCO_3, is illustrated.

in the basal-lateral membrane and/or by the HCO_3 reabsorptive (or proton secretory) pump in the apical membrane.

4. Since the (glucose + HCO_3)-binding component is assumed to be mobile and electrically charged in the loaded or unloaded state, it would have to account for some of the membrane conductance. Thus if the HCO_3 binding increased the electronegativity of the occupied (and mobile) component in relation to that of the unoccupied (and less mobile) component, this would account for some of the observed increase in the transepithelial conductance after adding HCO_3 to glucose-containing serosal fluids, with or without Na.[8, 15]

The other components of *System 1,* namely the HCO_3 secretory pump along with the HCO_3 reabsorptive (or H secretory) pump, can be assigned a location (or locations) in the apical membrane. It is possible that these two pumps are really different operational modes of a single HCO_3 pump (or a single proton pump), which is capable of reversing the direction of transport of the driven ion. (See the section on theoretical aspects of acidification and alkalinization).

System 2

One can assume that HCO_3 secretion and Cl reabsorption are obligatorily coupled by means of an active electroneutral Cl:HCO_3 exchanger in the basal-lateral membrane [16]; and that Na reabsorption is driven by an electrogenic pump located in the same membrane.[7, 17] The observed effect of HCO_3 on Cl reabsorption would have to be due to a direct interaction of HCO_3 with appropriate sites on the Cl:HCO_3 exchanger. Such an interaction of itself cannot change the electrical conductance or the Na transport across the bladder. However changes in both of these parameters have been found after addition of HCO_3 to or removal from the serosal fluid.[10, 15] In this system, the HCO_3-induced effect on Na reabsorption is qualitatively and topographically different from that on Cl reabsorption. The two kinds of HCO_3 interaction (that involving Na and that involving Cl transport) could occur on a single type or on two functional types of turtle bladder epithelial cells. The third component of this system, an electrogenic HCO_3 reabsorptive (or proton secretory) pump, is presumably located in the apical membrane.

In what follows, we present data in support of the electrogenicity of chloride reabsorption and of bicarbonate secretion as well as of HCO_3 reabsorption in the turtle bladder. Also presented is a model in which the net direction of the active transport of a single ion is inverted without requiring any fundamental change in the metabolic reaction that provides the driving force for this active ion transport. The application of this model, originally developed by Heinz,[18] to the observed process of HCO_3 reabsorption from and secretion into the mucosal fluid of the turtle bladder is discussed. The methods of procedure and techniques used have been described in detail elsewhere.[5]

The Mucosal Acidification Current

The short-circuiting current (I_{sc}) across ouabain-treated turtle bladders bathed on both surfaces by HCO_3rich (20 mM) media devoid of all exogenous Cl (SO_4 substitution), is thought to be carried solely by the net flow of HCO_3 ions from the mucosal to the serosal fluid (irrespective of whether this net charge translocation is driven by a primary HCO_3 reabsorption, or a primary proton secretion, or by both). This conclusion is based on the following lines of evidence:

1. The removal of HCO_3 (not CO_2) from the mucosal fluid under the aforementioned conditions is followed by a marked decrease in I_{sc}; and the subsequent removal of mucosal CO_2 by a smaller decrease in I_{sc}. After CO_2 was added back to the mucosal fluid, no change in I_{sc} was detected; but after

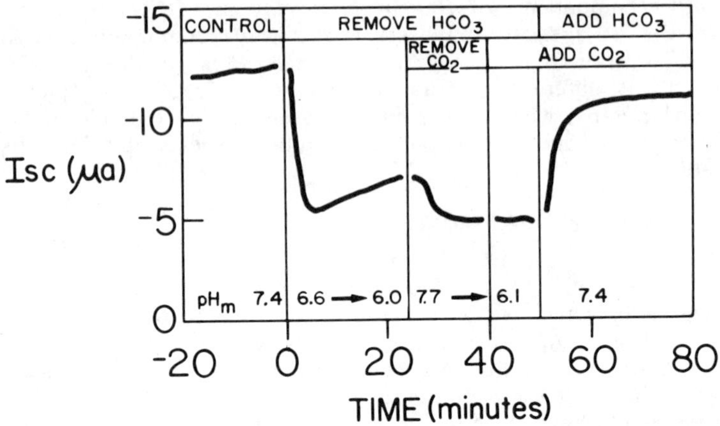

FIGURE 4. Effect on I_{sc} of isocapnic, nonisohydric additions and removals of HCO_3 and the effect of iso-bicarbonate, nonisohydric removals and additions of CO_2. Mucosal pH is indicated at the lower end of each panel. Bladders bathed on both surfaces by Na-free (choline), Cl-free (SO_4) Ringer's media.

HCO_3 was added back to the O_2-containing mucosal fluid, I_{sc} increased rapidly to the control level (FIGURE 4).

2. Isohydric additions of HCO_3 (and CO_2) to a mucosal fluid (pH, 7.2, initially devoid of exogenous HCO_3, CO_2, Na, and Cl) are followed by the development of increments in the transepithelial *PD* and I_{sc} in the presence of a fixed composition of the serosal fluid (pH, 7.2; HCO_3, 20 mM; CO_2, 2.0 mM; Cl-free and Na-free). The I_{sc}, under these isohydric conditions, is a Michaelis-like function of the mucosal HCO_3 and/or CO_2 (FIGURE 5); and a similar I_{sc} function can be generated by symmetrical additions of the same

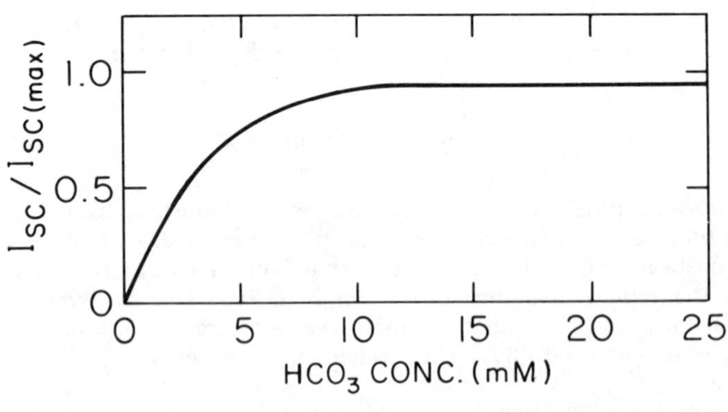

FIGURE 5. Relative I_{sc} or I_{sc} at any HCO_3 concentration, divided by maximal I_{sc} versus HCO_3 concentration in the mucosal fluid. Serosal HCO_3 concentration constant at 25 mM. Bladders bathed by Na-free (choline), Cl-free (SO_4) Ringer's media. (From Gonzalez and Schilb.[8])

quantities of HCO_3 and CO_2 to the mucosal and serosal fluids, which are therefore kept identical in composition before and after these additions.[8]

3. The net flux of SO_4 across bladders in these Na-free media is near-zero.[10]

4. In ouabain-treated bladders, bathed on both surfaces by identical Na-rich or Na-free media with or without Cl and devoid of exogenously added HCO_3 and CO_2, the magnitude of the I_{sc} is approximately equal to that of the chemically measured net rate of mucosal acidification. This original finding of the Steinmetz group [19] has been confirmed in our laboratory.[10]

Chloride Reabsorption and Bicarbonate Secretion

Electrogenicity of Cl Reabsorption

A tentative hypothesis holding that the active reabsorption of chloride by the turtle bladder is also an electrogenic process has been adduced from data on short-circuited bladders bathed on both surfaces by Cl-containing, isohydric (pH, 7.2) Na-free media under two conditions: 1. With 20 mM HCO_3 in both the mucosal and serosal fluids, which are therefore identical; and 2. with 20 mM HCO_3 in the serosal fluid, but with no exogenously added HCO_3 in the mucosal fluid, the pH of which is fixed at 7.2. The results of these experiments,[5] summarized in TABLE 1, are the following:

Under the first condition, the magnitude of the short-circuiting (I_{sc}) exceeds that of the isotopically measured net Cl reabsorption (i.e., $I_{sc} = I_r^{Cl} + I^x$), the excess (I^x) being attributable to the active mucosal acidification current (i.e., the net HCO_3 reabsorption or the net H secretion). Under the second condition (no mucosal HCO_3 present), the magnitude of the short-circuiting current approximates that of the net Cl reabsorption (i.e., $I_{sc} = I_r^{Cl}$), which would establish the electrogenicity of the active Cl reabsorption providing that the asymmetry in the composition of the bathing fluids does not induce transepithelial diffusion potentials of any significant magnitude.

Electroneutral Cl:HCO_3 Exchange

An alternative hypothesis, proposed by Leslie *et al.*[13] and others of the Steinmetz group,[14, 16] holds that the chloride reabsorption occurs via a one-for-one electroneutral and active (metabolically driven) exchange of chloride reabsorption for bicarbonate secretion. This hypothesis was adduced from the fact that the serosal addition of HCO_3 is followed by an alkali flow into an acidic mucosal fluid, together with a nearly equal increment of the unidirectional chloride flux from the mucosal to the serosal fluid (see above). The concomitant small changes in the *PD* and I_{sc} were attributed to the diffusional flow of HCO_3 from the serosal to the mucosal fluid through the intercellular channels.[13]

Disulfonic Stilbene Effects

Additional insight into the nature of Cl reabsorption and HCO_3 secretion has been gained from observations on the effects of serosal additions of certain disulfonic stilbenes (SITS, DIDS). These compounds, which inhibit the Cl:HCO_3 exchange in erythrocytes,[20] reduce the rate of mucosal acidification by turtle bladders in the presence and absence of exogenous HCO_3.[9, 10, 21] However, SITS-induced changes in Cl reabsorption in the turtle bladder apparently depend on the state of this function or on the state of the bladder when the disulfonic stilbene is administered.[10] Thus, the greater the control rate of Cl reabsorption, the greater is the degree of SITS-induced inhibition of this transport. However, the *PD* and I_{sc} are invariably reduced to zero after the serosal addition of SITS, while the rate of Cl reabsorption is decreased (but remains finite) in some bladders or remains constant in others (TABLE 2).

A closer inspection of these data reveals that the turtle bladders can be subdivided into two operationally defined groups: 1. Those with a SITS-sensitive Cl transport function, in which the control levels of the forward (m-to-s) Cl flux exceed 20 ua/cm² and the Cl flux ratio exceeds 6.0; and

TABLE 1

MEAN VALUES (± STANDARD ERRORS) OF *PD*, I_{sc} AND NET Cl REABSORPTION (I_{net}^{Cl}) IN TWO SETS OF BLADDERS *

In Mucosal Fluid	*PD* (mv)	*I*sc	I_{net}^{Cl} (ua)
HCO_3+Cl (*n*=19)	39.8±5.8	35.6±5.4	18.4±5.5
Cl (*n*=29)	20.7±2.8	19.6±2.7	21.3±3.9

* The bladders were bathed on the serosal surface by (HCO_3+Cl)-containing, Na-free (choline) Ringer's solution. The mucosal fluid of one set contained 80 mM Cl and 20 mM HCO_3 at a pH of 7.4 and was identical to the serosal fluid (upper row); that of the second set contained 80 mM Cl and no exogenous HCO_3 (SO_4) substitution) at a pH of 7.4.

2. those with a SITS-insensitive Cl transport function, in which the control levels of the same Cl flux parameters are less than 10 ua/cm² and 4.0, respectively (TABLE 2). Husted *et al.*[16] have also found SITS-insensitive Cl flux parameters under similar conditions, that is, when the control Cl flux ratio is less than 4.0.

It should be noted that the mean Cl flux ratio obtained from pairs of mated half bladders (or from pairs of unmated whole bladders) is the only legitimate parameter with which the net driving force for Cl transport (or the "state" of the Cl transport function) across bladders of one size can be compared with that across bladders of a different size, in the same or in another laboratory. This claim presupposes that the mean values of exposed area, dry mass and electrical parameters associated with the tissues in which the mean forward flux of ^{36}Cl is determined are approximately the same as those associated with the paired tissues in which the mean backflux of this isotope is determined.

In the SITS-treated state, of all of the aforementioned bladder groups,

TABLE 2

MEAN VALUES (± STANDARD ERRORS) OF *PD*, I_{sc}, AND THE MUCOSAL-TO-SEROSAL Cl FLUX (I_{ms}^{Cl}); AND MEAN VALUES (± STANDARD ERRORS) Cl FLUX RATIOS (I_{ms}/I_{sm}) IN 16 PAIRS OF MATED HALF BLADDERS: 32 PIECES OF TISSUE OF SIMILAR DRY WEIGHT AND IDENTICAL EXPOSED AREA *

Period	*PD* (mv)	I_{sc} (ua/cm²)	I_{ms}^{Cl} (ua/cm²)	Flux Ratio (I_{ms}/I_{sm})
		SITS-sensitive Cl transport ($n=8$)		
Control	50.1±6.4	26.2±5.9	21.0±5.7	7.8±3.3
SITS	0.2±1.2	0.3±0.73	11.3±2.5	2.5±1.2
		SITS-insensitive Cl transport (n=8)		
Control	31.8±8.2	18.8±5.0	8.6±1.7	2.3±1.1
SITS	0.1±0.5	0.0±0.3	7.4±1.3	2.0±1.1

* All 32 hemi bladders were bathed on both surfaces by identical Na-free (choline) Ringer's media containing: 20 mM HCO_3 at pH, 7.2; 25 mM Cl; and 27 mM SO_4. Values shown reflect the uniformly reproducible SITS-induced inhibition of *PD* and I_{sc} in both groups of paired half-bladders (upper and lower rows), the electrical and Cl flux parameters associated with a SITS-sensitive Cl flux (upper rows), and the same parameters associated with a SITS-resistant Cl flux (lower rows).

there remains a finite net reabsorption of Cl in an electrically silent system.[10] This would be the expected result if the Cl transport were driven by a SITS-resistant, electroneutral Cl:HCO_3 exchanger, such as that included in the general model shown in FIGURE 1. However, the electrical effects of SITS on ouabain-treated bladders bathed on both surfaces by identical Cl-free, HCO_3-rich media are *not* predicted by any electroneutral ion exchange mechanism. For example, the SITS-treated state of such bladders is characterized by the steady state maintenance of a finite *PD* and I_{sc} oriented in the direction of negative ions flowing from the serosal to the mucosal fluid (TABLE 3). Not shown in the table is the fact that the measured net sulfate flux is zero before and after the addition of SITS. That this I_{sc} is carried only by HCO_3 secretion is further suggested by obtaining the same effect of SITS after removal

TABLE 3

MEAN VALUES (± STANDARD ERRORS) OF THE *PD* AND I_{sc} BEFORE AND AFTER THE ADDITION OF SITS (10^{-4} M) TO THE SEROSAL FLUIDS OF NINE BLADDERS BATHED ON BOTH SURFACES BY IDENTICAL Cl-FREE (SULFATE SUBSTITUTION), NA-FREE (CHOLINE) MEDIA CONTAINING: 20 MM HCO_3 AT PH, 7.2 AND 40 MM SO_4*

Condition	*PD* (mV)	I_{sc} (ua/cm²)
Control	26.6±6.7	9.7±2.9
SITS	−7.3+0.3	−2.3+0.3

* Not shown is the doubling of these negative values after theophylline (10^{-3} M) was added to the mucosal and serosal fluids in four of the SITS-treated bladders. Data shown are summarized from those of Brodsky, Durham and Ehrenspeck.[10]

(with appropriate substitutions) of all non-HCO_3 ions that might be implicated as carriers of this I_{sc}.[10]

It can be concluded that there is an active electrogenic secretion of HCO_3 in the turtle bladder (as depicted in the general model of FIGURE 1); and that this process accounts for the *PD* and carries the I_{sc} across SITS-treated bladders bathed on both surfaces by identical Cl-free, HCO_3-rich media (TABLE 3).

Since the absence of Cl from the media is associated with an electrical state (TABLE 3), which differs from that seen in the presence of Cl (TABLE 2) in the SITS-treated bladders, it follows that part of the Cl reabsorption could also be an active electrogenic process (FIGURE 1), which was originally suggested by Gonzalez *et al.*[5] as discussed above.

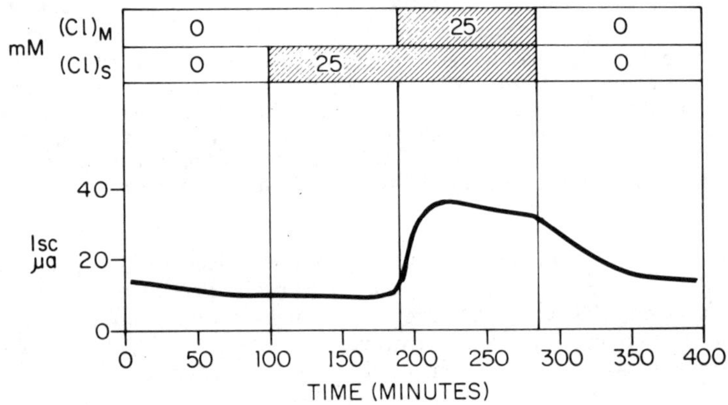

FIGURE 6. Responses of I_{sc} to additions and removals of mucosal and serosal Cl in bladders bathed by Na-free, HCO_3-containing media.

Cl-induced Electrical Changes

In keeping with the alleged electrogenicity of chloride reabsorption, the addition of Cl to, or its removal from, the mucosal and serosal fluids of a bladder, initially bathed on both surfaces by identical Cl-free (SO_4-containing), HCO_3-rich media, is followed by a significant increase or decrease in the electronegativity of the serosal fluid (FIGURE 6). Similar Cl-dependent changes in the Isc, found in three additional experiments of this design, occurred while the rate of HCO_3 reabsorption was kept constant at a maximal level. This maximal level was obtained by keeping the HCO_3 concentration of the media at 20–25 mM (see FIGURE 5). Not shown are findings that nearly identical Cl-induced changes in I_{sc} and *PD* were elicited in two other bladders in which the HCO_3 reabsorption was kept constant at a zero level. The zero level of HCO_3 reabsorption was obtained by fixing the mucosal HCO_3 and pH at 0.02 mM and 4.0, respectively, with continuous pH statting of the mucosal fluid, while the serosal HCO_3 and pH were fixed at 20 mM and 7.2, respectively.[22]

Acidification and Alkalinization (Theoretical Aspects)

Data cited here indicate that the turtle bladder epithelial cells have the capacity for alkalinizing as well as acidifying the urine. This dual transport function can be represented by the operations of the general model in FIGURE 2. In an all-electrogenic model, net acidification of the urine would occur whenever $I_r^{HCO_3}$ exceeds $I_s^{HCO_3}$ (or I_s^H exceeds I_r^H), and net alkalinization of the urine would occur with a reversal of these ion flow inequalities. In a part-electrical and part-electroneutral model (i.e., a zero level of electrical conductance for the exchange of Cl reabsorption for HCO_3 secretion), a similar set of ion flow inequalities can be adduced to account for the acidification of the urine seen under the same conditions and for the alkalinization of the urine seen under other conditions. In either model, each of two discrete ion pump mechanisms (coupled to one or two exergonic metabolic reactions) drives the acidification and the alkalinization process.

However, Heinz has recently shown how a single (ion specific) pump mechanism can reverse the direction of transport of a given ion in a model system.[18] This direction-reversing mechanism can be exploited to account for the active acidification and the active alkalinization of urine by the turtle bladder [10,13] or mammalian nephron,[23] as shown schematically in FIGURE 7. In this scheme, the active secretion or reabsorption of a given ion (a) is driven by the coupling of a single metabolic source of free energy (generated by the conversion of a substrate, S to products, P) to a single intramembrane reaction, namely the conversion of Y to X. The net direction of the transport of ion a can be made to depend only on the relative mobilities of either the loaded carriers (X_a and Y_a) or the unloaded carriers (X and Y); the latter case is described below.

The operations required for a secretory mode of the active transport of anions are shown in the upper panel. In this mode, the functionally-effective intramembrane translocations are those of the loaded carrier, X_a (from the cellular to the mucosal fluid boundary of the membrane) and the unloaded carrier, Y (in the reverse direction). These directional orientations are consequences of a low coefficient of "slippage" of the unloaded carrier, X (from left to right) and/or a relatively higher rate coefficient of back-flow of the unloaded carrier, Y (from right to left). Therefore, most of the X' (produced from Y' by coupling with exergonic reaction, $S \rightarrow P$) remains available for complexation with a at the cellular interface of the membrane while little of the Y'' (spontaneously forming from X'') remains available for complexation with a at the mucosal interface of the membrane; at this interface, most of the unloaded carrier (Y'') is translocated from right to left, so that the rate of back-leakage of the loaded carrier (Y_a'') is small.

In order to switch from the secretory to the reabsorptive mode (lower panel) of the active transport of a, it is only necessary to assume that the mobility coefficient of X increases and/or that of Y decreases. Under these new conditions, the functionally effective intramembrane translocations are those of the unloaded carrier, X (again, from left to right within the membrane) and the loaded carrier Y_a (again, from right to left in the membrane). Because of the relative inequality in the magnitude of the mobility coefficients ($L_x > L_y$), little of the unloaded X' is left for complexation with a at the inner membrane interface (most of the X' is translocated as rapidly as it is formed from Y'); and since little of the Y'' (forming spontaneously for X'' at the mucosal boundary) is translocated, most of this Y'' is available for complexation with a″. The subsequent rate of Y_a translocation (from right to left) is accelerated by coupling with the exergonic reaction ($S \rightarrow P$) which converts Y' to X', depletes Y', and accelerates the dissociation of Y_a' at the cellular interface of the membrane.

Because of the general nature of this model, the ion translocations repre-

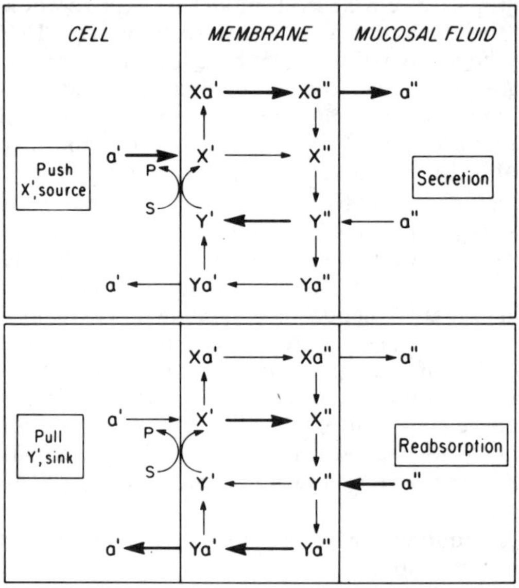

FIGURE 7. The Heinz model [18] for the induced reversal in direction of the active transport of a single solute driven by a single pump mechanism ($S \longrightarrow P$) coupled to a single intramembrane reaction ($Y' \longrightarrow X'$), as applied to the alkalinization or acidification of the urine in turtle bladder [10] or mammalian renal collecting tubule.[23] Upper panel denotes the net secretion of solute a, occasioned by a high mobility coefficient of Y relative to X and a consequent high transfer rate of X_a relative to that of Y_a as indicated by thick and thin arrows in each case. Lower panel denotes net reabsorption of the same solute occasioned by a single perturbation. When this occurs, the mobility coefficient of X becomes high relative to that of Y. See text for further details.

sented could occur through electrically conductive or nonconductive paths. Moreover, these paths could be adapted to fit a secretion of HCO_3 or protons (e.g., as shown in the upper panel) or a reabsorption of HCO_3 or protons (as shown in the lower panel).

SUMMARY

Although the electrogenicity of the active reabsorption of sodium and bicarbonate (or secretion of protons) has been well-established in the short-circuited turtle bladder preparation, the nature of the active reabsorption of chloride and secretion of bicarbonate has been controversial. These processes have been ascribed to the separate actions of two discrete electrogenic pumps or to the single action of a metabolically driven electroneutral anion exchange mechanism. The present report deals with these transport processes per se; with the relations among serosal bicarbonate, glucose, and the reabsorption of Na and Cl; and finally with the application of the Heinz model for inversion of an active transport as a tentative single mechanism for the alkalinization and acidification of the urine.

References

1. Brodsky, W. A. & T. P. Schilb. 1966. Ionic mechanisms for sodium and chloride transport across turtle bladders. Am. J. Physiol. **210:** 987–996.
2. Schilb, T. P. & W. A. Brodsky. 1966. Acidification of mucosal fluid by transport of bicarbonate ion in turtle bladders. Am. J. Physiol. **210:** 997–1008.
3. Schilb, T. P. & W. A. Brodsky. 1972. CO_2 gradients and acidification by transport of HCO_3 in turtle bladders. Am. J. Physiol. **222:** 272–281.
4. Schilb, T. P. 1978. Bicarbonate ion transport: a mechanism for the acidification of urine in the turtle. Science **200:** 298–299.
5. Gonzalez, C. F., Y. E. Shamoo & W. A. Brodsky. 1967. Electrical nature of active chloride transport across short-circuited turtle bladder. Am. J. Physiol. **212:** 641–650.
6. Gonzalez, C. F., Y. E. Shamoo, H. R. Wyssbrod, R. E. Solinger & W. A. Brodsky. 1967. Electrical nature of sodium transport across the isolated turtle bladder. Am. J. Physiol. **213:** 333–340.
7. Solinger, R. E., C. F. Gonzalez, Y. E. Shamoo, H. R. Wyssbrod & W. A. Brodsky. 1968. Effect of ouabain on ion transport mechanisms in the isolated turtle bladder. Am. J. Physiol. **215:** 249–261.
8. Gonzalez, C. F. & T. P. Schilb. 1969. Acetazolamide-sensitive short-circuiting current versus mucosal HCO_3-concentration in turtle bladders. Biochim. Biophys. Acta **193:** 419–429.
9. Ehrenspeck, G. & W. A. Brodsky. 1975. Effects of 4-acetamido-4′-isothiocyano-2,2′-disulfonic stilbene on ion transport in turtle bladders. Biochim. Biophys. Acta **419:** 555–558.
10. Brodsky, W. A., J. Durham & G. Ehrenspeck. 1979. The effects of a disulphonic stilbene on chloride and bicarbonate transport in the turtle bladder. J. Physiol. **287:** 559–573. London, England.
11. Brodsky, W. A., Z. I. Cabantchik, N. Davidson, G. Ehrenspeck, E. M. Kinne-Saffran & R. Kinne. 1979. Localization and characterization of transport-related elements in the plasma membrane of turtle bladder epithelial cells. Biochim. Biophys. Acta **556:** In press.
12. Wilczewski, T. & W. A. Brodsky. 1975. Effects of ouabain and amiloride on the Na pathways in turtle bladder. Am. J. Physiol. **228:** 781–790.
13. Leslie, B. R., J. H. Schwartz & P. R. Steinmetz. 1973. Coupling between Cl-reabsorption and HCO_3-secretion in turtle urinary bladder. Am. J. Physiol. **225:** 610–617.
14. Steinmetz, P. R. 1974. Cellular mechanisms of urinary acidification. Physiol. Rev. **54:** 890–956.
15. Gonzalez, C. F., Y. E. Shamoo & W. A. Brodsky. 1969. The accelerating effect of serosal HCO_3^- on Na^+ transport in short-circuited turtle bladders. Biochim. Biophys. Acta. **193:** 403–418.
16. Husted, R. F., L. H. Cohen & P. R. Steinmetz. 1979. Pathways for bicarbonate transfer across the serosal membrane of turtle urinary bladder: studies with a disulfonic stilbene. J. Membrane Biol. **47:** 27–37.

17. Schilb, T. P. & W. A. Brodsky. 1970. Transient acceleration of transmural water flow during inhibition of sodium transport in turtle bladders. Am. J. Physiol. **219:** 590–596.
18. Heinz, E. 1978. Mechanics and energetics of biological transport. *In* Molecular Biology, Biochemistry and Biophysics.: 29. Springer-Verlag, New York, N.Y.
19. Steinmetz, P. R., R. S. Omachi & H. S. Frazier. 1967. Independence of hydrogen ion secretion and transport of other electrolytes in turtle bladder. J. Clin. Invest. **46:** 1541–1548.
20. Cabantchik, Z. I. & A. Rothstein. 1972. The nature of the membrane sites controlling anion permeability of human red blood cells as determined by studies with disulfonic stilbene derivatives. J. Membrane Biol. **10:** 311–330.
21. Cohen, L. H., A. Mueller & P. R. Steinmetz. 1978. Inhibition of the bicarbonate exit step in urinary acidification by a disulfonic stilbene. J. Clin. Invest. **61:** 981–986.
22. Durham, J. H., W. Nagel, G. Ehrenspeck & W. A. Brodsky. 1979. Sodium-independent, active chloride reabsorption in the turtle bladder: an electrogenic process. Symposium on Chloride Transport. Fed. Proc. In press.
23. McKinney, T. D. & M. B. Burg. 1977. Bicarbonate transport by rabbit cortical collecting tubules. Effect of acid and alkali loads *in vitro* on transport *in vitro*. J. Clin. Invest. **60:** 766–786.

Discussion

W. McD. Armstrong: With respect to the data showing a turning on or an increase of bicarbonate transport by theophylline, it seems pertinent to note that John White recently reported a similar theophylline effect on bicarbonate transport in the intestine. Do you have any comment on that work in relation to your own?

W. Brodsky: I think that White found an increased reabsorption, rather than secretion, of bicarbonate, but I can't speculate on his data right now. We did, however, find that part of the bicarbonate secretion seen after SITS had to be electrogenic because of the presence of a finite *PD* (serosa positive) and maintenance of a short circuiting current in the absence of exogenous chloride. Not shown here are some of John Durham's data suggesting that part of the chloride reabsorption also had to be electrogenic in nature. Thus, by lowering mucosal pH to 4.0 in a system originally free of exogenous chloride, the bicarbonate reabsorption is stopped. Then chloride is added (sulfate substitution) to both fluids. This initiates a bicarbonate secretion, but also induces a large *PD* (serosa negative) in some cases. In our model, we had conceived of anion receptors with different affinities for chloride and bicarbonate. If these receptors were not saturated with anions, one could account for the observed independence of the chloride, bicarbonate and sodium reabsorption in the short-circuited preparation. The Heinz model is useful because it can account for both bicarbonate reabsorption and secretory secretion with one and the same element.

J. M. Salhany (*University of Nebraska Medical Center, Omaha, Neb.*): Have you tried DIDS as well as SITS in the turtle bladder? DIDS is presumably a more potent inhibitor of $Cl:HCO_3$ transport.

Brodsky: The effect of DIDS in the turtle bladder is almost indistinguishable from that of SITS at the same dosage level.

SECRETION AND REABSORPTION OF BICARBONATE IN SINGLE RENAL TUBULES OF THE RABBIT

Maurice B. Burg

Laboratory of Kidney and Electrolyte Metabolism
National Heart, Lung, and Blood Institute
Bethesda, Maryland 20205

Control of renal excretion of bicarbonate is important for the maintenance of acid-base homeostasis. Bicarbonate is present in the glomerular filtrate in approximately the same concentration as in plasma. Bicarbonate excretion is determined by the balance of its glomerular filtration and its transport across renal tubular epithelium. This paper is a summary of some recent studies of the mechanism and control of bicarbonate transport in isolated perfused rabbit renal tubules.

Each nephron consists of numerous different kinds of tubule epithelium arranged in series (FIGURE 1). The tubule segments differ from one another both in cellular ultrastructure and in transport function. In order to characterize the different transport functions, tubules have been dissected and perfused *in vitro* [1] (FIGURE 2).

In the studies to be presented, the transepithelial voltage was measured using the perfusion pipet as a bridge into the tubule lumen.[2] Fluid absorption was measured using ^{14}C inulin in the perfusate as a volume marker.[3] Bicarbonate was estimated from total CO_2, measured by microcalorimetry.[4] The principle of the total CO_2 measurement is illustrated in FIGURE 3.

When perfused with solutions containing 25 mM of bicarbonate, the different tubule segments transported bicarbonate at greatly differing rates *in vitro* (FIGURE 4). Sl proximal convoluted tubules (segment 1 in FIGURE 1) transported bicarbonate most rapidly.[5] This segment normally reabsorbs the bulk of the filtered bicarbonate. The concentration of bicarbonate in the tubule fluid normally falls to a relatively low level ($<$10 mM) by the end of Sl *in vivo*.[6] The rate of bicarbonate reabsorption was less in proximal straight tubules (segment 2 and segment 3 in FIGURE 1) than in proximal convoluted tubules, but the straight tubules were capable of maintaining bicarbonate at as low a concentration as the convoluted tubules.[7] The fact that bicarbonate transport is intrinsically slower in proximal straight tubules than in proximal convoluted tubules may reflect the smaller amount of bicarbonate reaching the straight tubules *in vivo*. Cortical thick ascending limbs did not transport bicarbonate (FIGURE 4).[8] This absence of bicarbonate transport in this segment had previously been inferred from renal clearance studies.[9] Finally, cortical collecting ducts transported bicarbonate at a slower rate than the proximal tubules (FIGURE 4).

The magnitude and direction of bicarbonate transport across cortical collecting ducts depended on the treatment of the rabbits from which the collecting ducts were obtained (FIGURE 5).[10] The rabbits were divided into three treatment groups: 1. Those given $NaHCO_3$ on the day before the experiment had mean urine pH of 8.1 at the time they were killed in preparation for the perfusion of their tubules. 2. Those not specially treated had mean urine pH

0077–8923/80/0341–0225 $01.75/0

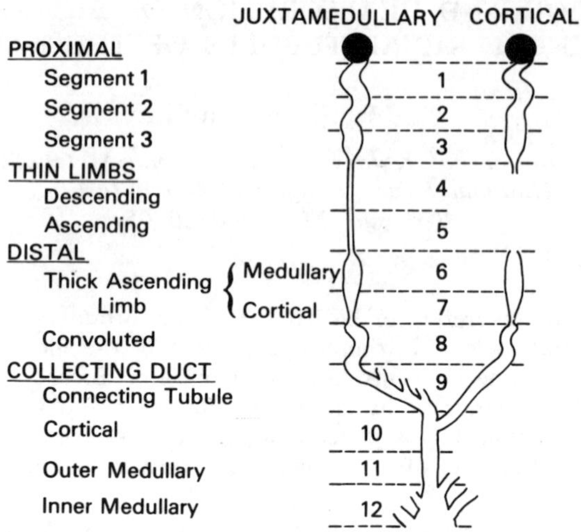

FIGURE 1. Anatomy of rabbit nephrons. (From Burg.[18])

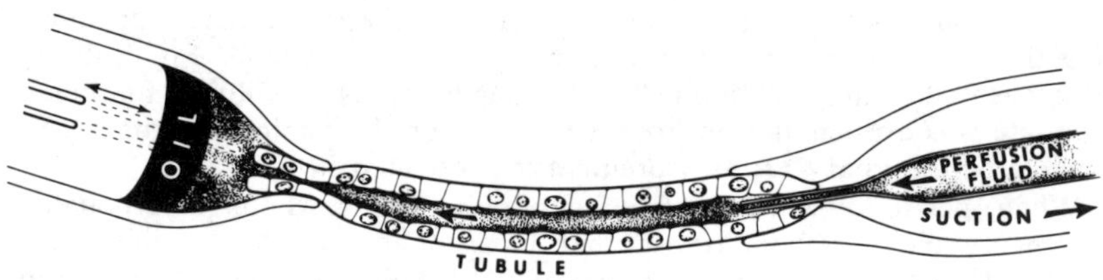

FIGURE 2. Method for perfusing renal tubules *in vitro*. (From Burg and Orloff.[19])

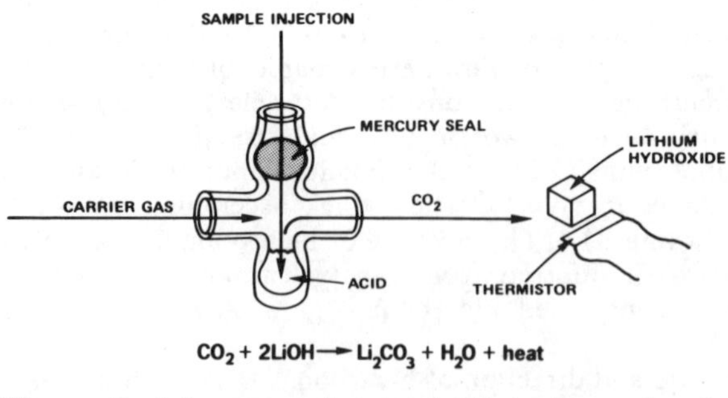

FIGURE 3. Micromethod for measuring total CO_2.[4] A sample of tubule fluid is delivered into acid. The bicarbonate (and also carbonate, and carbonic acid) in the sample are converted into CO_2. The CO_2 is carried by a stream of inert gas to a crystal of LiOH. The temperature increase caused by reaction of CO_2 with LiOH is measured by the thermistor and is directly proportional to the total CO_2 content of the tubule fluid sample.

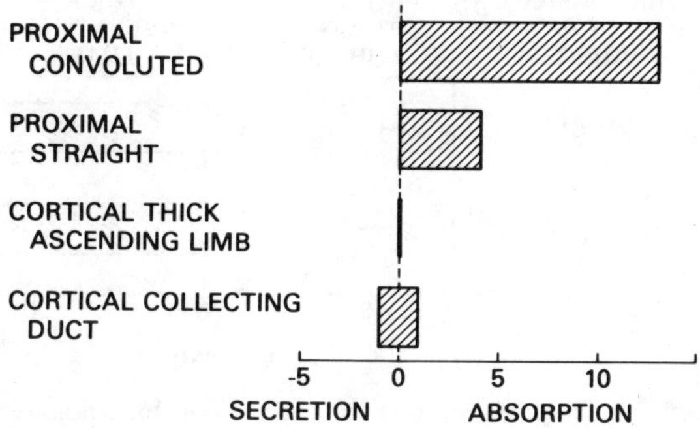

FIGURE 4. Rate of bicarbonate transport by rabbit nephron segments (pEq./cm·s). (From Burg and Green,[5] McKinney and Burg,[7] Iino and Burg,[8] and McKinney and Burg.[10])

of 7.4. 3. Those given NH_4Cl had mean urine pH of 5.0. Since the collecting ducts were perfused under identical conditions *in vitro,* the observed differences in bicarbonate transport must be due to conditioning of tubules by the treatment given to the rabbit. When the rabbits had been given $NaHCO_3$ and were excreting alkaline urine, their tubules secreted bicarbonate *in vitro.* When they had been given NH_4Cl and were excreting acid urine, their tubules absorbed

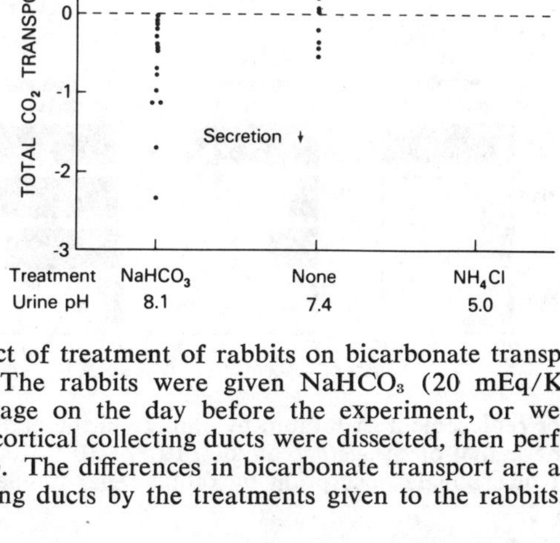

FIGURE 5. Effect of treatment of rabbits on bicarbonate transport by their cortical collecting ducts. The rabbits were given $NaHCO_3$ (20 mEq/Kg) or NH_4 Cl (15 mEq/Kg) by gavage on the day before the experiment, or were given no special treatment. Their cortical collecting ducts were dissected, then perfused under standard conditions *in vitro.* The differences in bicarbonate transport are a result of conditioning of the collecting ducts by the treatments given to the rabbits. (From McKinney and Burg.[10])

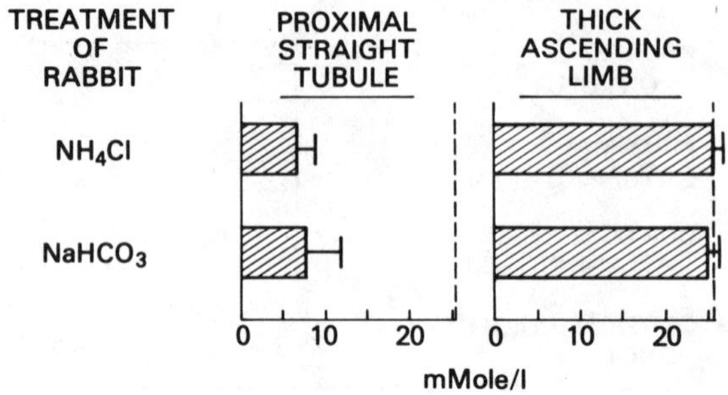

FIGURE 6. Lack of effect of treatment of rabbits on bicarbonate transport by their proximal straight tubules and thick ascending limbs of Henle's loop. The groups of rabbits received the same treatments as in FIGURE 5, but in contrast to the results with cortical collecting ducts, there was no evidence of conditioning of bicarbonate transport in the proximal straight tubules or thick ascending limbs (From Iino and Burg.[8])

bicarbonate *in vitro*. Presumably conditioning of the collecting ducts also affects bicarbonate transport *in vivo,* contributing to control of its excretion.

Conditioning of bicarbonate transport was limited to the collecting ducts. It was not observed in proximal straight tubules or thick ascending limbs (FIGURE 6).[8] Bicarbonate absorption by proximal straight tubules did not differ whether the rabbits had received $NaHCO_3$ or NH_4Cl. Also, there was no significant bicarbonate transport by cortical thick ascending limbs from either group.

In order to elucidate the mechanisms of bicarbonate transport, different baths and perfusates were tested. Sodium was found to be necessary for bicarbonate transport in proximal tubules, but not in cortical collecting ducts. When all of the sodium in the bath and perfusate was replaced by choline, tetramethyl ammonium, or lithium, bicarbonate absorption by proximal tubules was greatly inhibited[5,7] (FIGURE 7). Ouabain, added to the bath, had the same effect. Although the proximal tubules reabsorbed some bicarbonate in

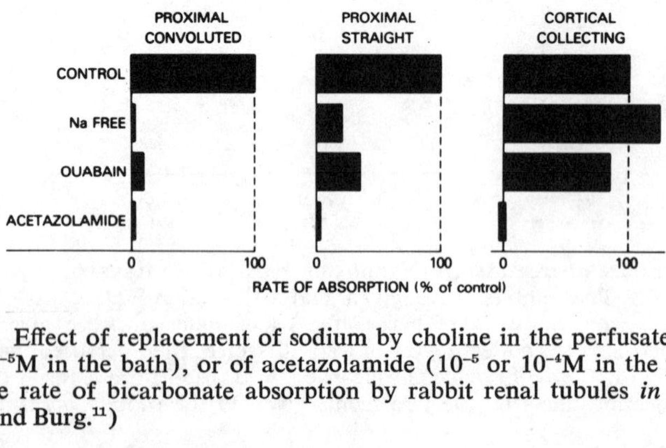

FIGURE 7. Effect of replacement of sodium by choline in the perfusate and bath, of ouabain (10^{-5}M in the bath), or of acetazolamide (10^{-5} or 10^{-4}M in the perfusate and bath) on the rate of bicarbonate absorption by rabbit renal tubules *in vitro*. (From McKinney and Burg.[11])

the absence of sodium, the magnitude of the remaining bicarbonate transport was small. Thus, at slow flow rates, bicarbonate concentration in proximal tubule fluid decreased by more than 15 mM in the presence of sodium, but only by 4 mM in its absence or when ouabain was added. The sodium-dependence of bicarbonate transport in proximal tubules is consistent with a model proposed by Pitts [12] (FIGURE 8). In this model, there is sodium-proton exchange at the lumen membrane that couples absorption of sodium into the tubule cells to secretion of protons into the lumen. In proximal tubules, the principal fate of secreted protons is reaction with bicarbonate in the tubule fluid to form CO_2. The CO_2 produced by this reaction diffuses into the cells in which it reacts to produce bicarbonate and more protons. There is facilitated transport of cellular bicarbonate across the peritubular membrane into the blood, completing its reabsorption. In support of this theory, a sodium-proton exchanger has been identified in vesicles prepared from lumen membrances of proximal tubule cells.[13] One important consequence of the theory is that, because the sodium concentration difference between the tubule fluid and cell is the principal driving force for proton secretion, the maximum proton concentration difference between cell and lumen is limited by the sodium concentration difference. That may explain why proximal tubule fluid generally does not become very acidic. (The observed pH generally is not less than approximately 6.7). A direct test of this consequence of the theory is not practical since it requires simultaneous measurement of cell and tubule fluid pH and sodium concentration. Based on the estimates that are available, however, it is plausible that the steady-state proton ratio between cell and lumen is approximately equal to the reciprocal of the sodium ratio, and that both ratios are less than ten. A further consequence of the theory is that introduction of a very acidic solution into the lumen of proximal tubules could cause the exchanger to "run backwards," leading to a rapid increase of tubule fluid pH to the steady state value. The latter result was observed in rat proximal tubules, but was ascribed to high transepithelial proton permeability [14] rather than proton entry into the cells.

In contrast to proximal tubules, when choline replaced sodium in the perfusate and bath of cortical collecting ducts, bicarbonate absorption was not inhibited (FIGURE 7).[11] In fact, the rate of bicarbonate absorption increased under those conditions. Also, ouabain did not inhibit bicarbonate absorption in collecting ducts, as it did in proximal tubules. Bicarbonate absorption evidently is not dependent on sodium-proton exchange in collecting ducts. Protons presumably are directly secreted, as in mitochondria and bacteria (or bicarbonate is directly absorbed).

In both proximal convoluted tubules and cortical collecting ducts, the carbonic anhydrase inhibitor acetazolamide greatly decreased bicarbonate absorption (FIGURE 7).[5, 7, 11] This result supports the theory that hydration of CO_2, catalyzed by cellular carbonic anhydrase, is the source of protons in all the different segments (FIGURE 8).[15] Carbonic anhydrase activity is present in the lumen as well as cells of proximal tubules. Inhibition of the luminal carbonic anhydrase activity slows the reactions in the tubule fluid (FIGURE 8) and causes protons to accumulate, resulting in a fall in pH. The role of the carbonic anhydrase activity in the lumen presumably is to prevent such a fall in pH under normal conditions, since the fall in pH would

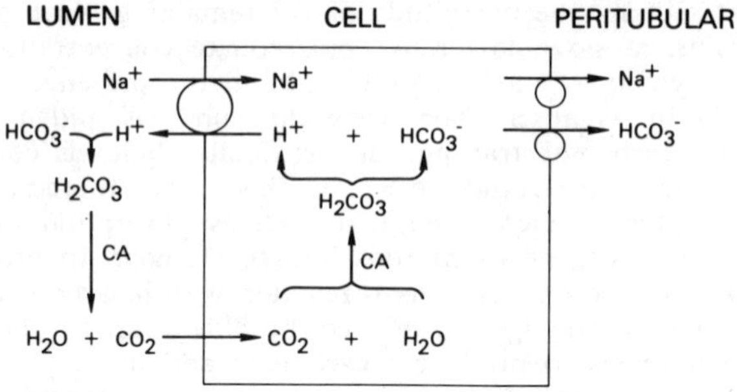

FIGURE 8. Model of bicarbonate transport by proximal tubules. CA is carbonic anhydrase (From Pitts.[12])

limit proton secretion as the maximum pH difference between cell and lumen was approached.

The effects of ion substitutions and drugs were also tested on cortical collecting ducts that were secreting bicarbonate as well as on the collecting ducts that were absorbing bicarbonate (FIGURE 9).[16] Acetazolamide inhibited bicarbonate secretion. Replacement of sodium by choline converted bicarbonate secretion to absorption. Ouabain had no significant effect on bicarbonate secretion. Amiloride inhibited bicarbonate absorption, but caused bicarbonate secretion to increase. In order to analyze these results, it would be helpful to know whether absorption and secretion of bicarbonate are two processes that occur simultaneously or whether there is only one process whose direction varies. Turtle bladders both absorb and secrete bicarbonate, re-

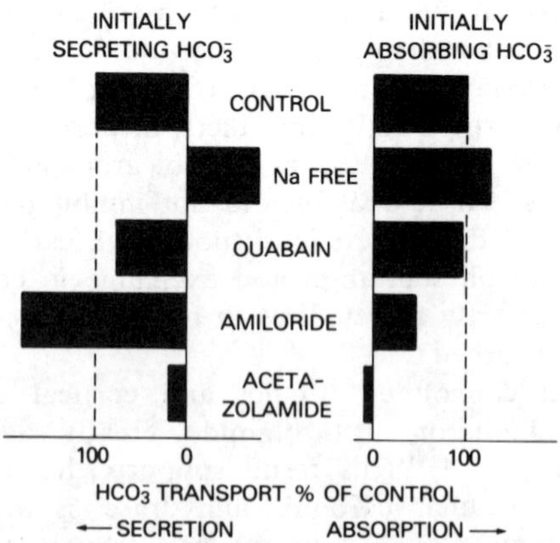

FIGURE 9. Effect of replacement of sodium by choline in the perfusate and bath, of ouabain (10^{-5}M in the bath), of amiloride (10^{-5}M in the perfusate), and of acetazolamide (10^{-4}M in the perfusate and bath) on the rate of bicarbonate transport by cortical collecting ducts perfused *in vitro*. The tubules are divided into two groups, depending on whether they were secreting or absorbing bicarbonate during the control periods (From McKinney and Burg.[16])

sembling cortical collecting ducts in that respect. It has been proposed [17] that there are two separate processes in the turtle bladder: 1. acidification by proton secretion and 2. direct bicarbonate secretion. If the same two processes occur in the cortical collecting ducts, the rate and direction of net bicarbonate transport must be determined by their balance, and changes in either or both could account for the results with choline, ouabain, and amiloride.

Secretion of bicarbonate by turtle urinary bladders was ascribed to the electrically-neutral exchange of bicarbonate for chloride that was absorbed. Although anion exchange may also be involved in bicarbonate secretion by cortical collecting ducts, there is no specific requirement for chloride. Bicarbonate secretion by cortical collecting ducts was unaffected when all of the chloride in the perfusate and bath was replaced by either nitrate or methyl sulfate.[16]

In summary, bicarbonate is transported by a number of different renal tubule segments. Both the rates of transport and the mechanisms involved differ between the segments. Although bicarbonate is absorbed by both proximal tubules and collecting ducts, bicarbonate secretion has been observed only in the collecting ducts. Bicarbonate secretion in cortical collecting ducts results from conditioning of that tubule segment, following administration of alkali to the animals.

References

1. Burg, M., M. Abramow & J. Orloff. 1966. Preparation and study of fragments of single rabbit nephrons. Am. J. Physiol. **210:** 1293–1298.
2. Burg, M. B., L. Isaacson, J. Grantham & J. Orloff. 1968. Electrical properties of isolated perfused rabbit renal tubules. Am. J. Physiol. **215:** 788–794.
3. Burg, M. & N. Green. 1973. Function of the thick ascending limb of Henle's loop. Am. J. Physiol. **224:** 659–668.
4. Vurek, G., D. Warnock & R. Corsey. 1975. Measurement of picomole amounts of carbon dioxide by calorimetry. Anal. Chem. **47:** 765–767.
5. Burg, M. B. & N. Green. 1977. Bicarbonate transport by isolated perfused rabbit proximal convoluted tubules. Am. J. Physiol. **233:** F307–314.
6. Malnic, G. & G. Giebisch. 1972. Mechanism of renal hydrogen ion secretion. Kidney Int. **1:** 280–296.
7. McKinney, T. D. & M. B. Burg. 1977. Bicarbonate and fluid absorption by renal proximal straight tubules. Kidney Int. **12:** 1–8.
8. Iino, Y. & M. Burg. Effects of acid-base status *in vivo* on bicarbonate transport by rabbit renal tubules *in vitro*. Submitted for publication.
9. Rosin, J. M., M. A. Katz, F. C. Rector, Jr. & D. W. Seldin. 1970. Acetzolamide in studying sodium reabsorption in diluting segment. Am. J. Physiol. **219:** 1731–1745.
10. McKinney, T. D. & M. B. Burg. 1977. Bicarbonate transport by rabbit cortical collecting tubules: effect of acid and alkali loads *in vivo* on transport *in vitro*. J. Clin. Invest. **60:** 766–768.
11. McKinney, T. D. & M. B. Burg. 1978. Bicarbonate absorption by rabbit cortical collecting tubules *in vitro*. Am. J. Physiol. **234:** F141–145.
12. Pitts, R. F. 1948. Renal excretion of acid. Fed. Proc. **7:** 418–426.
13. Murer, H., U. Hopfer & R. Kinne. 1976. Sodium/Proton antiport in brush-border membrane vesicles isolated from rat small intestine and kidney. Biochem. J. **154:** 597–604.

14. Cassola, A. C., G. Giebisch & G. Malnic. 1977. Mechanisms and components of renal tubular acidiflcation. J. Physiol. **267:** 601–624.
15. Rector, F. C. 1973. Acidification of the urine. *In* Handbook of Physiology. B. M. Brenner & F. C. Rector, Eds. Section 8: Renal Physiology. p. 431–454. Am. Physiol. Soc. Washington, D.C.
16. McKinney, T. D. & M. B. Burg. 1978. Bicarbonate secretion by rabbit cortical collecting tubules *in vitro*. J. Clin. Invest. **61:** 1421–1427.
17. Leslie, B. R., J. H. Schwartz & P. R. Steinmetz. 1973. Coupling between Cl^- absorption and HCO_3^- secretion in turtle urinary bladder. Am. J. Physiol. **225:** 610–617.
18. Burg, M. B. The renal handling of sodium chloride, water, amino acids, and glucose. *In* The Kidney. B. Brenner & F. Rector. Eds. 2nd edit. W. B. Saunders Co. Philadelphia, Pa. In press.
19. Burg, M. B. & J. Orloff. 1968. Control of fluid absorption in the renal proximal tubule. J. Clin. Invest. **47:** 2016–2024.

CHLORIDE ACTIVE TRANSPORT, MEMBRANE LIPIDS AND RECEPTORS IN THE CORNEAL EPITHELIUM *

José A. Zadunaisky, Barry E. Schaeffer, and Bruce Cherksey

Department of Physiology and Biophysics and
Department of Ophthalmology
New York University Medical Center
New York, New York 10016

The movement of salt across certain epithelia, whether by absorption or secretion, is the result of chloride active transport. The cornea of the eye, and the intestine are the best-known examples of this category of epithelia, in which the driving force is chloride secretion. In both tissues, the activation of the membrane enzyme adenylate cyclase, resulting in an increase in the concentration of cyclic AMP in the tissue, stimulates chloride secretion. The parallel between the corneal epithelium and the intestine is striking, and from knowledge of these two systems, it is possible to extrapolate to other less well-understood epithelia, such as the chloride-secreting rectal gland in sharks,[1] or the gill epithelium of fish.[2] All these systems utilize chloride secretion for osmoregulatory function, but the thick ascending loop of the mammalian kidney is a most typical example of chloride active transport.[3]

Here the corneal epithelium will be discussed. The following aspects will be described: A. A short introduction to corneal transport. B. New developments for a cellular model obtained from determinations of intracellular potential differences and intracellular ionic activities. C. The activation of specific membrane receptors and the effects on chloride transport of changes in membrane viscosity or fluidity. D. The kinetics of interaction between receptors and membrane bound proteins. The work represents recent results of this laboratory.

Corneal Chloride Transport

The corneal epithelium secretes chloride ions from the aqueous or endothelial side towards the tear or epithelial side. This movement of chloride occurs against an electrochemical gradient and fulfills all the basic criteria for active ion transport. Chloride transport is extremely sensitive to the concentration of sodium on the aqueous side.[4] This sodium activation, however, does not appear to act directly on the chloride translocating mechanism.[5] The function of chloride transport is to contribute to the dehydration of the stroma or connective tissue of the cornea that would otherwise tend to swell and become opaque. The osmotic pressure developed inside the stroma is counteracted by the secretion of ions and water, resulting in the maintenance of a minimum level of hydration. Any disturbance in this delicately balanced

* This work was supported by grants from the National Institutes of Health, Public Health Service Nos. EY01340, EY07009, EY03151 and GM 25002.

0077–8923/80/0341–0233 $01.75/0 © 1980, NYAS

transport mechanism in the endothelium and epithelium results in loss of corneal transparency.[6]

Classical inhibitors of ion transport, such as ouabain, markedly reduce the active chloride transport [6] and specific inhibitors of anion secretion, such as furosemide,[7] have a drastic effect on chloride transport. Activation of chloride transport, other than that which is sodium activated, also occurs when drugs or natural hormones are added to the cornea causing a resultant increase in cyclic AMP concentration in the epithelium.[8–10]

Epinephrine, theophylline, adenosine and prostaglandins are among these agents. FIGURE 1 illustrates our present knowledge of the relationship between corneal epithelial transport and the receptors on the cell membranes of the epithelium.[11]

Models for the Chloride Secretory Cell

There are two possibilities for the location and orientation of the ion pumps of the chloride secretory epithelium. One is that chloride penetrates into the cell, possibly from the stromal or aqueous side and a pump mechanism on the outside, or tear side membrane of the cell, takes care of the translocation against the electrochemical gradient. The second hypothesis, represented in FIGURE 2, is that the entry of chloride occurs by an active component

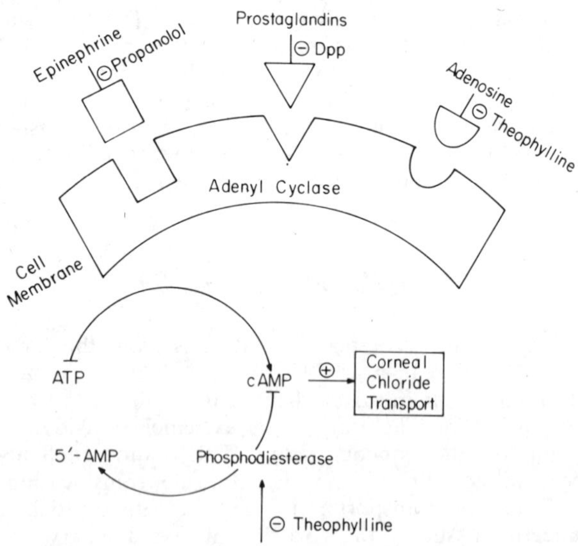

FIGURE 1. Membrane receptors affecting chloride secretion in the corneal epithelium. The specific agents acting on adenyl cyclase and their inhibitors are shown in a diagramatic manner, including the final effect on the production of cyclic AMP. The interactions between these receptors has been studied by means of electro-physiological methods, flux analysis and cyclic AMP determinations. (From Spinowitz and Zadunaisky.[11])

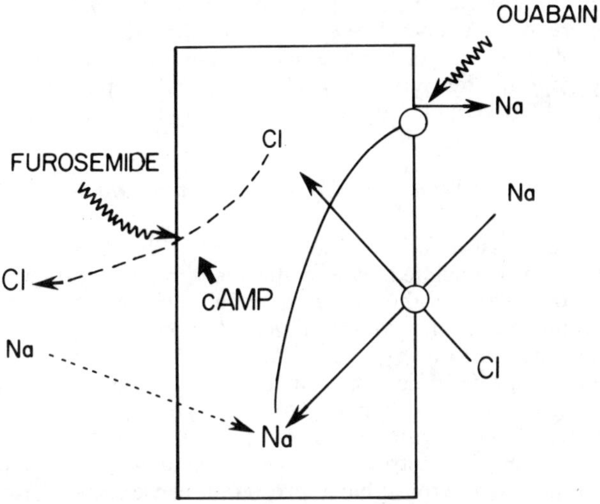

FIGURE 2. Cellular model for chloride secretion in the corneal epithelium. On the right side of the figure the events at the inner or stromal membrane of the epithelial cell are shown, and on the left side are shown the events at the outside or tear-facing membrane of the idealized cell. The sodium transport inhibited by ouabain keeps the sodium concentration low in the cell and an independent entry of Cl due to the sodium gradient keeps intracellular chloride concentration at approximately three times higher than predicted by the electrochemical gradient.[12] Diffusion or mediated exit of chloride occurs at the tear side membrane and is stimulated by cAMP, inhibited by furosemide.

in the inside or stromal portion of the cell and passive diffusion occurs at the tear or outside of the cell. The recent measurement of the intracellular activity of chloride ions in the cells of the corneal epithelium [12] tends to favor the model shown in FIGURE 2. The entry would consist of a sodium-coupled mechanism, with a low sodium concentration inside the cell maintained by the sodium pump located in the stroma side of the epithelial cell, and a higher chloride activity inside the cell that permits diffusion to the tear side. The chloride activity inside the cells of the corneal epithelium was found to be approximately three times higher than the activity predicted by the electrochemical gradient.[12] In fact, this description fits the model for the secretory epithelial cells of the intestine. There are other pieces of evidence that fall into place when this model is considered. For instance, the oxygen uptake of corneas incubated in appropriate solutions [13] or placed as membranes in chambers,[14] changes little when chloride is removed from the media bathing the cornea. Instead, a great change in oxygen uptake is observed when sodium is removed from the bathing media in incubated corneas [13] and especially from the inside or stromal side of the cornea.[14]

The action of stimulants and inhibitors of chloride transport of these epithelial cells can also be referred to this model of cotransport on the inside and passive diffusion on the outside. Epinephrine has been shown to affect the outside membrane of the cells [15] and this could explain the rapid effect of this hormone on chloride transport, with very little change in transcorneal potential difference. A complete picture of the action of stimulants and inhibitors is not available yet but the field is moving in the direction of localization of appropriate receptors in each side of the epithelial membrane by determination of intracellular electrical potentials and ionic activities.[12, 15]

Membrane Fluidity and Receptor-Enzyme Interactions

Membrane processes, such as the active chloride transport of the cornea described above, are mediated by specific membrane-bound proteins; that is, receptors and the enzymes they regulate. The cell membranes are in a state of dynamic equilibrium, and compositional changes exert profound influences on membrane function. For example, it has been shown that membrane lipids mediate changes in the fluid state of the membrane, thus controlling the capacity for membrane components to undergo topographic redistribution within the plane of the membrane. It is also well known that the immediate lipid microenvironment can strongly influence the activity of membrane-bound enzymes, and may, therefore, have profound consequences for membrane-mediated cell function. Thus, one can expect membrane lipid composition to exert an influence on chloride transport.

Recently it has been shown that the rate of active chloride transport across frog and rabbit cornea is affected by unsaturated fatty acid supplementations into corneal cell membranes.[16] Briefly, exogenous fatty acid supplementations are added to the medium bathing the corneas. As can be seen in TABLE 1, unsaturated fatty acids have the ability to stimulate the chloride secretion to varying extents, while saturated fatty acids have no effect. It would appear, also, that those unsaturated fatty acids having a longer chain length as well as a greater number of unsaturated bonds, have greater stimulatory ability (FIGURE 3).

The electrical response of the frog cornea following the addition of unsaturated fatty acids to the medium is quite rapid; it is initiated within 1–2 minutes, reaches a peak in approximately 10 minutes, and then gradually declines to a new baseline level (FIGURE 4). Similar levels of stimulation are observed for the rabbit cornea with unsaturated fatty acids, but the response is delayed in onset, probably due to the greater thickness of the cornea. When saturated fatty acids ranging in chain length from 16-C to 20-C were applied to mounted frog corneas in a similar manner, there were no observable effects.

^{14}C-linoleic acid was used as a representative fatty acid to check incorporation into cell membranes. Incorporation into whole cells after a 20 min. incubation in ^{14}C-linoleic acid (1 μCi/1 $\times$ 10^7 cells) was 27% of the applied label, equivalent to 1.5 μg of linoleic acid per 1 $\times$ 10^7 cells. Incorporation into the cell membrane fraction was found to be 49.7% of whole cell incorporation. Other studies have confirmed similar levels of incorporation into cell membranes of a large number of different saturated and unsaturated fatty acids.[17]

In order to confirm that the increase in current after the addition of unsaturated fatty acids was due to increased chloride transport across frog cornea, chloride-free Ringer's solution was substituted as the bathing medium. It was found that in the absence of chloride, only low currents could be obtained, and no stimulation could be elicited with fatty acid additions. Upon

TABLE 1

EFFECTS OF VARIOUS FATTY ACIDS ON THE SHORT-CIRCUIT CURRENT (SCC) OF ISOLATED FROG AND RABBIT CORNEAS *

	Unsaturated Fatty Acids	Chain Length: Unsaturations	% Increase in S.C.C.±S.E.M.	No. of Expts.
Frog cornea	Myristoleic	(14:1)	28±7.7	(5)
	Linoleic	(18:2)	44±8.9	(9)
	Linolenic	(18:3)	45±14.9	(5)
	Arachidonic	(20:4)	65±11.6	(7)
Rabbit cornea	Linoleic	(18:2)	69±9.1	(5)
	Saturated Fatty Acids	**Chain Length: Unsaturations**	**% Increase in S.C.C.±S.E.M.**	**No. of Expts.**
Frog cornea	Palmitic	(16:0)	0	(5)
	Stearic	(18:0)	0	(8)
	Arachidic	(20:0)	0	(4)

* Corneas were mounted in Ussing-type chambers, as described in the text, and fatty acids were added to the medium bathing both corneal surfaces with a final concentration of 100 μg/ml for the unsaturated fatty acids, and 16–80 μg/ml for the saturated fatty acids. The data is expressed as the mean percentage increase over control levels, ± the standard error. Note that for frog cornea, arachidonic acid (the most highly unsaturated fatty acid used) appears to have the greatest ability to increase SCC, while saturated fatty acids have no effect on SCC. (From Schaeffer and Zadunaisky.[16])

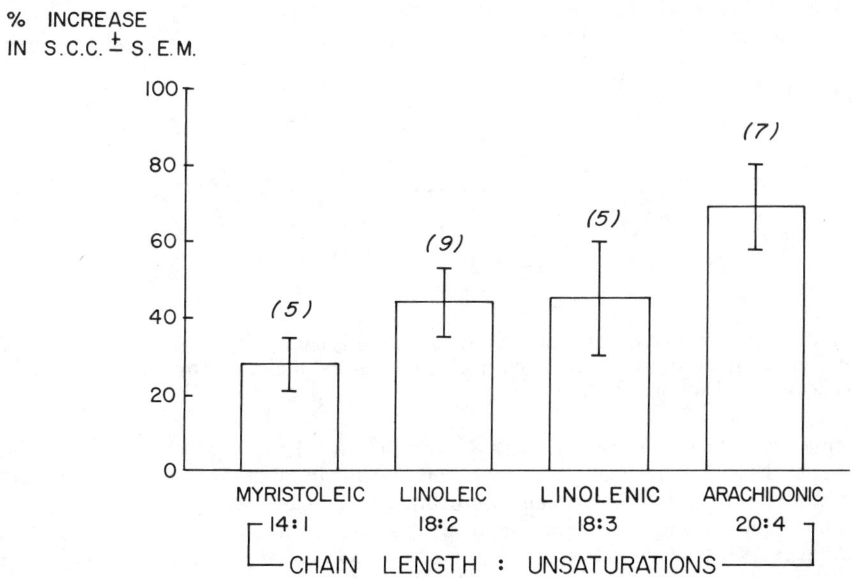

FIGURE 3. Effect of fatty acids of increasing unsaturation and chain length on SCC in frog cornea.

returning to chloride-containing Ringer's, typical responses were again observed (FIGURE 5).

In addition, the unidirectional ^{36}Cl flux increased by 56% ± 11.1 when linoleic acid was added to mounted corneas (FIGURE 6). Both of the above results appear to indicate that treatment with unsaturated fatty acids stimulate active chloride transport in the frog cornea.

Since the unsaturated fatty acid incorporation into corneal cell membranes results in the stimulation of chloride transport, we looked for changes in

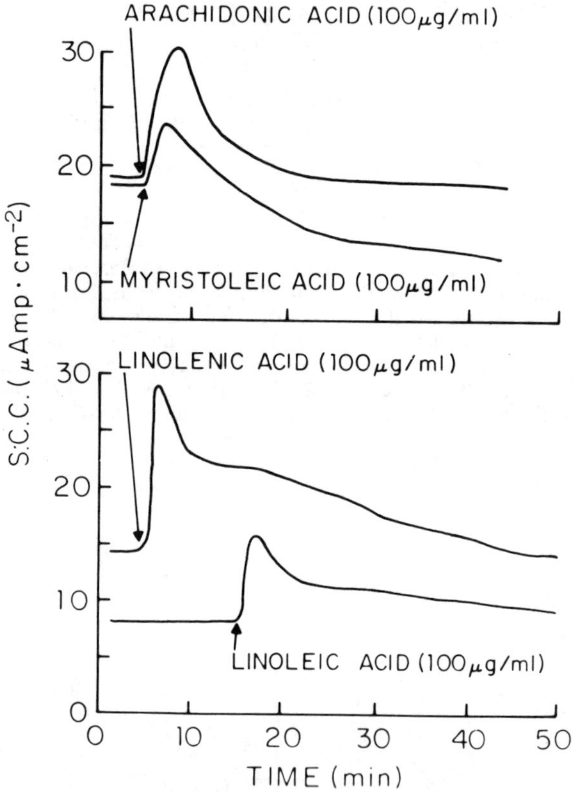

FIGURE 4. Stimulation of short-circuit current in isolated frog cornea by unsaturated fatty acid application to both corneal surfaces as indicated by arrows. (From Schaeffer and Zadunaisky.[16])

membrane fluidity using an assay described in detail elsewhere.[16] Isolated frog and rabbit corneal epithelial cells were incubated with linoleic acid (200 μg/ml) while control cells were untreated. Cells were then subjected to immunofluorescence labelling at 0 C, which results initially in a smooth, homogeneous ring fluorescent staining of the cell surface. After labelling, cells were given a further 15 minute incubation period at physiologic temperatures, during which time the labelled antigens diffused randomly in the plane of the membrane. Any collisions between antigens that occur at this time result in

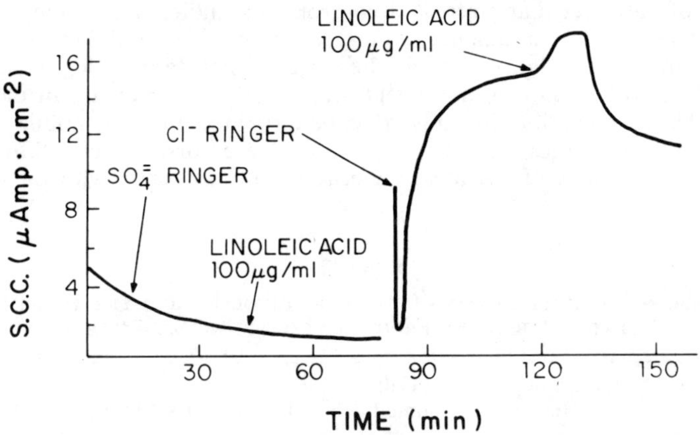

FIGURE 5. Isolated frog corneas bathed in chloride-free (sulfate) Ringer's show no response to linoleic acid application. When returned to standard (chloride-containing) Ringer's, the same corneas once again show typical response to linoleic acid treatment. Brief removal of fluid from corneal chambers during the wash out results in a discontinuity in the graph, followed by a rapid return of current to its former level. (From Schaeffer and Zadunaisky.[16])

cross-linking, and aggregates of fluorescent antibody begin to form that soon become visible as fluorescent patches in the membrane. Since the diffusion rate of membrane antigens is limited primarily by membrane lipid viscosity (the inverse of membrane fluidity), any observed change in the rate of fluorescent patching serves as an indicator of a change in membrane fluidity. We found that linoleic acid-treated corneal epithelial cells always have a higher

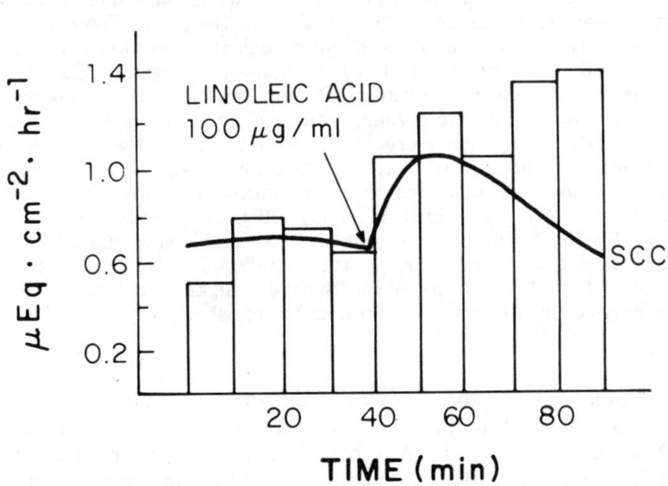

FIGURE 6. Addition of linoleic acid to mounted frog corneas results in the stimulation of 36-chloride active flux (bars) simultaneously with short-circuit current. (From Schaeffer and Zadunaisky.[16])

incidence of patching than untreated control cells, indicating an increased rate of membrane diffusion, and hence, an increase in membrane fluidity (TABLE 2).

Given the above data, it might be argued that fatty acid activation of chloride transport in cornea may result directly from an increase in membrane fluidity. The explanation for this may be provided by the mobile receptor hypothesis.[18] According to this hypothesis, receptors, which diffuse independently in the plane of the membrane, reversibly associate with the enzymes that they regulate. Receptor affinity for enzyme is greater when the receptor is occupied by the hormone. The level of enzyme activation must also be dependent upon the rate of encounter between receptor and enzyme, and should therefore be sensitive to changes in diffusion rate, which would increase the chances of receptor interaction with enzyme. In the corneal epithelium, the activation of adenylate cyclase to produce cAMP stimulates chloride transport (See FIGURE 1).

TABLE 2

MEMBRANE FLUIDITY CHANGES INDUCED BY LINOLEIC ACID TREATMENT OF ISOLATED RABBIT AND FROG CORNEAL EPITHELIAL CELLS *

	Experiment No.	No. Cells Counted	Rings	Patches	% Patches
Rabbit Cells					
Linoleic acid treated	1	67	26	41	61
	2	113	37	76	67
	3	199	72	127	64
Control	1	157	114	43	27
	2	85	46	39	46
	3	129	69	60	47
Frog Cells					
Linoleic Acid treated	4	205	118	87	42
Control	4	210	146	64	30

* Concentrated cell suspensions were prepared and incubated with linoleic acid (200 μg/ml) for 10 minutes prior to immunofluorescence labelling of the cell surface. Control cells were not treated with fatty acid. Labelling with fluorescent antibody is done on ice. Cells are then returned to physiological temperature for 15 minutes to allow for lateral diffusion of labelled membrane antigens. Collisions between polyvalent antibody molecules during diffusion result in cross-linking, producing antibody clusters within the membrane, which, when sufficiently large, may be seen as fluorescent patches by fluorescence microscopy. Since the rate of patching is limited primarily by the rate of antibody diffusion in the plane of the membrane, patching may be used as an indicator of membrane viscosity (fluidity). Each cell counted by microscopy is classified as having either ring fluorescence or patching of fluorescence. TABLE 2 shows the results of four such experiments, three of which were done on rabbit, and one on frog corneal epithelial cells. In each experiment the data indicate that linoleic acid-treated cells have a higher incidence of patching than untreated control cells. (From Schaeffer and Zadunaisky.[16])

The β-Receptor and Membrane Fluidity

α- and β-adrenergic agonists stimulate active chloride transport in the cornea[8, 9] and the presence of both α- and β-adrenergic receptors has been demonstrated in the rabbit cornea. The response of the chloride current to fatty acids strongly resembles a catecholamine response.[16] Therefore, the interaction between fatty acids and adrenergic receptors was tested.

Initially, both α and β receptors were blocked using the α-adrenergic antagonist phentolamine (10^{-5} M) and the β-adrenergic antagonist propranolol (10^{-5} M) by adding both to the Ringer's bathing the cornea. As shown in FIGURE 7, the initial response to the addition of the antagonists was a decrease in the short-circuit current. When the current had stabilized, about 30 min. after the addition of antagonists, arachidonic acid (600 μg/ml) was added to each side of the chamber. This caused an increase in the short-circuit current to only 8.5% above its base line value, compared to an increase of 65% in the control corneas. The combination of antagonists was sufficient to nearly eliminate the effects of fatty acids and the relative importance of α and β receptors in the stimulation of chloride transport by fatty acids was tested next. To this end, the antagonists were added individually to the Ringer's bathing the cornea, as shown in FIGURE 8. The lower recording is for a cornea treated with propranolol (10^{-5} M), i.e., β-adrenergic blocked. The upper recording is for a cornea treated with phentolamine (10^{-5} M) and was α-adrenergic blocked. Blocked corneas were then allowed to reach a stable short-circuit current, at which time arachidonic acid (600 μg/ml) was added to the Ringer's on both sides of the cornea. The α-blocked cornea exhibited an increase in the short-circuit current of 50% above the base-line value, not significantly different from the control increase of 64%. The effect of epinephrine (10^{-6} M) was significantly blocked in these corneas (FIGURE 8).

Corneas pretreated with the β-blocker propranolol exhibited an entirely different response. Arachidonic acid (600 μg/ml) produced only a 25% increase in the short-circuit current compared to the 64% control, which is a significant difference. In the β-blocked corneas, the effect of 10^{-6} M epinephrine was not significantly altered (TABLE 3).

It is apparent from these experiments that the increase in short-circuit current and chloride transport produced by administration of fatty acids involves the adrenergic receptors and that the effect is due to a mechanism involving mainly the β-adrenergic receptor.

Kinetic Models for Hormone-Receptor Interaction and Chloride Secretion

In order to gain a clearer understanding of the molecular events involved in the β-receptor-adenylate cyclase interaction, kinetic models have been developed and tested in turkey erythrocyte membrane fragments.[19] Two models are presented in FIGURE 9. The first model has been taken from the literature[20, 21] and the kinetic assumptions involved in its formulation may be found in the references cited. In this model, which is based on the mobile receptor hypothesis, the receptor (R) and enzyme (E) exist separately in the membrane. The initial step in activation is binding of Hormone (H) by the receptor unit. This is an equilibrium process characterized by a rate

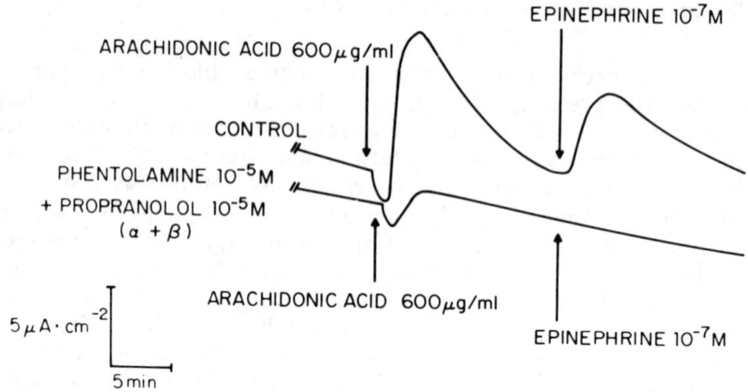

FIGURE 7. Representative responses of mounted frog corneas to adrenergic blocking agents, followed by arachidonic acid and epinephrine.

constant K_H. In the next step the hormone-receptor complex (HR) diffuses in the plane of the membrane colliding with E to form the hormone-receptor-enzyme complex (HRE). The rate of formation of HRE is a function of the fluidity of the membrane. In the next steps, the enzyme is activated (E) and dissociates from HR. Under appropriate conditions, this final step will be irreversible.[21] Kinetically, the observed rate constant for the model is first-order and a function of the collisional-coupling step.

Our system differs from the turkey erythrocyte in significant ways. First, epithelia are more complex tissues and the interactions can be occurring in

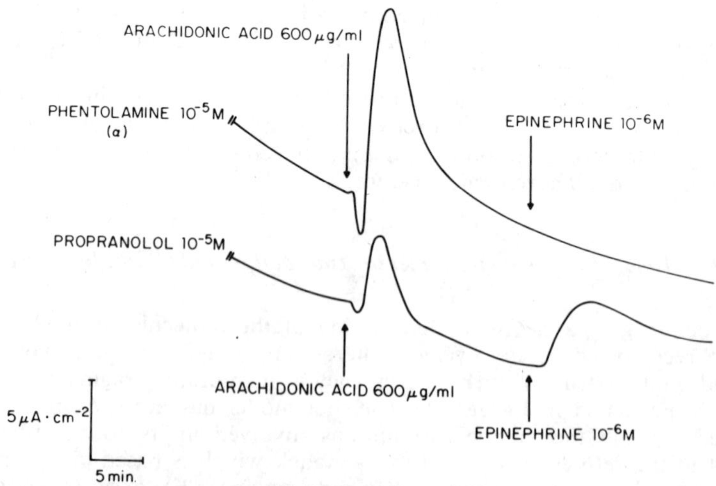

FIGURE 8. Representative responses of mounted frog corneas to adrenergic blocking agents, followed by arachidonic acid and epinephrine.

TABLE 3

EFFECT OF BLOCKING α- AND β-ADRENERGIC RECEPTORS ON SUBSEQUENT RESPONSE OF SCC IN FROG CORNEA *

	Percent Increase in SCC (Mean±SEM) Arachidonic Acid	Epinephrine
1. Phentolamine (10^{-5})	50±11.2 (3)	19±6.1 (6)
2. Propranolol (10^{-5})	23± 9.2 (4)	69±9.5 (6)
3. Phentolamine (10^{-5}) and Propranolol (10^{-5})	8.5± 1.5 (2)	0±0 (4)
4. Control	64± 8.1 (3)	104±39 (3)

* Numbers indicate percent increase in short-circuit current immediately following addition of arachidonic acid or epinephrine. Corneas were previously treated with appropriate antagonists as shown.

the outside facing or the inside facing membrane as discussed in the section on models for the corneal epithelial transport. Second, while the turkey erythrocyte exhibits no basal adenylate cyclase activity in the absence of hormone, the cornea is quite active without stimulation. The second kinetic model presented in FIGURE 9 allows for hormone-independent activity while preserving the details of the model for the turkey erythrocyte.

In our model in FIGURE 9, an activated receptor (*R′*) is postulated. The transition from inactive to active receptor would occur when *RH* was formed under equilibrium conditions. *R′* would then rapidly (10^{-13} to 10^{-16} secs) be formed. *R′* might then migrate in the plane of the membrane undergoing

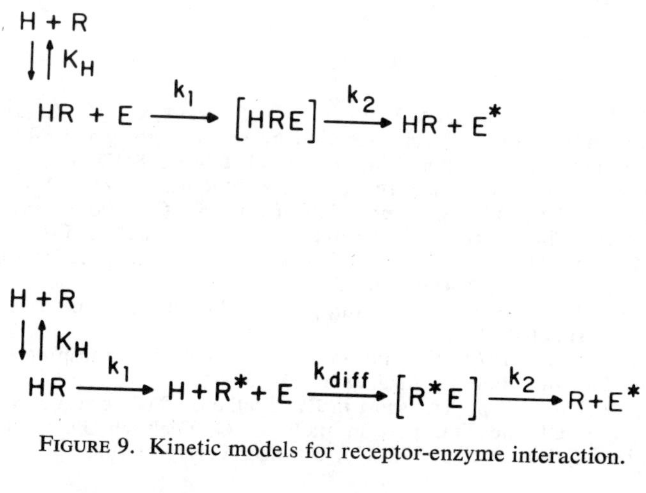

FIGURE 9. Kinetic models for receptor-enzyme interaction.

collisional coupling with E. The remainder of this model is the same as the first model.

A small population of active receptors would also exist in the absence of hormone if the transition from the inactive to the active state was due to a conformational change of the receptor. In this case, the transition activation energy (E_A) would be a function of the ground state vibrational energy levels.[23] A portion of the receptors would be expected to have energies above E_A.

This kinetic model also allows for the possibility that either the ground state vibrational level or E_A might shift due to alteration of membrane composition produced by incubation with the fatty acids. This would allow additional receptors (or the enzyme itself) to enter the active population.

This model follows first-order kinetics. The observed rate could be a function of a collisional-coupling step.

These results support the interpretation given for our previous data. If β-receptor sites are involved in fatty acid activation of chloride transport, then this would provide confirmation that fatty acids exert their influence at the level of the cell membrane. However, actions for fatty acids at other levels cannot be excluded at this time. According to the mobile receptor hypothesis, fatty acids might effect membrane-bound enzyme activity in several different ways. As was stated previously, an increase in membrane lipid fluidity would increase the diffusion rate of membrane-bound receptors and enzymes, and would have the effect of increasing the rate of enzyme catch by activated receptor. It is also known that membrane-bound enzyme activity is strongly influenced by lipid composition. Unsaturated fatty acids would tend to increase enzyme activity. Another possibility is that changes in membrane fatty acids might alter the conformation of receptor molecules that could effectively lower the activation energy required for receptor-enzyme interaction.

Acknowledgments

The efficient technical and secretarial assistance of Leon Garretson and Sally Wrigley are gratefully acknowledged.

References

1. Silva, P., J. Stoff, M. Field, L. Fine, J. N. Forrest & F. H. Epstein. 1977. Mechanism of active chloride secretion by shark rectal gland: Role of NaK ATPase in chloride transport. Am. J. Physiol. **233:** F298–F306.
2. Degnan, K. J., K. J. Karnaky Jr. & J. A. Zadunaisky. 1977. Active chloride transport in the *in vitro* opercular skin of a teleost (*Fundulus heteroclitus*), a gill-like epithelium rich in chloride cells. J. Physiol. **227:** 155–191.
3. Burg, M. & N. Green. 1973. Function of the thick ascending limb of Henle's loop. Am. J. Physiol. **224:** 659–668.
4. Zadunaisky, J. A. 1966. Active transport of chloride in frog cornea. Am. J. Physiol. **211:** 506–512.
5. Zadunaisky, J. A. 1972. Sodium activation of chloride transport in the frog cornea. Biochim. Biophys. Acta. **282:** 255–257.
6. Zadunaisky, J. A. 1978. Transport in Eye Epithelia: The cornea and crystalline lens. *In* Membrane Transport in Biology. G. Giebisch, D. C. Tosteson & H. H. Ussing, Eds. Vol. **3:** 307–335. Springer-Verlag. Berlin.

7. CANDIA, O. A. 1973. Short-circuit current related to active transport of chloride in frog cornea; effects of furosemide and ethacrynic acid. Biochim. Biophys. Acta. **298:** 1011–1014.
8. CHALFIE, M., A. H. NEUFELD & J. A. ZADUNAISKY. 1972. Action of epinephrine and other cyclic AMP-mediated agents on the chloride transport of the frog cornea. Invest. Ophthal. **11:** 663–650.
9. KLYCE, S. C., A. H. NEUFELD & J. A. ZADUNAISKY. 1973. The activation of chloride transport by epinephrine and Db cyclic-AMP in the cornea of the rabbit. Invest. Ophthalmol. **12:** 127–139.
10. BEITCH, B. R., I. BEITCH & J. A. ZADUNAISKY. 1974. The stimulation of chloride transport by prostaglandins and their interaction with epinephrine, theophylline, and cyclic AMP in the corneal epithelium. J. Membrane Biol. **19:** 381–396.
11. SPINOWITZ, B. S. & J. A. ZADUNAISKY. 1979. Action of adenosine on chloride active transport of isolated frog cornea. Am. J. Physiol. **237(2):** F121–127.
12. ZADUNAISKY, J. A., K. R. SPRING & T. SHINDO. 1979. Intracellular chloride activity in the corneal epithelium. Fed. Proc. **38:** 1059.
13. BENTLEY, P. J. & T. YORIO. 1978. The properties of active Cl transport across the cornea of the toad (*Bufo marinus*). Comp. Biochem. Physiol. **61A:** 635–641.
14. ZADUNAISKY, J. A. 1979. Characteristics of chloride secretion in some non-intestinal epithelia. *In* Mechanisms of Intestinal Secretion. H. J. Binder, Ed. : 53–64. Alan R. Liss. New York.
15. KLYCE, S. D. & R. K. S. WONG. 1977. Site and mode of adrenalin action on chloride transport across the rabbit corneal epithelium. J. Physiol. **266:** 777–799.
16. SCHAEFFER, B. E. & J. A. ZADUNAISKY. 1979. Stimulation of chloride transport by fatty acids in corneal epithelium and relation to changes in membrane fluidity. Biochim. Biophys. Acta. **556:** 131–143.
17. CURTIS, A. S. G., C. CHANDLER & N. PICTON. 1975. Cell surface lipids and adhesion. III. The effects on cell adhesion of changes in plasmalemmal lipids. J. Cell. Sci. **18:** 375–384.
18. CUATRECASAS, P. 1974. Membrane Receptors. Annu. Rev. Biochem. **43:** 169–214.
19. RIMON, G., E. HANSKI, S. BRAUN & A. LEVITZKI. 1978. Mode of coupling between hormone receptors and adenylate cyclase elucidated by modulation of membrane fluidity. Nature **276:** 394–396.
20. TOLKOVSKY, A. & A. LEVITZKI. 1978. Mode of coupling between the β-adrenergic receptor and adenylate cyclase in turkey erythrocytes. Biochemistry **17:** 3795–3810.
21. TOLKOVSKY, A. & A. LEVITZKI. 1978. Coupling of a simple adenylate cyclase to two receptors: adenosine and catecholamine. Biochemistry **17:** 3811–3817.
22. CUATRECASAS, P., V. BENNETT & S. JACOBS. 1975. Irreversible stimulation of adenylate cyclase activity of fat cell membrane by phosphoramidate and phosphonate analogs of GTP. J. Membrane Biol. **23:** 249–278.
23. BENSON, S. 1960. The Foundations of Chemical Kinetics. McGraw-Hill. New York.

CURRENT STATUS OF MEMBRANE-BOUND CARBONIC ANHYDRASE *

Thomas H. Maren

Department of Pharmacology and Therapeutics
University of Florida College of Medicine
Gainesville, Florida 32610

INTRODUCTION

Since the discovery of carbonic anhydrase 47 years ago, the main attention paid to its properties has been devoted to the cytoplasmic enzyme, not only in red cells but also in tissues. So far as techniques allow, it appears that this enzyme is not present in red cell membranes.[1] In the kidney, however, it gradually became apparent that there was a small amount of carbonic anhydrase in both mitochondria and microsomes, relative to that in the supernate. In terms of specific activity, however, the enzyme concentration in the particulate fractions was of the order of one-third that in the supernatant.[2-4]

With new techniques to separate membranes, carbonic anhydrase was found in both brush border and basolateral fractions, by free-flow electrophoresis [5] and gradient centrifugation.[6] In both of these studies, care was taken to exclude contamination from the soluble fraction, and to use appropriate markers for the membranes.

As new data and ideas developed in secretory physiology, and specifically in the field of renal acidification, it became clear that secretory function must depend on vectorial properties of cells, and that membranes must be able to maintain charge separation. Characteristics of the cytoplasm—pH, for example—are of less interest than pH of cell surfaces. Similarly, the properties of cytoplasmic carbonic anhydrase, which was the enzyme type almost universally analyzed until very recently, may not be directly relevant to physiological processes. It is thus essential to generate rate and equilibrium constants for the membrane carbonic anhydrase, and compare these to the cytoplasmic enzymes, particularly to those of the red cell, whose structure is known and is available in chemically pure form.

The present paper has two objectives: To give the kinetic properties and susceptibility to inhibition of the renal membrane carbonic anhydrase, based on previous and present work; and to show a scheme for the workings of membrane and cytosolic carbonic anhydrase in renal bicarbonate reabsorption and acidification.

METHODS

Dog kidneys were thoroughly perfused with saline through the renal artery, following removal. The fractions were prepared by a centrifugation

* Suppc :ed by NIH Grants EY 02227 and HL 22258

0077-8923/80/0341-0246 $01.75/0

scheme yielding crude microsomes, which were then washed 3 times with buffer before analysis. The crude microsomes were also subjected to selective precipitation yielding the antiluminal or basolateral membranes, and a supernate ultimately yielding a washed luminal fraction.[7] The kinetics of CO_2 hydration were studied as recently described,[8] except that a temperature of 0.5 C was used. In inhibition studies, enzyme, substrate and drug were mixed for one-four minutes to attain equilibrium before the reaction was started by the addition of buffer. It is important to note that this time varies both with drug and with enzyme preparation. Further details are given in text and figure legends. The chemistry of the drugs has been described.[9] They were obtained from Lederle Laboratories, Pearl River, New York.

Results of Present and Previous Investigations

Recovery of Enzyme from Kidneys

The whole kidney from dog has about 400 enzyme units per gram wet weight, and from rat about 250.[9] These enzyme units are defined specifically for the system then used (CO_2 hydration at 0 degrees using 100% gas, in a 7 ml volume, in carbonate buffer) where one enzyme unit doubles the uncatalyzed rate. However, for cytoplasmic carbonic anhydrases in secretory cells (except in the rumen of cattle), these units are convertible into molar equivalents by the relation one enzyme unit per gram tissue equals 17 nM. Thus, the whole dog kidney has about 7 μM, and the rat kidney about 4 μM. In terms of these enzyme units, only about 10–20 units per gram kidney are in the microsomes of dog or rat [3,5] and the same fraction, 3–5% of the whole kidney, is reported for human kidneys.[4,10] However, in terms of enzyme units/mg protein the value (for all species) is about three for homogenate and one or two for the microsomes.[3,4] The concentration of enzyme (units/mg. protein) in brush border and basolateral membranes are about the same.[5,6]

Physical and Immunological Properties of Renal Membrane Carbonic Anhydrase

The "standard carbonic anhydrase" is C (or II) from human red cells, which has been thoroughly characterized with respect to structure, activity, and immunological properties.[11] It is of high activity and sensitivity to sulfonamides, compared to human B (or I), or the very low activity enzyme in muscle and liver of male rats.[11–13] Whole human kidney yields an enzyme indistinguishable from C, in terms of amino acid composition, electrophoretic mobility, immunology, and kinetics.[10] Similarly, the soluble enzyme from rat kidney has nearly identical properties to the high activity form in rat red cells.[14] †

When membranes are separated either by differential centrifugation or electrophoresis, they yield activity in the assay system, even though they are not truly solubilized,[3–5] as noted above. However, further characterization

† It may be noted, however, that the rat kidney enzyme does have a two-fold lower k_{cat} and two-fold higher K_I against sulfanilamide than human red blood cell C.[14]

requires solubilization, and this has been achieved, in one case, by sodium dodecyl sulfate (which curiously inactivates the supernatant but not the microsomal carbonic anhydrase),[4] and in another by 0.2% Triton 100 at pH 9.7.[15] The data from these two papers are fragmentary, but clearly suggest that the microsomal carbonic anhydrase is different from the supernatant. It is suggested that the molecular weight is of the order of 66,000 or less; that of human red cell C and other mammalian carbonic anhydrases is 30,000 moles. The amino acid composition of human membrane-bound renal carbonic anhydrase is different from that of the cytoplasmic enzyme which, in turn, is the same as red cell C, and is also immunologically distinct.[15]

Kinetic Properties

There have been no previous data on the critical values, k_{cat}, or turnover number of any membrane bound carbonic anhydrase. Since $k_{cat} = V_{max}/E$ it is required that V_{max} be generated by Lineweaver-Burk or equivalent data, and that E is known. Except for red cell enzymes, purification has not been done to yield E in molar terms, but fortunately there is another route to finding E. This depends on the fact that there are inhibitors of the enzyme powerful enough to be active when present in about the same concentration as enzyme itself. Thus, the enzyme is titrated with a drug that binds it on an equimolar basis, and from the residual activity the molar concentration of E, and also the K_I of the drug-enzyme complex, is generated.[16] Thereafter, any level of enzyme activity can be translated into molar terms; that, in fact, is the basis of the conversion given above for whole kidney. In the system we are currently using,[8] the barbital assay at 0 degrees, one enzyme unit in the assay system $= 10^{-9}$ M of C, and 7×10^{-9} M of the B enzyme.

FIGURES 1 and 2 show lines to yield E, and K_I for chlorazolamide (CL 13,580) or benzolamide (CL 11,366) for the canine cytoplasmic and membrane carbonic anhydrase from kidney. The equation used[16] is

$$\frac{I}{i} = K_I\left(\frac{1}{1-i}\right) + E$$

where I is the concentration of drug, i is the fractional inhibition, K_I is the enzyme drug dissociation constant, and E is the concentration of enzyme present. All are molar terms except i, which is dimensionless. FIGURE 1 yields the molar values of E based on titration of the enzymes with chlorazolamide. This is the chemical of choice for such work, since it requires little or no equilibration time with enzyme, and is perhaps the most active of all the sulfonamides. E per enzyme unit is 1.5×10^{-9} M for the cytosolic enzyme (very close to the value for C, as noted above) and 3×10^{-9} M for microsomes. The differing values for K_I will be considered in the next section. FIGURE 2 shows a similar plot for benzolamide against microsomes, yielding the same value for E. (Benzolamide cannot be run accurately against cytosolic enzyme because the long incubation period yields some deterioration of enzyme.)

Luminal and antiluminal membranes yielded the same lines as shown in FIGURES 1 and 2 for crude microsomes.

Having precise molar values for enzyme concentrations, we can now find the critical kinetic parameters for k_{cat} and K_m. FIGURE 3 shows reciprocal plots of renal cytoplasmic and membrane carbonic anhydrase compared to human red cell B and C. It is clear that the supernatant renal enzyme is closely allied to C; the membrane renal enzyme has distinct properties from the other three. These relations are summarized in TABLE 1.

An unexpected finding of great interest was the high resistance of membrane enzyme to inhibition by I^- (TABLE 1) and Cl^-. The latter (not shown) had no effect at 0.5 M, while the K_I for cytosol enzyme was 63 mM. These may

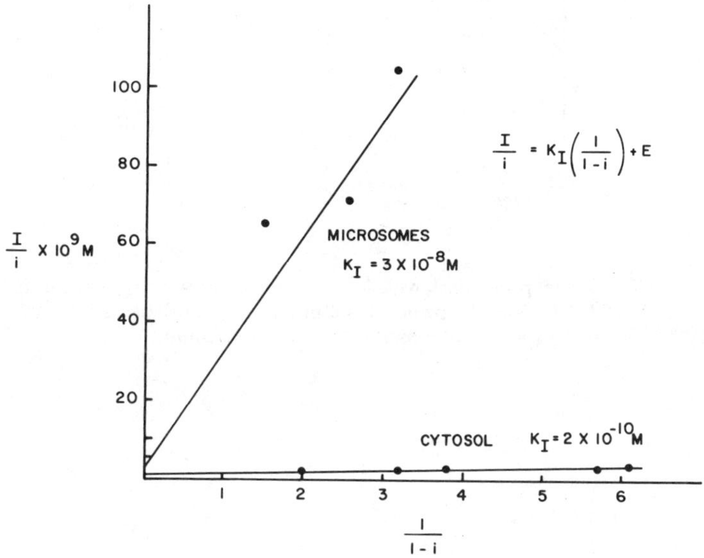

FIGURE 1. Easson-Stedman plot for chlorazolamide (CL 13,580, 2-o-chlorphenyl-thiadiazole-5-sulfonamide) inhibition of canine renal carbonic anhydrases. Hydration reaction at 0.5 C, using 70 mM CO_2 in barbital buffer. Uncatalyzed rate constant 0.0025 sec^{-1}. Enzyme, substrate and drug were mixed for one minute, and reaction was started by addition of buffer.[8] All experiments were carried out with/or data adjusted to one unit of enzyme, i.e., the amount that doubles the uncatalyzed rate. E (ordinal intercept) is 1.5×10^{-9} M for cytosol and 3×10^{-9} M for microsomes.

be compared to values for C (200 mM) and B (6 mM).[17] This is compelling physiologically, since membrane enzyme is probably exposed to higher Cl^- concentrations than the cytoplasmic; it may also be a very useful marker for distinguishing membrane from cytoplasmic carbonic anhydrase. This finding does recall our recent work on carbonic anhydrase from the shark rectal gland. This organ handles 0.5 M NaCl, and the K_I against I^- and Cl^- was about 10 times greater than against human red cell C, and nearly 1000 times greater than against human red cell B.[18]

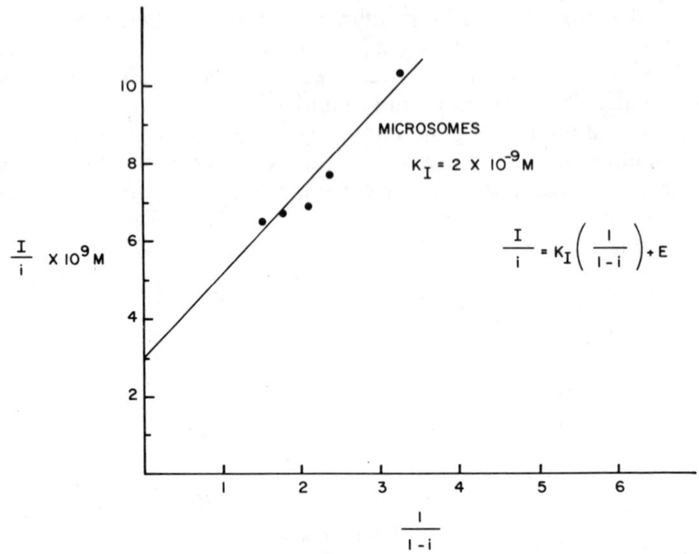

FIGURE 2. The same procedure was followed as in FIGURE 1, except for the use of benzolamide (CL 11,366, 2-benzenesulfonamido-1, 3, 4, thiadiazole-5-sulfonamide). $E = 3 \times 10^{-9}\, M$ for microsomes. Incubation time was five minutes.

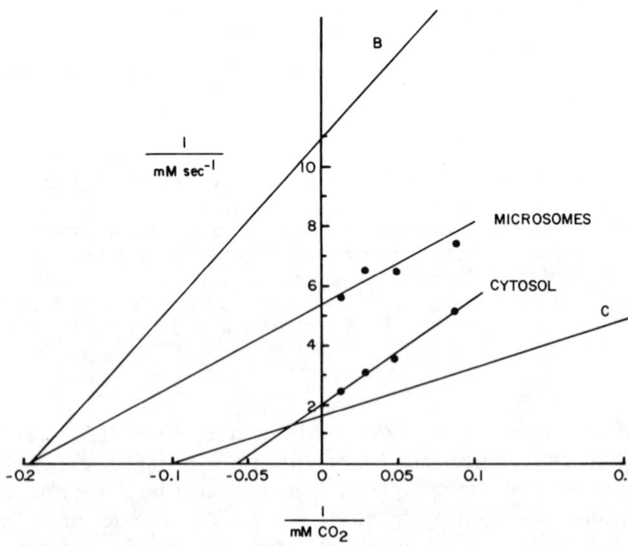

FIGURE 3. Plots of enzymic rate vs. substrate for the canine renal carbonic anhydrases. The lines for human B and C are taken from literature values[11] and current experiments. The same procedure as that given for FIGURE 1 was followed. Substrate variations were made by using different concentrations of gaseous CO_2.

Inhibition by Sulfonamides

TABLE 2 summarizes data on seven sulfonamides against four carbonic anhydrases: human red cell B and C, renal supernatant, and microsomes. Where luminal and antiluminal fractions were tested, results were equal and the same as for crude microsomes.

The first two columns show that the drugs vary widely in their ability to discriminate between the low activity (B) and high activity (C) isozymes in the red cells. It was at one time supposed that low activity enzyme was equivalent to relatively low sulfonamide inhibitory activity; this is indeed true for the first two drugs studied, sulfanilamide and acetazolamide. But methazolamide shows the same activity against human red cell isozymes B and C. For the other four drugs, there is probably a two–fivefold difference, but the K_I of these against human red cell C is only approximate, since the concentrations of I and E in the reaction are almost exactly equal.[16]

TABLE 1

KINETIC PROPERTIES OF RED CELL AND RENAL CARBONIC ANHYDRASES–CO_2 HYDRATION AT 0°

	k_{cat} * $sec^{-1} \times 10^{-4}$	K_m† mM	K_I KI mM
Human red blood cell			
B	3	5	0.5
C	22	10	20
Canine kidney			
Supernate	17	17	9
Microsomes	6	5	130

* $k_{cat} = V_{max}/E$ from FIGURE 3.
† K_m from FIGURE 3.

Turning to the renal enzymes, it is apparent that for the dog the drugs are more active against the supernatant than the microsomal fraction. The difference in K_I varies from two to 135-fold, depending on the drug and the study. In the main, this confirms our fiindings of twelve years ago, when we used cruder isolation techniques.[3] For the rat, no differences in K_I were found in a relatively small number of observations for acetazolamide and methazolamide.[6] For ethoxzolamide and CL 13,475, there was a four–fivefold difference in I_{50},[5] but since a microsomal enzyme has lower activity per mole (or greater E per unit) the larger amount of enzyme present in the microsomal experiment dictates that the K_Is calculated from $I_{50} - \frac{1}{2} E$ may be quite similar in the two fractions. (See TABLE 6 in Reference 9.) More data are needed here.

It must be emphasized that these lower values for K_I of the inhibitors against microsomes do not enter into the controversy, surprisingly still extant,[19] as to whether renal carbonic anhydrase is completely inhibited by 20 mg/kg acetazolamide. The basis for thinking that this *is* the case, which was clearly

TABLE 2

K_I OF SULFONAMIDES AGAINST RENAL CARBONIC ANHYDRASES COMPARED TO HUMAN RED CELL ISOZYMES B AND C, $(M \times 10^8)$ *

	B	C	Super-natant	Micro-somes	Species	Ref.
Sulfanilamide	5000	200	340	600	Dog	3
			190	850	Rat	5
				400	Man	15
Acetazolamide	20	1	3	12	D	3
			1	12	D	P
			2	2	R	6
Methazolamide	1	1	3	9	D	3
			2	12	D	P
			2	2	R	6
Benzolamide	0.2	0.04	0.08	0.2	D	P
CL 13,580	0.1	0.04	0.2	3.0	D	3
			0.02	2.7	D	P
Ethoxzolamide	0.1	0.02	0.2	0.8	R	5 (I_{50})
CL 13,475	0.2	0.04	0.2	1.0	R	5 (I_{50})

* For the first three drugs, $K_I = I_{50}$ since $I_{50} > E$. For the next four drugs, the K_I was calculated from the relation $K_I = I_{50} - \frac{1}{2} E_0$. For the last two drugs, only I_{50}s are available for the renal enzymes. Incubation times for equilibrium were two minutes or less for sulfanilamide, methazolamide, acetazolamide and CL 13,580. For ethoxzolamide and benzolamide, it is about five minutes. In the special case of microsomal enzyme, however, equilibrium appears complete for all drugs in $<$ one minute. P indicates present data.

stated in the first renal experiment ever done with this drug, is that the dose response curve becomes flat between 5–20 mg/kg.[20] The calculated fractional inhibition (i) of course depends on the K_I used; the value originally used was 6×10^{-8} M (taken from 38° data on whole kidney homogenate) yielding $i = 0.9998$ for the 20 mg/kg dose.[21] The present microsomal value (TABLE 1) of 12×10^{-8} M would yield $i = 0.9995$. More precisely, this must be determined at 37° (rather than 0°). I do not feel that these relatively small differences are of importance for understanding of the physiology.

The fractional inhibition (i) at which the dose response curve becomes flat is an observed and calculated value, and depends in the last analysis on how much residual enzyme is necessary for function. This may vary with the organ system [22]; or, for a single organ, will probably change with variations in function.

Roles of the Enzymes at the Several Sites in Kidney

The older diagrams in which CO_2 and H_2O and the products of their equilibria are depicted as if in a bag of fluid in cells, must give way to more realistic schemes in which pH gradients are generated within cells, and in which the role of carbonic anhydrase in the membranes is recognized.

FIGURE 4 is an attempt to do this. The essential features are: 1. The membranes on the luminal side are cathodic, and able to build up H^+ concentration, while membranes at the blood side are anodic and attract OH^- and HCO_3^- ions; 2. Cellular energy is transduced in part into the protolysis of water; and 3. Carbonic anhydrase is involved in the reactions shown at both membranes, the formation of H^+ at the luminal and of HCO_3^- at the blood side.

The first two of these are essential for the operation of the system, the third provides only an acceleration of the rate. The scheme is a general one, and applies to most secretory epithelia, with appropriate modifications as to orientation of acid or base with blood or lumen. A brief look at some other tissues is of interest with regard to the importance of the third or catalytic element. In the stomach, the acid and base gradients are so enormous that the uncatalyzed reactions are adequate to provide gastric acid; total inhibition of carbonic anhydrase reduces (by some 80%) but does not abolish acid formation.[9] In the case of aqueous humor secretion, it is likely that gradients in the ciliary processes are much lower, but rates of HCO_3 formation are also low; here complete inhibition reduces the rate by about 70%.[23] In both the stomach and eye, calculation of the uncatalyzed rates yields values close to what is the observed rate after complete inhibition.[9, 23]

In the kidney, the analysis is uniquely complicated by the fact that there is a second mechanism for HCO_3^- reabsorption (shown at the right of FIGURE 4), independent of catalyzed or uncatalyzed hydroxylation of CO_2 and of enzymology which we have been discussing. That is why in the normal intact mammal, the effect of complete inhibition (on HCO_3^- reabsorption)

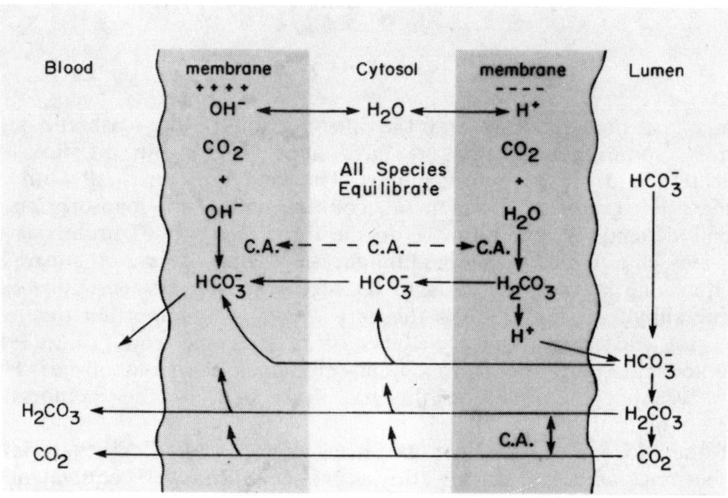

FIGURE 4. Scheme for proton formation and HCO_3^- reabsorption in the proximal tubule. The two processes may occur at separate or the same sites. C.A. = carbonic anhydrase. See text.

is not 70%–80%, as in other organs, but only 20%.‡ (I have discussed this[26] and do not wish to divert attention now from the scheme of FIGURE 4 with regard to CO_2 and water.) In the case of the kidney, the rates of CO_2-induced acidification controlling both H^+ and NH_4^+ output and HCO_3^- reabsorption are high, and the pH gradients low. Thus it is reasonable that the enzyme is a requirement for the observed rate; uncatalyzed rate is small. Attempts to calculate the uncatalyzed reaction yield—unlike other organs—rates less than 10% of the enzyme related process.[26]

The renal processes, despite their special elements of "HCO_3^- reabsorption as such" and negligible uncatalyzed rate, may be regarded as general with respect to protolysis of water, charged membranes and carbonic anhydrase catalysis. FIGURE 4 shows the formation of HCO_3^- at the blood side and formation of H^+ at the luminal side, without reference to the intricacies of the catalytic mechanism. (This is a separate matter, of much current interest and some dispute, but need not affect the present line of thought, since for any proposed mechanism, the initial and final products are the same.) There is abundant evidence to support such orientation; the actual pH in the membranes is not known but may be guessed at 8–9 at the blood and 5–6 at lumen. It may be noted that in this model the function of luminal enzyme is not the dehydration of H_2CO_3 (although this will be inevitable), but of formation of H^+. Nor does the term "luminal" mean that the enzyme is actually facing the lumen. These constraints do not appear necessary, since H_2CO_3 and HCO_3^- are highly permeable across the peritubular membrane.[27]

What is the role of the cytosolic enzyme? As indicated in FIGURE 4, it would allow rapid equilibration of all the species of CO_2 in cell water, stabilizing pH. Perhaps, also, cytosolic carbonic anhydrase may be a source or precursor for the membrane-bound enzyme. Does the cytosolic enzyme furnish pressure for production or binding of the membrane enzyme? Are the differences in properties due to the binding?

Tests of the Model

Since the time of Pitts and the introduction of the carbonic anhydrase inhibitors, some thousand papers have appeared on the question of renal acidification and HCO_3^- reabsorption. The data are consistent with a model in which H^+ gradients develop in concert with Na^+ reabsorption. These gradients depend, as noted above, on the three elements of protolysis of H_2O, membrane charge and carbonic anhydrase. Thus, it is not surprising that more than one means is at hand to reduce or abolish H^+ secretion. Although carbonic anhydrase inhibition is the best known, it is important not to neglect similar (at least qualitatively) effects of maleic acid and lithium[26] and 4-pentenoic acid.[28] In the stomach, an electrical charge abolishes H^+ secretion.[29] Whether these act on the protolysis step, or the membrane itself, remains to be found.

At the blood side, far less has been done. It is clear that peritubular fluid becomes alkaline during the renal acidification,[30] equivalent to the

‡ Micropuncture data taken from superficial proximal loops in the *in situ* kidney of the deeply anesthetized rat show that carbonic anhydrase inhibition causes rejection of nearly 100% of filtered HCO_3.[24, 25] The reason this is so different from findings in the intact animal has been widely discussed, but not yet explained.

alkaline tide of gastric secretion. It will be important to explain why the substitution of Cl^- or HCO_3^- in peritubular capillaries causes depolarization of the antiluminal membrane, and acetazolamide nearly abolishes the change.[27] Enzyme inhibition may alter charge or pH at membranes, due to decrease of the catalytic reactions shown in FIGURE 4.

Attention has also turned to the differential effects of the several inhibitors when added from the lumen, on HCO_3^- reabsorption and H^+ formation;[25] this may be due to varying rates of penetration, or different rates of achieving equilibrium with the two types of enzymes or different equilibrium values (TABLE 2).

This fragmentary summary of recent work is intended to show the significant shift in emphasis toward events at the membranes. The present data is a beginning toward definition of kinetics and inhibition of membrane-bound carbonic anhydrase, so that quantitative physiological experiments may follow.

SUMMARY

We have studied the kinetic properties, and susceptibility to inhibition, of cytoplasmic and membrane carbonic anhydrase from dog kidneys, and attempted to place the data in the context of earlier work on this subject.

The cytoplasmic enzyme thus far seems the same as human red cell carbonic anhydrase C, on the basis of kinetics, inhibition, amino acid composition and immunochemistry.

On the other hand, the membrane enzyme is quite a different protein from either the cytoplasmic, or human red cell B or C. This enzyme is found in both luminal (brush border) and antiluminal (basolateral) fractions, and there appear no differences between the two. The turnover number (k_{cat}) lies between those of B and C, and susceptibility to sulfonamide inhibition is two to 135-fold less than for the cytoplasmic enzyme, depending on the drug used. The usual difference is about fivefold. The K_I for acetazolamide against the membrane enzyme is 10^{-7} *M*, so that at the renal concentrations achieved at the usual *in vivo* doses (~20 mg/kg) or used in current *in situ* perfusion work (both 10^{-4} *M*) the enzyme is 99.9% inhibited. A striking difference between the membrane carbonic anhydrase and cytoplasmic or red cell B or C is its resistance to inhibition by halions. At 0.5 M chloride, there is no effect, whereas for the other three types inhibition ranges from 70%–99%. The membrane renal enzyme is also immunologically distinct from the other three types. The membrane enzyme has activity in its native state, but can be solubilized without loss of activity by treatment with Triton or sodium dodecyl sulfate.

The actions of the renal carbonic anhydrases are depicted in a scheme that takes into account the protolysis of water, the attraction of H^+ and of OH^- (and HCO_3^-) to the luminal and antiluminal membranes respectively, and the catalytic hydration and hydroxylation of CO_2.

ACKNOWLEDGMENTS

I thank Dr. Gautam Sanyal and Mr. Ness Pessah for carrying out with particular fidelity the present experiments reported here. A full account of this work will appear separately.

REFERENCES

1. RANDALL, R. F. & T. H. MAREN. 1972. Absence of carbonic anhydrase in red cell membranes. Biochim. Biophys. Acta **268:** 730–732.
2. KARLER, R. & D. M. WOODBURY. 1960. Intracellular distribution of carbonic anhydrase. Biochem. J. **75:** 538–543.
3. MAREN, T. H. & A. C. ELLISON. 1967. A study of renal carbonic anhydrase. Mol. Pharmacol. **3:** 503–508.
4. MCKINLEY, D. N. & P. L. WHITNEY. 1976. Particulate carbonic anhydrase in homogenates of human kidney. Biochim. Biophys. Acta **445:** 780–790.
5. WISTRAND, P. J. & R. KINNE. 1977. Carbonic anhydrase activity of isolated brush border and basal-lateral membranes of renal tubular cells. Pflügers Arch. **370:** 121–126.
6. EVELOFF, J., E. R. SWENSON & T. H. MAREN. 1979. Carbonic anhydrase activity of brush border and plasma membranes prepared from rat kidney cortex. Biochem. Pharmacol. **28:** 1434–1437.
7. KINSELLA, J. L., P. D. HOLOHAN, N. I. PESSAH & C. R. ROSS. 1979. Isolation of luminal and antiluminal membranes from dog kidney cortex. Biochim. Biophys. Acta **552:** 468–477.
8. MAREN, T. H. & E. COUTO. 1979. The nature of anion inhibition of human red cell carbonic anhydrases. Arch. Biochem. Biophys. **196:** 501–510.
9. MAREN, T. H. 1967. Carbonic anhydrase: Chemistry, physiology and inhibition. Physiol. Rev. **47:** 595–781.
10. WISTRAND, P. J., S. LINDAHL & T. WÅHLSTRAND. 1975. Human renal carbonic anhydrase. Purification and properties. Eur. J. Biochem. **57:** 189–195.
11. LINDSKOG, S., L. E. HENDERSON, K. K. KANNAN, A. LILJAS, P. O. NYMAN & B. STRANDBERG. 1971. Carbonic anhydrase. *In* The Enzymes. P.D. Boyer, Ed. 3rd edit. **5:** 587–665. Academic Press. New York.
12. REGISTER, A. M., M. K. KOESTER & E. A. NOLTMANN. 1978. Discovery of carbonic anhydrase in rabbit skeletal muscle and evidence for its identity with "basic muscle protein." J. Biol. Chem. **253:** 4143–4152.
13. KING, R. W., L. C. GARG, J. HUCKSON & T. H. MAREN. 1974. The isolation and partial characterization of sulfonamide-resistant carbonic anhydrases from the liver of the male rat. Mol. Pharmacol. **10:** 335–343.
14. WISTRAND, P. J. & T. WÅHLSTRAND. 1977. Rat renal and erythrocyte carbonic anhydrases. Purification and properties. Biochim. Biophys. Acta **481:** 712–721.
15. WISTRAND, P. J. 1980. Solubilization and preliminary characterization of membrane-bound carbonic anhydrase **85**. Uppsala J. Med. Sci.
16. MAREN, T. H., A. L. PARCELL & M. N. MALIK. 1960. A kinetic analysis of carbonic anhydrase inhibition. J. Pharmacol. Exp. Ther. **130:** 389–400.
17. MAREN, T. H., C. S. RAYBURN & N. E. LIDDELL. 1976. Inhibition by anions of human red cell carbonic anhydrase B: Physiological and biochemical implications. Science **191:** 469–472.
18. MAREN, T. H. & B. R. FRIEDLAND. 1978. Further studies on the phylogeny of vertebrate carbonic anhydrase in red cells and secretory organs. Bull. Mt. Desert Island Biol. Lab. **18:** 79–82.
19. WARNOCK, D. G. & F. C. RECTOR, JR. 1979. Proton secretion by the kidney. Ann. Rev. Physiol. **41:** 197–210.
20. BERLINER, R. W., T. J. KENNEDY & J. ORLOFF. 1951. Relationship between acidification of the urine and potassium metabolism. Am. J. Med. **11:** 274–282.
21. MAREN, T. H. 1963. The relation between enzyme inhibition and physiological response in the carbonic anhydrase system. J. Pharmacol. Exp. Ther. **139:** 140–153.
22. MAREN, T. H., J. R. HAYWOOD, S. K. CHAPMAN & T. J. ZIMMERMAN. 1977. The pharmacology of methazolamide in relation to the treatment of glaucoma. Invest. Ophthalmol. **16:** 730–742.

23. MAREN, T. H. 1976. The rates of movement of Na^+, Cl^- and HCO_3^- from plasma to posterior chamber: Effect of acetazolamide and relation to the treatment of glaucoma. Invest. Ophthalmol. **15:** 356–367.
24. LUCCI, M. S., D. G. WARNOCK & F. C. RECTOR, JR. 1979. Carbonic anhydrase-dependent bicarbonate reabsorption in the rat proximal tubule. Am. J. Physiol. **5:** F58–F65.
25. LUCCI, M. S., T. D. DUBOSE, L. R. PUCACCO, J. P. KOKKO & N. W. CARTER. 1979. Effect of carbonic anhydrase inhibitors of varying permeabilities on proximal tubular pH and total CO_2 reabsorptive rate (J_{tCO_2}). Clin. Res. **27:** 423A.
26. MAREN, T. H. 1974. Chemistry of the renal reabsorption of bicarbonate. Can. J. Physiol. Pharmacol. **52:** 1041–1050.
27. FRÖMTER, E., K. SATO & K. GESSNER. 1976. Electrical studies on the mechanism of H^+/HCO_3^- transport across rat kidney proximal tubule. *In* Proceedings of the 6th International Congress of Nephrology. S. Giovannetti, V. Bonomini & G. D'Amico, Eds. pp. 108–112. S. Karger. Basel.
28. KLEINMAN, J. G. & M. L. LEVIN. 1975. Effects of pent-4-enoic acid on renal free water, bicarbonate and phosphate excretion in the dog. Clin. Sci. Mol. Med. **49:** 409–417.
29. KIVILAAKSO, E., R. CRASS, J. G. ALLISON, S. ITO, D. FROMM & W. SILEN. 1979. Mechanism of the inhibitory action of electric current on H^+ secretion by frog gastric mucosa. Gastroenterology **76:** 795–803.
30. FILHO, E. M. G. & G. MALNIC. 1976. pH in cortical peritubular capillaries of rat kidney. Pflügers Arch. **363:** 211–217.

DISCUSSION

J. KOKKO: I don't think that these studies shed any light on the question of whether or not the bicarbonate ion is reabsorbed as such. You indicated that you must postulate the bicarbonate reabsorptive process per se because only 20% of the bicarbonate reabsorption is inhibited by diamox, whereas there may be an incomplete inhibition of bicarbonate reabsorption by the whole kidney. There is no question of the proximal tubular as you depicted on the slide. You, in fact, then inhibit bicarbonate reabsorption completely by various carbonic anhydrase inhibitors, but, I think it is incorrect to connect these data with the conclusion that the bicarbonate ion per se is being reabsorbed.

T. MAREN: I didn't make that connection, except as an idea in a footnote shown in a slide. The data are entirely kinetic and this model is an attempt to show how the membranes act. I put the bicarbonate moiety in, "as such" because I believe it exists. If you don't believe it exists, you can use that slide without it and I quite agree there is no direct connection between the present data and modes of HCO_3 reabsorption.

J. A. ZADUNAISKY: Some systems that are extremely sensitive to bicarbonate are not affected by any of the inhibitors of carbonic anhydrase. This is not the only one. There are other systems and we are perplexed.

MAREN: I believe, just as you are saying, that there are bicarbonate transport systems as such: Dr. Young's paper with respect to the salivary glands, for instance. I am speaking about the carbonic anhydrase system, and I think it is very important to show where the CO_2 plus OH reaction is formed to yield bicarbonate, examples of which are in the pancreas, in the aqueous humor and in the cerebral spinal fluid. The analysis of those data suggests that there is no bicarbonate, as such, moving in these three systems. On the other hand, there are systems in which the bicarbonate mechanism, as such, is dominant. In the kidney, as I just said to Dr. Kokko, both mechanisms exist. The salivary glands may be another.

M. BERRIDGE: In regard to the pH gradient, I believe that Thomas Zeuthen working with intestinal epithelial cells has, in fact, demonstrated a clear-cut gradient from one side of the epithelium to the other with a pH electrode. So I think it is conceivable that pH gradients could exist within the cells, including potassium and chloride gradients as well. It is my belief this work has not yet been confirmed by others.

E. FRÖMTER: On the question of intracellular gradients, we haven't been able to find any.

PART IV. TRANSPORT MECHANISM FOR PROTON SECRETION IN GASTRIC EPITHELIUM

THEORIES ON GASTRIC ACID SECRETION *

George W. Kidder, III

Department of Physiology
University of Maryland School of Dentistry
Baltimore, Maryland 21201

The acid secretory mechanism of the gastric mucosa probably holds the record for the production of a large ionic gradient, since the secreted solution is essentially an HCl solution isotonic to the plasma of the organism. With blood (or serosal solution) pH of 7.4, and a secreted fluid of 107 mM HCl (as in frog) with a pH of 0.97, the pH gradient (ΔpH) across the tissue is 6.4, which is a concentration ratio of 2.7×10^6. When compared to the ratios of 100 or so that are the limit for ions such as Na^+, this H^+ ratio is large indeed. (Even higher ratios can be expected in species with higher plasma osmolalities; thus, in mammals the secreted fluid is about pH 0.8 (ratio 4.0×10^6), while in elasmobranchs, whose blood is isotonic to seawater, the secreted fluid should be pH 0.57, with a concentration ratio of 6.8×10^6).[1] Since the cytoplasm of the secretory cell cannot be far from pH 7, most of the gradient, and most of the energy required to produce it, must be referred to the mucosal-facing membrane.

In the following discussion, I will be dealing largely with the gastric mucosa of the bullfrog, which seems to be as simple a system as available, and one of the best studied. This tissue can be readily isolated and mounted in an Ussing chamber, where it secretes acid into the mucosal bathing solution at a rate (J_H) of about 5μEq/cm^2·hr, which can be measured by titration. In doing so, it produces a transepithelial potential difference (*PD*) of about 30 mV, serosal side positive. If this potential be short-circuited, the resulting short-circuit current (J_{sc}, if expressed in flux units) is around 1 μEq/cm^2·hr, and it was shown long ago [2, 3] that the total chloride flux across this tissue (J_{Cl}), measured by tracer methods, is given by $J_H + J_{sc}$; H^+ and Cl^-, both transported from serosal to mucosal solution, appear to account for the active transmural ion movements in this tissue.

Although the mucosa is mounted in the chamber as a flat sheet, the secretory surface is far from flat, since the acid-secreting oxyntic cells line the depths of secretory tubules, with a consequent large increase in total area over chambered area.[4–7] This factor (130 in frog gastric mucosa [7]) accounts for the observation of a low electrical resistance (100–200 Ω·cm^2 chambered area) in a tissue that by all other criteria is a "tight epithelium," in the sense that ionic channels between the cells are virtually absent. This anatomy also produces a large unstirred layer on the mucosal surface of the cell, which has interesting consequences. For instance, FIGURE 1 shows a schematic representation of a single secretory pit, lined with the acid-secreting oxyntic cells, while the area between the pits is paved with surface epithelial cells. It is generally accepted [8] that as the oxyntic cells secrete HCl into the pit lumen, the resulting osmotic force extracts H_2O from the cells, and pro-

* Supported in part by National Science Foundation Research Grant PCM 77-03336.

0077–8923/80/0341–0259 $01.75/0 © 1980, NYAS

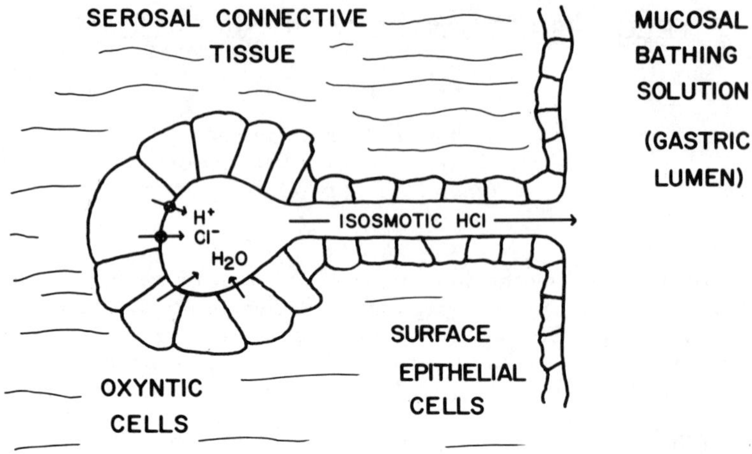

FIGURE 1. Schematic diagram illustrating the formation of isomotic HCl by oxyntic cell secretion of HCl and the consequent osmotic water flow. HCl then appears in the mucosal bathing solution as a conseqence of diffusion and bulk flow. While the pH at any intermediate point along the tubule will vary with bulk solution pH and secretory rate, the pH immediately adjacent to the oxyntic cells will be constant at that of primary secretion for any finite secretory rate.

duces a small pressure difference that drives fluid (the "primary secretion") from the pits as an HCl solution isotonic to the oxyntic cell cytoplasm and (by extension) to the blood plasma. To the extent that H^+ diffuses (as opposed to bulk flow) from the tubules in exchange for Na^+ from the solution, the pH at any point along the tubule will be higher than that of primary secretion, but the pH immediately adjacent to the oxyntic cell membrane will be the unmodified result of its own secretion. While this is easy to see in a system such as the gastric mucosa with an obvious anatomical unstirred layer, it should also apply in other systems secreting isomotic fluid within the unstirred layers that bind any tissue.

Thus, at any finite secretory rate, the luminal surface of the oxyntic cell is exposed to a 107 mM solution of HCl at about pH 1, quite independent of the pH of the bulk solution. It is not surprising that measurements of acid secretory rate show no dependence of J_H on mucosal solution pH down to about pH 2 (where measurement becomes impossible [9]), and little change in the electrical properties usually associated with secretion down to pH 1, which is as far as one can go with isotonic HCl. Any hypertonic solution, to the extent that it penetrates the tubules, will only dilute itself by pulling additional water from the cells, so one cannot test the potency of the pump by challenging it with a pH much greater than its normal operating load.[9] We conclude that the gastric H^+ pump is probably operating at near its maximum gradient at all times.

With a knowledge of the concentrations of the transported species on both sides of the tissue, one can calculate the free energy requirement for acid secretion. From the expression:

$$\Delta G = RT\ ln\ (C_2/C_1) + zF\ \Delta\psi$$

where C_1 and C_2 are the source and sink concentrations of the ion in question, z is its valence, and $\Delta\Psi$ is the electrical potential difference measured with the sink (here the mucosal side) taken as zero, one can calculate ΔG when R, T and F have their usual thermodynamic meanings. As shown at the top of FIGURE 2, when this is applied to the tissue as a whole, it shows that both H^+ and Cl^- are actively transported (ΔG is positive) under short circuit conditions, although only a minor amount of work is expended per mole of Cl^-. Removal of the short circuit, allowing the *PD* to rise to 30 mV serosal positive, will increase ΔG^{Cl} by 690 cal/mole, while decreasing ΔG^{H} by the same amount. The qualitative conclusion remains; most of the work is expended in transport of H^+.

These ions do not pass directly from serosal to mucosal solution. Rather, they pass through the cytoplasm of the cell, and we must therefore consider the energy requirements at both membranes as influenced by the cytoplasmic

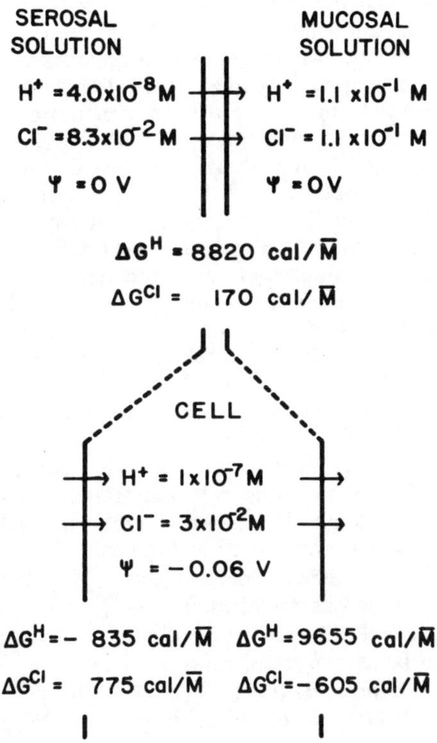

FIGURE 2. Summary of the energetics of H^+ and Cl^- secretion in the frog gastric mucosa under short circuit conditions. The top portion shows the tissue as a whole; both H^+ and Cl^- require energy for transport. With reasonable assumptions as to cell pH, chloride concentration and potential, individual membrane energy requirements can be calculated. As shown in the bottom part of the figure, ΔG^{Cl} for the mucosal membrane is negative (passive movement) while the serosal membrane has a positive energy requirement for this ion. See text for further discussion.

concentrations and potential. Knowing these values, we can easily calculate the free energies for each face of the cell; the problem is to make these estimates. I have assumed a cellular pH of 7; it surely cannot be far from this and still provide an appropriate biochemical milieu. If there is an error here, it is probably that the assumed pH is too low, since acid secretion causes the intracellular pH to shift toward more basic values.[10] Cell chloride and intracellular potential pose a greater problem. It has frequently been assumed (e.g.[11]) that cell chloride is low enough to allow passive entry across the serosal membrane at the cell potential that exists. Chemical measurements on frog gastric mucosa show values from 30–40 mM[12] to a high of 70 mM,[13] although one is never sure what cells are being measured. In other epithelia, direct measurements with the chloride microelectrode have given values between 25 and 70 mM,[14–17] this last for bullfrog intestine. To be conservative, I will guess that 30 mM is the appropriate chloride concentration for bullfrog gastric oxyntic cells.

The intracellular potential is likewise a problem. Villegas[18] measured —18 mV across the serosal membrane of oxyntic cells of bullfrog stomach in open circuit conditions, and similar values have been measured in bullfrog and *Necturus* by Shoemaker.[19] If the high-resistance membrane is the mucosal membrane, as we have reason to expect, then the value under short-circuit conditions would be similar to this across both membranes. This is a rather low potential: recent measurements give values between —24 and —52 mV for a range of epithelia,[14–17] and there are measurements of potentials of —70 mV in toad gastric mucosa.[40] All in all, and in consideration of the observation that improvements in technique generally have lead to more negative potentials,[20] I have chosen —60 mV for purposes of illustration.

With these values, the calculations give the figures shown in the bottom half of FIGURE 2. On the mucosal surface of the cell, ΔG_H is strongly positive, but ΔG_{Cl} is negative, indicating that the movement of Cl^- across this membrane is downhill along its electrochemical gradient. Correspondingly, the serosal surface of the cell shows a moderate positive ΔG_{Cl} and a moderate negative ΔG_H; an active process causes Cl^- to enter this face of the cell, while H^+ or its equivalent is passive here.

If these values are even close to accurate, we conclude that the active process for H^+ is located on the mucosal membrane, as usually assumed, but that active Cl^- transport is on the serosal membrane. This implies a clear separation of the H^+ and Cl^- transport mechanisms, although it would still be theoretically possible to have an energy-requiring system transporting Cl^- "downhill" across the mucosal membrane. Of course, the whole argument turns on the values for cell chloride and potential; if, for instance, at 30 mM $[Cl^-]$ the potential can be shown to be more positive than —37 mV, the model fails. If, on the other hand, cell $[Cl^-]$ is 70 mM, the potential need only be —11 mV. Reliable simultaneous measurements of these values would be most timely.

Thermodynamics treats of states, and not of specific mechanisms that move material between states. In particular, the serosal influx of H^+ shown in this diagram is probably to be understood as an outflux of bicarbonate ion, which would have the same effect and is known to occur.[21,22] Moreover, it has been shown that this HCO_3^- outflux is coupled to a portion of the Cl^- influx at the serosal membrane by a 1:1 Cl^-/HCO_3^- exchange mechanism.[23]

When we note that ΔG^{H} at this surface is more negative than ΔG_{Cl} is positive, it becomes attractive to speculate that the Cl^- active transport across the tissue is driven, at least in part, by a secondary active transport of Cl^- at the expense of the HCO_3^- gradient, with no need for a direct interaction with metabolic energy. The HCO_3^-, produced as a product of the neutralization of the OH^- that remains following H^+ secretion, is then the driving force for the Cl^- secretion, and only a single pump (that for H^+) would consume energy directly. This would also explain the inhibitory effect of removal of Cl^- on acid secretion, since this would inhibit HCO_3^- exit from the cell. However, there would have to be an additional mechanism for Cl^- entry at the serosal surface of the cell, to account for the observation that J_{Cl} is greater than J_H under steady-state, short-circuit conditions, and to carry the charge represented by J_{sc} across the serosal membrane.

From the magnitude of the energy required for the proton secretory event, and the energy that can be extracted from the hydrolysis of 1 mole of ATP at reasonable concentrations of ATP, ADP and phosphate, it is clear that while 1 H^+ could be transported by the energy from 1 ATP, 2 H^+/ATP would exceed the available energy. There are, however, acid secretory mechanisms such as those found in turtle bladder,[24, 25] that cannot produce gradients greater than $\Delta pH = 3$ or so; in these cases, not only is a 2 H^+/ATP ratio thermodynamically possible, but it would be expected, since it would be a more efficient mechanism, and might well have evolved in this direction. We therefore expect to find two sorts of H^+ pumps; the high-potency (gastric) H^+ pump, and a low-potency (kidney, bladder, etc.) H^+ pump. The latter may be common to many sorts of tissues, since it may well be part of the mechanism for regulation of cytoplasmic pH. There is no a priori reason to expect these pumps to be similar in any way, and some reasons to expect differences.

In considering the gastric acid pump, we do not know for sure what the transported species is. For the low-potency pumps, two mechanisms have been considered; transport of H^+ toward the lumen, and transport of HCO_3^- out of the lumen. In the stomach, the vanishingly small concentration of HCO_3^- in equilibrium with 5% CO_2 at pH 1 (8×10^{-9} M) makes the latter unlikely on these grounds. Having stated this, I must admit that the H^+ pump must function with a source concentration of 10^{-7} M H^+ or so, and perhaps this is not a compelling argument. However, there seems to be no good evidence against a proton pump per se in gastric mucosa, and this will be assumed.

There are many possible forms of proton pumps, with various properties and cotransport possibilities, and thermodynamics is little help in unraveling them. For years, there has been a discussion between proponents of two rather different views of the secretory apparatus. Impressed by the role of Cl^- in acid secretion, one group[26, 27] has maintained that H^+ is secreted across the mucosal face of the cell by an active mechanism that is constrained to transport one H^+ and one Cl^- together; that is, an electrically neutral or "unitary" HCl pump, as shown in FIGURE 3. To account for the generation of a *PD* and a short circuit current, it is necessary to include an additional, current-carrying Cl^- transport mechanism. This model readily explains the observed simultaneous changes in J_H and J_{Cl} under a variety of conditions. The other view[28] is that the pumps for H^+ and Cl^- are separate and therefore electrogenic, but coupled by electrical and/or biochemical events within

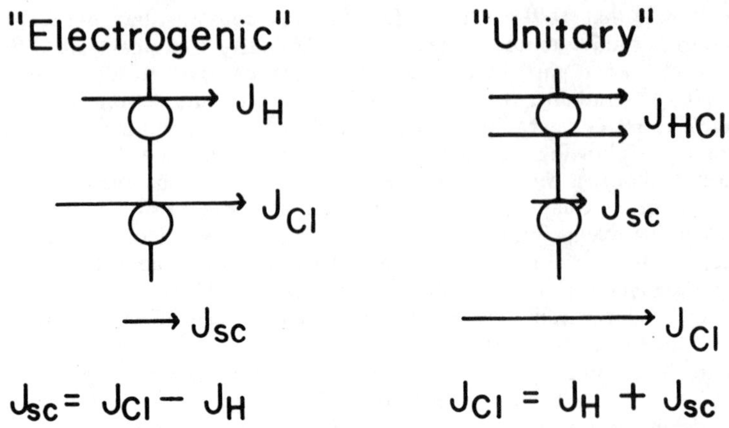

FIGURE 3. Schematic representation of the two extreme models for gastric acid secretion. Note that while the equations relating J_{sc}, J_{Cl} and J_H are identical in the two models, the identification of J_{sc} and J_{Cl} with pump activities is different.

the cell. The proponents of this viewpoint to the residual secretion of H^+ in the absence of solution Cl^-,[29] as well as to the effects of large imposed currents on acid secretion,[30, 31] which are most readily understood in terms of the electrogenic model. Clearly, the sort of pump postulated by these two models is very different, although, of course, both models must come to grips with the same set of facts.

In either case, three fluxes can be measured, and the relationship $J_{sc} = J_{Cl} - J_H$ will hold. What differs between the models is the meaning to be assigned to these fluxes. In the unitary pump model, J_H measures the rate of the HCl pump, J_{sc} measures the rate of the "additional" Cl^- pump, but J_{Cl} (the total Cl^- flux) does not have a corresponding mechanism, but is just the sum of J_H and J_{sc}. Conversely, in the electrogenic model, J_H measures the rate of the H^+ pump, J_{Cl} the rate of Cl^- pump, and it is J_{sc} that has no mechanism, being the difference between J_{Cl} and J_H. We might expect that fluxes that measure real pumps would show rational kinetic behavior with respect to external conditions, while fluxes that are combinations of other real events might be irrational if plotted in the expectation of simple kinetics. This provides a test of the two models.

In a series of bullfrog mucosae in the steady state, J_H and J_{sc} were measured [32] at various Cl^- concentrations (Cl^- varied in both solutions) and J_{Cl} was calculated as the sum of J_H and J_{sc}. These values were then plotted as 1/J vs. 1/[Cl^-] in the conventional Lineweaver-Burke manner, as shown in FIGURE 4. Note that J_H and J_{Cl} give reasonable straight lines with high correlation coefficients and reasonable Michaelis coefficients, while J_{sc} does not make any sense when examined in this way. From these data alone, one would conclude that J_{sc} does not represent a real pump, which would support the electrogenic hypothesis. We also note that K_m for J_H and K_m for J_{Cl} are different, which is allowed under the electrogenic hypothesis, but that the difference is not great.

In the presence of SCN^-, both J_H and J_{Cl} are inhibited, while J_{sc} is essentially unchanged, as shown in FIGURE 5. Choosing a concentration of thiocyanate (SCN^-), which gives significant but not total inhibition of J_H, we repeated the kinetics of the tissue's response to Cl^- concentration changes, as shown in FIGURE 6. Note that while J_H has its K_m dramatically increased by SCN^-, that for J_{Cl} is much less affected. This is consonant with an electrogenic hypothesis, but nearly impossible to fit into a unitary pump mechanism, where any factor that changes J_H should have an identical effect on the majority of the chloride transport rate as well, were the pump a coupled HCl pump. We conclude that H^+ and Cl^- are secreted at separate sites, and not by way of a unitary HCl pump.

We thus require a model in which the transport of H^+ and Cl^- are separated, yet share some common pathways. One such model, synthesized from the above considerations and having many features of previous models [33] is shown as FIGURE 7. A primary proton pump, located on the mucosal membrane, contributes the bulk of the free energy necessary for secretion, while Cl^-

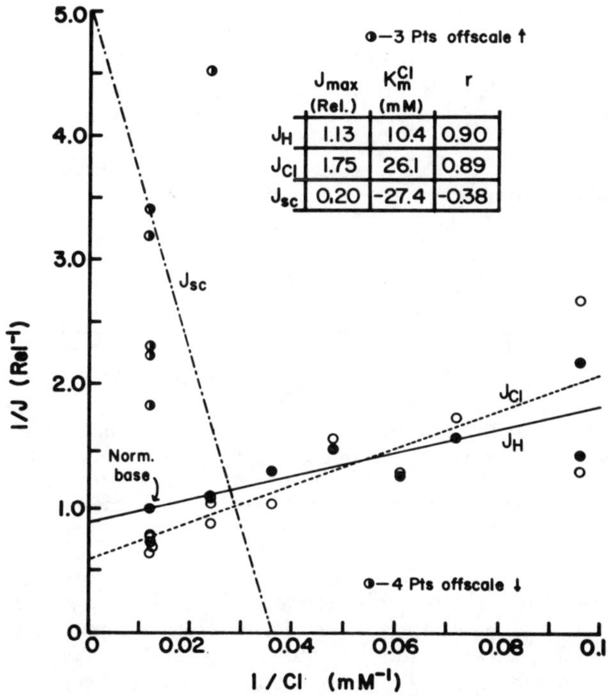

FIGURE 4. Double-reciprocal plot of $1/J_x$ vs. $1/[Cl^-]$, where x=H, Cl, or "sc" as previously defined. Chloride concentrations were varied in both bathing solutions simultaneously, by replacement with sulfate and sucrose, with the rates measured at the end of a one-hour equilibration period. All rates normalized to the J_H measured for that tissue under control conditions (82.6 mM Cl^-) to correct for variations in tissue activity. Note reasonable fits for J_H and J_{Cl}, but poor fit and impossible slope for the J_{sc} line.

moves passively across this membrane at a different site, driven largely by the cell potential. This negative potential is established, in this cell as in others, largely by a K^+ diffusion potential, aided by the H^+ pump activity. The necessary K^+ gradient is maintained by a conventional, primary Na^+/K^+ pump. Most of the Cl^- enters the cell in exchange for the HCO_3^- produced as a by-product of H^+ secretion; since the HCO_3^- gradient is directed outward across the serosal membrane, this is a secondary active transport process for inward Cl^- movement (a HCO_3^-/Cl^- antiport). However, under short-circuit conditions, more Cl^- is required than can be provided by this mechanism, and the balance (equal to J_{sc}) is carried into the cell by a NaCl symport, such as is found in some other epithelia, while the necessary charge transfer is

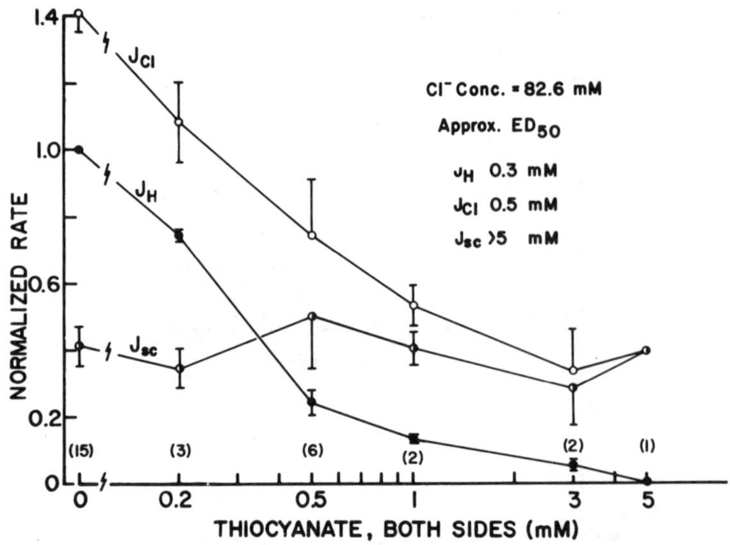

FIGURE 5. Effect of NaSCN on rate after one-hour exposure under full Cl^- conditions. Note lack of effect on J_{sc} over this concentration range. Tissue sensitivity to SCN^- is somewhat greater than usually reported; this is probably due to application of inhibitor to both sides and to use of 10% CO_2 in gas mixture, which gives higher initial rates.

achieved by K^+ outflux. Since this system is driven by the Na^+/K^+ pump, the necessity for these ions becomes clear. If the NaCl symport is a 1:1 system, an additional Cl^- pathway would be required to explain the Cl^- conductivity of the serosal membrane; if this symport can carry a charge, this might not be necessary.

This model is satisfying in that it explains many of the properties of the gastric mucosa by mechanisms that are found in many other tissues. Even the HCO_3^-/Cl^- antiport, while first demonstrated in gastric mucosa,[23] has been found in a variety of cell types, and may be considered ubiquitous. Since evolution presumably started from a common set of properties, one feels that models that do not radically depart from the norm are more likely to explain the results of such evolution. The one unique feature of the gastric secretory

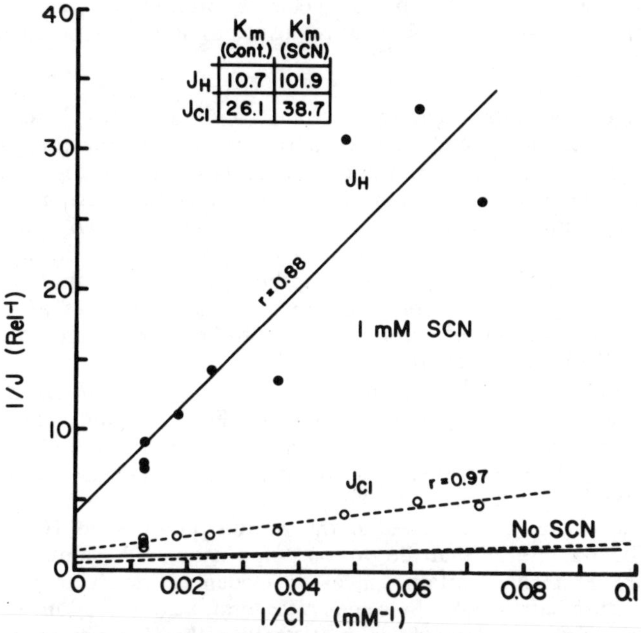

FIGURE 6. Double-reciprocal plot of J_H and J_{Cl} vs. [Cl^-] in the presence of 1 mM SCN^-, as compared to the corresponding controls (lower lines). Note large change in J_H slope with only moderate changes in J_{Cl}. As before, all values normalized to control J_H (full Cl^-, no SCN^-).

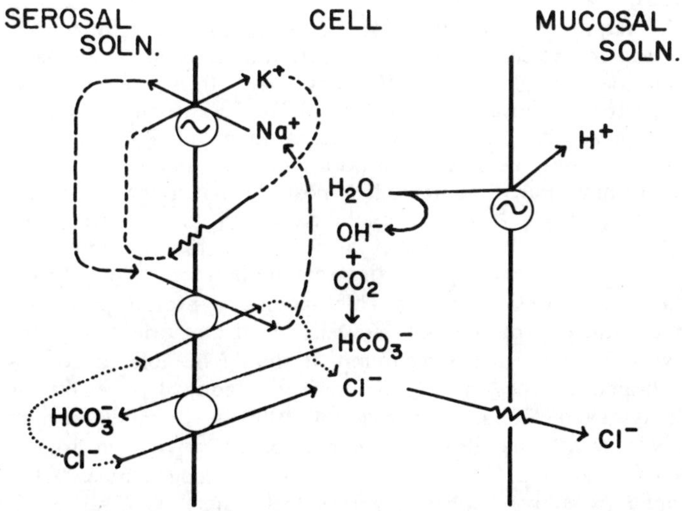

FIGURE 7. A model for gastric acid secretion in the frog. Circles represent carrier-mediated systems; those with tildes are metabolically-driven active transport systems, while plain circles are driven only by the gradients of transported species. The resistor symbol indicates diffusion without interaction with any other species, and may or may not be carrier-mediated. See text for details.

cell is the high-potency H^+ pump. This is presumably restricted to sites where a million-fold H^+ gradient is required, as it would be inefficient elsewhere.

A very interesting development of the last few years has been the isolation of membrane fragments, organized as spherical vesicles, which seem to consist of portions of cellular membrane containing active ion transport mechanisms, and which can produce concentration differences between the vesicle interior and the bathing solution. From the gastric mucosa, there have been two different preparations; a HCO_3^- -stimulated ATPase,[35] which is now generally thought not to be directly involved in acid secretion, and the current candidate, a K^+ -stimulated ATPase.[36, 37] Without minimizing the importance of studies on these vesicles, I would like to sound several notes of caution with respect to these purported proton pumps by discussing the properties we might expect from such preparations.

To isolate vesicles, one disrupts the mucosa and purifies the resulting soup to obtain a small amount of active material. If one assumes that the gastric H^+ pump is driven by ATP, one selects procedures and an assay suitable for an ATPase. However, there may be many such ATPases and an additional criterion is handy. At present, the property of stimulation by K^+ is used, which leads to the implication that the proton pump is an H^+/K^+ exchange system. The assumption that the proton pump is an ATPase may well be true, but it is by no means certain, since the evidence,[39] such as it is, does not support a close connection between ATP and acid secretion in the intact tissue. If this assumption be false, it is possible that the current preparations, which after all do consume ATP and secrete acid, are really samples of a low-potency H^+ pump, present in this tissue (as in others) for control of cellular pH. Such a pump would not produce large pH gradients, and could function at 2 H^+/ATP; both properties are consistent with current vesicle preparations.

The K^+ requirement is also suspect. Since the response to K^+ was used as a criterion for selection of this preparation, it is understandable that the final preparation should exhibit K^+ sensitivity. It is curious, however, that the frog gastric mucosa does not transport K^+ to any significant degree, although it is sensitive to the complete absence of K^+. Since mammalian gastric mucosa does have K^+ transport capacity, and since the vesicles are isolated from mammalian sources, it is possible that a species difference exists at the H^+ pump level. Another possibility, which seems worthy of consideration, is that the transformation of a plasma membrane into a vesicle has conferred upon the system properties that it did not have before.

Consider a theoretical gastric (high-potency) proton pump that has no other ionic cofactors, mounted in a vesicle wall and oriented to transport H^+ into the vesicle. Being a gastric pump, it should be able to do this until the vesicular lumen reaches pH 1, or until the equivalent electrical potential difference is reached. For a vesicle of 1000 Å diameter, which seems a characteristic size,[38] the lumen volume is 5×10^{-16} ml, in round numbers, which means that at pH 1 it contains 30,000 hydrogen ions. All of this H^+ was supplied by active transport across the vesicle wall, since at an initial (bathing solution) pH of 6.5, there would be only 1 H^+ for every 10 vesicles, on the average. Leaving aside the question of the meaning of pH under these circumstances, we can make several predictions based on the requirements for electroneutrality and osmotic equilibrium in such systems. First, it should be clear that as H^+ enters the vesicle, either some positive ion must leave, or an anion must enter, since otherwise the rapid buildup of positive charge in the vesicle lumen will prevent any appreciable H^+ entry. The most effective ions for this purpose will be those to which the vesicle is most permeable, so the rate of H^+ entry will be limited by the rate of co- or counter-ion flow. Now, if the charge were balanced by anion entry alone (say, for instance, Cl^-), the increase in HCl concentration within the vesicles to pH 1 would cause the osmolality of the vesicle contents to double if there were no osmotic water flow. However, one would predict from the properties of the tissue that the vesicles would be quite permeable to water, since they presumably represent just those parts of the membrane across which water movement occurs during secretion *in vivo*. Therefore, we expect the vesicles to maintain osmotic equilibrium by swelling, lowering the H^+ concentration and allowing further H^+ entry. In the limit, a volume increase to three times the initial volume would occur, which would increase the diameter by 44% and double the area of the membrane. If the membrane could withstand this, such an increase should be readily observable. More likely, the vesicles would burst, or be so severely stretched as to become very permeable to passive H^+ diffusion, which would prevent measurable H^+ accumulation. Electroneutrality maintained by anion cotransport would thus not be compatible with measurable H^+ accumulation by vesicles.

On the other hand, electroneutrality might be maintained by exchange of H^+ for a cation already in the vesicle, by passive diffusion of the cation in response to the electrical fields created by electrogenic H^+ entry. In this case, volume changes would be minimal, and no vesicle rupture would be expected. The initial concentration of the counterion within the vesicles would have to be equal to the final H^+ concentration, in this case 100 mM. If, for instance, K^+ were the most permeable cation, its presence in the medium (and therefore in the vesicles equilibrated with this medium) would be a requirement for rapid H^+ accumulation, and therefore (in the case of an ATP-driven pump) for high ATPase activity. Such a mechanism could give rise to an apparent requirement for K^+, even though the proton pump itself had no such requirements.

In the absence of H^+ backflux (leaks), the pump will stop when the pH gradient plus the electric potential difference reach thermodynamic equilibrium with energy source. If a leak exists that allows H^+ to leak from the vesicle, then the internal pH would be limited to that value at which the leak balanced the pump, and energy would thereafter be consumed at a steady rate. While modifications in the permeability of the membrane to K^+ (as, for instance, by valinomycin) would alter the initial rates, it would have no effect on the final equilibrium or steady state pH, since in neither of these conditions is there any K^+ movement.

If we now allow some anion permeability to exist as well (and we have every reason to expect permeability for Cl^-) the situation we observe will depend on the permeabilities for both K^+ and Cl^-, and their concentrations in the vesicle at the beginning. The lesser of these permeabilities will essentially determine the rate of KCl entry under conditions where the initial K^+ is not sufficient to balance the H^+ required to reach whatever pH limit is set by H^+ backflux. The ratio of permeabilities and the changing concentration gradients will determine, at any point in time, how much of the countercurrent is carried by Cl^-, which in turn will determine vesicle swelling. Swelling itself will probably cause changes in passive membrane properties, further altering the results expected. Furthermore, if variations exist in vesicle structure and function, the swelling of some vesicles but not others could selectively destroy part of the initial population. The point is that any practical situation, even a relatively simple one, is really quite complex when more than about two flows have to be considered.

The same proton pump, mounted in the tissue, has fewer constraints. Electroneutrality can be satisfied by Cl^- movement alone, and no osmotic damage will occur since the secreted fluid can flow away. Cation countertransport is therefore not required for the integrity of the system, and indeed (as one might expect from a pump that secretes its own sink solution), there is little effect on secretion of large changes in bulk mucosal fluid composition.

I would counsel caution, therefore, in translating the results of vesicle experiments to an interpretation of the properties of the H^+ pump. For the reasons given above, I remain doubtful that the current candidate for the primary proton pump will be more successful than the last contender. However, there seems little question that an ATP-driven membrane system has been isolated from gastric mucosa, which can transport H^+ against some gradient, and that is a very interesting observation in its own right. But until it can be shown that a reasonably high ΔpH is obtainable, and that a thermodynamically plausible H^+/ATP ratio exists, I feel justified in reserving judgement as to whether this, at last, is the gastric proton pump.

In conclusion, it would appear on purely thermodynamic grounds, based on some reasonable but unverified assumptions, that there is no need to postulate an active chloride transport mechanism at the mucosal surface of the oxyntic cell, but that the energy required for the uphill transport of chloride could be derived from a combination of a HCO_3^-/Cl^- antiport and a NaCl symport at the serosal membrane. Since HCO_3^- is a product of H^+ secretion, this would provide a mechanism, external to the mucosal membrane, to explain the coupling between H^+ and Cl^- transport, while retaining separate electrogenic transport for H^+ (an active pump) and Cl^- (diffusion), which can be demonstrated under suitable conditions. There would then exist only two primary pumps, that for H^+ on the mucosal membrane, and the Na^+/K^+ exchange pump. The requirement for K^+ of the vesicular membrane preparations could well be a consequence of vesicular organization, electroneutrality and osmotic balance, and not a requirement of the pump mechanism itself. It will be interesting to see if these speculations can survive experimental testing; models and calculations are, after all, merely targets for the slings and arrows of experimental science.

References

1. Kidder, G. W., III. 1977. Effect of mucosal pH on PD in dogfish gastric mucosa. Bull. Mt. Desert Island Biol. Labs **17:** 21–23.
2. Hogben, C. A. M. 1951. The chloride transport system of the gastric mucosa. Proc. Natl. Acad. Sci. **37:** 393–395.
3. Hogben, C. A. M. 1955. Active transport of chloride by isolated frog gastric

epithelium: Origin of the gastric mucosal potential. Am. J. Physiol. **180**: 641–649.

4. Canosa, C. A. & W. S. Rehm. 1958. Microscopic dimensions of the pit region of the dog's gastric mucosa. Gastroenterology **35:** 292–297.
5. Helander, H. F. & B. I. Hirschowitz. 1972. Quantitative ultrastructural studies on gastric parietal cells. Gastroenterology **63:** 951–961.
6. Helander, H. F. & R. P. Durbin. 1977. Secretory surface area and phosphatase activity of frog gastric mucosa. Am. J. Physiol. **232:** E48–E52.
7. Helander, H. F. 1977. An attempt to correlate functional and morphological data for the gastric parietal cells. Gastroenterology **73:** 956–957.
8. Durbin, R. P. 1979. Osmotic flow of water in isolated frog gastric mucosa. Am. J. Physiol. **236:** E63–E69.
9. Kidder, G. W., III & J. T. Blankemeyer. 1978. Cytochromes and gastric acid secretion. A reevaluation of mucosal acidification experiments. Biochim. Biophys. Acta **512:** 192–198.
10. Hersey, S. J. 1978. Intracellular pH and gastric acid secretion. Acta Physiol. Scand. Sp. Suppl. **:** 242–252.
11. Durbin, R. P. 1967. Electrical potential difference of the gastric mucosa. *In* Handbook of Physiology. C. F. Code, Ed. Sec. 6: Alimentary Canal. 879–888. Am. Physiol. Soc. Washington, D.C.
12. Davenport, H. W. & F. Alzamora. 1962. Sodium, potassium, chloride and water in frog gastric mucosa. Am. J. Physiol. **202:** 711–715.
13. Shanbour, L. L. & W. S. Rehm. 1971. H^+ secretion and tissue Cl^- content of frog gastric mucosa bathed in Cl^--free media. Proc. Soc. Exptl. Biol. Med. **136:** 1236–1239.
14. Armstrong, W. McD., W. Wojtkowski & W. R. Bixenmen. 1977. A new solid-state microlectrode for measuring intracellular chloride activities. Biochim. Biophys. Acta **465:** 165–170.
15. Duffey, M. E., K. Turnkeim, R. A. Frizzell & S. G. Schultz. 1978. Intracellular chloride activities in rabbit gallbladder: Direct evidence for the role of the sodium-gradient in energizing "uphill" chloride transport. J. Membr. Biol. **42:** 229–245.
16. Spring, K. A. & G. Kimura. 1978. Chloride reabsorption by renal proximal tubules of *Necturus.* J. Membr. Biol. **38:** 233–254.
17. White, J. F. 1977. Activity of chloride in absorptive cells of Amphiuma small intestine. Am. J. Physiol. **232:** E553–E559.
18. Villegas, L. 1962. Cellular localization of the electrical potential difference in frog gastric mucosa. Biochim. Biophys. Acta **64:** 359–367.
19. Shoemaker, R. L. 1978. Micropuncture studies using the amphibian fundic gastric mucosa, *in vitro*. Acta Physiol. Scand. Special Suppl. 173–180.
20. Armstrong, W. McD. 1978. Intracellular ionic activity measurements with ion-selective microelectrodes/some technical considerations. Arzneim-Forsch. **28:** 705.
21. Daves, R. E. 1951. The mechanism of hydrochloric acid production by the stomach. Biol. Rev. **26:** 87–120.
22. Teorell, T. 1951. The acid-base balance of the secreting isolated gastric mucosa. J. Physiol. London **114:** 267–276.
23. Sanders, S. S., J. O'Callaghan, C. F. Butler & W. S. Rehm. 1972. Conductance of submucosal-facing membrane of frog gastric mucosa. Am. J. Physiol. **22:** 1348–1354.
24. Schilb, T. P. & W. A. Brodsky. 1966. Acidification of mucosal fluid by transport of bicarbonate ion in turtle bladders. Am. J. Physiol. **210:** 997–1008.
25. Al-Awqati, Q. 1978. H^+ transport in urinary epithelia. Am. J. Physiol. **235:** F77–F88.
26. Hogben, C. A. M. 1968. Observations on ionic movement through the gastric mucosa. J. Gen. Physiol. **51:** 240S–249S.

27. FORTE, J. G. 1968. The effects of inhibitors of HCl secretion on the unidirectional fluxes of chloride across bullfrog gastric mucosa. Biochim. Biophys. Acta **150:** 136–145.
28. REHM, W. S. 1950. A theory of the formation of HCl by the stomach. Gastroenterology **14:** 401–417.
29. HEINZ, E. & R. P. DURBIN. 1959. Evidence for an independent hydrogen-ion pump in the stomach. Biochim. Biophys. Acta **31:** 246–247.
30. REHM, W. S. 1956. Effect of electric current on gastric hydrogen ion and chloride ion secretion. Am. J. Physiol. **185:** 325–331.
31. REHM, W. S. 1962. Acid secretion, resistance, short-circuit current, and voltage-clamping in frog's stomach. Am. J. Physiol. **203:** 63–72.
32. KIDDER, G. W., III & C. W. MONTGOMERY. 1974. The kinetics of chloride and thiocyanate effects on acid secretion in the frog gastric mucosa. J. Baltimore Coll. Dent. Surg. **29:** 52–61.
33. FORTE, J. G., H-C. LEE & H. BREITBART. Chloride transport by gastric mucosa and membranes isolated from gastric oxyntic cells. Fed. Proc. In press.
34. FRIZZELL, R. A., M. FIELD & S. G. SCHULTZ. 1979. Sodium-coupled chloride transport by epithelial tissues. Am. J. Physiol. **236:** F1–F8.
35. KASBEKAR, D. K. & R. P. DURBIN. 1965. An adenosine triphosphatase from frog gastric mucosa. Biochim. Biophys. Acta **105:** 472–482.
36. FORTE, J. A., A. L. GANSER & A. TANESAWA. 1974. The K^+-stimulated ATPase system of microsomal membranes from gastric oxyntic cells. Ann. N. Y. Acad. Sci. **242:** 255–267.
37. SCHACKMAN, R., A. SCHWARTZ, G. SACCOMANI & G. SACHS. 1977. Cation transport by gastric H^+:K^+ ATPase. J. Membr. Biol. **32:** 361–381.
38. LEE, H-C., H. BREITBART, M. BERMAN & J. G. FORTE. Potassium-stimulated APTase activity and hydrogen transport in gastric microsomal vesicles. Biochim. Biophys. Acta. In press.
39. KIDDER, G. W., III. ATP effects on intact gastric mucosa. *In* Gastric Secretion. G. Sachs, E. Heinz and K. J. Ullrich, Eds. pp. 273–292. Academic Press. New York.
40. NAGEL, W. Personal communication.

DISCUSSION

W. S. REHM: We made simple calculations for the dog stomach and the minimum free energy necessary to transport the chloride was a little over 20% of the minimum free energy necessary for the hydrogen transport. How does that fit in with your particular scheme?

G. KIDDER: If you calculate for transport across the entire epithelium, you find that the energy necessary for the hydrogen ion is less than that across the mucosal membrane while the energy for the chloride is more than that across the mucosal membrane. The sum of the energies across the two membranes, of course, adds up to the energy across the entire system. In the frog, it would seem that the difference between the energy requirements for transepithelial H^+ and Cl^- transport is somewhat greater than in the dog.

F. H. EPSTEIN: What is the oxygen cost of transporting a mole of hydrogen ion in the system that you are working with? Does that square with the one hydrogen ion per ATP that you are talking about?

KIDDER: Do you mean in theory?

EPSTEIN: As is actually measured.

KIDDER: There are always problems with doing these kinds of measurements. Dr. Forte, I believe, was one of the first people to attempt this sort of thing. The problem, basically, is a problem of baseline respiration. Does it, in fact, remain constant or does it go down? In either case, as I remember Dr. Forte's data, the excess oxygen consumption was sufficient to account for the energy that is expended at a one ATP per hydrogen. Am I right in my remembrance?

J. G. FORTE: If you take the total equivalence of transport, there were more than four equivalents of H^+ plus Cl^- transported per oxygen consumed.

KIDDER: But you are still within the possibility of a one H per ATP pump.

Dr. FORTE: Yes.

J. A. ZADUNAISKY: In your model, you postulate the sodium potassium transport and sodium chloride obligatory entry into the cell for the chloride secretion to occur.

KIDDER: That is one possibility. What I have proposed is that a sizable fraction of the chloride entry into the cell is mediated by a chloride bicarbonate exchange. This, however, would not allow the steady state secretion of chloride at a higher rate than the secretion of hydrogen ion that is observed under short circuit conditions. Therefore, it is necessary that there be another way to get chloride into the cell, and with the concentrations I have assumed, it is necessary that this be an active entry step.

J. A. ZADUNAISKY: But, in all of the systems that we have analyzed, including the cornea, we need the evidence to be sure that it matches. Do you agree that you need sodium for the chloride secretion to occur in the stomach?

KIDDER: Certainly, in the long term you need sodium for everything to occur.

ZADUNAISKY: Do you know the activity of chloride inside the cell?

KIDDER: This has yet to be measured.

FORTE: I think in answer to your first question, Terry Macken has measured the relationship between chloride transport and sodium concentration in the serosal fluid and found a relationship that's consistent. It requires sodium, and the amount of chloride transport is dependent upon the amount of sodium.

A. M. WEINSTEIN (*National Institutes of Health, Chevy Chase, Md.*): Earlier, Dr. Maren asked us to consider the possibility of substantial intracellular pH gradients which would, therefore, require the expenditure of energy in translocating protons from one side of the cell to the other in the cell itself. Could you comment on that and how would that model would fit in with these considerations?

KIDDER: I don't believe Dr. Maren was postulating a pH gradient across the cytoplasm of the cell. He was postulating a pH gradient across the membrane bounding the cell or within that membrane, which is very different. The problem with the pH gradient in the cytoplasm with an ion as mobile as hydrogen ion is that the amount of energy necessary just to maintain that gradient becomes enormous. I think one can effectively rule it out on that basis.

ATP METABOLISM IN ISOLATED GASTRIC GLANDS *

S. J. Hersey

Department of Physiology
Emory University
Atlanta, Georgia 30322

Introduction

The fundamental nature of hydrochloric acid production by the gastric mucosa has been the subject of intense investigation for over one hundred years. Among the various aspects of this problem has been the question of interaction between energy metabolism and proton translocation. The fundamental features of this interaction include the absolute dependence of H^+ transport on oxidative metabolism and the control of respiration by acid secretory activity.[1-4] These general observations strongly support a mutually dependent relationship between oxidative metabolism and acid production. The exact nature of this relationship remains an area of intense investigation and controversy. Arguments concerning the role of metabolism in H^+ transport have centered around a distinction between the so-called redox and ATPase models for proton translocation.[1,5] The fundamental distinction between these models involves the nature of the immediate energy source for proton translocation. However, beyond this distinction, the two models involve quite different mechanisms for the actual translocation mechanism and for fhe control of metabolism. Additionally, information related to this distinction can provide a basis for identification of the subcellular components responsible for acid production, e.g., K-H-ATPase, and nonmitochondrial cytochromes.

Among the approaches to identifying the energy source for H^+ transport, one of the most direct has been an attempt to demonstrate that acid secretion involves an increased utilization of ATP. Studies utilizing metabolic inhibitors have shown that inhibition of acid secretion is accompanied by a fall in ATP levels.[4,6,7] While this supports the idea that ATP is necessary, other studies [8] have shown that normal cellular ATP levels alone are not sufficient to support H^+ transport. Attempts to demonstrate changes in steady-state ATP levels associated with changes in acid secretion have been controversial.[5-7] Very small or no changes in ATP were found for both *in vitro* and *in vivo* preparations upon stimulation of acid secretion. Furthermore, the small changes in ATP/ADP that have been reported [7] do not appear capable of accounting for the control of respiration exerted by the acid secretory process.[13] Thus, previous studies have failed to clearly demonstrate that ATP is directly utilized by the acid secretory mechanism.

Recently we have begun studying a preparation of isolated gastric glands in our laboratory. This preparation, developed by Dr. Berglindh,[9] has been shown to respond to physiological stimulants and inhibitors [10,11] in terms of changes in respiration, cyclic AMP content, and the accumulation of the

* This work was supported in part by grants from USPHS Nos. AM14752 and AM21588.

0077–8923/80/0341–0274 $01.75/0 © 1980, NYAS

weak base aminopyrine (AP). The gastric glands appear to be a very suitable preparation for the study of biochemical parameters associated with changes in secretory activity. Consequently, we have undertaken a re-examination of the role of ATP utilizing this preparation and conventional biochemical analyses. The present communication represents some initial studies concerning the production and turnover of ATP. The results show that both oxidative metabolism and glycolysis contribute to ATP production, that ATP levels decline during acid secretory activity, but that the activity-related decline may be due to decreased production rather than increased utilization.

METHODS

The preparation of isolated gastric glands from rabbit stomach was obtained as previously described.[9, 10] The glands were incubated in 25 ml flasks or respirometer flasks, as appropriate, in medium containing, in mM: NaCl, 132.4; KCl, 5.4; $MgSO_4$, 1.2; $CaCl_2$, 1.0; HEPES, 10; glucose, 1.0; phenol red, 10 mg/ml, and 2 mg/ml rabbit albumin, pH 7.4. The flasks were gassed with room air and maintained at 37°C. Respiration was measured by standard techniques using a differential respirometer (Gilson) and rates were normalized on a dry weight-hour basis. Accumulation of the weak base aminopyrine (AP) was measured as previously described [10] using ^{14}C-AP (New England Nuclear). The accumulation of AP was expressed as the ratio of AP in intraglandular water to AP in extraglandular water. Adenine nucleotide measurements were performed on perchloric acid extracts of gland suspensions using conventional enzymatic methods.[12] ATP was measured via hexokinase-glucose-6-phosphate dehydrogenase and ADP was measured via pyruvate kinase-lactic dehydrogenase. In general, agents were added at the start of the experiment and incubations were for 60 min. In the case of timecourse experiments, a preincubation period of 20 min was employed prior to addition of the test agent. All experiments were performed on duplicate or triplicate samples of glands and the mean value used for subsequent statistical analysis. The *n* values refer to number of experimental animals.

RESULTS AND DISCUSSION

ATP Production

Freshly prepared or control glands exhibit a small AP ratio and a basal respiration rate. However, control glands are close to a resting state relative to the maximum stimulation that can be achieved (see below). Control glands thus represent a reasonable preparation with which to compare various metabolic and secretory states. Typical values for control adenine nucleotide levels are presented in TABLE 1. The mean ATP value of approximately 4 nmol/mg dry wt. is very similar to that found for *in vivo* dog [5] and *in vitro* amphibian [6–8] gastric mucosa. The ADP values and ATP/ADP ratios are also similar to those found in other preparations.

A relatively crude attempt to identify the metabolic pathways responsible for maintenance of the ATP levels was made by using selected metabolic inhibitors. Respiratory inhibitors including amytal, cyanide and azide were uniformly found to suppress the ATP levels and inhibit AP accumulation and respiration. Some results using inhibitors are given in TABLE 1 for comparison with control tissues. The substantial reduction of ATP levels with inhibitors indicates that oxidative metabolism is an important source of ATP production. Amytal appears to be more effective than azide as a respiratory inhibitor for the gastric glands; however, both agents are potent inhibitors of AP accumulation and produce equivalent reductions in ATP levels. Since the dose of amytal or azide employed does not completely suppress respiration, a portion of the residual ATP could be due to continued oxidative phosphorylation, particularly in resting glands where ATP utilization is presumed to be slow. Alternatively, the residual ATP may be due to glycolytic activity. To test the latter possibility the glycolytic inhibitor 2-deoxyglucose (DOG) was tested alone and in combination with azide. As shown in FIGURE 1, with glucose as the sole exogenous substrate, DOG is an effective inhibitor of AP accumulation. This agent also produces a significant suppression of ATP. The inhibitory actions of DOG are overcome by the addition of oxidative substrates such as the fatty acid, butyrate, or pyruvate (not shown). These results indicate that glucose metabolism can sustain, but is not essential for, ATP levels and secretory activity. Azide alone rather completely suppresses AP accumulation and reduces ATP levels by about 70%. The combination of azide plus DOG does not reduce the ATP levels to zero. Surprisingly, about 40% of the tissue ATP content remains in the presence of DOG plus azide. This high residual ATP level is not reduced by uncoupling agents, such as valinomycin (5μM) or FCCP (5μM). However, incubating the tissue extracts with commercial ATPase destroys the residual signal indicating that it is, in fact, ATP. Conceivably, the residual ATP represents a bound component of ATP that does not participate in the normal metabolic turnover. The source and significance of this component represent interesting questions but ones which cannot be addressed at this time. The metabolically labile component of ATP appears to be accounted for by a combination of oxidative and glycolytic activity. Of these two, oxidative metabolism appears capable of supporting completely normal ATP levels and secretory activity (DOG plus butyrate), whereas glycolysis alone (azide) appears inadequate to support either ATP levels or AP accumulation.

TABLE 1

EFFECTS OF RESPIRATORY INHIBITORS *

	n	ATP nmol/mg	ADP dry wt	ATP/ADP	Q_{o2} (%)	AP Ratio control
Control	6	3.7±0.6	0.9±0.1	4.9±1.1	100	100
NaN_3	6	1.0±0.4	1.4±0.2	0.8±0.2	56±5	7±4
Control	5	4.0±0.3	0.6±0.1	6.8±0.4	100	100
Amytal	5	1.7±0.3	0.8±0.1	2.2±0.6	26±5	11±6

* Values are mean±SEM; NaN_3 (5×10^{-3}) and amytal (2×10^{-3}); treatment for 60 min; all values for inhibitors are significantly different ($p < .01$) from controls by paired t-test.

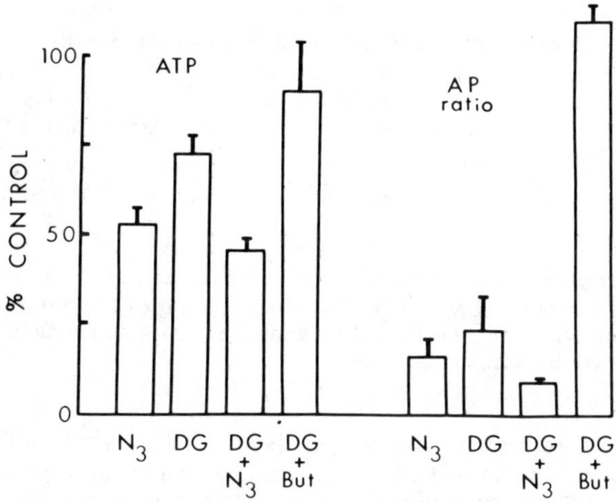

FIGURE 1. Influence of azide and DOG on ATP levels and AP accumulation. The levels of ATP and the AP ratio are presented as percent of control, i.e., no inhibitor, values. Measurements were made after 60 min. exposure to agents. The agents used were N_3, sodium azide (5 mM); DG, 2-deoxyglucose (20 mM); But, sodium butyrate (10 mM). All tissues were exposed to 1 mM glucose and treated with histamine (0.1 mM) plus isobutyl methyl xanthine (IMX, 0.01 mM) to obtain maximal AP ratios. Values represent the mean$\pm$SEM for $n=5$.

ATP Utilization

According to the ATPase model for acid secretion, proton translocation is accompanied by an increased hydrolysis of ATP. The products of this hydrolysis, ADP and P_i, in turn, would stimulate metabolic reactions that resynthesize the ATP. This general scheme accounts both for the energy supply for secretion and a mechanism for metabolic control. During various states of acid secretory activity, the steady-state level of ATP is predicted to be inversely related to the rate of secretion. The magnitude of the change in ATP will be a function of the "closeness" of coupling, but some finite change must occur in order to account for the maintained stimulation of metabolism. Previous attempts to demonstrate a change in ATP level have met with little success; however, in the present study we find substantial changes in ATP and ADP levels associated with secretory activity. Typical results for histamine stimulation are given in TABLE 2. Compared to control glands, those treated with histamine show a large increase in AP accumulation and approximately a doubling of respiration. This increased activity is associated with a significant decrease in ATP and a concomitant rise in ADP. The changes in nucleotide levels result in substantial changes in the calculated ratio, ATP/ADP. Similar results are obtained for stimulation by dibutyryl cyclic AMP, indicating that the results are not specific for histamine. Moreover, the nucleotide changes do not occur in the presence of thiocyanate, a specific inhibitor of acid secretion, nor in K^+ depleted glands.

The timecourse of ATP changes were measured to ascertain if they correlate

TABLE 2

COMPARISON OF RESTING AND STIMULATED STATES *

	ATP nmol/mg	ADP nmol/mg	ATP/ADP	Q_{o2} µl/mg hr	AP Ratio
Control	3.7±0.3	0.9±0.1	4.5±0.7	9.5±0.4	9.6±1.7
Histamine + IMX	2.6±0.3	1.3±0.2	2.4±0.4	18.6±0.7	42±4.3

* Values are mean±SEM for $n=9$; histamine (1×10^{-4}) and IMX (1×10^{-5}); incubations for 60 min; all values for stimulated glands are significantly different ($p<.01$) from controls by paired t-test.

with changes in secretory activity. As shown in FIGURE 2, ATP falls rapidly upon addition of histamine reaching a steady value at essentially the same time as does AP accumulation. Previous reports have shown a similar timecourse for changes in respiration.[10] Thus, the timecourse of ATP change is consistent with an association to secretory activity.

The steady-state decrease in ATP levels suggests an increased utilization during acid production. The concomitant increase in respiration further sup-

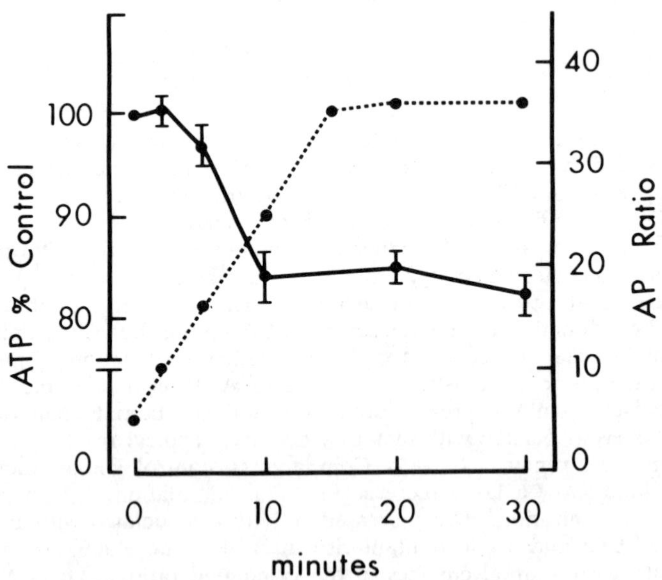

FIGURE 2. ATP timecourse following histamine. Glands were preincubated 20 min. and histamine (0.1 mM) plus isobutyl methyl xanthine (0.01 mM) was added at time zero. Samples were taken for ATP analysis at various times thereafter. The values represent mean±SEM for $n=3$. The dashed line is a representative timecourse for AP accumulation shown for comparison.

ports this idea, assuming that respiration is associated with ATP synthesis. This latter assumption represents an important element in the ATPase model and, if proven, would confirm that ATP is utilized during H^+ production. The alternative possibility is that the decline in ATP level is due to decreased synthesis without increased hydrolysis. Such an hypothesis would require that the respiration associated with secretory activity is required for reactions not involving ATP synthesis, such as a direct redox coupling. These two possibilities may be tested in the so-called ATP-sparing experiments. This test involves a rapid inhibition of ATP synthesis and comparison of the timecourse of ATP decline under conditions of rest and activity. If the resting condition requires less ATP hydrolysis than does the active state, a difference

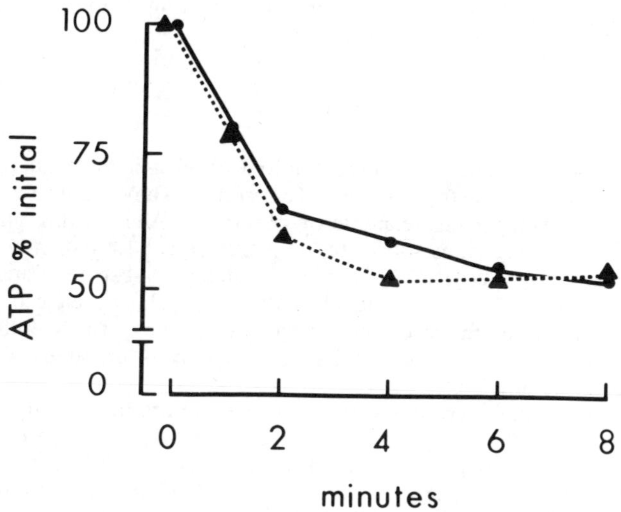

FIGURE 3. ATP timecourse following azide. Glands were preincubated 20 min. in the presence (dashed line) or absence (solid line) of histamine plus IMX. At time zero sodium azide (5 mM) was added and samples withdrawn at timed intervals for ATP analysis.

in timecourse should be detected. In the case of the gastric glands, the increment in respiration associated with stimulation indicates that the rate of ATP utilization should approximately double. This magnitude of change in ATP hydrolysis should be readily detectable in the timecourse comparison. Any such test depends on the metabolic inhibition being specific and rapid. Previous tests of this type have reported a "sparing" effect;[6] however, this was only observed over long time periods and the initial (15 min) phase showed no difference in ATP decline. We have repeated these studies in the gland preparation using azide as a rapid inhibitor. The timecourse of ATP decline for resting versus histamine-stimulated glands is shown in FIGURE 3. While the initial values for stimulated glands are lower, the rate of ATP decline following azide is not different from that of resting glands. The half-time for the loss of the labile component of ATP does not differ significantly under the two conditions. The half-times for ATP loss, obtained from semilog plots of fractional decline versus time, were found to be 1.6 ± 0.3 min and 1.8 ± 0.3 min. (mean ± SEM, $n = 4$) for control and histamine stimulated glands, respectively. Thus, there does not appear to be a sparing of ATP in nonsecreting glands. These results, subject to the specificity of azide action, would argue that the reduced steady-state level of ATP during secretory activity is due to reduced synthesis of ATP and the rate of hydrolysis remains essentially constant. The obvious implication is that during secretion, redox energy is shunted away from phosphorylation reactions and is coupled to other energy-dissipating reactions, presumably those associated with proton translocation. An additional consideration is that while the fall in ATP/ADP ratio observed could account for the increased respiration, it could also reflect an equilibrium with a reduced energy state of the mitochondria, resulting from rapid activity of the coupled energy-dissipating processes.

Conclusions

The studies presented here represent initial observations of adenine nucleotide levels in isolated gastric glands. The results show that both oxidative metabolism and glycolysis are capable of generating ATP in this tissue. When glucose is the only exogenous substrate, an intact glycolytic pathway is required for full H^+ secretory activity. However, oxidizable substrates, such as fatty acids, can maintain ATP levels and H^+ secretion in the absence of glycolysis. Thus, it is presumably the products of glycolysis that are essential and not the specific reactions; nor is it possible for glycolysis alone to sustain ATP levels or acid secretion.

Stimulation of acid secretion with histamine results in a significant decline in ATP levels and ATP/ADP ratio. These changes appear to be related to the increased secretory activity. Attempts to show increased ATP hydrolysis during activity failed. The reduced ATP levels observed during activity thus appear to be due to reduced synthesis of ATP rather than increased hydrolysis. Despite the observed changes in ATP levels associated with H^+ production, the results do not provide direct support for an ATPase model of acid secretion.

Acknowledgment

The author gratefully acknowledges the expert technical assistance of Ms. Linda Campbell.

References

1. Hersey, S. J. 1974. Biochim. Biophys. Acta **344:** 157–203.
2. Moody, F. G. 1968. Am. J. Physiol. **215:** 127–131.
3. High, W. L. & S. J. Hersey. 1974. Am. J. Physiol. **226:** 1408–1412.
4. Sachs, G., R. Shoemaker & B. I. Hirschowitz. 1967. Biochim. Biophys. Acta **143:** 522–531.
5. Sachs, G., E. Rabon, G. Saccomani & H. M. Sarau. 1975. Ann. N.Y. Acad. Sci. **264:** 456–475.

6. FORTE, J. G., P. H. ADAMS & R. E. DAVIES. 1965. Biochim. Biophys. Acta **104:** 25–38.
7. DURBIN, R. P., F. MICHELANGELI & A. NICKEL. 1974. Biochim. Biophys. Acta **367:** 177–189.
8. HERSEY, S. J. 1977. Biochim. Biophys. Acta **496:** 359–366.
9. BERGLINDH, T. & K. J. ÖBRINK. 1976. Acta Physiol Scand. **96:** 150–159.
10. BERGLINDH, T., H. F. HELANDER & K. J. ÖBRINK. 1976. Acta Physiol. Scand. **97:** 401–414.
11. CHEW, C. S., S. J. HERSEY, G. SACHS & T. BERGLINDH. 1979. Am. J. Physiol. In press.
12. BERGMEYER, H. U. 1965. Methods of Enzymatic Analysis. Academic Press. New York, N.Y.
13. WILSON, D. F., C. OWEN, L. MELA & L. WEINER. 1973. Biochem. Biophys. Res. Commun. **53:** 326–333.

DISCUSSION

E. RACKER: I recall a demonstration in the stomach of the fetal lamb in which gastric secretion could take place under anaerobic conditions. If that were the case, of course, it would eliminate your suggested mechanism or, at least, under those circumstances.

S. J. HERSEY: I do not recall the exact paper that you are referring to. There have been reports, not just in fetal tissues, but in adult tissues as well of residual acid secretion under presumed anaerobic conditions. Many of these experiments have later proved that, in fact, the anaerobic condition was not completely anaerobic or that the acid production that people were measuring was lactic acid and not hydrochloric acid; so most of the data in the literature concerning this is quite open to question. Careful studies have been published showing that in amphibian gastric mucosa there is zero acid secretion under anaerobic conditions and I believe that this is reasonably accepted at this time.

J. G. FORTE: In newborn piglet stomachs, which are sensitive to oxygen removal, the secretion goes to zero. We have observed that this stomach tissue is absolutely dependent upon substrates for secretion and seems to prefer carbohydrate sources to a much greater extent than fatty acids, in contrast to the frog gastric mucosa. How does this fit in with your hypotheses regarding the fatty acid pathways as primary energy sources of secretion?

HERSEY: I am not sure of the exact type of measurements that you made, but as a general rule, the type of substrate preferred by a tissue is going to depend on the enzyme activities associated with those particular pathways in that particular tissue. We have found in amphibian that the oxidation of fatty acids is much more rapid than that of carbohydrates. Given a choice of substrates, rabbit gastric glands prefer to oxidize lipids as opposed to carbohydrates. We have not worked with the newborn pig and I am not sure of the nature of your experiments, but I think the essence of what substrate is utilized is not so important as the nature of the final product formed from that substrate. This would be the direct energy source for the proton translocation.

FORTE: I wholly agree with your interpretation. It is just that I was referring to it relative to the hypothesis that cyclic AMP liberated fatty acids are the primary source of energy for gastric secretion.

HERSEY: I think that the hypothesis probably would not hold absolutely even in the rabbit gastric glands. Certainly the evidence from deoxyglucose would say that exogenous glucose can be an essential part of the substrate supplied for the proton translocation. So, we would have to say that in this preparation there is no absolute dependence on mobilization of lipid for the secretory process.

ENERGY PRODUCTION AND ENERGY CONSUMPTION IN FROG GASTRIC MUCOSA

Lazaro J. Mandel

Department of Physiology
Duke University Medical Center
Durham, North Carolina 27710

INTRODUCTION

This paper reviews the results obtained by two approaches used to study the relationship between oxidative metabolism and active transport in various epithelia, emphasizing proton secretion by the gastric mucosa. Both of these approaches utilize optical techniques to monitor the redox state of the various mitochondial respiratory enzymes, in conjunction with measurements of active transepithelial transport in the intact tissue. The first of these concerns the observation of changes in the redox state of the respiratory enzymes as a function of alterations in the rate of active transport by the tissue.

The predicted changes in the redox state of the mitochondrial enzymes in response to alterations in active transport could be most easily understood in terms of the mitochondrial transitions described by Chance and Williams.[1] In isolated mitochondria, these authors defined a series of experimental conditions under which mitochondria could undergo transitions from one state to another. Of special interest is the transition from passive (State 4) to active (State 3) and vice versa. This transition is obtained by initially bathing the mitochondria in a solution containing metabolic substrates, oxygen, and orthophosphate but lacking ADP to obtain a low, resting, respiratory rate. Upon addition of ADP, the respiratory rate increases rapidly and is maintained at this accelerated rate until the ADP is phosphorylated to ATP. This transition is accompanied by predictable changes in the redox state of the respiratory enzymes, as shown in TABLE 1, becoming more oxidized (less reduced) when approaching the active state and returning to a more reduced state when the respiratory rate decreases. This net oxidation is obtained because the rate at which electrons are oxidized (i.e., oxygen consumption), is increased relative to the rate at which they are supplied to the respiratory chain.

These changes in redox level may be detected by optical techniques because all the components listed in TABLE 1 have characteristic absorption peaks (except flavoproteins [FAD], which have a trough) associated with the reduced form of the enzyme. Thus, their redox state may be detected spectrophotometrically in the intact tissue.[2, 3] A more sensitive method involves the use of fluorometry to detect the fluorescent emission from NADH because its reduced form fluoresces in a wide band of visible light (420–500 nm) when stimulated with 340–370 nm light, whereas the oxidized form does not fluoresce. This technique has also been applied successfully to a number of intact tissues.[4, 5]

0077–8923/80/0341–0283 $01.75/0 © 1980, NYAS

TABLE 1

SOME CHARACTERISTICS OF THE RESPIRATORY ENZYMES IN MITOCHONDRIAL STATES 3 AND 4 *

State	ADP Level	Oxygen Consumption	Steady-state Percent Reduction NADH	FAD	cyt *b*	cyt *c*	cyt *a*
3	high	fast	53	20	16	6	4
4	low	slow	99	40	35	14	0

* (From Chance and Williams.[1])

Responses of Mitochondrial NAD to Alterations in the Work Output of Various Tissues

Various investigators have shown that alterations in work output cause corresponding changes in the redox state of NAD in intact tissues. Increases in work obtained by contractile activity in skeletal muscle [4] and electrical stimulation in brain cortex [6] lead to decreases in NADH fluorescence (oxidation of NAD). The opposite response, that is, a reduction of NAD is obtained in perfused rat kidney cortex when active transport is inhibited by ouabain.[7] Both of these type of responses are those expected for mitochondrial transitions from resting to active and vice versa, as described earlier.

The relationship between NADH fluorescence and active transport in the kidney has been specifically explored by Robert Balaban in my laboratory. The doubly perfused bullfrog kidney as described by Balaban and Mandel,[8] was used for microfluorometric monitoring of the redox state of NAD while simultaneously measuring the rates of water and solute reabsorption. Results demonstrated that all experimental conditions that primarily caused increases in transport work of the bullfrog kidney produced an oxidation of NAD, whereas conditions that decreased transport work caused reductions in NAD.[8]

These results with the bullfrog kidney demonstrate that the mitochondria in the intact kidney undergo state transitions in response to alterations in work similar to those displayed by isolated mitochondria subjected to varying rates of oxidative phosphorylation and oxygen consumption. Since alterations in the transport work of the kidney produce parallel changes in the oxygen consumption rate,[9] these mitochondrial state transitions suggest that a cellular signal exists that regulates tissue oxygen consumption and phosphorylation rate in response to the energetic needs of active transport. In conjunction with ample experimental evidence that ATP provides the energy for active transport in the kidney,[9] the observation that these mitochondrial transitions occur in the kidney suggests that ATP may be involved in this feedback regulation of the mitochondrial state and oxygen consumption rate in this tissue.[8]

Respiratory Chain Responses to Alterations in the Rate of Acid Secretion by the Frog Gastric Mucosa

In the gastric mucosa, transmission spectrophotometry of the redox state of the respiratory enzymes has been performed rather than fluorometry of

NADH emission. However, similar conclusions regarding mitochondrial transitions may be made from this measurement.[1] The gastric mucosa from the frog is mounted in a modified Ussing double chamber with optical windows to allow transmission spectrophotometry. As shown in FIGURE 1, this chamber also permits the simultaneous measurement of short circuit current and acid secretory rate.

Initiating the experiment in the resting, nonsecreting state, stimulation of the frog stomach to secrete acid with histamine, or any other secretagogue, causes a reduction of the entire cytochrome chain, as obtained by Hersey and Jobsis.[10] Such a stimulation of acid secretion is accompanied by an increase in oxygen consumption.[11] Inhibition of secretion with thiocyanate causes the opposite response; that is, an oxidation of the entire chain.[10] These responses are the *opposite* of those just described for the kidney, where increases in work cause oxidations of NAD. The observed redox responses of the gastric mucosa have been interpreted by Kidder [12] as being due to local anoxia, by Hersey and High,[13] as reflecting the properties of a redox pump and not anoxia, and by Hersey,[14] as describing a mitochondrial transition from a substrate-depleted (State 2) to an active (State 3) state. Thus, in contrast to the comparatively simple explanation given to the kidney results, the redox responses of the gastric mucosa are more complex, suggesting that the action of secretagogues is to initiate a series of events with multiple manifestations on the mitochondrial redox state. Unfortunately, these results have not provided clear information regarding the immediate energy source for acid secretion.

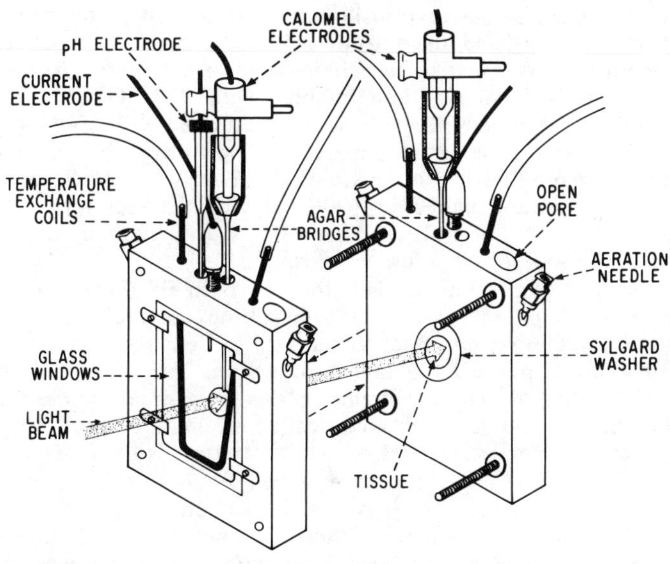

FIGURE 1. Schematic diagram of chamber used in present studies. This is a flattened version of Ussing double-chambers with clear windows to permit transmission spectrophotometry of intact tissue. The rate of acid secretion and the short current circuit could be measured simultaneously. (From Mandel and Riddle.[15] By permission of *American Journal of Physiology*.)

Kinetic Relationship between Energy Production and Consumption in Frog Gastric Mucosa

The second approach to be discussed in this paper concerns the use of spectrophotometry to monitor the rate of reduction of the terminal oxidase of the cytochrome chain, cytochrome $a(+a_3)$, during the transition to anoxia as described by Mandel and Riddle.[15] In this largely aerobic tissue, changes in the redox state of cytochrome oxidase are fairly well correlated with changes in the rate of oxygen consumption during the approach to anoxia [16, 17]; therefore, the reduction kinetics of this cytochrome closely represent the kinetics of energy-producing reactions. On the other hand, the kinetics of energy consumption may be assumed to follow the changes in the rate of acid secretion. Any delay between energy production and energy consumption would be due to energy storage in the tissue; quantitative knowledge regarding this storage would give important information regarding the immediate source of energy for acid secretion. For example, a long delay between the cytochrome kinetics and acid-secretory kinetics would be indicative of a storage pool of energy in the parietal cells that could be ATP, suggesting that this compound could be the direct source of energy for acid secretion, as proposed by various investigators.[18, 19] At the other extreme, the lack of any delay between these two processes would be consistent with either a redox source of energy [14] or an extremely small amount of energy stored in these cells that could be used for acid secretion. This delay between energy production and energy consumption was measured in the present experiments.

The experiments were performed in the same chamber as described earlier, one which allows simultaneous transillumination as well as the measurements of short-circuit current and the rate of acid secretion with a pH-stat.[15] The gastric mucosa was mounted in the chamber with a buffered Ringer solution (pH 7.4) on the nutrient side and bubbled with 95% O_2, 5% CO_2, along with an unbuffered secretory Ringer's solution buffered with 100% O_2 and pH-stated to 5.0. The transition to anoxia was performed by rapidly flushing the existing oxygen-bubbled solutions with large volumes of Ringer solutions of identical composition bathed in 100% N_2 for the secretory solution and 95% N_2, 5% CO_2 for the nutrient solution. In this manner, it was possible to change bathing solutions in a few seconds.

When the frog gastric mucosa is bathed on both sides with sulfate Ringer, the secretory potential reverses and the resulting short circuit current (I_{sc}) appears to be directly proportional to the rate of acid secretion.[15, 20, 21] Thus, the rate of change in the I_{sc} obtained in sulfate Ringer's could be used to monitor the kinetics of acid secretion. In this manner, it is possible to measure the kinetics of acid secretion electrically without the intervening diffusion delays inherent in a measurement of acid secretion with a pH-stat.

Results of a typical experiment are shown in FIGURE 2, in which anoxia was initiated at the time indicated. Visual inspection suggests that the kinetics of I_{sc} inhibition are very similar to those of cytochrome $a(+a_3)$ reduction. The average of six similar experiments, in which the initial rates were fitted to an exponential time constant are summarized in TABLE 2. The average initial rate for I_{sc} inhibition was 0.6 min slower than the cytochrome $a(+a_3)$ reduction rate, although these two values were not significantly different from each other. This slight delay could be totally explained by the fact that

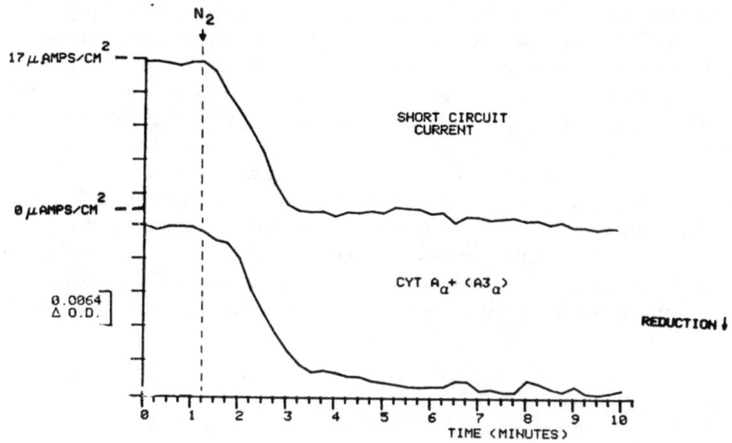

FIGURE 2. Simultaneous kinetics of short-circuit current and cytochrome $a(+a_3)$ reduction of gastric mucosa subjected to rapid anoxia in sulfate Ringer's solutions. The peak wavelength used was 605 nm, with references at 575 and 630 nm. Note the almost simultaneous kinetics of these parameters.

the cytochrome $a(+a_3)$ kinetics do not follow exactly the kinetics of oxygen consumption, although the latter is completely inhibited when 100% reduction is reached.[1] The actual relationship between cytochrome $a(+a_3)$ kinetics and oxygen consumption kinetics is variable, depending on various factors, which include the intracellular ATP and ADP levels, substrates utilized, and other variables. However, it appears that under most conditions oxygen consumption lags behind the cytochrome $a(+a_3)$ reduction rate. A calculation by Jobsis [22] could be used as an example to quantitate the kinetic relationship between these variables. Using these values, it may be calculated that a time constant of 0.9 min for cytochrome $a(+a_3)$ (TABLE 2) would result in a time constant of 1.5 min for the oxygen consumption rate, not significantly different from the I_{sc} time constant. Thus, it may be argued that no delay exists between the rate of inhibition of acid secretion and the rate of decrease of oxygen consumption, implying that no measurable storage of high energy compounds exists in this tissue that may be utilized for acid secretion.

TABLE 2

EXPONENTIAL TIME CONSTANTS (IN MINUTES) FOR CYTOCHROME $a(+a_3)$ AND I_{sc} KINETICS TO ANOXIA ($\pm$S.E.) N=6

Bathing media	cyt $a(+a_3)$	I_{sc}	Delay *
Sulfate	0.9±0.2	1.5±0.3	0.25
Chloride	1.4±0.2	3.8±0.3	1.85

* The delay was measured as the time between the onset of the cyt a response and the onset of the I_{sc} response.

When the same experiment is performed in chloride bathing media, the results are quite different because the I_{sc} does not measure the rate of acid secretion. Instead, when acid secretion is present, the I_{sc} is a sum of active Cl and active proton transport.[23] However, when acid secretion is inhibited, the I_{sc} measures solely the rate of active Cl transport. Results with secreting mucosae in chloride media show that I_{sc} decays much more slowly than the reduction kinetics of cytochrome $a(+a_3)$, as shown in FIGURE 3. The average results of six such experiments are also summarized in TABLE 2, demonstrating that a clear delay exists between these two kinetic parameters. As shown earlier, the kinetics of acid secretion are comparable to the cytochrome kinetics in the transition to anoxia; and, thus, the I_{sc} kinetics really represent the time course of inhibition for active nonacid Cl transport. The long delay between

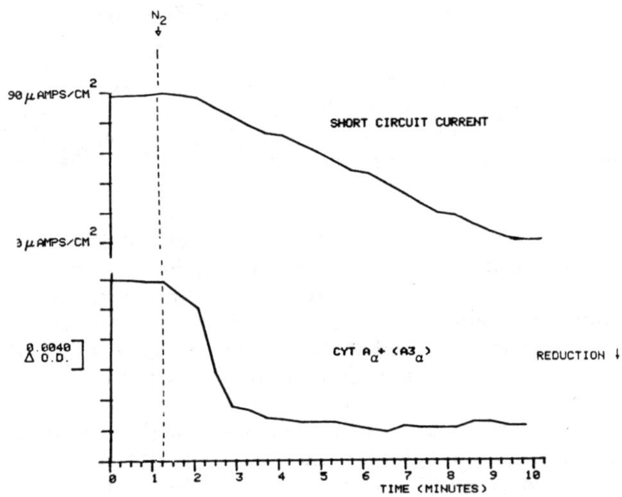

FIGURE 3. Simultaneous kinetics of I_{sc} and cytochrome $a(+a_3)$ reduction of gastric mucosa subjected to anoxia in chloride Ringer solutions. Peak and reference wavelengths used for spectrophotometry are as in FIGURE 2. Note the delay and slower inhibition of I_{sc} when compared to the cytochrome kinetics.

the cytochrome and I_{sc} kinetics seems to represent a period of time during which active transport utilizes energy stored within the cells because aerobic metabolism has ceased.

The contrast between the kinetic results in sulfate and chloride provide important information regarding the immediate energy sources for active proton and active nonacid chloride transport. The clear delay between the I_{sc} kinetics and cytochrome oxidase (oxygen consumption) kinetics to anoxia in chloride indicate that a stored form of energy exists in this tissue capable of supporting active Cl transport for a limited time period after oxygen consumption ceased. It may be speculated that this form of energy is also the immediate energy source under normal aerobic conditions. The best candidate for this role is ATP because the gastric mucosa maintains more than 60% of its normal ATP level after 10–30 minutes of anoxia.[18, 19]

In contrast to the results in chloride solutions, the kinetics to anoxia in sulfate indicate that no delay exists between the rate of oxygen consumption, which determines the rate of energy production from oxidative metabolism, and the rate of acid secretion, which should be directly related to the rate of energy consumption for this process. The lack of delay suggests that no stored energy measurable by the present techniques exists in this tissue in a form capable of maintaining acid secretion in the absence of oxidative metabolism. Total tissue ATP is clearly not the source of energy for acid secretion because it decays much more slowly than the oxygen consumption rate, as discussed earlier. The present results are consistent with either of the two main theories concerning the source of energy for acid secretion, the ATPase [18, 19] and redox [14] theories, without being able to differentiate between them. In terms of these theories, the results suggest at least the following two possibilities:

1. Acid secretion may be energized by a chemical form of energy, such as ATP, contained within a cellular compartment that becomes dissipated in about two minutes with anoxia. This is a very real possibility because it has been demonstrated that tissues highly dependent on oxidative metabolism, such as the kidney cortex, lose 90% of their ATP after only two minutes of anoxia.[24] In the two studies in which total tissue ATP was measured in the gastric mucosa during anoxia, the tissue lost 23% of its ATP in the first 15 minutes.[18, 19] This figure constitutes an average for all cell types, and, therefore, it is conceivable that most of the initial loss of ATP occurs in the parietal cells, which occupy 28% of the total cell mass in the frog.[25] If the parietal cells became rapidly depleted of ATP with anoxia, the kinetics of inhibition of acid secretion would be expected to be fast, too.

2. Electron transport through the cytochrome chain may be essential for acid secretion because the rate of inhibition of electron transport (which is identical with oxygen consumption) is not different from the rate of inhibition of acid secretion by anoxia. These results are consistent with those of Hersey,[26] showing that flow through the intact cytochrome chain is required for acid secretion. This author demonstrated that bypassing the cytochrome chain blocked by amytal restored tissue ATP levels but not acid secretion. Therefore, the present results cannot distinguish between the ATPase and redox theories, but they place constraints on the former theory by restricting the maximal size of the ATP compartment that could be utilized for acid secretion.

Kinetic Relationship between Oxidative Metabolism and Active K Transport in the Midgut of the Tobacco Hornworm

I would like to conclude this communication with a brief description of the results obtained when the same kinetic approach is applied to another transporting epithelium, the midgut of the tobacco hornworm, as described by Mandel, Riddle and Storey.[27] This tissue actively transports K from hemolymph to lumen by a process highly dependent on oxidative metabolism.[28, 29] The question of whether ATP is the immediate source of energy for the K transport has persisted for many years because of the inability of investigators to find any K-ATPase or Na,K-ATPase in the midguts of either the hornworm [30] or the silkworm.[31]

We have recently investigated this problem by measuring simultaneously the kinetics to anoxia of the active transport rate, the reduction level of cytochrome oxidase, and the ATP levels in the tissue.[27] The experiments were performed in the same chamber described earlier, allowing the simultaneous measurement of the short circuit current (I_{sc}) and the mitochondrial redox state. In this tissue, the I_{sc} measures net K transport.[29] The midgut was fastened by initial mounting on a plastic disc with a grooved ridge surrounding the central circular opening. The tissue was tied down with a thread loop that fit into the groove, making a tight seal with little edge damage. The disc with the tissue was, thereafter, placed in a recessed space flush against one of the half-chambers; a slight amount of grease was used to obtain a good seal between the plastic disc and the plastic chamber. The actual area of tissue in contact with the solutions was 0.33 cm^2.

After the appropriate experimental condition was achieved with the tissue mounted as described earlier, the chamber was quickly taken apart and excess Ringer shaken off the tissue. The plastic disc holding the tissue was placed over a plastic vial containing liquid nitrogen, and the central portion of the tissue punched out with a metal rod chilled in liquid nitrogen and pushed into the vial. The frozen samples were weighed and assayed for protein and ATP.

The transition to anoxia was obtained in a similar manner to that in the gastric mucosa by initially equilibrating the tissue in a K Ringer solution [32] bubbled with 100% oxygen; and, thereafter, rapidly flushing this solution with an excess of the same solution that had been preequilibrated with 100% nitrogen. A typical result is shown in FIGURE 4, displaying the clear difference

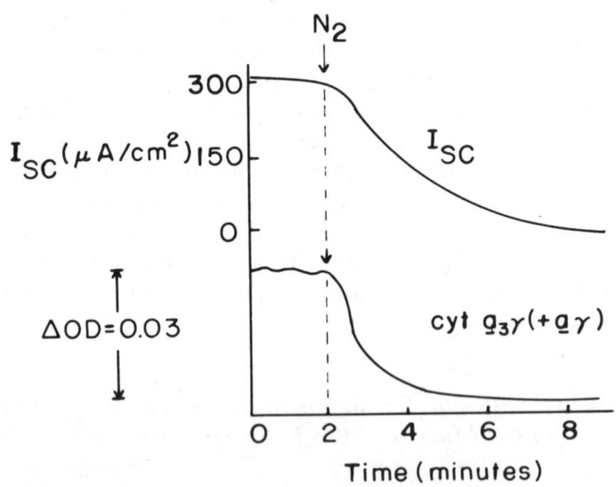

FIGURE 4. Kinetics to anoxia of I_{sc} and the reduction of cytochrome $a_3(+a)$ in the intact midgut of the tobacco hornworm. The peak wavelength used for spectrophotometry was 445 nm and references were at 400 and 455 nm. At the indicated time, solutions pre-equilibrated with 100% N_2 were rapidly exchanged for the solutions in the chamber. Note the delay and slower kinetics of I_{sc} in comparison with the cytochrome rate of reduction. (From Mandel *et al.*[27] By permission of *American Journal of Physiology*.)

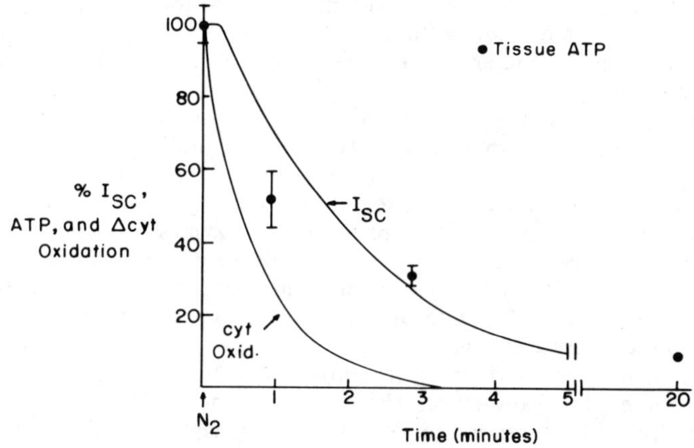

FIGURE 5. Simultaneous plot showing percentage of I_{sc}, percentage of change in cytochrome oxidation and percentage of ATP remaining in the tissue as a function of time after changing solutions in the transition to anoxia. (From Mandel *et al.*[27] By permission of *American Journal of Physiology*.)

between the kinetics of cytochrome a_3 and the I_{sc}. The I_{sc} displays an initial delay and slower kinetics, some current remaining after the cytochrome has become 100% reduced. This result is similar to that obtained earlier on the silkworm midgut.[33] In five such experiments, the average time constant for cytochrome a_3 reduction was 0.75 ± 0.05 min, whereas the time constant for I_{sc} was 2.10 ± 0.15 min after a delay of 0.25 min.

The ATP levels in the midgut were measured in another set of experiments at various times during the transition to anoxia. The initial ATP value obtained during equilibration in 100% O_2 was 1.19 ± 0.10 μmoles ATP/g wet weight in five experiments. The kinetics to anoxia of the ATP levels are plotted (FIGURE 5) in conjunction with the average calculated kinetic values for the cytochrome and the I_{sc}. This figure demonstrates that the cytochrome kinetics are faster than those of the ATP or I_{sc} inhibition. The delay existing between the cytochrome and I_{sc} kinetics demonstrates that an intermediate form of energy must exist that is capable of maintaining active transport during the delay. The kinetics of ATP are shown to be sufficiently slow as compared to the rate of I_{sc} inhibition for this nucleotide to be considered as the possible source of energy for active K transport in the midgut.

Very recently, Wolfersberger[34] has succeeded in identifying and partially purifying a K-modulated ATPase from a plasma membrane fraction from the midgut of the tobacco hornworm. This new finding complements in timely fashion the conclusion made on the basis of the kinetic experiments that ATP may be the source of energy for active K transport in this tissue.

SUMMARY

The present communication describes two approaches used to study the relationship between oxidative metabolism and active transport in various epithelia, as compared to the gastric mucosa. Mitochondrial transitions are observed in the intact kidney that respond to induced changes in transport work in the same manner as has been observed in isolated mitochondria stimulated from a resting to active transition and vice versa. Since other experiments suggest that ATP is the source of energy for active Na transport, it is possible that ATP is also involved (directly or indirectly) in the feedback regulation of mitochondrial state and oxygen consumption in the kidney. In contrast, the stimulation of acid secretion in the gastric mucosa by secretagogues causes a reduction of all respiratory enzymes, which is the opposite response to that observed in the kidney. This response indicates that secretagogues elicit a complex series of events that are not open to simple interpretation. Unfortunately, this type of experiment does not help in separating the redox from the ATPase theory.

Experiments were performed to simultaneously measure the rapid kinetics to anoxia of energy production (by oxidative metabolism) and energy consumption (by acid secretion and/or active Cl transport) in the gastric mucosa. Results show that active Cl transport (nonacid) continues after aerobic energy-producing reactions have stopped; this transport is probably powered by ATP. Conversely, acid secretion ceases with identical kinetics to those of oxidative metabolism, demonstrating that no form of stored energy exists in these cells that is capable of maintaining acid secretion. Although these results do not allow a differentiation between the redox and ATPase theories, the results place some constraints on the properties of this energy source, emphasizing its extremely close association with oxidative metabolism.

Similar kinetic experiments to anoxia in the K-transporting midgut of the tobacco hornworm, demonstrate that a delay exists between the kinetics of cytochrome oxidase reduction and those of active transport inhibition. Since some active transport remains after oxidative metabolism has ceased, some intermediate form of energy must be capable of powering the K transport. The best candidate for this energy is ATP, since its kinetics of inhibition are sufficiently slow to provide the necessary energy after the inhibition of oxygen consumption.

Acknowledgments

I would like to acknowledge the collaboration of Dr. Robert Balaban, Ms. Terry Riddle, and Ms. Janet Storey in individual aspects of the studies cited. The helpful comments of Dr. Robert Balaban in the preparation of this manuscript are gratefully acknowledged. My appreciation is also extended to Ms. Joyce Cannon for typing this manuscript.

References

1. Chance, B. & G. R. Williams. 1955. Respiratory enzymes in oxidative phosphorylation. J. Biol. Chem. **217:** 383–427.
2. Jobsis, F. F. 1963. Spectrophotometric studies on intact muscle. I. Components of the respiratory chain. J. Gen Physiol. **46:** 905–928.
3. Oshino, N., T. Sugano, R. Oshino & B. Chance. 1974. Mitochondrial function under hypoxic conditions: The steady states of cytochrome $a+a_3$ and their relation to mitochondrial energy states. Biochim. Biophys. Acta **368:** 298–310.

4. JOBSIS, F. F. & J. C. DUFFIELD. 1967. Oxidative and glycolytic recovery metabolism in muscle. Fluorometric observations on their relative contributions. J. Gen. Physiol. **50:** 1009–1047.
5. JOBSIS, F. F., V. LEGALLAIS & M. O'CONNOR. 1966. A regulated differential fluorometer for the assay of oxidative metabolism in intact tissues. Proc. I.E.E.E.B.M.E. **13:** 93–99.
6. ROSENTHAL, M. & F. F. JOBSIS. 1971. Intracellular redox changes in the functioning cerebral cortex. II. Effect of direct cortical stimulation. J. Neurophysiol. **34:** 750–762.
7. FRANKE, H., C. H. BARLOW & B. CHANCE. 1976. Oxygen delivery in perfused rat kidney: NADH fluorescence and renal functional state. Am. J. Physiol. **231:** 1082–1089.
8. BALABAN, R. S. & L. J. MANDEL. 1980. Coupling of aerobic metabolism to active ion transport in the kidney. J. Physiol. (London). In press.
9. COHEN, J. J. & D. E. KAMM. 1976. Renal metabolism: relation to renal function. *In* The Kidney. B. M. Brenner and F. C. Rector, Eds. W. B. Saunders. Philadelphia.
10. HERSEY, S. J. & F. F. JOBSIS. 1969. Redox changes in the respiratory chain related to acid secretion by the intact gastric mucosa. Biochem. Biophys. Res. Commun. **36:** 243–250.
11. FORTE, J. G. & R. E. DAVIES. 1963. Oxygen consumption and active transport of ions by isolated frog gastric mucosa. Am. J. Physiol. **204:** 812–816.
12. KIDDER, G. W., III. 1970. Unstirred layers in tissue respiration: application to studies of frog gastric mucosa. Am. J. Physiol. **219:** 1789–1795.
13. HERSEY, S. J. & W. L. HIGH. 1972. Effect of unstirred layers on oxygenation of frog gastric mucosa. Am. J. Physiol. **223:** 903–909.
14. HERSEY, S. J. 1975. Interactions between oxidative metabolism and acid secretion in gastric mucosa. Biochim. Biophys. Acta **344:** 157–203.
15. MANDEL, L. J. & T. G. RIDDLE. 1979. Kinetic relationship between energy production and consumption in frog gastric mucosa. Am. J. Physiol. **236(3):** E301–E308.
16. CHANCE, B., B. SCHOENER & F. SCHINDLER. 1964. The intracellular oxidation-reduction state. *In* Oxygen in the Animal Organism. F. Dickens & E. Neil, Eds. :367–388. Pergamon. London.
17. OSHINO, N., T. SUGANO & B. CHANCE. 1974. Mitochondrial function under hypoxic conditions: The steady states of cytochrome a $(+a_3)$ and their relation to mitochondrial energy states. Biochim. Biophys. Acta **368:** 298–310.
18. DURBIN, R. P., F. MICHELANGELI & A. NICKEL. 1974. Active transport and ATP in the frog gastric mucosa. Biochim. Biophys. Acta **367:** 177–189.
19. FORTE, J. G., P. H. ADAMS & R. E. DAVIES. 1965. Acid secretion and phosphate metabolism in bullfrog gastric mucosa. Biochim. Biophys. Acta **104:** 25–38.
20. HEINZ, E. & R. P. DURBIN. 1959. Evidence for an independent hydrogen-ion pump in the stomach. Biochim. Biophys. Acta **31:** 246–247.
21. REHM, W. S., T. L. DAVIS, C. CHANDLER, E. GOHMANN & A. BASHIRELAHI. 1963. Frog gastric mucosae bathed in chloride-free solutions. Am. J. Physiol. **204:** 233–242.
22. JOBSIS, F. F. 1972. Oxidative metabolism at low pO2. Federation Proc. **31:** 1404–1413.
23. HEINZ, E. & R. P. DURBIN. 1957. Studies of the chloride transport in the gastric mucosa of the frog. J. Gen. Physiol. **41:** 101–117.
24. NEEDLEMAN, P., J. V. PASSONNEAU & O. H. LOWRY. 1968. Distribution of glucose and related metabolites in rat kidney. Am. J. Physiol. **215:** 655–659.
25. HELANDER, H. F., S. S. SANDERS, W. S. REHM & B. I. HIRSCHOWITZ. 1972. Quantitative aspects of gastric morphology. *In* Gastric Secretion. G. Sachs, E. Heinz & K. J. Ullrich, Eds. :69–88. Academic Press. New York.

26. HERSEY, S. J. 1977. Influence of amytal and menadione on high-energy phosphates and acid secretion in frog gastric mucosa. Biochim. Biophys. Acta **496:** 359–366.
27. MANDEL, L. J., T. G. RIDDLE & J. M. STOREY. 1979. Role of ATP in respiratory control and active transport in tobacco hornworm midgut. Am. J. Physiol. In press.
28. HARVEY, W. R., J. A. HASKELL & K. ZERAHN. 1967. Active transport of potassium and oxygen consumption in the isolated midgut of *Hyalophora cecropia.* J. Exp. Biol. **206:** 235–248.
29. MOFFETT, D. F. 1979. Bathing solution tonicity and potassium transport by the midgut of the tobacco hornworm, *Manduca sexta.* J. Exp. Biol. **78:** 213–223.
30. KEYNES, R. D. 1973. Comparative aspects of transport through epithelia. *In* Transport Mechanisms in Epithelia. H. H. Ussing & N. A. Thorn, Eds. : 505–511. Munksgaard. Copenhagen.
31. JUNGREIS, A. M. & G. L. VAUGHAN. 1977. Insensitivity of lepidopteran tissues to ouabain: absence of ouabain binding and Na^+—K^+ ATPase in larval and adult midgut. J. Insect. Physiol. **23:** 503–509.
32. HARVEY, W. R. & K. ZERAHN. 1972. Active transport of potassium and other alkali metals by the isolated midgut of the silkworm. *In* Current Topics in Membranes and Transport. **3:** 367–410.
33. MANDEL, L. J., D. F. MOFFETT & F. F. JOBSIS. 1975. Redox state of respiratory chain enzymes and potassium transport in silkworm midgut. Biochim. Biophys. Acta **408:** 123–134.
34. WOLFERSBERGER, M. G. 1980. A potassium-modulated plasma membrane ATPase from the midgut of *Manduca sexta* larvae. Insect Biochem. In press.

DISCUSSION

UNIDENTIFIED SPEAKER: Was any correlation made between either delay or the rate constant with the degree of stimulation? If the tissue is stimulated more, you might expect that the delay would be shorter or the time constant faster in sulphate media. Did you do any correlation instead of maximally stimulating them?

L. J. MANDEL: No, we didn't, and part of the reason is that in sulphate the acid secretory rate is about 50% or so of what it is in chloride. Therefore, we always stimulated maximally in order to obtain the maximal separation. Actually, if there is little delay when you stimulate maximally, then I really don't understand what you would try to achieve if you stimulate less.

UNIDENTIFIED SPEAKER: I remember your data said there was about a quarter of a minute delay in sulphate when maximally stimulated and the question was whether that delay could be prolonged if you had stimulated less and, therefore, imply that there was. . . .

MANDEL: That there might be longer storage. No, we did not do those experiments.

T. P. SCHILB: What is the oxygen tension in your perfusing fluid?

MANDEL: It is about 550 millimeters of mercury. We bubble with 95% oxygen.

SCHILB: I thought that might be so and I was just wondering if you had ever tried lower oxygen tensions?

MANDEL: What would lower oxygen accomplish?

SCHILB: The rate of oxygen consumption might be constant until one reaches a fairly low oxygen tension. If one had the oxygen tension just above that point in perfusing fluid; then, when one stopped the flow to let it go anoxic, one might get a cleaner, faster, anoxia, a cleaner break in the cytochrome a_3 curve.

MANDEL: That is possible. Actually, with any kind of epithelium that is metabolizing rather quickly, local areas of anoxia are always a problem; therefore, we like to bubble it with as high an oxygen concentration as possible. It's amazing how fast oxygen is dissipated in a tissue like the gastric mucosa.

J. G. FORTE: At least in the presence of chloride bathing media, oxygen tensions below approximately 80% show decreases in secretory rate in the frog gastric mucosa. So, it is not as far off as you might think from other tissues.

W. S. REHM: Did you say that the delay in the short circuit current decline in chloride media was in favor of an ATP source? I thought you said that the delay in the decline of the short circuit current in chloride media could be taken as evidence that the source of energy must be ATP.

MANDEL: The source of energy for active nonacid chloride transport.

REHM: There is an alternative explanation: There is no active transport during this decline. But before the decline, the active transport process stored up ion gradients. This is consistent with the very recent picture of the mechanism of ion transport proposed by Forte, Machen, Kidder, Sachs and our group. So, there is a possibility that during the decline in the short circuit current, there is no more active transport. Active transport has established gradients and now these gradients are being dissipated.

MANDEL: For ten minutes?

REHM: Yes.

MANDEL: That is a very long time.

REHM: Saying it is a long time or a short time would have to depend upon experimental data concerning the resistance and other electrical parameters of the two membranes.

E. RACKER: I notice that all of you in this field are using glucose. Can you make and prepare these cells and keep them secreting if you use, let's say, just butyrate without glucose?

MANDEL: Yes, you can just use butyrate.

RACKER: But can you isolate the cell in butyrate without glucose and get good secretion?

MANDEL: Yes.

RACKER: You can?

MANDEL: Yes.

RACKER: I haven't seen this in the literature.

MANDEL: Most people just put glucose naturally in the Ringer. Dr. Hersey has performed a lot of experiments, has published some of his data and might be able to answer this question.

S. J. HERSEY: We have not published such findings for the gastric glands, but we have prepared gastric glands in several substrates including butyrate, β-hydroxy butyrate, pyruvate, and others. So prepared, most of them are perfectly functional glands. Their secretory activities in terms of aminopyrine accumulation and respiratory responses are the same as those in glucose.

There are a few substrates which are not so effective. Succinate is one which we believe doesn't get in rapidly; but every other substrate you might envision gets into the cells rapidly, and when the tissues have been prepared in any of them, they are perfectly viable.

Glucose is not essential; it is just traditional.

REHM: We have done many experiments on *in vitro* frog gastric mucosa with no exogenous substrates; and in many of those, there are perfectly normal secretory rates.

FORTE: For how long?

REHM: For quite some time; but the complete answer is a long story.

UNIDENTIFIED SPEAKER: Does acid secretion with sulfate Ringer's continue in the presence of uncoupling agents?

MANDEL: I don't think so.

UNIDENTIFIED SPEAKER: What about deriving the energy from electron transport?

MANDEL: As I said, my results cannot differentiate between the redox theory and the ATPase theory. Therefore, they are consistent with both possibilities as long as there is a small pool of energy.

THE FUNCTIONAL ROLE OF K^+-ATPASE IN PROTON TRANSPORT BY GASTRIC MICROSOMAL VESICLES

Hon Cheung Lee, Haim Breitbart, and John G. Forte

Department of Physiology-Anatomy
University of California
Berkeley, California 94720

INTRODUCTION

A K^+-stimulated ATPase activity has been shown to be associated with a low density, vesicular membrane fraction isolated from gastric epithelial cells.[1,2] A possible relationship for this unique enzyme to gastric H^+ secretion is suggested by 1. Its occurrence in the acid secreting region of gastric mucosa;[3] 2. An ontogenetic parallelism between enzyme and HCl secretion;[4] 3. The well-known requirement for K^+ by gastric secretion;[5] 4. Localization of fluorescent enzyme antibodies to oxyntic cells;[6] and 5. The demonstration of ATP and K^+-dependent H^+ transport by the isolated membrane vesicles.[7–10]

One interesting feature of the gastric K^+-stimulated ATPase is the action of various K^+ ionophores in further stimulating ATP hydrolytic activity.[2,11] For instance, K^+-ATPase activity is stimulated to about the same extent, by valinomycin, gramicidin and nigericin; however, specific effects of each ionophore on the ability of the vesicles to produce and/or sustain a pH gradient depends on the relative effects of the ionophore on H^+ permeability. We have developed a pump-leak model of the gastric membrane vesicle[12] in which the K^+-stimulated ATPase function as a H^+/K^+ exchange pump, as suggested by others;[7,8] and which shows both kinetic and gradient limitations based on the relative vesicular permeability to the principal ions, H^+, K^+ and Cl^-. We report here on a study of several detailed characteristics for activating vesicular K^+-ATPase activity and the H^+ pump system. We explore those conditions that best demonstrate coupling between utilization of ATP and H^+ transport. Finally, we examine the detailed nature of the K^+ requirement for gastric vesicular function and apply this to what is known for K^+ involvement in HCl secretion by intact gastric mucosa.

METHODS

Membrane Preparation

Membrane vesicles were harvested from homogenates of hog gastric mucosal scrapings as previously described.[2] The membranes were first recovered by differential centrifugation, and further purified by sucrose density gradient centrifugation. The low density membranes (1.08–1.11 gm/ml) prepared from gastric mucosa in this manner are rich in tubulovesicle membrane derived from oxyntic cells, and are relatively free of contamination by mitochondria and rough-surfaced endoplasmic reticulum.

0077–8923/80/0341–0297 $01.75/0

ATPase Measurements

ATPase activity was standardly assayed as the rate of inorganic phosphate liberation from ATP. Reaction tubes typically contained 1 mM ATP, 1 mM $MgSO_4$, 10 mM Pipes buffer, and various salts and other additions as described for individual experiments. The reaction medium was set at pH 6.7, the value established as optimum for K^+-stimulated ATPase and vesicular H^+ transport in previous studies.[12] The reaction was carried out either at 37° C or at room temperature 21–22° C). The latter conditions were employed as a convenience when we wished to carry out simultaneous or parallel measurements of H^+ transport using the spectrofluorometer.

H^+ Uptake

Vesicular transport of H^+ was assessed by the accumulation of fluorescent amine probes that are sensitive to the transvesicular pH gradient (ΔpH) by virtue of their characteritics as weak bases. Our earlier studies using 9-aminoacridine and acridine orange demonstrated that a quenching of fluorescence occurred that was proportional to the amount of probe accumulated in response to a ΔpH.[10] We showed that 9-aminoacridine distributed across the membrane in accord with the H^+ gradient, thus it behaves closely to the theoretical prediction of an ideal ΔpH probe. On the other hand, acridine organe showed characteristics of binding to membrane sites, in addition to accumulation in response to the H^+ gradient. Because of this binding, the effect of the ΔpH on acridine orange accumulation is magnified, thus it serves as a highly sensitive, although qualitative, H^+ gradient probe.

Measurements of fluorescence were made with a Perkin-Elmer MPF-44 spectrofluorometer. Wave lengths used were 422 → 455 nm (excitation → emission) for 9-aminoacridine and 493 → 530 nm for acridine orange.

Results and Discussion

Role of K^+ and Anions in Activating Gastric Microsomal ATPase and H^+ Transport

ATPase activity of typical and freshly isolated hog gastric microsomal vesicles is shown in Table 1. In the presence of Mg^{2+} as the only activator, there is a relatively small ATPase activity, which, by virtue of numerous criteria,[12, 13] is neither related to the K^+-ATPase nor the H^+ transport activities. This "basal" Mg^{2+}-stimulated ATPase will be subtracted from the ATPase activity measured in the presence of K^+, such as that shown in Table 1 as the K^+-stimulated ATPase activity. Addition of K^+-ionophores, such as valinomycin, produces a relatively large further stimulation of ATPase activity. We will refer to the ATPase activity that is specifically activated by valinomycin as the val-stimulated K^+-ATPase activity.

Figure 1 shows the influence of [K^+] in stimulating gastric vesicular ATPase in the absence and presence of valinomycin. The apparent K_a for K^+ is about 4 mM without K^+ ionophore. The val-stimulated K^+-ATPase, on the

TABLE 1

ATPASE ACTIVITIES OF GASTRIC MICROSOMES *

	ATPase Activity (μmole · mg^{-1} · L^{-1})	
	21–24° C (n=10)	37° C (n=6)
Mg^{2+} only	2.7±0.3	6.0±0.5
K^+-stimulated	4.8±0.5	17.6±2.4
K^+ plus val-stimulated	14.7±1.3	54.7±7.7
Val-stimulated	9.7±0.3	38.0±5.1

* Values are shown for rate of ATP hydrolysis at room temperature and 37° C. Basal ATPase activity for preparation is that measured in the presence of 1 mM $MgSO_4$. The additional stimulation of ATPase produced by 140 mM KCl (K^+-stimulated) and 140 mM KCl plus 10 μM valinomycin (K^+ plus val-stimulated) is also shown. The enhancement of K^+-ATPase specifically produced by valinomycin is shown as val-stimulated activity.

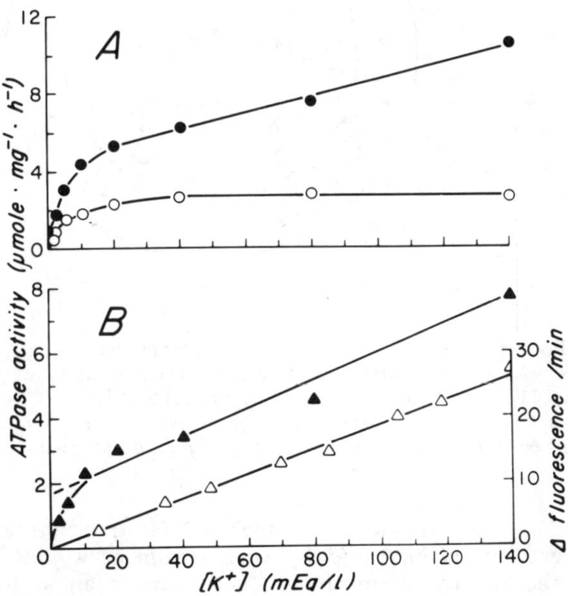

FIGURE 1. The effect of [K^+] on gastric vesicle ATPase activity and H^+ transport as measured by the initial rate of acridine orange fluorescence quenching. A. ATPase activity is shown in the absence (○) and presence (●) of 10 μM valinomycin. B. The val-stimulated K^+-ATPase (▲) calculated as the difference between ATPase activity measured in the presence and absence of valinomycin and shown along with parallel measurements for the initial rate of acridine orange fluorescence quenching (△). The ionic strength and [Cl^-] of the reaction medium was maintained constant with varying amounts of NaCl as the K^+ concentration was changed, as indicated.

other hand, does not show clear saturation kinetics, but continues to increase even at 140 mM KCl. A comparison of val-stimulated K^+-ATPase activity and H^+ transport rate, as monitored by the fluorescent dye uptake method, shows that both of these functions are directly dependent on [K^+], suggesting some degree of "coupling." Thus there is some basis for differentiating the K^+-stimulated ATPase from the val-stimulated K^+-ATPase activity: the association with H^+ secretion and the apparently different kinetics for K^+ activation.

In general, these results are consistent with a hypothesis whereby K^+ is required at some intravesicular, or intramembranous, site for activation of a H^+/K^+ exchange pump and associated (coupled) hydrolysis of ATP. A schematic representation of a gastric vesicle with an ATP-driven H^+/K^+ pump, and associated leak pathways for the major ions, H^+, K^+ and Cl^-, is given in FIGURE 2. K^+ accessibility would be enhanced by valinomycin, thus

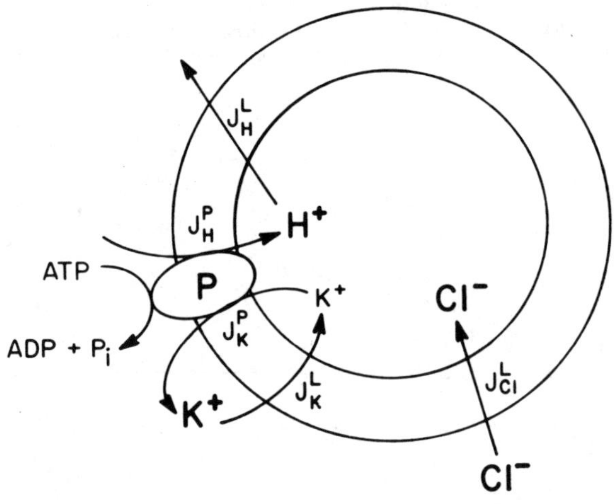

FIGURE 2. Proposed pathways for ion movements across gastric microsomal vesicles. The *J*s refer to ionic fluxes with the superscripts designating pump flux (P) or leak pathway (L). The model consists of an ATP driven H^+/K^+ exchange, in accord with earlier work [7–10] and passive leak pathways for the principal ions, K^+, H^+ and Cl^-. (From Lee *et al.*[12] By permission of *Biochimica Biophysica Acta.*)

activating the H^+ pump ATPase. Whether net H^+ transport and the transvesicular pH gradient is increased by such treatment would very critically depend upon the relative permeability of the major anion in the system. This notion was tested by monitoring ATPase activity and net H^+ transport in the presence of various anions.

As shown in FIGURE 3, the rate of H^+ accumulation measured in the presence of valinomycin by the fluorescence quench method is very sensitive to the nature of the anion; transport rates were highest for NO_3^-, Br^- and Cl^-, and were virtually zero with acetate, isethionate or sulfate. K^+-stimulated ATPase activity varied only slightly with the different anions tested as shown in FIGURE 4. However, val-stimulated K^+-ATPase was markedly increased in

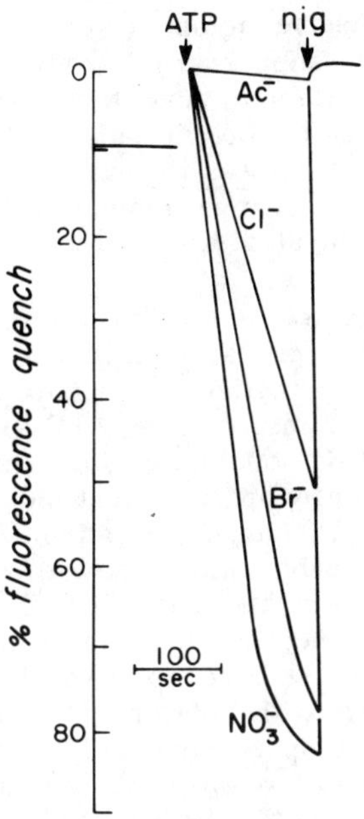

FIGURE 3. The effect of various anions on acridine orange uptake by gastric vesicles. The greater rate of fluorescence quench represents a greater uptake of acridine orange into vesicles, which indicates a lower pH inside the vesicles. Concentration of K^+ was kept constant at 150 meq/l, with the anion being as indicated. In addition, the reaction medium contained 10 mM Pipes buffer (pH 6.5), 1 mM $MgSO_4$, 60 μg/ml membrane protein, 5 μM acridine orange and 2.5 μM valinomycin. 1 mM ATP was added where indicated to start the H^+ uptake and the system was discharged (uncoupled) by addition of 10^{-6} M nigericin (nig). The use of isethionate or sulfate as the major anion (not shown) gave results identical to that shown for acetate (Ac^-). (From Lee *et al.*[12] By permission of *Biochimica Biophysica Acta*.)

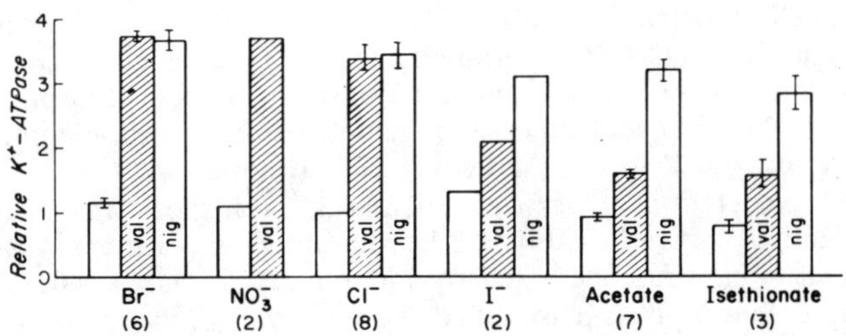

FIGURE 4. Effect of various anions on K^+-ATPase activity. Each of the indicated anions was present as the K^+ salt at 140 mM. Results for a number of experiments (shown in parentheses) on different preparations have been normalized as the relative ATPase activity, taking the activity measured in 140 mM KCl, with no ionophore, as 1.0 for any given experiment. Where indicated, ionophore concentrations were 10^{-5} M valinomycin (val) and 10^{-6} M nigericin (nig).

the presence of the permeable anions (NO_3^-, Br^-, Cl^-), while the larger anions (acetate, isethionate) supported relatively meager increases in ATPase by valinomycin. It is important to note that this restrictive effect of anion permeability on ATPase activity was virtually abolished when gastric vesicular permeability to both K^+ and H^+ was increased, e.g., valinomycin plus a protonophore, or the K^+/H^+ exchange ionophore, nigericin.[12] As shown in FIGURE 4, the nigericin-stimulated K^+-ATPase was relatively independent of the anion tested.

The results with various anions are thus consistent with a direct coupling between the val-stimulated K^+-ATPase and the vesicular H^+/K^+ pump activity. With no ionophore present, penetration of K^+ to its internal activating site is low; hence, little or no H^+ transport, low ATPase activity and independence from anions. Addition of K^+ ionophore will increase accessibility of K^+ to the pump site when matched by an equivalent movement of a counterion, i.e., anion in same direction or exchangeable cation (H^+). For valinomycin, the dependence upon the permeable anion showed analogous behavior for the val-stimulated K^+-ATPase and the enhanced H^+ accumulation. When H^+ permeability, as well as K^+ permeability, was increased (e.g., nigericin) the system was essentially uncoupled, i.e., passive H^+/K^+ exchange increased as pump and ATPase increased, with little or no anion dependence. These interdependencies between the pump enzyme (ATPase) and the ionic permeation pathways have been formally developed into a pump-leak model that predicts many characteristics of the gastric vesicle ATPase and H^+ transport activities.[12]

Modification of ATPase Activity and H^+ Transport

Another approach to assess correlative functional activities involves the use of selective activation and inhibition. Modification of chemical groups by certain reagents provides a range of effective inhibition of val-stimulated K^+-ATPase with only slight alteration in K^+-ATPase activity in the absence of ionophore. This is demonstrated for sulfhydryl group modification by *N*-ethylmaleimide and for amino group modification by ethylacetimidate in FIGURES 5 and 6, respectively. It is also clear from these data that there is a striking parallelism between inhibition of the val-stimulated ATPase and the H^+ transport activity. Other experiments that support a close correlation between val-stimulated K^+-ATPase and H^+ transport include pH optima (~6.7), K_m for ATP (~50 μM) and kinetics of temperature activation (similar activation energies and break in the Arrhenius plot at about 14° C).[12]

The close correlation of gastric vesicular H^+ transport with val-stimulated K^+-ATPase activity is well-established by the above experiments. But what of K^+-stimulated ATPase not mediated by ionophore? Is this a separate and distinct enzyme activity? Is it a modified form of val-ATPase? Is it simply the result of very leaky or broken vesicles?

K^+-stimulated ATPase can be enhanced, presumably at the expense of the val-stimulated K^+-ATPase, by a number of procedures or treatments, such as aging of the membrane vesicle preparations, lyophilization, and mild treatment with detergent.[11] These treatments might also render the vesicles leaky, thus the rate limitation for K^+ accessibility to its activating site might be relaxed.

We chose to examine this possibility for a particular kind of treatment that provides a very reproducible activation of K+-ATPase activity: preincubation in an alkaline environment. The phenomenological changes in ATPase activity for preparations preincubated at pH 8.5 are shown in FIGURE 7. Membrane preparations examined 10–20 minutes after pH 8.5 treatment were incapable of generating a significant pH gradient, with or without valinomycin, and there appeared to be no rate limitation for val-stimulated ATPase activity imposed by the nature of the anion (e.g., Cl^- vs. acetate). Although these afore-

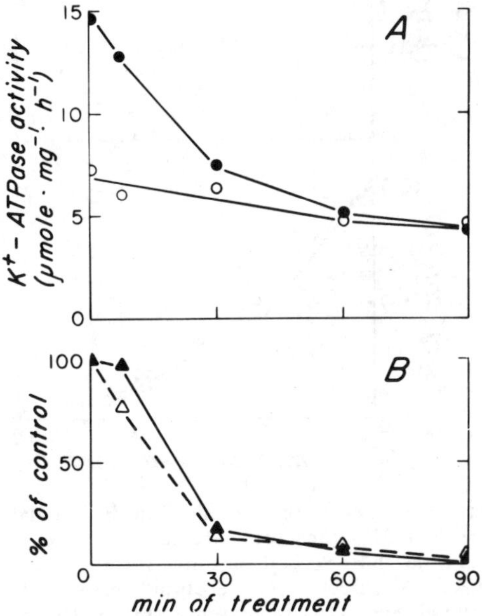

FIGURE 5. Effect of sulfhydryl group modification by *N*-ethylmaleimide on gastric vesicular K+-ATPase and H+ transport. Microsomal vesicles were preincubated with 10 mM *N*-ethylmaleimide (NEM) at room temperature for various periods of time, as indicated. The membranes were then diluted and measurements of ATPase activity and acridine orange uptake were carried out at room temperature (21° C). A. Effects of NEM on absolute rates of ATPase activity measured with 140 mM KCl in the absence (○) and presence (●) of 10 μM valinomycin. B. Time course of inhibition by NEM is shown as the percentage of control (i.e., no NEM treatment) for both the valinomycin-stimulated portion of K+-ATPase (▲), and the initial rate of acridine orange uptake (△).

mentioned observations might be interpreted in terms of a leakiness of the vesicles on exposure to high pH, direct measurements did not bear this out. Results of FIGURE 8 show that pH 8.5-treated vesicles could still sustain an artificially-imposed pH gradient as well as control membranes, and they remained subject to the same permeability alterations produced by valinomycin and protonophores. It is possible that such procedures as aging, and high pH incubation result in degenerative uncoupling of the ATP hydrolytic activity

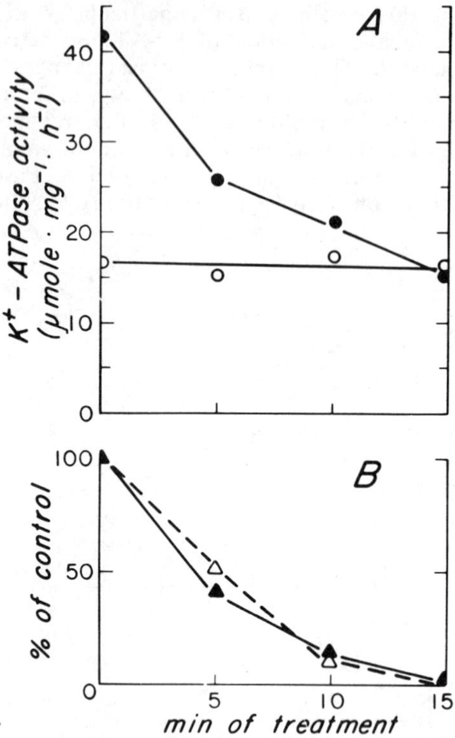

FIGURE 6. Effect of amino group modification by ethylacetimidate on gastric vesicular K^+-ATPase and H^+ transport. Membranes were preincubated at room temperature for the time indicated, then diluted for ATPase assay (37° C) and acridine orange uptake. A. Effect of ethylacetimidate on ATPase activity measured in 150 mM KCl in the absence (○) or presence (●) of 10^{-5} M gramicidin. B. Time course of inhibition by ethylacetimidate shown as percentage of control (i.e., no treatment) for both the gramicidin-stimulated portion of K^+-ATPase (▲) and the rate of acridine orange uptake (△).

from the transvesicular ion transport process. Thus ATPase activity might continue, and even be enhanced, with K^+ serving as the activator ($K_a \approx 4$ mM) at an external site. The logical extension of such a proposal is that there are two sites for K^+ activation in normally functioning vesicles: an "internal" site, with ionophore-limited access, that is the primary transport site; and an external site, in which K^+ participates in phosphoprotein phosphatase activity of the total enzyme cycle. Support for an external site for K^+ activation is available from experiments in which membrane phosphoprotein intermediates or model phosphatase substrates are rapidly dephosphorylated or hydrolyzed by K^+, with very little apparent permeability restrictions.[14] Evidence supporting an "internal" K^+ site was presented above in the characterization and correlation of the val-stimulated K^+-ATPase and vesicular H^+ transport. This system would be analogous to the one for K^+ participation in the ($Na^+ + K^+$)-ATPase.

Application of Gastric Microsomal ATPase and H^+/K^+ Pump to HCl Secretion

There are a number of questions that must be addressed and problems that must be solved for the vesicular ATPase and H^+/K^+ transport system before we can appreciate its precise involvement in the gastric HCl secretory process. Several of these, such as the maximum pH gradient, the stoichiometry between H^+ and K^+, the stoichiometry between H^+ and ATP, and the electrogenic behavior of the system, have been discussed elsewhere.[15, 16] One problem, which is especially apparent from the studies presented here, concerns the availability of K^+ to the internal transport site under physiological conditions. Assuming that the pump operates as an ATP-driven, forced K^+/H^+ exchanger, there must be an equivalent supply of K^+ to the apical secretory surface. What is the source of this K^+?

The K^+ requirement for gastric H^+ secretion has been established for some time.[5] Furthermore, in its support of H^+ secretion, K^+ is more effective when added from the serosal side than from the mucosal side.[17] For gastric H^+ secretion to be powered by the H^+/K^+ exchange pump, it is obvious that K^+ would have to be recycled from cytoplasm to transport site and back again. Because there are several species with relatively homogeneous oxyntic cell populations in their gastric glands, the K^+ recycling process would very likely have to be through the same apical membrane. Thus the vesicular model of

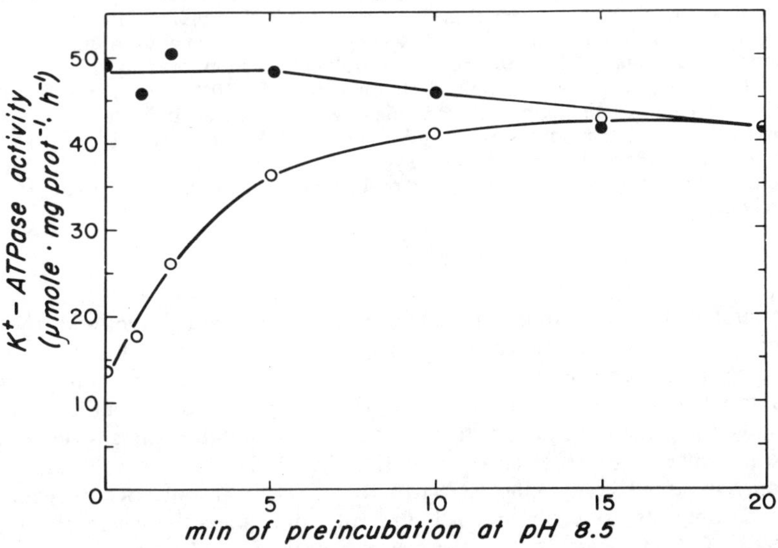

FIGURE 7. Effect of pretreatment at pH 8.5 on gastric vesicular ATPase activity. Membranes were pretreated at 37° C in 20 mM HEPES buffer, pH 8.5, for the indicated times, then diluted for the subsequent ATPase activity measurements with the typical conditions: 140 mM KCl, pH 6.7 and 37° C. K^+-ATPase activity is shown in the absence (○) and presence (●) of 10 μM valinomycin. With increasing time of preincubation at pH 8.5, stimulation by K^+ alone increases until valinomycin produces no further activation.

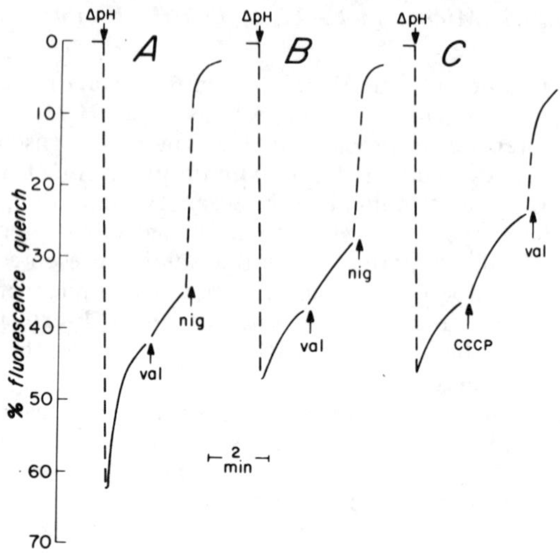

FIGURE 8. Effect of preincubation at pH 8.5 on the relative permeability of gastric vesicles to H^+ as measured by the ability to sustain an artificially induced transvesicular ΔpH. For induction of the pH gradient, vesicles were incubated for several min with 20 mM succinate buffer, pH 4.0, so as to allow the vesicular interior to equilibrate at this pH. At the time indicated by the arrow at ΔpH, Tris base was added to rapidly raise the pH of the external medium to about 7. The pH gradient was assessed by the intravesicular accumulation of acridine orange, shown as a quenching of fluorescence. Leakage of H^+ (diminution of gradient) is monitored by the restoration rate of fluorescence. Ionophores, e.g., valinomycin (2.5×10^{-6} M, val), protonophore (10^{-6} M, CCCP) or nigericin (10^{-6} M, nig), were added as indicated in order to demonstrate the relative importance of both H^+ and K^+ conductance pathways for rapid relaxation of the pH gradient. A. Typical preparation of gastric vesicles. B and C. Same vesicle preparation as in A, but preincubated at pH 8.5 for 20 and 15 min, respectively.

anion-limited K^+ accessibility to some internal activating site and the recycling of K^+ through the H^+/K^+ exchange pump would fit the criteria, in principle. However, the rate of K^+ access in the absence of ionophore is much too low to account for effective H^+ transport. A comparison of the capability for generating pH gradients is shown in FIGURE 9 for gastric vesicles in the presence of external or internal K^+, and in the absence or presence of valinomycin as a K^+ ionophore. Virtually no H^+ transport occurs with external K^+ only, unless valinomycin is present (and unless there is a permeable anion like Cl^-, cf., FIGURE 3). For vesicles preincubated with K^+, there is a relatively rapid initial rate of H^+ transport, but the maximum pH gradient is not as great as for those conditions in which valinomycin was present. The somewhat slower rate of net H^+ accumulation seen in these latter cases is not due to a slower pump rate, but rather due to the increased K^+/H^+ leak pathway. In fact, the limitation on the pumping rate in the presence of valinomycin is set by the anion permeation, i.e., no more H^+ can be accumulated than Cl^- uptake.

It is clear that the ionophore activated system is more effective in generating large pH gradients and stimulating related ATPase activity. However, for its application to the gastric HCl secretory system, the physiological equivalent of valinomycin has not been identified. We will propose the hypothetical scheme shown in FIGURE 10, in which a K^+-carrying compound or K^+ channel is activated by the events that stimulate HCl secretion. In order to explain the greater effectiveness of K^+ from the serosal or cytoplasmic side in the intact tissue [17] and isolated vesicles (FIGURE 9), the secretory responses to passing current in Cl^--free Ringers,[18] and some of the electrical activity of intact tissue,[16] we further propose that the K^+ channel substance is operative only from cytoplasm to some "interior" K^+ activating site. Thus K^+ accessibility would show an asymmetry through the membrane: a region of high K^+ conductance and one of low K^+ conductance. The system would be analogous to one in which two membranes with different properties were placed in series.

The scheme shown in FIGURE 10 will explain a number of the observations made on H^+ transport *in situ* and in isolated vesicles. A permeable anion is necessary for optimal H^+ transport, even when a high concentration of K^+ is provided at the outer mucosal surface.[18] If the K^+ conductance is higher through the inner membrane region than the outer one (i.e., $R_K < R_K'$), then externally applied current would be carried by K^+ and H^+ differentially in the inner and outer regions, respectively, and by the anion. Externally applied

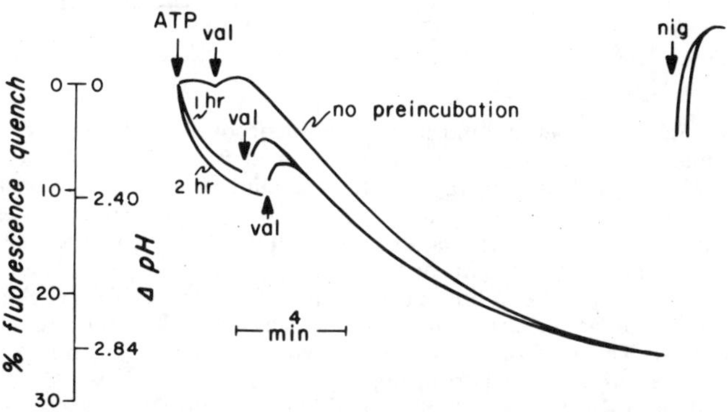

FIGURE 9. Effect of time of preincubation in KCl and of valinomycin on gastric vesicular acidification as measured by fluorescence quenching of 9-aminoacridine. Membranes were placed in 150 mM KCl at room temperature for (a) no preincubation, (b) 1 hr or (c) 2 hrs, as indicated. In addition reaction medium contained 10 mM Pipes buffer (pH 6.7), 1 mM $MgSO_4$ and 10 μM 9-aminoacridine. Dye uptake (denoted by percentage of fluorescence quench) was initiated by addition of 1 mM ATP. The calculated transvesicular H^+ gradient (ΔpH) is also shown. Vesicular acidification was virtually nonexistent for preparations that were not preincubated in KCl until the addition of 2.5 μM valinomycin (val). The K^+ ionophore also increased the pH gradient for preparations which had been preincubated in K^+. All H^+ gradients were dissipated after addition of 1 μM nigericin (nig).

current, enhancing or opposing net H^+ secretion, could serve to speed up or slow down the H^+/K^+-ATPase, and hence the rate of H^+ secretion. Effects of current flow on H^+ secretory rates are well established,[19, 20] and concomitant current induced changes on tissue O_2 uptake (and possibly ATP turnover) have been demonstrated.[21] Although such consistencies are intriguing, the

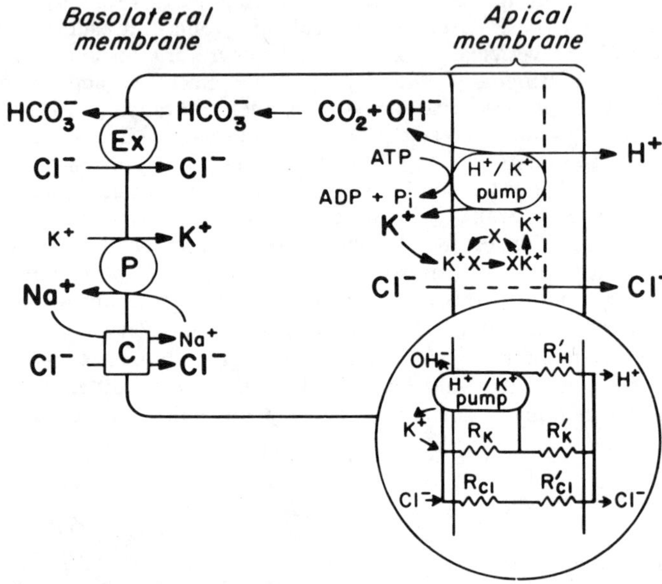

FIGURE 10. Schematic representation for the incorporation of the vesicular ATPase and H^+/K^+ pump into gastric HCl secretion. The pump is shown at the apical, secretory membrane with a pathway for K^+ recycling from the cytoplasm (perhaps facilitated by a specific channel or carrier, X) to the pump activating site then back through the H^+/K^+ exchange. An equivalent circuit is shown in the inset, in which resistance to K^+ movement might be different in the inner and outer membrane regions ($R_K' > R_K$). The recycling of K^+, coupled to the H^+/K^+ exchange, would provide the basis for an electrogenic pump which would be especially apparent in Cl^--free solutions. The role of Cl^- at the secretory surface could be to provide the return limb of the circuit, thus permitting the net flow of HCl, which might appear tightly coupled or open circuit. Under conditions of active HCl secretion Cl^- entry into the oxyntic cell at the basolateral surface is provided by the neutral Cl^-/HCO_3^- exchanger (Ex). The Na^+ pump (P) at that same surface would serve to maintain the cellular ionic milieu and K^+ balance. The Na^+ gradient, serosa to cell, could provide the driving force for moving Cl^-, via a coupled NaCl entry (C), into the cell against its electrochemical potential gradient. (From Forte *et al.* By permission of *Annual Review of Physiology*.)

proposed scheme for H^+ secretion depends heavily upon identification of the pathway and regulation of the passive K^+ flow in the recycling system. Recent studies have suggested calcium and cytoplasmic proteic components,[22, 23] as possible regulatory activators of the gastric K^+-ATPase system. Such studies must be pursued in greater depth to provide specific mechanistic detail for the putative activation processes for HCl secretion.

References

1. Ganser, A. L. & J. G. Forte. 1973. K^+-stimulated ATPase in purified microsomes of bullfrog oxyntic cells. Biochim. Biophys. Acta **387:** 169–180.
2. Forte, J. G., A. L. Ganser, R. Beesley & T. M. Forte. 1975. Unique enzyme of purified microsomes from pig fundic mucosa. Gastroenterology **69:** 175–189.
3. Forte, J. G., A. L. Ganser & A. S. Tanisawa. 1974. The K^+-stimulated ATPase system of microsomal membranes from gastric oxyntic cells. Ann. N.Y. Acad. Sci. **242:** 255–267.
4. Limlomwongse, L. & J. G. Forte. 1970. Developmental changes in ATPase and K^+-stimulated phosphatase of tadpole gastric microsomes. Am. J. Physiol. **219:** 1717–1722.
5. Davis, T. L., J. R. Rutledge, D. C. Keesee, F. J. Bajandas & W. S. Rehm. 1965. Acid secretion, potential and resistance of frog stomach bathed in K^+-free solutions. Am. J. Physiol. **209:** 146–152.
6. Sachs, G., H. Chang, E. Rabon, R. Schackmann, H. M. Saran & G. Saccomani. 1977. Metabolic and membrane aspects of gastric H^+ transport. Gastroenterology. **73:** 931–940.
7. Lee, J., G. Simpson & P. Scholes. 1974. An ATPase from dog gastric mucosa: Changes of outer pH in suspensions of membrane vesicles accompanying ATP by dialysis. Biochim. Biophys. Res. Comm. **60:** 825.
8. Sachs, G., H. H. Chang, E. Rabon, R. Schackmann, M. Lewin & G. Saccomani. 1975. A nonelectrogenic H^+ pump in plasma membranes of hog stomach. J. Biol. Chem. **251:** 7690–7698.
9. Chang, H., G. Saccomani, E. Rabon, R. Schackmann & G. Sachs. 1977. Proton transport by gastric membrane vesicles. Biochim. Biophys. Acta **464:** 313–327.
10. Lee, H. C. & J. G. Forte. 1978. A study of H^+ transport in gastric microsomal vesicles using fluorescent probes. Biochim. Biophys. Acta **508:** 339–356.
11. Ganser, A. L. & J. G. Forte. 1973. Ionophoretic stimulation of K^+-ATPase of oxyntic cell microsomes. Biochem. Biophys. Res. Comm. **54:** 690–696.
12. Lee, H. C., H. Breitbart, M. Berman & J. G. Forte. 1979. Potassium stimulated ATPase activity and H^+ transport in gastric microsomal vesicles. Biochim. Biophys. Acta **553:** 107–131.
13. Lee, H. C. & J. G. Forte. 1979. Chemical Modification of gastric microsomal K^+-stimulated ATPase. Biochim. Biophys. Acta. Submitted for publication.
14. Ray, T. K. & J. G. Forte. 1976. Studies on the phosphorylated intermediate of a K^+-stimulated ATPase from rabbit gastric mucosa. Biochim. Biophys. Acta **306:** 169–180.
15. Sachs, G., J. G. Spenney & M. Lewin. 1978. H^+ transport: Regulation and mechanism in gastric mucosa and membrane vesicles. Physiol. Rev. **58:** 106–173.
16. Forte, J. G., T. E. Machen & K. J. Öbrink. 1979. Mechanisms of gastric H^+ and Cl^- transport. Ann. Rev. Physiol. In press.
17. Sanders, S. S., D. H. Noyes, S. G. Spangler & W. S. Rehm. 1973. Demonstration of a barium-potassium antagonism on the lumen side of in vitro frog stomach. Am. J. Physiol. **224:** 1254–1259.
18. Rehm, W. S. & S. S. Sanders. 1977. Electrical events during activation and inhibition of gastric HCl secretion. Gastroenterology **73:** 959–969.
19. Crane, E. E., R. E. Davies & N. M. Longmuir. 1948. The effect of electric current on HCl secretion by isolated frog gastric mucosa. Biochem. J. **43:** 336–342.

20. REHM, W. S. 1950. A theory of the formation of HCl by the stomach. Gastroenterology **14:** 401–417.
21. FORTE, J. G. & R. E. DAVIES. 1964. Relation between hydrogen ion secretion and oxygen uptake by gastric mucosa. Am. J. Physiol. **206:** 218–222.
22. PROVERBIO, F. & F. MICHELANGELI. 1978. Effect of calcium on the H^+/K^+-ATPase of hog gastric microsomes. J. Membr. Biol. **42:** 301–315.
23. RAY, T. K. 1978. Gastric K^+-stimulated adenosine triphosphatase. Demonstration of an endogenous activator. F.E.B.S. Letters **92:** 49–52.

DISCUSSION

D. WARNOCK: I am interested in this chloride conductance path you described. Have you not shown that amino reactive groups inhibited the stimulation of your potassium ATPase in the presence of valinomycin? Do you think that is due to a decrease in the anion conductance? Have you tried a lipophilic anion like thiocyanate that presumably could cross the membrane, and, in that setting, restore the potassium ATPase activity? In other words, do you think that this amino modification changed the anion conductance in these vesicles?

J. G. FORTE: I don't believe so. The treatment may affect the anion conductive pathway; but, I don't think it is the major limiting effect here and the reason I don't is that the nigericin-stimulated ATPase is affected to the same extent. Nigericin stimulation of the ATPase should be independent, and as far as we know it is independent, of the anion and it, too, is inhibited by the amino group reagents.

In recent studies using some fluorescent amino reagents, we found that one of the particularly active amino groups for this site is an ATP binding site since we could protect it with ATP, with very minimal amounts of ATP. Whether there is an additional anion pathway, I don't yet know specifically, but I don't believe that it accounts for the decrease in ATPase activity that ethylacetimidate produces.

W. S. REHM: If I correctly understand your model, it is formerly equivalent to the old electrogenic model.

FORTE: It would be an electrogenic model. The double membrane puts two membranes in series with the ATP generating system providing the source of protons for the electrogenic pump.

REHM: In other words, there is a parallel pathway for chloride conductance from the lumen through the membrane to the cytoplasm. If we then trace the rest of the circuitry, potassium could conduct the current from the cytoplasm to calcium halfway through the membrane, and from that point out to the lumen, the conductance would be via protons. Is that correct?

FORTE: That's right.

RACKER: I also was puzzled about the effect of the amino modification—how that possibly could affect both valinomycin and nigericin. I have no possible explanation for how this amino group modification could inhibit the effect of valinomycin.

My second question is, what is your ratio of ATP to protons? Do you have any information on that.

FORTE: No, I don't have that information. Our best estimate is that it's considerably less than one. That's because in our system, there is a big leak that we had to account for. So, I don't have a direct estimate of the number of protons per ATP. The net proton accumulation, in our case, to ATP utilization is less than one. George Sachs may have additional information on this ratio.

With respect to your first question, you are puzzled by the fact that an amino reagent affects both the valinomycin-stimulated and the nigericin-stimulated ATPase activities. The amino group reagent may affect an anion conducting channel, but, also may directly affect the enzyme. If it affects the active site of the enzyme, then it could inhibit the enzyme independent of any ionophoric effects. I think we have independent evidence that it is affecting the active center of the enzyme.

It is quite true that it primarily affects the stimulation by valinomycin. Either the potassium ATPase is an absolutely and totally different enzyme than the valinomycin ATPase, if I may call it that, or that there has been some detailed alteration in the transport ATPase from its functional state, including both an intravesicular K^+ transporting site and an external phosphatase activating site to a degenerated state with only the phosphatase activating K^+ site. These two states might have different sensitivities to these inhibitors. That's the only alternative I could suggest.

ASPECTS OF PARIETAL CELL BIOLOGY: CELLS AND VESICLES *

G. Sachs, T. Berglindh, E. Rabon, H. B. Stewart, M. L. Barcellona,
B. Wallmark, and G. Saccomani

Laboratory of Membrane Biology
University of Alabama in Birmingham
University Station
Birmingham, Alabama 35294

INTRODUCTION

The purpose of this paper is to describe the current status of an interesting mammalian H^+ transport model, the gastric parietal cell. It was the first vital H^+ generator to be recognized, and concepts as to mechanism, many decades ago, made much use of redox-driven $\Delta\bar{\mu}_{H^+}$ pumps. Ironically, however, recent information has implicated an ATP-driven $\Delta\bar{\mu}_{H^+}$ pump, no evidence having been obtained as yet for a $\Delta\Psi$ component, whether redox or ATP driven. Here we shall outline some properties of the intact cell and of subcellular components identified as H^+ pump-related; discuss these components in terms of structure and transport functions (lipid, carbohydrate and protein), and enzyme mechanism; and, finally, present a working model that attempts correlation of enzyme mechanism and transport. The systems to be discussed are the isolated rabbit gastric gland and highly purified hog gastric plasma membrane vesicles.

Cellular Aspects of H^+ Pump Activity

This part of the discussion is based on certain unresolved questions that must be settled in order to correlate the vesicle results with data from the intact cell. Some of these questions are (a) the necessity of cAMP and/or Ca^{++} for H^+ pumping; (b) the precise site of acid secretion; (c) the cellular role of K^+; and (d) the location of the ATPase component of the H^+ pump.

cAMP/Ca^{++} Regulation

The histamine-induced secretory response of isolated rabbit glands is dependent on binding to an H_2 receptor.[1] The acid secretory response of gastric glands (monitored by the accumulation of ^{14}C-aminopyrine (AP),[2] a weak base, $pK = 5$) is correlated with increases in cAMP levels in the glands and adenylate cyclase activation in particulate fractions from the glands. There is a line of identity correlation between the ability of several histamine analogs to increase AP accumulation or O_2 consumption and to increase cAMP levels or adenylate cyclase activity (e.g., histamine, 4-methyl histamine, dimaprit, cimetidine, tolazoline, pyridylethylamine).[3]

* Supported in part by Grants NIH AM15878, AM21588, NSF PCM 78–09028 and PSCM 77–18951.

0077–8923/80/0341–0312 $01.75/0 © 1980, NYAS

The need for activation of the H_2 receptor, even by unrelated hormones, is demonstrated by the glandular response to gastrin. This occurs to a significant extent only in the presence of the phosphodiesterase inhibitor, isobutyl methyl xanthine. FIGURE 1 shows that the response of the glands to gastrin is blocked by low doses of the histamine antagonist, cimetidine. Addition of 0.1 U of histaminase also potently inhibits the gastric response (data not shown). Since the activation of the H_2 receptor is essential for the gastrin response, we conclude that cAMP levels must be increased for an H^+ secretory response

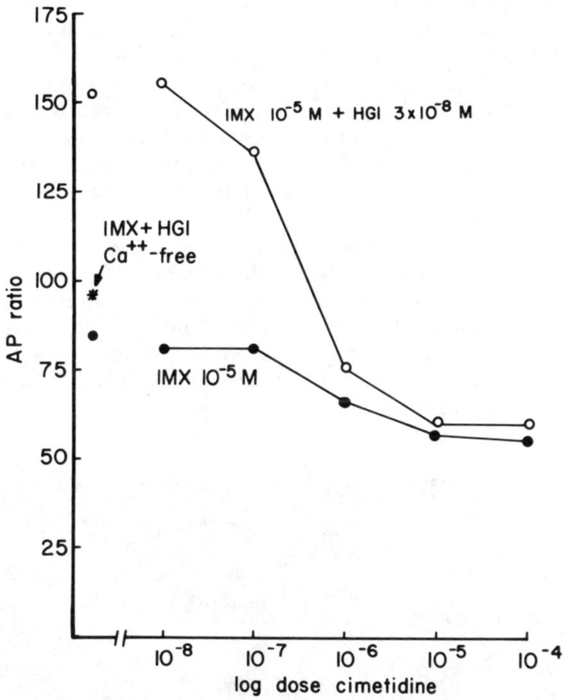

FIGURE 1. Stimulation of AP accumulation by human gastrin I in the presence of IMX (○) and by IMX 10^{-5} M alone (●) and the dose-dependent inhibition of these responses by cimetidine (○—○; ●—●). The star shows the drastically-reduced response to gastrin seen in the absence of Ca^{++}. The unstimulated AP ratio was 26.7.

to be observed to non-neurogenic stimuli of H^+ secretion, a finding consistent with most, but not all studies on amphibian mucosa.[4, 5]

Variations in extracellular Ca^{++} do not affect the histamine response.[6] However, as also shown in FIGURE 1, the response to gastrin is dependent on medium Ca^{++} levels. The cholinergic response is also sensitive to Ca^{++} levels in the medium.[6, 7] Hence, there is also evidence implicating medium [Ca^{++}] in the secretory response to some stimuli. No evidence has been obtained as yet for modulation of the histamine response by intracellular redistribution of Ca^{++}. We postulate therefore a cAMP or Ca^{++}-dependent kinase regulation of H^{+} pump activity.

Cell and Site of H^{+} Secretion

The cell responsible for acid secretion has been assumed to be the parietal cell.[8] It has also been assumed, in the mammal, that the subcellular site is the secretory canaliculus. The morphological transformation of the oxyntic[9] or parietal cell,[10] whereby intracellular tubulo-vesicles transform into the microvilli lining the secretory canaliculus (in the case of the parietal cell), has led to the invocation of a fusion-desegregation membrane process as crucial in providing pump sites at the cell surface.[11] In this section we will provide evidence defining the cellular site of H^{+} secretion and evidence arguing for a simpler macroscopic model for the morphological transformation.

Rabbit glands in the living state can be viewed by fluorescence or Nomarski optics separately or simultaneously. Acridine orange is one of several optical probes of H^{+} (chemical) gradients that have been used in organelles[12] and quantitated in gastric vesicles.[13, 14] At high concentration of the dye, the characteristic green fluorescence (530 nm) is quenched and a less intense red fluorescence (660 nm) appears. In the absence of stimulus, gastric glands show an uniform green fluorescence appearance with some red shift in peptic granules. With addition of 10^{-5} M histamine, a progressive intracellular expansion of a red fluorescent region is seen, corresponding to the secretory canaliculus (FIGURE 2), showing the region of low pH within the cells. This defines the cell and site of acid secretion.

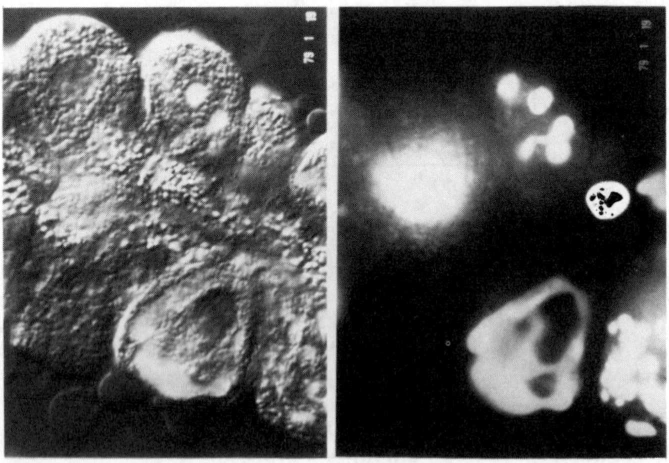

FIGURE 2. On the left is shown a combined Nomarski and fluorescent micrograph of stimulated gastric glands in the presence of AO 10^{-4} M, with regions of expanded secretory canaliculi evident in the two cells on the bottom right. On the right is shown the same gland, using fluorescence microscopy. The white regions correspond to regions of intense red fluorescence. The outline of an expanded secretory canaliculus is clearly seen in the cell at bottom center. (From Dibona *et al.*[3])

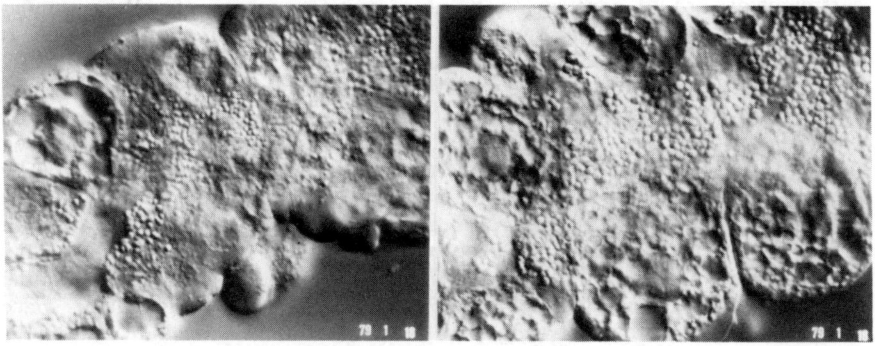

FIGURE 3. A Nomarski micrograph of living rabbit gastric glands in normal K^+ (5.4 mM left) and high K^+ (108 mM right) following the addition of 10^{-3} M AP. The vacuolization of the parietal cells is clearly visible, the major difference being the larger number of vacuoles seen with high medium K^+. (From Dibona *et al.*[6])

Aminopyrine is accumulated by gastric glands by 20–300-fold depending on various parameters, but the basal ratio of 20 is independent of secretagogue or Ca^{++}. The addition of K^+ above control (5.4 mM) levels alone increases the AP accumulation. The pH of the acid compartment is regulated by a balance between HCl influx and efflux. The former is determined by KCl influx coupled to an H^+:K^+ antiport pump. It follows that at AP concentrations in the medium, which, when accumulated, will be sufficient to buffer the compartment, the AP will continue to accumulate as the compartment is driven back towards the original pH. Given a 100-times accumulation of AP in 10% gland volume, the 10^{-4} M AP will reach 100 mM in this space and induce hypertonicity. Moreover, the hypertonicity will swell the acid compartment.

Such experiments, at normal and high K^+ (108 mM), done in living glands illustrate the vacuolization induced entirely because of the AP accumulation (FIGURE 3). This vacuolization is found also with histamine addition; and electron microscopy has shown this to be due to swollen secretory canaliculi lined by microvilli stretched to various degrees.[15] Two conclusions follow. Firstly, consonant with the acridine orange studies above, the parietal cell is the site of AP accumulation, hence of H^+ secretion. Secondly, the morphological transformation of the cell (loss of tubulo-vesicles, appearance of canaliculi) might be achieved by osmotic gradients. Hence, we propose that a fusion-desegregation hypothesis is unnecessary. In fact, one action of histamine and other secretagogues must be to provide the osmotic gradient necessary to expand the collapsed tubules, and to move solution out of the parietal cell and into the gastric gland lumen. This osmotic hypothesis does not exclude submicroscopic translocation of proteins or other more subtle processes.

Role of K^+

The seminal observation in this area was made over two decades ago[16]; acid secretion is absolutely dependent on medium K^+. Using gastric glands, it is straightforward to construct a K^+ dose response curve of AP accumulation ± histamine. When histamine is present, the apparent K^+ affinity is 1 mM. In the absence of histamine a similar apparent affinity is observed, but a second increment of AP accumulation is seen peaking, as with histamine, at 100 mM medium [K^+]. With ouabain present to allow cellular free K^+ levels to approach that of medium [K^+], inhibition of AP accumulation is seen at normal [K^+]. This inhibition is overcome by increasing medium [K^+]. The apparent $K_{0.5}$ is shifted to 32 mM in the presence of, and 58 mM in the absence of, histamine (FIGURE 4). (It should be noted that the Na^+ inhibition of K^+-ATPase activity or of ATPase phosphorylation,[17] is also found in the glands. Removal of both Na^+ and K^+ from the medium, allows the residual cell K^+ to maintain a low level of AP accumulation in the glandular space and this is reduced by increasing [Na^+]). From this we conclude that K^+ is necessary for activation of acid secretion, and that the activating site has a relatively low apparent affinity (32–58 mM) for this ion. Histamine increases the apparent affinity for K^+.

Localization of Gastric ($H^+ + K^+$)-ATPase

By a combination of differential, zonal gradient centrifugation and free flow electrophoresis, the vesicles containing the H^+:K^+ ATPase have been

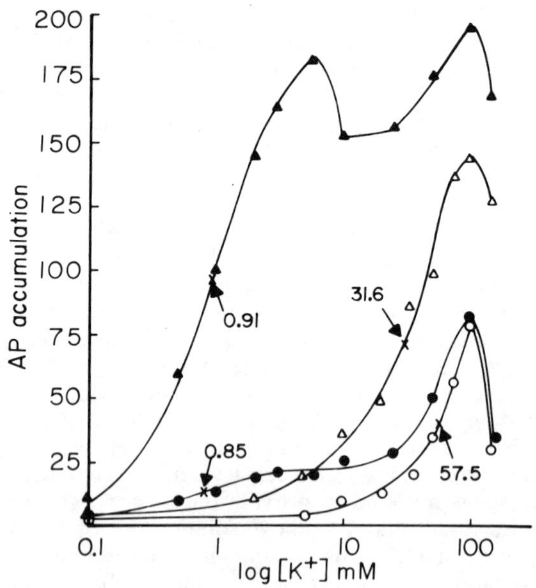

FIGURE 4. The AP accumulation ratio is shown as a function of medium K^+ under various conditions. In the absence of histamine (●—●) an initial $K_{0.5}$ of .85 mM and a secondary $K_{0.5}$ of 50 mM is obtained. In the presence of histamine (▲—▲) a $K_{0.5}$ of 0.91 mM is found. With ouabain a single $K_{0.5}$ of 57.5 mM is observed in the absence of histamine (○—○) and of 31.6 mM in the presence of histamine (△—△). The two phases seen in the curve labelled (●—●) probably correspond to activated and high K^+ recruited cells, respectively.

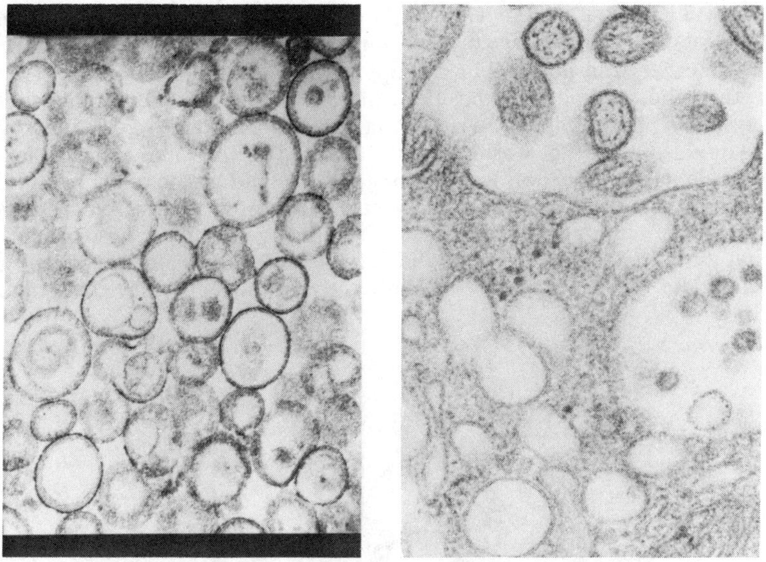

FIGURE 5. Left: Electron micrograph of purified gastric membrane vesicles fixed in 2% glutaraldehyde and 1% tannic acid. The outermost leaflet is almost always thicker than the inner ones, ×84,500. Right: Electron micrograph of a parietal cell in the gastric tissue fixed as above. The microvillus shows a protein distribution opposite to that of the isolated gastric vesicles, ×71,500.

purified,[18] showing that only a 105,000 daltons polypeptide region is present. The vesicles are highly asymmetric in terms of protein distribution, as revealed by tannic acid-glutaraldehyde fixation, and this distribution corresponds to the distribution of protein in the microvillus of the intact tissue; but apparently not to that of the collapsed tubule, a point which may become significant as our understanding of the pump process develops (FIGURE 5).

Antibodies to this polypeptide region (both inhibitory and noninhibitory), monospecific by rocket immunoelectrophoretic analysis,[19] bind selectively to the microvillus surface, as revealed by the peroxidase method [20] (FIGURE 6).

Thus, the K^+ requiring ATPase is localized to the membrane present at the site of acid secretion where the pH is expected to fall to about 0.8. The estimated pH of this region in the gland, at maximal AP accumulation is about 1.5.

Structural Features of $(H^+ + K^+)$*-ATPase*

The overall aminoacid composition of the K^+-ATPase is shown in TABLE 1. Tryptic hydrolysis, ATP or ADP protection therefrom, and simultaneous phosphorylation studies using [$\gamma = {}^{32}P$] ATP indicate that three peptide subunits of similar M_r may be present, a catalytic subunit (the active site present first on a 87,000 then 47,000 fragment) a glycoprotein (trypsin/insensitive) and a highly trypsin-sensitive subunit.[21] Isoelectric focussing also indicates possible heterogeneity of the ATPase complex. The carbohydrate composition of the K^+-ATPase is shown in TABLE 2, and accounts for 28% by weight of the protein. Notable is the absence of neuraminic acid.

As for other transport ATPases, lipid plays an important role. TABLE 3 shows the phospholipid composition of this class of membrane vesicles. Hydrolysis by phospholipase A_2 in intact vesicles can remove 50% of hydrolysable lipid. Following washing with defatted BSA there is a time- and hydrolase-dependent loss of K^+-ATPase activity and no loss of K^+-pNPPase activity (FIGURE 7).[22] ATPase activity can be restored to a varying degree by the addition of various phospholipids. The most effective phospholipids are phosphatidyl ethanolamine, phosphatidylserine and phosphatidylcholine Essentially complete reconstitution of ATPase activity can be achieved (TABLE 4).

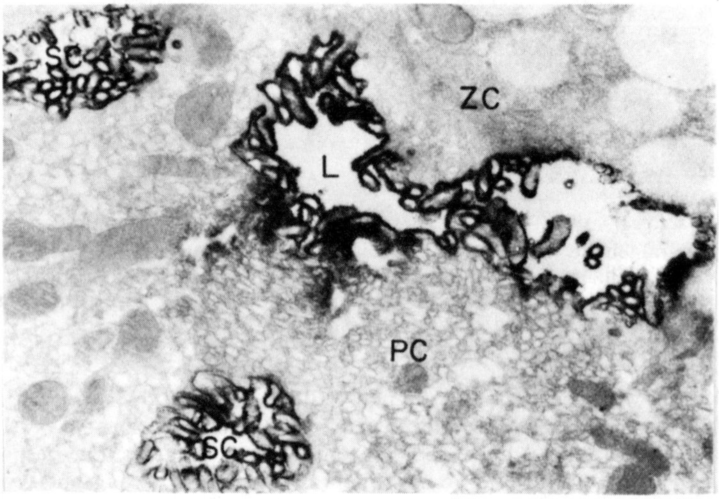

FIGURE 6. Ultrastructural appearance of gastric membrane K^+-ATPase localization with peroxidase procedure on a rat parietal cell. This section was stained with anti-K^+-ATPase antibody. The staining reaction is strong along the secretory surface of the parietal cell, both on the apical cell surface (facing the gland lumen, L) and on the surface that borders the secretory canaliculi (SC). There is no antibody reaction in the cytoplasm. (ZC)-zymogen cell, ×9,750.

The possible interaction between protein and phospholipid can be further assessed by a study of transition temperatures. For example, K^+-ATPase activity shows a transition at 28° C, whereas H^+ transport activity has a transition at 22° C. Probes of bulk lipid viscosity such as diphenylhexatriene [23] show a progressive increase of polarization with decrease in temperature but no definable transition. The above transitions therefore probably reflect protein or local protein lipid. On the other hand, differential scanning calorimetry shows two transitions, at 54° C and 66° C. These are apparently irreversible transitions, indicating that they are due to protein or protein-lipid complexes (FIGURE 8). Trypsinization [21] removes the transition at 54° C, suggesting that

TABLE 1

AMINO ACID COMPOSITION OF THE GASTRIC MEMBRANE *

Amino acid residue	Residues/100
Lysine	5.69
Histidine	1.88
Arginine	4.92
Aspartic acid	9.63
Threonine	6.32
Serine	8.40
Glutamic acid	10.80
Proline	6.36
Glycine	7.53
Alanine	8.25
Cysteine	1.59
Valine	5.43
Methionine	1.90
Isoleucine	4.51
Leucine	9.30
Tryosine	3.25
Phenylalanine	4.57
Tryptophan	N.D. †

* Gastric membranes were resuspended in 6 N HCl and hydrolyzed for 22 hrs at 106° C in closed ampules under vacuum. Amino acid analysis was performed in a Beckman Model 116 amino acid analyzer, average of 3 determinations. (From Sterner.[46])

† N.D., not determined.

TABLE 2

CARBOHYDRATES COMPOSITION OF GASTRIC MEMBRANES *

Monosaccharides	mg/100 mg protein †
Fucose	2.72
Mannose	2.95
Glucose	8.08
Galactose	5.76
N-acetyl glucosamine	4.95
N-acetyl galactosamine	3.03
Sialic acid	none
Inositol	0.56

* The monosaccharides were determined by gas-liquid chromatography as trifluoroacetates of methyl-glycosides. Gastric membrane protein was subjected to methanolysis with 1.5 M methanolic-HCl at 80° C for 24 hours. The liberated methyl-glycosides were derivatived with a mixture containing 10% trifluoracetic anhydride and 90% ethyl acetate. The trifluoroacetates of methyl-glycosides were separated in a glass column (180 cm × 2 mm diameter) packed with 2% XF-1105 on 100/120 mesh Gas Chrom P. The sugars were quantitated from their ratio to myo-inositol that was used as internal standard. (From Tomana.[47])

† Average of 6 determinations.

TABLE 3

PHOSPHOLIPID COMPOSITION OF GASTRIC MEMBRANES *

Phospholipid †	μ moles/mg prot.	% total
Total	0.832±0.02	100.0
Sphingomyelin	0.208±0.01	25.0
Glycerol-phospholipids	0.624±0.01	74.4
PC	0.273±0.02	32.8
PE	0.182±0.02	21.9
PS	0.094±0.01	11.3
PI	0.070±0.01	8.4

* Phospholipid composition of gastric membranes is the mean of four different preparations (±S.D.).

† PC=phosphatidylcholine; PE=phosphatidylethanolamine; PS=phosphatidylserine and PI=phosphatidylinositol.

this transition is due to the trypsin sensitive subunits, and not the glycoprotein component.

Another aspect of protein structure that has been studied is the overall helical structure, as revealed by circular dichroism (FIGURE 9) corrected for scattering artefacts.[24] The addition of ATP (but not Mg^{++} or K^{+}) results in a change of CD spectrum interpretable as an increase in α-helical structure.

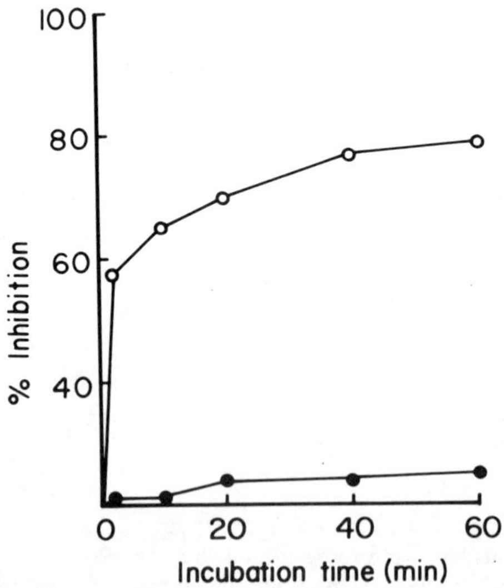

FIGURE 7. Effect of phospholipase A_2 treatment on the enzymatic activities of the gastric membranes (○—○); loss of K^+-ATPase and (●—●) K^+-pNPPase activities. Two mg of membrane protein were incubated with 100 μg of phospholipase A_2 at 22° C in 1 ml solution containing 2 mM $CaCl_2$ and 20 mM Tris-Cl (pH 7.4) for different periods of time. Reaction was stopped by diluting the incubation mixtures with ice-cold, fat-free 1% BSA.

TABLE 4

EFFECT OF PHOSPHOLIPID ADDITION ON THE Mg^{++}-ATPASE AND K^{+}-ATPASE ACTIVITIES OF PHOSPHOLIPASE A_2 TREATED AND BSA WASHED GASTRIC MEMBRANES *

Systems	Phospholipids added	Mg^{++}-ATPase μ moles Pi mg^{-1} hr^{-1}	K^{+}-ATPase † μ moles Pi mg^{-1} hr^{-1}
Intact vesicles (untreated)	—	7.4 ± 1.0	119.6 ± 2.3
Intact vesicles (treated)	—	26.4 ± 1.7	29.9 ± 4.3
Intact vesicles	PE (brain extract)	13.4 ± 0.7	104.1 ± 11.9
Intact vesicles	PC (soybean)	23.5 ± 1.5	102.5 ± 10.4
Intact vesicles	PS (bovine)	15.1 ± 2.0	78.4 ± 1.5
Intact vesicles	PI (bovine)	25.3 ± 0.9	25.9 ± 3.1

* Phospholipase A_2 treated membranes were incubated at 37° C for 15 min with phosphatidylethanolamine (PE), phosphatidylcholine (PC), phosphatidylserine (PS) and phosphatidylinositol (PI) at concentration ratios 3/1, 30/1, 3/1, 3/1, (w/w) respectively and ATPase assay was carried out immediately (Mean of 6 experiments ± S.D.).

† Activity in presence of 0.02 M KCl minus the "basal" Mg^{++}-ATPase rate. For intact vesicles (untreated) the K^{+}-ATPase activity was measured in presence of 10^{-5} M valinomycin.

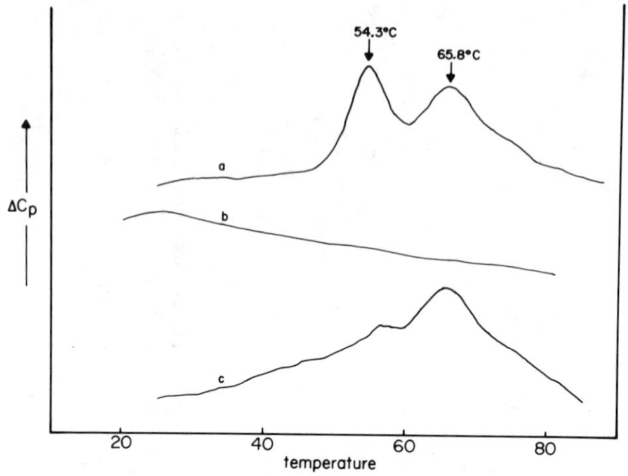

FIGURE 8. Representative differential scanning calorimetric profiles of the gastric membrane K^+-ATPase (scan rate 60° C hr^{-1}). (a) 3.2 mg/ml of gastric membrane protein in 0.25 M sucrose-5 mM Tris-acetate were seen as control showing a transition at 54.3° C and a transition at 65.8° C. (b) Sample from (a) was cooled to 0° C after completion of the first scan and rerun over the same temperature range. Of interest is the irreversibility of the two transitions. (c) 3.2 mg/ml of gastric membrane protein were preincubated at 30° C for 30 min with trypsin (protein: trypsin = 100:1). The reaction was stopped by addition of trypsin inhibitor to a final ratio of 5 to 1 inhibitor to trypsin. The calorimetric profile shows the almost complete disappearance of the 54.3° C transition while the 65.8° C transition remains. (From Segrest.[48])

These studies show that gastric vesicles contain a protein-carbohydrate complex that is asymmetric with respect to the lipid bilayer, undergoing thermal transitions in different parts of its structure, and capable of conformational changes as a function of one group of ligands, adenine nucleotides.

Kinetics of the K^+-ATPase

Steady State

This ATPase, as for the other plasma membrane ATPases [25] has several ligands and several partial reactions. The analogies to the other ATPases are quite striking; for example, the ligands are ATP, ADP, Mg^{++}, H^+ and K^+ and the partial reactions are ATP/ADP exchange, phosphorylation from [$\gamma = {}^{32}P$] ATP, conversion of phosphorylated intermediate, K^+-dependent dephosphorylation, K^+-dependent phosphatase and Mg^{++}-dependent phosphorylation from Pi.

FIGURE 10, a velocity and Eadie Hofstee plot as a function of ATP concentration shows that there are two ATP sites on the enzyme, with K_M of 74 μM and 1 mM at 37° C. ATP at high concentration (>2 mM) inhibits K^+-ATPase activity. The K_M for ATP decreases in the range from pH 7.4 to 6.1, and with decrease in temperature indicating interaction of H^+ with the enzyme or substrate protonatable groups.

Only one K^+ site is found in these membrane preparations with a $K_{0.5}$ of the order of 0.2–0.8. The affinity for K^+ is decreased by increasing concentrations of ATP (FIGURE 11) and the affinity for ATP is decreased by increasing K^+ concentrations showing interaction between the K^+ site and ATP.

The enzyme catalyzes an ATP/ADP exchange. The rate is less than the overall reaction. Cations ($K = Rb^+ > Na^+ > Li^+$) activate the exchange reaction, which might be expected if the bulk of the phosphorylated form of the enzyme was cation sensitive and ADP insensitive (TABLE 5). The adenylate kinase inhibitor diadenosine pentaphosphate only inhibits at very high concentration.[26]

The phosphorylation from ATP occurs in the absence of K^+, giving a value of 1600 pmol mg^{-1} protein. Assuming three subunits,[21] this is a ratio of 1.6 nmol/3.3 nmol catalytic subunit. This phosphorylated enzyme can be dephosphorylated by K^+ by 60%, or by NH_2OH, but is insensitive to ADP alone, which is surprising in view of the transphosphorylation reaction.

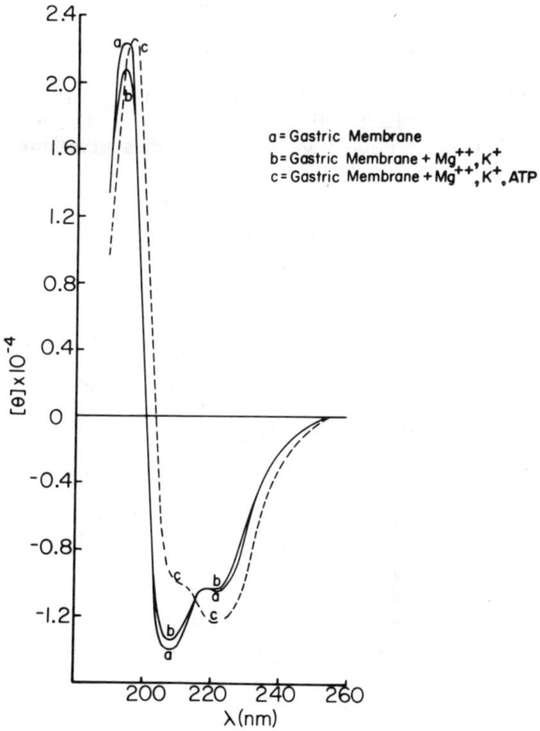

FIGURE 9. Corrected circular dichroism spectra. The gastric membrane preparation (a) at concentrations of 0.56 mg/ml were solubilized in 0.1% SDS-20% TFE and a cell of 0.211 mm path length was used; (b) as in (a) but in the presence of 2 mM Mg^{++} and 20 mM K^+; (c) as in (a) but in the presence of 2 mM ATP. (From Spisni.[40])

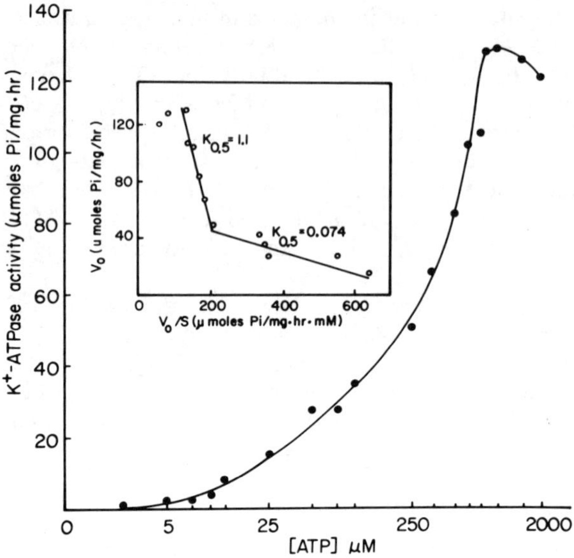

FIGURE 10. Activation of K^+-ATPase by different ATP concentrations. The enzyme was incubated with 2 mM $MgCl_2$, 20 mM KCl, in 40 mM Tris-Cl pH 7.4, for 15′ at 37° C. The ATP concentration was varied as indicated in the figure. The inset shows the Eadie-Hofstee plot.

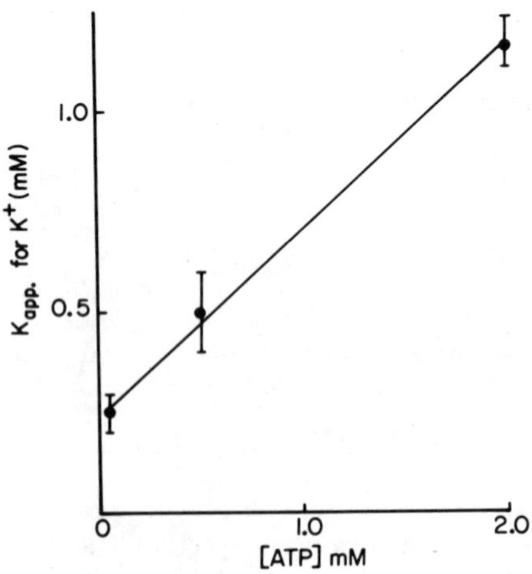

FIGURE 11. Dependence of K_m app. for K^+ of K^+-ATPase on the concentration of ATP.

The enzyme can be phosphorylated by ^{32}Pi, in the presence of Mg^{++} and the level of phosphorylation is 20% of the level reached with ATP.

The p-nitrophenyl phosphatase reaction can be separated from ATPase activity by phospholipase A_2 digestion,[22] and by treatment with the specific histidine reagent, diethyl pyrocarbonate (DEPC). This reagent blocks K^+-ATPase activity by reacting with 8–9 of the 18 histidine residues present in the enzyme complex. Its specificity for histidine in this system has been demonstrated by the pH dependence of the inhibition of ATPase activity and of the spectral changes observed at 240 nm. FIGURE 12 shows that the phosphatase activity is not inhibited, and also that ATP can protect against inactivation of the ATPase. Proton transport is also inhibited by DEPC. Histidine residues are therefore involved in the ATP reaction path of this enzyme complex, and in H^+ transport.

TABLE 5

ATP/ADP EXCHANGE *

Additions	μ moles mg^{-1} hr^{-1}
Control (Mg^{++} only)	6.65
K^+, Rb^+	25.2
Na^+	15.4
Li^+	9.38
.75 mM Di-Adenosine-5-Pentaphosphate	7.07

* In each experiment, exchange activity was measured in a 50 μl suspension containing 50 mM Tris-Cl pH 7.4, 5 μg enzyme; 1 mM ATP; 1 mM ADP; 2 mM $MgCl_2$ and 3000 cpm/n mole ^{14}C-ADP. For cation activation, 20 mM salt was added. Di-adenosine-5-pentaphosphate (.75 mM) was incubated 15 minutes at 25° C before initiation of the exchange reaction. Following a five minute, 25° C incubation, the reaction was stopped by addition of 5.5 μl of 50 mM CDTA.

(*b*) *Transient kinetics*

Data obtained, using a density gradient preparation of the K^+-ATPase, have been amplified using the electrophoretically purified enzyme and a similar flow-quench apparatus.[17]

FIGURE 13 shows the rate of phosphorylation at 22° C of the K^+-ATPase over the first 100–120 msec, when enzyme is mixed with Mg^{++} and then pulsed with ATP or enzyme-ATP mixed and then pulsed with Mg^{++}. The phosphorylation rate in the latter condition is considerably greater than in the former. This demonstrates an E-ATP complex. Either rate of phosphorylation, 4400 or 1400 min^{-1}, is considerably faster than the overall rate of 210 min^{-1}. A steady state of phosphorylation is reached at 130 msec, and if the enzyme is first phosphorylated for this 130 msec, and then mixed with K^+, two phases of dephosphorylation are seen: a fast phase with an initial rate constant of 4100 min^{-1} and a slow phase with a rate constant of 210 min^{-1} (FIGURE 14). The enzyme-Pi complex is only very slowly discharged by ADP. The phosphorylation rate, the possible interconversion of the enzyme-phosphate forms and the initial rate of discharge of the enzyme-phosphate by K^+ are con-

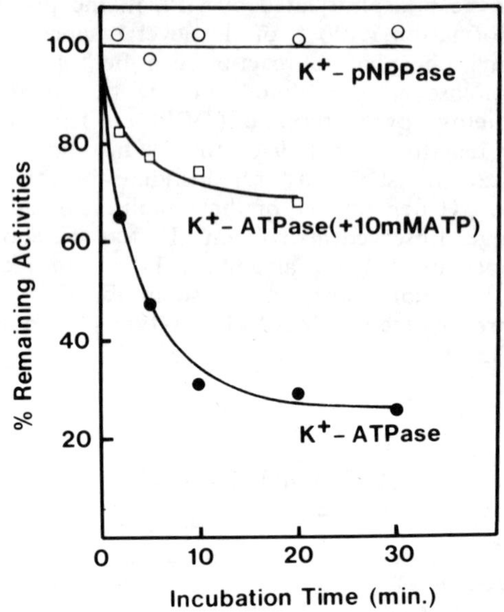

FIGURE 12. Effect of diethyl pyrocarbonate (DEPC) on the gastric membrane. 1.0 mg of protein was resuspended in 1.0 ml of 60 mM HEPES (pH 6.8) and incubated at 22° C for different periods of time in the presence of 0.24 mM. DEPC. The reaction was stopped by diluting the mixtures in ice-cold buffer and immediately afterwards enzyme assays were carried out. (●—●) K^+-ATPase; (○—○) K^+-pNPPase activities remaining after treatment with DEPC; (□—□) K^+-ATPase activity remaining after incubation of the membrane with 10 mM ATP following DEPC treatment.

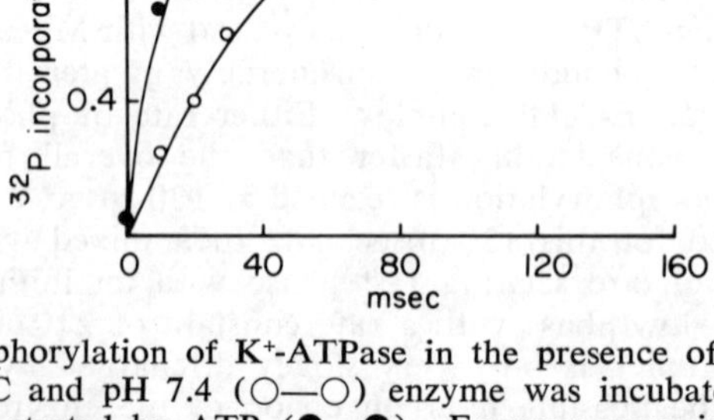

FIGURE 13. Phosphorylation of K^+-ATPase in the presence of 2 mM $MgCl_2$ and 5 μM ATP at 22° C and pH 7.4 (○—○) enzyme was incubated with $MgCl_2$ and the phosphorylation started by ATP (●—●). Enzymes were incubated with ATP 100 ms and phosphorylation started by $MgCl_2$.

siderably faster than the overall reaction. Hence, these can be intermediate steps in ATP hydrolysis. A possible rate-limiting step is the removal of ADP from the enzyme.

From a combination of steady state and transient kinetics, the enzyme sequence may be identified as

$$\left.\begin{array}{lcl} \text{ENZ} + \text{ATP} & \rightleftharpoons & \text{ENZ}_{\text{I}} - \text{ATP} \\ \text{ENZ}_{\text{I}} - \text{ATP} & \overset{\text{Mg}^{++}}{\rightleftharpoons} & \text{ENZ}_{\text{I}} - \text{P} - \text{ADP} \\ \text{ENZ}_{\text{I}} - \text{P} - \text{ADP} & \rightleftharpoons & \text{ENZ}_{\text{I}} - \text{P} + \text{ADP} \\ \text{ENZ}_{\text{I}} - \text{P} & \rightleftharpoons & \text{ENZ}_{\text{II}} - \text{P} \end{array}\right] \text{H}^{+}$$

$$\left.\begin{array}{lcl} \text{ENZ}_{\text{II}} - \text{P} + \text{K}^{+} & \rightleftharpoons & \text{ENZ}_{\text{II}} - \text{K}^{+} + \text{P} \\ \text{ENZ}_{\text{II}} - \text{K}^{+} + \text{ATP} & \rightleftharpoons & \text{ENZ}_{\text{I}} - \text{ATP} + \text{K}^{+} \end{array}\right] \text{K}^{+}$$

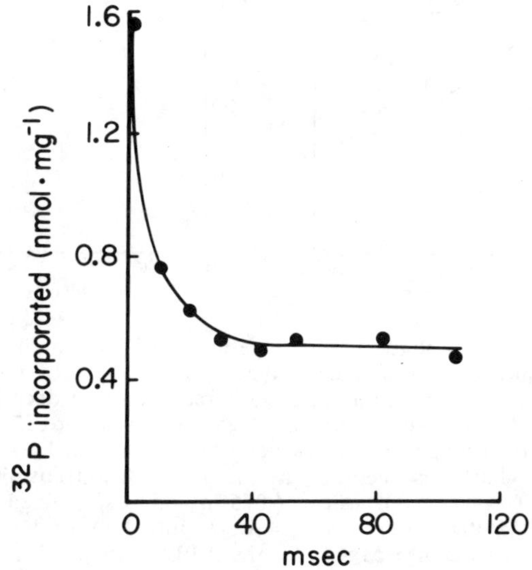

FIGURE 14. Dephosphorylation of phosphoenzyme of K^+-ATPase, by 20 mM KCl. The phosphoenzyme was formed in the presence of 5 μM ATP and 2 mM $MgCl_2$ for 100 ms before the addition of KCl.

This reaction is written in the monomeric form. If the enzyme is a dimer, with one-half site reactivity,[27, 28] the reaction scheme is more complex. ATP binding to the second (low affinity) site could be essential for loss of ADP from the E − Pi − ADP complex.

From these studies, we can conclude that the overall mechanism of the enzyme may be quite similar to the $(Na^+ + K^+)$-ATPase,[25] but that there is considerable variation in the details of the reaction pathway.

Passive Transport Properties of Gastric Vesicles

Various means are available for measurement of osmotic and diffusional flow of water, ranging from light scattering [29] for the former, to NMR [30] for the latter. Studying net flow of H_2O across gastric membrane vesicles, by applying an osmotic pulse of a K^+ salt, allows simultaneous measurement of the rate of water exit (vesicle shrinkage) and K^+ salt entry (vesicle reswelling). A typical result is presented in FIGURE 15. The right part of the figure, B, shows tracing of water exit, giving a $t_{1/2}$ on the order of 700 msec, and giving an apparent H_2O permeability of 10^{-5} cm sec^{-1}, with an E_A of 14 Kcal/mol. These values are independent of the nature of the K^+ salt. The high E_A strongly suggests a hydrophobic route for H_2O flow. In the other part of the

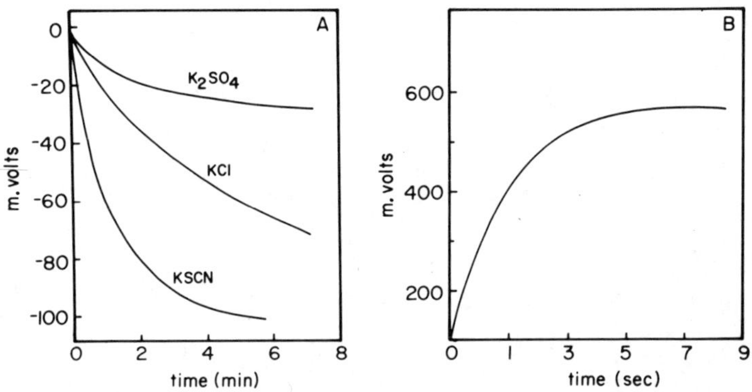

FIGURE 15. (A) Reswell phase of gastric vesicles (.037 mg ml^{-1}) that were partially collapsed by exposure to a 138 m.osm hypertonic gradient of the listed salt (i.e., K_2SO_4, KCl or KSCN). The volume change is recorded as a decrease in transmitted light 90° to incident. (B) Shrink phase of gastric vesicles (.037 mg ml^{-1}) due to H_2O loss resulting from exposure of vesicles to a 138 m.osm hypertonic gradient of KCl. The volume change is recorded as an increase in transmitted light 90° to incident. For both (A) and (B) vesicles (.075 mg ml^{-1}) suspended in .25 M sucrose, 2 mM $MgCl_2$, 2 mM PIPES/Tris pH 7.4 were pulsed (1:1) with a solution consisting of .25 M sucrose, 2 mM $MgCl_2$, 2 mM PIPES/Tris pH 7.4 and 150 mM salt.

figure, A, the rate of reswelling (to 20% initial value) is shown for KCl, K_2SO_4 and KSCN. The major fraction of the vesicles (90%) show, with KCl, a $t_{1/2}$ of 48 min and a consequent permeability of 10^{-9} cm sec^{-1}, and an E_A of 8.7 Kcal/mol. The salt entry in this group of vesicles does not discriminate between Na^+ and K^+ but is restricted when Cl^- is replaced by SO_4^{--} and barely influenced by SCN^-. A possible interpretation of this pathway is a cation-anion symport. Since the major active H^+ transport signal is derived from this group of vesicles, it is relevant to inquire whether the KCl entry rate is adequate to account for net H^+ flux in the isolated vesicles and intact tissue. The answer to the first proposition is "yes," to the second, "no"; implying the need for regulation of the KCl entry pathway, via a cAMP/Ca^{++}-

regulated protein phosphokinase.[31] Data showing an increased rate of Cl^- permeation [32] following ATP may be interpreted in this way.

The second group of vesicles has a faster rate of KCl entry, amplified by SCN^- substitution for Cl^- and again reduced when SO_4^{--} replaces Cl^-. This can be taken to imply a K^+ conductance step limited by an anionic conductance in this group of vesicles. Aging or mild trypsinization of the vesicles exaggerates the Cl^- conductance component as probed by pH dependent dyes in combination with ionophores.[33]

H^+ permeability can be measured by the rate of dissipation of a pH pulse in the presence of salt, protonophores and other ionophores using acridine orange as a sensing dye.[34] The H^+ pathways appear to be an HCl symport and a H^+:cation exchange, giving a permeability of 10^{-8} cm sec^{-1}, i.e., about 10 times that of the passive permeability for K^+. This passive permeability ratio, given an H^+:K^+ exchange pump, determines the steady state H^+ gradient that can be achieved by the gastric vesicle preparation when energized by ATP.

Transport Reactions of the K^+-ATPase

Major interest arose in this enzyme with the demonstration of its H^+ transport capability [35, 36] and several aspects of this have been studied.[13, 27, 37–41] The original hypothesis of a KCl leak coupled to an H^+:K^+ exchange pump that appears electroneutral has been confirmed recently,[42] and we will only summarize some of the salient features, before presenting some new data.

The enzyme is an electroneutral H^+:K^+ exchange pump. The K_M for H^+ and K^+ transport is similar to the high affinity site found in steady state kinetic analysis. The use of potential difference probes such as ANS^-, SCN^-, diethyloxacarbodicyanine$^+$ and dis-$SC_3(5)^+$ using dual wavelengths spectrophotometry or stop-flow spectrophotometry demonstrate the absence of a measurable potential, unless ionophores are added to induce a diffusion potential of a pumped ion (tetrachlorsalicylanilide for H^+, valinomycin for K^+). Various probes have been used to monitor loss of H^+ from the external medium, and appearance of H^+ in the internal medium. From these, the disappearance of H^+ from the external medium exceeds the appearance of H^+ internally by between 1 and 2 orders of magnitude, the maximal pH gradient being between 3 and 4. Hence, intravesicular buffering is present, and the maximum Δ pH expected of the pump (i.e., 6.6 units) is not achieved, since the H^+ leak exceeds the K^+ leak by one order of magnitude.

When ATP is added to pre-equilibrated vesicles and either a pH electrode or a dye is used to monitor either $\Delta\ pH_o$ or $\Delta\ pH_i$, there is a characteristic overshoot (FIGURE 16). The entire curve can be modelled using the following assumptions: (1) there is active electroneutral H^+:K^+ exchange; (2) there is an inward leak of KCl; (3) there is an outward leak of HCl; (4) transport is a saturable function of internal KCl; (5) isotonicity is maintained; (6) intravesicular buffering occurs, and (7) $P_{H^+}/P_{K^+} = 10$. It is of interest to follow the expected osmolarity changes. Thus, the exchange of H^+ for K^+, if electroneutral, maintains isotonicity. However, as internal H^+ exceeds the buffering capacity, there is an efflux or HCl until KCl influx equals the HCl efflux that determines the steady state. During the phase of HCl efflux, there is a loss of internal solute and the vesicles will shrink. FIGURE 17 shows,

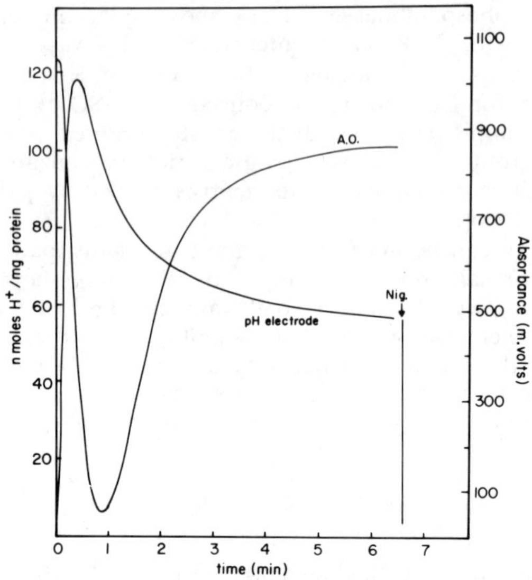

FIGURE 16. Gastric vesicles incubated for 2 hours at 25° C in 150 mM KCl, 2 mM $MgCl_2$ and 1 mM PIPES/Tris pH 6.1 and a .167 mg ml^{-1} protein. For proton transport measured by the pH electrode, .8 mM MgATP pH 6.1 was added to a 2.0 ml volume of the suspension at 25° C. The change in external pH was quantitated by comparison of the signal excursion to standard additions of 1.0 mM HCl. At steady state H^+ transport was uncoupled by addition of 1 μg nigericin/ml. The ΔpH dependent loss of absorbance of 10 μM acridine orange at 490 nm was observed in the same medium (except protein concentration diluted to .086 mg ml^{-1}) as that used for the pH electrode.

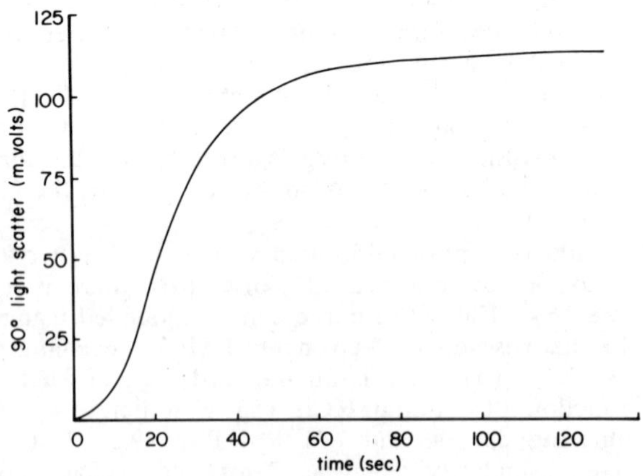

FIGURE 17. 90° light scatter produced by addition of .8 mM MgATP pH 7.4 to gastric vesicles equilibrated in a medium containing 150 mM KCl, 2 mM $MgCl_2$, 2 mM PIPES/Tris pH 7.4 and .086 mg ml^{-1} protein. The change in transmitted light at 500 nm 90° to incident is observed after addition of MgATP at time 0.

using a stop-flow scattering spectrophotometer, that there is indeed a delay phase during the initial H^+:K^+ exchange, the duration of this phase being determined by the buffering of the vesicle interior; this is followed by a shrinkage by about 15% of vesicle volume correlating with the HCl efflux phase. This shows that osmotic events can also be used to determine transport mechanisms as discussed elsewhere.[43]

A troublesome defect in the gastric vesicles model as isolated from hog gastric mucosa is the absence of the expected inhibition by SCN^-. However, SCN^- inhibits translocation of H^+ by frog gastric vesicles, without affecting the

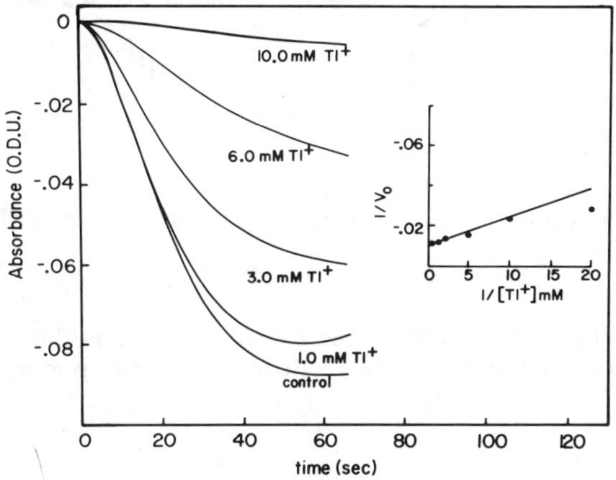

FIGURE 18. Gastric vesicles were incubated for 2 hours at 25° C in medium containing 150 mM KNO_3, 2 mM $MgNO_3$, 2 mM PIPES/Tris pH 7.4, .086 mg ml^{-1} protein and $TlNO_3$ ranging from 1.0 mM to 10.0 mM. The transport-dependent formation of a ΔpH is followed as the loss of absorbance of 10μM acridine orange at 490 nm after addition of .8 mM MgATP pH 7.4. The insert is a Lineweaver Burke plot of the gramicidin (1 μg ml^{-1}) + thallium-stimulated component of the gastric ATPase in the absence of K^+.

K^+ flux, i.e., SCN^- uncouples H^+ and K^+ movement.[44] Another means of uncoupling ATP turnover from H^+ transport in intact vesicles is the use of Tl^+. Althought Tl^+ has a higher affinity for the K^+ site of the ATPase, (0.13 mM) over a wide range of Tl^+ concentrations, no H^+ transport can be detected. Moreover, in the presence of 150 mM KNO_3, Tl^+ inhibits H^+ uptake in a concentration dependent manner (FIGURE 18). External rather than internal Tl^+ is required for this inhibition, but ATP turnover is stimulated by internal Tl^+ since gramicidin increases ATPase activity in intact vesicles (FIGURE 18). An explanation for these data would be that Tl^+ can exchange for H^+ at the pump, or elsewhere in the membrane as well as substitute for K^+ at the K^+ site. Hence, ATP turnover would result in no net flow of either H^+ or TI^+ but an initial TI^+ transient would, and does, occur. Consonant

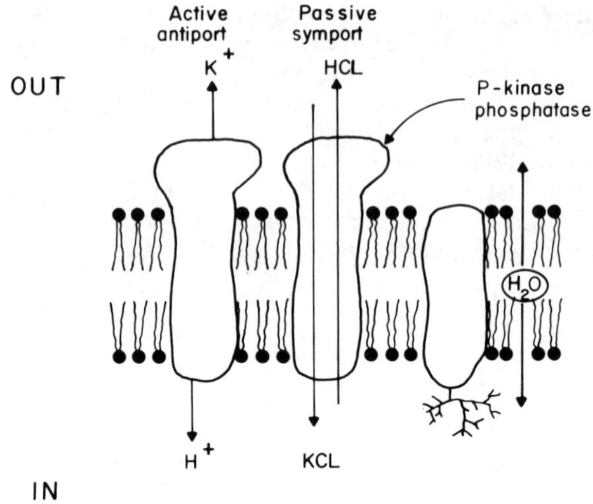

FIGURE 19. A representative model summarizing some of the properties of gastric membrane vesicles.

with these findings is the inhibition of H^+ secretion by rabbit gastric glands by 1 mM TI^+, (data not shown). In contrast TI^+ is transported by the $(Na^+ + K^+)$-ATPase.[45]

Model of ATPase Transport

FIGURE 19 is intended as a summary of some of the properties of this vesicular system that we have discussed. The catalytic subunit of the pump is modelled as an electroneutral H^+:K^+ antiport. The second subunit of the pump is considered responsible for the KCl and HCl leak components. Ca^{++}

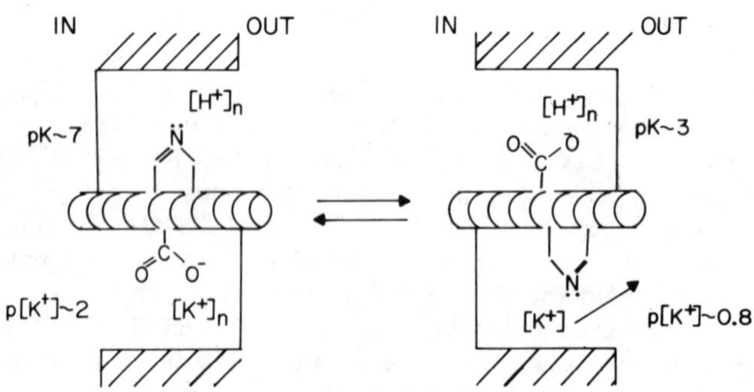

FIGURE 20. Schematic representation of the H^+:K^+ antiport component of the gastric pump.

or cAMP regulation is postulated to be by phosphorylation of this subunit increasing the KCl permeability. The third subunit is the glycoprotein. H_2O is shown as permeating phospholipid regions.

FIGURE 20 shows a conceptual model of the H^+:K^+ antiport component of the pump. The essential action of the ATP phosphorylation-dephosphorylation cycle is visualized as a change in H^+ and K^+ affinity reciprocally across the membrane; an alteration in sidedness of the H^+ and K^+ channels; and, an obligatory site occupancy, n, for these events to occur. The high affinity H^+ site is postulated to be histidine, and the low affinity site, the carboxyl group of glutamyl or aspartyl. In turn, the high affinity K^+ site is carboxyl and histidine the low affinity site. The stoichiometry of the process, n, appears to be four initially, but may dcrease as the Δ pH increases.[37]

SUMMARY

Many features of these gastric vesicles satisfy the requirements for the gastric H^+ pump. For example, we have: (a) K^+ requirement, (b) K_A for K^+ of about 30 mM; (c) identical cation sequence for tissue and vesicles, (d) similar anion sequence, (e) localization at the microvillus of the secretory canaliculus, (f) TI^+ inhibiting H^+ transport of both systems, and (g) the K^+ gradient satisfying the osmotic gradient requirement for HCl-flow out of the parietal cell. Points that require explanation are lack of SCN^- effects and regulation of KCl permeability.

REFERENCES

1. BLACK, J. W., W. A. M. DUNCAN, C. J. DURANT, C. R. GANELLIN & E. M. PARSONS. 1972. Nature **236:**385–390.
2. BERGLINDH, T. & K. J. ÖBRINK. 1976. Acta Physiol. Scand. **96:**150–159.
3. CHEW, C. S., S. J. HERSEY, G. SACHS & T. BERGLINDH. Am. J. Physiol. In press.
4. HARRIS, J. B. & D. ALONSO. 1965. Fed. Proc. **24:**1368–1376.
5. CHEW, C. S. & S. J. HERSEY. 1978. Am. J. Physiol. **235:**E140–E149.
6. TAKEGUCHI, N., G. SACHS & T. BERGLINDH. Am. J. Physiol. In press.
7. SOLL, A. H. 1979. Gastroenterology **75:**A–36.
8. REHM, W. S. 1972. Arch. Internal Med. **129:**270–278.
9. SEDAR, A. W. 1965. Fed. Proc. **24:**1360–1367.
10. HELANDER, H. F. & B. I. HIRSCHOWITZ. 1972. Gastroenterology **63:**951–961.
11. FORTE, T. M., T. E. MACHEN & J. G. FORTE. 1977. Gastroenterology **73:** 941–955.
12. DELL'ANTONE, P. R. COLONNA & G. F. AZZONE. 1972. Eur. J. Biochem. **24:** 553–561.
13. RABON, E., H. H. CHANG & G. SACHS. 1978. Biochemistry **17:**3345–3353.
14. LEE, H. C., A. QUINTAMHILHA & J. G. FORTE. 1976. Biochem. Biophys. Res. Comm. **72:**1179–1186.
15. DIBONA, D. R., S. ITO, T. BERGLINDH & G. SACHS. 1979. Proc. Nat. Acad. Sci. **76:**6689–6693.
16. HARRIS, J. B., H. FRANK & I. S. EDELMAN. 1958. Am. J. Physiol. **195:**499–504.
17. WALLMARK, B. & S. MÅRDH. J. Biol. Chem.
18. SACCOMANI, G., H. B. STEWART, D. SHAW, M. LEWIN & G. SACHS. 1977. Biochim. Biophys. Acta **465:**311–330.
19. SACCOMANI, G., H. F. HELANDER, S. CRAGO, H. H. CHANG, D. W. DAILEY & G. SACHS. 1979. J. Cell. Biol. **83:**271–283.

20. Sternberger, L. A. & J. P. Petrali. 1975. Cell Tissue Res. **162:**141–176.
21. Saccomani, G., D. W. Dailey & G. Sachs. 1979. J. Biol. Chem. **254:**2821–2827.
22. Saccomani, G., H. H. Chang, A. Spisni, H. F. Helander, H. L. Spitzer & G. Sachs. J. Supramol. Struct. In press.
23. Shinitsky, M. & M. Inbar. 1974. J. Mol. Biol. **85:**605–615.
24. Urry, D. W. 1972. Biochim. Biophys. Acta **265:**115–168.
25. Albers, R. W. 1976. *In* The Enzymes of Biological Membranes. A. Martonosi, Ed. **3:**283–301. Plenum Press, New York.
26. Lemhardt, G. 1973. J. Biol. Chem. **248:**1121–1124.
27. Chang, H. H., G. Saccomani, E. Rabon, R. Schackmann & G. Sachs. 1977. Biochim. Biophys. Acta **464:**313–327.
28. Sachs, G. 1977. Rev. Physiol. Biochem. Pharmacol. **79:**133–162.
29. Sha'afi, R. I., C. M. Gary Bobo & A. K. Solomon. 1971. J. Gen. Physiol. **58:** 238–258.
30. Fabry, M. E. & M. Eisenstadt. 1975. Biophys. J. **115:**1101–1110.
31. Kirchberger, M. A., M. Tada & A. M. Katz. 1974. J. Biol. Chem. **249:**6166–6173.
32. Soumarmon, A. & E. Racker. 1979. *In* Frontiers of Bioenergetics. A. R. Croft, A. Scarpa & J. Leigh, Eds. Academic Press, New York.
33. Rabon, E., I. Kajdos & G. Sachs. 1979. Biochim. Biophys. Acta. **556:**469–478.
34. Sachs, G., E. Rabon & G. Saccomani. 1979. *In* Cation Flux Across Biomembranes :53–66. Academic Press, New York. In press.
35. Lee, J., G. Simpson & P. Scholes. 1976. Biochim. Biophys. Res. Comm. **60:** 825–832.
36. Sachs, G., E. Rabon, G. Saccomani & H. M. Sarau. 1976. Ann. N.Y. Acad. Sci. **264:**456–475.
37. Sachs, G., H. H. Chang, E. Rabon, R. Schackmann, M. Lewin & G. Saccomani. 1976. J. Biol. Chem. **251:**7690–7698.
38. Schackmann, R., A. Schwartz, G. Saccomani & G. Sachs. 1977. J. Membr. Biol. **32:**361–381.
39. Lewin, M., G. Saccomani, R. Schackmann & G .Sachs. 1977. J. Membr. Biol. **32:**301–318.
40. Sachs, G., H. H. Chang, E. Rabon, R. Schackmann, H. M. Sarau & G. Saccomani. 1977. Gastroenterology **73:**931–940.
41. Rabon, E., H. H. Chang, G. Saccomani & G. Sachs. 1978. Acta Physiol. Scand. Special Suppl. pp. 409–426.
42. Lee, H. C., H. Breitbart, M. Berman & J. G. Forte. 1979. Biochim. Biophys. Acta **553:**107–131.
43. Rabon, E., N. Takeguchi & G. Sachs. J. Membr. Biol. In press.
44. Rabon, E., G. Saccomani, D. K. Kasbekar & G. Sachs. 1979. Biochim. Biophys. Acta **551:**432–447.
45. Skulskii, I. A., V. Maninnen & J. Järnefelt. 1978. Biochim. Biophys. Acta **506:**233–241.
46. Sterner, R. 1979. Department of Cell Biology. Rockefeller University, New York. Private communication.
47. Tomana, M. 1979. Division of Rheumatology. University of Alabama in Birmingham. Private communication.
48. Segrest, J. P. Department of Pathology. University of Alabama in Birmingham. Private communication.
49. Spisni, A. Laboratory of Molecular Biophysics. University of Alabama in Birmingham. Private communication.

TRANSPORT ATPases IN ANION AND PROTON TRANSPORT

S. L. Bonting, J. J. H. H. M. de Pont, J. M. M. van Amelsvoort,
and J. J. Schrijen

Department of Biochemistry
University of Nijmegen
Nijmegen, the Netherlands 6500 HB

Introduction

A first suggestion for a transport ATPase involved in anion and proton transport was made in 1965 by Durbin and Kasbekar.[1] They postulated that an anion-sensitive Mg-ATPase (anion ATPase) is involved in the gastric acid secretion process. This was based on the observation that the ATPase activity is stimulated by bicarbonate and that both the ATPase and acid secretion are inhibited by thiocyanate. The relatively high carbonic anhydrase activity in gastric mucosa,[2] and the inhibition of gastric acid secretion by the carbonic anhydrase inhibitor acetazolamide suggested that this enzyme is also involved in the secretory process. This led them to a model,[1] which was based on the mechanism of the Na-K ATPase cation transport system. Bicarbonate and chloride together would stimulate the enzyme, leading to phosphorylation by ATP and to inward bicarbonate and outward chloride transport. Bicarbonate plus a proton would result from hydration of CO_2 by the carbonic anhydrase. Extrusion of the proton with the chloride ion would then constitute HCl secretion.

Since then, many reports on anion ATPase activity in gastric mucosa and other tissues have appeared. The presence of both mitochondrial and non-mitochondrial activity in gastric mucosa was claimed,[3, 4] and the properties of these two activities strongly resembled each other.[5] On the other hand, Soumarmon *et al.*[6] could only detect a mitochondrial but not a plasma-membrane bound anion ATPase. Since mitochondria are abundant in the parietal cell,[7] there is a great risk of contamination of a microsomal or plasma membrane fraction with mitochondrial fragments. This led us to a critical study of the properties and location of anion ATPase in various tissues, the results of which will be reviewed in the first part of this paper (see also Bonting *et al.*[8]).

Our negative findings about anion ATPase in various tissues then led us to investigate the other candidate for a role in anion and proton transport. This is the K-activated Mg-ATPase (K-H ATPase), which was shown to be located in the gastric mucosal plasma membrane[9, 10] and was reported to be involved in acid secretion.[11] The results of our studies are described in the second part of this paper.

0077–8923/80/0341–0335 $01.75/0 © 1980, NYAS

Anion-Sensitive ATPase

Nature of the Anion Effects

Anion ATPase is stimulated not only by HCO_3^-, but also by various other oxyanions like borate, selenite, arsenite, arsenate and sulfite.[3, 4] The enzyme in lizard gastric mucosa appears to have only one anionic binding site,[12] rather than two, as suggested in the model of Durbin and Kasbekar.[1] The enzyme activity is inversely proportional to the logarithm of the affinity of the anion to the enzyme (FIGURE 1), although HCO_3^- gives about twice the

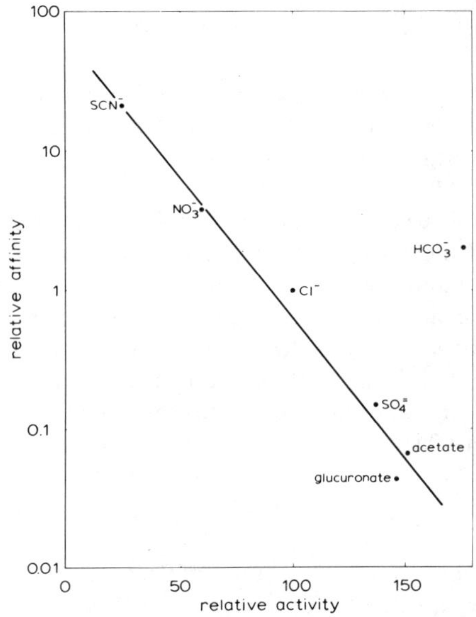

FIGURE 1. Relationship between relative anion affinity of Mg^{2+} ATPase and the relative enzyme activity in the presence of each anion. (From De Pont *et al.*[12])

expected activity. Thus, the enzyme activity actually seems to be inhibited rather than activated when an anion is bound to the anionic-binding site.

Hence, the anion effects on anion ATPase differ sharply from the cation effects on Na-K ATPase, which has two specific cationic binding sites: one for Na^+ and one for K^+, and which is activated upon binding of the two cations. In addition, the molar ratio of H^+ transported per ATP hydrolyzed by anion ATPase in lizard gastric mucosa was very low (0.06–0.17), compared to the universal Na^+/ATP ratio of three for Na-K ATPase.[12]

These findings suggested that the Durbin and Kasbekar model would at least need modification. Even more serious was the fact that the F_1-ATPase of mitochondria displays similar anion effects as the anion ATPase.[13, 14] Therefore, a critical study of the intracellular origin and characteristics of the anion ATPase in five tissues (gastric mucosa, gill, kidney, pancreas and erythrocytes) was carried out by us.[8, 15–18]

Distribution of Anion ATPase Activity in Gradient Subfractions

The tissues were homogenized under mild conditions (Potter-Elvehjem teflon-glass homogenizer, 4–5 strokes at 850 rev./min.[15] The initial centrifugal fractionation and the subsequent gradient subfractionation of the light mitochondrial and the microsomal fractions, as well as the various enzyme assays have been described elsewhere.[15]

A triple assay of anion ATPase activity was carried out routinely in a medium containing 100 mM Tris (pH 8, 4), 2 mM $MgCl_2$, 2 mM Na_2ATP, 0.1 mM ouabain and 50 mM Na-thiocyanate, NaCl or $NaHCO_3$. The presence of anion ATPase was concluded when the activity in the HCO_3 medium was higher and that in the thiocyanate medium was lower than that in the chloride medium, which was used as reference medium. The highest anion ATPase activity was always present in the heavy mitochondrial fraction (10 min, 10,000 × g), but activity was also found in the light mitochondrial (20 min, 20,000 × g) and microsomal (60 min, 100,000 × g) fractions.

The latter two fractions, which were most likely to contain plasma membrane were subfractionated by sucrose (30%–55% w/v) gradient centrifugation (16 h., 64,000 × g). In gastric mucosa, gill and pancreas three peaks (on protein basis) were found: peak I (low density), peak II (medium density) and peak III (high density). The findings for kidney and erythrocyte will be discussed separately. FIGURE 2 shows the distribution of anion ATPase and various markers over these peaks for gastric mucosa microsomal fraction. Peak II and III contain both anion ATPase (HCO_3^- stimulation and thiocyanate inhibition, latter not shown) and cytochrome *c* oxidase, but both are absent from peak I. The latter is rich in K^+-activated phosphatase, Na-K ATPase and 5′ nucleotidase, suggesting that it represents a plasma membrane fraction. Since peak III has the highest cytochrome *c* oxidase and monoamine oxidase activities, this peak appears to be of mitochondrial origin. The presence in peak I of considerable monoamine oxidase activity, a marker for mitochondrial outer membrane, and the absence of anion ATPase in this peak suggests that the latter enzyme is located in the mitochondrial inner membrane, which carries the F_1-ATPase.

Peak II combines a high anion ATPase activity with the presence of rather low cytochrome C oxidase and monoamine oxidase activities and with the absence of plasma membrane and endoplasmic reticulum markers. What does it represent?

Nature of Peak II

Electronmicroscopic observation of peak II material shows the presence of mitochondria, mostly lacking an outer membrane, as well as of lysosomes and some rough and smooth endoplasmic reticulum.[15] Peak III contains many small mitochondria and rough endoplasmic reticulum, while peak I consists mainly of vesicular structures indicative of plasma membrane.

Comparisons of the phospholipid composition of the three peak fractions [15]

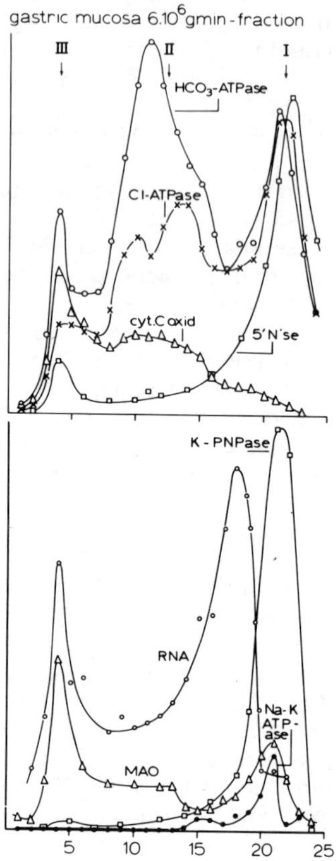

FIGURE 2. Enzyme distribution patterns after density gradient centrifugation of a microsomal fraction of rabbit gastric mucosa. (From Van Amelsvoort *et al.*[15])

show a much closer resemblance between peak II and III than between peak II and I (TABLE 1). The low level of phosphotidylcholine and the high levels of phosphatidylserine and sphingomyelin in peak I confirm its plasma membrane nature. The high cardiolipin level in peak II suggests a high content of mitochondrial inner membrane.

Anion ATPase activity in peak II can be solubilized by the detergent Triton X-100, optimally at a 3:1 ratio of detergent to protein.[8] Subjection of the solubilized fraction to SDS-gel electrophoresis shows two dominant bands at 64,200 and 61,600 daltons and a lesser band at 28,500 daltons,[8] which is very similar to the subunit pattern of mitochondrial F_1-ATPase from various tissues.[19]

Effects of Mitochondrial ATPase Inhibitors

The effects of mitochondrial inhibitors, like oligomycin, aurovertin, quercetin and DCCD, on the anion ATPase activity of various tissue fractions have been determined. TABLE 2 shows that for peak II and III of gastric mucosa [15] and pancreas,[18] the effects are very similar for each inhibitor. The same is true for the microsomal and mitochondrial fractions of rat kidney.[16] This strongly suggests a mitochondrial origin for the enzyme in all cases. The nearly complete inhibition by a combination of oligomycin and aurovertin in the case of gastric mucosa further supports this conclusion.

In the case of the pancreas, Simon *et al.*[20] had based their conclusion that there is a nonmitochondrial anion ATPase, in large part, on its oligomycin insensitivity. In our study of the pancreas,[18] we found anion ATPase activity only in the mitochondrial fraction, but not in the microsomal fraction. Upon subfractionation of the mitochondrial fraction, peak I was found to lack anion ATPase activity. Initially, when studying the effect of oligomycin on the anion ATPase activity in peak II and III, we also observed very little inhibition by oligomycin (maximal inhibition ca. 15%). It occurred to us that this might be due to the presence of (phospho)-lipase activity in the preparation. Since these enzymes require Ca^{2+}, we added 1 mM EDTA to the homogenization buffer and gradient medium. Now indeed a strong (ca. 86%) inhibition by oligomycin was observed,[18] which equals that found in gastric mucosa and kidney fractions (TABLE 2).

Causes for Incorrect Localization of Anion ATPase

How can other investigators in several cases find anion ATPase activity in what appears to be a plasma membrane fraction, while we find in gastric mucosa, gill, kidney and pancreas the enzyme only in subfractions that are clearly of mitochondrial origin?

Experiments involving recentrifugation, centrifugation by flotation and prolonged (65 hr instead of 16 hr) centrifugation of gastric mucosa microsomal fraction [15] indicate that in our subfractionation procedure equilibrium has been reached after 16 hr, and also that peak II anion ATPase does not originate from peak III by osmotic lysis of mitochondria in this subfraction during centrifugation.

TABLE 1

PHOSPHOLIPID COMPOSITION OF THREE SUBFRACTIONS OF RABBIT GASTRIC MUCOSA

	Subfraction		
Phospholipid	I (%)*	II (%)	III (%)
Phosphatidylethanolamine	28.0±1.0	28.6±0.5	24.4±0.7
Phosphatidylcholine	27.2±0.7	43.5±1.5	49.6±0.7
Phosphatidylinositol	4.7±0.3	6.0±0.4	6.6±0.3
Phosphatidylserine	13.1±1.0	5.3±0.4	4.2±0.2
Sphingomyelin	18.8±1.5	6.7±1.0	6.2±0.7
Lysophosphatidylcholine	2.5±1.6	—	0.9±0.4
Diphosphatidylglycerol (cardiolipin)	1.2±0.4	7.0±1.1	3.1±0.5
Unidentified	4.7±0.6	2.7±0.8	4.2±1.0

* Values are percentages of total phosphorus with standard error for four experiments. (From Van Amelsvoort *et al.*[15])

However, gradient centrifugation for only 8 hr instead of 16 hr presents a very different distribution pattern: peak III is in its normal position, but peak I and II are not separated and now contain both anion ATPase and cytochrome *c* oxidase.[15] This means that short gradient centrifugation times, as used by some authors, are insufficient to move down the very small (sub) mitochondrial particles normally found by us in peak II.

The choice of the mitochondrial marker enzyme is another critical point.

TABLE 2

EFFECT OF INHIBITORS ON ANION ATPASE OF DIFFERENT TISSUE FRACTIONS *

Inhibitor	pI_{50}	pI_{99}	Rest act. (%)	pI_{50}	pI_{99}	Rest act. (%)
			Gastric mucosa [15]			
	peak II			peak III		
Oligomycin	7.2	5.0	15	7.1	4.9	20
Aurovertin D	6.1	4.9	46	6.2	5.1	50
Olig. 20 μM + Aurovertin 35 μM	—	—	9	—	—	17
Quercetin	4.2	3.7	6	4.1	3.7	8
DCCD	6.6	5.8	16	6.7	6.0	16
Mitoch. inhib. protein (g/U)	3.7	—	—	3.4	—	—
Acetazolamide (10 mM)	—	—	93	—	—	93
			Pancreas (1 mM EDTA present)[18]			
	peak II			peak III		
Oligomycin	7.8	5.8	16	7.7	5.6	11
Aurovertin D (16 μM)	—	—	44	—	—	42
Mitoch. inhib. protein (g/U)	3.4	—	—	3.1–3.3	—	—
			Kidney [16]			
	microsomal fraction			mitochondrial fraction		
Oligomycin	7.8	4.8	26	7.0	4.7	10
Aurovertin D	6.3	5.1	48	6.2	5.0	51
DCCD	6.6	5.6	29	6.5	5.7	18

* pI_{50} is the negative logarithm of the inhibitor concentration (molar if not otherwise indicated) at half-maximal inhibition; pI_{99} is that giving 99% of the maximal inhibition; Rest act. is the percent of activity remaining at maximal inhibition. Activity represents total ATPase activity in the HCO_3^- medium. The figures are averages for 2–4 experiments.

Succinic dehydrogenase is relatively easily solubilized, especially at higher pH, and an insensitive assay method for this enzyme is often used. Monoamine oxidase is not suitable either, since it is located in the mitochondrial outer membrane, which is easily detached from the inner membrane containing the anion-sensitive F_1-ATPase. Cytochrome *c* oxidase appears to be the mitochondrial marker of choice.

Drastic homogenization of the tissue may be another cause for incorrect localization of the anion ATPase since this may disrupt the mitochondria to such an extent that even after 16 hr of centrifugation, peak II has reached its equilibrium position. We have demonstrated this by sonicating the light mitochondrial fraction of gastric mucosa before gradient centrifugation.[15] Anion ATPase activity has now disappeared from peak III and is increased in peak I and II, and cytochrome *c* oxidase has shifted in parallel fashion. In connection with an injudicious choice of mitochondrial marker, this would lead again to an incorrect localization of anion ATPase. At the same time, this observation stresses again the strong relationship between cytochrome *c* oxidase and anion ATPase activity.

Another example of the effect of homogenization is presented by the gill, for which we had to use a Polytron homogenizer in view of the cartilaginous nature of this tissue.[15] In contrast to gastric mucosa, kidney and pancreas, the highest anion ATPase activity is now found in the light microsomal fraction. The activity in the microsomal fraction is relatively higher than in any of these other tissues, while upon subfractionation, the anion ATPase and cytochrome *c* oxidase activities run strictly parallel.

So, too drastic homogenization, too short centrifugation times and the use of unsuitable mitochondrial markers (or no marker at all) may all cause incorrect conclusions about the location of anion ATPase.

Anion ATPase in Kidney Brush Border

In our initial study of the kidney,[15] we were unable to reach a definite conclusion about the location of anion ATPase because the mitochondrial and plasma membrane marker enzymes greatly overlapped in the subfractions of the light mitochondrial and microsomal fractions of rabbit kidney. In addition, there were reports about the presence of the enzyme in kidney brush border.[21, 22] This led us to a further investigation of the kidney.[16] The effects of mitochondrial inhibitors on anion ATPase activity in the mitochondrial and microsomal fractions of rabbit kidney (TABLE 2) leave little doubt about the mitochondrial origin of this activity.

A brush border membrane fraction, isolated as described by Liang and Sacktor,[22] resembled theirs with respect to electron microscopic appearance, anion ATPase activity and oligomycin inhibition. Bicarbonate stimulation of anion ATPase activity is strongly (64%–68%) inhibited by oligomycin (50 μM), aurovertin (22 μM and DCCD (3 μM). Upon purification by gradient centrifugation there is a single peak, in which alkaline phosphatase and γ-glutamyl-transpeptidase (thought to be marker enzymes for the brush border membrane) coincide with low anion ATPase, Na-K ATPase and cytochrome *c* oxidase activities (FIGURE 3). Bicarbonate stimulation, cytochrome *c* oxidase activity and oligomycin inhibition have decreased in parallel to about one fourth. This residual anion ATPase activity, in our opinion, derives from the mitochondria, which are trapped in brush border membranes on our electronmicrographs.

Recently, the presence of an anion-sensitive ATPase activity in the brush border of rat small intestine has been reported.[23] The substrate specificity, the anion stimulation profile, and its inhibition by L-phenylalanine, EDTA and L-cysteine suggest that this activity represents mainly alkaline phosphatase of the brush border membrane, as the authors themselves point out. Part of the activity may also represent Ca-Mg ATPase, since EGTA was omitted from the assay medium, and mitochondrial ATPase, since the authors have not properly excluded contamination by mitochondrial fragments.

Anion ATPase in Erythrocyte Membranes

A report [24] on the presence of a particulate anion ATPase activity in rabbit erythrocytes, which are devoid of mitochondria, led us to investigate

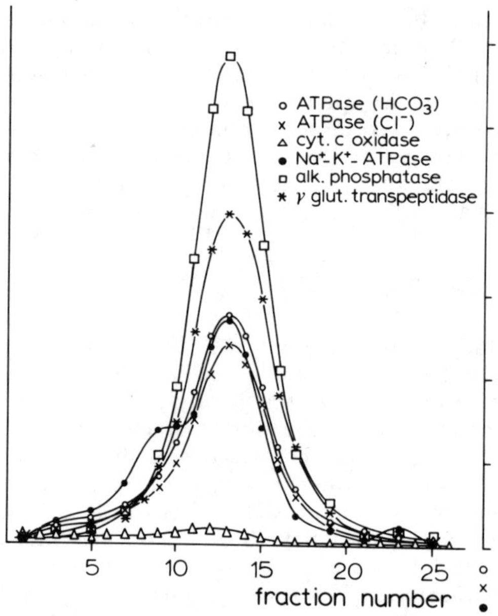

FIGURE 3. Enzyme distribution patterns after density gradient centrifugation of brush border membrane fraction of rabbit kidney cortex. (From Van Amelsvoort *et al.*[16])

this activitiy in more detail.[17] While the reported findings could be confirmed, the characteristics of the anion ATPase in erythrocyte membrane differ strikingly from those of the enzyme studied in other tissues. Its anion-stimulation profile is markedly different: e.g., sulfite inhibits the ATPase activity slightly instead of strongly activating it; thiocyanate has only a slight inhibitory effect on the activity. Calcium stimulates the erythrocyte activity, whereas EGTA, ruthenium red and chlorpromazine inhibit it. We are, therefore, forced to conclude that this enzyme is part of the Ca-Mg ATPase activity known to be located in the erythrocyte membrane.[25]

Conclusions

From the evidence reviewed here, we are forced to conclude that no proof for the existence of a plasma membrane located anion ATPase has been obtained in any of the tissues studied. In gastric mucosa, gill, kidney and pancreas only a mitochondrial anion ATPase, probably identical with the F_1-ATPase, could be demonstrated. In the erythrocyte membrane, anion ATPase activity appears to be part of its Ca-Mg ATPase system.

In our opinion, these findings appear to rule out a function of anion ATPase in the transport of anions and protons across the plasma membrane in these tissues.

(K-H) ATPase

Isolation and Properties

A vesicular membrane fraction is isolated from fresh pig gastric mucosa by means of the method described by Forte *et al.*[26, 27] The homogenate, obtained after mild homogenization (teflon-glass homogenizer, three strokes at 300–500 rpm) in a medium containing 113 mM mannitol, 37 mM sucrose, 0.2 mM EDTA and 5 mM Tris-HCl (pH 7.4) is centrifuged for 20 min at 20,000 g. The supernatant is centrifuged on 37% (m/v) sucrose, 25 mM Tris HCl (pH 7.4) for 30 min at 100,000 g. The membrane fraction at the interface of sucrose layer and supernatant (interface fraction) is collected. It is diluted with 25 mM Tris-HCl (pH 7.4) to 15% (w/v) sucrose, and layered on an 18–43% (w/v) sucrose, 25 mM Tris-HCl (pH 7.4) gradient, resting on a 55% (w/v) sucrose solution. After 16 hr centrifugation at 150,000 × g the fractions are collected and scanned at 254 nm. The major 254 nm peak represents the membrane fraction ($d = 1.11$–1.13). This fraction is diluted with sucrose-free buffer and centrifuged for 1 hr at 100,000 × g. The pellet is resuspended in 0.25 M sucrose to a protein concentration of 1 mg/ml.

The K-H ATPase assay is carried out in a medium containing 30 mM imidazole-HCl (pH 7.0), 5 mM Na_2ATP, 5 mM $MgCl_2$, 0.1 mM ouabain and 20 mM KCl or choline-Cl. In butanedione studies, the buffer is 125 mM Na-borate (pH 7.0), which stimulates the same activities. K-activated phosphatase is determined in the same medium with 5 mM p-nitrophenylphosphate instead of ATP. Standard assay procedures are used for cytochrome *c* oxidase, 5′ nucleotidase, Na-K ATPase and anion ATPase.[15]

The interface fraction is high in K-H ATPase and K-pNPPase activity. It is low in Mg-ATPase (⅕ of K-H ATPase) and cytochrome *c* oxidase (less than supernatant), indicating very low mitochondrial contamination. The final membrane fraction is 4 times as high in K-H ATPase as the interface fraction, and Mg-ATPase is now only 1⁄10 of K-H ATPase. The main Mg-ATPase, cytochrome *c* oxidase, and anion ATPase activities are found at higher densities. The membrane fraction has a specific K-H ATPase activity of 75–95 nmol P_i/mg protein/h, only slightly lower than the values obtained by Saccomani *et al.*[28] after free flow electrophoresis.

The K-H ATPase activity has a pH optimum of 7.0 ± 0.1, which is lower

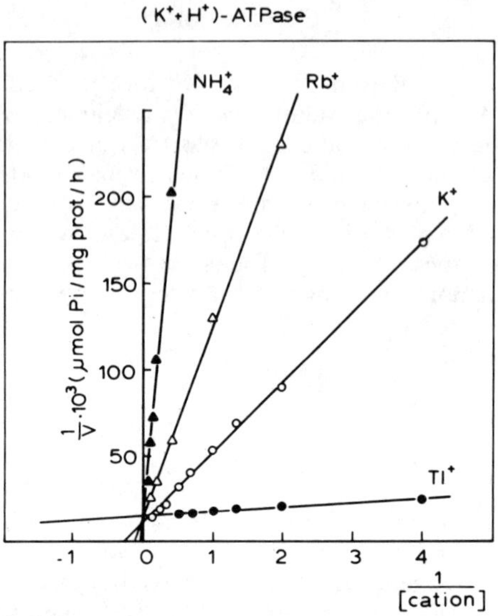

FIGURE 4. Effects of Tl^+, K^+, Rb^+ and NH_4^+ on gastric microsomal ATPase activity. Activity is determined as described in the text. Various concentrations of the indicated cations are added to the assay medium. Results are presented as Lineweaver-Burk plots.

than the value of 7.5 reported by Ganser & Forte,[9] but more recently they have obtained a value of 7.1.[39] Cation activation has been studied with various cations. Clear activation is shown by K^+, Tl^+, Rb^+ and NH_4^+, while Cs^+ has only a slight effect and Na^+, Li^+ and choline do not activate at all. From Lineweaver-Burk plots (FIGURE 4) Km values for the activating cations are obtained, increasing in the order $Tl^+ < K^+ < Rb^+ < NH_4$ (TABLE 3, 2nd col.). The same sequence has been reported for stimulation of proton uptake by the vesicles,[11] as well as for dephosphorylation of the phosphorylated enzyme.[26]

TABLE 3

ACTIVATING AND PROTECTIVE EFFECTS OF CATIONS ON K–H ATPASE *

Cation	K_m for activation mM	K_d for protection against butanedione 0 mM ATP mM	1 mM ATP mM
Tl^+	0.13	0.06	0.6
K^+	2.7	0.14	3.2
Rb^+	7.6	0.50	8.4
NH_4^+	26	0.94	—

* Gastric mucosa vesicle fraction used for determination of K_m (from data as shown in FIGURE 4) and K_d (from data as shown in FIGURE 10A by the Scrutton and Utter method).

The membrane preparation contains proteins and phospholipids. On SDS gel electrophoresis, a 100,000 protein, presumably the K-H ATPase, comprises approximately 70% of the total protein content. The phospholipid content is 1.07 μmol lipid P·mg protein^{-1}. The major phospholipids are: phosphatidylcholine (32%), phosphatidylethanolamine (27%), sphingomyelin (26%), phosphatidylserine (9%) and phosphatidylinositol (6%). Both phospholipid content and composition are qualitatively and quantitatively very similar to those of a highly purified Na-K ATPase preparation from rabbit kidney outer medulla.[29] The high sphingomyelin content is characteristic for a plasma membrane preparation. Removal of up to 50% of the phospholipids by phospholipase C(*Clostridiun welchii*), results in a similar reduction in K-H ATPase activity, suggesting that a phospholipid environment is essential for the activity of the enzyme.[40]

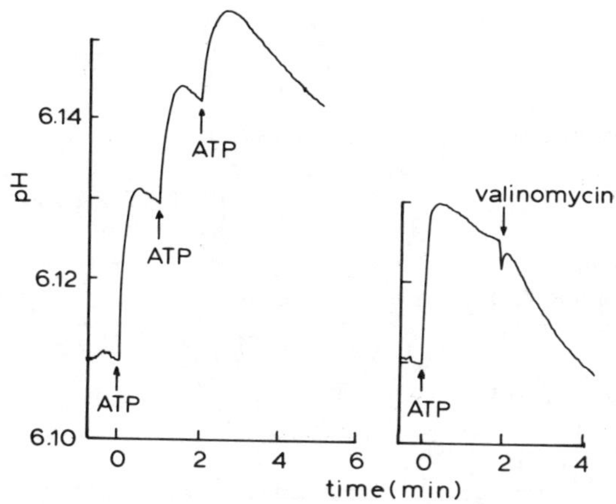

FIGURE 5. Evidence of ATP-dependent proton uptake and valinomycin-induced lowering of proton gradient in vesicle fraction from pig gastric mucosa.

Transport Function

The vesicles of the membrane fraction are capable of proton uptake. When 1 ml membrane fraction (4.4 mg protein/ml in 0.25 M sucrose, pH 6.11) is mixed with 5 ml medium containing 5 mM glycylglycine, 2 mM $MgCl_2$, 150 mM KCl (pH 6.11), the addition of 50 nmol ATP raises the pH by 0.031 unit maximally in 24 sec (FIGURE 5). This can be repeated a few times with slightly decreasing rate and extent of the pH shift.

Addition of 10 nmol valinomycin or nigericin after ATP depletion increases the rate at which the pH decreases. When valinomycin is added together with ATP there is an accelerated pH increase, i.e., an accelerated proton uptake. These findings agree with those of others,[11, 30, 31] indicating active proton uptake through a neutral $K^+ - H^+$ exchange system dependent on ATP hydrolysis. The addition of the ionophore makes the vesicle more permeable to K^+, leading to passive K^+ influx and similar proton efflux. In agreement with this, we find that valinomycin raises the V_{max} of the enzyme, but does not change its K_m value for K^+.

Modification of Arginine by Butanedione

Having thus established that the membrane fraction isolated from pig gastric mucosa has a high K-H ATPase activity and is capable of active proton transport, we have begun a study of the mechanism of the enzyme reaction. In first instance, we have investigated the presence of an essential arginine group, since this is generally a feature of enzymes reacting with anionic substrates such as ATP. Recently, we have established the presence of such a group in Na-K ATPase by means of modification with butanedione.[32]

The membrane fraction (0.5 mg protein/ml is preincubated with 0.5 mM butanedione in 125 mM Na-borate buffer (pH 7.0) containing 5 mM $MgCl_2$

TABLE 4

REVERSIBILITY OF THE INACTIVATION OF GASTRIC K–H ATPASE BY BUTANEDIONE *

Buffer (125 mM, pH 7,0)	K–H ATPase activity (percent of control) 0 min. incubation	90 min. incubation
	(%)	(%)
Na-borate	10	15
Imidazole-HCl	12	43
Tris-HCl	13	41
HEPES-NaOH	9	69

* Membranes, treated with butanedione, are incubated for 0 or 90 min at 37° C in the indicated buffers. Results are expressed as percent of K-H ATPase activity in the absence of butanedione treatment.

for 20 min. at 37° C. The membranes are then separated from the reaction mixture by gel filtration on a Sephadex G25 column, and are subjected to K-H ATPase assay.

There is a considerable inactivation of the enzyme, which is reversible upon incubation for 90 min at 37° C in various buffers, but not in borate buffer (TABLE 4). This observation indicates specific modification of one or more arginine groups. Butanedione inactivation follows pseudo first-order kinetics, as shown by the linear plots of log activity vs. time (FIGURE 6). The slope of a double-logarithmic plot of half-inactivation time vs. butanedione concentration will then yield the number of modified arginine groups necessary for inactivation.[33] The plot shown in FIGURE 7 has a slope of 0.94, indicating that modification of one arginine group leads to inactivation, as is also the case for Na-K ATPase.[32]

The apparent first-order rate constant for inactivation varies proportionally with the butanedione concentration (FIGURE 8), indicating that the reaction

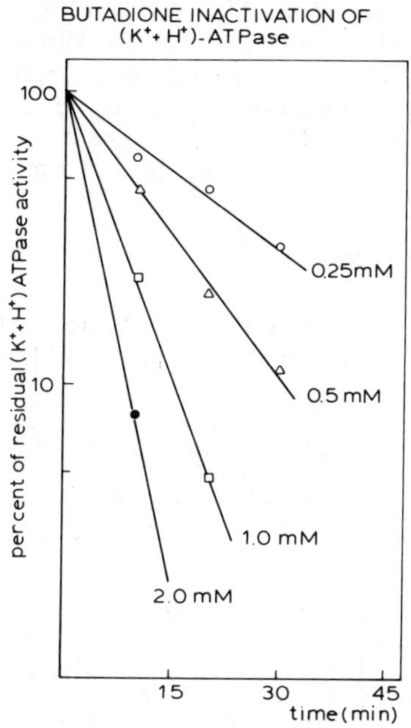

FIGURE 6. Inactivation by butanedione as a function of time. Gastric mucosa vesicle fraction (0.5 mg protein/ml) is incubated at 37° C during the indicated times with various concentrations of butanedione in 125 mM sodium borate (pH 7.0), 5 mM $MgCl_2$. K-H ATPase is determined as described in the text. Enzyme activity is expressed as a percentage of control activity without butanedione.

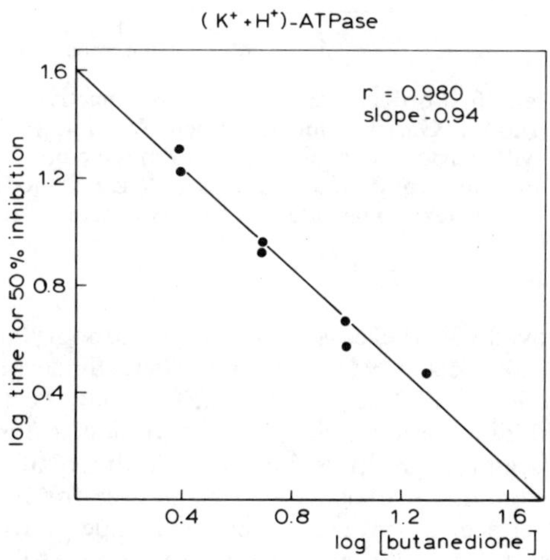

FIGURE 7. Double logarithmic plot of half-maximal inhibition time vs. butanedione concentration. Derived from experimental results as displayed in FIGURE 6. Slope indicates number of arginine groups involved in inactivation.

obeys true second-order kinetics. This relationship has been determined in the presence of 5 mM $MgCl_2$, as well as in the absence of Mg^{2+} and addition of 5 mM CDTA. In the first case, the second-order rate constant is 145 min^{-1} M^{-1}, and in the second case only 52 min^{-1} M^{-1}. Apparently the essential arginine group is more exposed in the presence of Mg^{2+} than in its absence, which suggests that Mg^{2+} induces a change of conformation in K-H ATPase.

Protective Effects of Nucleotides

In the presence of ATP, inactivation of the enzyme by butanedione is decreased (FIGURE 9b). This protective effect of ATP is considerably enhanced

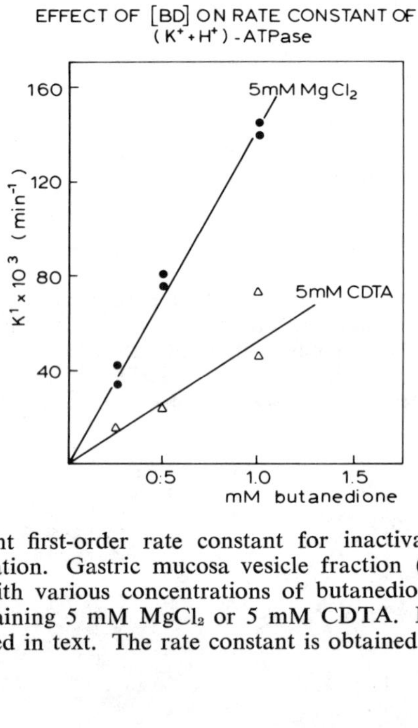

FIGURE 8. Apparent first-order rate constant for inactivation as a function of butanedione concentration. Gastric mucosa vesicle fraction (0.5 mg protein/ml) is incubated at 37° C with various concentrations of butanedione in 125 mM sodium borate (pH 7.0) containing 5 mM $MgCl_2$ or 5 mM CDTA. K-H ATPase activity is determined as described in text. The rate constant is obtained from plots of inactivation vs. time.

when Mg^{2+} is removed. In the case of the nonphosphorylating ATP-analogue AMPPNP there is no effect of Mg^{2+} (FIGURE 9a). Since Mg^{2+} is required for phosphorylation of K-H ATPase,[26] this indicates that the protective effect of ATP involves its binding rather than phosphorylation. This, in turn, suggests that the essential arginine group is involved in the binding of ATP and is located in the ATP binding center, as is also the case for Na-K ATPase.[32]

The protective effects of several other nucleotides have been determined (TABLE 5). Apart from ATP and AMPPNP, only ADP and deoxy-ATP show marked protective effects, which are again more pronounced in the absence of Mg^{2+} In the case of ADP, which must be released upon phosphorylation of the enzyme by ATP, this means that the Mg^{2+}-induced con-

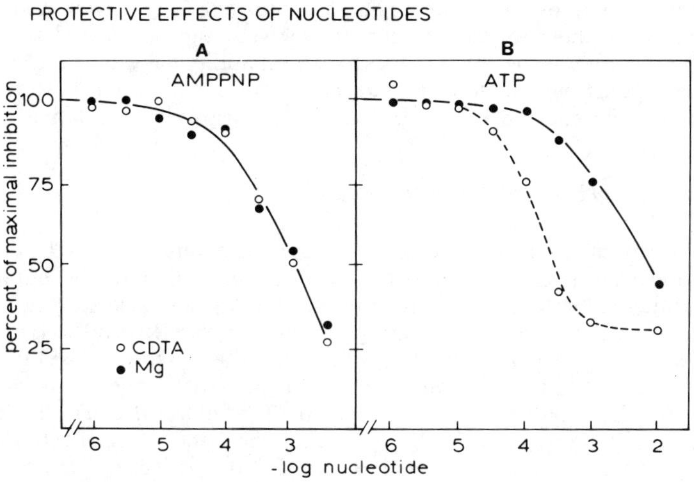

FIGURE 9. Protective effect of nucleotides on inactivation of K-H ATPase by butanedione. Gastric mucosa vesicle fraction (0.5 mg protein/ml) is incubated at 37° C in 125 mM sodium borate (pH 7.0), containing either 5mM $MgCl_2$ or 5 mM CDTA, and either ATP+1 mM butanedione (B) or AMPPNP+0.5 mM butanedione (A). Inhibition without nucleotide is set at 100%.

TABLE 5

PROTECTIVE EFFECTS OF NUCLEOTIDES ON INACTIVATION BY BUTANEDIONE *

Nucleotide added (1 mM)	Percent Inhibition 5 mM $MgCl_2$	5 mM CDTA
none	≡100	≡100
ATP	59±2	24±4
ADP	61±4	42±3
deoxy-ATP	75±2	43±2
AMPPNP	67±3	57±1
AMPPCP	92±2	87±2
AMP	92±2	100±4
CTP	85±3	91±3
GTP	83±3	89±2
ITP	86±2	89±1

* K-H ATPase is incubated for 20 min at 37° C in 125 mM Na borate buffer (pH 7.0) containing 0.5 mM butanedione, 1 mM nucleotide and either 5 mM $MgCl_2$ or 5 mM CDTA. Results are expressed as a percentage of the inhibition obtained in the absence of nucleotide, with standard error for two experiments.

formational change exposing the arginine group not only favors ATP binding and subsequent phosphorylation, but also lowers the affinity of the enzyme for ADP, and thus leads to its release. The inability of CTP, GTP and ITP to protect against butanedione inactivation agrees with their low activity as substrates for the enzyme (17%, 0% and 0%, respectively, relative to ATP).

Cation Effects on Butanedione Inactivation

The effects of monovalent cations on the inactivation of K-H ATPase by butanedione have been investigated by adding the cation to the butanedione-borate medium (1 mM CDTA, no Mg^{2+}). Choline (and also Li^+ and Cs^+) has no significant effect, but the activating cations Tl^+, K^+, Rb^+ and NH_4^+ show a protective effect (decreasing in this order) in the absence of ATP (FIGURE 10a). With an equation derived by Scrutton & Utter [34] the dissociation constants K_d for these cations can be calculated (TABLE 3, 3rd col.). These values are considerably lower than the K_m values for their activating effects (TABLE 3, 2nd. col.). These protective effects indicate that the activating cations induce a conformational change in the enzyme, which decreases the exposure of the essential arginine group.

When these experiments are repeated in the presence of 1 mM ATP, there is a striking reversal (FIGURE 10b). The inactivation is increased in the same order ($Tl^+ > K^+ > Rb^+ > NH_4$) as it was decreased in the absence of ATP.

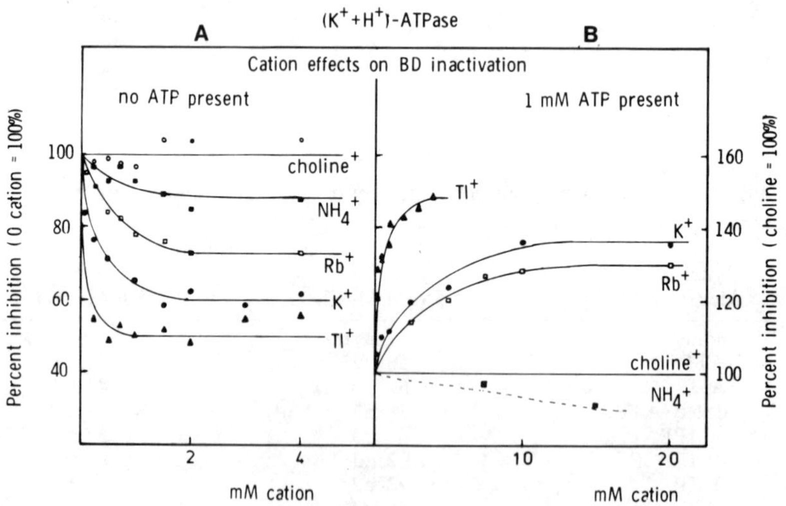

FIGURE 10. Effects of activating monovalent cations on inactivation of K-H ATPase by butanedione in absence (A) or presence (B) of ATP. Gastric mucosa vesicle fraction (0.5 mg protein/ml) is incubated for 15 min (A) or 40 min (B) at 37° C in 125 mM sodium borate (pH 7.0), 1 mM CDTA, 0,5 mM (A) or 2 mM (B) butanedione, activating cations in the stated concentrations, and either 0 mM (A) or 1 mM (B) ATP. Results are expressed as percent inhibition of that in the absence of activating cation (=100%).

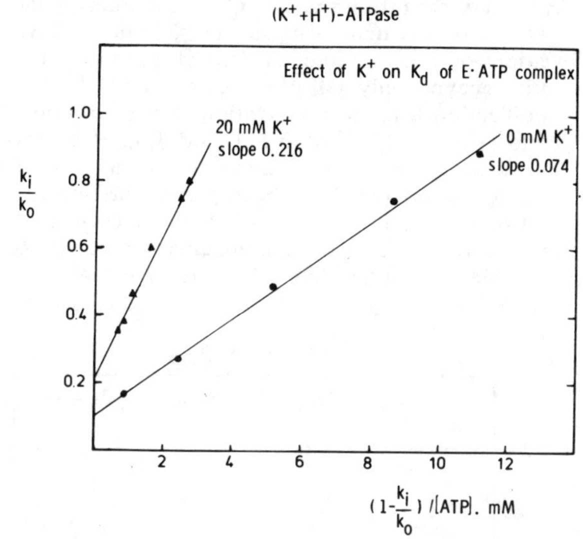

FIGURE 11. Determination of dissociation constant K_d for the enzyme-ATP complex in the presence or absence of 20 mM K^+. Data obtained from experiments as shown in FIGURE 9 B. k_i and k_o are apparent first-order inactivation rate constants in the presence or absence of ATP, respectively. K_d is determined from the slope of the plot.

NH_4^+ shows in this experiment slightly lower values than choline. The dissociation constants for the cations calculated from these data (TABLE 5, 4th col.) are close to the K_m values for activation. The interpretation of these findings appears to be that the cations lower the protection by ATP by decreasing the exposure of the essential arginine group and thus the binding of ATP to it. In order to test this conclusion, we have determined the dissociation constant K_d for ATP by observing butanedione inactivation in the presence and absence of 20 mM KCl at different concentrations of ATP (FIGURE 11). The value of K_d can then be calculated by means of the Scrutton and Utter method.[34] In the presence of 20 mM K^+ a K_d value of 216 μM and in its absence of 74 μM is found, which confirms our conclusion.

Modification of Sulfhydryl Groups by DTNB

In analogy to our previous studies of the sulfhydryl groups of Na-K ATPase,[35] we have also started an investigation of these groups in K-H ATPase. The modifying reagent DTNB (5,5′dithiobis(2-nitrobenzoic acid) reacts with sulfhydryl groups through a disulfide exchange reaction under release of 5′thio(2-nitrobenzoic acid), which has a characteristic absorption peak at 412 nm. The extent of the reaction can thus be determined by measuring the 412 nm absorption. The enzyme preparation (0,5 mg protein/ml) is typically incubated in a solution of DTNB in 25 mM Tris-HCl (pH 7.5), 2 mM CDTA for 60 min at 37° C. Incubation of the enzyme with DTNB in the presence of the denaturing agent sodium dodecylsulfate (SDS) leads to modification of 9 ± 1 mole sulfhydryl groups per 100,000 gram protein. In the native enzyme only half of these are modified.

Sulfhydryl modification leads to inactivation of the enzyme, which is however never complete (FIGURE 12). At 0.1–1 mM DTNB a maximal inactivation of only 85% is obtained. Half-maximal inactivation is obtained at 10 μM DTNB. The inactivation reaction obeys pseudo first-order kinetics. Calculations show that only one sulfhydryl group is involved in inactivation.

There are again cation effects on the inactivation rate. Replacement of CDTA by Mg^{2+} doubles the initial rate of inactivation. Na^+, Li^+ and choline

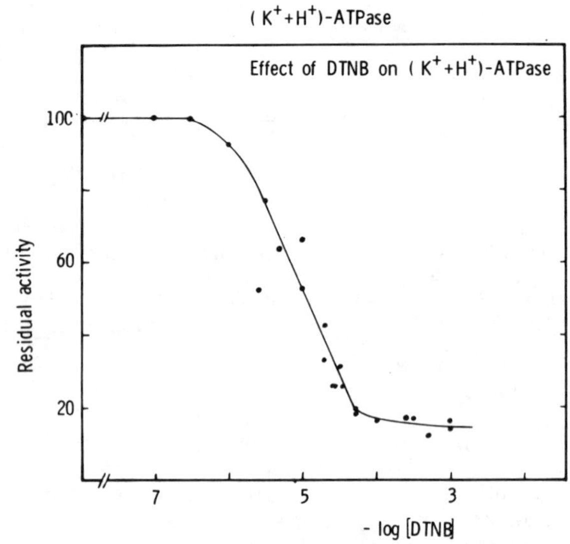

FIGURE 12. Inhibition of K-H ATPase by varying concentrations of DTNB. Gastric vesicle fraction (0.5 mg protein/ml) is incubated in 25 mM Tris-HCl (pH 7.5), 2 mM CDTA plus DTNB for 60 min. at 37° C. K-H ATPase activity is determined as described in the text.

have no effect. K^+ increases the rate of inactivation, which effect is opposite to that of K^+ on inactivation by butanedione in the absence of ATP (FIGURE 10).

ATP also protects against inactivation by DTNB. A dissociation constant $K_d = 34$ μM for the enzyme-ATP complex is calculated by means of the Scrutton & Utter [34] method, which value is in the same order of magnitude as the value $K_d = 74$ μM found in the butanedione experiments. The inactivation rate is maximally inhibited by 90% by ATP, as is also found for butanedione (FIGURE 12).

The role of the "essential" sulfhydryl group is not quite clear. The opposite effect of K^+, as in the case of butanedione, makes it unlikely that the sulfhydryl group is directly involved in the binding of ATP, such as has been concluded for the essential sulfhydryl group in Na-K ATPase.[36] Possibly, the protective effect of ATP is due to its induction of a conformational change in the enzyme, which makes the sulfhydryl group less accessible. Further studies of this problem are obviously required.

CONCLUSIONS

The gradient-purified vesicle fraction prepared from pig gastric mucosa has a high K-H ATPase activity (75–95 μmol Pi/mg protein/h, excl. <10% K-insensitve Mg-ATPase) and a potent proton uptake capacity. It is thus comparable to similar preparations from various species reported by other investigators. The enzyme has a pH-optimum of 7.0, and is stimulated by Tl^+, K^+, Rb^+ and NH_4^+ (in order of decreasing effect), slightly by Cs^+ and not by Na^+, Li^+ and choline. The vesicle membrane has the phospholipid content and composition characteristic of plasma membranes. The phospholipid environment is essential for the activity of the enzyme.

The proton uptake by the vesicles in the presence of K^+ in the medium requires ATP and is accelerated by valinomycin added together with ATP, while ionophore addition after ATP depletion removes the proton gradient. This confirms the presence of the neutral $K^+ — H^+$ exchange transport system previously demonstrated by other investigators. This explains our previous finding of an indirect role of Na-K ATPase in gastric acid secretion,[37] involving the maintenance of a high K^+ content in the parietal cell by this enzyme located mainly on the serosal side of the cell.[38] The K^+ ions must, together with Cl^-, passively cross the tubulovesicular membrane, fused with the microvillar membrane in the secreting cell, in order to permit proton transport across this membrane to the lumen by means of the K-H ATPase located in this membrane.

The presence of an essential arginine group has been demonstrated by treatment with butanedione. Conditions and kinetics of the reaction indicate that a single arginine group is essential for activity of the enzyme, and that this group is more exposed in the presence of Mg^{2+} than in its absence.

Studies of the protective effects of ATP and some other nucleotides indicate that the essential arginine group is involved in binding of ATP, as is also the case for Na-K ATPase. They also confirm the conformational change induced by Mg^{2+}, which leads to exposure of the arginine group, thus favoring binding of ATP, subsequent phosphorylation and release of ADP.

Another conformational change is induced by the monovalent activating cations. In the absence of ATP they protect the arginine group against butanedione ($Tl^+ > K^+ > Rb^+ > NH_4^+$), but in the presence of ATP they show the reverse effect (in the same order of effectiveness). This indicates that the cation-induced conformational change decreases exposure of the arginine group and its binding of ATP, and probably favors dephosphorylation of the enzyme.

The enzyme also possesses an essential sulfhydryl group accessible to DTNB, but its modification never completely inactivates the enzyme. $Mg,^{2+}$ as well as K^+, increase the rate of inactivation, while ATP has a protective effect. These findings suggest that both conformational changes induced by Mg^{2+} and by K^+ increase the exposure of the sulfhydryl group, while ATP has the reverse effect. The sulfhydryl group does not appear to be directly involved in ATP binding.

Summary

Studies in our laboratory have shown that the anion-sensitive Mg-ATPase is located in mitochondria, but not in the plasma membrane of rabbit gastric mucosa, trout gill, rabbit kidney and rat pancreas; whereas in rabbit erythrocyte membrane, it is part of the Ca-Mg activated ATPase system. These findings appear to rule out a function of the anion-sensitive ATPase in the transport of anions and protons across the plasma membrane in these tissues.

On the other hand, the K-activated ATPase in a gradient-purified vesicle fraction of pig gastric mucosa mediates proton uptake in exchange for K^+ in the presence of ATP, in agreement with earlier findings of other investigators. The enzyme requires a phospholipid environment for its activity. Studies of arginine modification with butanedione in the presence or absence of ATP and its analogues, and of activating cations indicate that the enzyme contains an essential arginine group involved in ATP binding; and that K^+ induces a conformational change, which leads to decreased ATP binding and probably coincides with enzyme dephosphorylation. Similar studies of sulfhydryl modification with DTNB indicate that the enzyme contains an essential sulfhydryl group, which does not appear to be directly involved in ATP binding, but rather that ATP binding may induce a conformational change which makes the sulfhydryl group less accessible.

Acknowledgments

These studies were funded in part by the Netherlands Organization for Basic Scientific Research (ZWO) through the Netherlands Biophysics Foundation.

References

1. Durbin, R. P. & D. K. Kasbekar. 1965. Fed. Proc. **24:** 1377–1381.
2. Bundy, H. F. 1977. Comp. Biochem. Physiol. **57 B:** 1–7.
3. Blum, A. L., G. Shah, T. St. Pierre, H. F. Helander, C. P. Sung, V. D. Wiebelhaus & G. Sachs. 1971. Biochim. Biophys. Acta **249:** 101–113.
4. Sachs, G., G. Shah, A. Strych, G. Cline & B. I. Hirschowitz. 1972. Biochim. Biophys. Acta **266:** 625–638.
5. Sachs, G., V. D. Wiebelhaus, A. L. Blum & B. I. Hirschowitz. 1972. *In* Gastric Secretion. G. Sachs, E. Heinz & K. J. Ullrich. Eds., pp. 321–343. Academic Press, New York.
6. Soumarmon, A., M. Lewin, A. M. Cheret & S. Bonfils. 1974. Biochim. Biophys. Acta. **339:** 403–414.
7. Ito, S. 1967. *In* Gastric Secretion, Mechanism and Control. T. K. Shnitka, J. A. L. Gilbert & R. C. Harrison. Eds., :3–24. Pergamon Press. Elmsford, New York.
8. Bonting, S. L., J. M. M. van Amelsvoort & J. J. H. H. M. de Pont. 1978. *In* Gastric Ion Transport. K. J. Öbrink & G. Flemström. Eds., :329–340. Acta Physiol. Scand., Special Suppl.
9. Ganser, A. L. & J. G. Forte. 1973. Biochim. Biophys. Acta **307:** 169–180.
10. Forte, J. G., A. L. Ganser & A. S. Tanisawa. 1974. Ann. N.Y. Acad. Sci. **242:** 255–267.

11. Sachs, G., H. H. Chang, E. Rabon, R. Schackman, M. Lewin & G. Saccomani. 1976. J. Biol. Chem. **251:** 7690–7698.
12. De Pont., J. J. H. H. M., T. Hansen & S. L. Bonting. 1972. Biochim. Biophys. Acta **274:** 189–200.
13. Mitchell, P. & J. Moyle. 1971. Bioenergeties **2:** 1–11.
14. Moyle, J. & P. Mitchell. 1975. FEBS Letters **56:** 55–61.
15. Van Amelsvoort, J. M. M., J. J. H. H. M. de Pont & S. L. Bonting. 1977. Biochim. Biophys. Acta **466:** 283–301.
16. Van Amelsvoort, J. M. M., J. J. H. H. M. De Pont, A. L. H. Stols & S. L. Bonting. 1977. Biochim. Biophys. Acta **471:** 78–91.
17. Van Amelsvoort, J. M. M., P. M. K. B. Van Hoof, J. J. H. H. M. De Pont & S. L. Bonting. 1978. Biochim. Biophys. Acta **507:** 83–93.
18. Van Amelsvoort, J. M. M., J. W. C. M. Jansen, J. J. H. H. M. De Pont & S. L. Bonting. 1978. Biochim. Biophys. Acta **512:** 296–308.
19. Senior, A. E. 1973. Biochim. Biophys. Acta **301:** 249–277.
20. Simon, B., R. Kinne & G. Sachs. 1972. Biochim. Biophys. Acta **282:** 293–300.
21. Kinne-Saffran, E. & R. Kinne. 1974. Proc. Soc. Exp. Biol. Med. **146:** 751–753.
22. Liang, C. T. & B. Sacktor. 1976. Arch. Biochem. Biophys. **176:** 285–297.
23. Humphreys, M. H. & L. Y. N. Chou. 1979. Am. J. Physiol. **236:** E 70–76.
24. Duncan, C. J. 1975. Life Sci. **16:** 955–966.
25. Bonting, S. L. & J. J. H. H. M. De Pont. 1977. *In* Mammalian Cell Membranes. G. A. Jamieson & D. M. Robinson. Eds. **4:** 145–183 Butterworth. London.
26. Ray, T. K. & J. G. Forte. 1976. Biochim. Biophys. Acta **443:** 451–467.
27. Lee, H. C. & J. G. Forte. 1978. Biochim. Biophys. Acta **508:** 339–356.
28. Saccomani, G., H. B. Stewart, D. Shaw, M. Lewing & G. Sachs. 1977. Biochim. Biophys. Acta **465:** 311–330.
29. De Pont, J. J. H. H. M., A. Van Prooyen & S. L. Bonting. 1978. Biochim. Biophys. Acta **508:** 464–477.
30. Ganser, A. L. & J. G. Forte. 1973. Biochem. Biophys. Res. Commun. **54:** 690–696.
31. Scholes, P. & J. Lee. 1978. *In* Gastric Ion Transport. K. J. Öbrink & G. Flemström, Eds. :427–434. Acta Physiol. Scand., Special Suppl.
32. De Pont, J. J. H. H. M., B. M. Schoot, A. Van Prooyen & S. L. Bonting. 1977. Biochim. Biophys. Acta **482:** 213–227.
33. Levy, H. M., P. D. Leber & E. M. Ryan. 1963. J. Biol. Chem. **238:** 3654–3659.
34. Scrutton, M. C. & M. F. Utter. 1965. J. Biol. Chem. **240:** 3714–3723.
35. Schoot, B. M., J. J. H. H. M. De Pont & S. L. Bonting. 1978. Biochim. Biophys. Acta **522:** 602–613.
36. Patzelt-Wenczler, R., H. Pauls, E. Erdmann & W. Schoner. 1975. Eur. J. Biochem. **53:** 301–311.
37. Hansen, T., S. L. Bonting, J. F. G. Slegers & J. J. H. H. M. De Pont. 1972. Pflügers Arch. **334:** 141–153.
38. Hansen, T., J. F. G. Slegers & S. L. Bonting. 1975. Biochim. Biophys. Acta **382:** 590–608.
39. Lee, H. C., H. Breitbart, M. Berman & J. G. Forte. 1979. Biochim. Biophys. Acta **553:** 107–131.
40. Bonting, S. L. *et al.* Unpublished results.

Discussion

Unidentified speaker: Similar experiments have been done on DTNB inhibition of Na-K ATPase. There different ligands alter the effect of DTNB.

Did you get different end effects with the various ligands for which you showed that some increased and some decreased the modification by DTNB?

S. L. BONTING: We have so far looked at the effects on the rate of inactivation rather than on the final level of modification. Further work is in progress.

UNIDENTIFIED SPEAKER: What kind of animal did you use to isolate your vesicles?

BONTING: Pig.

J. G. FORTE: Your proposed model has potassium with chloride presumably leaking into the lumen. In my model, I have proposed a tight coupling because it is difficult to explain the fact that potassium added to the luminal side is not as good a stimulant to acid secretion as potassium added to the serosal side in intact mucosa. Moreover, your proposed recycle, with potassium getting out into the lumen, and there serving as an activator of the ATPase, is not compatible with the *in vivo* data of gastric secretion, which indicate that the potassium must be coupled tightly.

BONTING: Our model is not particularly authoritative on this point, because we have no evidence from our studies on how the intact system functions. In your model however, it remains possible that potassium can stay within the membrane, which goes against my intuitive feelings. On the other hand, with all of the canalicular tubular spaces, there could be an accumulation of potassium ions close to the pump sites and you would still obtain a close coupling. However, I have no hard data either supporting or denying what you said.

G. SACHS: If you compare your site-active agents in transporting and nontransporting modes of the pump in vesicles or broken membranes, do you see any difference? If you have a closed vesicular system where you know there is an asymmetry and proton transport going on, can you distinguish a DTNB or butanedione effect there from that in broken membranes, in which no gradients are being generated?

BONTING: Our modification experiments have been done so far without any relation to transport events. It is simply a vesicular preparation which has been treated by the modifying agents. All that I can say is that the vesicles, before they were mistreated by us, showed fairly normal transport characteristics; we haven't yet done parallel studies on the transport after the modification.

SACHS: I was rather asking the question whether transport modified the modification.

BONTING: I cannot answer that question either at this stage.

Part V. Biochemical Aspects of Anion and Proton Transport Mechanisms in the Erythrocyte

ELECTRICAL RESISTANCE OF THE RED CELL MEMBRANE AND THE RELATION BETWEEN NET ANION TRANSPORT AND THE ANION EXCHANGE MECHANISM *

J. F. Hoffman, J. H. Kaplan,† T. J. Callahan ‡ and J. C. Freedman §

Department of Physiology
Yale University School of Medicine
New Haven, Connecticut 06510

This paper concerns the transport pathway through which net movement of anions occurs across the red cell membrane. This pathway is also referred to as the conductive or current carrying pathway to distinguish it from the facilitated mechanism that mediates the self-exchange of anions such as Cl^- or $SO^=_4$ and the physiologically important heteroexchange, the Cl^--HCO^-_3 shift. (For a recent comprehensive review of this extensive literature see, in addition to the papers in this volume, Knauf.[1]) Anion flow through the exchange pathway is rapid and known to occur electroneutrally, with an exchange stoichiometry of 1:1. Anion flow through the net pathway is slower than transfer by exchange and is sensitive to and a determinant of the membrane potential. It is not yet known whether the net and exchange pathways represent separate mechanisms or are two aspects of the same system. Net transfer could occur by "slippage" in the exchange pathway, that is when the 1:1 stoichiometry breaks down and an unloaded carrier crosses the membrane. Evidence supporting a common mechanism is based on the fact that certain inhibitors, such as DIDS, inhibit anion transport through both pathways to about the same extent; but the case is not conclusive since it is difficult to rationalize the effects of other inhibitors (e.g., pronase and phlorizin) and certain detailed differences in the kinetic characteristics (selectivity, pH, temperature) with the transport of anions taking place through a common membrane pathway.[2] On the other hand, measurements of membrane resistance (the reciprocal of conductance) may provide a way of sorting out the relationship between the two pathways by seeing to what extent the membrane resistance is affected by agents which alter anion exchange.

The electrical resistance of a membrane, R_m, is of interest because it provides a measure of the permeability of that membrane to ions. It was known early on from the studies of Höber and Fricke [3] on the properties of direct and alternating current flow through red cell suspensions, that the resistance of red cell membranes is high compared to the red cell's interior; but, as pointed out by Cole,[4] methodological limitations could only set a minimum value for

* Supported by USPHS National Institutes of Health Grants AM–17433 & HL–09906.

† Current address: Department of Physiology and Biophysics, University of Iowa, Iowa City, Iowa 52242.

‡ Current address: F.D.A., Bureau of Medical Devices, Silver Spring, Maryland.

§ Current address: Department of Physiology, Upstate Medical Center, Syracuse, New York 13210.

0077–8923/80/0341–0357 $01.75/0 © 1980, NYAS

R_m, about 3 ohm cm^2. More recent estimates, summarized below, place the actual value of R_m of red cells several orders of magnitude higher (10^3 to 10^6 ohm cm^2) depending upon the method used and the type of red cell studied.[5]

Direct measurement of the membrane potential and the membrane resistance have been made on the red cells of the salamander, *Amphiuma,* because these cells are large and can be impaled with conventional microelectrodes.[5] Thus, estimates obtained by means of diphasic current pulses passed through the microelectrode indicated that R_m was no less than 1000–2000 ohm cm^2, and probably higher because of current leakage that occurred around the microelectrode during the measurement. Rathlev and Lassen[5] suspended single *Amphiuma* red cells between two micropipettes and after dipping the cell into oil measured the transverse resistance. By this technique they found that the maximum value of R_m averaged 2.3×10^5 ohm cm^2.

It is also possible to estimate the membrane resistance indirectly by measuring in different ways the red cell's permeability to ions such as K^+, Cl^- or $SO_4^=$. Permeabilities, such as P_K, P_{Cl} or P_{SO_4}, and ionic conductances are usually determined from the constant field equation,[6, 7] with the membrane potential either measured directly or approximated in some indirect fashion (e.g., flux ratio equation, fluorescent probes). To a first approximation, and certainly for the purposes of this paper, we can assume that the measurement of a membrane's conductance to Cl^- (or its reciprocal, R_{Cl}) can be taken as a measure of that membrane's total conductance (or R_m). Thus estimates of R_{Cl} in *Amphiuma* red cells range[5, 8] from 10^3–10^5 ohm cm^2 to 2.6×10^6 ohm cm^2. R_{Cl} in sheep red cells has been reported[9] to be about 10^6 ohm cm^2. There is general agreement in the estimates of P_{Cl} in human red cells determined by a variety of means[2, 5, 10] and the values of R_{Cl} computed from these various estimates range between 10^5 and 10^6 ohm cm^2. Since P_{Cl} is about 4 times larger than P_{SO_4},[2] and since under the conditions of these experiments conductance is approximately equivalent to permeability times the valence squared, the ratio of R_{SO_4} to R_{Cl} should follow in proportion to the ratio of the concentration of Cl^- to $SO_4^=$. Thus, this discussion serves to emphasize that not only is the R_m of the red cell high but that because of its magnitude another method can be devised to help evaluate the relation between the net and exchange modes of transport for anions.

This other method is based on the fact that the coupled Na:K pump, which moves Na out and K in, is electrogenic in red cells.[11] Thus, where Post and Jolly[12] found that the stoichiometry of net transport by the pump was 3 Na^+_{in} for 2 K^+_{out}, it required an electrical measurement to show that this exchange was accompanied by a net transfer of charge. Electrogenicity of the Na:K pump was demonstrated in *Amphiuma,* human and other types of red cells by utilizing the fluorescent dye technique that had been previously shown to provide an accurate measure of membrane potentials (E_m) in red cells.[13] R_m could therefore be estimated (with an accuracy of $\pm 20\%$) from the net current flow through the pump (assumed to be one-third the ouabain-sensitive Na efflux) together with the pump-dependent change in E_m. The validity of this approach was substantiated by the finding that R_m calculated from the DIDS-resistant or insensitive efflux of $SO_4^=$ (where the total flux of $SO_4^=$ was inhibited by more than 99% by DIDS), parallelled estimates of R_m obtained by using Ohm's Law. Thus, R_m for different species of red cells was found to be about 1×10^6 ohm cm^2 for those from man, 3×10^5 ohm cm^2 for sheep, 5×10^5 ohm cm^2 for duck and 3×10^5 ohm cm^2 for *Amphiuma*. These values apply to red cells that had been treated so that Na_{in} was increased (to increase the pumping rate and therefore the net current flow), that had $SO_4^=$ substituted for Cl^- inside and outside (to decrease the membrane's permeability to anions since, as mentioned before, P_{SO_4} is about four times smaller than P_{Cl}), and that had been exposed to DIDS (in order to maximally inhibit P_{SO_4}). While these changes in the experimental conditions would appear to optimize the measurement of the electrogenic component, they could result in an over-estimation of R_m in the various types of red cells. Therefore, R_m was evaluated in human red cells, in which, except for containing high Na_{in}, the conditions were made as physiological as possible. It was found that R_m in $SO_4^=$-loaded and DIDS-treated cells was 1.1×10^6 ohm cm^2 and fell to 4.3×10^5 ohm cm^2 in the absence of DIDS; for Cl^--containing cells suspended in a Cl^- medium, R_m was found to be 6.3×10^5 ohm cm^2 when DIDS-treated but 3.6×10^5 ohm cm^2 without DIDS.

The significance of this latter result, namely, that R_m in the normal cell (Cl^-, no DIDS) is only one-third the R_m in the $SO_4^=$-loaded, DIDS-inhibited cell, lies in its implication regarding the relation between the anion pathway used for net flow and the anion exchange mechanism. This result indicates that the transport pathway through which net movement of anions can occur can be separated from the exchange transport pathway because of the marked difference in the sensitivity of the two types of anion flow to DIDS. This view is supported by the recent results of Knauf and Law [14] who found that the net movement of anions was only partially inhibited by DIDS, the DIDS-insensitive component being about one-third the total net flow; and by Kaplan, *et al.*[15] who found that the DIDS-insensitive outward rate constant of K (and, therefore, presumably P_{Cl}) from valinomycin-exposed red cells also comprised about one-third the total K efflux. It would appear then that there are two ways that net flow of anions can take place across the red cell membrane: one way is associated with the anion exchange mechanism (the slippage pathway), the other way is evidently an independent pathway.

References

1. Knauf, P. A. 1979. Erythrocyte anion exchange and the Band 3 protein; transport kinetics and molecular structure. *In* Current Topics in Membranes and Transport. F. Bronner & A. Kleinzeller, Eds. **12:** 249–363. Academic Press. New York, New York.
2. Knauf, P. A., G. F. Fuhrmann, S. Rothstein & A. Rothstein. 1977. The relationship between anion exchange and net anion flow across the human red cell membrane. J. Gen. Physiol. **69:** 363–386.
3. Höber, R. 1945. Physical chemistry of cells and tissues. The Blakiston Co. Philadelphia, Pa.
4. Cole, K. S. 1940. Permeability and impermeability of cell membranes for ions. *In* Cold Spring Harbor Symposia on Quantitative Biology. **8:** 110–121.
5. Lassen, U. V. 1977. Electrical potential and conductance of the red cell membrane. *In* Membrane Transport in Red Cells. J. C. Ellory & V. L. Lew, Eds. :137–172. Academic Press. New York, New York.
6. Goldman, D. E. 1943. Potential, Impedance and Rectification in Membranes. J. Gen. Physiol. **27:** 36–60.

7. HODGKIN, A. L. & B. KATZ. 1949. The effect of sodium ions on the electrical activity of the giant axon of the squid. J. Physiol. **108:** 37–77.
8. HOFFMAN, J. F. & P. A. KNAUF. 1973. The mechanism of the increased K transport induced by Ca in human red blood cells. *In* 2nd International Symposium on Metabolism and Membrane Permeability of Erythrocytes, Thrombocytes and Leukocytes. E. Gerlach, K. Moser, E. Deutsch & W. Wilmanns, Eds. :66–70. George Thieme. Vienna.
9. TOSTESON, D. C., R. B. GUNN & J. O. WIETH. 1973. Chloride and hydroxyl ion conductance of sheep red cell membrane. *In* 2nd International Symposium on Metabolism and Membrane Permeability of Erythrocytes, Thrombocytes and Leukocytes. E. Gerlach, K. Moser, E. Deutsch & W. Wilmanns, Eds. : 62–66. George Thieme. Vienna.
10. SACHS, J. R., P. A. KNAUF & P. B. DUNHAM. 1975. Transport through red cell membranes. *In* The Red Blood Cell. D. MacN. Surgenor, Ed. **2:** 613–703. 2nd Edit. Academic Press. New York, New York.
11. HOFFMAN, J. F., J. H. KAPLAN & T. J. CALLAHAN. 1979. The Na:K pump in red cells is electrogenic. Fed. Proc. **38:** 2440–2441.
12. POST, R. L. & P. C. JOLLY. 1957. The linkage of sodium, potassium and ammonium active transport across the human erythrocyte membrane. Biochim. Biophys. Acta **25:** 108–128.
13. HOFFMAN, J. F. & P. C. LARIS. 1974. Determination of membrane potentials in human and *Amphiuma* red blood cells by means of a fluorescent probe. J. Physiol. (London) **239:** 519–552.
14. KNAUF, P. A. & F. Y. LAW. 1980. Relationship of net anion flow to the anion exchange system. *In* Membrane Transport in Erythrocytes, U. V. Lassen, J. O. Wieth & H. H. Ussing, Eds. Alfred Benzon Symposium **14**. Munksgaard, Copenhagen. In press.
15. KAPLAN, J. H., M. PRING & H. PASSOW. 1980. Concentration dependence of chloride movements that contribute to the conductance of the red cell membrane. *In* Membrane Transport in Erythrocytes, U. V. Lassen, J. O. Wieth & H. H. Ussing, Eds. Alfred Benzon Symposium **14**. Munksgaard, Copenhagen. In press.

ANION TRANSPORT ACROSS THE RED BLOOD CELL MEMBRANE AND THE CONFORMATION OF THE PROTEIN IN BAND 3

H. Passow, H. Fasold,* E. M. Gärtner, B. Legrum, W. Ruffing and L. Zaki

Department of Cell Physiology
Max-Planck-Institut für Biophysik and
** Biochemisches Institut der Universität*
D-6000 Frankfurt am Main 71, West Germany

Anion Transport and the Protein in Band 3

For several decades, anion transport across the red blood cell membrane was believed to proceed via aqueous channels that are lined with positive fixed charges. These charges were assumed to repel cations, to accumulate anions, and thus to account for the well-known anion-cation discrimination of the membrane.[1] An extensive reinvestigation of the kinetics of anion transport in our laboratory confirmed many of the predictions of the fixed-charges hypothesis, including the very general prediction of competition between Cl and SO_4.[2] Other predictions of the hypothesis were disproved. In particular, it was demonstrated that penetration through pores could not be the rate-limiting step for anion translocation through the membrane.[3] Pursuing this work, we and others observed that a number of covalently binding agents that had, as a common property, the capacity of reacting with amino groups were able to produce inhibition of anion transport.[3–5, 46] Since certain proteolytic enzymes also exerted a strongly inhibitory effect, it was suggested that amino groups of membrane protein were involved in anion transport.[6, 46] These suggestions formed the basis for subsequent attempts to identify a specific transport protein. First pieces of evidence concerning the nature of this protein were obtained by the use of the amino reagents, H_2DIDS (4,4′-dihydroisothiocyano-2,2′-stilbene disulfonate), SITS (4-acetamido-4′-isothiocyano stilbene-2,2′-disulfonic acid), N_2ph-F, (1-fluoro-2,4-dinitrobenzene) and a noncovalently binding analogue of H_2DIDS called DAS (4,4′-diacetamido-2,2′-stilbene disulfonate). It could be shown that the capacity of these agents to inhibit anion transport was related to their ability to bind to a glycoprotein of a molecular weight of 96,000 daltons that migrates in Band 3 of SDS polyacrylamide gel electropherograms of the red cell membrane.[7–10] The evidence was, however, not quite conclusive since it remained open whether or not small quantities of some other membrane protein inside or outside Band 3 had escaped detection, and that modification of such protein was responsible for the observed inhibition.

Attempts were made, therefore, to study anion transport in vesicles containing the purified Band 3 protein. *In situ*-purification was first attempted by incorporation of trypsin into resealed ghosts[8, 11] and later by treatment of ghosts with a detergent.[12, 13] Trypsin treatment resulted in a high yield of resealable vesicles that contained primarily the 55 K fragment (a doublet) of the Band 3 protein and only one additional major band, which, however, was

0077–8923/80/0341–0361 $01.75/0 © 1980, NYAS

incapable of SITS binding. Anion transport in these vesicles was close to normal and responded to a number of representative inhibitors as in ghosts that had not been treated with trypsin. The detergent-treated membranes yielded a fairly pure preparation of Band 3-containing vesicles. An unknown fraction of the vesicles in these preparations could be made to reseal. The anion transport in these resealed vesicles was rather similar to the transport in intact red cells. However, it remained open whether or not the resealing was confined to a small fraction of vesicles that had not been as successfully purified as the bulk. Incorporation of partially [14, 39, 40] or more completely [16] purified Band 3 protein into liposomes leads to an increase in anion transport. However, an increased flux is not a sufficient criterion for a successful reconstitution of the specific anion transport system as indicated, for example, by the finding that incorporation of lysozyme into lipid vesicles also enhances anion transport drastically.[31] In liposomes containing more [16] or less [39, 40] purified Band 3 preparations, phloretin,[39] certain local anesthetics,[40] and APMB [16] were able to produce inhibition, as in intact red cells, while certain other inhibitors failed to be effective.[16] It remained unclear, therefore, whether the observed increase in transport represented a partial restoration of the normal function of the transport protein or an artifact.

Although the described attempts to achieve *in situ* purification or reconstitution yielded interesting results, they failed to provide conclusive evidence for a participation of Band 3 in anion transport.* For this reason, the assumption that the integral Band 3 protein constitutes the anion transporter is still largely based on circumstantial evidence. However, this evidence comprises a large variety of diverse observations that are most conveniently and consistently explained on the assumption that Band 3 constitutes the anion transporter. Three examples (for a review of other examples, see Cabantchik *et al.*[15]) may serve to illustrate this type of evidence. Firstly, there exists a linear relationship between inhibition of anion transport and the number of Band 3 molecules that are modified by H_2DIDS, DIDS, SITS, or DAS. At complete inhibition of transport, the number of modified molecules is equal to the number of Band 3 molecules in the membrane.[8, 9, 17–19] Secondly, the pK value of the sites for covalent binding of H_2DIDS to the protein in Band 3 is virtually identical with the pK value of the sites involved in irreversible inhibition of anion transport by this agent.[20, 21] Finally, the anion transport inhibitor APMB [2-(4′-aminophenyl-6-methylbenzenethiazol-3′,7-disulfonate] reduces the rate of anion transport and the availability of the sites for the binding of externally applied, nonpenetrating H_2DIDS in a parallel manner, regardless of whether the inhibitor is applied to the inner or outer membrane surface. Such unusual behavior is unlikely to be accidental and suggests a carrier-like role of the protein in Band 3.[22] On the basis of such evidence, it seems reasonable to base further work on the assumption that Band 3 does play an essential role in anion transport. It should be noted, however, that the presently available evidence does not preclude the possibility that Band 3 cooperates with other, as yet unidentified, membrane proteins [9, 23, 24] and that the failure to achieve satisfactory *in situ* purification or reconstitution may not be simply related to deficiencies in the techniques employed, but may indicate the absence of such auxiliary protein(s).

* See, however, the paper of Cabantchik *et al.* in this volume.

Anion Transport and the Conformation of the Anion Transport Protein

The problem at hand may be characterized as follows:[21, 25] Anion transport in red cells does not contribute to the conductance of the membrane.[37, 38] This implies that the transport protein does not simply form an aqueous pore. It also does not act like a diffusible carrier since the large molecule that traverses the lipid bilayer is firmly attached by its carbohydrate moiety to the other membrane surface, and by a hydrophilic segment of the peptide chain to the inner surface. Thus, anion translocation across the membrane can neither be brought about by translational nor by rotational diffusion of the transport protein.[36] Instead, the transport process is likely to constitute a sequence of conformational changes that are associated with anion binding, translocation and release at the other membrane surface.

The peptide chain of the Band 3 molecule is large enough to traverse the membrane several times. One may suspect, therefore, that only a small fraction of the molecule would need to undergo these changes, while the rest of the molecule would primarily serve to maintain a stable stationary environment in which the transport related conformational changes take place. This view is supported by the observation that the removal of the hydrophilic segment of the peptide chain mentioned above has little, if any, effect on anion transport.[11, 26] Nevertheless, even though the transport-related region of the transport molecule may seem to be small as compared to the transport molecule as a whole, it is likely that the transport-related conformational changes involve many sites and allosteric interactions between many of these sites.[21]

From the study of the kinetics of anion transport, it has been inferred that there exist several anion transport-related sites on the transport protein: a transfer site that binds the substrate anions, and one or several modifier sites, which, after combination with another substrate anion or other chemical agents, reduce or enhance the rate of transport. Each of these sites can be assumed to represent a number of amino acid residues that cooperate in the binding of the substrates or modifiers, and each of these sites can be expected to be allosterically linked to other portions of the protein molecule, including one or several of the sites that can be identified as binding sites for modifying agents.

Conformation Probing by Measuring the Rate of Dinitrophenylation of a Specific Lysine Residue at the H_2DIDS Binding Site of the Band 3 Molecule

The present paper deals with the conformation of the Band 3 protein near a specific lysine residue that resides in the outward facing portion of the Band 3 molecule and that can be identified by its capacity to react covalently with the anion transport inhibitor H_2DIDS. This site will be called *a*. For the present purpose, it is important that *a* can be easily and fairly selectively dinitrophenylated. This is due to the fact that *a* is more accessible and has a lower pK value than most other ϵ-amino groups on the Band 3 protein.[21] Accessibility and reactivity of *a* depend on the microenvironment in which the reaction takes place, and, hence, are functions of protein conformation near the selected amino acid residue. Thus, measuring the rate of dinitrophenylation of *a* yields information about the conformation of the peptide chain at the H_2DIDS-binding site. Dinitrophenylation leads to an irreversible modification of the "reporter group" *a*. However, the information derived from the determination of the rate of dinitrophenylation pertains to the conformation that existed before that group is irreversibly modified. Since the structural factors that determine the absolute rates of dinitrophenylation of site *a* are unknown, it is only possible to use the rate of dinitrophenylation as an indicator for the discrimination between different conformations. In spite of the exceptionally high reactivity, the rate of dinitrophenylation of *a* is too low to allow one to follow the time course of transport-related conformational changes. We are restricted, therefore, to comparisons of conformations in different steady states.

The rate of dinitrophenylation of *a* can be easily followed by measuring the reduction of the capacity of Band 3 to bind $^3H_2DIDS$ covalently after exposure of the red cell membrane to N_2ph-F for a defined length of time. Since each Band 3 molecule carries only one H_2DIDS binding site, we can equate the decrease of the number of sites that are available for *covalent* H_2DIDS binding with the number of lysine residues *a* that have been blocked by dinitrophenylation.

For an understanding of the data in figures and tables, it is essential to keep the following details of the experimental procedure in mind: Red cell ghosts are exposed to N_2ph-F in media of varying Cl concentration, or at a fixed Cl concentration of 122.5 mM, in the presence or absence of reversibly binding modifiers of anion transport. After termination of the reaction by removal of the unreacted N_2ph-F, they are reequilibrated in media that have the same composition as the media in which dinitrophenylation had been performed, except that neither N_2ph-F nor the other modifying agents are present. Each patch of ghosts is then subdivided into two portions. One is used for measuring anion flux, the other is exposed to a large excess of $^3H_2DIDS$ for a fixed length of time under conditions in which reaction with lysine residue *b* can be neglected. The membranes of the $^3H_2DIDS$-treated ghosts are subsequently isolated, dissolved in SDS, subjected to polyacrylamide gel electrophoresis, and the capacity of Band 3 for covalent $^3H_2DIDS$ binding is determined. Thus, regardless of the conditions under which dinitrophenylation took place, the effects on transport and the modification of lysine residue *a* are determined under identical conditions. It should be noted that a high degree of modification of lysine residue *a* by N_2ph-F is reflected by a low capacity for $^3H_2DIDS$ binding and vice versa.

The following pages are devoted to a description of the effects of a substrate anion, Cl, and a range of reversibly acting inhibitors of anion transport on the conformation near lys (lysine) *a* as determined by the described method of conformation probing. The material to be presented will be subdivided into four sections: 1. The relationship between the reporter group *a* and anion transport; 2. The effects of reversibly binding inhibitors of anion transport on the conformation near *a*; 3. Transmembrane effects, and 4. Conclusions.

Lysine Residue a *and Anion Transport*

The lysine residue *a* is a constituent of the H_2DIDS binding site of the Band 3 molecule (FIGURE 1). This site constitutes a superficially located niche that is formed by segments of the peptide chain in contact with the external medium. It is composed of amino acid residues that convey to it a partially hydrophobic quality, and the capacity to exchange electrons with the benzene rings of the H_2DIDS molecule.[27] The amino acid residues include two lysine residues called *a* and *b*. They are located on the 60K and 35K segments, respectively, of the peptide chain at a distance that can be bridged by an H_2DIDS molecule (about 20 Å). Thiourea bond formation between the two isothiocyanate groups of an H_2DIDS molecule and the two lysine residues leads to intramolecular crosslinking.[25] The rate of reaction with *a* exceeds that with *b*, indicating that *a* is more accessible or reactive than *b*.[34]

a and *b* form part of a cluster of at least five basic amino acid residues.[34] Due to the high charge density in the immediate vicinity, their pK values are much lower (8.5 and 8.8, respectively,[21, 34] than the pK of normally dissociating ϵ-amino groups (about 10.5). As has already been pointed out, the low pK of *a* is partially responsible for the high reactivity of this group and for the selectivity of the reaction with amino reagents that, in principle at least,

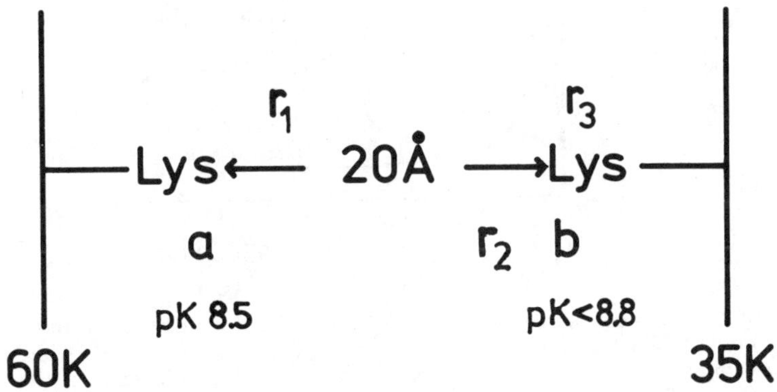

FIGURE 1. The H_2DIDS binding site. r_1, r_2, r_3 represent positively charged amino acid residues. r_1 and r_2 are supposed to be involved in noncovalent H_2DIDS binding. (From Passow *et al.*[34])

should also be capable of reacting with the many other lysine residues on the Band 3 molecule.

Exposure to N_2ph-F reduces the capacity for covalent H_2DIDS binding to *a* to a much larger extent than for the noncovalent binding that precedes the formation of the covalent bond (FIGURE 2). This indicates that *a* is more susceptible to dinitrophenylation than the other amino acid residues at the H_2DIDS-binding site and that *a* is not directly involved in noncovalent H_2DIDS binding.[20]

Noncovalent binding of H_2DIDS to Band 3 is associated with an inhibition of anion transport. Cl competes with H_2DIDS [28] (or the H_2DIDS analogue DAS, unpublished observations from this laboratory). From the competition studies, it was possible to calculate an apparent dissociation constant for Cl-binding to the common binding site with H_2DIDS. This constant was close to the apparent K_m value that was derived from the saturation kinetics of Cl transport. Assuming that the apparent constants derived from the two types of measurements represent affinities, it was suggested that Cl and H_2DIDS

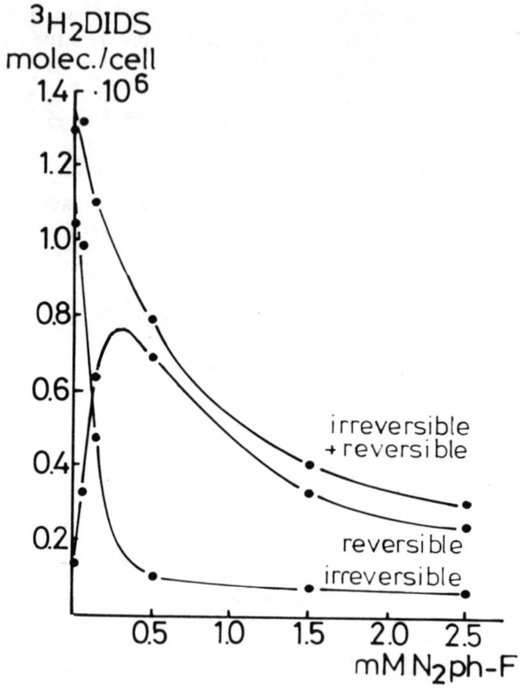

FIGURE 2. Effect of dinitrophenylation of the red cell membrane on reversible and irreversible H_2DIDS binding. Prior to the exposure to $^3H_2DIDS$ (3 μM), the red cells had been dinitrophenylated at the N_2ph-F concentrations indicated on the abscissa for 30 min at 37° C, pH 7.4. After exposure to $^3H_2DIDS$ (37° C, pH 7.4) for 5 min, the cells were separated, and covalent and noncovalent H_2DIDS binding was determined from the decrease of $^3H_2DIDS$ in the supernatant and from the amount of $^3H_2DIDS$ that remained attached to the cells after washes with BSA and isolation of the membranes. Note: the reduction of covalent binding ("irreversible") is partially compensated by an increase of noncovalent binding ("reversible"). This means that the covalent bond formation with lys *a* is reduced while noncovalent binding is still possible. (From Passow.[20])

compete for the transfer site.[28] In terms of the structure of the H_2DIDS binding site suggested in FIGURE 1, this would imply that several or all of the amino acid residues r_1–r_3 that are adjacent to lys *a* participate in complex formation with the substrate anions and the subsequent translocation across the membrane.

Although lys *a* does not seem to be directly involved in transport, it interacts with the side chains that form the transfer site. This follows from the finding that dinitrophenylation of *a* is always associated with a proportional decrease of the rate of anion transport. This is observed regardless of the conditions (pH, ion composition of the medium) under which dinitrophenylation is performed. It indicates a 1:1 relationship between the modification of *a* and the inhibition of transport. Conversely, increasing the Cl concentration in red cell ghosts enhances the rate of dinitrophenylation of *a* (as indicated by a decrease of the capacity for covalent H_2DIDS binding to Band 3) and the concomitant irreversible inhibition of anion transport (FIGURE 3). This finding is a very strong piece of evidence for the participation of Band 3 in anion transport. It supports the previous inference that lysine *a* is not involved in the complex formation with substrate anions: If *a* were a constituent of the complex between transfer site and substrate anion, complex formation with Cl would render *a* less available for reaction with N_2pH-F. This is not observed. Instead we find the described enhancement of the rate of dinitrophenylation.

It should be noted that the curves in FIGURE 3 cannot be used directly for the determination of the Cl concentration at which acceleration becomes half maximal. This is due to the fact that dinitrophenylation is an irreversible process. The curve represents, therefore, only a certain stage of the reaction on the way to the maximal degree of modification that can be brought about by the amounts of N_2ph-F that are present in the various samples. These

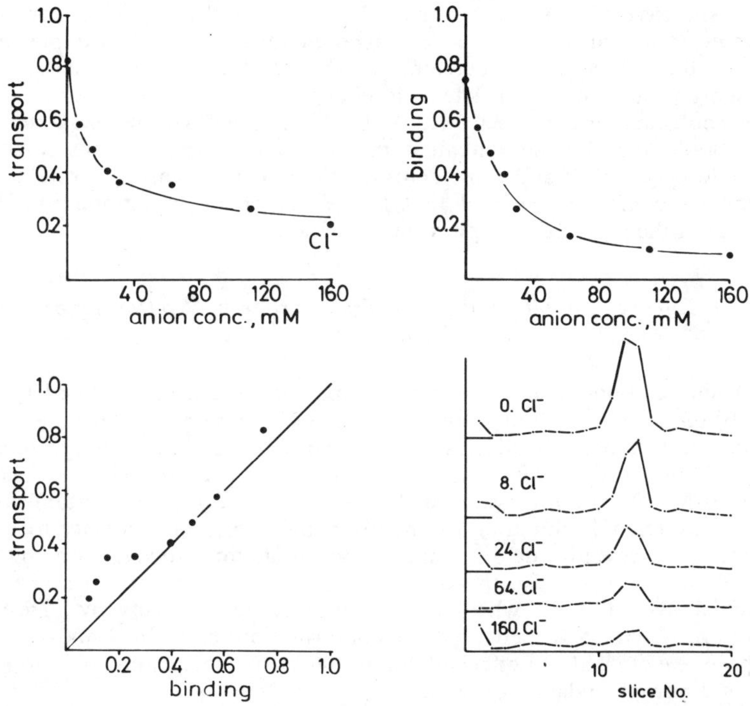

FIGURE 3. Effect of Cl on inhibition of sulfate equilibrium exchange by N_2ph-F (upper left), and on dinitrophenylation of H_2DIDS binding sites on Band 3 (lys *a*) (upper and lower right). The peaks at the lower right represent H_2DIDS binding to Band 3 on the SDS polyacrylamide gel electropherograms on which the curve in the upper right is based. Lower left: transport vs. H_2DIDS binding sites (lys *a*) that did not react with N_2ph-F. Transport and binding are expressed as fractions of parallel runs at each Cl concentration with ghosts that had not been dinitrophenylated. We assured ourselves that for each Cl concentration the volumes of controls and dinitrophenylated ghosts were the same. The Cl^- concentrations indicated at the lower right are in mM.

amounts are always equal and sufficient to produce virtually complete inhibition at infinite reaction time, regardless of the Cl concentration. However, by a mathematical treatment of the data that takes these considerations into account (unpublished), it is possible to show that half maximal acceleration is achieved at a Cl concentration of about 45 mM, i.e., at a concentration that is close to the $K_{1/2}$ value for Cl transport. If one assumes that $K_{1/2}$ has the meaning of a Michaelis-constant, and thus represents the affinity of Cl to the transfer site, one can conclude that the conformational change that manifests itself by the acceleration of the rate of dinitrophenylation of *a* is due to Cl-binding to the transfer site.

Cl-binding is associated with a decrease of the positive charge of the transport protein. This should increase the pK value of the lysine residue *a* and thus reduce its susceptibility for dinitrophenylation. The observed enhancement of the rate of dinitrophenylation does not agree with this expectation and has to be attributed, therefore, largely to a conformational change that leads to an increase of the accessibility of site *a* and that possibly consists of a movement of that site from a less exposed to a more exposed position.

For the subsequent discussion, we shall call those conformations of the transport protein in which site *a* is either more or less exposed the *cis* and *trans* conformations, respectively. With this terminology, we may reformulate our observations by the statement that Cl binding to the transfer site shifts the ratio between *cis* and *trans* conformations of the transport protein in favor of the *cis* conformation, and that this shift can be monitored by following changes of the rate of dinitrophenylation of *a*.

Effect of Inhibitors of Anion Transport on the Conformation at Lysine Residue a

Anion transport across the red cell membrane can be inhibited by a large variety of reversibly binding chemical agents. These agents include molecules of the most diverse chemical structure and show, as the only common property, a high degree of amphiphilicity (for review, see Deuticke).[35] The recent work of Motais and his associates has demonstrated that for certain groups of chemically related inhibitors, the hydrophobicity and the capacity to exchange electrons are essential determinants of the inhibitory power.[43] In the present section, we wish to raise the question whether the inhibitors can be roughly classified according to whether they produce their effects by arresting the transport system in the *cis* or *trans* conformation, as defined above. For this purpose, we studied the effect of the presence of the various inhibitors on the rate of dinitrophenylation of lys *a*.

Reversibly binding inhibitors of anion transport could affect the rate of dinitrophenylation of a reporter group by direct combination with that site, or by combination with a site that is allosterically linked to the reporter group. In the first case, the reporter group would be occupied by the reversibly acting inhibitor part of the time period that is allotted for reaction with N_2ph-F. This would lead to a reduction of the rate of dinitrophenylation, as compared to a control that is dinitrophenylated in the absence of an additional inhibitor. In the second case, conformational changes could reduce or increase the rate of dinitrophenylation of the reporter group. Thus, measuring the change of the rate of dinitrophenylation of the reporter group represents a simple and generally applicable means of demonstrating that the effects of a noncovalently-binding inhibitor of anion transport are associated with a modification of the protein in Band 3. In addition, in cases in which direct binding to the reporter group can be excluded, information can be obtained about allosteric interactions between the binding site of the inhibitor and the site at which the reporter group is located.

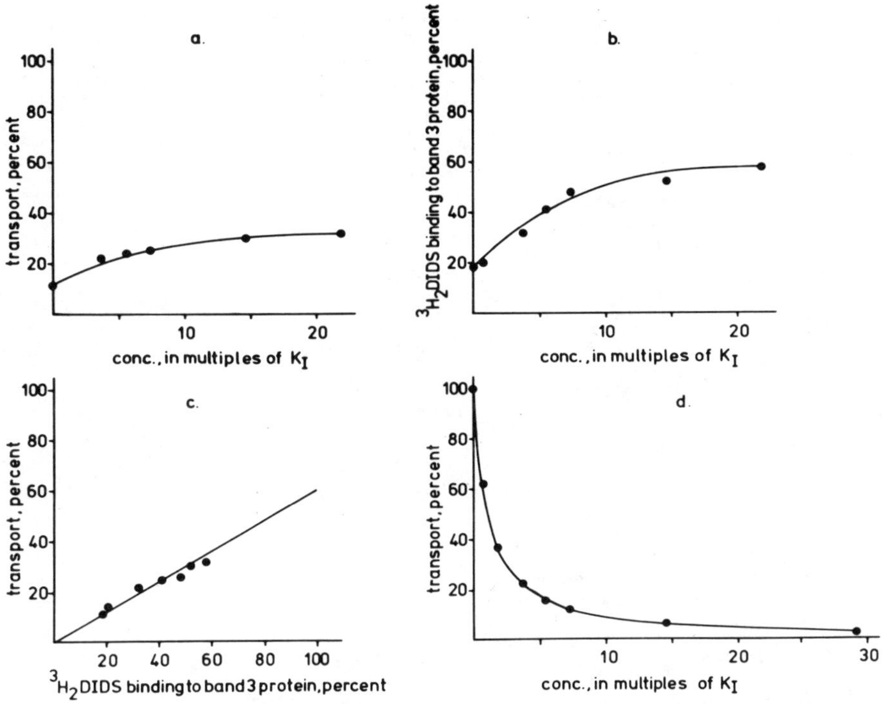

FIGURE 4. (a)–(d): Effect of persantin on dinitrophenylation of H_2DIDS binding sites on Band 3 (lys *a*) and inhibition of anion transport. Transport and H_2DIDS binding are expressed as percents of corresponding values in red cells that had not been exposed to N_2ph-F. (d) Effect of persantin on SO_4 equilibrium exchange, as measured in red cells that had not been dinitrophenylated. All concentrations are expressed in multiples of the K_I value given in TABLE 1.

The basic idea of the experimental arrangement is illustrated in FIGURE 4. FIGURE 4d shows that increasing concentrations of persantin produce inhibition of sulfate equilibrium exchange with a K_I value of 1.4 μM (TABLE 1, compound No. 9). To test the effect of persantin on the rate of dinitrophenylation of *a*, red cells were incubated at a fixed concentration of N_2ph-F (0.15 mM, pH 7.4) and a range of persantin concentrations that, according to FIGURE 4d, produce more or less inhibition. After a specified time (30 min), the reaction is interrupted. After removal of unreacted N_2ph-F and persantin by several

TABLE 1

EFFECT OF REVERSIBLY ACTING INHIBITORS ON RATE OF SULFATE EQUILIBRIUM EXCHANGE (K_I=CONCENTRATION AT WHICH 50% INHIBITION IS PRODUCED), ON DINITROPHENYLATION OF H_2DIDS BINDING SITES ON BAND 3 (LYS *a*) AND ON IRREVERSIBLE INHIBITION BY N_2PH-F OF ANION TRANSPORT *

			Δ binding, %		Δ transport, %	
			$\frac{[I]}{K_I}$ 2.0	6.0	$\frac{[I]}{K_I}$ 2.0	6.0
1	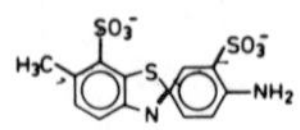	K_I = 1.127 mM	-28	-47	-20	-32
2	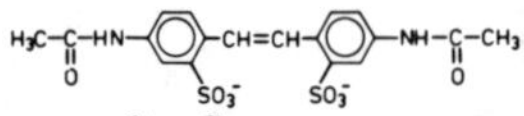	K_I = 0.335 mM	-25	-44	-27	-37
3	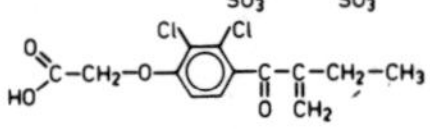	K_I = 0.102 mM	-32	-50	-8	-14
4	$^{-}O_3S-S-S-SO_3^{-}$	K_I = 3.030 mM	-19		-45	
5	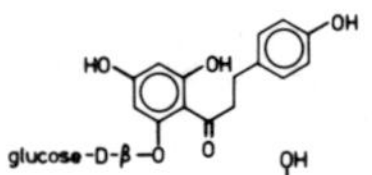	K_I = 1.020 mM	+9		+8	
6	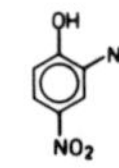	K_I = 0.367 mM	-34	-53	-17	-34
7	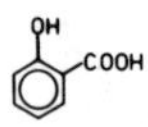	K_I = 2.900 mM	-22		-11	
8	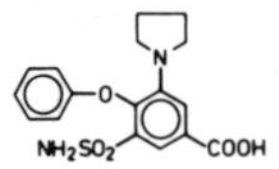	K_I = 0.422 mM	-27	-35	-15	-25
9	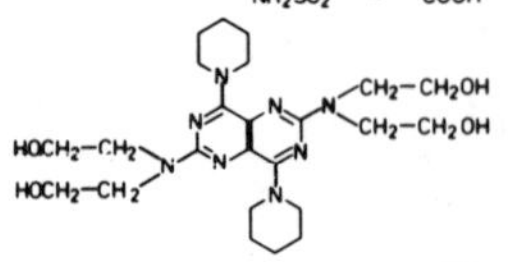	K_I = 0.0014m	-23	-32	-11	-20
I	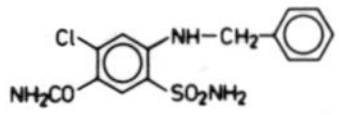	K_I = 0132 mM	+14		+9	
II	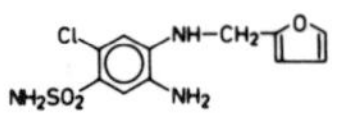	K_I = 0134 mM	+17		+8	
III	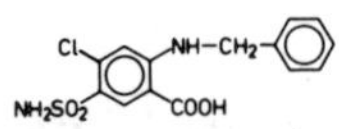	K_I = 0083 mM	-12	-29	-9	-17
IV	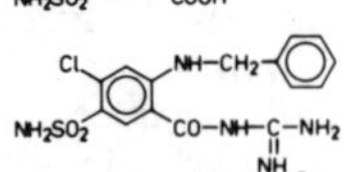	K_I = 0032 mM	+8		+7	
V	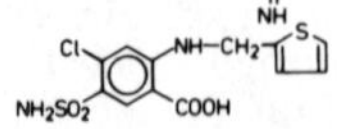	K_I = 0148 mM	-10	-22	-7	-14
VI (Furosemide)	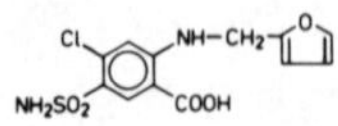	K_I = 0247mM	-26	-44	-13	-21
VII	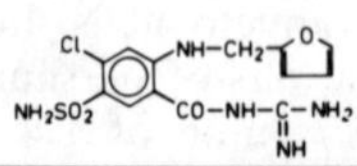	K_I = 0.150 mM	-9		± 0	

* K_I=concentration at which 50% inhibition is produced. [I]=concentration of inhibitor exposure to N_2ph-F at pH 7.4, 122.5 mM Cl, 37° C, for 30 min, intact red cells, hct 2%. The concentrations of the reversibly acting inhibitors are 2.0 and 6.0 times K_I. Δ binding, Δ transport= change of capacity to bind H_2DIDS and change of rate of SO_4 transport, as percent of control that had not been exposed to N_2ph-F. All measurements were made after removal of excess N_2ph-F and of the reversibly acting inhibitor. Formulas 1 and 2 represent, respectively, APMB and DAS. (From Zaki *et al.*[29])

washes, each batch of modified cells is divided into two. One is used for measuring anion transport, the other for the determination of the $^3H_2DIDS$ binding capacity of the Band 3 protein. The results are plotted against the ratio $[I]/K_I$, at which dinitrophenylation had been carried out ($[I]$ = inhibitor concentration). The data points in FIGURES 4a and 4b at $[I]/K_I = 0$ indicate the effect of N_2ph-F alone. They show that under the specified reaction conditions, dinitrophenylation produces about 80%–90% inhibition of transport and H_2DIDS binding to Band 3. The inhibition decreases (or, as plotted in the figure, transport and H_2DIDS-binding capacity increase) with increasing persantin concentration. A plot of transport versus binding yields a straight line that passes through the origin, indicating that transport and H_2DIDS binding capacity of Band 3 (i.e., lys *a*) were affected proportionally.

In analogous experiments, we studied the effects produced by seven sulfonamide derivatives (including three with negative, three with positive and one with no electrical charge) and eight other compounds. All agents produced an inhibition of anion transport that could be described by the equation:

$$^{o}k_s = {}^{o}k_{s0}/(1 + [I]/K_I)$$

in which $^{o}k_s$ and $^{o}k_{s0}$ represent the rate constants for sulfate transport at the inhibitor concentrations $[I]$ and zero, respectively; K_I is a constant that is specific for each inhibitor.

TABLE 1 shows that all inhibitors modify the rate of dinitrophenylation of lys *a*. This represents another piece of circumstantial evidence for the participation of Band 3 in anion transport.

In contrast to the inhibition of anion transport, the effects of the various inhibitors on the rate of dinitrophenylation of lys *a* cannot be represented as a unique function of the ratio $[I]/K_I$. Some agents like dipyridamol, and the negatively charged sulfonamide derivatives, reduce the rate of dinitrophenylation (indicated by a minus sign in TABLE 1). Others, like the nonpenetrating phlorizin and the positively charged sulfonamide derivatives produce an enhancement (indicated by a plus sign). Perhaps, even within each group, each agent exerts some individual effect.

As has already been stated, inhibition of anion transport could be brought about by a direct combination of the inhibitor with lys *a* or by binding to some other site that is allosterically linked to lys *a*. In the latter case, inhibition could be associated with conformational changes of the transport protein that render lys *a* either more or less susceptible to dinitrophenylation. For those inhibitors (phlorizin and positively charged furosemide derivatives) that enhance the rate of dinitrophenylation, inhibition by direct combination with lys *a* as the sole cause for the effect on transport can be excluded, and allosteric effects must be involved. The transport protein is arrested in a conformation where lys *a* is exposed. In analogy with the terminology used in the discussion of the effects of the substrate anion Cl on the susceptibility of lys *a* for dinitrophenylation, we shall call this state of the transport protein the *cis* conformation. When an inhibitor reduces the rate of dinitrophenylation of lys *a*, it is not immediately clear whether this is due to a protection by direct combination with that amino acid residue or to an allosteric effect. Additional information is required to decide between the two possibilities. It is most likely that non- or slowly penetrating compounds like APMB, DAS and H_2DIDS, which compete with Cl for the transfer site, will convert the transport system into the *cis* conformation and that the reduction of the rate of dinitrophenylation is related to the physical obstacle that the binding of the inhibitor to the transfer site creates in the immediate vicinity of lys *a*. For all the other inhibitors that reduce the rate of dinitrophenylation of *a*, it is impossible to make suggestions concerning their mode of action solely on the basis of the data in TABLE 1: whether they stop the transport protein in the *cis* conformation by direct combination with *a* or in *trans* conformation by an allosteric effect (TABLE 2).

There exists, however, one case where it could be demonstrated with a reasonable degree of certainty that inhibition is associated with a shift of lys *a* from an exposed to a buried state, rather than with a direct combination with the inhibitor. The case in point is the inhibition of anion transport by Ca^{++}, first demonstrated by Low.[32] At high Ca^{++} concentrations, inhibition reaches maximally about 60%. Inhibition takes place only if Ca^{++} is present inside the ghosts while external Ca^{++} is ineffective.[32, 33] The absence of inhibition by external Ca^{++} indicates that this cation is incapable of inducing inhibition by combination with the H_2DIDS-binding site. Nevertheless, we have observed that internal Ca^{++} reduces the rate of dinitrophenylation of the lysine residue *a* at the H_2DIDS binding site in the outer surface in proportion to its capacity to inhibit anion transport by interaction with an internally located binding site (FIGURE 5). In other words, an ion that obviously cannot serve as a substrate for the anion transport system produces an allosteric effect that is transmitted across the membrane and that leads to a shift of *a* from an exposed to a buried state. The transport system is arrested in the trans-conformation.

Transmembrane Movements of Lys a?

It would be most interesting to know whether the removal of lys *a* from the outer surface in the course of the *cis-trans* transition of the transport protein leads to the exposure of that amino acid residue at the inner membrane surface. The study of the sidedness of action of two disulfonic acids, DAS and APMB, provides some information on this important question. Both

TABLE 2

CONFORMATION OF TRANSPORT SYSTEM AFTER INHIBITION

cis	Protection or *trans*	*trans*
phlorizin positively charged furosemide derivatives	$APMB_{out}$ * DAS * H_2DIDS * tetrathionate * salicylate 2,4 DNP negatively charged furosemide derivatives	Ca^{++}_{in} $APMB_{in}$

* Most likely protection with arrest in *cis* position.

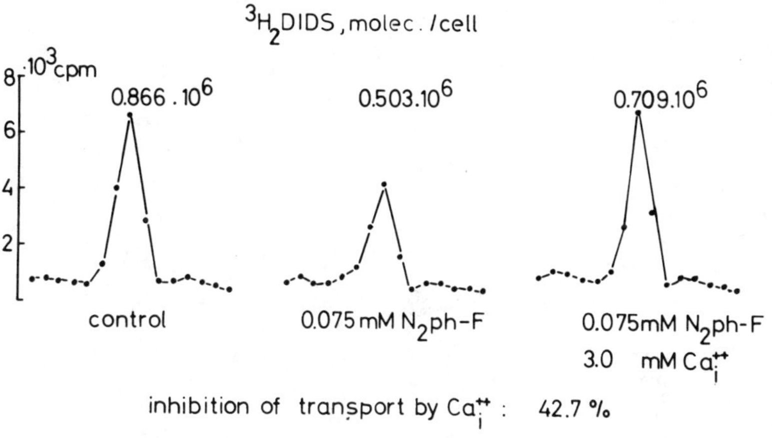

FIGURE 5. Effect of internal Ca^{++} (3 mM) on dinitrophenylation of H_2DIDS binding sites (lys *a*) on Band 3. pH 7.4, 37° C, 122.5 mM Cl, 30 min.

agents are fairly effective inhibitors (TABLE 2) that penetrate about two orders of magnitude more slowly than the sulfate ions that are used for the measurement of anion flux. The dimensions of the two inhibitor molecules are rather similar. In particular, the distances between the sulfonate groups differ by no more than 1–1.5 Å.

As has been shown above, if applied at the outer cell surface, both agents reduce the rate of dinitrophenylation of *a* (FIGURE 6) and produce a concomitant protection of the anion transport system against inactivation. This is exemplified for DAS in FIGURE 7. The cells are dinitrophenylated at a constant concentration of N_2ph-F but at varying concentrations of DAS, subsequently washed to remove excess N_2ph-F and DAS, and then subjected to flux measurements. One observes that the higher the DAS concentration, the less the irreversible inhibition by dinitrophenylation (FIGURE 7a). This is accompanied by a parallel protection of lys *a* against modification, as determined from measurements of the H_2DIDS-binding capacity of Band 3 (FIGURE 7b). If one plots transport against the number of H_2DIDS-binding sites that had not been dinitrophenylated, one obtains a straight line relationship with a slope of 45°, as one would expect if there was a one-to-one correlation between the binding of DAS and the protection against dinitrophenylation of the lys *a* on the H_2DIDS binding site of the Band 3 protein (FIGURE 7c). Similar results were obtained with APMB.

With the described information in mind, it is now possible to approach the problem of transmembrane effects on *a*. For this purpose, we consider the sidedness of action of APMB and DAS.

If a nonpenetrating agent exerts on inhibitory effect at one surface only, the situation is rather simple (FIGURE 8). On the basis of such observations, one would infer that the binding sites that combine with the inhibitor are immobile and confined to that surface of the membrane at which the inhibitor exerted its effect. One can also conclude that there are no functionally significant allosteric interactions between the inner membrane surface and the sites

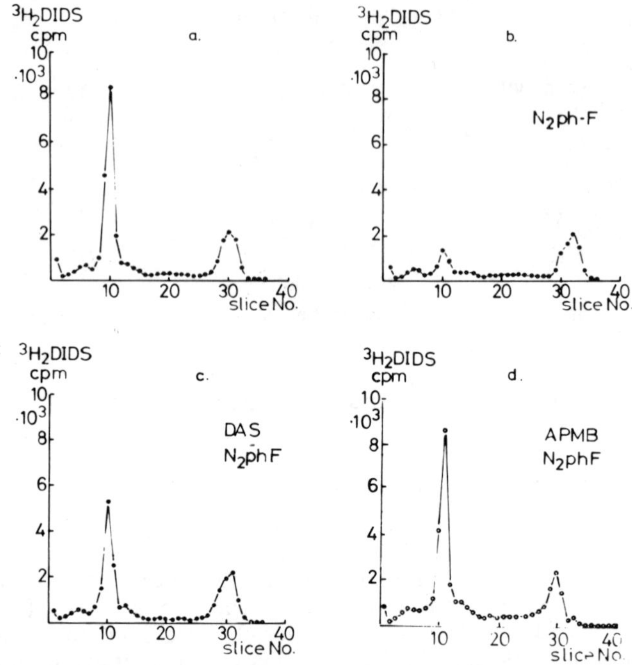

FIGURE 6. $^3H_2DIDS$ binding to red cell ghosts (90 min, 37° C, pH 7.4). Prior to the exposure to $^3H_2DIDS$, the ghosts had been treated with N_2ph-F with or without DAS (2 mM) or APMB (5 mM) present. N_2ph-F concentration (a) 0 mM, (b, c) 0.15 mM, (d) 0.2 mM. pH 7.4, Cl-concentration 122.5 mM, 37° C, time of incubation 20 min.

at the outer surface. The experimental findings with DAS would agree with such a picture (FIGURE 9): The agent inhibits if applied at the outer membrane surface of red cell ghosts, but not if incorporated into the ghosts.[9]

The behavior of APMB is quite different from that of DAS.[8, 9] This agent produces inhibition at either membrane surface (FIGURE 9). The interpretation of this result is ambiguous. Inhibition at either surface would be compatible with three different explanations: 1. Binding to fixed sites at either surface; 2. Combination with mobile sites that move to that surface at which they can react with the inhibitor, and 3. An allosteric effect that is transmitted across the membrane (FIGURE 8).

To discriminate between these possibilities, we made use of the fact that APMB is capable of preventing the dinitrophenylation of *a*. N_2ph-F penetrates across the membrane. Hence, it should be capable of reacting with mobile sites or with fixed sites at both surfaces. If we are dealing with fixed sites at both surfaces, internal APMB should be able to protect internal binding sites against dinitrophenylation, but not external sites. Correspondingly, external APMB should only protect external sites, but not the internal ones. In both cases, APMB would be incapable of protecting the anion transport system against dinitrophenylation. If, however, we are dealing with mobile sites, APMB should be able to prevent dinitrophenylation, regardless of whether it is present at the inner surface or outer surface.

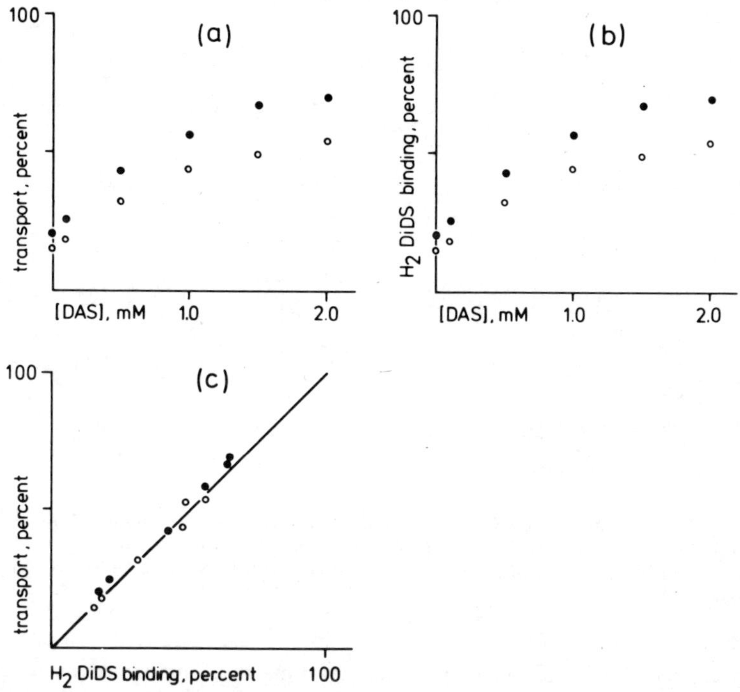

FIGURE 7. Action of DAS on dinitrophenylation of Band 3 (lys *a*) (b) and inhibition of sulfate equilibrium exchange by N_2ph-F (a) in red cell ghosts; (c) plot of (a) vs. (b). Time of dinitrophenylation 30 min, 37° C, pH 7.4, 122.5 mM Cl. Concentration of N_2ph-F: 100 μM (open circles) and 200 μM (filled circles). (From Passow.[20])

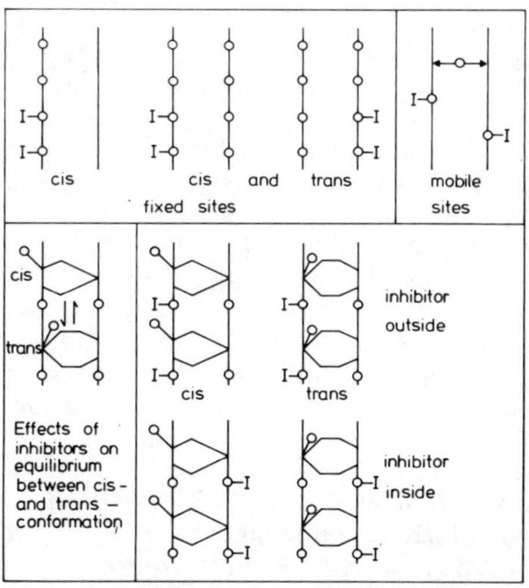

FIGURE 8. Schematic representation of possible modes of actions of inhibitors.

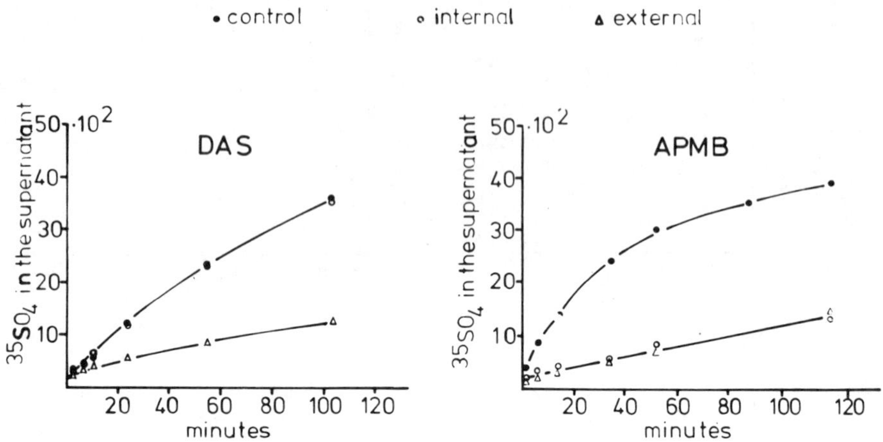

FIGURE 9. Sidedness of action of DAS and APMB. "Control" "internal" and "external" refer, respectively, to ghost suspensions without internal or external inhibitor, with internal inhibitor only, with external inhibitor only. (From Zaki *et al.*[9] and Passow.[20])

An experiment based on these considerations was performed as follows:[22] One batch or ghosts was used as a control. Another was exposed to N_2ph-F in the absence of APMB. A third batch and a fourth batch contained internal or external APMB, respectively, in the presence of N_2ph-F. After thirty minutes at 37° C, the ghosts were washed, rehemolyzed to remove internal APMB, and then resealed again. Anion flux was measured in the twice-sealed ghosts. TABLE 3 shows the result: both internal as well as external APMB protect the binding sites against dinitrophenylation to about the same extent.

This result points to a reaction with mobile sites. We are now raising the question: Could such mobile sites be identical with the lysine residue *a* that is protected against dinitrophenylation by external APMB and DAS?

To answer this question, we determined the H_2DIDS-binding capacity of the ghosts that had been dinitrophenylated in the presence or absence of internal or external APMB, as described above. The result is also represented in TABLE 3. Like external APMB, internal APMB protects against dinitrophenylation not only the transport system but also the site that is responsible for covalent binding of externally applied $^3H_2$DIDS. This finding shows that the protected binding site is in fact identical with lys *a*.

In contrast to APMB, DAS inhibits anion transport and protects the lys *a* against dinitrophenylation only if applied to the outer membrane surface. This finding is not easily reconcilable with the assumption that internal APMB exerts its effect by direct combination with an inward facing lysine residue *a* since that residue should also be accessible to internal DAS. It seems unlikely that two different sets of H_2DIDS binding sites exist on the protein in Band 3 with different relative affinities for DAS and APMB. Moreover, there is experimental evidence which strongly suggests that this is not so:

Ghosts were loaded with APMB and dinitrophenylated. The transport system, as studied after removal of APMB and unreacted N_2ph-F, was found to be largely protected by internal APMB against irreversible modification. In these ghosts, after removal of internal APMB by rehemolysis and a second sealing procedure, external DAS was still able to inhibit to the same extent as before. In other words, internal APMB protected DAS binding sites in the outer cell surface against irreversible modification by N_2ph-F. This confirms, therefore, that the binding sites for APMB and DAS in the outer cell surface are the same.

This led us to consider an explanation for the differences of the sidedness of the effects of DAS and APMB that do not involve the assumption that the common binding sites are mobile. The ghost membrane is not completely impermeable to APMB or DAS. Thus, diffusing inhibitors could conceivably react with the fixed site somewhere inside the membrane. Even though the external medium may be sufficiently large to dilute the inhibitor that leaves the ghosts to the point where an inhibition at the outer surface becomes unlikely—inside the membrane—at the location of such a fixed site, the concentration could be high enough to produce inhibition. The effect on the inhibitory site would depend on the ratio between that local concentration and the association constant for the binding to the fixed site.

Experimentally, we found that both APMB and DAS penetrate. We adjusted the internal concentrations of DAS and APMB to the values at which external APMB and DAS produce 80% inhibition (about 5 mM for APMB and 1.8 mM for DAS). The ratio of the two concentrations is reasonably close to the ratio of the affinities of the two compounds to the inhibitory sites on the outer cell surface. Under these conditions, the penetration rate for DAS is a little less than one half the penetration rate of APMB. This implies that the permeabilities of the two agents are rather similar and that the concentration of DAS inside the membrane should be high enough to produce at least some inhibition. Yet DAS, in striking contrast to APMB, does not inhibit.

This result suggests that both APMB and DAS react at the same site in the outer membrane surface, and that this site is identical with the site for reversible H_2DIDS-binding that is adjacent to lys *a*. This site does not become accessible to the inner membrane surface since internal DAS neither produces inhibition nor "protection" against dinitrophenylation of *a*.

TABLE 3

SIDEDNESS OF ACTION OF APMB ON RATE OF DINITROPHENYLATION OF H_2DIDS BINDING SITE (LYS *a*) AND RATE OF ANION TRANSPORT *

Condition	${}^{o}k_s \times 10^2$ min^{-1}	H_2DIDS $\times 10^{-6}$ molec/cell
Control	1.78	1.48
N_2phF + ext. APMB	1.13	0.73
N_2phF + int. APMB	1.05	0.72
N_2phF no APMB	0.48	0.29

* Ghosts dinitrophenylated in the presence or absence of APMB (external or internal) washed, rehemolysed, sealed, divided into two batches, one for flux measurements, the other for measurements of H_2DIDS binding capacity. N_2ph-F concentration: 0.125 mM. APMB concentration: 5 mM. Control: no dinitrophenylation. (From Passow and Zaki.[22])

We assume, therefore, that the inhibition observed with internal APMB is due to APMB-binding to a site at the inner membrane surface that is not a DAS-binding site and consequently not identical with the site for reversible H_2DIDS-binding.

Thus we conclude that lys *a* is not capable of alternating between the two membrane surfaces and that APMB binds to some other site at the inner membrane surface. This APMB binding then leads to a shift of the transport protein from the *cis* to the *trans* conformation (FIGURE 8) and a concomitant movement of the amino acid residue *a* from an exposed to a buried state. This conformational change should translocate the transfer site from an outward facing to an inward facing position, exactly as one would expect after a direct combination of the transfer site with APMB. However, the absence of an inhibition by internal DAS suggests that the conformation at the inward facing transfer site is different from that of the outward facing site. In other words, the properties of the transfer site seem to undergo a profound change when the transport protein undergoes the transition from the *cis* to the *trans* conformation.

CONCLUSIONS

Anion transport can be modified by a surprising variety of chemically unrelated compounds. The compounds include agents of positive, negative and no net charge, aromatic and aliphatic compounds [35] and even inorganic cations such as calcium.[32, 33] The present paper shows that the inhibition produced by all of a number of selected representatives of the different classes of agents is associated with a conformational change of the Band 3 protein. It must remain open whether or not these changes are induced by binding to the protein directly, to other proteins that interact with the Band 3 protein, or to the lipid bilayer. In any case, our result is compatible with the view that all inhibitors used produce their effects on transport by modifying the Band 3 protein.

The various inhibitors could be grouped into two classes that produced different conformational changes at equal degrees of inhibition of transport. The accessibility of the reporter group *a* was either increased or reduced (FIGURE 8, lower part).

The increase must be clearly attributed to a conformational change of the transport protein that leads to an exposure of *a*. Phlorizin and positively charged furosemide derivatives produce the increase by their action on the outer membrane surface. The decrease may be explained in at least two different ways: either as a conformational change that leads to a burial of site *a* somewhere inside the protein structure, or as a "protection" through combination of the inhibitor with lys *a,* or with an adjacent site that controls the access to lys *a.* In a few cases, it is possible to make a reasonable guess. DAS, APMB, tetrathionate, i.e., compounds that are known to react with the outwardly directed transfer site (presumably sites r_1 and r_2 in FIGURE 1) probably block the access to lys *a.* In the case of internal APMB and Ca^{++}, the situation is different. Binding is unlikely to take place at the inwardly directed transfer site. Nevertheless, *a* is removed from the outer membrane surface. Hence, it is likely that allosteric effects are involved. Altogether, the parallel study of the effects of the various inhibitors on anion transport and on the susceptibility of the transport-related lysine residue *a* for dinitrophenylation indicates that there exist at least five, possibly six, different identifiable sites on the transport protein that are allosterically linked to lys *a* (FIGURE 10). Usually allosteric effects are not simply unidirectional. It is most likely, therefore, that all of these sites are mutually interacting and thus participate in those conformational changes of the Band 3 molecule that are associated with anion transport.

The two positions of site *a* have been attributed to two different conformations of the transport protein: *cis* and *trans*. In the absence of substrate anions

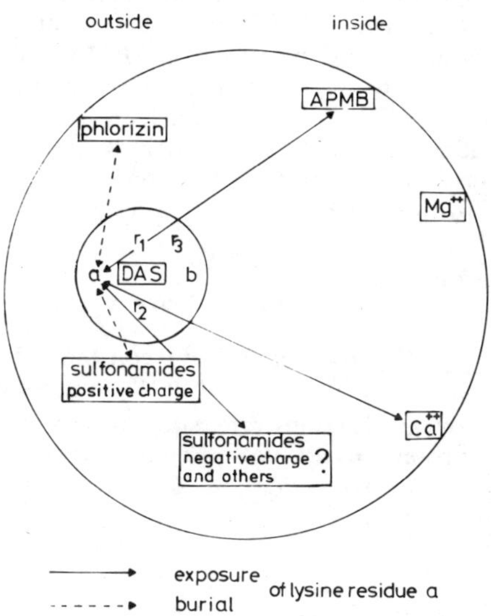

FIGURE 10. Interactions between the transport-related lysine residue *a* and various reversibly acting inhibitors of anion transport. Mg^{++} has not been discussed in the text. It is included in the figure since it has been shown that it inhibits anion transport by combination with an endofacial site that is not identical with the Ca^{++} binding site.[32] It should be noted that the allosteric interactions depicted in this figure are not necessarily confined to the Band 3 molecule itself, but may also involve binding of the inhibitors to the lipid bilayer or adjacent protein molecules.

and inhibitors, lys *a* is buried and the protein is in the *trans* conformation. Addition of Cl exposes lys *a,* indicating a change of the ratio *cis/trans* to higher values. This change is a saturable process with a half saturation constant that is close to the half saturation constant for Cl transport, which is measured under conditions of equilibrium exchange. It is likely, therefore, that the change of this ratio indicates a shift of an equilibrium between the two conformations of the transport protein toward the *cis* conformation. For the time being it must remain open whether or not "extroverting" inhibitors like phlorizin or "introverting" inhibitors like Ca^{++} act simply by shifting the normally existing equilibria between *cis* and *trans* form of the protein further towards the extremes or if they produce additional conformational changes near a that change the rate of dinitrophenylation to values below or beyond those observed after deprivation of/or saturation with Cl. If one accepts our interpretation of the sidedness experiments with APMB and DAS, one has to postulate those portions of the transport protein that are relevant for the binding of these inhibitors and for the substrate anions are quite different in *cis* and *trans* conformations.

The conclusions derived above have important consequences for the kinetics of anion transport that can be illustrated by a simple mathematical model. This model is based on the assumption that, unlike carrier molecules in a homogeneous phase, the transfer site within the transport protein will undergo a conformational change when it traverses the energy barrier in the course of anion translocation. This would imply that the affinity of the transfer site to the substrate anions and to inhibitors (like APMB and DAS) that bind to the substrate site are likely to be different at the two sides (1 and 2) of the energy barrier, and that the rates of transition from *cis* to *trans* (k_{21}) and vice versa (k_{12}) are not identical. If, for simplicity's sake, it is stipulated that substrate binding is fast as compared to the conformational changes involved in substrate transfer, for the flux from 1 to 2, one arrives at the expression:

$$\text{flux} = k_{12} \cdot \frac{a \cdot \alpha \cdot \overline{RS}}{a + \alpha K_1 + (1 - \alpha) K_2 + \alpha \cdot \frac{K_1}{K_3} \cdot I} \qquad \textbf{(1)}\dagger$$

where $\alpha = \dfrac{k_{21}}{k_{12} + k_{21}}$, K_1, K_2 = mass law constants for substrate binding at surfaces 1 and 2; K_3 = mass law constant for the binding of the inhibitor at the substrate site on surface 2, and $\overline{RS}$ = total number of transport protein molecules per cell. If no inhibitor is present (inhibitor concentration $I = 0$), equation 1 is formally identical with the Michaelis Menten equation:

$$\text{flux} = \frac{a \cdot V_{max}}{a + K_S} \qquad \textbf{(2)}$$

(a = concentration of substrate anion, V_{max} = efflux at $a \gg K_S$, K_S = half saturation constant)
where the coefficients assume the following meaning:

$$V_{max} = k_{12} \cdot \alpha \cdot \overline{RS} \qquad \textbf{(3)}$$

$$K_S = \alpha \cdot K_1 + (1 - \alpha) \cdot K_2 \qquad \textbf{(4)}$$

For competitive inhibition, one can also obtain an expression that is formally identical with that for competitive inhibition in enzymes. However, the significance of K_I is different:

$$K_I = \frac{1}{(1 - \alpha)} \cdot \frac{K_3}{K_1} [a + \alpha K_1 + (1 - \alpha) K_2] = \frac{1}{(1 - \alpha)} \frac{K_3}{K_1} (a + K_S) \qquad \textbf{(5)}$$

The symbols have the same meaning as above. K_3 is the mass law constant for inhibitor binding to the *trans* side (surface 2) of the transporter.

† The model discussed above is related to the ping-pong mechanism of enzyme kinetics. The special case for $I = 0$ of this equation is implicit in a very general treatment of noncarrier transport published in 1957 by Patlak.[42]

The equation for the relationship between transport and substrate concentration reduces to an expression in which the constant K_S assumes the meaning of an equilibrium constant when $k_{12} \gg k_{21}$ or $k_{21} \gg k_{12}$. In the former case $K_S = K_1$.

So far as Cl transport in the red cells is concerned, it is important that the determination of K_I values for reversible H_2DIDS binding at different Cl concentrations yielded a K_S value that agreed well with the K_S value as determined from the saturation kinetics of the uninhibited transport system. This suggests that the Cl binding to the transfer site is fast as compared to the conformational change associated with the transfer across the membrane, and that K_S represents an equilibrium constant that must be equated to the equilibrium constant for the Cl binding to one of the two conformations of the transport protein.

The mode of action of those inhibitors that do not compete for the substrate binding site of the transport protein (e.g., phlorizin, the positively charged furosemide derivatives, Ca^{++}) may be quite complex. The shift of the ratio between *cis* and *trans* conformation may be related to a preferential binding to either the *cis* or the *trans* conformation and be associated with a change of the stability of one of these forms. Alternatively, or in addition, a change of the ratio k_{12}/k_{21} could be induced. A more detailed analysis of the data is, however, outside the scope of the present paper and will be presented elsewhere.[44]

It should be noted that considerations essentially similar to those presented above about the kinetics of protein-mediated anion transport in the red cell should apply to many transport processes that involve the participation of transport proteins with asymmetric structures at the rate-limiting barrier.

Summary

Measuring the rate of dinitrophenylation of a specific lysine residue (called *a*) that is allosterically linked to the transfer site, it could be demonstrated that the anion transport protein may exist in two different conformational states, designated *cis* and *trans*. In the *cis* conformation *a* is easily accessible for reaction with dinitrofluorobenzene; in the *trans* conformation, *a* is less accessible. In the presence of the substrate anion Cl, the equilibrium between the *cis* and *trans* conformation is towards the *cis* conformation. Reversibly acting inhibitors of anion transport arrest the transport system, either predominantly in the *cis* or in the *trans* conformation. Phlorizin and certain positively charged derivatives of furosemide produce arrest in *cis* conformation, internal 2-(4′-aminophenyl)-6-methylbenzenethiazol-3′,7-disulfonate (APMB) and Ca^{++} in *trans* conformation. Within this frame of reference, the different susceptibilities of the transfer site to internal and external 4,4′ diacetamido-2,2′-stilbene disulfonate (DAS) are interpreted on the assumption that the conformation of the transfer site changes during the transition of the transport protein from the *cis* to the *trans* conformation, so that in the *trans* conformation a reaction with DAS is no longer possible.

ACKNOWLEDGMENTS

We thank Dr. R. Motais (Nice) for a stimulating discussion and Dr. Muschaweck (Hoechst AG, Frankfurt/Main) for his help with the selection of the furosemide derivatives. The derivatives were kindly supplied by Dr. Schorr and Dr. Bartmann of Hoechst AG.

REFERENCES

1. MOND, R. 1928. Pflügers Arch. Ges. Physiol. **219:** 467–480.
2. PASSOW, H. 1969. *In* Progress in Biophysics and Molecular Biology. J. A. V. Butler & D. Noble, Eds. :425–476. Pergamon Press. Elmsford, New York.
3. LEPKE, S. & H. PASSOW. 1971. J. Membrane Biol. **6:** 158–182.
4. PASSOW, H. & K. F. SCHNELL. 1969. Experientia **25:** 460–468.
5. ZAKI, L., C. GITLER & H. PASSOW. 1971. Proc. Int. Union Physiol. Sci. IX, XXV Intern. Congr. Munich. Abstr. No. 1835.
6. PASSOW, H. 1971. J. Membrane Biol. **6:** 233–258.
7. CABANTCHIK, Z. I. & A. ROTHSTEIN. 1974. J. Membrane Biol. **15:** 207–226.
8. PASSOW, H., H. FASOLD, L. ZAKI, B. SCHUHMANN & S. LEPKE. 1975. *In* Proc. 9th FEBS Meeting, Budapest 1974. G. Gárdos & I. Szász, Eds. Biomembranes: Structure and Function. **35:** 197–214.
9. ZAKI, L., H. FASOLD, B. SCHUHMANN & H. PASSOW. 1975. J. Cell Physiol. **86:** 471–494.
10. ROTHSTEIN, A., Z. I. CABANTCHIK & P. A. KNAUF. 1976. Fed. Proc. **35:** 3–10.
11. LEPKE, S. & H. PASSOW. 1976. Biochim. Biophys. Acta **455:** 353–370.
12. CABANTCHIK, Z. I., J. M. WOLOSIN, H. GINSBURG & O. ZEMEL. 1977. *In* Biochemistry of Membrane Transport. FEBS Symp. No. 42, G. Semenza & E. Carafoli, Eds. :306–315. Springer-Verlag, Berlin.
13. WOLOSIN, S. M., H. GINSBURG & Z. I. CABANTCHIK. 1977. J. Biol. Chem. **252:** 2419–2427.
14. ROTHSTEIN, A., Z. I. CABANTCHIK, M. BALSHIN & R. JULIANO. 1975. Biochem. Biophys. Res. Commun. **64:** 144–150.
15. CABANTCHIK, Z. I., P. A. KNAUF & A. ROTHSTEIN. 1978. Biochim. Biophys. Acta **515:** 239–302.
16. ROSS, A. H. & H. M. MCCONNELL. 1977. Biochem. Biophys. Res. Commun. **74:** 1318–1325.
17. LEPKE, S., H. FASOLD, M. PRING & H. PASSOW. 1976. J. Membrane Biol. **29:** 147–177.
18. SHIP, S., Y. SHAMI, W. BREUER & A. ROTHSTEIN. 1977. J. Membrane Biol. **33:** 311–324.
19. HALESTRAP, A. P. 1976. Biochem. J. **156:** 193–207.
20. PASSOW, H. 1978. *In* Cell Membrane Receptors for Drugs and Hormones: A Multidisciplinary Approach. R. W. Straub & L. Bolis, Eds. :203–218. Raven Press, New York City.
21. RUFFING, W., E. GÄRTNER, B. LEGRUM, S. LEPKE, L. ZAKI, H. FASOLD & H. PASSOW. 1979. Submitted for publication: Biochim. Biophys. Acta.
22. PASSOW, H. & L. ZAKI. 1978. *In* Molecular Specialization and Symmetry in Membrane Function. A. K. Solomon & M. Karnovsky, Eds. :229–252. Harvard University Press. Cambridge, Mass.
23. PASSOW, H. 1977. Acta Biol. Med. Ger. **36:** 817–821.
24. GRINSTEIN, S., L. MCCULLOCH & A. ROTHSTEIN. 1979. J. Gen. Physiol. **73:** 493–514.
25. JENNINGS, M. L. & H. PASSOW. 1979. Biochim. Biophys. Acta **554:** 498–519.
26. GRINSTEIN, S., S. SHIP & A. ROTHSTEIN. 1978. Biochim. Biophys. Acta **507:** 294–304.

27. Barzilay, M., S. Ship & Z. I. Cabantchik. 1979. Membrane Biochem. *2*, 227–254.
28. Shami, Y., A. Rothstein & P. A. Knauf. 1978. Biochim. Biophys. Acta **508:** 257–363.
29. Zaki, L., W. Ruffing, E. Gärtner, R. Motais & H. Passow. 1977. 12th Meeting Fed. Eur. Biochem. Soc. Copenhagen. Abstr. No. A4–17–671.
30. Kaplan, J. H., K. Scorah, H. Fasold & H. Passow. 1976. FEBS Letters **62:** 182–185.
31. Kaplan, J. H. 1972. Biochim. Biophys. Acta **290:** 339–347.
32. Low, P. S. 1978. Biochim. Biophys. Acta **514:** 264–273.
33. Gunn, R. B., O. Fröhlich, J. D. MacIntyre & P. S. Low. 1979. Biophys. J. **25:** Abstr. No. M–PM–P049.
34. Passow, H., H. Fasold, M. Jennings & S. Lepke. 1979. Fed. Proceedings. In press.
35. Deuticke, B. 1977. Rev. Physiol. Biochem. Pharmacol. **78:** 2–97.
36. Cherry, R. J., A. Bürkli, M. Busslinger & G. Schneider. 1977. *In* Biochemistry of Membrane Transport. G. Semenza & E. Carafoli, Eds. :86–95. Springer-Verlag, Berlin.
37. Harris, E. J. & B. C. Pressman. 1967. Nature **216:** 918–920.
38. Scarpa, A., A. Cecchetto & G. F. Azzone. 1970. Biochim. Biophys. Acta **219:** 179–188.
39. Gasko, O. D., A. F. Knowles, H. G. Schertzer, E. M. Suolinna & E. Racker. 1976. Anal. Biochem. **72:** 57–65.
40. Feinstein, M. B., M. Volpi, S. Perrie, A. Makriyannis & R. I. Sha'afi. 1977. Mol. Pharmacol. **13:** 840–851.
41. Deuticke, B. 1977. Rev. Physiol. Biochem. Pharmacol. **78:** 1–97.
42. Patlak, C. S. 1957. Bull. Math. Biophys. **19:** 209–235.
43. Motais, R. & J. L. Cousin. 1978. *In* Cell Membrane Receptors for Drugs and Hormones. A Multidisciplinary Approach. R. W. Straub & L. Bolis, Eds. : 219–225. Raven Press, New York City.
44. Passow, H., L. Kampmann, H. Fasold, M. Jennings & S. Lepke. 1980. *In* Membrane Transport in Erythrocytes. U. Lassen, H. H. Ussing & J. O. Wieth, Eds. :1–23. Munksgaard.
45. Gunn, R. B. & O. Fröhlich. 1979. J. Gen. Physiol **74:** 351–374.
46. Knauf, P. A. & A. Rothstein. 1971. J. Gen. Physiol. **58:** 190–223.

THE KINETICS OF THE TITRATABLE CARRIER FOR ANION EXCHANGE IN ERYTHROCYTES *

R. B. Gunn and O. Fröhlich

Department of Pharmacological and Physiological Sciences
University of Chicago
Chicago, Illinois 60637

The transport of small inorganic anions across the erythrocyte membrane is linked with a protonation reaction and proton transport. The site of this reaction and its precise relation to chloride and sulphate transport remains unknown. The transport of $^{36}Cl^-$ increases as pH increases from pH 5.5 to pH 7.8 in the extracellular phase (o) whose pH (pH_o) is in equilibrium with the intracellular pH (pH_i).[1] Since $pH_i - pH_o = \log (Cl_i/Cl_o)$, and since Cl_i is approximately Cl_o, there is little difference in pH in the two phases. The transport of ^{35}S-sulphate in exchange for sulphate has a pH dependence opposite to that of chloride.[2, 3] Sulphate exchange *declines* with increasing pH values above 6.0. In addition to these opposite trends for chloride and sulphate fluxes between pH 6 and 7.8, higher external pH values cause the chloride flux to decline because the intracellular anion concentration declines as the net hemoglobin charge becomes more anionic.[4] The opposite relations of chloride and sulphate fluxes between pH 6 and pH 7.8 were initially explained by a titration of the carrier transport mechanism.[5] This explanation supposes that the mechanism in low pH solutions is protonated and as a result can react with divalent anions SO_4^{-2} and HPO_4^{-2} while at high pH values without the proton, the common mechanism reacts only with monovalent anions: Cl^-, HCO_3^-, and Br^-. The purpose of this report is to delineate some of the complexities and errors in this simple explanation.

Dalmark [6] reported an extensive series of chloride self-exchange measurements in which the erythrocyte chloride concentration was varied by the nystatin technique.[7] He found that the apparent half-saturation coefficient, $K_{½}^{Cl}$, *increased* with increasing solution pH values if the flux was graphed against external chloride concentration. But since hemoglobin ionization changes with pH value, and since the internal chloride concentration is not a fixed fraction of the external chloride concentration at different pH values, the same data showed that $K_{½}$ *decreased* with increasing pH values if the flux was graphed against intracellular chloride concentration. Dalmark also reported that the apparent pK_a of the site which diminished chloride flux was not altered by increasing chloride from 150 mM to 600 mM on both sides of the membrane and, therefore, protons were noncompetitive inhibitors of chloride self-exchange flux. The titratable carrier model predicted that protons should be competitive inhibitors of the chloride flux since they remove chloride transporters through a reaction that converts monovalent transport sites into divalent transport sites. FIGURE 1 shows the pH dependence of chloride self-exchange that may reflect proton reactions with the exchange mechanism from either or

* This work was supported in part by USPHS grants HL-20725, HL-20365 and a Career Development Award to R. B. Gunn, HL-00208.

0077–8923/80/0341–0384 $01.75/0 © 1980, NYAS

both sides of the cell membrane. For example, the decline in flux above pH 8 is primarily due to the lower intracellular chloride concentration, as mentioned above. This decline in flux is consistent with a constant maximum flux over this pH range and $K_{½}^{Cl} = 65$ mM on the inside. The inhibition of the self-exchange flux by protons between pH 8 and pH 5.8 could, in principle, be due to competitive, uncompetitive or noncompetitive reactions (FIGURE 2) on either or both sides of the membrane. The data of Dalmark suggest non-

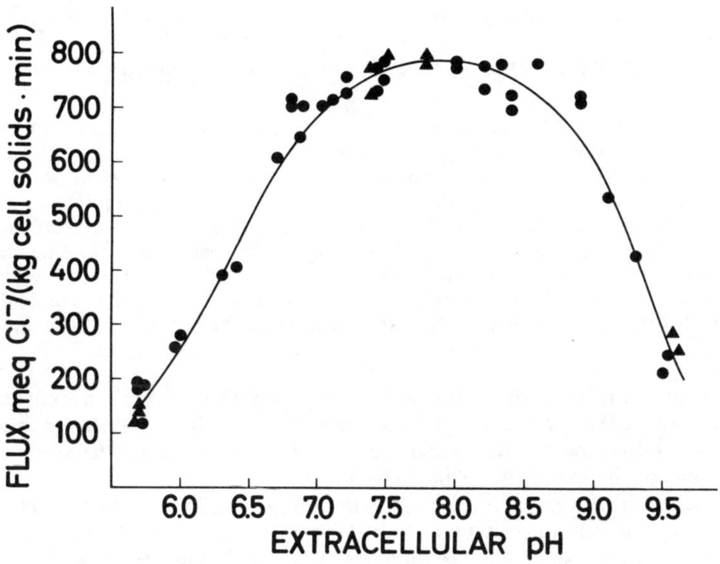

FIGURE 1. Chloride self-exchange flux as a function of extracellular pH. ^{36}Cl efflux was measured while total chloride, pH and water were at equilibrium at 0° C. The decline in flux at high pH values is attributable to two factors, the decline in intracellular chloride concentration that accompanies removal of protons from hemoglobin and the $K_{1/2-in}^{Cl}$ which is 65 mM. The decline in flux at low pH values is attributable to a protonation of the transport mechanism. (From Gunn *et al.*[1] By permission of *Journal of General Physiology*.)

competitive inhibition but do not indicate whether that inhibition is on the inside or outside or both sides of the membrane.

Protons are involved in sulphate transport. Jennings [8] showed that when chloride-containing cells were placed in unbuffered sulphate media, in which all CO_2 and bicarbonate had been purged by prior bubbling with N_2, the external pH became alkaline and approximately 0.8 protons entered with each sulphate for each chloride that exited. The time course of this exchange was slow, like sulphate exchange, and the dependence of the rate on the initial external pH was most easily explained by $H^+ — SO_4^{-2}$ cotransport in exchange for chloride rather than $OH^- — Cl^-$ cotransport in exchange for SO_4^{-2}. In a subsequent paper, Jennings also showed that there could be a very small $H^+ — Cl^-$ cotransport on the anion exchange mechanism.[9] This flux might be

1) $C + Cl \rightleftharpoons CCl$

2) $C + H \rightleftharpoons CH$

3) $CCl + H \rightleftharpoons CHCl$

reactions 1+2	competitive
reactions 1+3	uncompetitive
reactions 1+2+3	noncompetitive

FIGURE 2. Possible surface reactions of chloride and protons with the anion transporter, C. The first reaction is necessary for chloride transport and must occur between intrinsic membrane components and chloride from both solutions. The second and third reactions are possible mechanisms for proton inhibition of chloride flux through formation of alternative complexes. The CCl complex formed in the first reaction on each side can transport Cl to the opposite side. Whether the second and third reactions occur on both sides and whether the complexes are transported is argued in the text. It is concluded that reactions two and three occur but only to a significant extent on the inside of the erythrocyte in the absence of sulphate.

by the same mechanism as the H_2Cl_2 cotransport suggested by earlier studies at very low pH.[10] In order to integrate these different findings and appropriately modify the titratable carrier model, it is necessary to understand the conditions of the experiments in these reports.

Except for the experiments of Jennings, chloride transport was measured as the self-exchange flux when chloride was in equilibrium and approximately equal on the two sides of the membrane. It is implicit that such a measurement lumps a number of kinetic parameters that need further dissection. Recently, we have made steady state chloride flux measurements in which the concentration of chloride was independently varied on the inside and outside of the cells.[11] From these experiments a new perspective on the exchange mechanism has been obtained. The mechanism appears to involve a family of single sites that reciprocate between the two solutions from which they can complex or decomplex anions (ping-pong mechanism). In addition, these sites have very asymmetric environments or asymmetric intrinsic properties when they face the two different sides of the membrane. This asymmetry is striking for both chloride and protons.

First, we will discuss the asymmetry of chloride transport. FIGURE 3 is a schematic fiction separating the steps involved in chloride exchange. There are two half reactions in this ping-pong scheme. The efflux half reaction consists of: 1. internal $^{36}Cl^-$ binding to the transporter; 2. the transition of the complex from having access to the internal solution to having access to the external solution; and, 3. the decomplexation of the $^{36}Cl^-$ into the outside solution. The second half reaction is a similar series of steps for the influx of chloride.

There are four kinetic parameters that can be measured for each anion: $K_{1/2\text{-out}}$, the concentration of anion in the outside solution which gives half-maximal flux at some fixed internal chloride concentration; $V_{max\text{-out}}$, the maximal flux obtained by increasing the external chloride concentration; $K_{1/2\text{-in}}$, the half-saturation concentration of anion on the inside; and $V_{\text{max-in}}$, the maximal flux measured by increasing the inside chloride concentration at a fixed external chloride concentration.

The principle of dissecting these parameters is shown in FIGURE 4 in which bromide rather than $^{36}Cl^-$ was initially inside cells and in which we measured its efflux into different external chloride concentrations. The value of $K_{1/2\text{-out}}^{Cl}$ was 1 mM and $V_{\text{max-out}}$ was 300 mEq/(kg hemoglobin·min). This latter value is twice the maximal bromide self-exchange flux. The self-exchange carries bromide ions on both limbs of the cycle. Since bromide transport is eightfold slower than chloride transport, its steps will be rate-limiting compared to those of chloride. The overall reaction will be approximately doubled if chloride replaces bromide on half of the reaction cycle. The system appears to behave as if $1/V = 1/V_{in} + 1/V_{out}$ where V is the observed overall flux and V_{in} and V_{out} are the component unidirectional fluxes if they were not restricted by exchange flux. We have measured the apparent $K_{1/2\text{-out}}$ for chloride in eleven series of experiments, each with a different fixed internal chloride concentration. The results are shown in FIGURE 5, which shows the dependence of the apparent external $K_{1/2}$ for chloride on the *trans* or internal chloride concentration. From these data and others, we have concluded that the anion exchange mechanism behaves much as the ping-pong model in FIGURE 3 would predict.

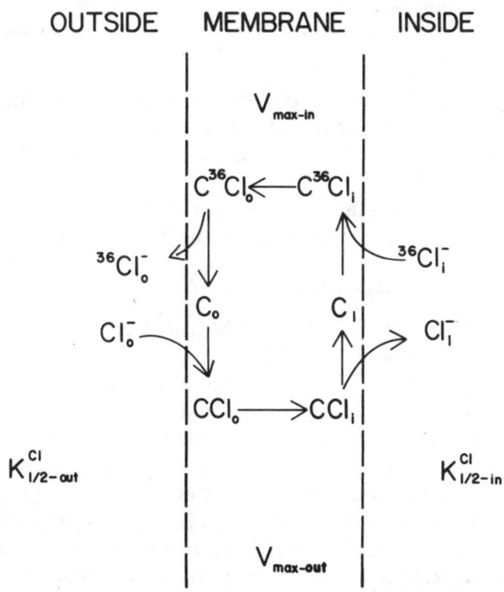

FIGURE 3. A scheme for a ping-pong anion exchange mechanism. ^{36}Cl on the inside reacts with an inward facing transporter, C_i, and the resulting complex $C^{36}Cl_i$ is transported to the outside where ^{36}Cl can decomplex into the outside solution. If Cl_i concentration is increased the flux is a hyperbolic function of Cl_i and is half-maximal at $K_{½\text{-in}}^{Cl}$ with a maximal flux, $V_{\text{max-in}}$. Analogous considerations for the influx partial reaction and increasing external chloride concentrations give the parameters $K_{½\text{-out}}^{Cl}$ and $V_{\text{max-out}}$.

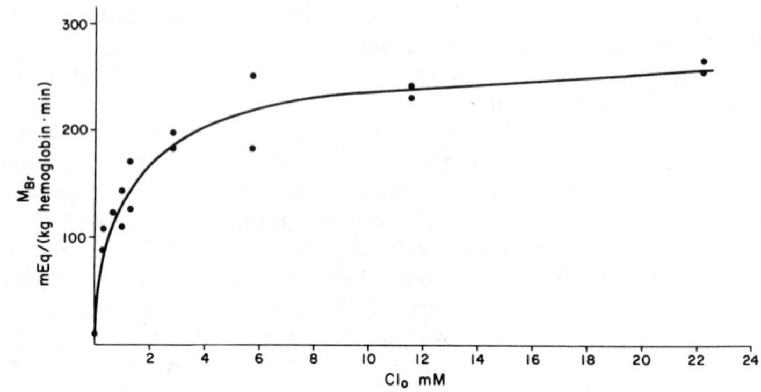

FIGURE 4. Stimulation of anion exchange by external exchangeable anions. ^{82}Br efflux from cells with 118 mEq Br/liter cell water was measured into isotonic, isoionic solutions of chloride, citrate and sucrose, pH 7.8 at 0° C. The activation of the anion exchange by external chloride was half-maximal at 0.84±0.14 mM. This value of $K_{½\text{-out}}^{Cl}$ is a function of internal anion concentration and of which anion is being exchanged for chloride. The $V_{max\text{-out}}$ measured here for chloride-bromide exchange is more than three times smaller than V_{max}^{max} for chloride exchange for chloride. This reflects the slower overall transport of bromide on the efflux limb of the exchange cycle. (From Gunn and Fröhlich.[11] By permission of *Journal of General Physiology*.)

The importance of FIGURE 5 for the present argument is the order of magnitude of $K_{1/2\text{-out}}$ on the ordinate. The maximal value is only 3 mM, so that the external sites are half saturated at 3 mM and completely saturated above 20 mM.

In contrast, the self-exchange flux and the internal $K_{1/2}$ for chloride are about 65 mM.[11, 12] For a ping-pong mechanism it can be shown that

$$K_{1/2\text{-self exchange}} = K_{1/2\text{-out}} + K_{1/2\text{-in}}$$

If $K_{1/2\text{-out}}$ is 3 mM and $K_{1/2\text{-in}}$ is 60 mM, the $K_{1/2\text{-self exchange}}$ will be 63 mM. Both the $K_{1/2}$ for self-exchange and $K_{1/2\text{-in}}$ have approximately these values. In contrast, the V_{max} values for the two sides are approximately equal.[11] These data show that at fixed pH (7.8 at 0° C), the entry of chloride is half-saturated at about 3 mM chloride concentration on the outside while the exit of chloride is half-saturated at 65 mM chloride on the inside. Because the exchange system is highly coupled and influx nearly equals efflux, the efflux step which is not saturated at physiological chloride concentrations (110 mM) is the rate-limiting step under these conditions. In the complete exchange cycle of chloride entry and chloride exit, it is the reactions of the internal facing carrier that complexes with the chloride, then moves it out and decomplexes the chloride, which are rate limiting. We, therefore, conclude that the chloride transport mechanism is asymmetric with respect to its reactions with chloride and it is the efflux half-reaction with the much greater $K_{1/2}$ that is rate limiting.

It is interesting now to reinterpret Dalmark's finding that the $K_{1/2}$ of self-exchange increased or decreased with increasing pH values depending on whether the external or internal chloride concentration was used. We propose that the internal chloride concentration is the correct independent variable for the following reason. $K_{1/2\text{-self-exchange}}$ is the sum of $K_{1/2\text{-out}}$, which is small and little affected by pH, and $K_{1/2\text{-in}}$, which is 15 times larger. The decrease in $K_{1/2\text{-self-exchange}}$ with increasing pH values primarily reflects a decrease in $K_{1/2\text{-in}}$ as the proton concentration falls as expected if protons are competitive inhibitors on the inside. But the inhibition is not simply competitive because the maximum flux declines as well, nor is it noncompetitive or uncompetitive since the $K_{1/2\text{-in}}$ increases with increasing internal protons when no change or a decrease in $K_{1/2\text{-in}}$ would be expected by an inhibitor of these types. The data of Dalmark suggest that if protons have little or no external effect on chloride transport, then on the inside they have a complex mode of inhibition. We will now show that external protons have little effect on chloride transport.

The second interesting asymmetry of the anion exchange mechanism is that the inhibition of chloride flux that occurs at low pH does not involve a titration reaction on the outside of the cell, but, by exclusion, must occur only on the inside of the cell. In order to show on which side of the membrane the inhibition of chloride occurs, we measured chloride efflux from cells with

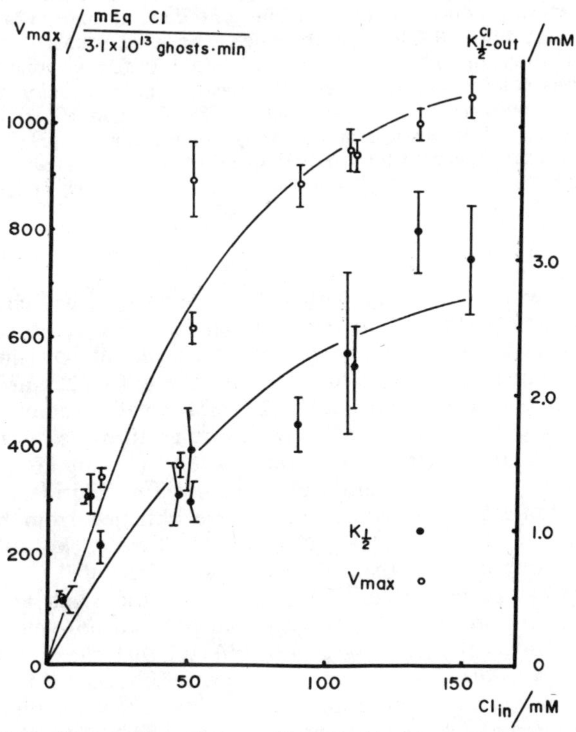

FIGURE 5. $K_{½\text{-out}}^{Cl}$ and $V_{max\text{-out}}$ for chloride-chloride exchange as a function of intraerythrocyte chloride concentration, Cl_{in}. The solid lines are the best hyperbolic curves that fit a ping-pong scheme. $K_{½\text{-out}}^{Cl}$ has a maximum value of about 3 mM and is half-maximal at Cl_{in} equal 60 mM. All experiments were performed at pH 7.8, 0° C with chloride-citrate-sucrose solutions. (From Gunn and Fröhlich.[11])

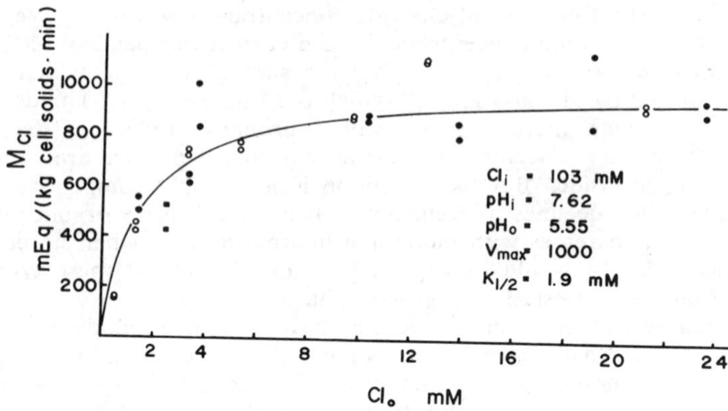

FIGURE 6. Chloride efflux into low pH media. Cell samples from one suspension, loaded with ^{36}Cl, were effluxed into different external chloride solutions at pH 5.55 at 0° C. The solid circles are from experiments in which the cells were placed into 25 mM gluconate with increasing chloride concentrations and decreasing sucrose to maintain isotonicity. The open circles are from experiments in which the efflux media had constant sucrose concentration and constant ionic strength (25 mM) by substituting gluconate for chloride, one for one. The solid square symbols are from experiments in which the efflux was measured into 160 mM gluconate with 2.5 mM Cl_o. The increased ionic strength does not appear to have a major effect. The best hyperbolic curve is shown with its parameters, $K_{½\text{-out}}^{Cl}$ and $V_{max\text{-out}}$. $V_{max\text{-out}}$ was 1000 mEq/(kg cell solids·min), approximately the same as at pH 7.8. But FIGURE 1 shows the self-exchange with low pH values on either side has about one-tenth this flux at pH 5.55. The $K_{½\text{-out}}$ is lower than $K_{½\text{-out}}=5.0$ mM in similar media at pH 7.8.

internal pH = 7.6 into solutions with pH 5.55 and different chloride concentrations. This was performed in two different ways, as shown in FIGURE 6. The solid points are the results when cells were placed into media with different chloride concentrations, different ionic strengths (25 mM to 50 mM) and constant gluconate concentrations (25 mM) made isotonic with sucrose. The open circles are the results after the same erythrocytes were placed into media with different chloride concentrations but constant ionic strength (25 mM)—made up by the substitution of gluconate for chloride and again with sucrose for isotonicity. The same results were obtained from both methods. The $K^{Cl}_{1/2\text{-out}}$ was 1.9 mM and $V_{max\text{-out}}$ was 1000 mEq/(kg cell solids·min). The $K^{Cl}_{1/2\text{-out}}$ value is lower than if $pH_{out} = 7.8$ where $K^{Cl}_{1/2} = 5.0$ mM with 160 mM gluconate substitution. The $V_{max\text{-out}}$ is the same as for fluxes at $pH_{out} = 7.8$. Therefore, lowering the external pH had no effect on the maximum transport rate. It is possible, but unlikely, that the proton reaction is slow compared with the time course of these experiments (2–10 seconds) and was, therefore, not seen. A more likely possibility is that inhibition by protons can only occur from the inside solution, which had an initial pH of 7.6 in these experiments.

Against the notion that no titration reaction occurs on the outside is the observation of Jennings [8] that protons and sulphate are cotransported by the anion pathway from outside to inside in exchange for intracellular chloride.

Protons, therefore, react with the carrier on the outside when sulphate is present, but the external proton alone does not inhibit chloride fluxes. If only a few external sites compared with the total number of chloride exchange sites were involved in sulphate transport, this difference in the effect of external protons might be explained. However, two sets of data are against this. First, titration with the irreversible inhibitor DIDS yields approximately the same number of sites for 100% inhibition of chloride transport and sulphate transport [13, 14]; and second, the fractional inhibition is the same for sulphate and chloride transport for several reversible inhibitors.[15] Therefore, in terms of the reactivity with DIDS and other inhibitors, sulphate transport sites behave as chloride transport sites. Second, sulphate competes with chloride on the external sites as if they were a single homogeneous class of sites on the transporters.

In FIGURE 7, the efflux of chloride into isotonic, isoionic chloride-gluconate solutions at pH 5.55 is shown when 1.2 mM sulphate is present. The $K^{\mathrm{Cl}}_{1/2\text{-out}}$ is 9.6 ± 1.6 mM, approximately five times the value in the absence of sulphate. From this value and two other experiments at different sulphate concentrations, we conclude that sulphate is competitive with chloride on the outside with a K_I around 0.5 mM at pH 5.55. These data are against sulphate transporters being a small subclass of the chloride transporters. Contrast the observations that protons do not react with the outward-facing site to reduce chloride fluxes, and are cotransported with sulphate from the outside to the inside, with the observation that sulphate and chloride compete for a common pool of external transport sites!

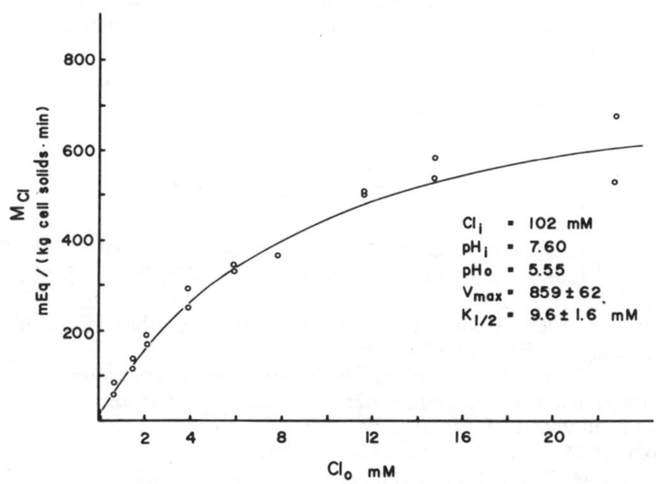

FIGURE 7. Chloride efflux into low pH media with 1.2 mM sulphate at different chloride concentrations. These experiments were similar to those in FIGURE 6 except gluconate and chloride were substituted one for one at a higher ionic strength (160 mM) and sulphate was present in the solutions. Sulphate greatly increased the $K_{1/2\text{-out}}^{\mathrm{Cl}}$ under these conditions although sulphate transport is nil at 0° C. While protons alone do not increase $K_{1/2\text{-out}}^{\mathrm{Cl}}$, sulphate does increase $K_{1/2\text{-out}}^{\mathrm{Cl}}$ especially at low external pH values.

The simplest resolution of these apparently contradictory data is made by assuming that the inhibition of chloride self-exchange by protons occurs only at an inside site in the absence of sulphate (or phosphate), and that sulphate binding to an external-facing transport site changes the pK of an external site, which then can bind protons; and this binding then permits the conformational change of proton-sulphate influx. We are, therefore, suggesting an ordered addition of a sulphate ion and then a proton to the outside site since we have shown that without sulphate protons do not form significant numbers of complexes with the outside-facing transport sites.

Another simple resolution of these data would be that protonation does occur on the outside but it does not affect the reaction with external chloride or its transport. That is, the CCl and CHCl complexes of the carrier, C, are equally well-transported. This would imply that the proton flux could be almost as great as the chloride flux via CHCl complexes; however, Jennings[9] has shown that proton flux is 10^4 slower than chloride, even in the absence of sulphate. We, therefore, exclude this possibility and conclude that there is almost no protonation reaction on the outside in the absence of sulphate.

In summary, the anion exchange mechanism is asymmetric with respect to its reaction with anions and with respect to protons. The outside site and the entry reaction ($K_{1/2\text{-out}}$) is half-saturated at 3 mM Cl^- whereas the internal site and efflux reaction ($K_{1/2\text{-in}}$) is half-saturated at 65 mM Cl^-. Protons in the outside solution do not diminish maximal chloride exchange or increase $K^{Cl}_{1/2\text{-out}}$. Therefore, the reactions $C + H \rightleftharpoons CH$ and $CCl + H \rightleftharpoons CHCl$ do not occur on the outside to a significant extent for chloride exchange, but the inhibition of chloride exchange by protons must occur from the inside. We suggest that an ordered complexation of a sulphate and then a proton precedes sulphate-proton cotransport in order to explain the competition between sulphate and chloride for a common set of transporters on the outside, and to explain the failure of protons on the outside to inhibit chloride transport in the absence of sulphate.

References

1. Gunn, R. B., M. Dalmark, D. C. Tosteson & J. O. Wieth. 1973. Characteristics of chloride transport in human red blood cells. J. Gen. Physiol. **61:** 185–206.
2. Passow, G. 1969. Passive ion permeability of the erythrocyte membrane. Prog. Biophys. Mol. Biol. **19:** 425–446.
3. Schnell, K. F., S. Gerhardt & A. Schoppe-Fredenburg. 1977. Kinetic characteristics of the sulfate self-exchange in human red blood cells and red blood cell ghosts. J. Membr. Biol. **30:** 319–350.
4. Funder, J. & J. O. Wieth. 1976. Chloride transport in human erythrocytes and ghosts: A quantitative comparison. J. Physiol. **262:** 679–698.
5. Gunn, R. B. 1972. A titratable carrier model for both mono- and divalent anion transport in human red blood cells. *In* Oxygen Affinity of Hemoglobin and Red Cell Acid Base Status. M. Rørth & P. Astrup, Eds. : 823. Mungsgaard. Copenhagen.
6. Dalmark, M. 1975. Chloride transport in human red cells. J. Physiol. **250:** 30–64.
7. Cass, A. & M. Dalmark. 1973. Equilibrium dialysis of ions in nystatin-treated red cells. Nature New Biol. **244:** 47–48.

8. JENNINGS, M. L. 1978. Proton fluxes associated with erythrocyte membrane anion exchange. J. Memb. Biol. **28:** 187.
9. JENNINGS, M. L. 1978. Characteristics of CO_2-independent pH equilibration in human red blood cells. J. Membrane Biol. **40:** 365.
10. GUNN, R. B., J. O. WIETH & D. C. TOSTESON. 1975. Some effects of low pH on chloride exchange in human red blood cells. J. Gen. Physiol. **65:** 731–749.
11. GUNN, R. B. & O. FRÖHLICH. 1979. Asymmetry in the mechanism for anion exchange in human red cell membranes: Evidence for reciprocating sites that react with one transported anion at a time. J. Gen. Physiol. **74:** 351–374.
12. BRAZY, P. C. & R. B. GUNN. 1976. Furosemide inhibition of chloride transport in human red blood cells. J. Gen. Physiol. **68:** 583–599.
13. LEPKE, S., H. FASOLD, M. PRING & H. PASSOW. 1976. A study of the relationship between inhibition of anion exchange and binding to the red blood cell membrane of 4,4′-diisothiocyano stilbene-2,2′-disulfonic acid (DIDS) and its dihydro derivative (H_2DIDS). J. Membrane Biol. **29:** 147–177.
14. FUNDER, J., D. C. TOSTESON & J. O. WIETH. 1978. Effects of bicarbonate on lithium transport in human red cells. J. Gen. Physiol. **71:** 721–746.
15. KU, C-P., M. L. JENNINGS & H. PASSOW. 1978. A comparison of the inhibitory potency of reversibly acting inhibitors of anion transport on chloride and sulfate movements across the human red cell membrane. Biochim. Biophys. Acta **553:** 132–141.

DISCUSSION

J. M. SALHANY: In your experiments about the asymmetry in K_m, did you use an impermeable anion substitute for chloride?

R. B. GUNN: Yes. At high pH, we used citrate and at low, we used gluconate.

SALHANY: Under the assumption that those don't interact with anion transport?

GUNN: That's correct. Citrate at low pH, of course, becomes a divalent anion and it does have some inhibitory effect; but gluconate, to the best of our current knowledge, has very little effect. Perhaps, there could be five percent inhibition.

SALHANY: These pH dependent experiments were done with intact erythrocytes; have you tried resealed ghosts in which the hemoglobin concentration is lower because there is a pH dependence to hemoglobin binding to the cytoplasmic side of Band 3?

GUNN: No we haven't, but that point should be pursued.

TRANSPORT AND INTERACTIONS OF ANIONS AND PROTONS IN THE RED BLOOD CELL MEMBRANE

Jens Otto Wieth,* Jesper Brahm * and Jørgen Funder †

* *Department of Biophysics*
† *Institute of Medical Physiology, Dept. A*
The Panum Institute
University of Copenhagen
DK-2200, Copenhagen N

INTRODUCTION

The purpose of this paper is 1. To analyze the interactions of protons with the anion transport system of the red cell membrane; 2. To review two ways in which the anion transport system can mediate the transfer of hydroxyl ions or protons, and 3. To discuss the mode of operation of two classes of lipid soluble artificial ion carriers, inducing conductive proton transfer or electrosilent hydroxyl/anion exchange in red cell membranes and in lipid bilayers.

Interactions of Hydrogen Ions with Anion Transport

This section deals first with the unspecific high anion permeability observed below pH 5, and thereafter with the effect of protons on the specific anion exchange system at higher pH values.

Unspecific Membrane Permeability at Low pH Values

The chloride self-exchange flux in red blood cells has a maximum, which at 0° C is broad and located between pH 7 and 8 (FIGURE 1). The flux decreases with increasing hydrogen ion concentration and reaches a minimum at pH 5.1–5.3, both in red cells and in resealed erythrocyte ghosts (FIGURE 2). At even lower pH values, an increased chloride permeability is observed (FIGURE 2). The kinetic properties of chloride transport at and below pH 5 differ fundamentally from the properties found at higher pH-values, suggesting that a different transport mode becomes significant at hydrogen ion concentrations above 10^{-5} M.[1] As shown in FIGURE 3, a gradual increase of chloride permeability with time can already be observed, when pH is lowered from 5.8. The chloride permeability increased about 25% per hour at pH 5.1. The work of Bjerrum[2,3] has made it clear that the permeability increase is correlated to a sequence of morphological changes in the cell membrane, initiated by the aggregation of cytoskeleton proteins. The mobility of integral membrane proteins is restricted by a fibrillar network of spectrin molecules located on the cytoplasmic side of the membrane. The hydrophobic integrity of the membrane depends on hydrophobic interactions between lipids and the integral proteins penetrating the hydrophobic membrane core. Aggregation of spectrin

0077–8923/80/0341–0394 $01.75/0

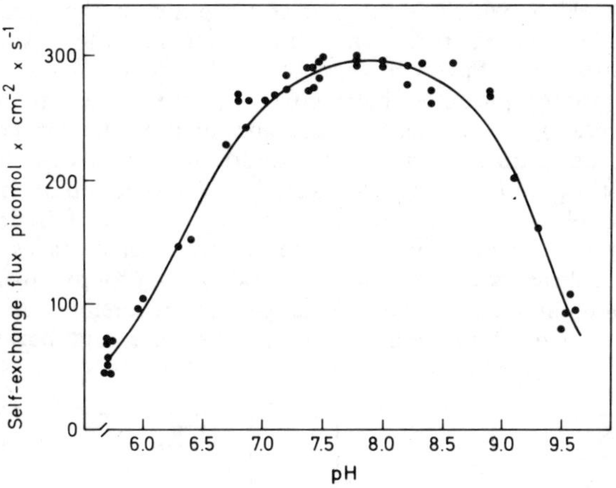

FIGURE 1. The pH dependence of the chloride self-exchange flux in human erythrocytes at 0° C. A membrane area of one cm^2 corresponds to 7×10^5 cells with a mean area of 1.42×10^{-6} cm^2. (From Gunn *et al.*[16])

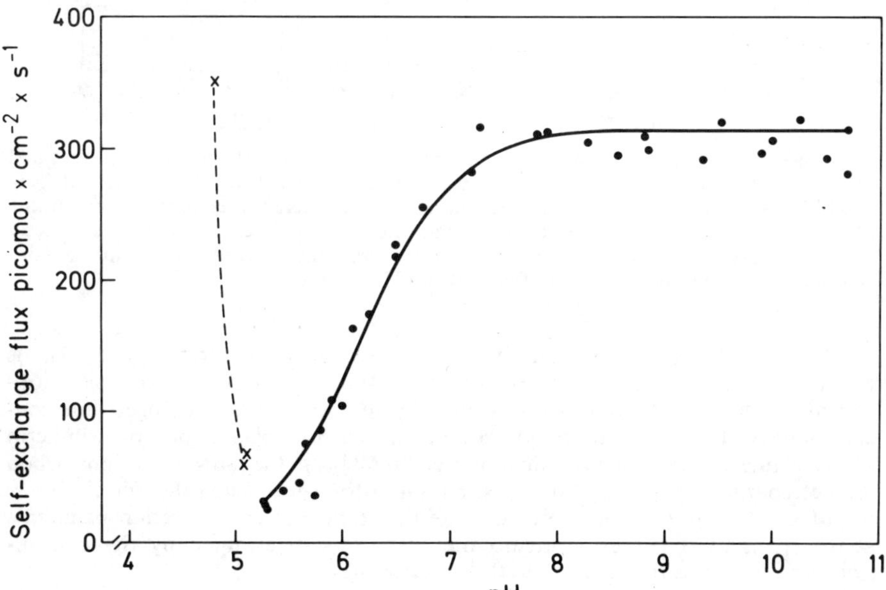

FIGURE 2. The pH dependence of the chloride self-exchange flux in resealed human erythrocyte ghosts at 0° C. Intracellular chloride concentration of ghosts does not vary with pH in contrast to the decrease found in erythrocytes when pH is increased. A considerable unspecific membrane permeability to chloride developed below pH 5.1. (From Funder and Wieth.[9] By permission of the *Journal of Physiology*.)

taking place at pH 5, which is the isoelectric point of the spectrin molecule, suggests that electrostatic forces are important for the maintenance of a normal microstructure. Spectrin aggregation is accompanied by aggregation of the intramembrane particles, believed to represent integral membrane proteins. The morphological changes are accompanied by the formation of hydrophilic channels permitting unspecific transport of small hydrophilic molecules and ions, but still narrow enough to exclude the larger inulin molecule.[3] It is likely that hydrophilic channels are formed, when areas of hydrophobic lipid-protein contact are replaced with protein-protein contact faces, involving the formation of salt-bridges between ionized and polar groups on the surfaces of individual protein molecules. These changes are reversible. Membranes that are retitrated to neutral pH regain normal structure and permeability.[3]

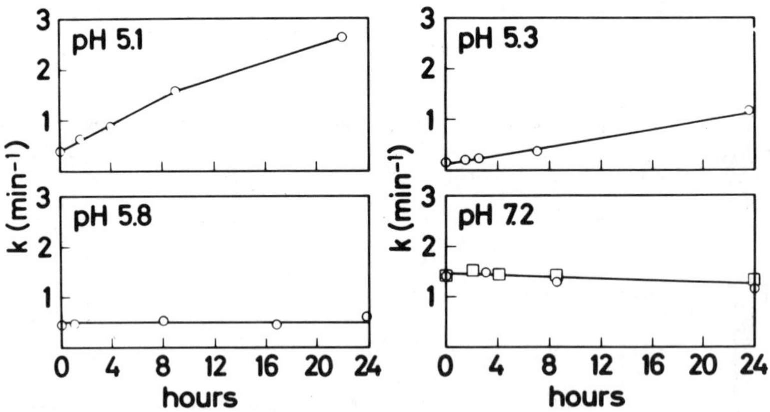

FIGURE 3. Time dependence of the rate of chloride self-exchange in resealed human erythrocyte ghosts between pH 5.1 and 7.2 (0° C). The increased permeability with time observed below pH 5.8 can be correlated to changes of membrane ultrastructure.[2,3] The rate coefficient of chloride self-exchange (k, min^{-1}) is directly proportional to the chloride self-exchange flux because intracellular chloride concentration and cell volume does not vary with pH.[9]

The net charge of spectrin becomes positive when titration with H^+ is carried beyond the isoelectric point. The unspecific chloride permeability observed below pH 5 increases proportionally to the squared hydrogen ion concentration.[1] It has been demonstrated that a monolayer of spectrin-actin behaves like an ion-selective membrane, facilitating the flow of anions when the net charge is positive, and impeding anion transport when the net charge is negative.[4] One might, therefore, surmise that the local concentration of anions at the shear of the interior membrane surface is determined by the concentration of fixed positive charges in the spectrin mesh.

Chloride Exchange between pH 5 and 7

The specific anion transport system, which performs a tightly coupled 1:1 exchange diffusion of anions across the red cell membrane, can be studied above pH 5. There is much evidence that the transport is mediated by an integral membrane protein with a molecular weight of $\sim 10^5$ dalton, as reviewed in the section on biochemical aspects of anion transport in the erythrocyte.[5–7] The exchange flux decreases steeply below pH 7 both in erythrocytes and in resealed ghosts. (FIGURES 1 and 2). The shape of the graphs relating flux to pH suggests that the transport system is inactivated when one proton is bound to a group in the transport system with a pK around 6. It has been maintained that the inhibition is effected by the binding of two hydrogen ions.[8] However, the evidence was based on few data, and the pH studies were not carried out below pH 5.9, where an extremely steep decrease of flux should have been found if the supposition was correct. Other investigators have found that the inhibition by H^+ as shown in FIGURES 1, 2, and 4 is adequately described at 0 and 38° C by assuming that transport is inhibited by protonation of a single group.[9, 10] The effect of hydrogen ions on anion transport is fully reversible.[16]

The titratable group is not identical with the anion binding site of the transport protein. Early interpretations of the effects of protons on transport of monovalent anions assumed that hydrogen and chloride ions compete for the site.[11] If this was so, the apparent pK should shift with increasing chloride concentration, a prediction that has not been verified. On the contrary, the apparent pK is insensitive to variations of chloride concentration between 150 and 600 mM,[8] demonstrating that the inhibition of chloride transport by hydrogen ions is of a noncompetitive nature.

The titratable group seems to be located near the protoplasmic face of the membrane. This conclusion is based on experiments in which cells with an intracellular pH of 7.3 were suspended in electrolyte media with pH-values from 5.6 to 7.4. Determinations of chloride self-exchange showed that the transport is relatively insensitive to the extracellular pH (TABLE 1). If the hydrogen ion binding group was exposed directly to the extracellular medium, one would anticipate a reduction of transport similar to the inhibition found in preequilibrated cells (FIGURES 1 and 2). Reduction of the extracellular pH from 7.4 to 5.6, which caused a 75% reduction of chloride transport in steady state experiments (FIGURES 1 and 2), only reduced the chloride self-exchange by 25% in the unsteady state experiments shown in TABLE 1. The interpretation of the small decrease is uncertain. The experiments demonstrate clearly that the pH-sensitive group is not directly exposed to the extracellular medium, but it was impossible to prevent transport of H^+ through the membrane, during the experiments which lasted 10–20 sec. As dealt with in the following section, equilibration of pH can take place by two mechanisms. One is the transfer of OH^- across the membrane through the Jacobs-Stewart cycle. This path was efficiently eliminated in the present experiments by the inhibition of carbonic anhydrase with acetazolamide. The second mechanism is the HCl/Cl^- exchange described by Jennings.[12] HCl transport via this mechanism has been found to increase tenfold between pH 7 and 5.5, and, therefore, becomes increasingly important as a source of error, the more the extracellular pH is lowered in experiments of the type shown in TABLE 1. Because HCl transport is mediated by the anion transport system itself, we have not been able to circumvent the experimental problems in deciding whether the titratable group senses the intracellular pH, or whether it is buried deep in the membrane and senses a hydrogen ion concentration that is slightly higher than that of the cytoplasm.

TABLE 1

CHLORIDE SELF-EXCHANGE—EXPOSURE OF NORMAL ERYTHROCYTES TO ACID MEDIA *

Extracellular pH	Rate Coefficient of Chloride Self-Exchange (s^{-1})	Self-Exchange Flux picomol $\times cm^{-2} \times s^{-1}$
7.4	0.040	259
7.0	0.038	246
6.75	0.038	246
6.5	0.036	233
6.25	0.035	227
6.0	0.032	207
5.8	0.030	194
5.7	0.029	188
5.6	0.029	188

* Chloride self-exchange at 0° C of human erythrocytes equilibrated at pH 7.4, 0° C after loading with acetazolamide, 1 mM, for 30 min at 38° C. The flux medium contained 165 mM KCl, 2 mM Tris, 0.5 mM KH_2PO_4, and 1 mM acetazolamide. It was titrated to the pH indicated in this TABLE with small amounts of HCl. The decrease of flux with decreasing extracellular pH is considerably smaller than the decrease found in red cells, which have been prerequilibrated at a low pH (FIGURE 1). The experiments were performed in duplicate. The coefficient of variation of the determination of the rate coefficient of chloride exchange was 2.1%, $n=10$.

The chemical nature of the inhibitory proton-binding group is not known. It has been suggested that it becomes positively charged, when it binds a hydrogen ion,[11, 13] and that the transport system, at the same time, is converted from one form carrying monovalent anions to another mediating the much slower transfer of phosphate [14] and sulphate,[15] which are both transported at increasing rates, when pH is lowered from 7 to 6. It is likely that the titratable group is located in the transport protein. The number of candidates is limited by the pK value around 6. The apparent pK of the transport regulating acid group changes very little (0.1–0.2 pH unit) between 0 and 38° C (FIGURE 4). This excludes an imidazole group of histidine as a likely candidate, and attracts attention to groups with a sufficiently low ionization enthalpy as, e.g., carboxyl groups of aspartic or of glutamic acid. As to the nature of the noncompetitive inhibition by H^+, one may suppose that the transport molecule can only mediate exchange diffusion of monovalent anions when it is stabilized in a configuration by a salt linkage involving the negatively charged carboxyl group. The role of this group would thus be similar to that of a charged group localized in the vicinity of the active center of an enzyme, stabilizing the molecular conformation needed for catalytic function, although the group itself is not involved in catalysis.

Chloride Exchange above pH 7

The shape of the graph depicting chloride self-exchange in red cells versus pH (FIGURE 1) has been taken as evidence for the existence of a titratable group in the alkaline pH-range with a pK of 8.8.[11] It has been assumed that the transport system can only carry out the exchange diffusion if the alleged group is protonated.[11, 13] Today it seems clear that there is no functional group in the accessible pH range that can be titrated into a nonfunctioning form with hydroxyl ions. It was realized at an early state that desaturation of the transport system due to decreasing intracellular anion concentrations at alkaline pH-values may invalidate conclusions based on data like those of FIGURE 1. Dalmark [8] showed that the decrease of flux with increasing pH disappeared when intracellular chloride concentration was raised to 400–600 mM. It could not be excluded that a shift of apparent pK was induced at high chloride concentrations, but results obtained with resealed ghosts (FIGURE 2) showed that chloride transport at 0° C was constant up to pH 10.8 under conditions in which chloride concentration was 0.16 M and the ionic strength was in the physiological range.[9]

The insensitivity of the transport system to OH^- was also reflected in unsteady state experiments (TABLE 2), in which red cells with a normal intracellular chloride concentration and pH were exposed to electrolyte media with pH values between 7 and 10. The large variations of extracellular hydroxyl ion concentration did not have any effect on the magnitude of the self-exchange flux. These results demonstrate that the transport system is insensitive to high concentrations of hydroxyl ions at the external side of the membrane. By performing experiments of this type in weakly buffered media, it was also observed that hydroxyl ions are not transported by the anion transport system.

At 0° C, anion transport in ghosts could be examined up to pH 11 (FIGURE 2), the highest pH value tolerated by the membranes. The interpretation of results obtained with ghosts at 38° C is less clear. Brahm [10] found that the flux decreased gradually with about 15% per pH unit between pH 7

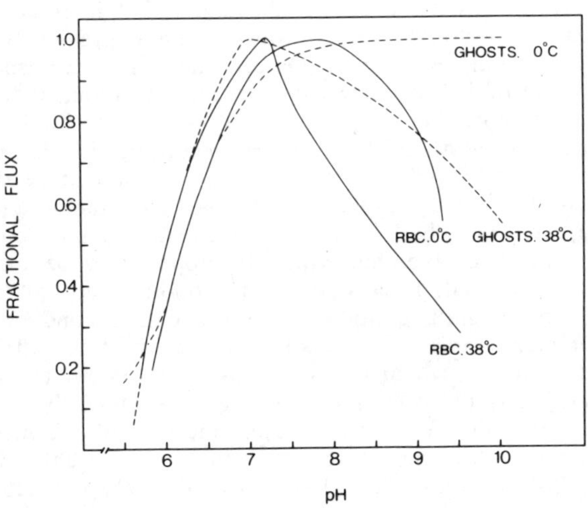

FIGURE 4. The fractional flux as a function of extracellular pH at 0° C and 38° C in erythrocytes and resealed ghosts at 0 and 38° C. (From Brahm.[10] By permission of the *Journal of General Physiology*.)

TABLE 2

CHLORIDE SELF-EXCHANGE—EXPOSURE OF NORMAL ERYTHROCYTES TO ALKALINE MEDIA *

Extracellular pH	Rate Coefficient of Chloride Self-Exchange (s^{-1})	Self-Exchange Flux picomol $\times$ $cm^{-2} \times s^{-1}$
7.0	0.043	278
7.6	0.047	304
8.0	0.043	278
9.0	0.044	285
9.5	0.044	285
10.0	0.047	298

* Chloride self-exchange at 0° C of human erythrocytes equilibrated at pH 7.4, 0° C after loading with acetazolamide, 1 mM, for 30 min. at 38° C. The flux medium contained 165 mM KCl, 2 mM Tris, 0.5 mM KH_2PO_4 and 1 mM acetazolamide. It was titrated to the pH indicated in this TABLE with small amounts of HCl or KOH. The self-exchange was insensitive to extracellular alkaline pH values.

and 10 at 38° C (FIGURE 4). The gradual decrease of transport with increasing pH bears no resemblance to that caused by the deprotonation of a single group with a well-defined pK. It would be worthwhile to examine whether the decrease of anion exchange with increasing pH found in ghosts at 38° C is reversible. If it is not, it would confirm our suspicion that the decrease of transport with increasing pH is due to a deteriorating effect of alkalinity on the transport system, unrelated to the functional state of a specific transport site.

Why does chloride transport decrease above pH 7–8 in intact red cells, if it is not due to the deprontonation of a transport site as previously assumed? Two factors have been held responsible: Desaturation of the transport system caused by the lowering of intracellular chloride concentration, and a redistribution of unpaired transport sites (S^+), which are assumed to be distributed in introvert (S_i^+) and extrovert (S_o^+) positions according to the membrane potential.[8, 10] The unidirectional flux of chloride is assumed to depend on the concentration of transport complexes (ClS) at each membrane surface. Under steady-state conditions, both anions and free sites should be distributed according to the membrane potential, which is supposed to be equal to the chloride equilibrium potential, because chloride conductance—although it is low—is still orders of magnitude higher than the cation conductances. As pointed out by Dalmark,[8] this would lead to a steady state distribution of anions and unloaded sites at which $Cl_i^-/Cl_o^- = S_o^+/S_i^+$. When the membrane potential at alkaline pH values becomes increasingly negative, the ratio S_o^+/S_i^+ decreases. Accumulation of sites at the cytoplasmic side of the membrane is supposed to compensate for the decrease of intracellular chloride, so the concentration of $(ClS)_i$ decreases less than it would if the carriers were not redistributed when membrane potential is altered. However, a comparison between the results of model calculations [8] and experimental data [8, 16] shows that the model predicts a smaller decrease of flux with pH, than has been found experimentally. It seems possible that the decrease of transport with increasing pH is due solely to the desaturation of the transport system, caused by the decrease of intracellular chloride concentration with pH. We have repeated the determinations of anion exchange at alkaline pH values (FIGURE 1). The flux was reduced by about 50% at pH 9.5, at which point the intracellular chloride concentration is about 30 mM when red cells are equilibrated in a 150 mM KCl medium at 0° C ($Cl_i^-/Cl_o^- = 0.2$). This is the same chloride concentration in which self-exchange is half-maximal when chloride concentration is varied symmetrically on both sides of ghost-membranes at neutral pH.[9] The studies of Gunn & Fröhlich [17] suggest that the affinity of the transport system for chloride differs on the two sides of the membrane. Their experiments showed that the affinity on the inside of the membrane was low, and one might, therefore, anticipate that the self-exchange flux would be halved at an intracellular chloride concentration of 20–30 mM, even in the absence of the alkaline intracellular pH.

Red cells have a negative membrane potential at alkaline pH-values. However, we have found that the exchange mechanism is not sensitive to variations of the potential. FIGURE 5 shows the results of two experiments, in which the influx of $^{36}Cl^-$ into ghosts with an intracellular KCl concentration of 165 mM was determined in an isotonic potassium citrate-sucrose medium [17] containing only 0.2 mM KCl. Variations of membrane potential were evoked by

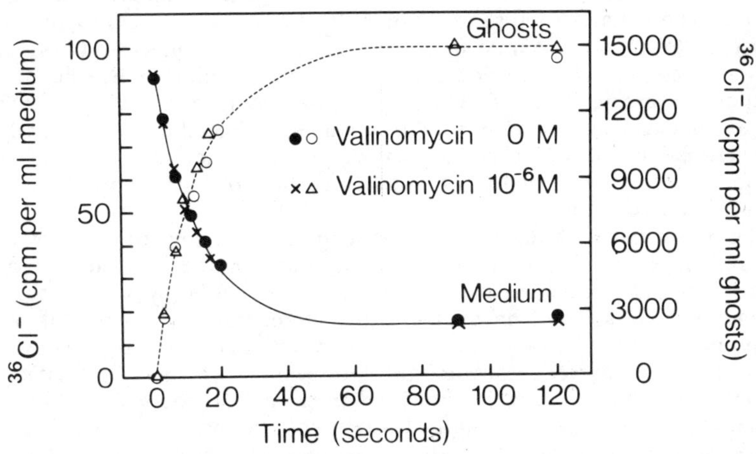

FIGURE 5. Lack of effect of membrane potential on the uptake of radioactive chloride into 165 mM KCl ghosts from a medium containing 0.2 mM chloride and 75 mM potassium in the absence and presence of 10^{-6} M valinomycin. The uptake of $^{36}Cl^-$ was monoexponential (k=0.08 sec^{-1}) both in the absence and presence of valinomycin. More than 80% of the radioactivity was taken up by the ghosts although the cytocrit was only 0.5%. After 90 and 120 sec the distribution of $^{36}Cl_{(ghosts)}/^{36}Cl_{(medium)}$ was 825. The steady state distribution ratio is stable, when bicarbonate transport via the Jacobs-Stewart cycle is prevented by incorporating acetazolamide into the ghosts.[18] Unidirectional chloride fluxes were 150 picomol× $cm^{-2}\times s^{-1}$ in both experiments. The isotonic medium contained potassium citrate 25 mM, sucrose 200 mM, acetazolamide 1 mM (pH 8.7, 10° C). Signatures: ●, x radioactivity in the medium in the absence and presence of 10^{-6} M valinomycin. Corresponding values for ghosts are designated ○ and △.

performing the influx experiment in the presence and absence of 10^{-6} M valinomycin. The figure shows the disappearance of $^{36}Cl^-$ from the medium (fully drawn curve, left hand ordinate) and the simultaneous accumulation of tracer in the ghosts (dotted line, right hand ordinate). Both the rates of chloride exchange and the steady state distribution ($^{36}Cl_i^-/^{36}Cl_o^- = 825$) were identical in the two experiments. It should be noted that more than 80% of the radioactivity was taken up by the ghosts, although the cytocrit was only 0.5%, simply because more than 80% of the chloride was located in the intracellular phase. The identical rates of chloride transport in the absence and presence of valinomycin demonstrate that the transport mechanism does not react to large variations of membrane potential. The exact values of the membrane potential are not known, but their magnitude can be estimated, and were confirmed by control experiments, using the fluorescent dye di-I-C_3 (5).[19] In the absence of valinomycin the membrane potential approaches the chloride equilibrium potential of 160 mV (inside positive), because the chloride conductance is about two orders of magnitude larger than the potassium conductance.[26] It was estimated that the membrane potential was close to zero in the presence of valinomycin at a concentration that increases potassium permeability (and conductance) 10^3 times. A similar insensitivity of the rate of chloride exchange to variations of membrane potential was found in experiments (not shown), where the rate of $^{36}Cl^-$ influx was determined in 0.3 M sucrose media containing 0.2 mM KCl. Under these conditions, the membrane potential varies more than 200 mV, from about 160 mV in the absence of to about —75 mV in the presence of valinomycin. These experiments showed that the chloride transport is barely affected by the dramatic reduction of extracellular ionic strength, and confirmed that the exchange mechanism does not react to sudden changes of membrane potential.

Investigations of the chemical nature of the anion binding site of the transport system is the object of another section of this symposium, so we will only add a few comments.

The insensitivity of the transport system to increasing pH argues against the direct involvement of charged amino-groups in the transport of anions. We have recently developed a method for the study of bicarbonate self-exchange in ghosts[18] and have shown that anion transport is not inhibited in membranes exposed to a carbon dioxide pressure of 700 mm Hg,[20] when care is taken to prevent concomitant intracellular pH-changes. This shows that anion transport is not inhibited when positively charged amino groups are transformed to negatively charged carbamino-groups.

It has been suggested that the functional transport group might be a guanidino-group of an arginine residue.[9] The guanidino group has a high ionization enthalpy. Therefore, the pK will shift towards less alkaline values with increasing temperature, but the pK will still be outside the range accessible to transport studies, if it, e.g., changes from 13 to 11 when temperature is increased from 0 to 38° C.

We have attempted to block membrane guanidino-groups by treating intact membranes with 1,2 cyclohexanedione,[21] without finding any effect on chloride transport. However, it is obvious that the reagent may not have gained access to the functional group. Treatment of red cell ghosts with pyridine-2,6-dicarboxylic acid in a concentration that was sufficient to remove all zinc ions from carbonic anhydrase,[41] had no effect on chloride and bicarbonate transport. Again the negative result only allows the conclusion that if the transport molecule is a metallo-protein, the metal ion is not readily accessible to the complexing reagent: a divalent anion that rapidly converts carbonic anhydrase into its apoenzyme by removing the zinc ion of the native enzyme.[41]

Diisothiocyanostilbene disulfonic acid (DIDS) binds covalently to the red cell anion transport protein, and appears to be a specific inhibitor of anion exchange in red cells. It has been established in several laboratories[18, 22, 23] that complete inhibition of anion transport is obtained after binding of 10^6 DIDS molecules per cell membrane, a number corresponding surprisingly well to the number of molecules of the purported anion transport protein. FIGURE 6 relates membrane binding of DIDS to the fractional inhibition of chloride

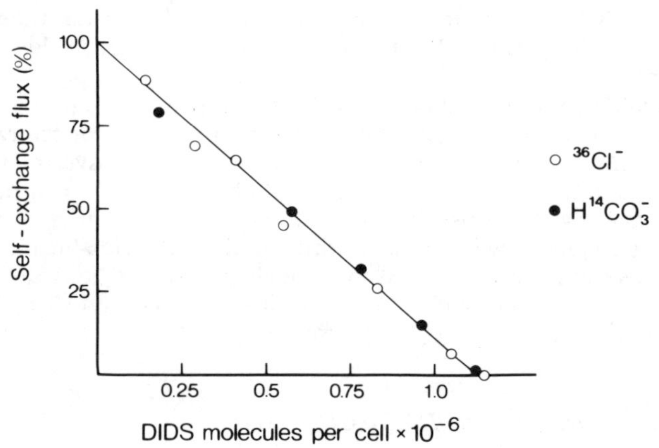

FIGURE 6. Inhibition of chloride (○) and of bicarbonate (●) self-exchange fluxes in ghosts prepared from DIDS-treated human red cells. The fractional flux is shown as a function of the number of DIDS-molecules irreversibly bound per cell. The ordinate represents the fractional self-exchange flux in percentage of the value found in control samples of untreated ghosts. (From Wieth.[18] By permission of the *Journal of Physiology*.)

and bicarbonate exchange. It is this type of documentation that suggests that a single group in the membrane protein may be responsible for anion binding and transport function. However, it may be fruitful to keep in mind that anion binding does not necessarily involve a single group, just as no single carbonyl group is more responsible than another for the binding of a potassium ion to a valinomycin molecule. It seems possible that more than a single group in the anion transport protein is involved in the change of molecular configuration that leads to exchange diffusion of anions through the protein. The molecular events may turn out to be similar to the mechanism of "induced fit" in enzyme kinetics,[24] in which the interaction between substrate and the active site leads to breakage of salt linkages, bringing the protein into an optimal conformation for catalysis. In the case of the exchange diffusion, the induced change of configuration of the transport protein is assumed to permit the facilitated diffusion of anions.

Mechanisms of Proton and Hydroxyl Ion Transfer

Hydrogen and hydroxyl ions can be transported through the red cell membrane by at least three mechanisms to be considered in this section. Dissipation of an induced pH-gradient proceeds extremely slowly, if the anion transport system of the cells has been inhibited. This low residual proton and hydroxyl ion permeability of DIDS-treated red cells is dealt with in the last part of the section. Before that we will consider the ability of the anion exchange mechanism to equilibrate a hydrogen ion gradient across the red cell membrane. There is no direct way to determine whether pH-equilibration is mediated by the transport of protons, hydroxyl ions or by transport of both ion species. As a first thought one might anticipate that OH^- can exchange with Cl^- or with other monovalent anions by exchange diffusion. However, it appears that the permeability to hydroxyl ions is extremely low in the absence of bicarbonate.[12]

In addition the data obtained with ghosts at pH 11 (FIGURE 2) show that chloride self-exchange is not inhibited by hydroxyl ion concentrations of about 1 mM. This does not mean that the anion exchange system is not involved in proton and hydroxyl ion transfer. Equilibration of pH can proceed through at least two mechanisms that depend on an intact anion exchange system. One is the Jacobs-Stewart cycle,[25] which utilizes chloride-bicarbonate exchange as one of several steps involved in smuggling hydroxyl ions through the membrane. The second mechanism is a cotransport of H^+ and Cl^-, which according to the findings of Jennings[12] requires an intact anion transport system.

Transfer of OH^- by the HCO_3/CO_2 Cycle

Jacobs & Stewart[25] were first to realize that transport processes involving the transfer of acid or base equivalents across the red cell membrane are enormously facilitated in the presence of bicarbonate. In a lucid set of experiments, they demonstrated that the pH-equilibration also depends on the presence of carbonic anhydrase, permitting the conclusion that transfer of OH^- through the membrane is "catalyzed" by a series of reactions initiated by the exchange, e.g., of intracellular chloride with extracellular bicarbonate, followed by the intracellular dissociation of HCO_3^- into OH^- and CO_2. The hydroxyl ion is taken care of by the buffers in the cell, and CO_2 leaves the cell readily by diffusion. The catalytic transfer of OH^-, proceeding without consumption of bicarbonate, is completed, when extracellular carbon dioxide ultimately is rehydrated to re-form an extracellular bicarbonate ion. The absence of carbonic anhydrase in plasma makes hydration of CO_2 rate limiting for pH-equilibration across the membrane in the capillary blood. Forster and collaborators[40] have shown that this is physiologically important, because the brief capillary transit times of an erythrocyte (0.3–1.0 s) are not long enough to ensure completion of pH-equilibration between red cells and plasma.

The role of intra- and extracellular chloride for the steady state distribution of OH^- and H^+ is illustrated in FIGURES 7a and 7b. Since the original work of Warburg,[26] it has repeatedly been confirmed that chloride, hydroxyl, and hydrogen ions establish partial ionic equilibria across the membrane, where the ratios:

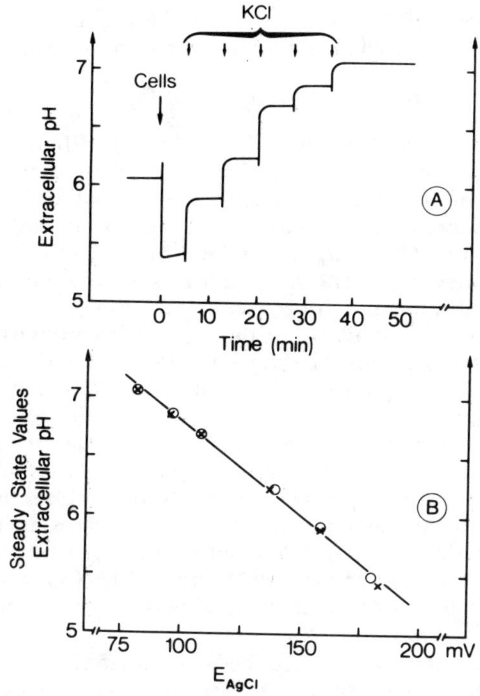

FIGURE 7. Hydroxyl ion equilibration via the Jacobs-Stewart cycle.[25] KCl-washed red cells with an intracellular pH of 7.6 were suspended at a hematocrit of 2% in a 0.3 M sucrose medium, initially containing 0.3 mM KCl (15° C). FIGURE 7A shows that addition of cells to the medium caused a fall of extracellular pH to 5.4 where $Cl_i^-/Cl_o^- = OH_i^-/OH_o^- = H_o^+/H_i^+ = 160$ (equation 1). Subsequent additions of increasing volumes of 150 mM KCl made the extracellular pH change to new steady state levels satisfying equation 1. Intracellular chloride concentration and pH are constant during the experiment. Extracellular chloride and hydrogen ion activities were determined with pH- and Ag/AgCl electrodes. The relation between the steady state signals of the two electrodes from duplicate experiments is shown in FIGURE 7B. Both potentials were measured with a calomelelectrode as a reference. Steady state extracellular pH was a linear function of E_{AgCl}: $pH = -0.16\ E_{AgCl} + 8.41$ (correlation coefficient 0.997). The slope of the relation averaged −57.7 mV/pH (S.D. 2.6 mV) in a series of 12 expts. at 15° C. The ideal slope is −57.2 mV/pH.

$$Cl_i^-/Cl_o^- = OH_i^-/OH_o^- = H_o^+/H_i^+ \quad \textbf{(1)}$$

are equal, when the anion distribution between the intracellular phase (i) and the extracellular phase (o) is in steady state. Until the nature of the exchange diffusion was discovered, it was assumed that chloride and hydroxyl ions were transported by electrodiffusion, implicating that the distributions expressed by equation 1 represent electrochemical ionic equilibria.[27] It is now clear that the quantitatively important route of pH equilibration is via the Jacobs-Stewart cycle, and that a normally functioning anion exchange is a prerequisite for the validity of equation 1. With our present knowledge about anion transport in red cells, it is more correct to state that the electrosilent chloride and bicarbonate exchange rapidly establishes a situation where:

$$Cl_i^-/Cl_o^- = HCO_{3(i)}^-/HCO_{3(o)} \quad \textbf{(1b)}$$

because the two anions exchange readily if their distribution ratios differ.[18] Thus, the equality of the ion distribution ratios expressed in equation 1 is a simple consequence of the situation described by equation 1b, if equilibrium has been established between the various derivatives of carbonic acid. The steady state distributions are solely determined by the anion concentrations of cells and media. They represent a chemical steady state, rather than an electrochemical equilibrium, and are consequently insensitive to large variations of the membrane potential, similarly to the voltage insensitivity of the electrosilent exchange diffusion of anions shown in FIGURE 5. Under conditions in which the membrane conductance is dominated by the chloride conductance, the membrane potential of the red cell will be equal to the chloride equilibrium potential, which can be calculated from the Nernst equation. This is the case in the intact red cell, where the chloride conductance is about 10^2 times higher than that of sodium plus potassium.[28] By treating the membranes with 1 μM valinomycin, the potassium conductance increases to a value that is about tenfold higher than the chloride conductance.[28] Nevertheless, the same chemical ionic equilibria, determined by the ratio Cl_i^-/Cl_o^- can be established in intact and in valinomycin-treated cells, although the membrane potentials may differ by 100 mV or more.

The steady state distributions described by equations 1 and 1b are easy to demonstrate experimentally by a variation of a classical experiment,[29] that illustrates the relation between chloride and hydrogen ion distribution when red cells are suspended in an isotonic sucrose medium with a low chloride content. FIGURE 7 shows the results of an experiment, which we have now used for 10 years as a students' practical to demonstrate the validity of equation 1. Packed red cells were suspended at a hematocrit of 2% in an unbuffered 0.3 M sucrose solution, initially containing 0.3 mM potassium chloride. The extracellular pH rapidly fell to 5.4, the pH value at which equation 1 is satisfied. After the initial fall of pH, the extracellular chloride concentration was increased stepwise by adding increasing volumes of 0.15 M KCl to the suspension. The extracellular pH and chloride concentration were both monitored continuously with glass and Ag/AgCl electrodes. Every time the KCl concentration in the medium was increased, pH rose to a new steady state level as shown in FIGURE 7A. The perfect accordance between chloride, hydroxyl, and hydrogen ion distribution was demonstrated by relating steady state values of extracellular pH after each KCl-step to the potential measured by the AgCl electrode. The results are shown in FIGURE 7B, which demonstrates that a tenfold increase of chloride concentration (altering the potential of the silver chloride electrode by —60 mV) induced the predicted change of extracellular pH by one unit (indicating a tenfold decrease of hydrogen ion concentration). The experiment of FIGURE 7 was carried out with KCl washed cells, and with solutions that were equilibrated with room air, having a pCO_2 of ~0.2 mm Hg. The intracellular pH of the packed, KCl-washed cells used for the experiment varied, at most, 0.02 units during the experiment because the medium was unbuffered. It is easy to demonstrate that the net hydroxyl ion transfer between cells and medium is mediated by the Jacobs-Stewart cycle: 1. The rates of equilibration are accelerated by adding small amounts of carbonic anhydrase to the extracellular phase, catalyzing the hydration of CO_2 to HCO_3^- and H^+ during the initial pH-fall, and the subsequent formation of CO_2 from HCO_3^- and H^+ during the stepwise removal of hydrogen ions from the extracellular phase. 2. The rates of pH equilibration are further increased by addition of micromolar concentrations of bicarbonate. 3. The equilibration of pH does not occur if exchange diffusion of anions is inhibited (as in DIDS treated cells). 4. The rate of pH equilibration is significantly retarded if intracellular carbonic anhydrase is inhibited by preloading the cells with the slowly permeating carbonic anhydrase inhibitor, acetazolamide, or if the experiment is carried out in CO_2-free solutions. The insensitivity of the cycle, and therefore of the ionic equilibria, to the membrane potential is evident from the finding that both the rate of pH decrease and the steady state value of pH are unchanged in short-lasting experiments using cells treated with sufficient valinomycin to increase potassium permeability 1,000 times.

Cotransport of H^+ and Cl^-

Jennings [12] has recently shown that there exists an alternative pathway of pH equilibration in CO_2-free systems, which also depends on the anion transport system of the red cell membrane. In experiments in which bicarbonate and carbon dioxide had been removed with utmost care, the dissipation of an induced pH gradient took place at a rate which was more than 25 times faster in intact cells than in DIDS-treated red cells.

The hypothesis that the pH-equilibration resulted from OH^-/Cl^- exchange was rejected on grounds of the pH-dependence. The pH equilibration was 12 times faster at pH 5.8 than at pH 7.5, and Jennings concluded that a cotransport of H^+ and Cl^- was most consistent with the experimental data. The process is electrically silent, i.e., induced changes of membrane potential have only minimal effects on the rate of pH equilibration. The rate of the presumed H^+ transport depends strongly on the halide anion of cells and medium. The rates of pH equilibration are 6 and 50 times slower in bromide and iodide, respectively, than in chloride media, reminding one of the relative rates of halide self-exchange.[12] Also, the activation energy of the proton transport bears a close resemblance to that of chloride self exchange,[10] being lower above 15° C (16 kcal/mole) than at low temperatures (27 kcal/mole). Together with a pronounced effect of anion exchange inhibitors, this led to the conclusion that pH equilibration, also in the absence of bicarbonate, can be mediated by the anion exchange system. It should be noted that the transport rate of HCl cotransport is only about one thousandth of the rate of chloride self-exchange. Moreover, the pH dependence of cotransport is different, being minimal above pH 7 and increasing gradually to the most acid pH values of pH 5.5 subjected to investigation.[12] The findings show that HCl cotransport is activated by H^+ in the same range stimulating the transport of SO_4^{--} and $H_2PO_4^-$ and in which inactivation of chloride self-exchange by H^+ takes place (FIGURES 1 and 2). There is strong evidence that sulphate is transported together with a proton.[30] It is thus possible that sulphate transport and HCl cotransport are both stimulated by the protonation of a group in the transport protein. The pK value of the stimulating effects seems, furthermore, to be similar to the pK of the inhibitory effect of H^+ on chloride self-exchange (FIGURES 1, 2 and 4). The same chemical group, therefore, might be involved both in stimulation and inhibition—a hypothesis that is supported by studies of the mutual effects of chloride and sulphate on anion transport rates.[13] As mentioned above, it is not likely that the inhibitory titratable group is identical with the binding site for monovalent anions. One must rather envision a group with proton-accepting and proton-donating capabilities, located in the vicinity of the putative binding site for monovalent anions. In agreement with the titratable carrier model,[11, 13] one single group may thus be responsible for both stimulatory and inhibitory effects. From the available data, it is not possible to decide if this is actually the case.

Residual Proton and Hydroxyl Permeability of DIDS-Treated Cells

The residual hydrogen and hydroxyl ion permeability of the red cell is extremely small after inhibition of anion exchange by DIDS-treatment of the

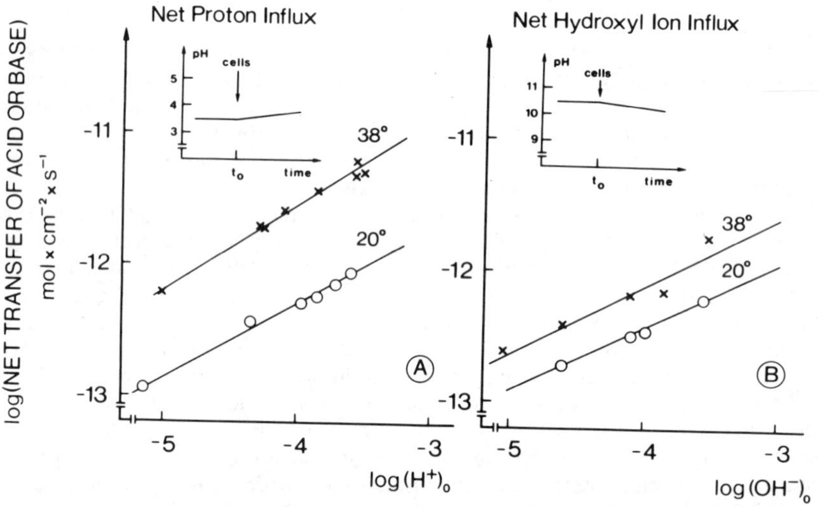

FIGURE 8. Net influx of protons (FIGURE 8A) and of hydroxyl ions (FIGURE 8B) into DIDS-treated red blood cells. Ordinates (logarithmic scales): Net movements of H^+ and OH^-. Abscissae (logarithmic scales): Extracellular concentrations of H^+ and of OH^- at the time of adding the cells (t_o). The experimental procedure is illustrated in the insets: Unbuffered media containing 140 mM KCl and 10 mM NaCl were made acid or alkaline by the addition of small amounts of 0.1 M HCl or KOH. After addition of the DIDS-treated cells (to a final hematocrit of 2%) the pH initially changed linearly with time. The rate coefficient (k, s^{-1}) of acid or base transfer was calculated from the initial slope of the pH change vs. time. The influx at t_o was calculated from the equation $J = k \times (V/A) \times C$ ($\text{mol} \times cm^{-2} \times s^{-1}$), V/A (cm) being the extracellular volume per cm^2 of cell membrane, and C the extracellular hydrogen or hydroxyl ion concentration (mol/cm^3) at t_o. Intracellular pH values were 7.2 (38° C) and 7.4 (20° C) so the low intracellular concentrations of H^+ and of OH^- could be neglected in the calculation of net fluxes. It was controlled that the addition of the washed DIDS-treated cells did not change the buffer capacity of the extracellular medium. The slopes of the relation log J vs. log C were within the range 0.5—0.6 in the four experimental series.

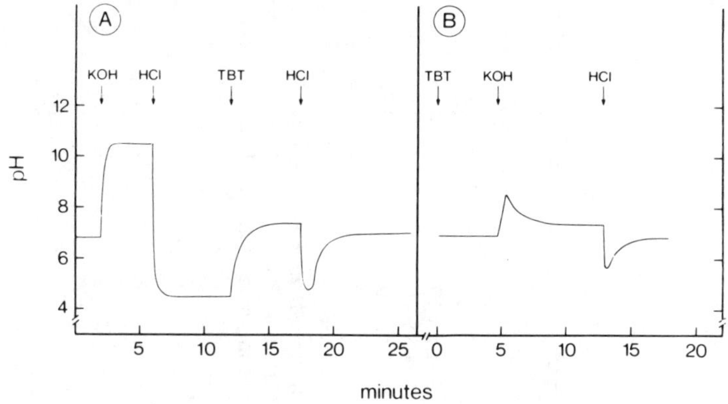

FIGURE 9. pH equilibration in DIDS-treated red blood cells before and after the addition of tributyltin 4×10^{-6} M (FIGURE 9A). Note that the DIDS-treated cells are virtually impermeable to hydrogen and hydroxyl ions at 0° C in the absence of the hydroxyl ion carrier. When organotin had been added before the cells, the suspension responded to pulses of acid or base with small pH excursions, which relaxed to a new equilibrium value in the course of a few minutes. (From Wieth and Tosteson.[36] By permission of the *Journal of General Physiology*.)

membrane. Equilibration of pH may take hours after exposing such cells to an acid or base pulse in an unbuffered medium at low temperatures (FIGURE 9 of the following section). FIGURE 8 shows the hydrogen ion influx at 20° C and 38° C as a function of the extracellular pH after addition of HCl to suspensions of DIDS-treated cells in unbuffered KCl medium. The net influx of hydrogen ions was calculated from the linear slope dpH/dt immediately following an acid pulse. With a buffer capacity of 50 mmol (cells $\times$ pH)$^{-1}$ and a membrane area of 1.5×10^7 cm^2 (cells)$^{-1}$, the intracellular pH of red cells will only change 0.1 unit in five minutes at a proton or hydroxyl net flux of 1 picomol $\times$ cm^{-2} $\times$ s^{-1}. FIGURE 8 shows that net influx of protons increased with decreasing pH, the slope of the logarithm of the flux vs. pH (i.e., the negative logarithm of the hydrogen ion concentration) being about −0.6 at both temperatures. The flux was a linear function of hydrogen ion concentration, but it increased only about four times ($10^{0.6}$) for a tenfold increase of hydrogen ion concentration. Similar experiments were carried out in the alkaline pH range, adding pulses of KOH to suspensions of DIDS-treated cells. The net influx of hydroxyl ions was smaller, but showed a similar relation to hydroxyl ion concentration, as found for the proton influxes (FIGURE 8). The net hydroxyl ion influx increased 3–4 times when the hydroxyl ion concentration was raised tenfold, reaching a value of 0.6 picomol $\times$ cm^{-2} $\times$ s^{-1} at the highest pH value examined (pH 10.6, 20° C). Thus, there was no apparent difference between the modes of pH equilibration when extracellular pH was either decreased below, or increased above, the equilibrium pH of the cell suspension. The flux increased by a factor of 5 when the temperature was increased from 20 to 38° C, corresponding to an apparent activation energy of about 16 kcal/mole (FIGURE 8). We cannot decide whether the pH equilibration across the DIDS-treated membrane is due to the transport of protons, hydroxyl ions, or both. The results of TABLE 3

TABLE 3

THE EFFECT OF MEMBRANE POTENTIAL ON NET HYDROGEN ION INFLUX IN DIDS TREATED CELLS (20° C)*

Expt. No.	Medium KCl mM	Medium NaCl mM	Valinomycin M	$pH_{t=0}$	C_{H^+} mol/cm^3	k s^{-1}	V/A cm	Net Proton Influx picomol $\times cm^{-2} \times s^{-1}$
1	140	10	0	3.58	2.6×10^{-7}	7×10^{-4}	4.7×10^{-3}	0.86
2	0	150	0	3.51	3.1×10^{-7}	6.5×10^{-4}	4.5×10^{-3}	0.91
3	140	10	10^{-6}	3.52	3.0×10^{-7}	6.5×10^{-4}	4.4×10^{-3}	0.86
4	0	150	10^{-6}	3.55	2.8×10^{-7}	9.0×10^{-4}	4.7×10^{-3}	1.18

* Net hydrogen ion influx into DIDS-treated cells with an intracellular pH of 7.39 exposed to unbuffered media with pH 3.5–3.6. The rate coefficient of hydrogen ion influx (k, s^{-1}) was calculated from the initial slope (dpH/dt) that was linear with time. Extracellular volume per cm^2 membrane area (V/A, cm) was calculated from the number of cells per ml, mean cell area 1.42×10^{-6} cm^2. $J_{H^+} = k \times (V/A) \times C_{H^+}$ was only slightly sensitive to large changes in membrane potential.

show that the net flux of hydrogen ions at an extracellular pH of 3.5 only increased from 0.9 to 1.2 picomol $\times$ cm^{-2} $\times$ s^{-1}, when the membrane potential was changed from a value near zero (experiment 3) to a value around −100 mV (experiment 4) by adding valinomycin to KCl and NaCl media. This observation suggests that the transport is largely electroneutral, as is the case in artificial lipid bilayers in which both chloride/hydroxyl exchange and cotransport of H^+ and Cl^- have been suggested [31, 32] as the main pathways of pH equilibration. Toyoshima and Thompson [32] observed a small net transport of $^{36}Cl^-$ across single-walled lecithin vesicles, which they ascribed to HCl cotransport. It is interesting to note that the pH dependence of the alleged HCl transport was identical with that shown in FIGURE 8, the logarithm of the chloride flux increasing linearly with pH with a slope of −0.5. We have not been able to find data for the permeability of artificial lipid bilayers to hydrogen ions under an imposed pH gradient. It would be worthwhile to compare such data with data for the permeability of the red cell membrane after inhibition of anion transport to decide if, e.g., the pathways allowing an osmotic flow of water convey a measurable permeability of H_3O^+ to the red cell membrane over that of a lipid bilayer. There is evidence that red cells treated with inhibitors of specific transfer mechanisms for ions and hydrophilic molecules reach ground state permeabilities, which are quite comparable to those of a lipid membrane,[33] and it is not unlikely that this also applies to the proton and hydroxyl ion permeabilities of DIDS-treated cells.

Carrier-Mediated Transfer of Hydroxyl Ions and Protons

The low permeability of the DIDS-treated red cell to protons and hydroxyl ions makes the membrane an excellent object for the study of the functional characteristics of artificial, lipid-soluble H^+ and OH^- carriers. The purpose of this section is to review the characteristic properties of two classes of carriers. The first group is represented by lipophilic monovalent organo-tin cations, which can mediate an electrosilent chloride/hydroxyl exchange across the red cell membrane.[34] The second class is exemplified by a number of weak acids, often used as uncouplers of oxidative phosphorylation in mitochondria, which induce proton conductance, both in biological and in artificial membranes.[35]

Electrosilent Hydroxyl Exchange Induced by Organotin Compounds

Monovalent trialkyltin derivatives can mediate a chloride-hydroxyl exchange across the membranes of mitochondria, erythrocytes, and liposomes.[34] It has recently been demonstrated that the organotin-induced exchange diffusion of anions, both in erythrocytes [36] and in lipid bilayers,[37] are perfect examples of the mechanism of electrosilent exchange diffusion, originally described by Ussing,[38] a transport process whereby ions can permeate a membrane rapidly by a tightly coupled exchange process, without increasing the electric conductance of the membrane.

The trialkyltin derivatives are efficient exchange carriers of hydroxyl ions. This is shown in Figure 9A, which also demonstrates that the DIDS-treated red cell membrane is virtually impermeable to hydrogen and hydroxyl ions at 0° C. Thus addition of KOH or HCl to a 1% suspension of DIDS-treated red cells caused equally large changes of pH in the unbuffered medium in the absence and presence of cells, and FIGURE 9A shows that the extracellular pH of the 1% cell suspension was completely stable with time. The addition of tributyltin (to make a concentration of 4×10^{-6} mol/l suspension) caused a rapid relaxation of pH to the equilibrium value, and a subsequent pulse of HCl caused a minor transient change of extracellular pH. The pH excursions were even smaller, and subsided in the course of a few minutes, when tributyltin had been added prior to the acid or base (FIGURE 9B).

Tributyltin is an efficient carrier of hydroxyl ions, and reestablishes rapidly

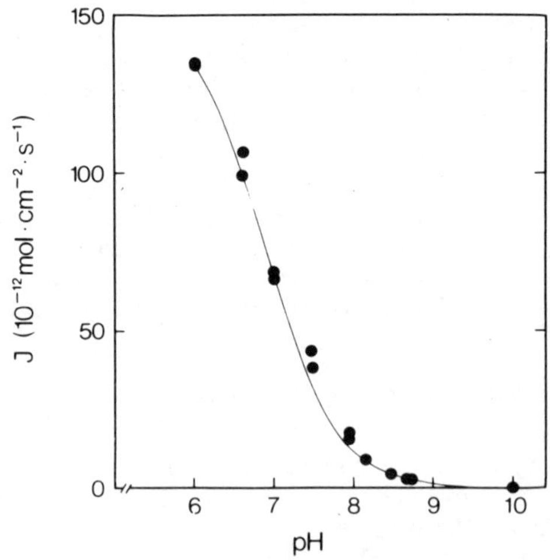

FIGURE 10. pH dependence of tributyltin mediated chloride exchange in DIDS-treated resealed ghosts (0° C, cytocrit 0.5%, tributyltin 5.2×10^{-6} mol/l suspension). The fully-drawn line corresponds to a dissociation constant for membrane-bound tributyltinhydroxide of 10^{-8} M at a chloride concentration of 165 mM. (From Wieth and Tosteson.[36] By permission of the *Journal of General Physiology*.)

a partial equilibrium where the ratio $Cl^-_{(i)}/Cl^-_{(o)} = OH^-_{(i)}/OH^-_{(o)}$. In contrast to the natural transport system, tributyltin was not able to perform chloride/bicarbonate exchange.[36] Moreover the high affinity of hydroxyl ions for tributyltin was apparent from the inhibition of tributyltin-mediated chloride exchange with increasing pH (FIGURE 10), differing markedly from the insensitivity of the natural anion exchange system to alkaline pH-values (FIGURE 2).

It has been demonstrated that the organotin-induced anion exchange both in red cells [36] and in artificial lipid bilayers [37] is electrosilent. TABLE 4 shows results obtained with black lipid membranes. A considerable permeability to the tracer chloride developed after addition of 6 μM tributyltin to the KCl solution bathing the membrane. The chloride permeability was not accompanied by any significant increase of membrane conductance, a finding that was substantiated in a large series of experiments.[37] The results of TABLE 4 also demonstrate that the unidirectional flux of chloride was not affected by an

TABLE 4

ELECTROSILENT EXCHANGE—DIFFUSION OF CHLORIDE THROUGH A BLACK LIPID MEMBRANE *

Tributyl-tin μM	Time after addition of $^{36}Cl^-$ min	V_m $\psi_2-\psi_1$ mV	Unidirectional chloride flux J_{1-2} picomole/$cm^2 \cdot$ sec	P $cm(\text{sec})^{-1}$	G_m S/cm^2	"G_{Cl}" † S/cm^2	"G_{Cl}"/G_m
0	0–15	0	not detectable		1.5×10^{-8}		
6	16–22	0	970	1.0×10^{-5}	3.0×10^{-8}	3.6×10^{-3}	1.2×10^{5}
6	23–32	0	1546	1.6×10^{-5}	5.0×10^{-8}	5.8×10^{-3}	1.2×10^{5}
6	33–37	0	1992	2.0×10^{-5}	5.0×10^{-8}	7.5×10^{-3}	1.5×10^{5}
6	38–41	−60	1864	1.9×10^{-5}	6.0×10^{-8}	7.1×10^{-3}	1.2×10^{5}
6	42–46	−60	2309	2.3×10^{-5}	6.0×10^{-8}	8.6×10^{-3}	1.4×10^{5}

* Membrane is bacterial phosphatidylcholine; area: 6.7×10^{-3} cm^2; electrolyte solutions: 0.1 M KCl, pH 7.0, 23° C.
† "G_{Cl}"$=J_{1-2}\cdot F^2/R\cdot T$

opposing membrane potential of —60 mV. The second last column of the table shows the chloride conductance that should have been found if chloride ions were transported by electrodiffusion. This calculated anion conductance was five orders of magnitude larger than the actual conductance measured, as shown in the last column of the table. This calculation even represents an underestimate because it assumes that chloride is totally responsible for the membrane conductance, which is not the case.

The suggested mechanism of the organotin-mediated exchange-diffusion is depicted in FIGURE 11. The electroneutral ion pair (R_3SnCl) can permeate

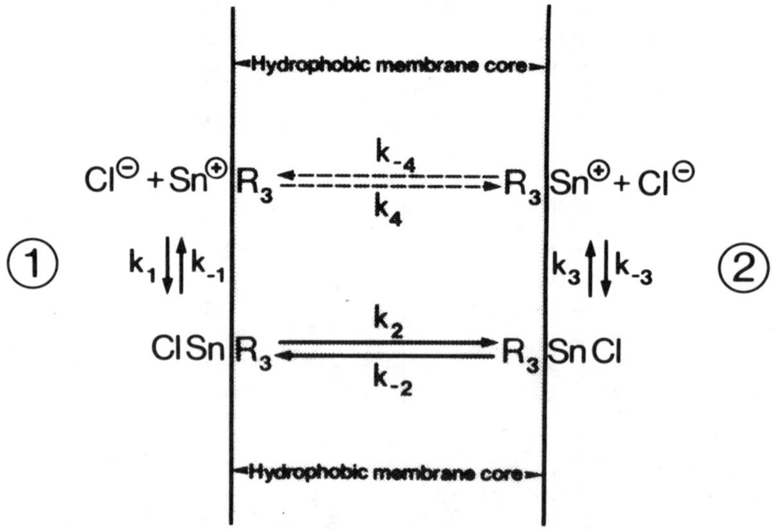

FIGURE 11. Schematic representation of the exchange diffusion of anions mediated by trialkyl organotin compounds. The electroneutral ion pair (R_3SnCl) can permeate the membrane, whereas the membrane permeability to the cation R_3Sn^+ is several orders of magnitude lower. (From Wieth and Tosteson.[36] By permission of the *Journal of General Physiology*.)

the membrane in contrast to the cation (R_3Sn^+), which is excluded from the hydrophobic core of the membrane. It is clear that the electrosilent nature of the transport process is not only due to the formation of electroneutral ion pairs. The transport process would be conductive if the unloaded carrier could permeate the membrane, thus mediating a net transfer of ions that represent a net transfer of electric current. Such behavior is exemplified by the proton carriers dealt with in the next section.

Conductive Proton Transport Induced by Weak Acids

Carbonylcyanide *m*-chlorophenylhydrazone (CCCP) is a weak acid, which is an uncoupler of oxidative phosphorylation in mitochondria. CCCP induces proton conductance in lipid bilayers. LeBanc [35] showed that CCCP makes the lipid membrane a perfect pH electrode in the range between pH 6 and 10. The membrane potential was 59 mV for a one unit pH difference of the two chamber solutions. Experimental evidence agreed favorably with a model, assuming that the weak acid uncoupler (HA) and its anion (A^-) both permeate the membrane with permeabilities of 11 and 2×10^{-3} cm $\times$ s^{-1}, respectively. The fundamental difference between the performance of tributyltin and CCCP as carriers of hydroxyl ions and protons can, thus, be ascribed to the differing permeabilities of the membrane to the unloaded carrier-species, R_3Sn^+ and A^-. The impermeability to R_3Sn^+ forces tributyltin to catalyze exchange diffusion of anions (FIGURE 11), whereas the permeability of the CCCP-anion (A^-) enables CCCP to mediate a net flow of protons (i.e., a proton current) through the membrane.

Macey *et al.*[39] have used the induced proton conductance to monitor changes of the red cell membrane potential. It was assumed that hydrogen ions will attain electrochemical equilibrium across the membrane, and that changes of membrane potential can be monitored by determining variations of extracellular pH in suspensions of erythrocytes in unbuffered media. The experiments were performed under conditions in which the intracellular pH may be assumed to be constant because proton fluxes causing large variations of extracellular pH in the unbuffered solution are still small compared to the buffer capacity of the cells. A precise estimate of the membrane potential is only obtained if the steady-state distribution of H^+ between medium and cells represents an electrochemical equilibrium. We will demonstrate that this is not the case in a membrane in which two pH equilibrating mechanisms operate in parallel: one representing a true proton conductance, the other being an electrosilent exchange mechanism, which tends to bring the ratio $OH^-_{(i)}/OH^-_{(o)}$ in accordance with the chloride distribution ratio regardless of the membrane potential. The experiment is shown in FIGURE 12. DIDS-treated red cells, having an intracellular pH of 7.36, were suspended in a 150 mM NaCl medium containing 1 μM valinomycin. A considerable negative membrane potential developed under these conditions; but, due to the low proton permeability of the DIDS-treated cells, no movement of H^+ took place until the permeability to protons was increased by adding CCCP to the medium. Extracellular pH then rapidly changed to a steady state value of 9.6. Assuming that hydrogen ions were in electrochemical equilibrium across the cell membrane, the pH difference between cells and medium represented a membrane potential of −130 mV. It appears likely that this value is correct because our examinations of the proton permeability of DIDS treated cells suggest that the CCCP-mediated proton conductance was the only pH-equilibrating mechanism of quantitative significance.

We next examined how the hydrogen ion distribution reacted to a carrier that mediates an electrosilent Cl^-/OH^- exchange across the red cell membrane. FIGURE 12 shows that the pH-gradient across the membrane collapsed immediately when tributyltin was added to the suspension. After a few minutes extracellular pH had established a new stable steady-state value of 7.85. Tributyltin introduced a hydroxyl ion exchange system that tended to equilibrate hydroxyl ions to a "chemical equilibrium," in which the hydroxyl ion distribution ratio corresponds to the chloride distribution ratio.

The membrane potential of the red cells is not affected by the addition of tributyltin.[36] Therefore, the exchange carriers established a hydroxyl ion distribution that is far from electrochemical equilibrium. The steady-state pH after the addition of tributyltin (7.85) corresponds to a ratio ($OH^-_{(i)}/OH^-_{(o)}$) of 0.3. The chloride distribution ratio of the cell sample was 0.62. The experiment thus demonstrates the differing modes of pH equilibration by proton conductance and by exchange diffusion of hydroxyl ions. The steady-state distribution of hydrogen ions after addition of tributyltin in the experiment of FIGURE 12 is a compromise between the tendency of CCCP to distribute

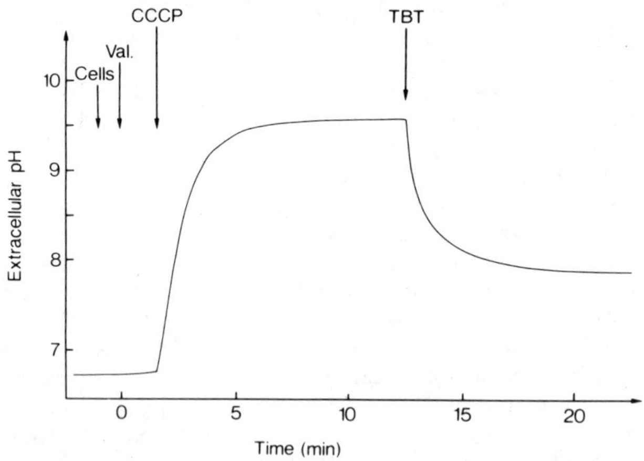

FIGURE 12. Evaluation of the membrane potential of DIDS-treated erythcocytes by means of the steady state distribution of protons between red cells and an unbuffered 150 mM NaCl medium after induction of proton conductance in the red cell membrane with CCCP as proposed by Macey *et al.*[39] There was no change of pH after addition of valinomycin (10^{-6} M) to a 2% cell suspension in an unbuffered NaCl medium (20° C). After addition of 10^{-5} M CCCP, extracellular pH rose to a steady state value of 9.61, 2.25 pH units higher than the intracellular pH of 7.36. This pH difference corresponds to a membrane potential of −130 mV (inside negative), assuming electrochemical equilibrium of protons. Addition of tributyltin (5 μM), an electroneutral exchange carrier of Cl^- and OH^-,[36, 37] caused a rapid fall of pH to a steady state value of 7.85. As described in the text, this extracellular pH is a compromise between the tendency of CCCP to equilibrate H^+ according to membrane potential, and the tendency of tributyltin to equilibrate OH^- according to the chloride distribution. A more detailed analysis is presented by Wieth.[42]

hydrogen ions according to the membrane potential, and the tendency of tributyltin to distribute hydroxyl ions in accordance with the chloride distribution ratio. It should be noted that the resultant steady-state distribution of H^+ and OH^- represents neither a chemical equilibrium (where $OH^-_{(i)}/OH^-_{(o)} = Cl^-_{(i)}/Cl^-_{(o)}$), nor an electrochemical equilibrium (where $H^+_{(o)}/H^+_{(i)} = \exp(zFVm/RT)$. Therefore, the experiment underlines that caution is needed when the red cell membrane potential is evaluated with the help of proton-conducting carriers, as analyzed more thoroughly elsewhere.[42]

ACKNOWLEDGMENTS

We wish to express our gratitude to Dr. Leon Pape, Zoophysiological Laboratory B, The August Krogh Institute, University of Copenhagen, for performing the measurements of membrane potential with fluorescent dye. The experiment with tributyltin in TABLE 4 was made in collaboration with Dr. M. T. Tosteson, Department of Physiology, Harvard Medical School. Our thanks are also due to Tove Soland and Lise Mikkelsen for expert technical assistance.

REFERENCES

1. GUNN, R. B., J. O. WIETH & D. C. TOSTESON. 1975. J. Gen. Physiol. **65:** 731–749.
2. BJERRUM, P. J. & J. TRANUM-JENSEN. 1977. Proc. Int. Un. Physiol. Sci. **13:** 80. Abs. 221.
3. BJERRUM, P. J., J. TRANUM-JENSEN & K. MØLLGAARD. 1980. *In* Membrane Transport in Erythrocytes. Alfred Benzon Sympos. 14. U. V. Lassen, H. H. Ussing & J. O. Wieth, Eds. Munksgaard. Copenhagen. In press.
4. BLANK, M., L. SOO & R. E. ABBOTT. 1978. Biophys. J. **21:** 11a.
5. CABANTCHIK, Z. I. 1980. This volume.
6. ROTHSTEIN, A. 1980. This volume.
7. PASSOW, H. 1980. This volume.
8. DALMARK, M. 1975. J. Physiol. **250:** 39–64.
9. FUNDER, J. & J. O. WIETH. 1976. J. Physiol. **262:** 679–698.
10. BRAHM, J. 1977. J. Gen. Physiol. **70:** 283–306.
11. GUNN, R. B. 1973. *In* Erythrocytes, Thrombocytes, Leucocytes. E. Gerlach, K. Moser, E. Deutsch & W. Wilmanns, Eds.: 77–79. Georg Thieme Verlag. Stuttgart.
12. JENNINGS, M. 1978. J. Membr. Biol. **40:** 365–391.
13. GUNN, R. B. 1978. *In* Membrane Transport Processes. J. F. Hoffman, Ed. **1:** 61–77. Raven Press. New York.
14. DEUTICKE, B. 1970. Naturwissenschaften **57:** 172–179.
15. SCHNELL, K., S. GERHARDT & A. SCHÖPPE-FREDENBURG. 1977. J. Membr. Biol. **30:** 319–350.
16. GUNN, R. B., M. DALMARK, D. C. TOSTESON & J. O. WIETH. 1973. J. Gen. Physiol. 185–206.
17. GUNN, R. B. & O. FRÖHLICH. 1979. J. Gen. Physiol. **74:** 351–374.
18. WIETH, J. O. 1979. J. Physiol. **294:** 521–539.
19. SIMS, P. J., A. S. WAGGONER, C. H. WANG & J. F. HOFFMAN. 1974. Biochemistry **13:** 3315–3330.
20. FUNDER, J. & J. O. WIETH. 1980. *In* Membrane Transport in Erythrocytes. Alfred Benzon Sympos. 14. U. V. Lassen, H. H. Ussing & J. O. Wieth, Eds. Munksgaard. Copenhagen. In press.
21. PATTHY, L. & E. L. SMITH. 1975. J. Biol. Chem. **250:** 557–564.
22. LEPKE, S., H. FASOLD, M. PRING & H. PASSOW. 1976. J. Membr. Biol. **29:** 147–177.
23. SHIP, S., Y. SHAMI, W. BREUER & A. ROTHSTEIN. 1977. J. Membr. Biol. **33:** 311–323.
24. KOSHLAND, D. E. 1973. Sci. Amer. **229:** 52–64.
25. JACOBS, M. H. & D. R. STEWART. 1942. J. Gen. Physiol. **25:** 539–552.
26. WARBURG, E. J. 1922. Biochem. J. **14:** 153–340.
27. FUNDER, J. & J. O. WIETH. 1966. Acta Physiol. Scand. **68:** 234–245.
28. KNAUF, P. A., G. F. FUHRMANN, S. ROTHSTEIN & A. ROTHSTEIN. 1977. J. Gen. Physiol. **69:** 363–386.

29. Wilbrandt, W. 1940. Pflügers Arch. **243:** 537–556.
30. Jennings, M. 1976. J. Membr. Biol. **28:** 187–205.
31. Nicholls, P. & N. Miller. 1974. Biochim. Biophys. Acta. **356:** 184–198.
32. Toyoshima, Y. & T. E. Thompson. 1975. Biochemistry **14:** 1525–1531.
33. Brahm, J. & J. O. Wieth. 1977. J. Physiol. **266:** 727–749.
34. Selwyn, M. J., A. P. Dawson, M. Stockdale & N. Gains. 1970. Eur. J. Biochem. **14:** 120–126.
35. LeBlanc, O. H. 1971. J. Membr. Biol. **4:** 227–251.
36. Wieth, J. O. & M. T. Tosteson. 1979. J. Gen. Physiol. **73:** 765–788.
37. Tosteson, M. T. & J. O. Wieth. 1979. J. Gen. Physiol. **73:** 789–800.
38. Ussing, H. H. 1948. Cold Spring Harbor Symp. Quant. Biol. **13:** 193–200.
39. Macey, R. I., J. S. Adorante & F. W. Orme. 1978. Biochim. Biophys. Acta **512:** 284–295.
40. Forster, R. E. & E. D. Crandall. 1975. J. Appl. Physiol. **38:** 710–718.
41. Hunt, J. B., M.-J. Rhee & C. B. Storm. 1977. Anal. Biochem. **79:** 614–617.
42. Wieth, J. O. 1980. *In* Membrane Transport in Erythrocytes. Alfred Benzon Sympos. **14**. U. V. Lassen, H. H. Ussing & J. O. Wieth, Eds. Munksgaard. Copenhagen. In press.

Discussion

R. B. Gunn: If you had nonDIDS-treated red cells in a CO_2 free medium and added enough CCCP, would the red cell membrane become a pH electrode?

J. O. Wieth: Yes. The pH equilibration with CCCP will also take place in intact red cells. But it is absolutely essential to remove even traces of CO_2 in such experiments. If CO_2 and bicarbonate were present in the system, hydroxyl ions will be shunted indirectly via the Jacobs-Stewart cycle,[25] and the proton distribution across the membrane will not be a correct measure of the membrane potential, because you induce an electrosilent hydroxyl ion transport, just as in the tributyltin experiments.

ARRANGEMENT OF THE RED CELL ANION TRANSPORT PROTEIN IN THE RED CELL MEMBRANE: INVESTIGATION BY CHEMICAL LABELING METHODS *

Kurt Drickamer †

The Biological Laboratories
Harvard University
Cambridge, Massachusetts 02138

Introduction

The experiments described in this paper concern the structure of the Band 3 [1] polypeptide from human erythrocyte membranes. Affinity labeling methods first implicated this 95,000-dalton polypeptide in anion transport across the erythrocyte membrane.[2,3] The involvement of Band 3 in anion transport has since been substantiated by reconstitution experiments in which purified Band 3 has been shown to catalyze anion transport across artificial lipid bilayers.[4] The structure of Band 3 is, therefore, of interest because this is one of the most readily accessible membrane transport proteins.

A considerable amount is now known about the structure of another transport protein, bacteriorhodopsin from *Halobacterium halobium.* Studies of this protein have employed physical methods such as X-ray diffraction analysis and electron microscopy.[5–7] Investigations of these types are possible because this protein in its native membrane forms an extensively ordered array in two dimensions. Unfortunately, although it would be of interest to know the detailed structures of other membrane transport proteins, such as Band 3, there has been very little progress toward obtaining ordered arrays of any membrane protein besides bacteriorhodopsin.

Because this barrier currently prevents detailed physical investigation of membrane proteins, an alternative "chemical" approach has been taken here. The experiments described here establish the location of portions of the erythrocyte Band 3 polypeptide with respect to the lipid bilayer. Identification of regions on each side of the membrane is possible because labeling reagents are available that cannot penetrate the lipid bilayer, due to their size or charge. The polypeptide segments whose locations have been established are relatively large (on the order of 100 amino acids). Thus, by analogy to physical methods, one might say that the arrangement of the polypeptide has been established at a "resolution" of 100 amino acids.

* This research was supported by Grant HL-08893 from the United States Public Health Service and Grant BMS73-06752 from the National Science Foundation to Guido Guidotti, Harvard University.

† Present address: Cold Spring Harbor Laboratory, P.O. Box 100, Cold Spring Harbor, New York 11724.

0077–8923/80/0341–0419 $01.75/0 © 1980, NYAS

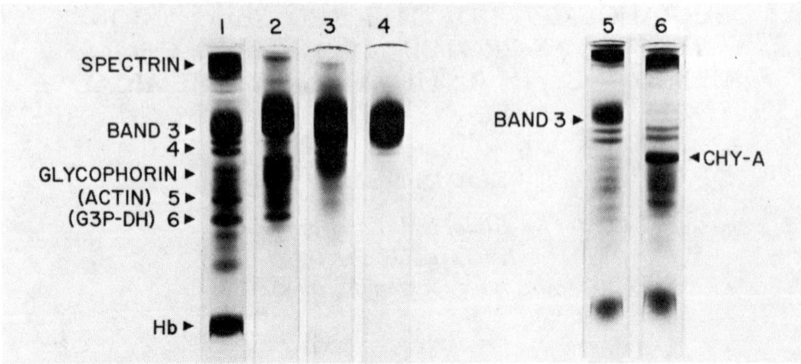

FIGURE 1. Polyacrylamide-sodium dodecyl sulfate (SDS) gels showing purification of Band 3 and proteolytic digestion of intact cells. Gels 1–4, successive stages in the purification of Band 3 from erythrocyte ghosts by the procedure of Drickamer.[8, 9] Gel 1, starting ghosts; gel 2, pellet after extraction with lithium diiodosalycilate; gel 3, pure Band 3 after chromatography on Sepharose 4B in the presence of Triton X-100; gel 4, Band 3 further purified by chromatography on Sepharose 6B in the presence of sodium dodecyl sulfate. Gels 5 and 6, ghosts prepared from cells first treated with trypsin (gel 5) or chymotrypsin (gel 6). Gel electrophoresis and staining with Coomassie blue was performed as described in Drickamer.[8] Each sample contained material from 10 μl of ghosts.

RESULTS

Purification of Band 3

FIGURE 1 summarizes the purification of Band 3 employed here. This method exploits the fact that when erythrocyte membranes are extracted with perturbants, such as lithium diiodosalycilate, the major proteins remaining associated with the lipid bilayer are Band 3 and glycophorin. Subsequent chromatography steps after detergent solubilization serve to remove lipid and minor protein contaminants. The Band 3 and glycophorin polypeptides are separated by destroying the glycophorin in a trypsin pretreatment of the intact cells. This treatment does not affect the Band 3 polypeptide.[10]

Cleavage of Band 3 and Arrangement of Fragments

In order to establish which portions of the Band 3 polypeptide are labeled by the various methods discussed in the following sections, it was necessary to create fragments of the molecule. For the experiments described here, it was desirable to divide the 95,000 dalton polypeptide into roughly ten regions; a combination of limited proteolysis and chemical cleavage was employed to achieve this.

Limited Proteolysis

Treatment of intact red cells with several different proteases has been shown to cleave Band 3, leaving only one obvious fragment ($M_r = 60{,}000$).

Among the proteases known to effect this cleavage are chymotrypsin,[10] thermolysin,[11] and pronase.[12] Chymotrypsin was employed in the present studies because its action can be conveniently terminated. The major cleavage fragment, designated Chy-A (FIGURE 1), was purified by a procedure similar to that employed for the intact polypeptide. The fate of the remaining portion of Band 3 after chymotrypsin cleavage is unclear; this problem will be discussed in subsequent sections.

Chemical Cleavages

The chemical methods investigated were limited to those that fragment at relatively rare amino acid residues. The most useful reagents employed were 2-nitro-5-thiocyanobenzoic acid (NTCB), *N*-bromosuccinimide (NBS), hydroxylamine, and cyanogen bromide; the results are summarized in FIGURE 2.

The amino acid composition of Band 3 indicates that NTCB, which cleaves specifically at cysteine residues, should attack the polypeptide at about five sites. In the event, most reaction is seen at three sites, with cleavage at one of these predominating. Similarly, the composition indicates that, on a statistical basis, one to three Asn-Gly bonds, sensitive to hydroxylamine, should be present. Two have actually been identified. Although the tryptophan content of Band 3 suggests that *N*-bromosuccinimide should cleave the polypeptide at about five sites, a single cleavage, in fact, predominates, producing two useful fragments. Finally, although the methionine content of Band 3 is relatively high (about 23 residues), analysis of fragments produced by BrCN indicates that many of these residues are clustered together, so there are only three large fragments produced by this reagent.

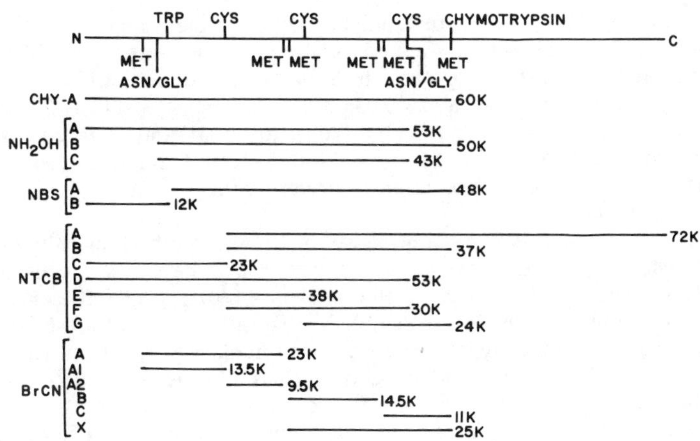

FIGURE 2. Summary of fragments produced by chemical and enzymatic cleavage of the Band 3 polypeptide. Number by each fragment indicates the apparent molecular weight on SDS gels. Location of amino acid residues indicated at the top is inferred from the cleavage patterns. Reagents used for cleavage are discussed in the text. (From Drickamer.[9] By permission of the *Journal of Biological Chemistry*.)

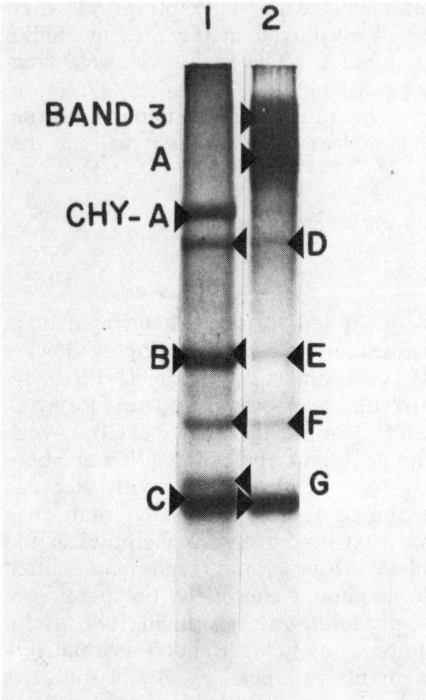

FIGURE 3. Example of chemical cleavage of the Band 3 polypeptide. SDS gels of NTCB-treated Chy-A (lane 1) and Band (lane 2). Major fragments from Chy-A are labeled B and C; major fragments from Band 3 are A and C. (From Drickamer.[9] By permission of the *Journal of Biological Chemistry*.)

Arrangement of Fragments

The arrangement of fragments shown in FIGURE 2 was established by cleaving by two methods in succession, and by labeling specific sites and comparing labeled fragments produced by different cleavages.

An example of the first method is shown in FIGURE 3. NTCB produces cleavage predominantly at one site in either intact Band 3 or in Chy-A. Either starting peptide produces a 23,000-dalton fragment (NTCB-C), suggesting that intact Band 3 and Chy-A share one terminal and that cleavage occurs at a cysteine (Cys 1 in FIGURE 2) 23,000 daltons from this terminal. Similar arguments were employed to align hydroxylamine fragments and some of the BrCN fragments.[8, 9]

The NBS fragments were aligned by the second method, as follows. Labeling studies discussed in the next section reveal that when intact cells are treated with ^{125}I and lactoperoxidase, all the label in Chy-A is near the site of chymotrypsin digestion; this result was established from the distribution of radioactivity in the already-aligned hydroxylamine fragments. Since removal of NBS-B ($M_r = 12{,}000$) from one end of Chy-A does not result in loss of this label, the sensitive tryptophan must be located near the opposite end of the polypeptide from the label. This type of argument was also used to align the remaining CNBr and NTCB fragments.[9]

Absolute Orientation

The absolute orientation of the fragments was determined in three ways. First, ^{14}C-NTCB was shown to label the newly-produced N-terminals of

NTCB-A and NTCB-B but not NTCB-C, which indicates that NTCB-C must come from the N-terminal of the original polypeptides.[8] Second, Chy-A and intact Band 3 were shown to have distinct C-terminals, indicating that their common end must be the N-terminal.[8] Third, the same blocked N-terminal peptide (Acetyl-Met-Glu-Glu) was isolated from intact Band 3, Chy-A, and NTCB-C.[13] All of these results indicate that the polypeptide orientation is as shown in FIGURE 2.

Amino Acid Compositions

The amino acid compositions of several of the fragments have been determined.[9] Although compositions of such large fragments are relatively uninformative, the hydrophobic amino acid contents of various regions are summarized in FIGURE 4.

Portions of Band 3 Exposed at the Cell Surface

Three chemical methods were employed to label the portions of Band 3 exposed at the cell surface: radioiodination of intact cells with ^{125}I and lactoperoxidase, reaction of cells with *p*-nitrophenyl-*N,N,N*-trimethyl[^{125}I]iodotyrosinate, and reaction of cells with diazotized [^{35}S]sulfanilate.

Representative results for the first method are given in FIGURES 5, 6, and 7. The results in FIGURE 5 show that cleavage with chymotrypsin removes no detectable Band 3-associated label (comparing Band 3 in FIGURE 5A and B with Chy-A in FIGURE 5C), although either trypsin or chymotrypsin treatment removes all of the glycophorin-associated label. In FIGURE 6, the radioactivity in Chy-A is shown to be restricted to NTCB-C and NTCB-G, indicating that

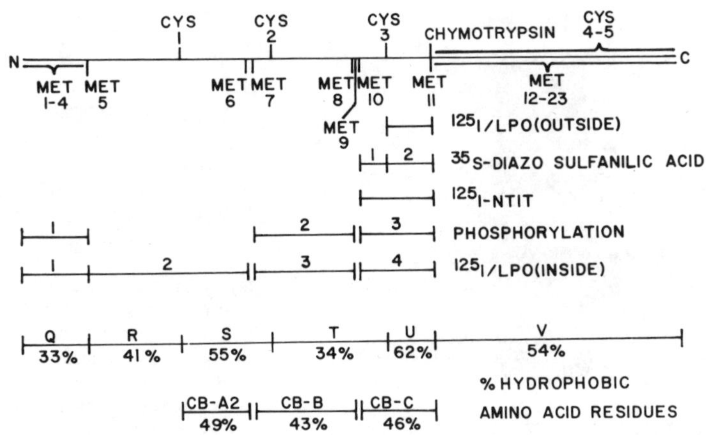

FIGURE 4. Summary of the location of cysteine and methionine residues and the distribution of various radioactive labels and of hydrophobic amino acids in the Band 3 polypeptide. LPO, lactoperoxidase; NTIT, *p*-nitrophenyl-*N,N,N*-trimethyl-iodotyrosinate. (From Drickamer.[9] By permission of the *Journal of Biological Chemistry*.)

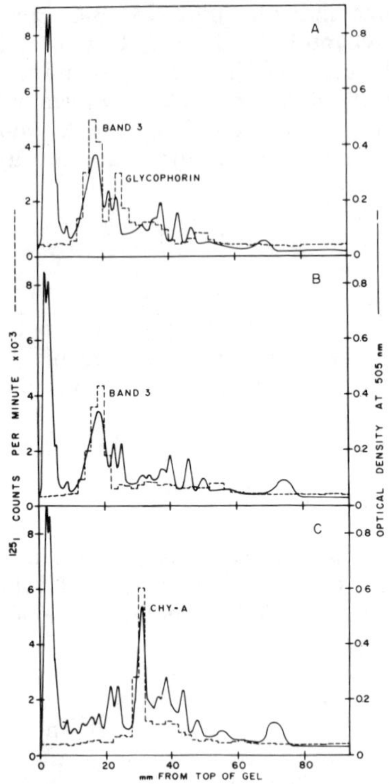

FIGURE 5. Proteolytic fragmentation of Band 3 from cells labeled at the external surface with ^{125}I and lactoperoxidase. SDS gels of ghosts from A cells not treated with proteases; B, cells treated with trypsin; and C, cells treated with chymotrypsin. All of the label in Band 3 moves to the Chy-A position on chymotrypsin treatment. (From Drickamer.[9] By permission of the *Journal of Biological Chemistry*.)

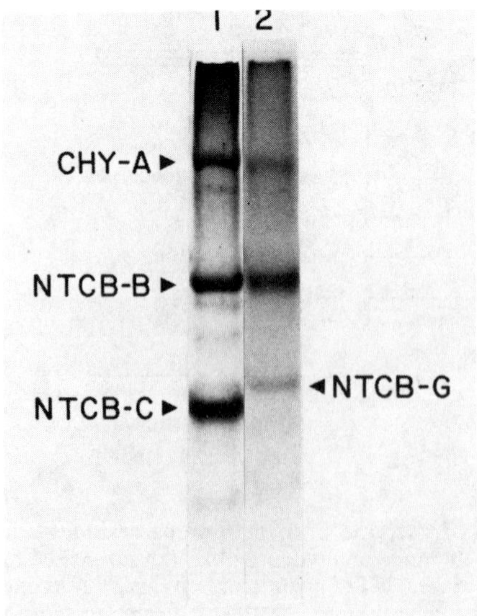

FIGURE 6. NTCB cleavage of Chy-A from cells labeled at the external surface with ^{125}I and lactoperoxidase. SDS gels of NTCB-treated Chy-A shown by Coomassie blue stain (in lane 1) and by autoradiography (in lane 2). Of the major fragments B and C, only B is labeled; of the minor fragments, G is also labeled. Treatment with NTCB and gel electrophoresis were performed as in Drickamer.[9]

the region of Band 3 that is exposed to radioiodination at the cell surface is limited to the segment indicated in FIGURE 4. Finally, only one of the major BrCN fragments (CB-C) is labeled by this method (see FIGURE 7). This is expected from the NTCB results.

FIGURE 4 also includes results for the other cell exterior labels. The conclusions are essentially similar, except that the diazotized [^{35}S]sulfanilate labels an additional segment not accessible to radioiodination, indicating a further region that must be exposed at the cell surface.

It has been shown that all of the carbohydrate in the erythrocyte membrane is arrayed at the surface of the cell.[14] This suggests that the sites of carbohydrate attachment to polypeptides such as Band 3 are another "cell

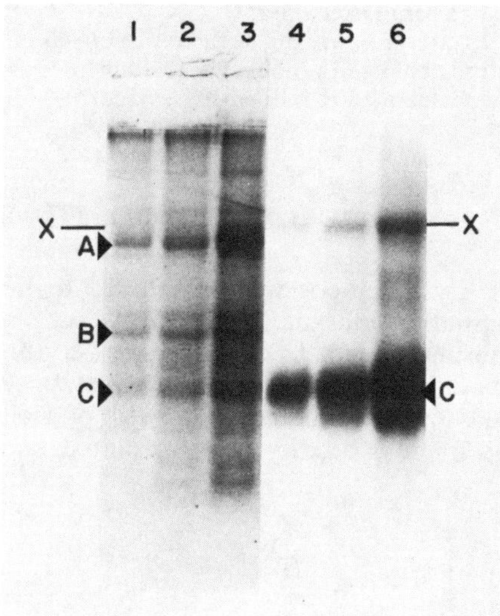

FIGURE 7. Cyanogen bromide cleavage of Chy-A from cells labeled at the external surface with ^{125}I and lactoperoxidase. SDS-urea gel of BrCN products shows three major fragments (A, B, and C) by Coomassie blue stain (lanes 1, 2, and 3). Of these, only C contains label detected by autoradiography (lanes 4, 5, and 6). X is a minor fragment resulting from incomplete cleavage between fragments B and C. (From Drickamer.[9] By permission of the *Journal of Biological Chemistry*.)

exterior marker." The carbohydrate associated with Band 3 was therefore investigated. It is attached to a single site on the polypeptide. This was established by showing that all of the Band 3 carbohydrate is present as a single high molecular weight glycopeptide in pronase digests of the purified protein (FIGURE 8). The composition of this glycopeptide matches the carbohydrate content of the intact polypeptide (TABLE 1). On the other hand, the carbohydrate content of Chy-A preparations is low and is probably due to contamination.[13] This indicates that the carbohydrate is attached in the C-terminal portion of Band 3, i.e., that portion which is not in Chy-A. This result is substantiated by the failure of Chy-A to become significantly labeled by the galactose oxidase-[^{3}H]$NaBH_4$ method for labeling carbohydrate (FIGURE 9). This means that at least part of the C-terminal 35,000 daltons of Band 3 must be outside the cell.

Finally, it should be mentioned that the site of chymotrypsin digestion in intact cells must itself, of course, be at the cell exterior.

TABLE 1

APPARENT CARBOHYDRATE COMPOSITION OF BAND 3 POLYPEPTIDE AND DERIVED FRAGMENTS *

Sugar	mol sugar residue/mol peptide		
	Intact Band 3 †	Glyco-peptide ‡	Chy A §
N-Acetylglucosamine	11.7	11.8	3.5
N-Acetylgalactosamine	0	0	0.7
Galactose	10.3	9.4	3.2
Mannose	2.9	2.3	1.2
Fucose	1.9	1.0	0.3
Sialic acid	2.3	1.9	0.8

* (From Drickamer.[14])
† Calculated in mol sugar residue/800 mol amino acids.
‡ Calculated in mol sugar residue/mol aspartic acid.
§ Calculated in mol sugar residue/550 mol amino acids.

Portions of Band 3 at the Cytoplasmic Side of the Membrane

Two methods were employed to label sites at the cytoplasmic side of the plasma membrane. In the first, use was made of the fact that Band 3 is phosphorylated by endogenous erythrocyte membrane protein kinases.[15, 16] These kinases are found exclusively at the cytoplasmic side of the membrane.[17] Cells were incubated with $^{32}P_i$ (or isolated membranes with γ-[^{32}P]ATP) and the Band 3 or Chy-A was isolated and subjected to various of the cleavages

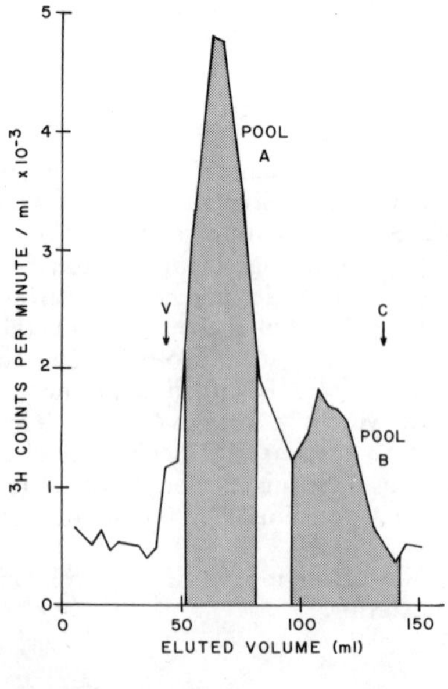

FIGURE 8. Gel filtration on BioGel P-10 of a pronase digest of Band 3 labeled in terminal galactose residues by the galactose oxidase-[^{3}H]NaBH$_4$ method. Examination by hydrolysis and paper chromatography reveals that radioactivity in Pool B is not carbohydrate-associated; direct sugar analysis confirms that there is no carbohydrate in Pool B. Thus the carbohydrate in Band 3 is exclusively in the form of the high molecular weight glycopeptide in Pool A. (From Drickamer.[13])

discussed above. Several regions of Band 3 were labeled (FIGURE 4), although the largest amount of label is found nearest the N-terminal.

In a second approach, sealed, inverted membrane vesicles were prepared from erythrocyte membranes. The external surface of these vesicles, which represents the cytoplasmic side of the original membranes,[18] was labeled with ^{125}I and lactoperoxidase. The distribution of this label is similar to but even more extensive than the labeling with ^{32}P (FIGURE 4).

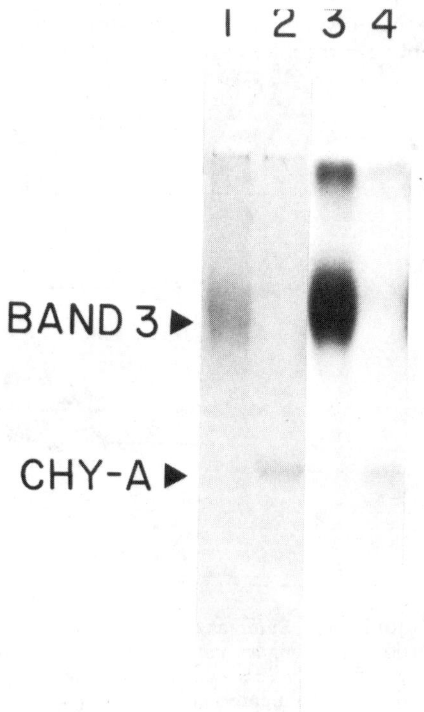

FIGURE 9. SDS gels of Band 3 and Chy-A polypeptides labeled in terminal galactose residues by the galactose oxidase-[^3H]$NaBH_4$ method. Lanes 1 and 3, Band 3 polypeptide; lanes 2 and 4, Chy-A. Lanes 1 and 2 show Coomassie blue, while lanes 3 and 4 show photofluorography of the same gel. (From Drickamer.[13])

Anion Binding Site

Finally, the same methods were employed to identify the site to which the reagent *p*-isothiocyanobenzene[^{35}S]sulfonate binds. This reagent is a covalent, competitive inhibitor of anion transport and is thus an affinity label for the anion transport "active site."[3] The pattern of labeling with this reagent is identical to the results for radioiodination of intact cells; that is, the label is located exclusively within 7,000 daltons of the C-terminal of Chy-A.

DISCUSSION

The Structure of Band 3

The results of the labeling studies just described are summarized at the top of FIGURE 10. A diagrammatic interpretation of the findings is given at the bottom of the figure. The following observations are suggested by this picture.

The Cytoplasmic Amino-Terminal Portion

One of the most striking features of the structure shown in FIGURE 10 is the very large portion of the Band 3 polypeptide that is located on the cytoplasmic side of the membrane: virtually the entire amino-terminal half of the polypeptide is in the cytoplasm. Thus, a large part of the Band 3 polypeptide is probably arranged like most soluble cytoplasmic proteins.

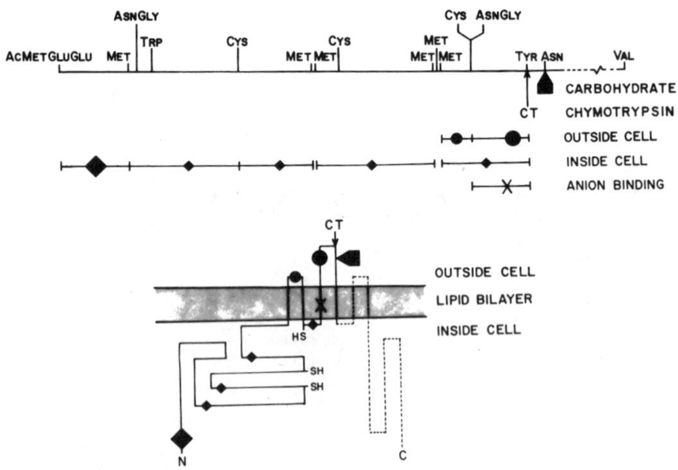

FIGURE 10. Summary of the location of various segments of Band 3. The upper portion of the figure indicates the linear order of various sites along the polypeptide. Larger symbols (diamonds and circles) represent more heavily labeled sites. Lower portion of the figure indicates a proposed arrangement of the Band 3 polypeptide in the lipid bilayer consistent with this labeling pattern. See text for details.

Noteworthy aspects of the "cytoplasmic domain" include the following features:

• The amino terminal residue of the polypeptide is acetylated, which is typical of cytoplasmic proteins.

• The overall amino acid composition of this half of the molecule is not extremely hydrophobic; in fact, the region nearest the amino terminal is quite polar.

• Sites along this entire part of the polypeptide are accessible to macromolecular reagents (lactoperoxidase and protein kinases), but the most accessible sites are near the amino terminal. Therefore, the bulk of the cytoplasmic material is shown folded like a globular protein, with the amino-terminal region extended away from the rest of the molecule.

Two other lines of evidence are consistent with this proposed arrangement.

Steck *et al.*[19] have identified a 41,000-dalton fragment of Band 3 that can be released by gentle proteolysis of the cytoplasmic surface of the erythrocyte membrane; this fragment corresponds nicely to the "cytoplasmic domain" identified here. Also, recent experiments demonstrate that the five cysteine residues in Band 3 are present as free sulfhydryl groups on the cytoplasmic side of the membrane.[20] The three cysteine residues that have been located in the primary structure of the polypeptide may therefore be taken as additional cytoplasmic markers. This is reflected in FIGURE 10, in which the polypeptide has been folded to place these three cysteines on the cytoplasmic side of the bilayer.

The Transmembrane Region

Potentially the most interesting part of the molecule is the region near the site accessible to chymotrypsin in intact cells. The arrangement of sites shown in FIGURE 10 is consistent with the available labeling data and indicates that the polypeptide is probably folded to traverse the membrane at least four times. This region contains a high concentration of hydrophobic amino acids and the anion binding site as would be expected if this part of the polypeptide is involved in forming an anion-selective pore through the lipid bilayer. This transmembrane region presumably corresponds to the 19,000-dalton "membrane domain" obtained by proteolytic digestion at both sides of the membrane.[19]

The Carboxyl-Terminal End

Although the structure of the carboxyl-terminal one third of the polypeptide is unresolved by the present method, it is known to contain the site of carbohydrate attachment. In FIGURE 10 this site is drawn very close to the site of chymotrypsin digestion. This location is suggested by evidence that the carbohydrate moiety is very heterogeneous, so that fragments containing the site of carbohydrate attachment appear as diffuse bands on dodecyl sulfate gels.[13] Difficulty finding discrete chemical fragments of the intact Band 3 polypeptide that overlap the chymotrypsin-sensitive site suggests that these overlap fragments contain the heterogeneous carbohydrate structure and are thus not readily observed on dodecyl sulfate gels.

In addition to the carbohydrate location, the dotted portion of FIGURE 10 is drawn to indicate that little of the carboxyl-terminal end of the Band 3 polypeptide is outside the cell. This is suggested by the insensitivity of the entire polypeptide to trypsin at the external surface of cells. Indeed, even cleavage with chymotrypsin occurs only at very high concentrations of enzyme. Some of the carboxyl-terminal end of the polypeptide is shown distributed in the membrane, a speculation that rests on the relatively high hydrophobic amino acid content of this end of the polypeptide.

Quaternary Structure

The disulfide bond that can be formed to link two copies of the Band 3 polypeptide is formed between one or more of the three cysteine residues whose specific location in the polypeptide have been established (i.e., those in Chy-A).[9, 19] In the absence of further information about the specific identity of the "cross-linking sulfhydryl(s)" all three of the cysteine residues are shown in FIGURE 10 positioned near the center of the molecule so that any or all of them could be at a subunit interface in a dimer.

Potential Refinements of the Structure

There are several ways in which the "resolution" of this method could be improved: increasing the density of cleavage sites and increasing the density and type of labeled sites.

The current results suggest that identifying more precisely the portions of the Band 3 polypeptide that are inside and outside the cell could be achieved by using the same reagents employed up to now to label extracellular and cytoplasmic portions of the polypeptide, but using methods to fragment the polypeptide more completely. Suitable fragments, which would average about 10 amino acids, would be those produced by complete trypsin digestion. Separation of all the tryptic peptides from Band 3 (about 75 are expected altogether) and establishing their order would be a formidable task. However, since the present "lower resolution" picture establishes that only a small portion of the total polypeptide is likely to interact directly with the lipid bilayer, the detailed investigation can be concentrated on a smaller region. Thus, it might be useful to examine the tryptic peptides generated from cyanogen bromide peptide *C* that contains most of the known membrane-spanning region.

A second method of increasing the resolution of this method would be to use additional reagents with a broader range of reactivity to identify sites on each side of the membrane. Reagents with just such a general reactivity have recently been developed; they contain photo-generated groups that can undergo insertion reactions with many side groups in proteins.[21, 22]

A third way to extend the labeling studies is to use reagents that partition preferentially into the hydrocarbon of the lipid bilayer and label the "intramembrane" portions of the polypeptide. These reagents have been successfully employed to label the intramembrane portions of erythrocyte glycophorin, lymphocyte histocompatibility antigens, viral glycoproteins, and renal Na^+, K^+-Adenosine-triphosphatase.[23-25] Recent results with Band 3 are consistent with the folding pattern shown in FIGURE 10.[26] All of the new labeling reagents just mentioned could be most advantageously applied after a higher resolution fragment map has been constructed.

Comparison of Band 3 with Other Membrane Proteins

TABLE 2 summarizes several differences between human erythrocyte membrane glycophorin and Band 3. The conclusions about glycophorin structure have now also been shown to be true of human lymphocyte histocompatibility antigens.[29-31]

The comparison between Band 3 and these other membrane proteins emphasizes that Band 3 is a representative of a class of membrane proteins distinct from those whose structures have been studied up to now. This result can be summarized by saying that while glycophorin and histocompatibility

TABLE 2

COMPARISON OF STRUCTURAL CHARACTERISTICS OF THE MAJOR HUMAN ERYTHROCYTE MEMBRANE GLYCOPROTEINS *

Characteristic	Band 3	Glycophorin
1. Traversal of bilayer	4 times or more	1 time only
2. Carbohydrate	Unique	Like serum glycoproteins
3. Amino terminal	Acetylated cytoplasmic	Unblocked extracellular

* Characteristics for Band 3 indicated are based on results discussed in text. Information about glycophorin was obtained from Tomita and Marchesi [27] and Winzler.[28]

antigens appear to be "partially secreted" proteins, Band 3 is much like a cytoplasmic protein that contains a small region specialized to form a transmembrane pore.

ACKNOWLEDGMENTS

I thank Guido Guidotti, in whose laboratory the experiments described above were performed. Cordie Grimm provided useful comments on the manuscript.

REFERENCES

1. FAIRBANKS, G., T. L. STECK, & D. F. H. WALLACH. 1971. Biochemistry **10:** 2606–2617.
2. ROTHSTEIN, A. & Z. I. CABANTCHIK. 1974. J. Membrane Biol. **15:** 207–226.
3. HO, M. K. & G. GUIDOTTI. 1975. J. Biol. Chem. **250:** 675–683.
4. ROSS, A. H. & H. M. MCCONNELL. 1978. J. Biol. Chem. **253:** 4777–4782.
5. HENDERSON, R. 1975. J. Mol. Biol. **93:** 123–138.
6. BLAUROCK, A. E. 1975. J. Mol. Biol. **93:** 139–158.
7. HENDERSON, R. & P. N. T. UNWIN. 1975. Nature **257:** 28–32.
8. DRICKAMER, L. K. 1976. J. Biol. Chem. **251:** 5115–5123.
9. DRICKAMER, L. K. 1977. J. Biol. Chem. **252:** 6909–6917.
10. TRIPLETT, R. B. & K. L. CARRAWAY. 1972. Biochemistry **11:** 2897–2903.
11. BOXER, D. H., R. E. JENKINS & M. J. A. TANNER. 1974. Biochem. J. **137:** 531–534.
12. BRETCHER, M. S. 1971. J. Mol. Biol. **59:** 351–357.
13. DRICKAMER, L. K. 1978. J. Biol. Chem. **253:** 7242–7248.
14. STECK, T. L. & G. DAWSON. 1974. J. Biol. Chem. **249:** 2135–2142.
15. PALMER, F. B. ST. C. & J. A. VERPOORTE. 1971. Can. J. Biochem. **49:** 337–347.
16. RUBIN, C. S. & O. M. ROSEN. 1973. Biochem. Biophys. Res. Commun. **54:** 494–500.
17. RUBIN, C. S., R. D. ROSENFELD & O. M. ROSEN. 1973. Proc. Natl. Acad. Sci. U.S.A. **70:** 3735–3738.
18. STECK, T. L. 1974. Methods Enzymol. **31:** 172–180.
19. STECK, T. L., B. RAMOS & E. STRAPAZON. 1976. Biochemistry **15:** 1154–1161.

20. RAO, A. 1979. J. Biol. Chem. **254:** 3503–3511.
21. KNOWLES, J. E. 1972. Acc. Chem. Res. **5:** 155–160.
22. STARROS, J. V. & F. M. RICHARDS. 1974. Biochemistry **13:** 2720–2726.
23. GOLDMAN, D. W., J. S. POBER, J. M. WHITE & H. BAYLEY. 1979. Nature. **280:** 841–843.
24. ABU-SALAH, K. M. & J. B. C. FINDLAY. 1977. Biochem. J. **161:** 223–228.
25. KARLISH, S. J. D., P. L. JORGENSON & C. GITLER. 1977. Nature **269:** 715–717.
26. GOLDMAN, D. W. Personal communication.
27. TOMITA, M. & V. T. MARCHESI. 1975. Proc. Natl. Acad. Sci. U.S.A. **72:** 2964–2968.
28. WINZLER, R. J. 1969. *In* Red Cell Membrane Structure and Function. G. A. Jamieson & T. J. Greenwalt, Eds.: 157–171. Lippincott, Philadelphia.
29. TERHORST, C., P. PARHAM, D. L. MANN & J. L. STROMINGER. 1976. Proc. Natl. Acad. Sci. U.S.A. **73:** 910–914.
30. PARHAM, P., B. N. ALPERT, H. T. ORR & J. L. STROMINGER. 1977. J. Biol. Chem. **252:** 7555–7567.
31. ROBB, R., C. TERHORST & J. L. STROMINGER. 1978. J. Biol. Chem. **253:** 5319–5324.

PROTEIN STRUCTURE IN RELATION TO ANION TRANSPORT IN RED CELLS *

A. Rothstein, M. Ramjeesingh, S. Grinstein, and P. A. Knauf

Research Institute
The Hospital for Sick Children
Toronto, Ontario, M5G 1X8, Canada

Band 3, an abundant protein of 95,000 daltons, has been identified as the component involved in anion transfer across the red cell membrane on the basis of localization of inhibitory probes, functional reconstitution, and retention of function in vesicles enriched in Band 3 by "negative" purification.[1–4] The most potent inhibitor of anion transport, the disulfonic stilbene, 4,4′-diisothiocyanostilbene-2,2′-disulfonic acid (DIDS) interacts with the inhibitory sites of Band 3 in a two-step reaction.[5, 6] A rapid reversible binding is followed by a slower, temperature-dependent, irreversible reaction. The former involves a chelation of the sulfonic acid groups of the probe, presumably with positively-charged groups in the membrane, and the latter involves a covalent reaction of one of its isothiocyano-groups.[2, 3] The second isothiocyano-group is not covalently linked to the inhibitory site, but under certain conditions it can interact with externally added ferritin,[7] or with another adjacent segment of Band 3.[8]

The amount of DIDS-binding to Band 3 follows a linear relationship with respect to the extent of irreversible inhibition, with virtually complete effect when 1.1 to 1.3×10^6 molecules of probe are bound per cell.[2, 5, 6] Because the number of monomers of Band 3 is also about 10^6 per cell, it seems that each monomer has one inhibitory site and can therefore bind one DIDS molecule. This conclusion has recently been confirmed by the finding that DIDS bound to the inhibitory site can be quantitatively cross-linked to another segment of each Band 3 monomer.[8]

The nature of the DIDS-inhibition has been assessed by a kinetic analysis of the reversible effects of H_2DIDS, a reduced form of DIDS that seems to react with the same membrane sites.[5, 6] At low temperatures, the covalent reaction of H_2DIDS is slow, so that the kinetic analysis can be performed when the interaction is essentially all reversible.[9] The inhibition data, plotted according to Hunter-Downs, are illustrated in FIGURE 1. The kinetic information suggests that H_2DIDS and Cl^- compete for a common binding site. The K_i for H_2DIDS binding is 0.1 μM (the intercept on the y-axis) and the apparent dissociation constant for Cl^- at the same site is 62 mM (the intercept on the x-axis), about the same as reported values for the apparent dissociation constant of Cl^- at its transport site.[10, 11] It can therefore be concluded that H_2DIDS and DIDS apparently compete with Cl^- for the anion transport site.†

* These studies were supported by Grants MT 4665 and MT 5149, Medical Research Council, Canada.

† Under conditions of heteroexchange,[12, 13] or of unequal Cl^- concentrations,[3] the anion transport system behaves asymmetrically, with affinities that are different on the two sides of the membrane. Nevertheless, for a competitive inhibitor acting at

0077–8923/80/0341–0433 $01.75/0 © 1980, NYAS

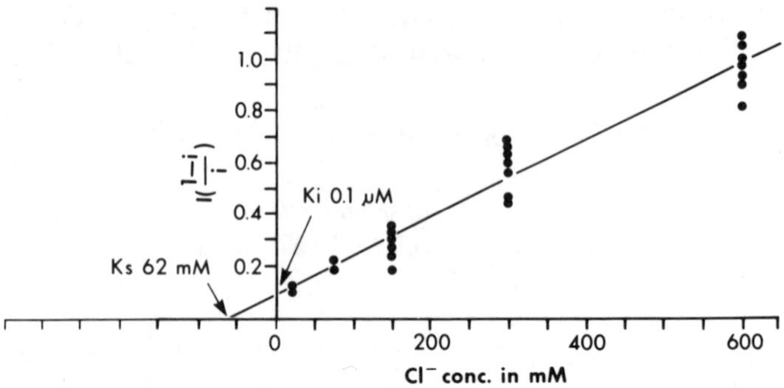

FIGURE 1. The kinetics of inhibition of Cl^- transport by H_2DIDS using the Hunter and Downs plot. *I* represents the H_2DIDS concentration and *i* the fractional inhibition. A noncompetitive inhibition would result in a horizontal line. The line of positive slope indicates competition between H_2DIDS and Cl^- for an inhibitory site. The dissociation constant of H_2DIDS for the site (K_i) is 0.1 M and of Cl^- (K_s), 62 mM. (From Shami *et al.*[9])

The kinetic characteristics of anion transport are consistent with a mobile carrier model [2, 3, 10, 14] in which the "loaded" carrier or transport site (*T*) crosses the membrane, allowing release of the anion on the other side, as in FIGURE 2. The "unloaded" carrier cannot cross, so that it can only return if accompanied by an anion. Then the cycle can be repeated. This model can account for the behavior of the anion transport system as a 1 for 1 electrically silent exchange system. The site labelled *M* in FIGURE 2 has been called the modifier.[10] Its interaction with anions results in a noncompetitive inhibition of transport, presumably by a reduction in the rate of translocation of the loaded carrier across the membrane. The modifier is assumed to be exposed only to the outside on the basis of reported asymmetries in its accessibility to the inhibitor, *N*-(4-azido-2-nitrophenyl)-2-aminoethylsulfonic acid (NAP-taurine),[2–4, 15] and to chloride.[13]

The model of FIGURE 2 involves a "sequential" exchange (sometimes called "ping pong") inasmuch as the transport site alternately interacts with anions at opposite sides of the membrane. Another option would involve a "simultaneous" exchange in which sites at opposite sides of the membrane are both loaded, followed by a simultaneous translocation. The kinetic data, however, favor the sequential model.[3, 12] A more direct test involves the use of DIDS in combination with NAP-taurine. DIDS is a nonpenetrating probe that only inhibits when applied to the outside surface of the membrane.[16] NAP-taurine, on the other hand, behaves as a slowly transported substrate.[17] In the dark, NAP-taurine is a reversible inhibitor. From the inside, by kinetic analysis similar to that presented in FIGURE 1 for DIDS, it also appears to inhibit by interaction with the transport site.[15] After exposure to light, NAP-taurine reacts covalently, largely at the same sites to which it was bound reversibly

the outer surface, such as H_2DIDS, the system will behave as if Cl^- competes with the inhibitor with a single affinity of Cl^- for the transport site, corresponding to the value found when its concentrations on the two sides of the membrane are equal.

in the dark.[18] When presented to the cytoplasmic face of the membrane in inside-out (IO) vesicles "stripped" of extrinsic proteins,[19, 20] NAP-taurine reacts largely with Band 3.[16] If the cells are first treated with DIDS to "recruit" the sites to the external face of the membrane (prior to formation of IO vesicles), the amount of binding of NAP-taurine to Band 3 from the cytoplasmic side is substantially reduced (FIGURE 3). The amount of reduction is consistent with the conclusion that each DIDS bound to the external side prevents the binding of one NAP-taurine to the cytoplasmic side. This finding supports the "ping-pong" model of FIGURE 2, based on the concept that the transport site is alternately exposed to outside and inside, but would be difficult to reconcile with a "simultaneous" model with noninteracting sites. It also provides further evidence that DIDS can be used as a marker for locating the mobile transport site within the Band 3 molcule. It should be pointed out, however, that the inhibitory action of DIDS is associated with the reversible interaction of the sulfonic acid groups with positively charged ligands of the transport site, whereas the covalent reaction is associated with the interaction of one of the isothiocyano groups with an adjacent ligand.[2] Thus, in localization experiments, DIDS marks a ligand that is spatially close to the transport site in undenatured Band 3, but that could possibly be some distance away in the primary structure of the protein. For example, under certain circumstances (prolonged incubation at high pH) DIDS can form a cross-link between two sites located in different parts of Band 3.[8]

Band 3 is a large transmembrane protein of 95,000 daltons. The location of the DIDS binding site within the primary structure has been determined by a series of proteolytic and chemical cleavages. Proteolysis from the outside results in a splitting of Band 3 into membrane-bound segments of about 65,000 and 35,000 daltons, with the DIDS largely localized in the 65,000 dalton segment.[21] Proteolysis from the inside (in resealed ghosts or in IO vesicles) releases a soluble cytoplasmic segment of Band 3 of about 40,000 daltons.[22] The DIDS is localized in membrane-bound segments of about

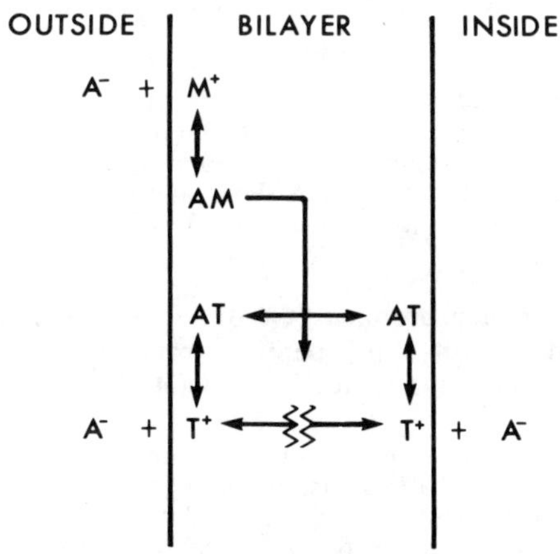

FIGURE 2. A mobile carrier model for anion transport. *T* is the carrier or transport site, and *M*, the modifier site, as described in the text. (From Shami *et al.*[9])

50,000 daltons.[23] After proteolysis from both outside and inside, the DIDS is highly localized in a transmembrane segment of 15,000 to 17,000 * daltons.[24]

Because the 15,000 dalton segment is highly resistant to proteolysis, its further dissection has been accomplished by chemical cleavage. The procedure is summarized in the diagram of FIGURE 4. The 15,000 dalton segment was purified by sodium dodecyl sulfate (SDS)-urea-acrylamide gel electrophoresis.[25]

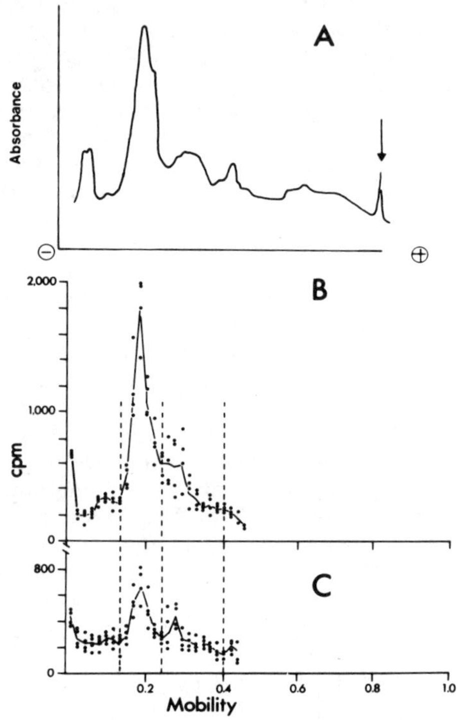

FIGURE 3. The labelling of membrane proteins by (^{3}H)NAP-taurine applied to the cytoplasmic side of IO vesicles from normal (B) and from DIDS-treated (C) cells. The proteins were separated by SDS-acrylamide gel electrophoresis and stained with Coomassie blue (as in the scan, A) or sliced and counted for (^{3}H)NAP-taurine content (as in B and C). The major peak is Band 3. (From Grinstein *et al.*[16])

It was treated with cyanogen bromide (CNBr) in formic acid. Two cleavages occurred, with production of three peptide fragments of 7,000, 4,000 and 4,000 daltons, as illustrated in FIGURE 5. The DIDS was found in one of the

* With chymotrypsin cleavage from the outside in intact cells, and trypsin cleavage from the inside in IO vesicles, the transmembrane segment containing DIDS has been reported as 17,000 daltons.[24] With chymotrypsin treatment of leaky ghosts, the transmembrane segment is also reported to be 17,000 daltons.[22] In our hands, the segment released by chymotrypsin treatment of leaky ghosts appears to be 15,000 daltons. The differences can probably be attributed to different techniques of SDS-acrylamide gel electrophoresis, and of calibration of the gels.

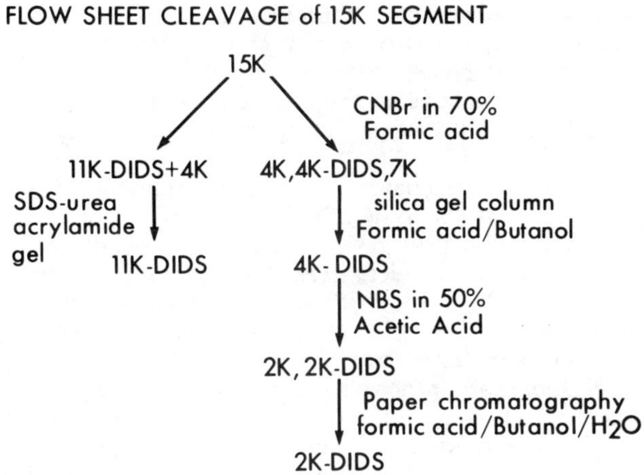

FIGURE 4. Flow sheet for the procedures used in the chemical cleavages to determine the location of the DIDS-binding site in the 15,000 dalton transmembrane segment of Band 3.

4,000 dalton fragments. CNBr cleaves at methionine, leaving a detectable lactone on the newly produced *C*-terminal ends of the segments.[26] The lactone was found in both 4,000 dalton segments, but not in the 7,000 dalton segment. Thus, the 7,000 dalton segment must be located on the C-terminal end of the 15,000 dalton peptide. The DIDS containing 4,000 dalton fragment appears to be in the middle of the 15,000 dalton segment, and the second 4,000 dalton fragment at the *N*-terminal end (FIGURE 5), based on partial CNBr cleavages and on *N*-terminal amino acid analyses.

The 4,000 dalton DIDS-containing segment was purified by chromatography on silica gel. It was then cleaved with *N*-bromosuccinimide (NBS) to produce

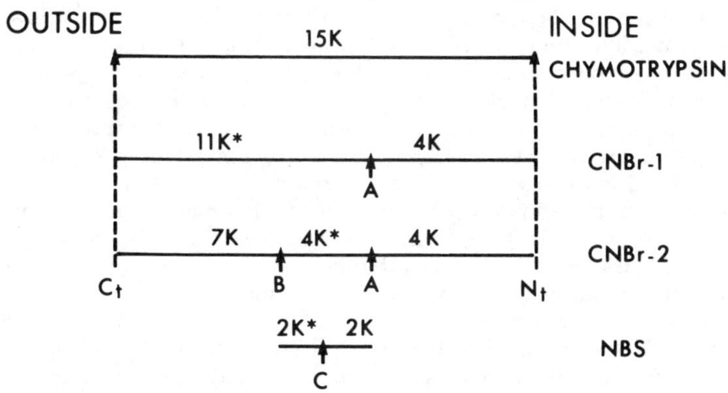

FIGURE 5. A peptide map illustrating the location of the DIDS-binding site in the 15,000 dalton transmembrane segment of Band 3.

two fragments of approximately 2,000 daltons, one of which contained the DIDS. They were separated by paper chromatography. The DIDS-labelled 2,000 dalton fragment contains the lactone produced by the previous CNBr cleavage, whereas the unlabelled fragment does not. Thus the former must be located at the *C*-terminal end of the 4,000 dalton fragment, as illustrated in FIGURE 5.

In the amino acid sequenator, the DIDS-containing 2,000 dalton segment stops after cleavage of the *N*-terminal amino acid (leucine). This may be due to the fact that the next residue is the DIDS-conjugate, although other explanations cannot be excluded. Tentatively then, the DIDS-site is located one residue to the left of the NBS cleavage point (*C* in FIGURE 5), about 9,000 daltons from the *C*-terminus and 6,000 daltons from the *N*-terminus of the 15,000 dalton transmembrane segment of Band 3.

Two other inhibitory probes have previously been localized in Band 3, based on chemical cleavages of a 65,000 dalton product of the chymotrypsin cleavage (at C_t in FIGURE 5) of the intact membrane.[27, 28] The compound 1-isothiocyano-4-benzene sulfonic acid (IBS) has been localized within 7,000 daltons of the external chymotrypsin site, based on cleavage with hydroxylamine.[28] Diazo-sulfanilic acid has been largely localized in a segment between 7,000 and 11,000 daltons from the chymotrypsin site, with perhaps some binding to the 7,000 dalton C-terminal segment, based on cleavages with CNBr, hydroxylamine and 2-nitro-5-thiocyano-benzoic acid (NTCB).[27] The reported CNBr cleavage at 11,000 daltons is presumably the same as the cleavage A in FIGURE 5. The cleavage with NTCB (reported as occurring 7000 daltons from the *C*-terminus) is not far from cleavage B. The major labelled site with DASA seems, therefore, to be located in the same 4,000 dalton segment as DIDS. The 7,000 dalton *C*-terminal segment that is a minor site of DASA labelling, and which is the only site of labelling with IBS,[27, 28] is not labelled by DIDS.

Although the 15,000 dalton transmembrane segment behaves as a hydrophobic peptide (based on the finding that it can only be dissolved out of the membrane with detergents or with strong organic acids), it has only about 40% hydrophobic residues.[29] Smaller segments of the peptide are also reported to contain about the same proportion of hydrophobic residues.[27, 30] The fragments produced in the present study, although insoluble in water, range in frequency of hydrophobic residues from 25–46 (TABLE 1). Thus, their hydrophobic behavior seems not to be due to long runs of hydrophobic residues, but rather to some particular tertiary arrangement with respect to hydrophilic and hydrophobic residues. In considering the specific arrangement of the 15,000 dalton segment, two complicating factors must be taken into account. Firstly, Band 3 probably exists in the membrane as a dimer,[31, 32, 33] although the presence of the dimer does not seem to influence the operation of the transport system. This latter conclusion is based on the findings that the kinetics of Cl^- transport and of its inhibition show no evidence of cooperativity, and that the relationship between DIDS binding and inhibition indicates the existence of one transport site per monomer of Band 3.[1–6] Secondly, another 35,000 dalton portion of Band 3 (derived from chymotrypsin cleavage at the outer face of the membrane) is also associated with the bilayer,[2, 3, 8, 21, 29] and it may also play a role in transport.[8] It seems to be closely associated with the 17,000 dalton segment, based on the finding that the two segments can be cross-linked with DIDS at high pH.[8] Under certain conditions of proteolysis, only a 9,000 dalton portion of the 35,000 dalton segment remains in the membranes, in a 1/1 mole ratio with the 15,000 dalton segment.

Based on the considerations outlined above, it seems premature to speculate about the specific organization of Band 3 within the bilayer, but in the most general sense it can be suggested that the 15,000 and 9,000 dalton segments from two monomers of Band 3 form a spherical or cylindrical structure within the bilayer, accounting for the fact that Band 3 is a major component of the intramembranous particles seen by freeze fracture electron microscopy.[34] Within the structure, the hydrophobic residues are presumably located on the outside facing the fatty acid side chains of the phospholipids, and the hydrophilic residues on the inside, forming a central core, or in contact with the aqueous phases at one side or the other of the bilayer.

TABLE 1

THE FRACTION OF HYDROPHOBIC AMINO ACID RESIDUES IN SEGMENTS OF THE 15,000 DALTON TRANSMEMBRANE FRAGMENT OF BAND 3 *

15,000 dalton segment	41 [29]
11,000 dalton (C_t-A)	36 [27]
7,000 dalton (C_t-B)	45 †
4,000 dalton (A-B)	27
2,000 dalton (B-C)	29
2,000 dalton (C-A)	25
4,000 dalton (A-N_t)	46 ‡

* Refer to FIGURE 5 for location of various segments. Hydrophobic residues are taken as Val, Met, Ile, Leu, Tyr, Phe, and Trp.

† Calculated by subtracting the amino acid composition of the 4,000 dalton (A-B) segment from that of the 11,000 dalton (C_t-A) segment.

‡ Calculated by subtracting the amino acid composition of the 11,000 dalton (C_t-A) segment from that of the 15,000 dalton segment.

From a topological point of view, certain sites in the 15,000 dalton transmembrane segment are accessible from the outside or from the inside. Thus the sites labelled C_t and N_t (FIGURE 5) are at the outside and inside respectively, based on their accessibility to proteolytic enzymes. As a nonpenetrating agent, DIDS must have accessibility to its binding site from the outside and, as pointed out previously, it cannot reach the site from the inside.[24] Furthermore, DIDS attached covalently to the site is capable of reaction with the large protein ferritin.[7] Thus the DIDS-binding site (the transport site, based on kinetic analysis [9]) seems to be located closer to the outer surface of the membrane than to the inner surface. Thus, the site, located in the middle of the 15,000 dalton segment (FIGURE 5) is geographically near the outside, indicating that the segment is probably folded within the bilayer. An asymmetric location of the transport apparatus is also suggested by studies with another inhibitory anion, 2-(aminophenyl)-6-methylbenzenethiazol-3,7′-disulfonic acid (APMB). When attached to a high molecular weight dextran, APMB is an effective inhibitor of anion transport if applied to the outside, but not from the inside. If, however, it is attached to the dextran by a long spacer (45 Å), it is also effective from the inside surface.[2]

In another sense, the question of the topological location of the DIDS-binding site is complicated by the fact that, on kinetic grounds, it appears to be the anion transport site. Although it can be "seen" by DIDS only from the outside, it has already been pointed out that it can also be "seen" by another probe, NAP-taurine, from the inside [15] and that, in fact, it behaves functionally as though it is alternately exposed to outside and inside.[12, 16]

A general model to account for the available information has been discussed previously.[1-4] It is shown in diagrammatic form in FIGURE 6. Although the membrane bound segments of Band 3 are proposed to exist as a cylinder with hydrophobic groups arranged on its outer surface facing toward the lipid components of the bilayer, and its hydrophilic groups forming an internal aqueous channel, it cannot be an open channel throughout its length because the anion transfer behaves as a nonconductive, one-for-one

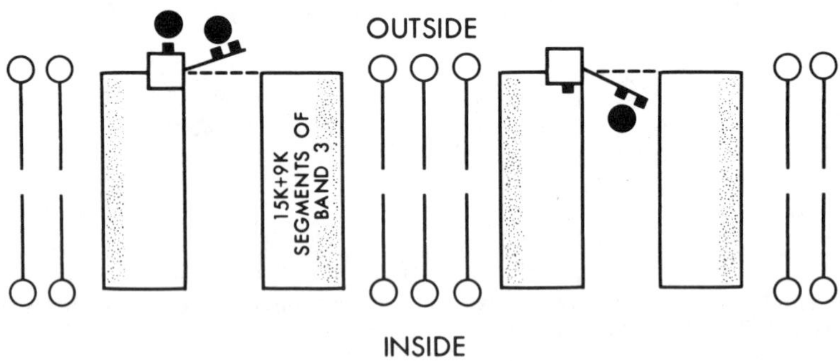

FIGURE 6. A model to explain the structure and function of Band 3 (adapted from Reference 4). (From Rothstein *et al.*[4])

exchange with high electrical resistance. Free anion diffusion occurs at a very low rate, whereas the rapid exchange behaves like a carrier with a "mobile" transport site, as in FIGURE 2. In the model, the peptide channel is assumed to be blocked to diffusion by a barrier located near the outer surface of the membrane. It is proposed that the anions can only pass the barrier when bound to the transport site, which is represented by a gate. The rapidity of turnover of anions through the system suggests that the transport site (gate) only moves a small distance during the transfer process. The movement of the loaded transport site (gate) across the barrier is therefore assumed to result from a local, spontaneous, conformational change. The two conformations (one with the anion binding site facing outward and the other facing inward) can rapidly equilibrate with each other provided that the anion site is occupied. Evidence suggesting that such a conformational change in Band 3 may occur have been compiled elsewhere.[2, 3] The model can account for the kinetics of anion transport, its inhibition, its asymmetrical arrangement in the membrane, and the structure of Band 3 within the membrane.

REFERENCES

1. ROTHSTEIN, A., Z. CABANTCHIK & P. KNAUF. 1976. Mechanism of anion transport in red blood cells: Role of membrane proteins. Fed. Proc. **35:** 3–10.
2. CABANTCHIK, Z. I., P. A. KNAUF & A. ROTHSTEIN. 1978. The anion transport system of the red blood cell: The role of membrane protein evaluated by the use of "probes." Biochim. Biophys. Acta **515:** 239–302.
3. KNAUF, P. A. 1979. Erythrocyte anion exchange and the Band 3 protein: transport kinetics and molecular structure. Curr. Top. Membr. Trans. In press.
4. ROTHSTEIN, A., S. GRINSTEIN, S. SHIP & P. A. KNAUF. 1978. Asymmetry of functional sites of the erythrocyte anion transport protein. Trends Biochem. Sci. **3:** 126–128.
5. LEPKE, S., H. FASOLD, M. PRING & H. PASSOW. 1976. A study of the relationship between inhibition of anion exchange and binding to the red blood cell membrane of 4,4′-diisothiocyano-2,2′-disulfonic acid and its dihydro derivative (H_2DIDS). J. Membr. Biol. **29:** 147–177.
6. SHIP, S., Y. SHAMI, W. BREUER & A. ROTHSTEIN. 1977. Synthesis of tritiated 4,4′-diisothiocyano-2,2′-stilbene disulfonic acid ([^{3}H]DIDS) and its covalent reaction with sites related to anion transport in red blood cells. J. Membr. Biol. **33:** 311–324.
7. ROTHSTEIN, A. & Z. I. CABANTCHIK. 1974. Protein structures involved in the anion permeability of the red blood cell membrane. *In* Comparative Biochemistry and Physiology of Transport. L. Bolis, K. Bloch, S. E. Luria & F. Lynen, Eds. 354–362. North Holland Publishing Co., Amsterdam.
8. JENNINGS, M. L. & H. PASSOW. 1979. Anion transport across the red cell membrane, *in situ* proteolysis of Band 3 protein, and crosslinking of proteolytic fragments by 4,4′-diisothiocyano dihydrostilbene-2-2′-disulfonate (H_2DIDS). Biochim. Biophys. Acta. **554:** 498–519.
9. SHAMI, Y., A. ROTHSTEIN & P. A. KNAUF. 1978. Identification of the Cl^- transport site of human red blood cells by a kinetic analysis of the inhibitory effects of a chemical probe. Biochim. Biophys. Acta **508:** 357–363.
10. DALMARK, M. 1976. Effects of halides and bicarbonate on chloride transport in human red blood cells. J. Gen. Physiol. **67:** 223–234.
11. BRAZY, P. 1976. Furosimide inhibition of chloride transport in human red blood cells. J. Gen. Physiol. **68:** 583–599.
12. GUNN, R. B. & C. FROHLICH. 1978. Evidence for an asymmetric, single site, ping-pong mechanism for anion transport in human red blood cells. J. Gen. Physiol. **74:** 351–374.
13. SCHNELL, K. F., E. BESL & A. MANZ. 1978. Asymmetry of the chloride transport system in human erythrocyte ghosts. Pflüg. Arch. Ges. Physiol. **375:** 87–95.
14. GUNN, R., M. DALMARK, D. TOSTESON & J. WIETH. 1973. Characteristics of chloride transport in human red blood cells. J. Gen. Physiol. **61:** 185–206.
15. KNAUF, P. A., S. SHIP, W. BREUER, L. MCCULLOCH & A. ROTHSTEIN. 1978. Asymmetry of the red cell anion exchange system: different mechanisms of reversible inhibition by NAP-taurine at the inside and outside of the membrane. J. Gen. Physiol. **72:** 607–630.
16. GRINSTEIN, S., L. MCCULLOCH & A. ROTHSTEIN. 1979. Transmembrane effects of irreversible inhibitors of anion transport in red blood cells: Evidence for mobile transport sites. J. Gen. Physiol. **73:** 493–514.
17. CABANTCHIK, Z., P. KNAUF, T. OSTWALD, H. MARKUS, L. DAVIDSON, W. BREUER & A. ROTHSTEIN. 1976. The interaction of anionic photoreactive probe with the anion transport system of the human red blood cell. Biochim. Biophys. Acta **455:** 526–537.
18. KNAUF, P. A., W. BREUER, L. MCCULLOCH & A. ROTHSTEIN. 1978. NAP-taurine as a photoaffinity probe for identifying membrane components containing the modifier site of the human red blood cell anion exchange system. J. Gen. Physiol. **72:** 631–649.

19. Steck, T. L. & J. A. Kant. 1974. Preparation of impermeable ghosts and inside-out vesicles from human erythrocyte membranes. Methods Enzymol. **32:** 172–180.
20. Yu, J. & Steck, T. L. 1975. Associations of Band 3, the predominant polypeptide of the human erythrocyte membrane. J. Biol. Chem. **250:** 9176–9184.
21. Cabantchik, Z. I. & A. Rothstein. 1974. Membrane proteins related to anion permeability of human red blood cells. II. Effects of proteolytic enzymes on disulfonic stilbene sites of surface proteins. J. Membr. Biol. **14:** 227–248.
22. Steck, T. L., B. Ramos & E. Strapazon. 1976. Proteolytic dissection of Band 3, the predominant transmembrane polypeptide of the human erythrocyte membrane. Biochemistry **15:** 1154–1161.
23. Lepke, S. & H. Passow. 1976. Effects of incorporated trypsin on anion exchange and membrane proteins in human red blood cell ghosts. Biochim. Biophys. Acta **455:** 353–370.
24. Grinstein, S., S. Ship & A. Rothstein. 1978. Anion transport in relation to proteolytic dissection of Band 3 protein. Biochim. Biophys. Acta **507:** 294–304.
25. Swank, R. T. & K. D. Munkres. 1971. Molecular weight analysis of oligopeptides by electrophoresis in polyacrylamide gel with sodium dodecyl sulfate. Anal. Biochem. **39:** 462–477.
26. Spande, T. F., B. Witkop, Y. Degani & A. Patchornik. 1970. Selective cleavage and modification of peptides and proteins. Advances in Protein Chem. **24:** 98–260.
27. Drickamer, L. K. 1977. Fragmentation of the Band 3 polypeptide from human erythrocyte membranes. Identification of regions likely to interact with the lipid bilayer. J. Biol. Chem. **252:** 6909–6917.
28. Drickamer, L. K. 1976. Fragmentation of the 95,000 dalton transmembrane polypeptide in human erythrocyte membranes. Arrangement of the fragments in the lipid bilayer. J. Biol. Chem. **251:** 5115–5123.
29. Steck, T. L., J. J. Koziarz, M. K. Singh, G. Reddy & H. Kohler. 1978. Preparation and analysis of seven major, topographically defined fragments of Band 3, the predominant transmembrane polypeptide of human erythrocyte membranes. Biochemistry **27:** 1216–1222.
30. Koziarz, J. J., H. Kohler & T. L. Steck. 1978. Chemical characterization of a membrane-spanning segment of Band 3. J. Supramol. Struct. Suppl. **2:** 215.
31. Steck, T. L. 1974. The organization of proteins in the human red cell membrane. A review. J. Cell. Biol. **62:** 1–19.
32. Reithmeier, R. A. F. 1979. Fragmentation of the Band 3 polypeptide from human erythrocyte membranes. Size and detergent binding of the membrane-associated domain. **254:** 3054–3060.
33. Nigg, E. & R. J. Cherry. 1979. Dimeric association of Band 3 in the erythrocyte membrane demonstrated by protein diffusion measurements. Nature **277:** 493–494.
34. Pinto da Silva, P. & G. L. Nicolson. 1974. Freeze etch localization of concanavalin A receptors to the membrane intercalated particles of human erythrocyte ghost membranes. Biochim. Biophys. Acta **363:** 311–319.

Discussion

J. F. Hoffman: What is the relationship between the binding site for DIDS and the transport site for an anion? If you recruit sites to one side or the other with the DIDS or the NAP-taurine, that says something about those sites, but what does it say about the transport sites?

A. ROTHSTEIN: Kinetically, the NAP-taurine on the inside and DIDS on the outside behave as though they are competing for the transport site and they, therefore, both mark the transport site. The complication in this conclusion is that the sulphonic acid groups of the probe may interact with the transport site; but the covalent reaction is at the other end of the molecule so that when we determine the location of the probe in the primary structure, we are localizing the covalent reaction site. The transport site could be located in another loop of the protein that is a close neighbor to the covalent site, but that is some distance away in the primary structure. This problem is inherent in using covalent probes to identify sites because the affinity site is not the covalent reaction site.

RECONSTITUTION OF THE ERYTHROCYTE ANION TRANSPORT SYSTEM: *IN VITRO* AND *IN VIVO* APPROACHES *

Z. Ioav Cabantchik,[†,‡] D. J. Volsky,[†] H. Ginsburg,[†] and A. Loyter [†]

† *Department of Biological Chemistry*
Institute of Life Sciences
Hebrew University, Jerusalem, Israel

‡ *Laboratory of Kidney and Electrolytes*
National Heart, Lung, and Blood Institute
National Institutes of Health
Bethesda, Maryland 20014

The assignment of red blood cell anion transport to the 100,000 dalton membrane polypeptides (commonly referred to as Band 3, according to their electrophoretic mobility) rested primarily on the use of chemical probes, which are, generally, inhibitors of the transport function.[1,2] With the aid of probes, it was possible to explore the chemical nature of the anion recognition sites, to specifically label the relevant functional membrane components and to study their arrangement in the membrane matrix. Probes were essential for delineating all the above properties, primarily because the anion transport system has virtually no obvious intrinsic features that could serve as chemical or functional markers throughout isolation procedures. However, the designation of Band 3 (B3) as the putative anion transport protein requires, ultimately, its isolation in a reasonably pure and demonstrably functional form. The latter is important for assessing the modulatory role of lipids and other protein constituents as well as for gaining further insights into the architecture of functional sites and the molecular mechanism of anion transport.

Previous attempts to isolate functional B3 polypeptides included a procedure of "negative purification" (NP) that consisted of selective removal of nonrelevant membrane proteins, either by a combination of ionic stresses and moderate treatment with detergents [3,4] or by proteolytic excision of external segments of polypeptides.[5,6] With either treatment, the highly hydrophobic B3, or parts thereof, remained *in situ* in their native lipid milieu. The resulting membranous fragments, which vesiculated either upon sonication [4] or spontaneously,[5] provided an experimental object amenable to transport studies. These studies revealed qualitative similarities between the functional properties of isolated vis-a-vis the native transport system.[4–6] However, other studies emphasized also the possibility that relatively minor components other than B3 might have contributed to the observed anion transfer of sulfate.[5] Similar features to those obtained with the B3 vesicles isolated by NP were also observed with B3 vesicles obtained by a partly selective *solubilization* of intrinsic

* This work was supported in part by grants from the Israel National Academy of Sciences and Humanities, Basic Research Branch (ZIC), The Hebrew University Central Fund (ZIC), the USA-Israel Binational Science Foundation (ZIC) and the German GSHmBH (München) (AL).

0077–8923/80/0341–0444 $01.75/0

membrane proteins and their *reconstitution* into lecithin vesicles either prior [7,8] or after B3 purification by Concanavalin-A affinity chromatography.[9] It should be stressed that with all these procedures a substantial fraction of the total population of isolated B3 was apparently nonfunctional, either because of B3 deterioration, leakiness of B3 vesicles, or unsuccessful choice of phospholipids. The remaining functional fraction, displayed some characteristic properties of erythrocyte sulfate exchange system, including the susceptibility to covalent binding probes, such as $[^3H]H_2$DIDS(4,4′-diisothiocyano-2,2′-ditritiostilbene disulfonate), which correlated with the specific labeling of B3 polypeptides.[8]

Recent Reconstitution Studies of Functional B3

In order to provide a more firm basis for the putative role of B3 in anion transport, we developed reconstitution procedures that permitted us to test and evaluate the alleged B3 mediated transfer of sulfate as well as test and evaluate the physiologically relevant chloride anion. We shall attempt to demonstrate here, that while B3 incorporated into liposomes of defined lipid composition (*in vitro* reconstitution) might express only partial capacity to transfer sulfate, B3 implanted into plasma membranes of living cells (lacking Cl-transfer properties) displays Cl-transport properties commensurate with those observable in intact erythrocytes. The method of *in vivo* reconstitution makes use of the fact that solubilized and reconstituted *Sendai* virus envelopes have demonstrable capacity to fuse with and integrate in plasma membrane of living eukaryotic cells.[10] The envelopes can be coreconstituted with membrane proteins and thus be used as vehicles for inserting these proteins into recipient cells.[11] Thus, the method provides a useful tool for assaying isolated components of membrane transport systems, particularly in those cases in which *in vitro* reconstitution is not easily attainable, either because of lipid composition unfavorable for reconstitution, leakiness of vesicles, or simply because of technical limitations in assaying rapid transport processes across systems of large surface/volume properties (e.g., liposomes). The rationale of this work rests on the assumption that plasma membranes of intact cells can provide to exogenous proteins a matrix suitable for transport measurements and an appropriate environment to favorably express their putative functions. The question that would require further study concerns itself with the more distant fate of the exogenous proteins implanted in plasma membranes of living cells.

Isolation Procedures

The basic procedure for isolation of B3 is given schematically in FIGURE 1. It follows, with minor modifications, the general lines for obtaining B3-rich residues [3,4] followed by detergent extraction (i.e., solubilization) of B3.[8] Solubilized B3 was reconstituted into B3 vesicles together with exogenous phospholipids (PL) (synthetic DL lecithin, i.e., 1 octadec-9-enyl, 2 hexadecyl ether and phosphatidyl serine, 94:6, w/w) (protein: PL ratio of 1:4, w/w) or alternatively, it was coreconstituted into vesicles with lipids and glycoproteins of *Sendai* virus envelopes (VG), essentially as recently published.[10] The "hybrid" B3VG, displays fusogenic properties, and as will be shown later, it

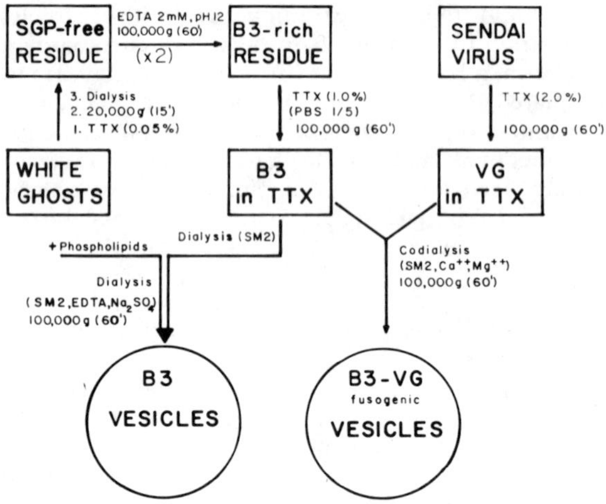

FIGURE 1. Schematic representation of Band 3 (B3) and viral glycoprotein (VG), isolation and reconstruction. Band 3 was isolated from human erythrocyte white ghosts (phenylmethyl sulfonyl fluoride-treated) as B-3 enriched vesicles (B3-rich residue), and subsequently extracted by a modification of a previously published method.[3, 8] B3 was solubilized in Triton X-100 (TTX) and, either mixed with TTX-solubilized VG or dialyzed at 4° C (Spectrapor-2) against SM2 beads, or bovine serum albumin in an ethylene diamine tetraacetic acid (EDTA) and Na_2SO_4-containing medium, mixed with phospholipids (synthetic DL-lecithin: phosphatidylserine, 94:6, w/w) and further dialyzed to yield B3 vesicles (TTX$<$0.02%). Solubilized VG, either alone or mixed with B3 in TTX, (B3 was from neuraminidase-treated cells) were dialyzed at 4° C (Spectrapor-2), either against SM2 beads or bovine serum albumin (to sequester TTX) to yield either VG vesicles[10] or B3VG vesicles.[11]

can be used as a vehicle for implanting B3 into recipient cells such as Friend erythroleukemia cells (FELC). The protein composition of the various membrane preparations are given in FIGURE 2 in terms of SDS-acrylamide-gel electrophoretograms. The latter show the high enrichment of reconstituted B3 (RB3) relative to the intact ghost as well as the protein profile of reconstituted viral glycoprotein (RVG). Most importantly, it shows that [^{3}H]H_2DIDS reaction with erythrocytes lead to the specific labeling of B3 which, as shown before,[1] could be fully correlated with the inactivation of either sulfate or chloride exchange.

Functional Characterization

Reconstitution in vitro

The B3 vesicles prepared, according to FIGURE 1, were of ca. 0.1 μM radius, monolamellar, and had a protein/lipid ratio of 0.25 (w/w), similarly to values computed for whole erythrocyte membranes (i.e., ghosts). While the main protein constituent was B3 (FIGURE 2), the majority of the lipid (80%)

was comprised of the synthetic DL-lecithin (77% of total lipid) and of phosphatidyl serine (3% of the total lipid). The remaining 20% were residual lipids from the erythrocyte ghost, primarily phosphatidyl choline. These B3 vesicles displayed sulfate transfer rates 50–100-fold faster than phospholipid vesicles, or displayed tenfold faster rates than B3 vesicles prepared from DIDS-treated red blood cells (FIGURE 3). The results indicate a specific, B3-mediated, sulfate transfer across vesicles, inasmuch as DIDS inhibits the anion exchange system of red blood cells and labels specifically the B3 polypeptides.[1] An additional feature displayed by RB3 vesicles is a symmetric (random) transmembrane disposition of B3. This is shown by the symmetric effects of DNDS, a nonpenetrating, reversibly acting, analog of DIDS that inhibits anion exchange across intact membranes only when exofacially applied.[12] In B3 vesicles, DNDS was effective from either membrane surface (FIGURE 3), although its overall inhibitory potency in B3 was less pronounced (by a factor of 10) than in intact cells or resealed ghosts.[12] The latter might indicate either a different affinity of each individual B3 for DNDS, or, more likely, a heterogeneous population of B3 polypeptides with either full or partial susceptibility to DNDS. In either case, the results with DNDS suggest that some, or all, of the critical anion transport sites on B3 vesicles differ from those in intact membranes, not just in their transmembrane disposition, but also in their tertiary structure. Other demonstrable differences between RB3 (FIGURE 4) and RBC-sulfate transfer properties,[1, 13] are based on the following observations obtained with RB3 vesicles. RB3 vesicles display:

A. Relatively lower and qualitatively different pH-dependence of sulfate transfer than in intact cells (FIGURE 4, A);

B. Relatively lower (30%) Q_{10} values for sulfate transfer than in intact cells (FIGURE 4, B);

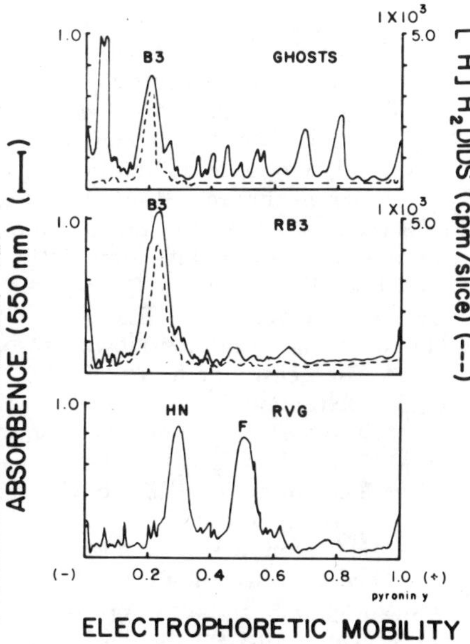

FIGURE 2. SDS-acrylamide (7.5%) gel electrophoresis of the various membrane preparations. Ghosts were prepared either from untreated or from [^{3}H]H_2DIDS-treated red blood cells[19] and electrophoresed as previously described.[20] Reconstituted Band 3 (RB3) was prepared either from unlabeled or labeled cells[19] and reconstituted viral glycoproteins (HN: hemagglutinin and neuraminidase, F: fusion factor) were prepared as previously described.[11] Ghosts were loaded at 75 μg protein (or 4×10^4 cpm [^{3}H]H_2DIDS labeled protein); RB3 were loaded at 15 μg protein (or 1.7×10^4 cpm [^{3}H]$_2$DIDS labeled protein). RVG were loaded at 30 μg protein. Gels were either stained with Coomassie Brilliant Blue and scanned at 550 nm (——) or sliced and counted for radioactivity (-----) as previously described.[19]

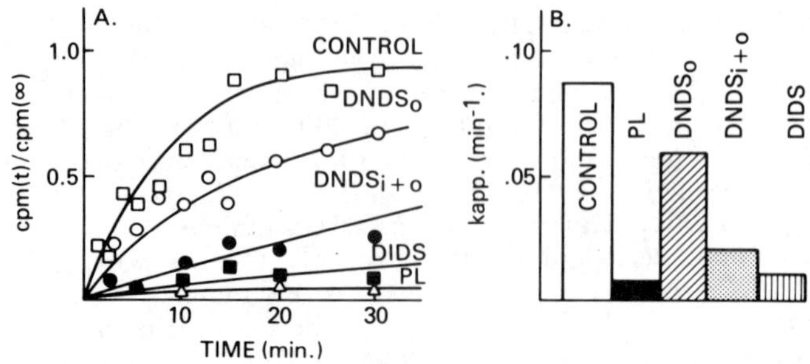

FIGURE 3. Anion transport of reconstituted B3 vesicles. B3 vesicles (70 μg/ml protein, 1 mg/ml phospholipids) were obtained in 50 mM Na_2SO_4, 0.2 mM NaN_3, 0.15 mM EDTA and 10 mM HEPES pH 7.0 (control). Similar vesicles were obtained from ghosts of DIDS-treated red blood cells.[19] Phospholipid vesicles (PL) were obtained by dialysis of TTX mixed with PL as shown for B3 in FIGURE 1. Sulfate ingress (25° C) was followed after addition of [^{35}S]Na_2SO_4 (200 μCi/ml) to the above vesicle preparations. Ingress was terminated by cooling 100 μl alliquots to 0° C, sequestration of extravesicular [^{35}S] on albumin-treated Dowex (Cl^-)/AG 1×8 (succinate form 20–50 mesh) columns (2 cm length×0.5 cm diameter), and elution of retained vesicles with isotonic succinate at 0° C.[21] Alternatively, vesicles were separated from medium on Sephadex G-50 columns.[4] A. [^{35}S] associated with vesicles is given as the fraction of cpm at time *t* relative to cpm at ∞ i.e., cpm at isotopic transmembrane equilibrium (ca. 6 hours at 37° C for all systems except DIDS and PL for which *t* at ∞ was 18 hours). Systems labeled $DNDS_{i+o}$ and $DNDS_o$ represent sulfate fluxes in the presence of externally added DNDS (500 μM). In $DNDS_{i+o}$, DNDS (500 μM) was incorporated into the vesicles during the TTX dialysis steps. B. Apparent rate constants k_{app} of sulfate ingress shown in A.

C. No demonstrable saturation and self-inhibition of sulfate transfer with increasing sulfate concentration (FIGURE 4, C), and

D. Small sulfate transfer capacity per B3 polypeptide (*vide infra*).

While observation B can be explained by the markedly different protein and lipid compositions that govern the fluidity and plasticity properties of the respective membranes,[14] and thus modulate their transport activities, the other observations (A, C and D) indicate more subtle changes either in B3 itself or in its relationship with other RBC membrane components absent in RB3 vesicles.[8] In the case of D, a comparison between the sulfate transfer capacity per B3 polypeptide of RB3 vesicles and intact membranes was carried out, taking into account the following three points:

- The density of B3 per lipid in B3 vesicles is comparable to that of intact membranes.
- The surface/volume ratio of vesicles relative to ghosts r^g/r^v (the respective radii of ghosts and of vesicles) is ca. 30, and
- The sulfate transfer activity of vesicles relative to cells is approximately 5 (k_{app} in vesicles is 0.095 min^{-1} (FIGURE 3) and k_{app} in cells or ghosts is 0.02 min^{-1}, at 50 mM sulfate, pH 7.0, 25° C).

This analysis shows that r^g/r^v is sixfold higher than $k^v{}_{app}/k^g{}_{app}$, thus indicating that B3 transport capacity is sixfold higher in cells as compared to vesicles. This substantial difference suggests either: that only 15% of B3 present in nonleaky vesicles are fully functional, or that each B3 functions at

15% the efficiency displayed in intact cells. However, at present, it would be premature to state whether the apparently lower efficiency in RB3 vesicles results from a genuine alteration, or even a deterioration, of B3 in the course of its preparation, from an unfavorable lipid environment or from the arbitrarily "wrong" choice of temperature, pH and sulfate concentration used for the above comparison. In that respect, it is possible to obtain substantially higher "efficiencies" for RB3 simply by carrying out the comparison with data obtained at 15° C and pH 7.5, in which case the "apparent efficiency" increases

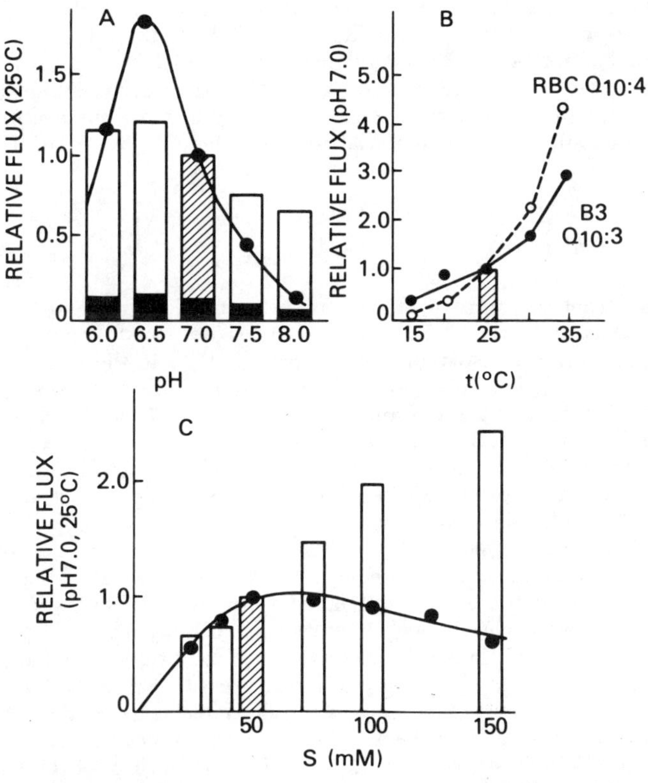

FIGURE 4. Kinetic properties of sulfate transport across reconstituted B3 vesicles (B3) and red blood cells (RBC). B3 vesicles were prepared at different H^+ and Na_2So_4 concentrations as indicated in FIGURE 1, and sulfate influx was performed, as detailed in legends to FIGURE 3. Sulfate fluxes in red blood cells (RBC) were performed as detailed elsewhere.[12, 13] Unless stated otherwise, standard conditions were: 50 mM Na_2SO_4, pH 7.0, 25° C. All influx values are given relatively to those observed in standard conditions (slashed columns). A. pH profile of sulfate in RBC (lines) and influx in B3 vesicles isolated from either untreated (columns) or DIDS-treated ghosts (black areas). B. Temperature t (° C) profile of sulfate influx. Q_{10} is the temperature coefficient equivalent to $k_{(t+10° C)}/k_{(t)}$. C. Sulfate concentration (S) dependence of sulfate influx in RBC (lines) and in B3 vesicles (columns).

to 25%–30%. This clearly underscores the fact that strict quantitative estimations of the efficiency of transport systems in the reconstituted state, relative to the native state, might be misleading. Nonetheless, using all the available sulfate transfer data for a variety of experimental conditions (FIGURE 4), it was obtained that B3 capacity in B3 vesicles ranged from 1/3 to 1/8 the capacity observed in intact ghosts. Finally, the implication of B3 in anion exchange based on reconstitution studies, must also be tempered by the fact that only 1/4 of the vesicle preparation were found to be tightly-sealed vesicles, whereas the rest were ion-leaky vesicles (not shown). This indicates that a large fraction of the chemically-detectable B3 (FIGURE 2), actually represented B3 associated with apparently unfunctional (i.e., nonassayable) B3 vesicles. The possibility exists that the observed sulfate transfer activity is mediated by minor factors other than, or in addition to, B3. While B3 cannot be fully excluded, it cannot be supported by presently available analytical techniques either. These techniques did not reveal in B3 vesicle preparation any discrete membrane protein component that, apart from B3, constituted more than 5% of Coomassie Blue stainable material and was labeled by $[^3H]H_2$DIDS (FIGURE 2).

Reconstitution in vivo

In an attempt to improve the above reconstitution procedure and further assess the suggested role of B3, we examined the possibility of assaying the functional properties of isolated and reconstituted B3 in the plasma membrane domain of living cells. We assumed that implantation of B3 into plasma membranes should allow the functional expression of the isolated, and perhaps otherwise inactive, B3. We have chosen to coreconstitute B3 with VG (glycoproteins of *Sendai* virus envelopes) in order to provide B3 vesicles with fusogenic potential (i.e., B3VG vesicles). These vesicles undergo fusion with Friend erythroleukemia cells (FELC) just as well as VG vesicles,[11] thus leading to the insertion of VG and B3 into their plasma membranes.[11] The latter was corroborated by immunological tests that revealed the location of both B3 and VG in the above plasma membranes.[11] FELC were selected as the B3VG recipient cells since they do not display the characteristic Cl-transport features of red blood cells,[15] and since they fully regain their permeability barrier, which is modified by the fusion stress.[16] Furthermore, the B3VG-fused FELC were shown to be demonstrably viable cells.[11] The experimental system has the advantage of being amenable to studies of Cl-transport performed at 0° C using standard techniques for measuring Cl-transport in red blood cells (RBC).[15] The various FELC preparations used for studying Cl-ingress are given in FIGURE 5. These represent essentially three types of cells (FIGURE 5): 1. Unfused FELC (a, b, c); 2. FELC fused with VG (d, e) and, 3. FELC fused with B3VG (f). All these cell preparations were tested for their Cl-transport ability, as shown in FIGURE 6. Whereas FELC showed very poor Cl-transport ability at 0° C, RBC, as expected, were very fast in transporting Cl. More importantly, however, FELC fused with B3VG showed a marked increase in Cl-transport relative to FELC. That this increase resulted from the newly implanted B3 is shown by the fact that neither VG alone, B3 alone, mixtures of vesicles containing either VG or B3 separately, nor B3VG vesicles derived from DIDS-treated cells, succeeded in conferring upon FEL cells the high Cl-transport capacity. Furthermore, implanted B3 displayed only partial susceptibility to DNDS, suggesting that only a fraction of the total implanted B3 retained the original asymmetric B3 orientation found in intact cells.[1]

An important aspect of the above experiments is the yield of B3 implantation versus the Cl-transport activity expressed by B3 in the FELC plasma

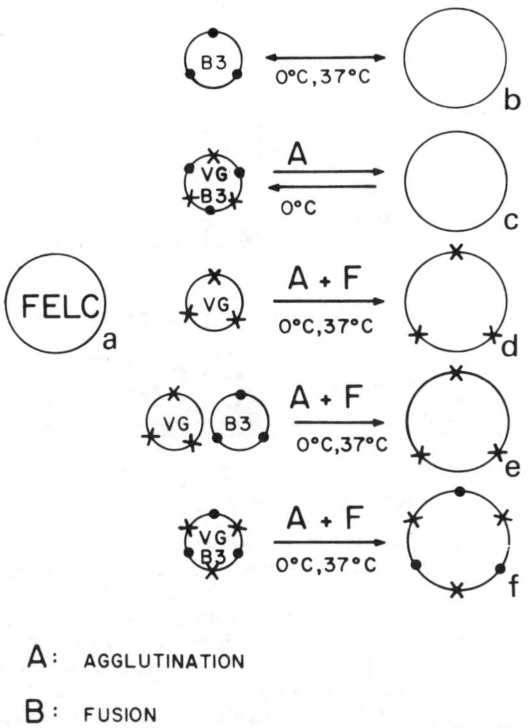

FIGURE 5. Schematic representation of interactions between various types of vesicles and Friend erythroleukemia cells (FELC). FELC were incubated at 0° C with reconstituted vesicles containing either viral glycoproteins (VG), Band 3 (B3), or both B3 and VG (B3VG), or with a mixture of B3 vesicles and VG vesicles. At this temperature, only agglutination (A) occurs.[10, 11] During subsequent incubations at 37° C, fusion (F) can also be observed in some of the systems, with concomitant appearance of viral and B3 antigens in the plasma membrane of FELC.[11]

membranes. Using [^{3}H]H_2DIDS, and assuming that all FELC contained B3 polypeptides, we estimated an average of 2×10^5 copies of B3 introduced per FELC. However, from FIGURE 6 we learn that only 25% of the total FELC population displayed high Cl-transport capacity, indicating that each FELC-B3VG harbored 8×10^5 B3 copies—a value fortuitously similar to that found in RBC.[1] However, despite the fact that FELC, and FELC-B3VG, each has a membrane surface considerably larger than that of RBC, and thus a smaller B3 density, the rate constant of Cl-ingress across FELC-B3VG (FIGURE 6, dotted line) was about twofold larger than in RBC. This clearly indicates that the implanted B3 is highly active in the FELC plasma membrane milieu, although an exact estimation of B3 activity in FELC-B3VG cannot be given. Evidently, factors such as membrane fluidity, and plasticity can possibly modulate B3 activity in FELC, as recently demonstrated for B3 reconstituted

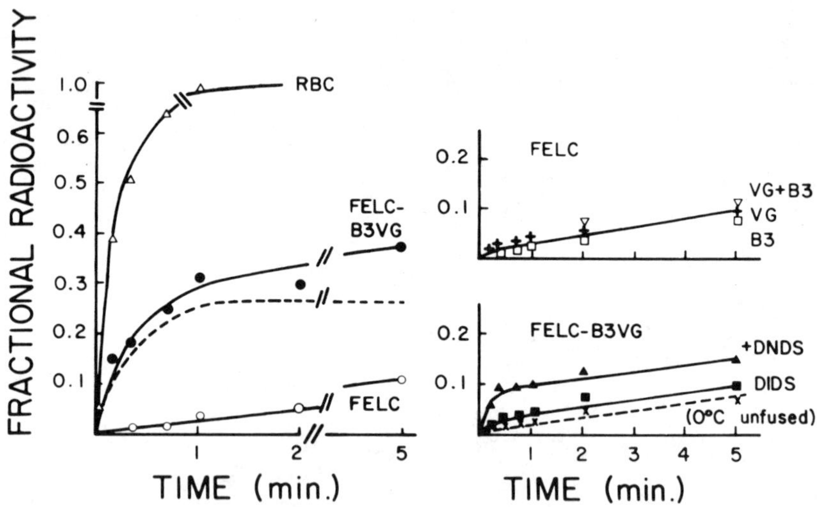

FIGURE 6. Functional properties of B3 implanted into living FELC. Fusion of the various vesicle preparations with FELC were as shown diagrammatically in FIGURE 5 and as detailed elsewhere.[11] The results are given as fractional radioactivity, which is equivalent to cpm t/cpm at infinity ∞ as shown for $^{35}SO_4^{-2}$ in FIGURE 4. The following symbols represent: RBC (△), red blood cells; FELC (○) Friend erythroleukemia cells. FELC-B3VG, FELC fused with B3VG and fluxes measured in the absence (●) or presence of 1 mM DNDS (▲); DIDS(O) denotes FELC-BVG containing B3 isolated from DIDS (0.1 mM) treated ghosts; 0° C-unfused (X), denotes FELC incubated with B3VG only at 0° C (i.e., agglutinated but not fused) and then cultivated. FELC+VG (+), FELC fused with VG. FELC+VG+B3 (▽), Friend cells fused with a mixture of separately reconstituted VG and B3 (see also FIGURE 5). The profile of Cl ingress in FELC-B3VG consisted of a fast transport component in ca. 1/4 of the cell population (or Cl-accessible compartment) that is attributed to fused cells (----, broken lines) and of a slow transport component in the remaining 3/4 of the cell population that is attributed to unfused cells.

into model lipid membranes.[8] These factors could conceivably provide B3 with an environment more favorable for expression of its functional properties than the RBC membrane itself. It is also conceivable that the above membrane environment might not be totally alien to B3, inasmuch as FELC resembles, in many respects, precursors of mouse erythrocytes and, thus, is used as a model for erythroid maturation.[17,1] However, further work will have to be done in order to verify whether the successful implantation of B3 into FELC could also be extended to other cell types, and, conversely, whether membrane proteins of other cell sources could be implanted in a highly functional state into plasma membranes of FELC. Finally, the activity of the implanted B3 in the plasma membrane of living cells might also alter with time and fade with membrane protein degradation uncompensated with B3 biosynthesis. This point, however, requires further investigation.

Conclusions

In the present work, we evaluated two novel procedures for isolation and reconstitution of functional B3 polypeptides. We first demonstrated the incorporation of purified and solubilized B3 into phospholipid vesicles. The *in vitro*-reconstituted system showed only partial expression of its sulfate transfer capacity and also showed low response of RB3-mediated sulfate transfer to pH, temperature, sulfate concentration and susceptibility to DNDS, a specific anion transport inhibitor. Reconstitution of B3 led to a symmetric transmembrane disposition of B3 polypeptides.

While *in vitro* reconstitution of B3 might not have provided highly functional polypeptides, thus leaving somewhat open the question of B3 involvement in anion transport, the *in vivo* reconstitution method, by allowing the apparently inactive B3 to display high Cl-transport capacity, has reinforced the alleged role of B3. It is, therfore, concluded that plasma membranes of living cells can provide an appropriate matrix for harboring isolated membrane transport proteins, for assaying their putative functions and for evaluating their role in physiological processes.

References

1. Cabantchik, Z. I., P. A. Knauf & A. Rothstein. 1979. Biochim. Biophys. Acta (Reviews) **515:** 239–302.
2. Passow, H. & L. Zaki. 1978. Studies on the molecular mechanism of anion transport across the red blood cell membrane. *In* Molecular Specialization and Symmetry in Membrane Function. A. K. Solomon & M. Karnovsky, Eds. 229–250. Harvard University Press. Cambridge, Mass.
3. Cabantchik, Z. I., J. M. Wolosin, H. Ginsburg & O. Zemel. 1977. Structural and functional properties of the anion transport system of human erythrocyte membranes. *In* FEBS Symp. No. **42:** Biochemistry of Membrane Transport. G. Semenza & E. Carafoli, Eds. 328–345. Springer Verlag, Berlin.
4. Wolosin, J. M., H. Ginsburg & Z. I. Cabantchik. 1977. J. Biol. Chem. **252:** 2419–2427.
5. Lepke, S. & H. Passow. 1976. Biochim. Biophys. Acta **455:** 353–370.
6. Grinstein, S., S. Ship & A. Rothstein. 1978. Biochim. Biophys. Acta. **507:** 294–304.
7. Rothstein, A., Z. I. Cabantchik, M. Balshin & R. L. Juliano. 1975. Biochem. Biophys. Res. Commun. **64:** 144–150.
8. Wolosin, J. M. 1977. Ph.D. Dissertation. Hebrew University. Jerusalem, Israel.
9. Ross, A. H. & H. M. McConnell. 1978. J. Biol. Chem. **253:** 4777–4782.
10. Volsky, D. J. & A. Loyter. 1978. FEBS Lett. **92:** 190–194.
11. Volsky, D. J., Z. I. Cabantchik, M. Beigel & A. Loyter. 1979. Proc. Natl. Acad. Sci. USA. **76:** 5440–5444.
12. Barzilay, M. & Z. I. Cabantchik. 1979. Membrane Biochem. **2:** 255–281.

13. SCHNELL, K. F., S. GERHARDT & A. SCHOPPE-FRIEDENBERG. 1977. J. Membrane Biol. **30:** 319–350.
14. GITLER, C. 1972. Ann. Rev. Biophys. Bioeng. **1:** 51–92.
15. HARPER, P. A. & P. A. KNAUF. 1979. J. Cell Physiol. **98:** 347–358.
16. USHIDA, T., J. KIM, M. YAMAIZUMI, Y. MIYAKE & Y. OKADA. 1979. J. Cell Biol. **80:** 10–20.
17. PATULEIA, M. C. & C. FRIEND. 1967. Cancer Res. **27:** 726–730.
18. HARRISON, P. R. 1976. Nature **262:** 353–356.
19. CABANTCHIK, Z. I. & A. ROTHSTEIN. 1974. J. Membrane Biol. **15:** 207–227.
20. FAIRBANKS, G., T. L. STECK & D. F. H. WALLACH. 1971. Biochemistry **10:** 2606–2617.
21. GASKO, O. D., A. F. KNOWLES, H. G. SHERTZER, E. M. SUOLINNA & E. RACKER. 1976. Anal. Biochem. **72:** 57–65.

DISCUSSION

J. M. SALHANY: I was intrigued by the pH dependence in your experiments. Is the pK of the ascending branch in the reconstituted system the same as that in the cell? How close is it?

Z. I. CABANTCHIK: Since the pH profile of sulfate transfer is considerably less pronounced in the reconstituted system than it is in intact cells, it is difficult to establish precisely the pK of the ascending branch. From visual inspection, it appears to be close to 7.5, or about 1 pH unit higher than that in intact cells.

J. WIETH: In your control experiments, did you or did you not determine that the sialoglycoprotein alone would not affect the anion transport?

CABANTCHIK: It appears that the sialoglycoprotein alone can lead to some increase (ca. 20% to 50%) in sulfate transfer relative to control liposomes. However, this increase is probably nonspecific inasmuch as it is insensitive to either DNDS or DIDS. When both Band 3 and the sialoglycoprotein were present, a tenfold increase in sulfate transfer activity was observed. This increase was not observed in similar experiments carried out with DIDS-treated samples.

WIETH: Is anything known about how sialoglycoprotein mediates fusion?

CABANTCHIK: In *Sendai* virus envelopes, there are two major proteins: the HN (hemagglutinin neuraminidase) protein and the fusion protein. Whereas the first is involved in the initial attachment of virus to receptors (sialoglycoproteins or sialoglycolipids), the fusion factor is apparently involved in subsequent steps of membrane fusion. The present view is that the two factors operate in tandem; however, the molecular mechanism of fusion is not well understood.

M. MORRISON (*St. Jude Childrens Research Hospital, Memphis, Tenn.*): Why didn't you fuse your vesicles with human intact red cells, for instance?

CABANTCHIK: Unlike Friend erythroleukemic cells, red blood cells do not regain their permeability barrier after fusion. Perhaps we should say that we have not yet developed the means to obtain fused erythrocytes that are also sealed to small ions.

STRUCTURE OF THE ERYTHROCYTE ANION TRANSPORT PROTEIN *

Michael J. A. Tanner, David G. Williams and Robert E. Jenkins

Department of Biochemistry
University of Bristol
Bristol BS8 1TD, England

Introduction

The transport of anions across the erythrocyte membrane is mediated by the major integral membrane protein of the erythrocyte.[1] This protein migrates on sodium dodecyl sulphate gel electrophoresis with an apparent molecular weight of 100,000 and has been designated Band 3[2] or polypeptide 3[3]. It represents about 25% of the Coomassie Blue-stainable proteins of the erythrocyte membrane, which is equivalent to about 10^6 copies per erythrocyte.[2,4] The protein is glycosylated[5,6,7] and contains receptors for concanavalin A and other lectins,[8,9] including those from *Phaseolus vulgaris* and *Ricinis communis.* This carbohydrate is *N*-glycosidically linked to the polypeptide. It has been suggested that a single oligosaccharide chain is present in the molecule,[10] but other workers have provided evidence for the presence of two oligosaccharide chains in the molecule.[11]

Structure and Folding of the Anion Transport Protein in the Membrane

The general approach to investigate the structure of the anion transport protein used by ourselves and other workers[12–16] has been to fragment the protein with proteases or chemical cleavage methods and to use markers on the polypeptide chain that have a known location, either at the extracellular or intracellular surface of the membrane.

We have used fragmentation of the native protein at a known side of the membrane in erythrocyte membrane preparations, together with lactoperoxidase radioiodination of tyrosine residues[17,18] in the protein to mark the intracellular or extracellular location of regions of the polypeptide that contain tyrosine residues accessible to the labelling agent. The intact anion transport protein was radioiodinated in both extracellular and cytoplasmic domains by using lactoperoxidase in leaky erythrocyte ghosts; and the protein was then isolated, digested with thermolysin, and the labelled peptides separated on peptide maps. Fourteen major radioactive peptides were obtained, each of which have a characteristic location on the peptide map (FIGURE 1). When the radioiodination was done on intact erythrocytes or resealed erythrocyte ghosts, so that only the extracellular domain of the protein was accessible to and labelled by the lactoperoxidase, only 5 of the peptides (numbers 1–5 in FIGURE 1) were found to contain radioactivity after this procedure.[19] This suggested that the remain-

* This work was supported in part by grants from the Medical Research Council and the Wellcome Trust.

0077–8923/80/0341–0455 $01.75/0

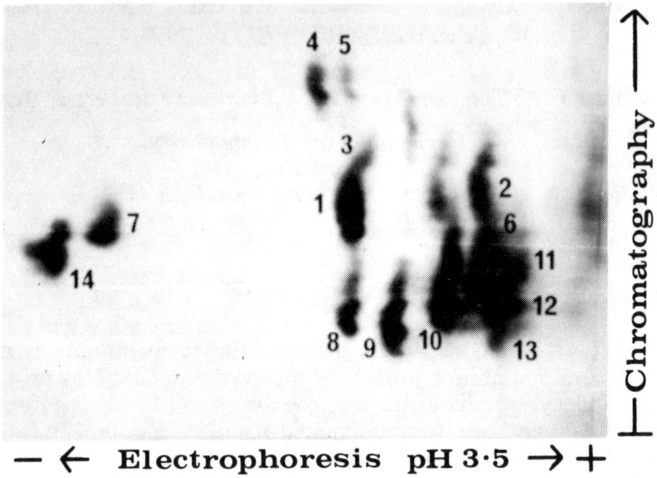

FIGURE 1. Autoradiograph of thermolysin peptide map of lactoperoxidase radioiodinated anion transport protein. The anion transport protein was isolated from lactoperoxidase-labelled erythrocyte ghosts, digested with thermolysin and the peptides separated on a peptide map.[19] The peptides numbered 1 through 5 are derived from tyrosine residues located at the extracellular surface of the membrane and peptides 6 through 14 are from residues located at the cytoplasmic surface of the membrane.

ing radioactive peptides (6–15, FIGURE 1) were derived from tyrosine-containing sequences in the cytoplasmic domain of the protein. Confirmation that this was the case came from similar experiments in which only the cytoplasmic domain of the protein was accessible to lactoperoxidase. These experiments used resealed erythrocyte ghosts containing internal lactoperoxidase and "inside-out" membrane vesicles. Only peptides 6–14 were radioactively labelled in this case.[19]

We have studied the structure of the protein by treating the lactoperoxidase radioiodinated protein with proteases in erythrocyte membrane preparations. The fragments obtained were isolated, thermolysin digested and the radioactive peptides separated on peptide maps. Knowing the radioactive peptide map peptides obtained from each fragment, it was possible to deduce the location of the fragment in the linear amino acid sequence of the protein and its topographical situation in the membrane. The labelled peptides obtained in thermolysin peptide maps of the anion transport protein are quite characteristic of the protein and distinct from those obtained from the major sialoglycoprotein, the other major intrinsic membrane protein present in the erythrocyte. The presence of the characteristic peptides of the anion transport protein in the peptide maps of a fragment also defined their origin from this protein, even if they were minor fragmentation products.

Trypsin Cleavage of the Protein

The extracellular region of the anion transport protein resists digestion by high concentrations of trypsin under iso-osmotic ionic strength conditions, although it is more susceptible to cleavage under hypo-osmotic ionic strength conditions.[3] Digestion of leaky erythrocyte ghosts with trypsin results in cleavage in the cytoplasmic portion of the anion transport protein, even when low concentrations of trypsin are used. Cleavage with trypsin occurs more readily at hypo-osmotic ionic strength than at iso-osmotic ionic strength, but under both conditions a large membrane-bound fragment of the protein (fragment T1) is obtained, which migrates as a broad band with an apparent molecular weight of 50,000–60,000.[3] Fragment T1 contained the extracellular-labelled peptide map peptides 1, 2, 3, 4, 5 and the intracellular peptides 8, 9, 10.[3] T1 represents the C-terminal portion of the protein, since the isolated fragment did not have a C-terminal basic residue.[20] This fragment had N-terminal glycine and contained the lectin receptors present in the intact protein.[20] Under certain conditions of gel electrophoresis, fragment T1 could be resolved into two bands. Peptide mapping of the two bands suggested that they represented the same polypeptide chain, but the two bands showed different abilities to bind concanavalin A and the lectin from *R. communis*.[20] The two bands probably differ only in carbohydrate composition.

When the protein is digested with low concentrations of trypsin, a number of water-soluble fragments are released from the membrane.[14, 15] Fragment T2s overlaps the smaller fragments T3s and T4s (FIGURE 2) and T2s and T3s share the blocked N-terminus [10] of the protein.[15, 21] T3s contained the cytoplasmically-located lactoperoxidase-labelled peptide map peptides 6, 7, 11, 12, 13 clustered within a small region of the protein,[22] while T4s contained the

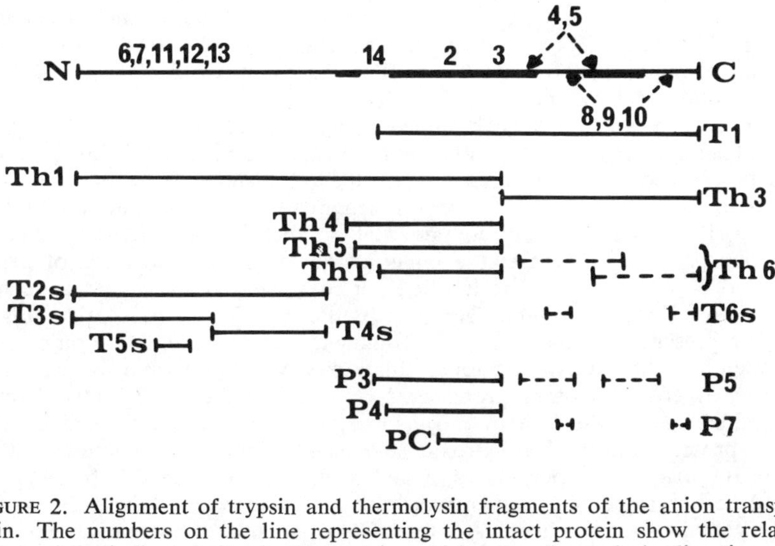

FIGURE 2. Alignment of trypsin and thermolysin fragments of the anion transport protein. The numbers on the line representing the intact protein show the relative positions of the iodinated sites in the protein, which, after thermolysin digestion, give rise to the radioactive peptides with corresponding numbers on peptide mapping (FIGURE 1). The thickened portions of the line represent extracellular regions of the protein. The position of the thermolysin (Th), pepsin (P) and trypsin (T) fragments of the protein are shown. The suffix *s* is used to indicate water soluble fragments of the protein. The broken lines indicate alternative positions for certain fragments and labelled sites. (From Williams *et al.*[22] and Tanner *et al.*[24])

sulphydryl group involved in the facile oxidative dimerization of the protein.[14, 15] Fragment T2s is located entirely in the cytoplasmic domain of the protein.[15, 21] A very small soluble peptide, T6s, which contained the cytoplasmic lactoperoxidase labelled peptide map peptides 8, 9, 10, and was not fixed by the Coomassie Blue protein staining method, could be obtained from the protein under these conditions.[22] Digestion with trypsin under more vigorous conditions, such as those used to produce fragment T1[3], yielded small soluble peptides that were poorly fixed after staining of acrylamide gels.[22]

In our earlier work, we obtained a membrane-bound band T2 on trypsin digestion of erythrocyte ghosts, which appeared to be complementary to the major fragment T1.[3] More recent studies using a higher resolution gel electrophoresis system to separate the fragments suggest that the T2 band contained more than one polypeptide.[22] We postulated a model for the structure of the protein that contained a duplicated set of extracellular sites and had the N-terminus of the protein located in the extracellular environment on the basis of the assumption that band T2 represented a homogenous polypeptide.[3] The model based on our present experiments (FIGURE 4), which will be discussed in more detail later, does not require the above features, and shows the N-terminus of the protein located at the cytoplasmic side of the membrane in agreement with the structures proposed by other workers.[15, 16]

Thermolysin Cleavage of the Protein

Cleavage at the extracellular region of the protein by thermolysin treatment of intact erythrocytes gave rise to several membrane-bound fragments[22] (FIGURE 2). The major fragment Th1 (approx. mol. wt. 60,000) contained the N-terminal region of the protein.[20, 22] This fragment had a blocked N-terminus like the intact protein and had a C-terminal residue of tyrosine.[20] A very similar fragment (Chy-A) has been obtained using chymotrypsin[12, 15] and this also had C-terminal tyrosine. Although we found that preparations of fragment Th1 contained carbohydrates,[20] these may well have resulted from the presence of contaminating glycolipids[10] since the fragment did not contain any lectin-binding sites. This fragment contained all of the lactoperoxidase-labelled peptide map peptides present in the intact protein with the exception of peptides 4, 5 and 8, 9 and 10.[22] The labelled peptide 3 was found to be close to the C-terminus of this fragment. The part of the protein not present in fragment Th1 is represented by the membrane fragment Th3 (FIGURE 2). Fragment Th3 migrated as a diffuse band of approximate mol. wt. 40,000 and has been found after treatment with other proteases.[14, 23] This fragment originated from the C-terminal end of the protein and, as expected, contained the lectin receptors on the protein, and the lactoperoxidase-labelled peptide map peptides not present in fragment Th1 (peptides 4, 5 and 8, 9, 10).[22] Since this fragment contained both extracellular and intracellular iodinatable sites, the polypeptide in this region of the protein must extend from one side of the membrane to the other.

Other minor fragments of the protein can be isolated when thermolysin treatment is restricted to the extracellular domain of the protein[22] (FIGURE 2). Fragments Th4 and Th5 are subfragments of Th1 and had the same C-terminus as Th1. These fragments contained the intracellular lactoperoxidase-labelled peptide map peptide 14 and the extracellular labelled peptides 1, 2 and 3. A further minor fragment (Th6) was obtained by digestion under these conditions. This is a subfragment of Th3 and contained only the extracellular labelled peptide map peptides 4 and 5. It migrated as a sharp band on gel electrophoresis that suggested it was probably not glycosylated.

Thermolysin and Chymotrypsin Digestion of Erythrocyte Ghosts

When erythrocytes were treated with thermolysin to yield fragments Th1 and Th3, lysed, and the resulting ghosts redigested with trypsin, a major fragment ThT was obtained [22] (FIGURE 3). Fragment ThT resulted from the cleavage of Th1 in the intracellular region of the protein by trypsin. ThT contained only the labelled peptide map peptides 1, 2 and 3, and has been shown to be coextensive with the C-terminus of fragment Th1.[22]

Digestion of ghosts from untreated erythrocytes with thermolysin or chymotrypsin yielded the major products shown in FIGURE 3. Digestion at iso-osmotic ionic strength with thermolysin yielded a fragment (GTh) of apparent mol. wt. 20,500 (FIGURE 3d). Under similar conditions chymotrypsin (FIGURE 3c) yielded a major fragment of the same molecular weight (GCh1) and small amounts of a fragment with an apparent molecular weight 500 daltons less (20,000, GCh2). This difference in molecular weight would represent 4–5 amino acids. When the digestion was done at low ionic strength chymotrypsin yielded only the smaller 20,000 mol. wt. band (FIGURE 3b). Inspection of the gels suggested that this band contained more than one component. Thermolysin yielded mainly the 20,500 mol. wt. component with small amounts of a lower molecular weight component (FIGURE 3a). Radioiodinated fragments were prepared by treating lactoperoxidase-labelled erythrocyte ghosts at iso-osmotic ionic strength with thermolysin or chymotrypsin. Although the two chymotrypsin fragments, GCh1 and GCh2, were just resolved on analytical gel electrophoresis (FIGURE 3), it was not possible to resolve them by electrophoresis on a preparative scale. Peptide maps of these fragments (in this case derived from the protein radioiodinated in both intracellular and extracellular domains) showed that GTh contained peptides 1, 2 and 3, while the mixture of GCh1 and GCh2 contained peptide 2 with small amounts of peptides 1 and 3 (TABLE 1). The labelled peptide map peptides 1 and 3 have been shown to be overlapping peptides, produced during the thermolysin digestion used for the preparation of peptide maps, and they contain the same radioiodinated tyrosine residue in the protein sequence.[24] Fragment GTh probably differs from the chymotrypsin fragment of corresponding mol. wt. in the presence of the tyrosine residue that gives rise to the labelled peptides 1 and 3. Fragment GTh appears to be very similar to fragment ThT.

Pepsin Digestion of Erythrocyte Ghosts

Digestion of erythrocyte membranes in 1 M acetic acid with pepsin yielded a number of membrane-bound fragments [24] (FIGURE 2). Fragments P3, P4 and PC (apparent mol. wt. 21,000; 18,500 and 13,500, respectively) all contained the lactoperoxidase-labelled peptide map peptides 1, 2 and 3, and were overlapping fragments that differed only at their N-terminal ends. They had common C-terminal ends that terminated very close to the C-terminus of fragment

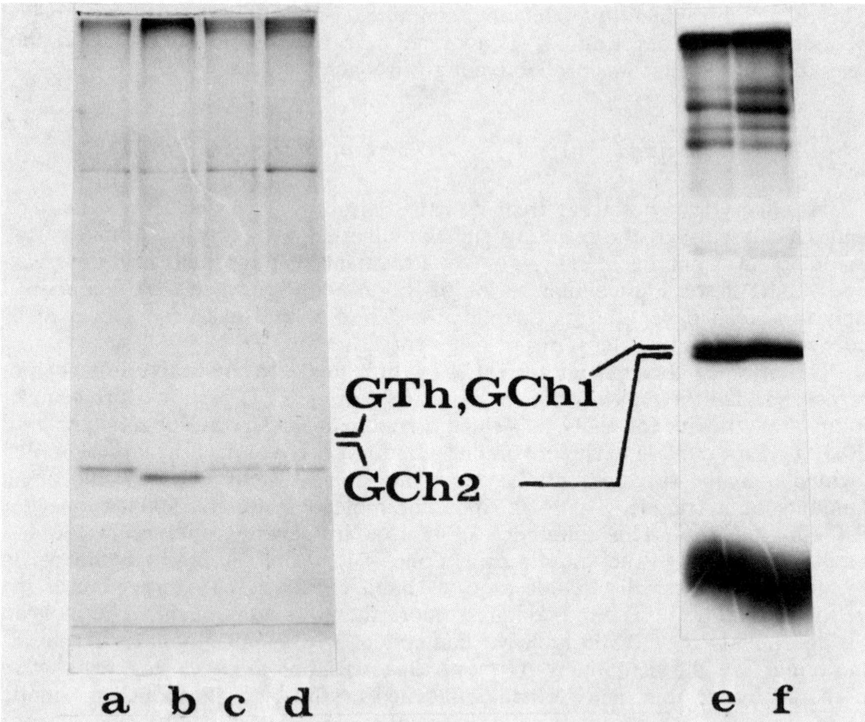

FIGURE 3. Digestion of erythrocyte ghosts with thermolysin and chymotrypsin at iso-osmotic and low ionic strengths. (a-d) A 50% suspension of packed erythrocyte ghosts were digested either at low ionic strength (in 5 mM sodium phosphate buffer pH 8.0) or at iso-osmotic ionic strength (in 0.15 M NaCl containing 5 mM sodium phosphate buffer pH 8.0). Digestion at low ionic strength was done for 1 hr at room temperature using 0.2 mg/ml thermolysin or chymotrypsin, while the reaction at iso-osmotic ionic strength was done for 1 hr at room temperature using 1 mg/ml of each protease. After digestion 1 μl of di-isopropyl-phosphofluoridate was added to the chymotrypsin digests and the thermolysin digests were made 2 mM in EDTA pH 8.0. After standing at room temperature for 10 min, the membranes were washed once with ice cold 0.1 M NaOH containing 10 mM 2-mercaptoethanol and twice with 15 imOsm sodium phosphate buffer pH 7.4, and dissolved in the sample buffer for gel electrophoresis.[22] Slab gel electrophoresis was done using the discontinuous buffer system of Laemmli [26] using a stacking gel containing 5% (w/v) acrylamide and a separating gel containing a concentration gradient of 8%–25% (w/v) acrylamide.[22] The gel was stained for protein with Coomassie Blue.

a) Thermolysin digestion at low ionic strength; **b)** Chymotrypsin digestion at low ionic strength; **c)** Chymotrypsin digestion at iso-osmotic ionic strength; **d)** Thermolysin digestion at iso-osmotic ionic strength; **e-f)** Erythrocyte ghosts were radioiodinated using lactoperoxidase [19] and digested under iso-osmotic ionic strength conditions with thermolysin or chymotrypsin, as described above. SDS gel electrophoresis was done, as described above, using a separating gel containing an acrylamide gradient of 10%–30% (w/v) acrylamide. An autoradiograph of the dried down gel is shown. **e)** Thermolysin digest; **f)** Chymotrypsin digest.

Th1. Fragment P3 appeared to be very similar to the fragments ThT, GTh and GCh 1. Two other membrane-bound fragments, P5 and P7, were also obtained.[24] These had apparent mol. wts. of 12,500 and less than 4,000, respectively. P5 contained only the extracellular-labelled peptide map peptides 4 and 5, while P7 contained only the intracellular-labelled peptide map peptides 8, 9 and 10. These two fragments were derived from the C-terminal region of the protein that is contained in fragment Th3.

Folding of the Polypeptide Chain in the Membrane

The alignment of the fragments of the protein (FIGURE 2), their contents of lactoperoxidase-labelled peptide map peptides, and the extracellular or intracellular location of the proteolytic cleavages leading to the termini of the fragments, suggested the model for the folding of the protein in the membrane

TABLE 1

LACTOPEROXIDASE-LABELLED PEPTIDE MAP PEPTIDES IN FRAGMENTS RESULTING FROM THERMOLYSIN AND CHYMOTRYPSIN DIGESTION OF ERYTHROCYTE GHOSTS *

	Relative radioactivity in peptide number					
	1	2	3	4	5	1+3
Intact protein	1.9	1.0	0.5	0.7	0.7	2.4
Thermolysin fragment	0.8	1.0	1.4	—	—	2.2
Chymotrypsin fragments	0.2	1.0	0.1	—	—	0.3

* Thermolysin peptide maps were prepared from radioiodinated GTh, and GCh1 and GCh2 (these two fragments were not resolved) obtained, as described in FIGURE 3, together with a sample of the original labelled intact protein, as described.[3, 19] The radioactivity in the labelled peptides in each peptide map was normalized so that the radioactivity in the labelled peptide 2 was put equal to 1.0.

shown in FIGURE 4.[22] This is the simplest model which is consistent with our experimental results and because of the limitations of the techniques we have used further folds of the polypeptide chain may not have been detected.

Since the large N-terminal domain of the protein is located on the cytoplasmic side of the membrane and the termini of fragments Th4 and Th5 result from extracellular cleavage, the polypeptide chain must cross the membrane on the C-terminal side of fragment T2s. The polypeptide then returns to the cytoplasmic side of the membrane to form loop A, as the N-terminus of fragment ThT is on this side of the membrane. ThT also lacks the intracellular peptide map peptide 14 that is present in fragments Th4 and Th5. Thus, the site giving rise to peptide 14 must be on loop B on the N-terminal side of the cleavage giving rise to ThT. The C-termini of fragments Th1, Th4, Th5 and ThT all derive from the same extracellular cleavage on the C-terminal side of the lactoperoxidase labelled site, which gives rise to the labelled peptide map peptides 1 and 3. Thus, the polypeptide chain again penetrates the membrane to form loop B. Fragment PC is the smallest membrane-bound fragment we observed that contained the peptide map peptides 1, 2 and 3. Thus, less than

8,000–12,500 daltons of polypeptide separate the point of emergence of the polypeptide from the membrane to form loop C and the point of cleavage by thermolysin in loop C that yields fragment Th1.[24]

The C-terminal region represented by fragment Th3 also contains regions that are accessible at both sides of the membrane. The relative order of the two groups of sites giving rise to the labelled peptide map peptides 4 and 5, and 8, 9 and 10 has not yet been defined. The occurrence of fragment Th6, which is derived from extracellular cleavage and contains only the extracellular peptide

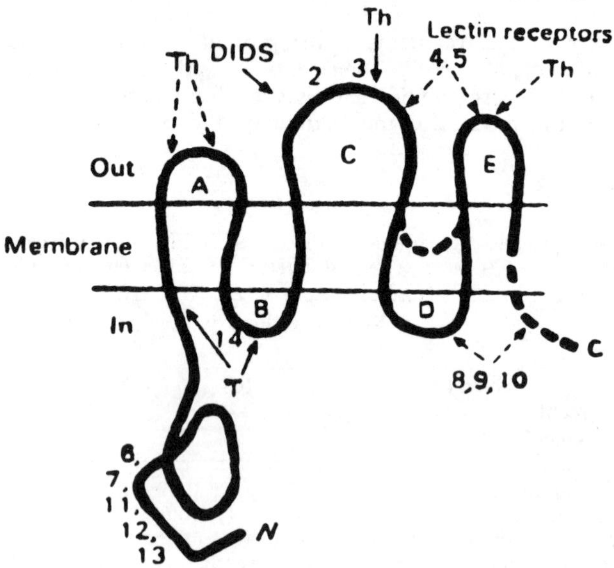

FIGURE 4. Schematic diagram of the folding of the polypeptide chain of the anion transport protein in the membrane. The broken lines show alternative possibilities for the topography of the C-terminal region of the protein, and alternative positions for the radioiodinated sites in this region. The major (solid arrows) and minor (broken arrows) sites of cleavage by trypsin (T) and thermolysin (Th) are shown, together with the location of the regions that bind the anion transport inhibitor, DIDS, and lectins. (From Williams *et al.*[22] By permission of the *Biochemical Journal.*)

map peptides 4 and 5, makes it necessary to postulate that a further fold in the polypeptide chain is present in this part of the molecule. There are two possible structures for the protein in this region of the molecule (represented by the solid and broken lines in FIGURE 4). In one structure (solid line in FIGURE 4), labelled peptide map peptides 8, 9 and 10 are present in the intracellular loop D and the peptide map peptides 4 and 5 are present in the extracellular loop E. Th6 would be the C-terminal peptide of the protein and would result from a minor cleavage in loop E on the N-terminal side of the sites giving rise to the labelled peptides 4 and 5. In this case, the C-terminus of the molecule need only be imbedded in the membrane, rather than penetrating through to the cytoplasmic side of the membrane. Alternatively, the sites yielding the labelled peptides 8, 9 and 10 could be present on a C-terminal intracellular "tail" of the protein (broken line in FIGURE 4). In this case, the region of the protein forming loop D need not penetrate right through the membrane but could simply be imbedded in it. The sites yielding the labelled peptide map peptides 4 and 5 could either be on loop C or on loop E on the N-terminal side of the point of cleavage by thermolysin that yields fragment Th6. The topographical location of the C-terminus of the molecule is not yet known.

The overall features of this model are similar to those proposed by Grinstein *et al.*[16] and Steck *et al.*[15] Our results clearly show that both the N- and C-terminal regions of the protein contain segments of polypeptide that are located at the intracellular side of the membrane. The detailed model in FIGURE 4 is based on the assumption that the cleavages leading to the minor fragments Th4, 5 and 6 occurred on the extracellular side of the membrane. It is difficult to be absolutely sure that this was indeed the case. They could conceivably result from intracellular cleavage by thermolysin because of the presence of a small proportion of leaky membranes in our preparations. However, no fragments corresponding to Th4, 5 and 6 are obtained on thermolysin treatment of leaky erythrocyte ghosts, and the peptide map of fragment GTh, the major product in this case, differs from that of Th4 and 5 in that it lacks the labelled peptide 14. This suggests that Th4, 5 and 6 are not the result of intracellular cleavage of the protein, and their termini are derived from extracellular proteolytic cleavage.

The occurrence, in high yield, of the membrane-bound pepsin fragments P5 and P7 that contain only the extracellular labelled sites 4, 5 and intracellular sites 8, 9, 10, respectively, further confirms that C-terminal end of the protein forms loop D at least (FIGURE 4). No fragment containing both the groups of sites 4, 5 and 8, 9, 10 were obtained on digestion with pepsin, which would be expected if fragments P5 and P7 were overlapping partial digestion products.

The model in FIGURE 4 shows that the protein can be divided into two structurally different domains. The N-terminal domain probably differs little in its structural characteristics from soluble proteins,[4] while the C-terminal domain is membrane-bound and can carry out anion transport.[25] The polypeptide in this region of the protein traverses the membrane at least 4–6 times, thus probably giving rise to the structural organization that allows it to act as a channel for the passage of anions across the erythrocyte membrane.

ACKNOWLEDGMENTS

We are grateful to Diana Barker, Sally Birch and Simon Peake for their assistance at different stages of this work. D.G.W. and R.E.J. were the recipients of Science Research Council Studentships.

REFERENCES

1. CABANTCHIK, Z. I., P. A. KNAUF & A. ROTHSTEIN. 1978. Biochim. Biophys. Acta **515:**239–302.
2. STECK, T. L. 1974. J. Cell. Biol. **62:**1–19.
3. JENKINS, R. E. & M. J. A. TANNER. 1977. Biochem. J. **161:**131–138.
4. STECK, T. L. 1978. J. Supramol. Struct. **8:**311–324.
5. HO, M. K. & G. GUIDOTTI. 1975. J. Biol. Chem. **250:**675–683.

6. Yu, J. & T. L. Steck. 1975. J. Biol. Chem. **250:**9170–9175.
7. Tanner, M. J. A., R. E. Jenkins, D. J. Anstee & J. R. Clamp. 1976. Biochem. J. **155:**701–703.
8. Findlay, J. B. C. 1974. J. Biol. Chem. **249:**4398–4403.
9. Tanner, M. J. A. & D. J. Anstee. 1976. Biochem. J. **153:**265–270.
10. Drickamer, L. K. 1978. J. Biol. Chem. **253:**7242–7248.
11. Gahmberg, C. G., G. Myllyla, J. Leikola, A. Pirkola & S. Nordling. 1976. J. Biol. Chem. **251:**6108–6116.
12. Drickamer, L. K. 1976. J. Biol. Chem. **251:**5115–5123.
13. Drickamer, L. K. 1979. J. Biol. Chem. **252:**6909–6917.
14. Steck, T. L., B. Ramos & E. Strapazon. 1976. Biochemistry **15:**1154–1161.
15. Steck, T. L., J. J. Koziarz, M. K. Singh, G. Reddy & H. Kohler. 1978, Biochemistry **17:**1216–1222.
16. Grinstein, S., S. Ship & A. Rothstein. 1978. Biochim. Biophys. Acta **507:** 294–304.
17. Phillips, D. R. & M. Morrison. 1971. Biochemistry **10:**1766–1771.
18. Hubbard, A. L. & Z. A. Cohn. 1972. J. Cell. Biol. **55:**390–405.
19. Boxer, D. H., R. E. Jenkins & M. J. A. Tanner. 1974. Biochem. J. **137:**531–534.
20. Jenkins, R. E. & M. J. A. Tanner. 1977. Biochem. J. **161:**139–147.
21. Fukuda, M., Y. Eshdat, G. Tarone & V. T. Marchesi. 1978. J. Biol. Chem. **253:**2419–2428.
22. Williams, D. G., R. E. Jenkins & M. J. A. Tanner. 1979. Biochem. J. **181:** 477–493.
23. Cabantchik, Z. I. & A. Rothstein. 1974. J. Membrane Biol. **15:**227–248.
24. Tanner, M. J. A., D. G. Williams & D. Kyle. 1979. Biochem. J. **183:**417–427.
25. Lepke, S. & H. Passow. 1976. Biochim. Biophys. Acta. **455:**358–370.
26. Laemmli, U. K. 1970. Nature **227:**680–682.

PART VI. ANION AND PROTON TRANSPORT MECHANISMS IN SINGLE MEMBRANE (NONEPITHELIAL CELLS); AXONES, MUSCLES, ERYTHROCYTES, TUMORS AND BACTERIAL CELLS

THE CATION SPECIFICITY FOR THE MELIBIOSE TRANSPORT SYSTEM OF *E. COLI* *

T. Hastings Wilson, Tomofusa Tsuchiya,† Jane Lopilato and Kathleen Ottina

Department of Physiology
Harvard Medical School
Boston, Massachusetts 02115

Chemical energy or light may be converted into an ion gradient across a biological membrane, and this potential energy is then utilized by a variety of membrane-related events, including nutrient accumulation, osmoregulation, action potential, ATP synthesis and motility. The proton is used as the currency for membrane energetics in microorganisms and mitochondria while the sodium ion is utilized by the plasma membrane of animal cells. It has been proposed [1–3] that during evolution the proton economy was utilized by membranes of the most primitive cells, while the use of sodium ions developed much later in time. Let us consider a possible series of events in the evolution of ion pumps.

In the most primitive anaerobes, ATP derived from substrate-level phosphorylation in glycolysis probably drove a proton-extrusion mechanism via a membrane-bound proton-ATPase (FIGURE 1A). A simple proton pump of this type has been described in *Neurospora* [4] and in yeast.[5] In photosynthetic organisms such as *Halobacterium halobium,* light provided the energy to drive another type of proton transport system [6] (FIGURE 1B). Finally, the most complex and most recently evolved mechanism is proton translocation by oxidation-reduction reactions [1, 7–9] (FIGURE 1C). The electrochemical potential difference for protons across the membrane (or protonmotive force) provides the energy for a variety of membrane transport processes.[8–11]

According to this view of evolution, a time arrived when selective advantage was obtained by cells that gave up the proton economy in favor of a sodium economy. Perhaps this was useful to provide a more stable pH environment for the cell. One simple type of sodium pump is widely distributed in microorganisms [12, 13] (FIGURE 2A). The inward movement of protons down their electrochemical potential gradient provides the energy for the extrusion of sodium ions out of the cell up their potential gradient. Present-day microbial halophiles show a partial preference for the sodium economy.[14] While the primary proton pumps are retained by the cells, sodium ion gradients are produced by the Na^+-H^+ exchange protein and the electrochemical potential gradient for the Na^+ is used to drive a large number of nutrient transport systems (substrate-Na^+ cotransport). The sodium economy is seen most highly developed in the animal cell plasma membrane where the Na-K-ATPase drives the Na^+ pump (FIGURE 2B). In the plasma membrane of these cells, a pure sodium economy is maintained.

* This work was supported in part by U.S. Public Health Service Grant No. AMO5736 from the National Institute of Arthritis, Metabolism and Digestive Diseases and the National Science Foundation (No. PCM-78–00859).

† Present address: Department of Microbiology, University of Okayama, 1–1–1 Tsushima-Naka, Okayama-shi, F700 Japan.

0077–8923/80/0341–0465 $01.75/0 © 1980, NYAS

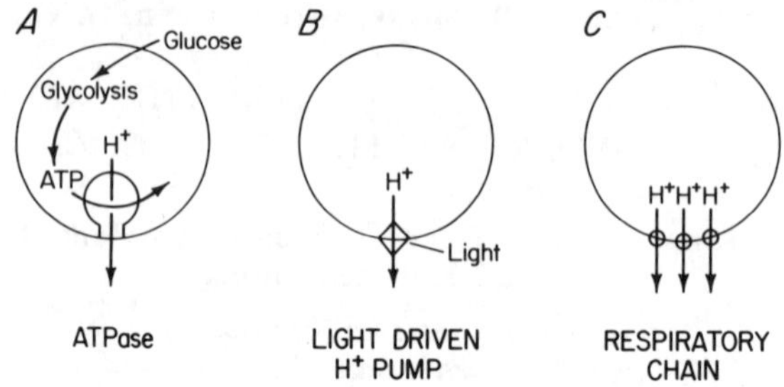

FIGURE 1. Proton pumps of microorganisms.

Associated with a change in cation economy of the membrane during evolution, there must have been a corresponding change in those processes that utilize the ion gradients as a source of energy. One important membrane function that must have changed its cation requirement was the nutrient-cation cotransport system. In the operation of this class of transport systems [15] there is obligatory coupling between cation entry and substrate entry with the inwardly directed electrochemical potential difference of the cation as the driving force (FIGURE 3). The ion gradient is maintained in the steady state by an independent cation extrusion pump. In most microorganisms, proton-substrate cotransport is the primary mechanism for this class of translocating devices (the prominent exception to this generalization being the extreme halophiles mentioned above [14]). The lactose carrier of *E. coli,* for example, utilizes protons for cotransport [16] and other cations such as Na^+ or Li^+ are completely without effect. Only the deuterium ion could substitute for protons.[17] In animal cells, on the other hand, sodium-substrate cotransport is universally utilized. While lithium ion may sometimes substitute for sodium ion,[18] protons have not been shown to interact with any of the cotransport systems of animal cells.

From an evolutionary point of view it would be interesting to find a cotransport system that was poised between the proton economy and the sodium economy (a "missing link" as it were) being able to use either protons or sodium ions for energy coupling. Such a transport system might also show a different cation preference for one substrate compared with a second substrate.

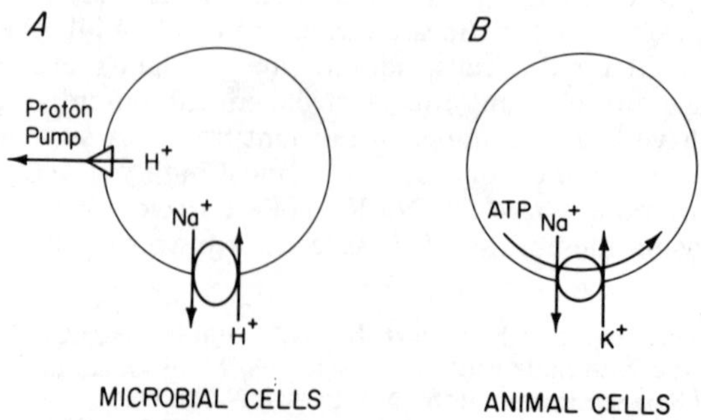

FIGURE 2. Sodium pumps.

We have found one such example of this type in the melibiose-cation cotransport system of *E. coli*. This communication describes the properties of this system (TABLE 1).

H^+-Melibiose Cotransport

In order to study the transport of the natural substrate, melibiose, it was necessary to utilize a mutant (RA11)[19] that lacks the hydrolytic enzyme α-galactosidase. This cell also carries a lac Y-gene deletion so that the lactose transport protein could not obscure the results with galactosides used in this study. The melibiose transport protein is induced by growth of the cells on amino acids in the presence of 5 mM melibiose.

The first series of experiments were designed to test the possibility of melibiose-H^+ cotransport. If there were obligatory couplings between the

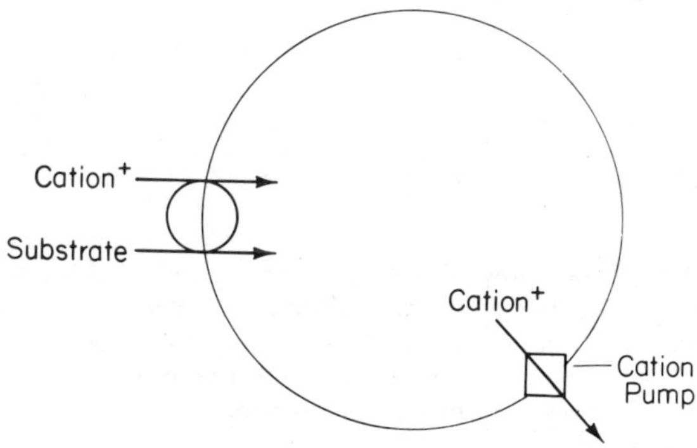

FIGURE 3. Cation-substrate cotransport.

inward movement of melibiose and protons, then addition of the disaccharide should result in the uptake of H^+ by the cells and an alkalinization of the external medium. In such experiments, it is essential to block the proton pumping of the respiratory chain by incubating the cells anaerobically as suggested by West.[16] The results of one such experiment[20] are shown in FIGURE 4. The addition of melibiose (5 mM) to anaerobic cells resulted in a prompt alkalinization of the external medium (an upward deflection in the figure). A similar proton movement resulted from the addition of α-methyl-galactoside, an analog of melibiose. On the other hand, there was no proton movement whatsoever with the addition of three other substrates, β-methyl-galactoside, β-thiomethylgalactoside and thiodigalactoside. Lactose, which is not a substrate for this system, was inactive. Thus, two substrates of the melibiose transport system show proton-substrate cotransport while three other substrates do not.

Consistent with the notion of melibiose-H^+ cotransport are a series of experiments showing that addition of this disaccharide facilitated proton entry when an inwardly directed protonmotive force was already established.[21] The sudden addition of HCl to anaerobic cells resulted in a very rapid acidification of the

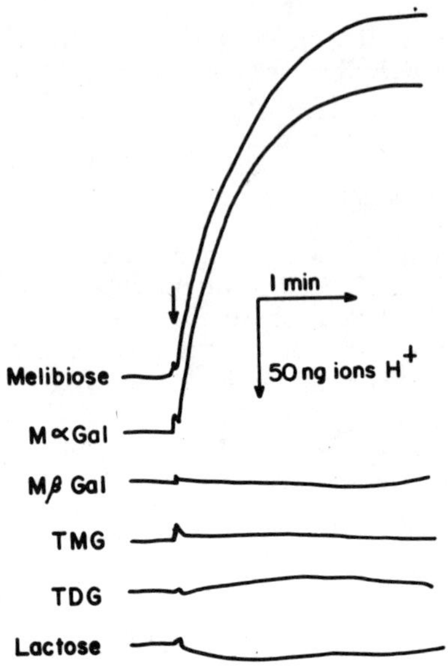

FIGURE 4. Proton movement induced by the addition of galactosides. A 20 μl sample of anaerobic 0.5 M galactoside was added to a 2 ml anaerobic cell suspension of RA11 (70 mg dry wt of cells/ml) in 90 mM KCl plus 30 mM KSCN. An upward deflection represents a rise in the pH of the medium. (From Tsuchiya and Wilson.[20])

medium followed by a slow entry of protons into the cell and a rise in external pH. The presence of melibiose greatly stimulated proton entry under these conditions (FIGURE 5). This experiment suggests that protons cannot enter the cell via the melibiose transport protein unless melibiose is present. This is consistent with the obligatory coupling between proton movement and melibiose movement on the membrane carrier.

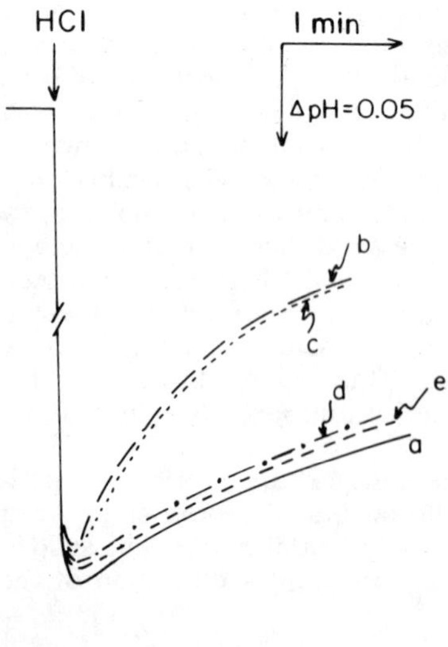

FIGURE 5. Stimulation of the rate of H^+ entry by melibiose and α-methyl-galactoside. Cells of RA11 were induced by growth in the presence of 5 mM melibiose. Two milliliters of cells (5 mg dry wt/ml) were incubated anaerobically in a vessel with a combined pH electrode. HCl (100 n moles) was added to the suspension at the time indicated by the arrow. Cells were preincubated with: **a**, no additions (control); **b**, melibiose (5 mM); **c**, α-methylgalactoside (5 mM); **d**, thiomethylgalactoside (5 mM); and **e**, lactose (5 mM).

Na^+-Melibiose Cotransport

The role of Na^+ was investigated by adding melibiose to anaerobic cells in the presence of 10 mM Na^+ and measuring the pH.[20] In contrast to the earlier experiment (FIGURE 4), the presence of Na^+ resulted in an exit of protons when the disaccharide was added (FIGURE 6). This result is consistent with melibiose-Na^+ cotransport and exit of protons in response to the positive potential produced by Na^+ entry.

This hypothesis was confirmed by the direct measurement of Na^+ entry into the cell. In the next experiment, Na^+ concentrations were monitored in the incubation medium with a Na^+ electrode.[20] Cells were incubated anaerobically in the presence of 25 μM Na^+ until the medium reached a steady pNa (-log Na^+). Melibiose was added and the change in pNa measured (FIGURE 7). There was a prompt fall in the Na^+ concentration of the medium on addition of the disaccharide. A similar stimulation of Na^+ uptake by induced cells was observed for the following substrates of the carrier: β-thiomethylgalactoside, thiodigalactoside, α-methylgalactoside, β-methylgalactoside and galactose.

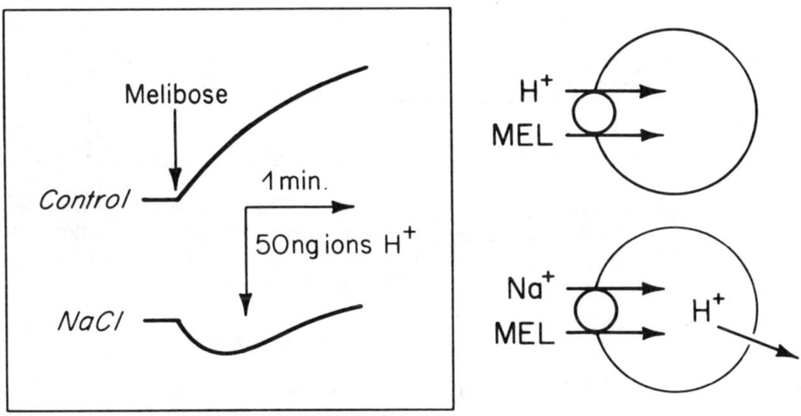

FIGURE 6. Proton movement induced by the addition of melibiose. Cells of RA11 were incubated anaerobically in the presence of 90 mM KCl and 30 mM KSCN with or without 10 mM NaCl. After preincubation for 45 min at room temperature, 20 μl of anaerobic 0.5 M melibiose was added to 2 ml of cell suspension and pH changes monitored. An upward deflection represents a rise in the pH of the medium. (From Tsuchiya and Wilson.[20])

Further confirmation of Na^+-melibiose cotransport comes from experiments on the accumulation of radioactive disaccharide by a transport positive, α-galactosidase negative, mutant.[22] Washed cells were incubated in the presence of ^{3}H-melibiose in the presence or absence of Na^+ (FIGURE 8). In the absence of added Na^+, the sugar accumulated within the cell to a concentration 13 times higher than the external medium. The addition of 5 mM Na^+ resulted in an intracellular concentration of melibiose 212 times that in the external medium.

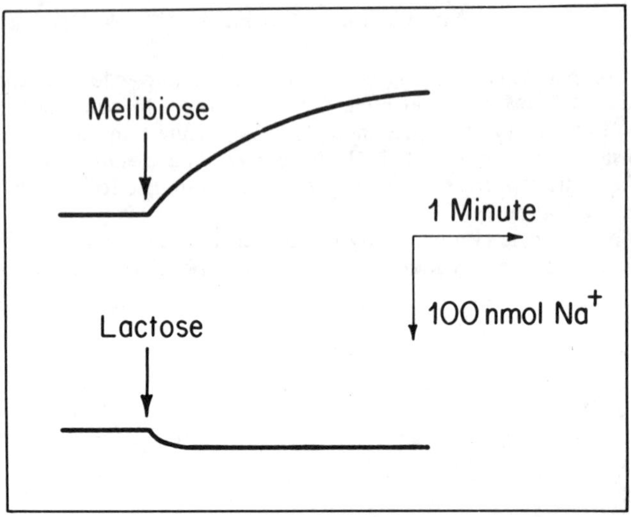

FIGURE 7. Sodium entry in response to addition of disaccharides. Melibiose or lactose was added to RA11 at the time indicated by the arrow to give a final concentration of 5 mM. An upward deflection indicates a fall in Na^+ concentration in the incubation medium. (From Tsuchiya and Wilson.[20])

Effect of Lithium

Lithium stimulation of melibiose transport was first observed in *Salmonella typhimurium* by Stock and Roseman [23] and by Tokuda and Kaback.[24] Lithium-TMG cotransport has also been demonstrated in *E. coli*.[25] The transport of α-methylgalactoside is also stimulated by this cation.[26] Two substrates, raffinose and galactose, are unaffected by Li^+. An unexpected finding was that melibiose (and α-p-nitrophenylgalactoside) transport was strongly inhibited by lithium.[26] The inhibition is so severe that cells will not grow on melibiose in the presence of 10 mM LiCl. Several Li^+-resistant mutants have been isolated. These mutants show a much reduced inhibitory effect of Li^+ on transport suggesting an alteration in the cation recognition site.

TABLE 1

THE EFFECT OF CATIONS ON THE TRANSPORT OF DIFFERENT GALACTOSIDES BY THE MELIBIOSE CARRIER OF *E. Coli*

Substrate	H^+	Na^+	Li^+
Melibiose	stimulates	stimulates	inhibits
α-Methylgalactoside	stimulates	stimulates	stimulates
β-Thiomethylgalactoside	no effect	stimulates	stimulates

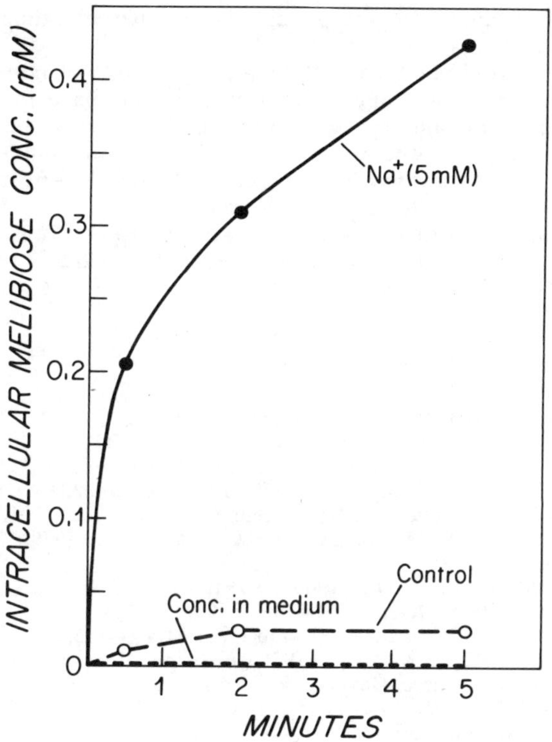

FIGURE 8. Stimulation of melibiose accumulation by Na^+. Cells of RA11 were grown on amino acids ("Tryptone") plus 10 mM melibiose and thiamine. Washed cells (0.25 mg dry wt cells/ml) were incubated in 0.1 M K PO_4 buffer pH 7.0 with or without 5 mM NaCl in the presence of ^{3}H-melibiose (2 μM). Samples (0.2 ml) were filtered on 0.65 μm pore size Millipore filters and washed once with 5 ml of buffer. The filter (containing cells) was dissolved in Brays solution and counted.

DISCUSSION

The melibiose transport system of *E. coli* is unusual in its cation requirement; it is capable of utilizing H^+, Na^+ or Li^+ for cotransport depending on the substrate. The natural substrate melibiose can undergo cotransport with either protons or Na^+. In the absence of added Na^+, protons are utilized; in the presence of 25 μM Na^+, this metal cation is used in preference to H^+.

From an evolutionary point of view, it is interesting to find a cotransport system capable of utilizing either H^+ or Na^+ as a source of energy. According to the hypothesis discussed earlier, the proton membrane economy of primitive cells must have given way at some point in time to the sodium economy of cells with more specialized needs. Presumably, the biochemical evolution of the membrane proteins of cotransport involved an alteration of the cation recognition site of the polypeptide. Consistent with this hypothesis is the discovery that the melibiose transport system can utilize both H^+ and Na^+. It is also of interest that the same transport protein shows different cation specificities for different substrates.

These views would be further strengthened by the isolation of mutants with altered cation recognition. Preliminary experiments have indicated that it is possible to isolate mutants resistant to the inhibitory effect of lithium on cells growing on melibiose. Studies with a few of these mutants indicate that the lithium inhibition of transport has been greatly reduced. Apparently, the cation recognition site in the carrier has been altered.

It is hoped that with this and other selective procedures mutants can be isolated that show the two extremes of cation specificity: affinity for Na^+ greater than H^+ and vice versa.

References

1. Mitchell, P. 1970. Soc. Gen. Microbiol. Symp. **20:**121–166.
2. Raven, J. A. & F. A. Smith. 1976. J. Theor. Biol. **57:**301–312.
3. Wilson, T. H. & P. E. Maloney. 1976. Fed. Proc. **35:**2174–2179.
4. Bowman, B. J. & C. W. Slayman. 1979. J. Biol. Chem. **254:**2928–2934.
5. Dufour, J. & A. Goffeau. 1978. J. Biol. Chem. **253:**7026–7032.
6. Stoeckenius, W., R. H. Lozier & R. A. Bogomolni. 1979. Biochim. Biophys. Acta **505:**215–278.
7. Mitchell, P. 1961. Nature (London) **191:**144–148.
8. Mitchell, P. 1963. Biochem. Soc. Symp. **22:**142–168.
9. Mitchell, P. 1966. Biol. Rev. (Cambridge) **41:**445–502.
10. Rosen, B. P. & E. R. Kashket. 1978. *In* Bacterial Transport. Barry P. Rosen, Ed. : 559–620. Marcel Dekker, New York.
11. Harold, F. M. 1978. *In* Bacteria. I. C. Gunsalus, Ed. Vol. 6:463–521 Academic Press, New York.
12. Harold, F. M. & D. Papineau. 1972. J. Membr. Biol. **8:**45–62.
13. West, I. C. & P. Mitchell. 1974. Biochem. J. **144:**87–90.
14. Lanyi, J. K. 1977. J. Supramol. Struc. **6:**169–177.
15. Crane, R. K. 1977. Rev. Physiol. Biochem. Pharmacol. **78:**99–159.
16. West, I. C. 1970. Biochem. Biophys. Res. Commun. **41:**655–661.
17. Linker, C. & T. H. Wilson. Unpublished observations.
18. Christensen, H. N. & M. E. Handlogten. 1977. J. Membr. Biol. **37:**193–211.
19. Lopilato, J., T. Tsuchiya & T. H. Wilson. 1978. J. Bacteriol. **134:**147–156.
20. Tsuchiya, T. & T. H. Wilson. 1978. Membr. Biochem. **2:**63–79.
21. Tsuchiya, T. & T. H. Wilson. 1979. Membr. Biochem. In press.
22. Ottina, K., J. Lopilato & T. H. Wilson. Unpublished observations.
23. Stock, J. & S. Roseman. 1971. Biochem. Biophys. Res. Commun. **44:**132–138.
24. Tokuda, H. & H. R. Kaback. 1977. Biochemistry **16:**2130–2136.
25. Tsuchiya, T., J. Raven & T. H. Wilson. 1977. Biochem. Biophys. Res. Commun. **76:**26–31.
26. Tsuchiya, T., J. Lopilato & T. H. Wilson. 1978. J. Membr. Biol. **42:**45–59.

MECHANISM OF LACTOSE TRANSLOCATION IN MEMBRANE VESICLES FROM *ESCHERICHIA COLI*

Gregory J. Kaczorowski,* Dan E. Robertson † and H. Ronald Kaback

Laboratory of Membrane Biochemistry
Roche Institute of Molecular Biology
Nutley, New Jersey 07110

The chemiosmotic hypothesis proposed by Peter Mitchell [1-5] has stimulated widespread interest in the role of the "protonmotive force" in bioenergetic processes. According to this hypothesis, energy derived from respiration or photochemical reactions is transformed into a transmembrane electrochemical gradient of protons ($\Delta\bar{\mu}_{H^+}$) * that represents the immediate driving force for the synthesis of ATP, active transport, and certain other energy-dependent processes.[6] In 1963, Mitchell postulated explicitly that $\Delta\bar{\mu}_{H^+}$ drives the accumulation of β-galactosides in *Escherichia coli* and that the active transport of these substrates occurs via coupled movements with protons (i.e., symport).[7] By this means, a β-galactoside-specific membrane protein (the product of the *lac y* gene) translocates substrate with protons, the substrate moving against and the proton(s) with their respective electrochemical gradients.

E. coli membrane vesicles retain the same configuration as the plasma membrane in the intact cell,[8-10] and this *in vitro* system has contributed increasingly to an understanding of chemiosmotic phenomena, particularly with respect to active transport.[11-15] Recent experiments have clarified the relationship between the electrical and chemical components of $\Delta\bar{\mu}_{H^+}$ and the accumulation of specific transport substrates.[16-20] Moreover, the concept of proton/substrate symport has been supported by the demonstration that active lactose accumulation leads to partial collapse of the electrical potential ($\Delta\Psi$) [21] and the pH gradient (ΔpH) [18] across the vesicle membrane, by studies of proton/substrate stoichiometry,[19] and by monitoring *lac* carrier function in response to $\Delta\Psi$ and ΔpH with fluorescent and photoreactive probes.[22-23]

Several laboratories have investigated proton/β-galactoside symport in intact *E. coli* by studying substrate-induced proton fluxes. West and Mitchell [24-26] demonstrated that addition of lactose to de-energized cells causes alkalinization of the medium, and Wilson's laboratory showed that transient accumulation of methyl-l-thio-β-D-galactopyranoside (TMG) can be coupled to an artificially-generated $\Delta\Psi$ (interior negative) or ΔpH (interior alkaline) in starved cells.[27] Similarly, Flagg and Wilson [28] observed that TMG efflux can drive proline uptake due to the ability of the *lac* carrier to catalyze proton translocation as TMG moves down its concentration gradient. Although these studies provide strong evidence for proton/substrate symport, the mechanism of the reaction is unknown.

* Fellow of the Helen Hay Whitney Foundation.

† Present address: Department of Pathology, Hahnemann Medical College, Philadelphia, Pennsylvania 19104. The abbreviations used in this paper are as follows: $\Delta\bar{\mu}_{H^+}$, the electrochemical gradient of protons; $\Delta\Psi$, the electrical potential; ΔpH, the pH gradient; CCCP, carbonylcyanide-*m*-chlorophenylhydrazone.

0077-8923/80/0341-0473 $01.75/0 © 1980 NYAS

One limitation in studying passive processes in intact cells is difficulty in achieving the completely de-energized state that is essential for quantitation of passive flux data. In view of this limitation, we have studied proton/β-galactoside symport in membrane vesicles since they are totally de-energized unless appropriate electron donors are added exogenously. Thus, in the experiments to be described (cf. Kaczorowski and Kaback [29] and Kaczorowski *et al.*, 1979 [23] for more detail), passive lactose movements were used to drive turnover of the *lac* carrier protein in order to study the symport reaction and to probe the mechanism of translocation.

Effect of pH on Efflux and Exchange

When concentrated *E. coli* membrane vesicles containing the *lac* carrier protein are equilibrated with 10 mM [1-^{14}C]lactose [a concentration that is approximately fivefold over the K_m for passive, carrier-mediated efflux [23]] and then diluted 200-fold into media devoid of lactose, the rate of efflux is first-order and increases about threefold from pH 5.5 to pH 7.5 [the time required for loss of 50% of the intravesicular lactose (i.e., the $t_{1/2}$) is about 45s, 27s, and 15s at pH 5.5, 6.6, and 7.5, respectively]. Moreover, the protonophore carbonylcyanide-*m*-chloro-phenylhydrazone (CCCP) does not influence the rate of efflux at any of the pH values tested, demonstrating that the generation of a $\Delta\Psi$ or ΔpH is not rate-limiting for efflux under these conditions.

Kinetically, efflux consists of a minimum of 4 steps (FIGURE 1): Binding of substrate to the carrier on the inner surface of the membrane, translocation across the membrane, release of substrate, and, return of the unloaded carrier. In order to determine whether return of the unloaded carrier is affected by pH, loss of intravesicular [1-^{14}C]lactose was studied under conditions in which the external medium contained equimolar concentrations of unlabeled lactose (i.e., under conditions of exchange). Thus, vesicles equilibrated with 10 mM [1-^{14}C] lactose were diluted 200-fold into media containing 10 mM ^{12}C-lactose, and loss of labeled sugar from the intravesicular pool was monitored. The rate of loss under these conditions (i.e., exchange) is very fast ($t_{1/2} < 2$s) relative to the rate of efflux and exhibits no dependence on pH from pH 5.5 to 7.5. The results imply that there are different rate-limiting steps for efflux and exchange, and that a step involving the return of the unloaded carrier is rate-determining for efflux, since the rate of translocation is markedly enhanced when the carrier is occupied by substrate on the external surface of the membrane.

If the *lac* carrier protein catalyzes the simultaneous translocation of protons and lactose, an increase in proton efflux should be induced by the movement of lactose down its concentration gradient, and this should be reflected by the generation of a $\Delta\Psi$ (interior negative) and/or a ΔpH (interior alkaline). When vesicles prepared in a sodium phosphate are equilibrated with 10 mM lactose, treated with excess valinomycin, and then diluted 200-fold into sodium phosphate containing ^{86}RbCl, transient ^{86}Rb uptake is observed at pH 5.5, 6.6, and 7.5, demonstrating that a $\Delta\Psi$ (interior negative) is indeed established.[21] Moreover, at each pH studied, the time course of the reaction reflects the rate of lactose efflux. That is, the rates of ^{86}Rb uptake, the maximum levels of accumulation, and the rates of decay increase with pH in a manner that approximates the increase in lactose efflux with pH. In addition, lactose efflux-induced ^{86}Rb uptake at each pH tested is abolished by CCCP, and importantly, by addition of lactose to the medium. Finally, ^{86}Rb uptake under these conditions is not observed with vesicles that are treated with sulfhydryl reagents that block *lac* carrier function. It is apparent, therefore, that downhill translocation of lactose via the *lac* carrier protein leads to the generation of a $\Delta\Psi$ (interior negative). On the other hand, all attempts to measure pH changes directly using a sensitive, rapidly-responding pH electrode were negative.

Calculation of $\Delta\Psi$ values from the time-points at which the transient Rb concentration gradients are maximal, reveals that the magnitude of $\Delta\Psi$ increases with pH from a value of about −31 mV at pH 5.5 to about −44 mV at pH 6.6 to about −51 mV at pH 7.5. Similar values are obtained under the same conditions with vesicles that are treated with N,N′-dicyclohexylcarbodiimide [30] and when the experiments are carried out in a weakly permeant buffer system, indicating that the $\Delta\Psi$ values obtained are not limited by passive membrane

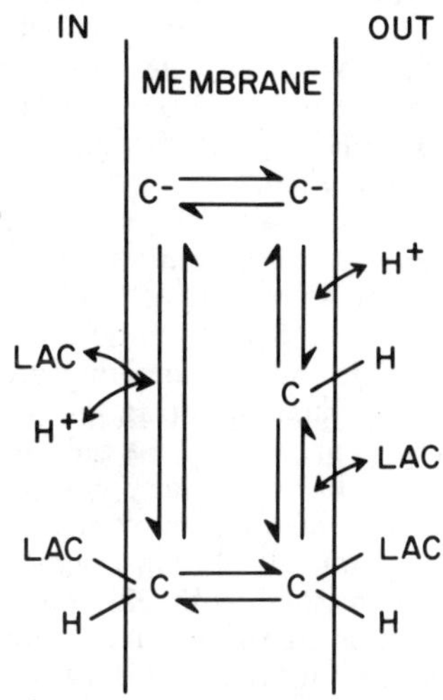

FIGURE 1. Schematic representation of reactions involved in lactose efflux and exchange. C represents the *lac* carrier protein. The order of substrate binding at the inner surface of the membrane is not implied.

permeability to protons or other ions. It is also noteworthy that similar results were obtained at pH 6.6 with two other probes that respond to $\Delta\Psi$ (interior negative)—[^{3}H]tetraphenylphosphonium$^+$, a permeant lipophilic cation, and 3,3′-dipentyloxacarbocyanine iodide, a fluorescent cyanine dye—and with lactose efflux-induced proline uptake.[29]

The ability of external lactose to inhibit the generation of $\Delta\Psi$ (interior negative) induced by lactose efflux is related to the apparent affinity of the carrier for substrate on the external surface of the membrane and not to diminution of the lactose concentration gradient across the membrane. When the maximal $\Delta\Psi$ generated is studied as a function of the external lactose concentration, 50% inhibition is observed at an external lactose concentration of about 0.3 mM and complete inhibition occurs at about 1 mM lactose. Thus, the ability of lactose to inhibit the generation of $\Delta\Psi$ under these conditions correlates reasonably well with the apparent K_m of the carrier (0.2 mM) and not with the lactose gradient across the membrane (i.e., even at 1 mM lactose, there is a tenfold concentration gradient across the vesicle membrane).

Clearly, the transient generation of $\Delta\Psi$ (interior negative) during carrier-mediated lactose efflux provides strong evidence for coupled movement of a charged species with lactose. Although it has not been possible to elicit lactose efflux-induced changes in external pH, it seems highly likely that the charged species involved is hydrogen ion since the $\Delta\Psi$ created during efflux is abolished by CCCP and independent of the ionic composition of the medium. In addition, this conclusion is consistent with studies in intact cells.[24-26, 28, 31] On the other hand, it should be emphasized that none of these studies can differentiate between proton/lactose symport and hydroxide/lactose antiport, a matter that is of obvious importance when considering the functional groups that may be involved in charge translocation.

Comparison of efflux and exchange rates demonstrates that the exchange process is very rapid, at least 10-times faster than efflux. Thus, the rate-determining step for this process must involve a reaction that corresponds to the return of the unloaded carrier to the inner surface of the membrane because this is the only step by which efflux and exchange differ (FIGURE 1). Assuming that the charged species translocated with lactose is protons and that loss of both substrates from the carrier is necessary for reinitiation of an efflux cycle, external pH might be expected to influence the rates of turnover in either of two ways. First, deprotonation of the carrier could be slow and thereby limit overall efflux rates in a pH-dependent fashion. Although proton transfers between accessible amino acid residues and water in soluble enzyme systems are typically very fast,[32] little is known about such processes with proteins embedded in a hydrophobic milieu. Second, pH could alter the equilibrium between protonated and unprotonated forms of the carrier (i.e., two different conformations), favoring the presence of the unprotonated form at higher pH. Since it is presumed that only the deprotonated carrier can recycle, the rate of efflux would be partially controlled by external pH, and the rate-limiting step might then involve "movement" of the unloaded carrier to the inner surface of the membrane. The observation that the rate of lactose efflux increases with pH is consistent with either of these possibilities. On the other hand, if deprotonation of the carrier is not obligatory for exchange (FIGURE 1), protons might remain bound to the carrier during this reaction, rendering it insensitive to pH. Given the ordered mechanism shown in FIGURE 1 whereby the carrier releases lactose first, followed by loss of a proton, deprotonation and/or return of the unloaded carrier could be slow and appear as the limiting step for carrier turnover during efflux. Further evidence for the ordered nature of the efflux reaction is given below.

Effect of pH on Counterflow

When intact *E. coli* containing the β-galactoside transport system are loaded with appropriate transport substrates and subsequently diluted into media containing the same substrate in radioactive form, transient influx of the radioactive substrate is observed.[33-34] The phenomenon is called "entrance counterflow," and its efficiency, in kinetic terms, is due in part to the frequency with which the carrier returns from the outer to the inner surface of the membrane in the loaded versus the unloaded form. When vesicles containing the β-galactoside transport system are equilibrated with 10 mM lactose and diluted 200-fold into media containing 0.4 mM [1-^{14}C]lactose, a concentration two times higher than the K_m for influx under energized conditions,[35, 23] rapid, transient uptake of radioactive substrate is observed at pH 5.5, 6.6, and 7.5. Under these conditions (i.e. saturating external lactose concentrations), the initial rate and peak level of radioactivity achieved are independent of pH, and it can be calculated that the intravesicular concentration of [1-^{14}C]lactose at the peak height is approximately 10 mM at each pH studied (i.e., the coupling efficiency between efflux of unlabeled lactose and influx of radioactive lactose is 1:1).

In contrast, when counterflow is repeated under identical conditions, but with an external [1-^{14}C]lactose concentration about threefold lower than the K_m for influx under energized conditions, the results are strikingly different. The "overshoot" phenomenon is now markedly sensitive to pH, the peak level of radioactivity decreases with pH from pH 5.5 to pH 7.5, and the coupling efficiency is reduced to 0.4:1 at pH 5.5, 0.27:1 at pH 6.6, and 0.13:1 at pH 7.5.

These results can be interpreted in the following manner (FIGURE 1): As efflux is initiated, lactose and protons bind to the carrier on the inner surface of the membrane and are translocated to the exterior. Lactose is released from the carrier, but in the presence of excess labeled substrate, rebinding and influx occur rapidly before deprotonation can occur. Thus, under these circumstances, proton release is infrequent and pH has no effect on the overall phenomenon. When [^{14}C]lactose is limiting externally, however, rebinding of radioactive substrate is less frequent, allowing deprotonation and return of the unloaded carrier. As a result, influx of [^{14}C]lactose is diminished and counterflow is inhibited. Moreover, when the pH is increased, deprotonation and return of the unloaded carrier are enhanced, resulting in further diminution of counterflow. Inhibition of lactose-efflux-generated $\Delta\Psi$ formation by external lactose is now readily understood. When substrate is present externally at saturating concentrations, release of lactose from the carrier and rebinding of substrate occur rapidly before deprotonation can occur, and the ability of the system to generate a $\Delta\Psi$ (interior negative) is abolished.

Effect of $\Delta\Psi$, ΔpH, and $\Delta\Psi + \Delta pH$ on Carrier-Mediated Lactose Efflux and Exchange

The observations discussed above indicate that the rate-determining step for lactose efflux down a concentration gradient is involved with the return of the unloaded carrier to the inner surface of the membrane. Thus, conditions that perturb this step should influence the rate of efflux without affecting exchange. In order to pursue this line of reasoning, we investigated the effects of artificially imposed $\Delta\Psi$s and/or ΔpHs on carrier-mediated efflux and exchange. When vesicles are equilibrated with 10 mM [1-^{14}C]lactose and 100 mM potassium phosphate (pH 7.0), treated with valinomycin, and then diluted 200-fold into media containing either potassium phosphate (no $\Delta\Psi$) or sodium phosphate ($\Delta\Psi$, interior negative), it is apparent that $\Delta\Psi$ (interior negative) retards efflux, and the $t_{1/2}$ is increased from about 12 sec to 21 sec. The corollary experiment performed with sodium-loaded vesicles diluted into sodium phosphate (no $\Delta\Psi$) or potassium phosphate ($\Delta\Psi$, interior positive) demonstrates that generation of $\Delta\Psi$ (interior positive) enhances the rate of efflux, and the $t_{1/2}$ is decreased from 12s to 10s (the effect of $\Delta\Psi$ in the latter experiment is less pronounced presumably because the potassium flux is directed into rather than out of the small intravesicular compartment).

In an analogous series of experiments, the effects of ΔpH, interior alkaline or acid, on lactose efflux were monitored. For this purpose, vesicles were first equilibrated with 10 mM [1-^{14}C]lactose, 5 mM potassium phosphate (pH 7.0), and 100 mM potassium acetate and then diluted 200-fold into media containing equimolar concentrations of either acetate (no ΔpH) or gluconate (ΔpH, interior alkaline).‡ In the absence of an imposed ΔpH, the $t_{1/2}$ for efflux is about 11s. When ΔpH (interior alkaline) is imposed, the rate of efflux is reduced, and $t_{1/2}$ increases to 17s. Conversely, when the acetate diffusion gradient is directed inward in order to generate a ΔpH of opposite polarity (i.e., interior acid), the rate of efflux is enhanced, and $t_{1/2}$ decreases from about 11s to 9.5s.

The effect of simultaneous imposition of outwardly directed potassium and acetate diffusion gradients on lactose efflux was also investigated. Vesicles were equilibrated with 100 mM potassium phosphate (pH 7.0), 100 mM potassium acetate and [1-^{14}C]lactose, treated with valinomycin, and diluted into media appropriate for the generation of ΔΨ (interior negative), ΔpH (interior alkaline), or ΔΨ + ΔpH (interior negative and alkaline). As expected, imposition of ΔΨ (interior negative) or ΔpH (interior alkaline) independently, decreases the rate of efflux. Moreover, when both gradients are imposed concurrently, leading to generation of ΨΔ + ΔpH (interior negative and alkaline), the rate of efflux is diminished even further, and the effects of ΔΨ and ΔpH appear to be additive.

In marked contrast to the experiments presented above, outward or inward directed fluxes of potassium and/or acetate have no discernible effect on the loss of intravesicular lactose under exchange conditions. In these experiments, vesicles were equilibrated with 10 mM [1-^{14}C]lactose and appropriate salts, treated with valinomycin, and diluted 200-fold into media containing equimolar concentrations of unlabeled lactose. In addition, the composition of the media was manipulated in order to establish potassium and/or acetate diffusion gradients of defined polarities. Under all of the conditions tested, the rates of exchange were essentially identical.

Clearly, these results support the prediction that the rate-determining step for lactose efflux down a concentration gradient involves a reaction associated with the return of the unloaded carrier to the inner surface of the membrane. Considering the normal polarity of $\Delta\bar{\mu}_{H^+}$ across the vesicle membrane, efflux is altered in the expected manner. Imposition of ΔΨ (interior negative) or ΔpH (interior alkaline) alone or simultaneously decreases the rate of efflux, while imposition of ΔΨ (interior positive) or ΔpH (interior acid) has the opposite effect. Importantly, moreover, a kinetic analysis of the alterations in rate measured under these conditions shows that the effects are due to changes in V_{max}, an observation that is commensurate with the notion that ΔΨ and ΔpH affect the limiting step in the overall reaction.

Although the rate of efflux is demonstrably altered by imposed ΔΨ and ΔpH,

‡ Since acetate is permeant in its protonated form only (i.e., as acetic acid), acetate diffusion gradients directed outward or inward lead to a net loss or gain of protons, respectively, giving rise to a ΔpH of appropriate polarity. In order to generate ΔpH in this fashion in the absence of undesired osmotic effects, vesicles equilibrated with acetate were diluted into media containing equimolar concentrations of gluconate or *vice versa.* Since gluconate is polyhydroxylated and induction of a specific transport system is required to observe uptake in the vesicles, it is presumably much less permeant than acetate.

these parameters exert no effect whatsoever on the loss of intravesicular lactose under exchange conditions. In addition to providing further support for some of the conclusions discussed, the findings taken together have important implications with regard to the translocation mechanism. It has been suggested[36-37] that an unloaded symporter carries a net negative charge, while the loaded porter (i.e., the ternary complex between the carrier, substrate and protons) is neutral, and evidence has been presented[18-19] that is most easily interpreted in light of this notion. On a simplistic level, the effects of imposed $\Delta\Psi$ and ΔpH on efflux and exchange are also consistent with this hypothesis. Given the observation that lactose translocation under exchange conditions is unaffected by the imposition of $\Delta\Psi$, ΔpH, or $\Delta\Psi + \Delta$pH, it seems unlikely that any of the species formed during this reaction carries a net charge. Moreover, since $\Delta\Psi$, in particular, decreases the rate of efflux when the polarity is interior negative and enhances the rate of efflux when interior positive, it is reasonable to suggest that the unloaded carrier may carry a net negative charge. If the unloaded carrier were positively charged, the opposite effects would be expected. On the other hand, it should be apparent that these considerations shed little light on the mechanism by which ΔpH alters the rate of efflux which highlights the importance of determining whether $\Delta\Psi$ and ΔpH affect the same or different steps in the translocation process.

Deuterium Solvent Isotope Effects

Given the ordered efflux mechanism depicted in FIGURE 1 in which a step(s) involving the loss of protons from the carrier and/or return of the unloaded carrier is rate-limiting for the overall efflux reaction, it is apparent that one means by which to further investigate the proposed mechanism is to search for a solvent deuterium isotope effect. This stratagem has been useful for investigating the mechanism of certain enzymatic reactions in which the rate-determining step involves proton transfer.[38] If the mechanism shown in FIGURE 1 is generally correct, replacement of protium in the reaction mixtures with deuterium might be expected to influence the rate of carrier-mediated efflux since proton loss from the carrier must be accomplished before the unloaded carrier can return to the inner surface of the membrane to reinitiate another efflux cycle. Alternatively, replacement of protium with deuterium should have no effect on exchange because the translocation reaction catalyzed under these conditions does not appear to involve proton loss.

Preliminary experiments have yielded very promising results in this regard. At equivalent pH and pD (i.e., pD = pH + 0.4), the rate of efflux is approximately 2- to 2.5-times slower in deuterated media (with >95% of the protium replaced with deuterium) relative to the rate observed under control conditions in protium. Importantly, moreover, the rate of exchange is identical in the presence of protium and deuterium.

Referring again to FIGURE 1 and to the previous discussion (cf. Effect of pH on Counterflow), when counterflow is studied under conditions where external lactose is below the apparent K_m of the carrier, the protonated form of the carrier (CH) partitions between two pathways: Loss of the proton and return of the unloaded carrier to the inner surface of the membrane, i.e., efflux; and rebinding of lactose and return in the loaded form, i.e., counterflow. Since deuterium would be expected to favor the CD form of the carrier, relative to the CH form, through an effect on either the rate of loss of deuterium, relative to protium, or on the equilibrium between the unloaded and deuterated forms of the carrier (i.e., a pK_a effect), the effect of deuterium on counterflow was investigated. Interestingly enough, under conditions where the external lactose concentration is below the apparent K_m of the carrier under energized conditions (cf. above), the magnitude of the peak height observed during the overshoot is greater in deuterium relative to protium.

Although the results taken as a whole provide interesting independent support for the proposed mechanism, it should be emphasized again that the effect of deuterium could be mediated by either a rate effect (i.e., an effect on the rate of proton loss from the carrier) or an effect on the pH-dependent equilibrium between two conformations of the carrier (i.e. a pK_a effect). Moreover, it should be recognized that these experiments do not allow distinction as to whether loss of proton(s) or return of the unloaded carrier to the inner surface of the membrane is rate-determining for efflux.

References

1. Mitchell, P. 1961. Nature (London) **191:**144.
2. Mitchell, P. 1966. Biol. Rev. Cambridge Philos. Soc. **41:**445.
3. Mitchell, P. 1968. *In* Chemiosmotic Coupling and Energy Transduction. Glynn Research. Bodmin, England.
4. Mitchell, P. 1973. J. Bioenerg. **4:**63.
5. Mitchell, P. 1977. FEBS Lett. **78:**1.
6. Hinkle, P. C. & R. E. McCarty. 1978. Sci. Am. **238:**104.
7. Mitchell, P. 1963. Biochem. Soc. Symp. **22:**142.
8. Owen, P. & H. R. Kaback. 1978. Proc. Nat. Acad. Sci. U.S.A. **75:**3148.
9. Owen, P. & H. R. Kaback. 1979. Biochemistry **18:**1413.
10. Owen, P. & H. R. Kaback. 1979. Biochemistry **18:**1422.
11. Kaback, H. R. 1974. Science **186:**882.
12. Kaback, H. R. 1976. J. Cell. Physiol. **89:**575.
13. Kaback, H. R., S. Ramos, D. E. Robertson, P. Stroobant & H. Tokuda. 1977. *In* Molecular Aspects of Membrane Transport. D. L. Oxender & C. F. Fox, Eds. **22:**473.
14. Harold, F. M. 1976. Curr. Top. in Bioenerg. **6:**83.
15. Konings, W. N. & J. Boonstra. 1977. Curr. Top. in Membr. and Transp. **9:**177.
16. Ramos, S., S. Schuldiner & H. R. Kaback. 1976. Proc. Nat. Acad. Sci. U.S.A. **73:**1892.
17. Ramos, S. & H. R. Kaback. 1977. Biochemistry **16:**848.
18. Ramos, S. & H. R. Kaback. 1977. Biochemistry **16:**854.
19. Ramos, S. & H. R. Kaback. 1977. Biochemistry **16:**4271.
20. Tokuda, H. & H. R. Kaback. 1977. Biochemistry **16:**2130.
21. Schuldiner, S. & H. R. Kaback. 1975. Biochemistry **14:**5451.
22. Schuldiner, S. & H. R. Kaback. 1977. Biochim. Biophys. Acta **472:**399.
23. Kaczorowski, G. J., D. E. Robertson & H. R. Kaback. 1979. Biochemistry **18:**3697.
24. West, I. C. 1970. Biochem. Biophys. Res. Commun. **41:**655.
25. West, I. C. & P. Mitchell. 1972. J. Bioenerg. **3:**445.
26. West, I. C. & P. Mitchell. 1973. Biochem. J. **132:**587.
27. Flagg, J. L. & T. H. Wilson. 1977. J. Membrane Biol. **31:**233.
28. Flagg, J. L. & T. H. Wilson. 1978. Membrane Biochem. **1:**61.
29. Kaczorowski, G. J. & H. R. Kaback. 1979. Biochemistry **18:**3691.
30. Patel, L., S. Schuldiner & H. R. Kaback. 1975. Proc. Nat. Acad. Sci. U.S.A. **72:**3387.

31. Bentaboulet, M., A Robin & A. Kepes. 1979. Biochem. J. **178:**103.
32. Eigen, M. 1963. Angew. Chem. Int. Ed. **3:**1.
33. Wong, P. T. S. & T. H. Wilson. 1970. Biochim. Biophys. Acta **196:**336.
34. Bentaboulet, M. & A. Kepes. 1977. Biochim. Biophys. Acta **471:**125.
35. Barnes, E. M. & H. R. Kaback. 1970. Proc. Nat. Acad. Sci. U.S.A. **66:**1190.
36. Schuldiner, S., G. Kerwar, R. Weil & H. R. Kaback. 1975. J. Biol. Chem. **250:**1361.
37. Rottenberg, H. 1976. FEBS Lett. **66:**159.
38. Schowen, R. L. 1977. *In* Isotope Effects on Enzyme Catalyzed Reactions. W. W. Cleland, M. H. O'Leary & D. Northrop, Eds.: 64–99. University Park Press. Baltimore.

TRANSPORT OF ANIONS IN EHRLICH ASCITES TUMOR CELLS: EFFECTS OF DISULFONIC ACID STILBENE IN RELATION TO TRANSPORT MECHANISM *

Charles Levinson

Department of Physiology
The University of Texas
Health Science Center at San Antonio
7703 Floyd Curl Drive
San Antonio, Texas 78284

There is abundant evidence to support the view that in erythrocytes a single class of intrinsic membrane glycoproteins is involved in the regulation of anion permeability.[1, 2] This protein, which spans the membrane, has an apparent molecular weight of 95,000 and is often referred to as Band 3 from its position on SDS polyacrylamide gel electrophoretograms.[3] Involvement of Band 3 proteins has been inferred primarily from studies of the interaction of a group of substituted disulfonic acid stilbenes, specific inhibitors of anion transport with intact erythrocytes or ghosts. One of the substituted stilbenes, DIDS (4,4′-diisothiocyano-stilbene-2,2′-disulfonic acid) and its dihydroderivative, H_2DIDS (4,4′-idisothiocyano-1,2-diphenyl ethane-2,2′-disulfonic acid) have been particularly useful. These agents do not enter the red blood cell but have the capacity to bind both reversibly and irreversibly to the membrane, and, as a consequence, inhibit anion transport. The extent of transport inhibition is linearly related to the number of molecules bound to Band 3 proteins.[4, 5]

The effects of H_2DIDS on anion transport in Ehrlich ascites tumor cells are more complex. For example, when H_2DIDS is present in the medium, both chloride and sulfate transport are inhibited. However, if H_2DIDS treated cells are washed and resuspended in H_2DIDS-free medium, only sulfate transport is inhibited.[6] Recently, the tritiated derivative of H_2DIDS ($^3H_2$DIDS) was utilized to explore in more detail the interaction of this agent with the tumor cell.[7] Data from these experiments show that $^3H_2$DIDS binds both reversibly and tightly, and, as a consequence, inhibits transport. Binding of 0.02–0.03 nmoles/mg dry cell mass or less did not effect transport. However, at greater than 0.03 nmoles and up to 0.15 nmoles, the relationship between tightly bound $^3H_2$DIDS and transport inhibition is linear. A reasonable explanation for this dose-response relationship is that the tumor cell membrane possesses a large number of nontransport-related $^3H_2$DIDS binding sites in addition to sites concerned with the regulation of sulfate permeability.

In the present experiments, we have directed our efforts toward identification of the membrane components that irreversibly bind $^3H_2$DIDS in an attempt to determine which are involved in anion transport. To this end, we have attempted to correlate the rate of $^3H_2$DIDS binding to specific membrane proteins with the development of transport inhibition.

* This study was supported by Grant CA 10917, U.S. Public Health Service, National Cancer Institute.

0077–8923/80/0341–0482 $01.75/0

Materials and Methods

Cells

Ehrlich-Lettré ascites tumor cells (hyperdiploid strain) maintained in Ha/ICR male mice were removed from unanesthetized animals by peritoneal aspiration and washed free of ascites fluid by gentle centrifugation and resuspension. The medium used for washing and incubation had the following composition: 135 mM NaCl, 10 mM Na_2SO_4, 6 mM KCl, and 10 mM HEPES-NaOH (pH 7.2–7.3; 295–302 mOsm).

Labeling with $^3H_2DIDS$

$^3H_2DIDS$ was prepared as described previously.[7] Binding was assessed by adding $^3H_2DIDS$ to cell suspension maintained at either 0° C or 37° C. The interaction between $^3H_2DIDS$ and the cells was stopped by the addition of an ice-cold medium containing 0.5% bovine serum albumin (Sigma Chemical Co.). The cells were separated from the medium by centrifugation at 1500 × g for one minute. The packed cells were then washed once in ice-cold albumin wash solution, twice in ice-cold medium and finally resuspended in medium at 21–23° C.

General Procedures

Samples of cell suspension were removed for the determination of wet and dry weight,[8] protein,[7] and for the measurement of $^{35}SO_3^{-2}$ transport.[7, 9] Radioactivity of the $^3H_2DIDS$-labeled cells was determined by dissolving 20 μl of whole cell suspension in 10 ml of scintillation cocktail.[8] In some experiments, additional aliquots of cell suspension were used to isolate the plasma membrane. The Brunette and Till[10] procedure as modified by Nachbar *et al.*[11] was used except that the $ZnCl_2$ concentration was reduced from 1 mM to 0.50 mM.[18]

Gel Electrophoresis

Membrane samples were heated in a boiling water bath for 5 min with 1% sodium dodecyl sulfate (SDS), 1% β-mercaptoethanol and 0.5 mM EDTA before electrophoresis. SDS-polyacrylamide slab gels (both 5% and 8%) with stacking gel (4% and 5%) were prepared and ran according to Laemmli.[12] Gels were stained with Coomassie Blue[3] and passively destained 24 hours in 7.5% acetic acid. Gels with radioactively-labeled proteins were fractionated on an electric gel slicer (Hoefer Scientific Instruments, San Francisco) into 90 1 mm fractions. In some experiments, a single edge razor blade was used to remove the prominently-stained bands. The gel fractions were collected directly into glass scintillation vials. A mixture[13] containing 0.2 ml 60% perchloric acid and 0.40 ml 30% H_2O_2 was added and the vials tightly capped. After incubation for 14–18 hours at 58° C, 10 ml of scintillation cocktail[8] were added to the dissolved gel slices. Radioactivity was measured with a Searle Analytic refrigerated liquid scintillation spectrometer.

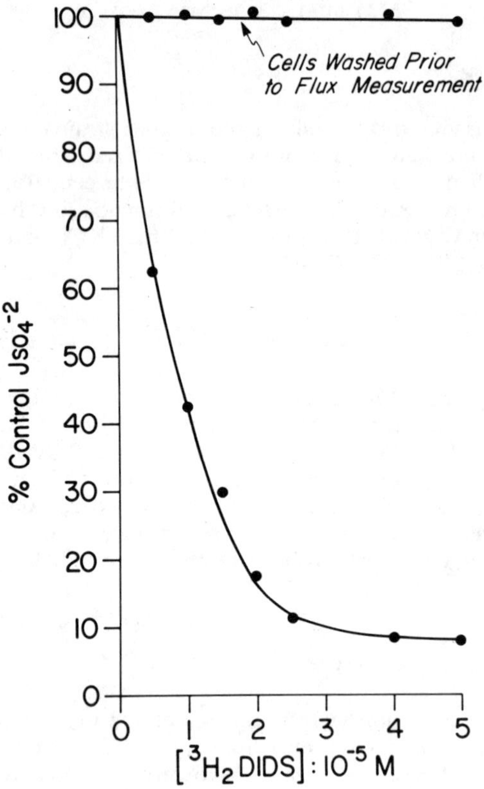

FIGURE 1. Inhibition of SO_4^{-2} transport ($J_{SO_4^{-2}}$, expressed as percent of control) by $^3H_2$DIDS when present in the medium. Cell suspensions, equilibrated for 60 min at 21° C, were added to medium containing $^3H_2$DIDS. Final concentration of reagent was from 5–50 μM. (^{35}S)SO_4^{-2}-uptake was then measured at 21° C each min for the next 5 min and the flux determined. Reversibility of $^3H_2$DIDS Inhibition: Cell suspensions were added to medium containing $^3H_2$DIDS, and, after 3 min, the cells were washed twice with medium. The flux of SO_4^{-2} was then measured.

RESULTS AND DISCUSSION

In Ehrlich ascites tumor cells, the steady state SO_4^{-2} exchange flux (control), measured at 21° C, is 1.02 ± 0.011 (S.E.) mmoles/Kg dry wgt. min^{-1}. However, as shown in FIGURE 1, in the presence of increasing concentrations of $^3H_2$DIDS in the medium SO_4^{-2} transport is reduced. The inhibitory effect of this agent is almost completely relieved when cells, exposed to $^3H_2$DIDS for 3 min or less, are washed twice with medium. Recently, it has been shown in erythrocytes that prolonged incubation with $^3H_2$DIDS leads to irreversible binding to the membrane.[4, 5, 14] This, in turn, results in the inhibition of anion transport. Since reversible inhibition in the tumor cell suggests direct interaction of $^3H_2$DIDS with the transport system, it was of interest to determine whether prolonged incubation resulted in irreversible inhibition. The data displayed in FIGURE 2 show that incubation (30 min, 37° C), with increasing concentrations of $^3H_2$DIDS followed by extensive washing, results in a dramatic inhibition of SO_4^{-2} transport. This established "bound" $^3H_2$DIDS as a potent SO_4^{-2} transport inhibitor in the tumor cell. The rate of development of irreversible inhibition of transport is shown in FIGURE 3. This process, which follows first order kinetics at 37° C, has a rate constant equal to 0.087 ± 0.009 (S.E., $n = 4$) minute^{-1}. In contrast to incubation at 37° C, incubation at 0°–3° C for up to 30 minutes does not result in significant inhibition of transport. This suggests that at this temperature $^3H_2$DIDS does not irreversibly bind to the membrane site(s) responsible for the regulation of SO_4^{-2} transport.

The relationship between SO_4^{-2} transport and $^3H_2$DIDS binding to both intact cells and the membrane isolated from these cells is shown in FIGURE 4. Binding of between 0.012–0.030 nmoles $^3H_2$DIDS per mg cellular protein was without detectable affect on transport. However, at greater than 0.03 nmoles the relationship between $^3H_2$DIDS binding and inhibition of transport is linear. This confirms a previous observation from our laboratory.[7] A similar relationship is observed for $^3H_2$DIDS binding to the plasma membrane. For example, when 0.07 nmoles $^3H_2$DIDS were bound per mg membrane protein, transport was not inhibited (i.e., 100% of control). However, at greater than 0.07 nmoles, and up to 0.33 nmoles, a linear relationship exists between transport inhibition and $^3H_2$DIDS binding.

The finding that a fraction of bound $^3H_2$DIDS was without detectable effect on transport suggests that the membrane possesses a population of binding sites unrelated to those concerned with the regulation of SO_4^{-2} permeability. Recently, it has been shown that these cells spontaneously release more than 50%

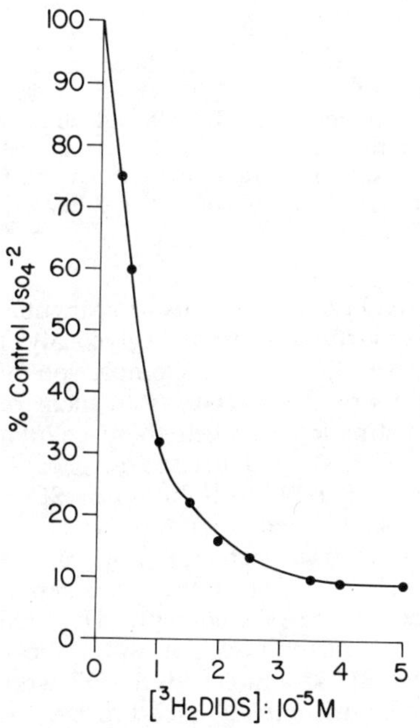

FIGURE 2. Inhibition of SO_4^{-2} transport by bound $^3H_2$DIDS. Cell suspensions, equilibrated for 60 min at 21° C, were mixed with a small volume of medium containing $^3H_2$DIDS. After incubation for 30 min at 37° C under a 100% oxygen atmosphere, the cells were washed twice with cold medium containing 0.5% serum albumin and twice with cold medium. (^{35}S)SO_4-uptake was then measured at 21° C each min for the next 5 min and the flux determined.

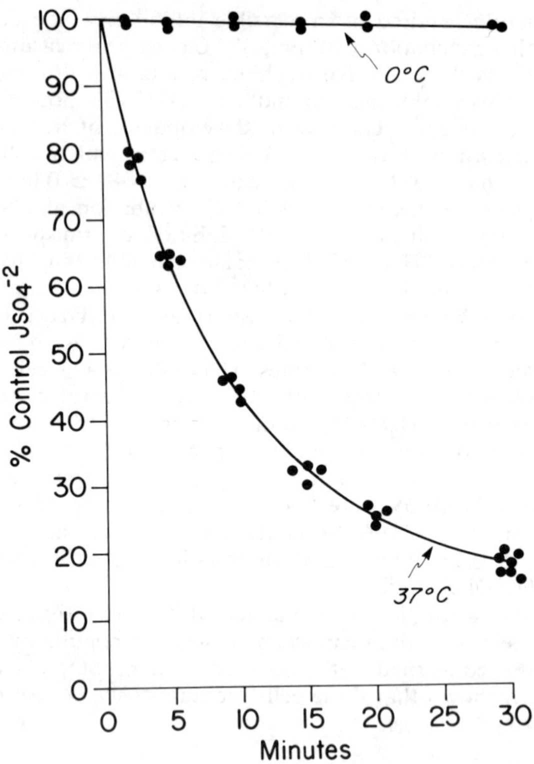

FIGURE 3. Development of inhibition of SO_4^{-2}-transport. Cells, incubated in room air at 21° C for 60 min, were transferred either to 37° C under a 100% O_2 atmosphere or to a 0° C ice-bath. After a 5 min equilibration period, $^3H_2$DIDS was added to a final concentration of 25 μM. Periodically during the next 30 min, aliquots of cell suspension were removed and the cells washed, as described in FIGURE 2. At each sampling time cell suspension was removed for the measurement of wet and dry cell weight and SO_4^{-2} transport at 21° C.

of the surface-iodinated proteins from lactoperoxidase-labeled cells after incubation in physiological salt solution.[15,16] This surface material (glycocalyx) contained a variety of glycoproteins and glycosaminoglycans. Comparison of the enzymatic activity, protein and sialic contents of the glycocalyx to those of the plasma membrane demonstrated a clear distinction between these cellular fractions. We have previously shown that $^3H_2$DIDS can label the glycocalyx fraction[17] and that its spontaneous release does not relieve inhibition of SO_4^{-2} transport.[7] Consequently, the observation that a substantial amount of $^3H_2$DIDS binds to the membrane and yet does not result in transport inhibition can be explained in terms of binding to the glycocalyx. This is illustrated by a representative experiment shown in FIGURE 5. Cells were labeled with $^3H_2$DIDS (30 min, 0° C), washed and resuspended at 37° C in the presence and absence of trypsin (1 μg/ml cell suspension). Although 0.045 nmoles $^3H_2$DIDS were bound per mg dry cell wgt., SO_4^{-2} transport was not inhibited (see FIGURE 4).

Upon incubation for 60 min at 37° C greater than 50% of the bound $^3H_2$DIDS was spontaneously released by the cells. The half-time ($t_{1/2}$) for this process was 5.4 min. Trypsin accelerated both the rate of $^3H_2$DIDS release ($t_{1/2} = 3.9$ min) as well as the total amount released in 60 min (73%). These observations are consistent with the view that $^3H_2$DIDS binds rapidly to superficial cell surface proteins that are then released into the medium. The factors responsible for this spontaneous "shedding" are, as yet, unknown.

FIGURE 6 demonstrates that spontaneous release of glycocalyx, prior to the analysis of the relationship between $^3H_2$DIDS binding to intact cells (or isolated membrane) and inhibition of transport, leads to reduction in nontransport-related binding. This is reflected by a shift in the binding curve toward the 100% intercept (FIGURE 4). Consequently, 25% to 40% of bound $^3H_2$DIDS can be removed without altering inhibition of SO_4^{-2} transport. The possibility

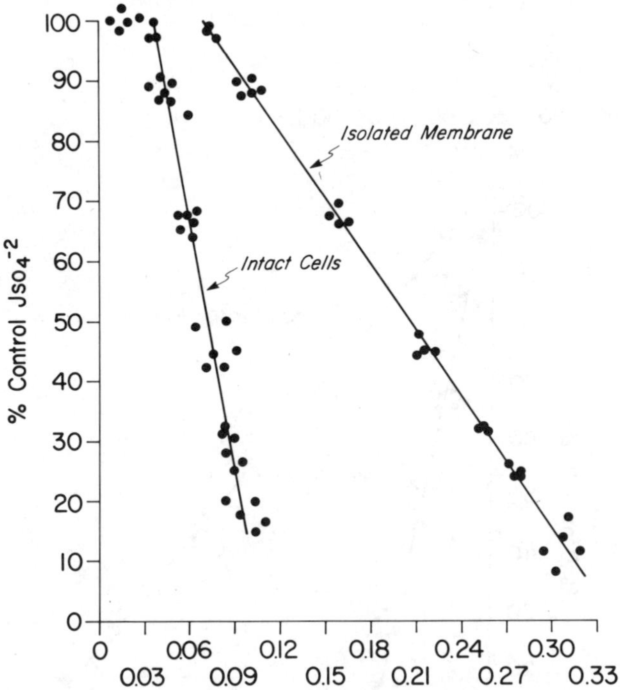

FIGURE 4. Relationship between inhibition of SO_4^{-2}-transport and $^3H_2$DIDS bound to intact cells and to plasma membrane. Cells, incubated in room air at 21° C for 60 min, were then transferred to 37° C under a 100% O_2 atmosphere. After a 5 min incubation period, $^3H_2$DIDS was added to a final concentration of 25 μM. Periodically, during the next 30 min, aliquots of cell suspension were removed and the cells washed, as described in FIGURE 2. At each sampling time, cell suspension was removed for the measurement of: wet cell weight, dry cell weight, cellular protein, SO_4^{-2}-transport at 21° C, and bound $^3H_2$DIDS. In some experiments, additional aliquots of cell suspension were removed and used for the preparation of plasma membranes.

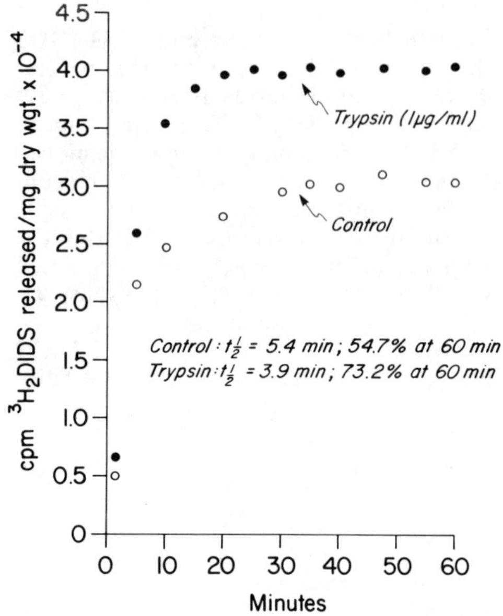

FIGURE 5. Time course of glycocalyx release. Cells, incubated at 0° C for 30 min with 25 μM $^3H_2DIDS$ were washed and resuspended in $^3H_2DIDS$-free medium at 37° C with and without trypsin (1 $\mu g/ml$). Periodically, 0.20 ml of cell suspension were removed and centrifuged one min at 12,000×g. Supernatant (50 μl) was then assayed for radioactivity.

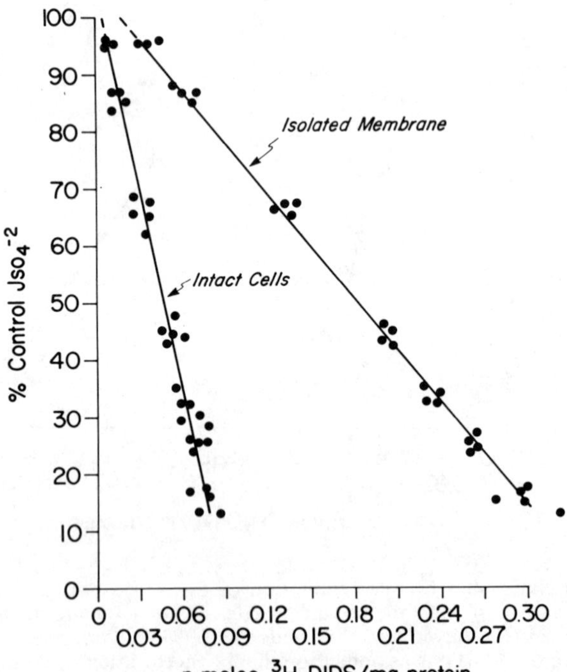

FIGURE 6. Relationship between inhibition of SO_4^{-2}-transport and $^3H_2DIDS$ bound to intact cells and to plasma membranes after glycocalyx release. The experimental details, with one important exception, are identical to those described in FIGURE 4. After washing to remove loosely adherent and extracellular $^3H_2DIDS$, the cells were incubated for 60 min at 37° C to permit loss of the labeled glycocalyx. At each sampling time cell suspension was removed and processed as described in FIGURE 4.

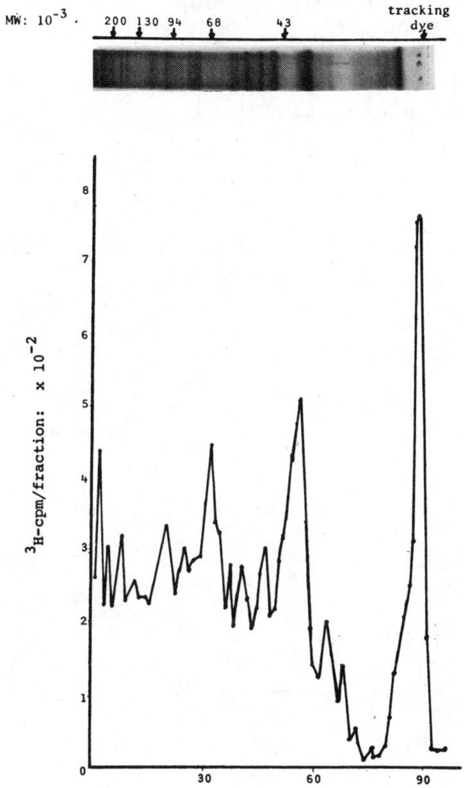

FIGURE 7. The staining and labeling profile of plasma membrane proteins isolated from $^3H_2$DIDS-labeled tumor cells. Cells were labeled with $^3H_2$DIDS (25 μM; 30 min at 37° C), washed and incubated 60 min at 37° C to permit release of glycocalyx proteins. The membrane was isolated and analyzed for protein and label distribution by SDS-polyacrylamide slab-gel (8%) electrophoresis. Proteins were stained with Coomassie Blue. Gels were fractionated, digested and analyzed for the distribution of $^3H_2$DIDS. Molecular weight markers were: myosin (200 K), β-galactosidase (130 K), phosphorylase A (94 K), bovine serum albumin (68 K), and ovalbumin (43 K) run in parallel with samples on the same slab gel.

of identifying transport-related binding sites associated with the membrane is thereby enhanced.

In an attempt to identify specific $^3H_2$DIDS-binding components of the membrane, cells were labeled with $^3H_2$DIDS and permitted to spontaneously release cell surface materials. The cell membrane was then isolated. FIGURE 7 shows the pattern of the membrane proteins separated by SDS-polyacrylamide gel electrophoresis and stained with Coomassie Blue. The lower portion of the figure describes the distribution of radioactivity in the gel. Clearly, no individual component is preferentially labeled by $^3H_2$DIDS. Lack of preferential labeling suggests that there is no specific $^3H_2$DIDS binding site unique to the anion transport protein. This result was surprising in view of the apparent specific interaction between $^3H_2$DIDS and Band 3 proteins of erythrocytes.[4, 5, 14] A possible explanation for the difference in labeling pattern could be the presence of radioactive contaminants in our preparation of $^3H_2$DIDS, which nonspecifically interacts with membrane proteins. However, the data shown in FIGURE 8 discount this possibility. Erythrocytes were labeled with $^3H_2$DIDS, washed and ghosts prepared according to the procedures described by Lepke *et al.*[5] $^3H_2$DIDS labeling was specific for (but not restricted to) Band 3 proteins. Based on 1.95×10^9 cells/mg ghost protein,[5] we find 1.75×10^6 molecules $^3H_2$DIDS-bound per cell and about 80% of the total bound $^3H_2$DIDS

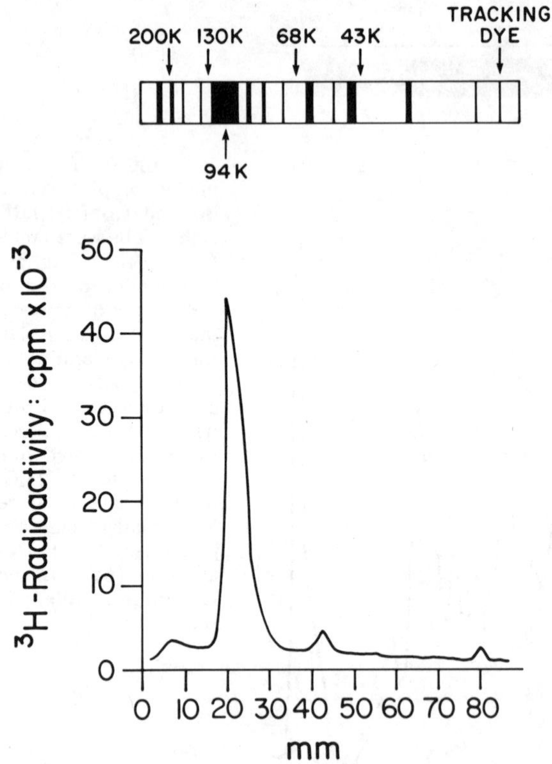

FIGURE 8. Representation of the staining and labeling profile of ghost proteins isolated from $^{3}H_2$DIDS-labeled erythrocytes. Fresh human erythrocytes were labeled with $^{3}H_2$DIDS (25 μM; 30 min at 37° C) and washed, as described in FIGURE 2. Ghosts were prepared according to the procedures described in REFERENCE 5. Ghost proteins were separated by SDS-polyacrylamide slab-gel (8%) electrophoresis, stained with Coomassie Blue and processed, as described in FIGURE 7.

associated with Band 3 proteins. These values are in good agreement with those reported by Lepke *et al.*[5]

Because of the lack of preferential labeling, we have adopted a kinetic approach aimed at indirectly identifying membrane proteins concerned with the regulation of SO_4^{-2} permeability. Cells were exposed to $^{3}H_2$DIDS for from 2 minutes to 30 minutes at 37° C and subsequently incubated for 60 min at 37° C to permit the spontaneous release of $^{3}H_2$DIDS-labeled glycocalyx proteins. For each time of exposure, transport inhibition was determined and the cell membrane was isolated from the same population of cells. The membrane proteins were separated by gel electrophoresis and stained with Coomassie Blue. Each gel was then sliced and the protein bands analyzed for radioactivity. This procedure permits evaluation of the rate constants describing labeling of the individual proteins as well as the rate constant describing the development of transport inhibition. These data are shown in FIGURE 9. The rate constant for development of transport inhibition (0.087 min^{-1}) is shown by the broken horizontal line. Two of the membrane proteins (apparent molecular weight: 99,000 and 93,000) label with rate constants closely approximating this value. The similarity of the kinetics of protein labeling and transport inhibition is suggestive of participation by these proteins in the regulation of SO_4^{-2} transport. However, $^3H_2$DIDS associated with these proteins accounts for only 1.8 ± 0.3% (SEM, $n = 3$) and 1.4 ± 0.3% (SEM, $n = 3$) of the total radioactivity of the membrane, respectively. When SO_4^{-2} transport is inhibited by 88%, approximately 1.3×10^7 molecules of $^3H_2$DIDS are bound per cell. If the distribution of bound $^3H_2$DIDS in the intact cell is the same as that in the isolated membrane, then in each cell 230,000 $^3H_2$DIDS molecules are associated with the 99,000 MW protein and 185,000 molecules are associated with 93,000 MW protein.

Previously, we have shown that although H_2DIDS when irreversibly bound to the tumor cell inhibits SO_4^{-2} transport, the steady-state exchange flux of Cl^- is unaffected. This led to the postulation that both anions are transported by a common mechanism that possesses two reactive sites.[6] Both sites, C_1 (transport) and C_2 (modifier), are necessary for SO_4^{-2} transport while only the transport site is required for Cl^- transport. It is tempting to speculate that either one or both of these $^3H_2$DIDS-labeled proteins is an integral component of the trans-

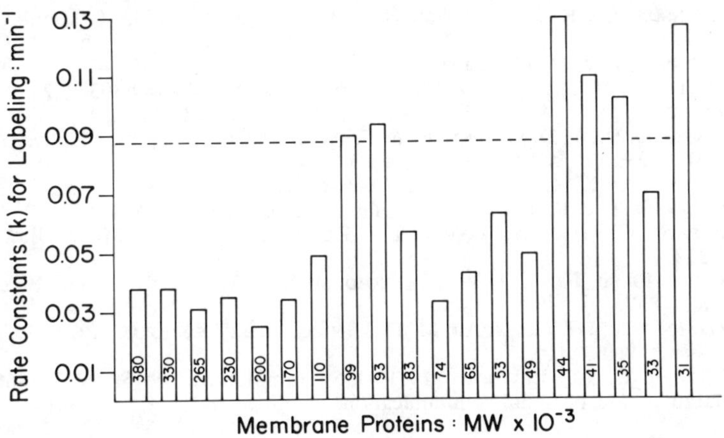

FIGURE 9. Rate constants describing the binding of $^3H_2$DIDS to individual membrane proteins. The experimental protocol for these experiments is identical to that described in FIGURE 6. Cells labelled with $^3H_2$DIDS (25 μM, 37° C), for from 2 min to 30 min, were subsequently incubated 60 min at 37° C to permit loss of glycocalyx proteins. Plasma membrane was then prepared from these cells at each time of exposure to $^3H_2$DIDS. Following slab-gel electrophoresis (FIGURE 7), radioactivity was determined in each of the protein bands. Rate constants were determined assuming first order labeling from the relationship: $1n\left[1-\frac{cpm(t)}{cpm(\infty)}\right]=-kt$, where $cpm(t)$ and $cpm(\infty)$ represents the radioactivity in each protein band at time t and at equilibrium, respectively; k is the rate constant (min^{-1}) describing the $^3H_2$DIDS labeling of the individual proteins. The data were curve fit using maximization of the regression coefficient r as the criterion for best fit. For all protein bands, r exceeded 0.97. These data represent mean values from three separate experiments.

port mechanism. However, a more unequivocal identification and quantitation of membrane proteins involved in the regulation of SO_4^{-2} permeability must await the development of agents that are more specific in their interaction with the tumor cell anion transport system.

Acknowledgments

The author is indebted to his colleague, Dr. Thomas C. Smith, for advice and help during the course of these studies and to R. J. Corcoran for expert technical assistance.

References

1. Rothstein, A., Z. I. Cabantchik & P. Knauf. 1976. Fed. Proc. **35:**3–10.
2. Passow, H. 1977. Acta Biol. Med. Germ. **36:**817–821.
3. Fairbanks, G., T. L. Steck & D. F. H. Wallach. 1971. Biochemistry **10:** 2606–2617.
4. Cabantchik, Z. I. & A. Rothstein. 1974. J. Memb. Biol. **15:**207–226.
5. Lepke, S., F. Fasold, M. Pring & H. Passow. 1976. J. Memb. Biol. **29:**147–177.
6. Levinson, C. 1978. J. Cell. Physiol. **95:**23–32.
7. Levinson, C., R. J. Corcoran & E. H. Edwards. 1979. J. Memb. Biol. **45:** 61–79.
8. Levinson, C. & M. L. Villereal. 1975. J. Cell. Physiol. **85:**1–14.
9. Villereal, M. L. & C. Levinson. 1976. J. Cell. Physiol. **89:**303–312.
10. Brunette, D. M. & J. E. Till. 1971. J. Memb. Biol. **5:**212–224.
11. Nachbar, M., J. D. Oppenheim & F. Aull. 1976. Biochim. Biophys. Acta **419:** 512–529.
12. Laemmli, U. K. 1970. Nature **227:**680–685.
13. Malin, D. T. & R. T. Lofberg. 1966. Anal. Biochem. **16:**500–509.
14. Ship, S., Y. Shami, W. Breuer & A. Rothstein. 1977. J. Memb. Biol. **33:**311–323.
15. Rittenhouse, H. G., J. W. Rittenhouse & L. Takemoto. 1978. Biochem. **17:** 829–837.
16. Rittenhouse, H. G., D. Ar, M. D. Lynn & D. K. Denholm. 1978. J. Supramol. Struct. **9:**407–419.
17. Smith, T. C. & C. Levinson. 1979. J. Supramol. Struct. **9:**(Suppl. 3):466.
18. Nachbar, M. Personal communication.

Discussion

D. G. Warnock (*University of California, San Francisco*): You showed that trypsin pretreatment or trypsin treatment enhances the release of the glycocalyx, or, at least, accounts for what you found. Can you tell me what trypsin treatment does to sulfate uptake in these cells?

C. Levinson: We've done very little with this but in contrast to red cells, even one microgram of trypsin per ml at 37° will simulate sulfate transport. I don't know if this is due to increasing the membrane's permeability to sulfate or due to exposing new sites. We haven't pursued it.

T. H. Wilson (*Harvard Medical School, Boston, Mass.*): Is this the same transport system that transports chloride and bicarbonate in these cells?

Levinson: Most likely it is, but there are some differences. The bound H_2DIDS clearly inhibits sulfate transport in these cells but does not inhibit chloride transport.

Wilson: What about bicarbonate?

Levinson: We haven't looked at it. As far as I know, no one has studied bicarbonate transport.

J. F. Hoffman: Did you calculate the turnover number for the sulfate flux of the two proteins that you identified to be associated with the transport, that is the number of sites per second?

Levinson: Based on about 400,000 transport molecules per cell and a maximal sulfate flux of 6 millimoles per kilo dry weight per minute, there would be between 5 and 10 sulfate molecules transported per site per second.

THE THERMODYNAMIC EFFICIENCY OF COTRANSPORT MECHANISMS WITH SPECIAL REFERENCE TO PROTON AND ANION TRANSPORT IN YEAST

A. A. Eddy, A. Seaston, D. Gardner and C. Hacking

Department of Biochemistry
University of Manchester
Institute of Science and Technology
Manchester M60 1QD, England

Despite the widespread acceptance of the concepts underlying the ion gradient hypothesis of energy coupling in the active transport of carbohydrates, amino acids and other solutes, our understanding of the energetics of these processes is rather limited. This statement applies both to the eukaryotic systems that form the subject of the present essay and to various prokaryotic systems, including the much studied mechanism of β-galactoside transport in *Escherichia coli.*

One important problem is the well-known dependence of the solute accumulation ratio ([cellular solute concentration]/[extracellular solute concentration] or $[S]_i/[S]_o$), observed in the steady state of solute accumulation, on two factors; namely, the magnitude of $[S]_o$ and the nature of S itself.[1,2] For instance, different substrates of the same carrier system may accumulate to very different extents. In general, it seems that a yeast or bacterial cell that concentrates a solute 10^3 times from a 1 μM solution is unlikely to concentrate it by the same factor from a 1 mM solution. However, if the mechanism of the pump ensures that a certain fixed amount of free energy is consumed per mole of S absorbed reversibly, the ratio $[S]_i/[S]_o$ will eventually reach a constant value, irrespective of the actual magnitude of $[S]_o$, unless other factors intervene. The solute gradient would thus tend to reach thermodynamic equilibrium with the energy source actuating the pump. Furthermore, analogues of the substrate S would accumulate to the same extent, though perhaps at a faster or a slower rate. A living cell operating on the above principle might be protected from osmotic explosion by the presence of leak pathways allowing the solute to escape. A continuous expenditure of energy would then be needed to maintain the gradient of S in a steady state.[3-5] Alternatively, a process of feedback regulation might limit the amount of solute accumulated to osmotically tolerable levels by controlling the working of the solute pump. The absorption of amino acids by yeast is believed to be regulated by such a process, but the mechanism of this so-called transinhibition is at present unknown.[2,6] In the present work, we have examined by means of specific mathematical models some of the properties of the pump and leak systems envisaged above, with the object of showing whether such schemes are relevant to the problem of explaining the extent of solute accumulation in yeast and mouse ascites tumor cells. There is now much evidence implicating protons and sodium ions, respectively, as cosubstrates in the solute pumps of these organisms.[2,7]

0077–8923/80/0341–0494 $01.75/0

The Flow of Ions Envisaged in Gradient Coupling

The absorption of protons along with various amino acids or phosphate will serve as an example of the principles involved.[8] FIGURE 1 shows the net uptake of protons and the net efflux of K^+ ions that occurred when the indicated, relatively small, amounts of glycine or arginine were added to washed cell suspensions of *Candida utilis* under conditions in which energy metabolism was deliberately prevented. It will be seen that about one equivalent of protons was

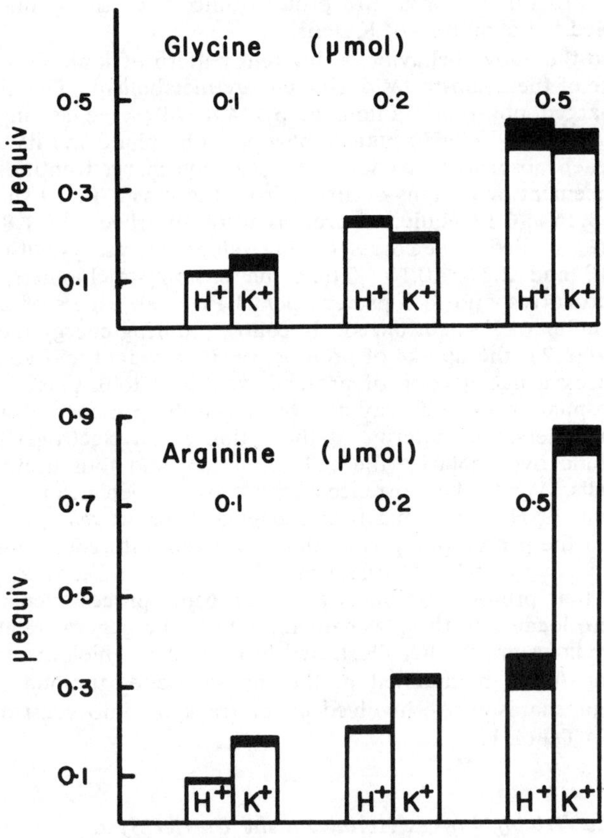

FIGURE 1. The absorption of protons and ejection of potassium ions caused by the addition of graded amounts of glycine or arginine to preparations of *Candida utilis* (N.C.Y.C. 193).* The indicated amounts of the amino acid were added to portions (50 mg dry weight) of the yeast depleted of ATP. The yeast was cultivated in a chemostat in a mineral salts solution with 2 mM L-glutamate as the limiting source of nitrogen. The washed cells were then assayed at 30° C, at about pH 4.8, in the presence of antimycin (15 μg) and 5 mM deoxyglucose. The ionic movements were assayed by means of ion-selective electrodes. The observations are mean values based on 3–6 assays.[8] The shaded areas correspond to the standard error of the mean. At least 95% of the added amino acid was absorbed when 0.1 μ mole or 0.2 μ mole was provided.
* British National Collection of Yeast Cultures, strain No. 193.

absorbed with glycine and one equivalent of K^+ was expelled from the yeast. Whereas a similar number of protons were absorbed with arginine, about two equivalents of K^+ were expelled, a circumstance that would allow electroneutrality to be maintained during the uptake of the arginine cation. The above behavior indicates that the amino acid carrier functions in the absence of energy metabolism and leads to the symport of protons. A number of circumstances, which we have discussed elsewhere, strongly indicate that the movement of K^+ ions is not an intrinsic part of the mechanism of amino acid uptake, but is simply the means whereby the energy-depleted yeast neutralizes the influx of positive charge.[2] In other words, the proton influx with the amino acid is electrically coupled to the outflow of K^+ ions.

In contrast to the above behavior, a different pattern of ionic flows accompanies the uptake of these substrates during energy metabolism. The absorption of the singly-charged phosphate anion, at pH 4.8 will serve as an example (FIGURE 2). Traces *d* and *e* of FIGURE 2 show how phosphate and its analogue, arsenate, were each absorbed; in effect, as positively charged entities since a significant displacement of K^+ ions occurred from the yeast. In 10 such assays with phosphate 3.26 ± 0.13 equiv. of protons were absorbed and 1.85 ± 0.14 equiv. of K^+ were expelled. The corresponding values in 7 assays with arsenate were 3.18 ± 0.10 and 2.23 ± 0.10. Thus, the proton stoichiometry of this system appears to correspond to 3 protons per $H_2PO_4^-$, an excess of 2 positive charges that the efflux of K^+ neutralized. In contrast, during energy metabolism (traces *a*, *c*, FIGURE 2), the uptake of protons was less evident and an influx of K^+ accompanied by a net ejection of protons eventually took place. It seems possible that phosphate primed energy metabolism in the presence of air. Such behavior can be understood in terms of the action of an electrogenic proton pump, which would hyperpolarize the cell membrane and thus draw K^+ ions into the yeast cells. While the existence in yeast of a mechanism exchanging equivalent amounts of H^+ for K^+ has been recognized for 40 years, the electrogenic character of the proton pump is a relatively recent inference which needs to be confirmed.[2] In the related organism *Neurospora crassa,* there is compelling evidence that proton ejection is an electrogenic process dependent on an ATPase system located in the plasmamembrane.[9] The general relationships between the ionic flows are further illustrated in FIGURE 3, which also describes the corresponding fluxes maintained in the mouse ascites tumour cells.[10, 11] Proton symport mechanisms are involved in the transport into yeast of all the substrates listed in TABLE 1.

The Driving Forces Acting on the Carrier System

On the likely assumption that the steady state distribution of K^+ across the yeast plasmalemma is a measure of the prevailing membrane potential, the magnitude of the proton electrochemical gradient is clearly large (TABLE 2).[2] Thus, the membrane potential in yeast strain 240 corresponds to a factor of about 10^3 and the pH gradient to a factor of at least 10, the proton gradient thus being of the order of 10^4, or the equivalent of 240 mV. Our work indicates that similar factors apply in the yeast strain 74 that concentrates glycine about 3×10^4-fold. However, the system involved is the general amino acid permease which absorbs 2 equiv. of protons per mole of glycine absorbed (TABLE 1).

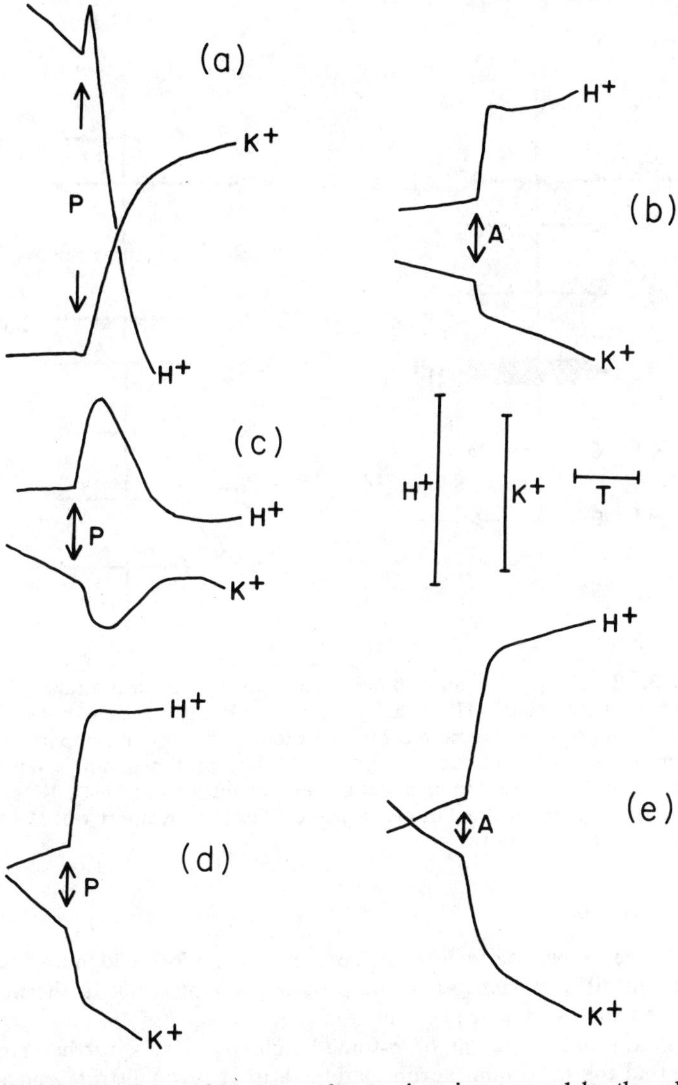

FIGURE 2. The net flows of proton and potassium ions caused by the addition of phosphate or arsenate to preparations of *Candida utilis*. The yeast strain N.C.Y.C. 193 was grown with 0.1 mM phosphate as the limiting nutrient in a chemostat. Portions of the washed yeast cells (50 mg dry weight) were fed with either 0.1 μ mole of arsenate **A** or phosphate **P**, as shown. The traces show how the recordings from the ion-specific electrodes varied with time. In traces **(a)** and **(b)** air was present. In trace **(c)** air was present with antimycin (15 μg) to inhibit respiration. In traces **(d)** and **(e)** both antimycin (15 μg) and 10 mM-iodoacetamide were present. The system was calibrated by the addition of 0.5 μ equiv. of HCl or KCl. The horizontal bar **T** represents 0.5 minutes.

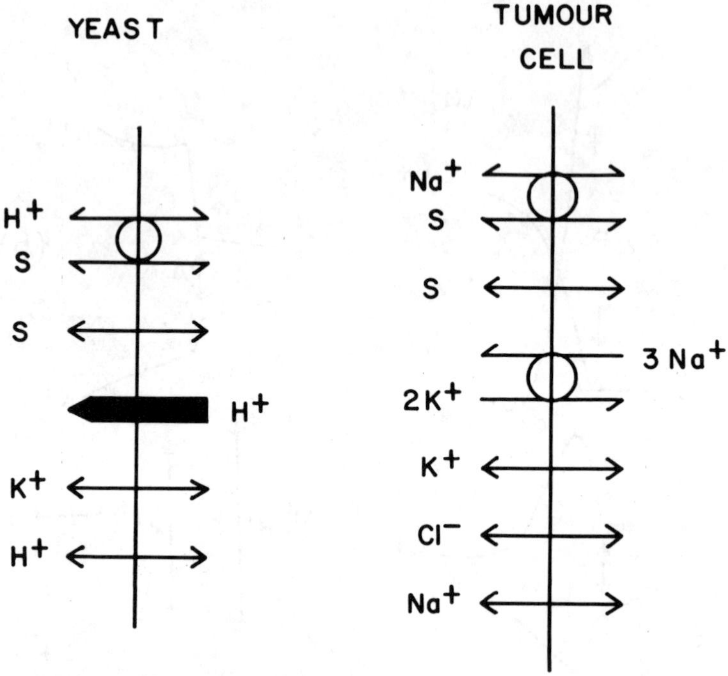

FIGURE 3. The respective ionic flows across the plasma membrane of yeast and mouse ascites tumor cells. The yeast plasma membrane appears to be highly permeable to K^+ and relatively impermeable to protons. The proton symport for substrate *S* functions in parallel with the leak pathway. The proton pump is represented as electrogenic. In the mouse tumor system, the relevant ions are Na^+, K^+ and Cl^- and the sodium pump is represented as electrogenic. The Na^+ symport with *S* functions in parallel with the leak pathway.

Clearly, if the above arguments are correct, the amino acid pump is working at only about 50% efficiency and the proton gradient is not in thermodynamic equilibrium with the glycine gradient.

The steady-state gradient of α-thioethylglucoside is a further striking example in that the maximum accumulation ratio at low substrate concentrations (<100 μM) does not exceed about 200-fold. This leads to the question of whether such small ratios arise because of the presence of a relatively large leak pathway bypassing the proton symport. We have recently obtained some preliminary evidence in favor of this interpretation.[16] The relevant observations show that, under appropriate physiological conditions, the steady-state accumulation ratio for the thioglycoside varied systematically, from about 3 to 200, with the activity of the proton symport. One interpretation of this behavior is that, under the relevant conditions, a graded activity of the pump operated against a fixed leak.

TABLE 1

PROTON STOICHIOMETRY OF VARIOUS YEAST CARRIERS *

System		Proton Equivalents
Saccharomyces		
General amino acid permease [12]	glycine	2
	L-methionine	2
	L-lysine	2
	L-citrulline	2
	L-phenylalanine	2
	L-leucine	2
Specific amino acid permease [12]	L-lysine	1
	L-methionine	1
	L-proline	1
	L-glutamate	2 or 3
Other permeases[12-15, 35]	maltose	1
	phosphate	3
	hypoxanthine	1
	sulphate	3
Candida [8]		
	glycine	1
	L-lysine	1
	L-arginine	1
	L-glutamate	2
	phosphate	3

* The number of equivalents of protons absorbed with each mole of substrate is shown.

Formulation of the Kinetic Steady State of Substrate Absorption

In order to define the flows of substrate through the leak pathway and the yeast symport mechanism, the scheme shown in FIGURE 4 was used. Let the substrate concentrations outside and inside the cells be $[S]_o$ and $[S]_i$ respectively, the corresponding hydrogen ion concentrations being $[H]_o$ and $[H]_i$. The free carrier E on either side of the membrane was assumed to reach equilibrium with protons on the same side, the protonated carrier, in turn, coming to equilibrium with the corresponding substrate pool. In other words, binding of the cosubstrate preceded binding of the substrate. Also the translocation steps were assumed to be rate limiting. The effect of the electric field on the carrier functioning has been represented in the literature in two contrasting ways that, in principle, can be combined. 1. The translocation step is regarded as being assisted or hindered by the electric field, dependent on the one hand, on whether the carrier or its substrate complexes are positively or negatively charged, and, on the other hand, on the direction of the field.[10, 17, 18] The translocation step is assumed, in effect, to sense the whole of the electric field. 2. Alternatively, the carrier is arbitrarily supposed to be in the center of the membrane, between two channels through which the cosubstrate penetrates, from either side of the membrane, under the influence of the electric field.[19] The latter thus raises or lowers the effective concentration of protons (relative to the concentrations

TABLE 2

THE STEADY-STATE GRADIENTS OF GLYCINE (GLY), α-THIOETHYLGLUCOSIDE (TEG), PROTONS AND POTASSIUM IONS MAINTAINED ACROSS THE YEAST PLASMAMEMBRANE DURING ENERGY METABOLISM AT ABOUT pH 4.5 *

Strain N.C.Y.C. 74 of *Saccharomyces uvarum* [23]

$[Gly]_o$	$[Gly]_i/[Gly]_o$
2 μM	3.7×10^4
11 μM	3.3×10^4

Strain N.C.Y.C. 240 of *Saccharomyces cerevisiae* †

$[TEG]_o$	$[TEG]_i/[TEG]_o$	$[K]_o$	$[K]_i/[K]_o$	pH_o	pH_i
0	—	85 μM	1941	4.5	6.03
0.1 mM	198	80 μM	2025	4.5	6.07
10 mM	25	140 μM	962	4.5	5.58
20 mM	10	—	—	—	—

* Intracellular pH was assayed by the propionate distribution method.[20]

† The values shown were obtained in one representative set of assays from a series of four.

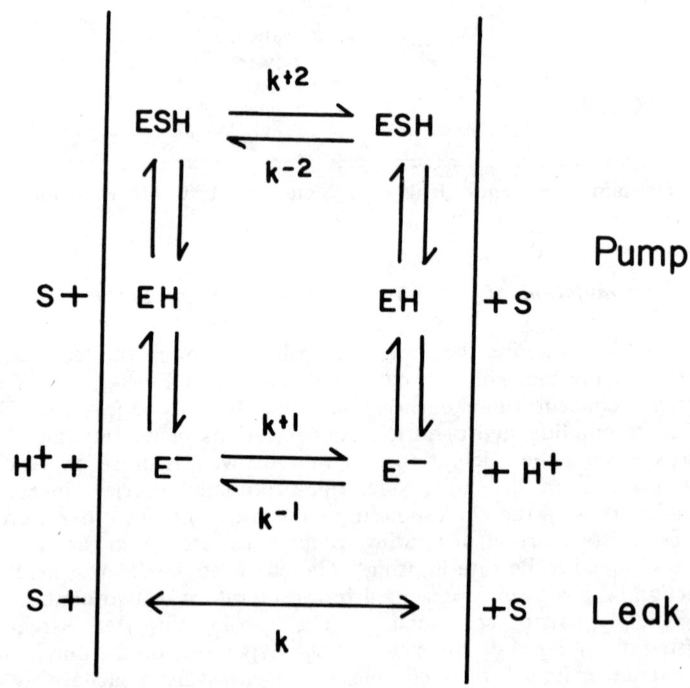

FIGURE 4. A specific carrier model of the proton symport for substrate S acting in parallel with a leak pathway. The "unloaded" carrier bears a single formal negative charge, the protonated carrier being uncharged. The electric field influences the magnitude of k^{+1} and k^{-1} but not the magnitude of k^{+2} or k^{-2}. Both H^+ and S equilibrate with the carrier E in an ordered process. Thus carrier translocation is rate determining.

bathing the two faces of the cell membrane) that participate in the binding of the substrate S. The translocation step is assumed to sense a negligible fraction of the electric field.

If the unloaded carrier is negatively charged, procedure 1 above can be represented as $k_{-1} = k'e^{-\frac{1}{2}VF/RT}$ and $k_{+1} = k'e^{\frac{1}{2}VF/RT}$, where k' is the rate constant when $V = 0$. In contrast, procedure 2 above means that $[\mathrm{H}]_o$ is raised by the factor $e^{-\frac{1}{2}VF/RT}$ at the external membrane surface and by the factor $e^{\frac{1}{2}VF/RT}$ at the inner membrane surface. In practice, we found that the kinetics of thioethylglucose transport were consistent with the former but not with the latter procedure and seemed to require, further, that the translocation of the unprotonated carrier, rather than that of the protonated carrier, responded to the electric field.

Two useful derivations concern first the magnitude of $[S]_i$ in the steady state reached at any given value of $[S]_o$ and, second, the magnitude of the leak coefficient k in terms of the other parameters of the model.

The steady state of the substrate S is defined by the equation $k([S]_i - [S]_o) = k_{+2}[EHS]_o - k_{-2}[EHS]_i$.
We then define the equilibrium binding constants K_1 and K_2 as,

$$K_1 = \frac{[E]_o[\mathrm{H}]_o}{[E\mathrm{H}]_o} = \frac{[E]_i[\mathrm{H}]_i}{[E\mathrm{H}]_i}$$

$$K_2 = \frac{[E\mathrm{H}]_o[S]_o}{[E\mathrm{H}S]_o} = \frac{[E\mathrm{H}]_i[S]_i}{[E\mathrm{H}S]_i}$$

the subscripts denoting the respective outer and inner membrane faces. It is convenient to make the substitution $\theta = e^{-\frac{1}{2}VF/RT}$ where V is the membrane potential and F, R and T have their conventional significance. We then find that the K_m value with respect to $[S]_o$ when $[S]_i$ is zero is given by

$$K_m = \frac{K_1K_2[K_1\,(1+\theta^2) + \theta^2[\mathrm{H}]_o + [\mathrm{H}]_i]}{\theta[\mathrm{H}]_o\,[K_1\,(1+\theta) + [\mathrm{H}]_i]}$$

and that

$$V_m = \frac{\theta\, K_1\, k'E_t}{K_1\,(1+\theta) + [\mathrm{H}]_i}$$

where E_t is the sum of the concentrations of the different forms of the carrier.

Moreover, the value of $[S]_i$ in the steady state is the positive real root of the equation

$$a\,[S]_i^2 + b\,[S]_i + C = O \text{ where}$$

$$a = [\mathrm{H}]_i\left[\theta\,[\mathrm{H}]_o\,(2\,[S]_o + K_2) + K_1K_2\,(1+\theta)\right]$$

$$b = \Big[(K_2[\mathrm{H}]_i + K_1K_2 - [\mathrm{H}]_i[S]_o)\,(\theta[\mathrm{H}]_o[S]_o + K_1K_2) + \theta\,([\mathrm{H}]_o[S]_o + K_2[\mathrm{H}]_o + K_1K_2)\,(\theta K_1K_2 - [\mathrm{H}]_i[S]_o) + \frac{k'E_t}{k}K_1K_2[\mathrm{H}]_i\Big]$$

$$c = -[S]_o\Big[\theta^2K_1K_2\,([\mathrm{H}]_o[S]_o + K_2\,[H]_o + K_1K_2) + (K_2\,[\mathrm{H}]_i + K_1K_2)\,(\theta[\mathrm{H}]_o[S]_o + K_1K_2) + \theta^2\,\frac{k'E_t}{k}K_1K_2\,[\mathrm{H}]_o\Big]$$

Further the leak coefficient k is given by

$$k = \frac{k'E_t\, K_1K_2\, (\theta^2\, [\mathrm{H}]_o[S]_o - [\mathrm{H}]_i[S]_i)}{([S]_i - [S]_o)\left[\theta\, ([\mathrm{H}]_o[S]_o + K_2[\mathrm{H}]_o + K_1K_2)\, ([\mathrm{H}]_i[S]_i + \theta K_1K_2) + ([\mathrm{H}]_i[S]_i + K_2[\mathrm{H}]_i + K_1K_2)\, (\theta[\mathrm{H}]_o[S]_o + K_1K_2)\right]}$$

Similar procedures were used to model the amino acid transport system of the mouse ascites tumor cells, although, in this case, the mathematical model

TABLE 3

THE CARRIER PARAMETERS DESCRIBING THE ACCUMULATION OF α-THIOETHYLGLUCOSIDE BY THE YEAST

Stage 1. K_1, K_2 and $k'E_t$ were calculated from the observed values of K_m and V_m at $\mathrm{p[H]}_o=5.0$. The experimentally determined values of K_m, V_m, $\mathrm{p[H]}_i$ and θ used for this purpose were, respectively, 4.0 mM, 31.0 nmole mg^{-1} min^{-1}, 6.08 and 31. The value of K_m at $\mathrm{p[H]}_o=7$ was required and was found to be 30.3 mM. $\mathrm{p[H]}_i$ was assumed also to be 7 under these conditions.

The conclusion was $K_1 = 6.8238 \times 10^{-4}$
$K_2 = 4.01184$
$k'E_t = 33.2183$

Stage 2. The value of the leak coefficient k was computed on the basis that at $\mathrm{p[H]}_o=5$ $[S]_i/[S]_o=216$ when $[S]=0.1$ mM.

The conclusion was that $k=0.02837$.

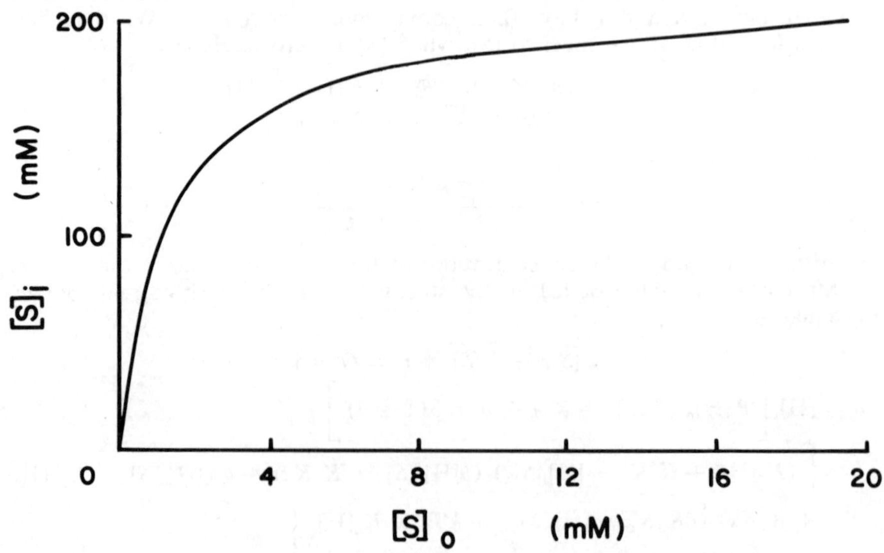

FIGURE 5. The predicted variation of $[S]_i$ with $[S]_o$ in the steady state of S. The parameters used are those listed in TABLE 3. In effect, the observed magnitude of $[S]_i/[S]_o$ when $[S]_o=0.1$ mM is used to predict the way the pool size varies with $[S]_o$ when the membrane potential, the pH gradient and the parameters of the model defined by K_m and V_m remain fixed.

was simplified by omitting the notion of an ordered binding to the carrier. Instead the dissociation of the ternary carrier complex ENaS was represented by a single constant. The free carrier was assumed to bear zero formal charge.[10, 17]

The Flow of Electric Charge in the Steady State

The magnitude of the leak coefficient k (TABLE 3), deduced from the mean value of the accumulation ratio of 216 observed at 0.1 mM thioethylglucoside, was used to predict how the cellular pool of the thioglucoside would increase with the substrate concentration when the proton gradient was constant (FIGURE 5). The prediction is based on the assumption that the cellular pH and the membrane potential remained constant. In fact, TABLE 2 shows that the cellular pH and the gradient of K^+ both fell when 10 mM, as opposed to 0.1 mM, thioethylglucoside was used. Nevertheless, the model leads to values of $[S]_i$ that are roughly consistent with those observed.

FIGURE 3 shows that, in the steady state of substrate accumulation, a net flow of cosubstrate ion occurs through the carrier system because the proportion of S leaving the cells through the leak pathway is greater than enters the yeast through the same route. Recent work in these laboratories[16] has verified that the expected flow of protons does occur in the steady state of accumulation in the presence of 8 mM α-thioethylglucoside. It was detected by assaying the proton flow in the period shortly after the proton pump had been inhibited. The magnitude of this steady state leak current is $k([S]_i - [S]_o)$. A possible interpretation of the small, but statistically significant, changes in the ionic balance in the presence of 10 mM thioethylglucoside is, therefore, that they result from the partial depolarization of the cells due to an accelerated uptake of protons. However, other factors may have to be taken into account. For instance, the yeast would swell in the presence of the thioglucoside, which might secondarily affect the ionic balance.

In other microbial systems, the accumulation of solutes in the steady state can be associated with a higher rate of respiration and with complex changes in the rate of dissipation of heat.[5, 21, 22] Both phenomena are consistent with the idea of energy dissipation through a leak pathway, although it has been suggested that a regulatory mechanism in the proton symport is also involved in one such instance.[22, 25, 26]

Amino Acid Transport in the Mouse Ascites Tumor Cells

The tumor cells provide an interesting test of the above concepts in that they accumulate various amino acids to different extents in a concentration-dependent fashion. For instance, TABLE 4 shows the driving forces that operate on the sodium symport for glycine in the steady state of amino acid accumulation, maintained with either 1 mM or 10 mM glycine. It is important to note that the sodium electrochemical gradient was somewhat larger than the glycine gradient in both instances.

Using our earlier estimates[24] of the magnitude of the leak coefficient for glycine in this system, we examined the way in which the accumulation ratio

TABLE 4

THE IONIC GRADIENTS MAINTAINED BY THE MOUSE ASCITES TUMOR CELLS DURING ENERGY METABOLISM IN THE PRESENCE OF 1 MM OR 10 MM GLYCINE *

Glycine (mM)	$[Na]_o$ (mM)	$[Na]_i$	$-V$ (mV)	$[Gly]_i/[Gly]_o$
1	155	38	61	26
10	155	40	48	9

* V=membrane potential. The system was assayed in the steady state of glycine accumulation maintained in glucose Ringer solution. Mean values from 4 separate assays are shown. The change in V and in the glycine accumulation ratio with the glycine concentration were highly significant ($p < 0.01$).

was expected to vary with such factors as the apparent K_m exhibited towards different amino acids (FIGURE 6) and the amino acid concentration itself (FIGURE 7).

FIGURE 6 shows that, when the other parameters of the system are constant, the largest amino acid gradients are produced when the amino acid concentra-

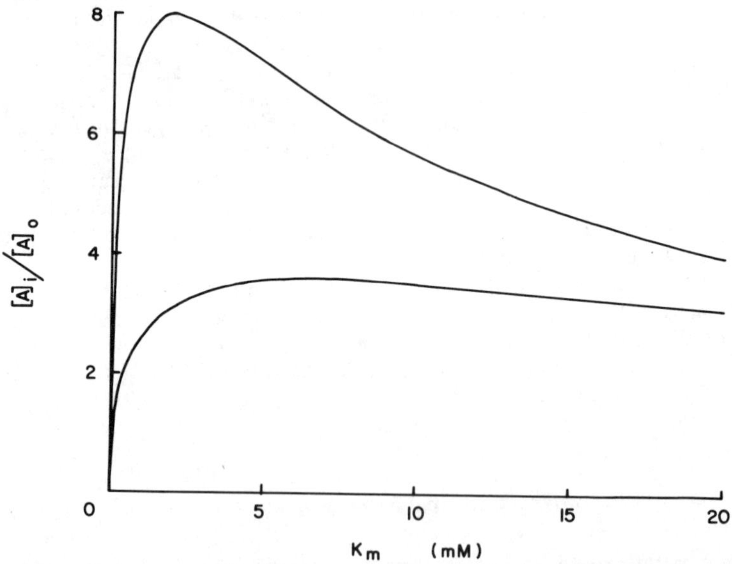

FIGURE 6. The variation of the amino acid accumulation ratio ($[A]_i/[A]_o$) with the apparent K_m of the carrier at two amino acid concentrations. The amino acid was assumed to bind to the carrier with Na^+ in a concerted process. The corresponding dissociation constant was systematically varied to give the K_m value shown when $[Na]_o$ was 0.155 M, V was -45 mV and V_m was constant at 8.575×10^{-5} mol 1^{-1} sec^{-1}. $[Na]_i$ was 0.039 M. The leak coefficient k was 0.001 sec^{-1}. The values of V_m, k, V and $[Na]_i$ are similar to those found in the steady state of glycine accumulation when $[A]_o$=10 mM. The upper line in the figure corresponds to $[A]_o$=2 mM and the lower line to $[A]_o$=10 mM.

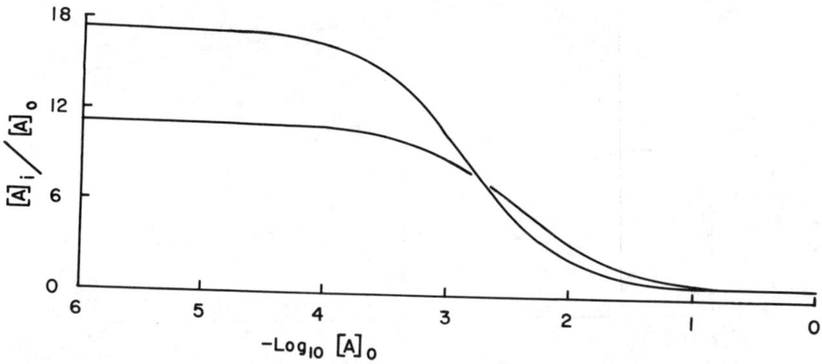

FIGURE 7. The variation of the amino acid accumulation ratio ($[A]_i/[A]_o$) with the amino acid concentration ($[A]_o$) at two K_m values. The parameters used are those listed in FIGURE 6. The upper line on the left hand side of the diagram corresponds to $K_m = 1$ mM, the lower line to $K_m = 4$ mM.

tion is near the K_m selected. The explanation appears to be that, at relatively low values of the K_m, the pump is merely circulating the amino acid, both inwards and outwards at roughly the same rate, whereas at relatively large values of the K_m, the net influx through the pump is small. The optimum accumulation thus occurs at some intermediate value of the K_m at which point the net flow through the pump is largest. As FIGURE 6 indicates, there are, in general, two values of the K_m that maintain a given accumulation ratio when $[A]_o$ is fixed. The intersection of the two lines shown in FIGURE 7 arises from that circumstance. One noteworthy point is that at very small amino acid concentrations the accumulation ratio is independent of $[A]_o$. However, the actual magnitude of the accumulation ratio depended on the K_m and, in both instances, it was smaller than what corresponds to thermodynamic equilibrium between the glycine and sodium ion gradients. As the influence of the leak pathway thus persists at very small solute concentrations, the observation of a constant accumulation ratio at different small solute concentrations is not evidence for tight coupling with the Na gradient.

Turning now to the sodium current through the leak, as depicted in FIGURE 8, this, too, varies with the K_m, reflecting indeed the tendencies exhibited in FIGURE 6. The availability of a procedure for assaying the membrane potential of the tumor cells with a fluorescent probe,[10] led us to attempt to detect the depolarization of the tumor cells that high concentrations of either glycine or leucine were expected to cause. Both amino acids were known to cause a fluorescence change of the potential-sensitive dye of about 25% (arbitrary units), under the standard conditions of the assay, when they are first added to the glycolysing tumor cells in sodium Ringer solution.[10, 11] A similar response is exhibited in the presence of L-alanine (FIGURE 9). The magnitude of the depolarization observed in the steady state was expected to be much smaller. This proved to be so (FIGURE 10). The observations summarized in TABLE 5 show, however, that a definite increase in fluorescence occurred in the presence of either amino acid. We attribute this behavior to the factors illustrated in FIGURE 8.

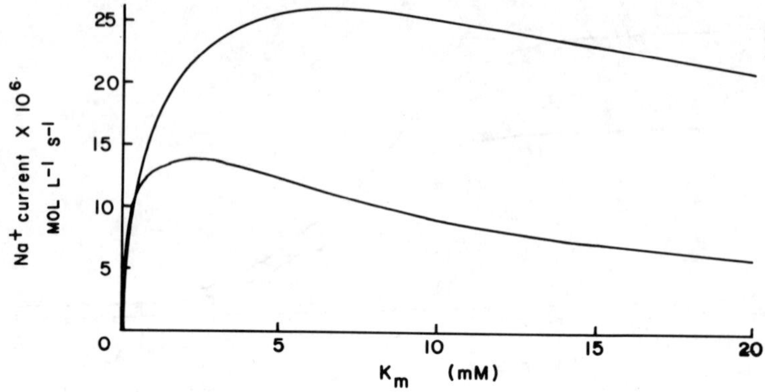

FIGURE 8. The predicted net sodium ion current through the pump in the steady state of amino acid accumulation, show as a function of K_m at two amino acid concentrations. The parameters used in connection with FIGURE 6 were again employed. The upper line corresponds to $[A]_o = 10$ mM, the lower line to $[A]_o = 2$ mM.

Significance in Relation to the Mechanism of Energy Coupling[27]

The general arguments outlined above, though preliminary in scope, show that the systems illustrated in FIGURE 3 have a number of characteristics resembling the behavior of the yeast and the tumor cells towards α-thioethylglucoside and glycine, respectively. The expected flow of the cosubstrate ions in the steady state of solute accumulation, which is implicit in FIGURE 3, does indeed appear to occur. The small depolarization of the tumor cells under

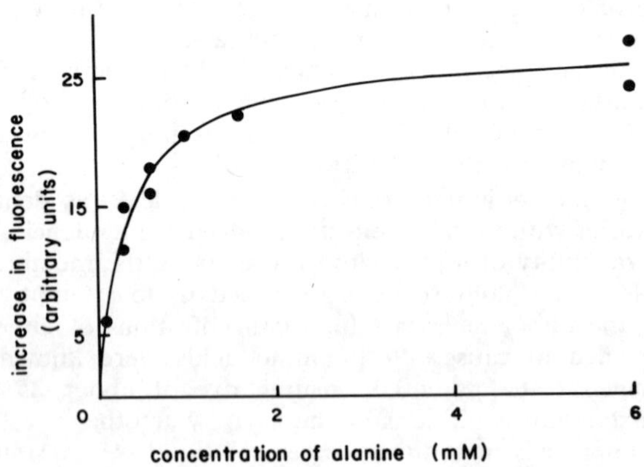

FIGURE 9. The fluorescence increase (depolarization) caused by graded additions of L-alanine to the mouse ascites tumour cells metabolizing glucose in sodium Ringer solution. The potential sensitive dye dipropyloxadicarbocyanine was used under conditions suppressing mitochondrial interference with the signal.[10, 11] Fluorescence was excited at 578 nm and recorded at 620 nm.

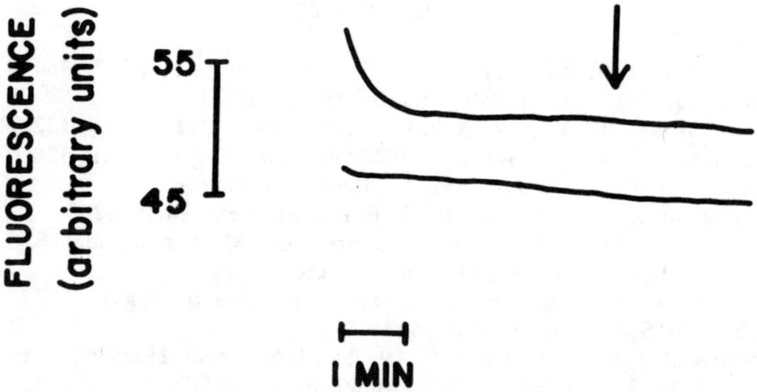

FIGURE 10. The assay of the fluorescence increase associated with the steady state of accumulation of 10 mM glycine. The upper line represents the fluorescence changes with time when the tumor cells had reached a steady state in the presence of glycine. The lower line is the corresponding control lacking added glycine. The arrow shows the time at which the readings are taken in order to infer the magnitude of the membrane potential from the standard calibration curve.[10]

these conditions is consistent with very early observations [34] that glycine absorption displaced potassium ions from the tumor cells during incubations of extended duration. Furthermore, more recent claims that amino acid uptake, while remaining Na^+-dependent, can be uncoupled from the flow of sodium ions into the tumor cells [32, 33] are inconsistent both with the depletion of cellular K [34] and the depolarization we believe we have now detected. The alleged uncoupling of the amino acid pump from the sodium gradient, together with other findings, has led to interesting speculations about the possible role of proton gradients,[28] or specific molecular conformations,[29] within the cell membrane [28] as the ultimate factor driving the system.[30-33] Some reassessment of these speculations now seems to be required.

TABLE 5

THE FLUORESCENCE INCREASE ASSOCIATED WITH MEMBRANE DEPOLARIZATION OF THE MOUSE TUMOR CELLS IN THE STEADY STATE OF AMINO ACID ACCUMULATION *

Amino acid	Fluorescence increase (%)
10 mM leucine	7.1±2.2 (23)
10 mM glycine	5.9±1.9 (14)

* The assays were done as indicated in the legend to FIGURE 10. The mean fluorescence increase (± standard deviation) was assayed on the indicated number of occasions.

References

1. Kotyk, A. & R. Stružinský. 1977. Biochim. Biophys. Acta **470:**484–491.
2. Eddy, A. A. 1978. Curr. Top. Membr. Transp. **10:**279–360.
3. Chevallier, M. R., R. Jund & F. Lacroute. 1975. J. Bacteriol. **122:**629–641.
4. Maloney, P. C. & T. H. Wilson. 1973. Biochim. Biophys. Acta **330:**196–205.
5. Kepes, A. 1970. Curr. Top. Membr. Transp. **1:**101–134.
6. Crabeel, M. & M. Grenson. 1970. Eur. J. Biochem. **14:**197–204.
7. Eddy, A. A. 1977. *In* Intestinal Permeation. M. Kramer and F. Lauterbach, Eds.: 332–347. Excerpta Medica, Amsterdam.
8. Eddy, A. A., R. Philo, P. Earnshaw & R. Brocklehurst. 1977. Fed. Eur. Biochem. Soc. Symp. **42:**250–260.
9. Bowman, B. J. & C. W. Slayman. 1977. J. Biol. Chem. **252:**3357–3363.
10. Philo, R. D. & A. A. Eddy. 1978. Biochem. J. **174:**801–810.
11. Philo, R. D. & A. A. Eddy. 1978. Biochem. J. **174:**811–817.
12. Seaston, A., C. Inkson & A. A. Eddy. 1973. Biochem. J. **134:**1031–1043.
13. Serrano, R. 1977. Eur. J. Biochem. **80:**97–102.
14. Cockburn, M., P. Earnshaw & A. A. Eddy. 1975. Biochem. J. **146:**705–712.
15. Reichert, U. & M. Forêt. 1977. FEBS Lett. **83:**325–328.
16. Eddy, A. A., A. Seaston, D. Gardner & C. Hacking. 1979. *In* Function and Molecular Aspects of Biomembrane Transport. F. Palmieri, S. Papa, M. Klingenberg and E. Quagliariello, Eds.: 219–228. Elsevier/North Holland Biomedical Press. Amsterdam.
17. Heinz, E. & P. Geck. 1978. *In* Membrane Transport Processes. J. F. Hoffman, Ed. **1:**13–30. Raven Press. New York.
18. Villereal, M. L. & J. S. Cook. 1978. J. Biol. Chem. **253:**8257–8262.
19. Mitchell, P. 1977. Symp. Soc. Gen. Microbiol. **27:**383–423.
20. Seaston, A., G. Carr & A. A. Eddy. 1976. Biochem. J. **154:**669–676.
21. Decker, M. & W. Tanner. 1972. Biochim. Biophys. Acta **266:**661–669.
22. Long, R. A., W. G. Martin & H. Schneider. 1977. J. Bacteriol. **130:**1159–1174.
23. Indge, K., A. Seaston & A. A. Eddy. 1977. J. Gen. Microbiol. **99:**243–255.
24. Morville, M., M. Reid & A. A. Eddy. 1973. Biochem. J. **134:**11–26.
25. Wilson, T. H., E. R. Kashket & M. Kusch. 1972. *In* The Molecular Basis of Biological Transport. J. F. Woessner, Jr. and F. Huijing, Eds. **3:**219–245. Academic Press. New York.
26. Lagarde, A. & A.-R. Prévot. 1974. C. R. Acad. Sc. Paris. **279:**97–100.
27. Garland, P. B. 1977. Symp. Soc. Gen. Microbiol. **27:**1–21.
28. Williams, R. J. P. 1975. *In* Electron Transfer Chains and Oxidative Phosphorylation. E. Quagliariello, S. Papa, F. Palmieri, E. C. Slater and N. Siliprandi, Eds.: 417–422. North-Holland. Amsterdam.
29. Boyer, P. D. 1974. Biochim. Biophys. Acta Library **13:**289–301.
30. Christensen, H. N. & M. E. Handlogten. 1975. Proc. Nat. Acad. Sci. U.S.A. **72:**23–27.
31. Garcia-Sancho, J., A. Sanchez, M. E. Handlogten & H. N. Christensen. 1977. Proc. Nat. Acad. Sci. U.S.A. **74:**1488–1491.
32. Christensen, H. N., C. De Cespedes, M. E. Handlogten & G. Ronquist. 1973. Biochim. Biophys. Acta **300:**487–522.
33. Christensen, H. N. 1979. Adv. Enzymol. **49:**41–101.
34. Christensen, H. N. & T. R. Riggs. 1952. J. Biol. Chem. **194:**57–68.
35. Roomans, G. M., G. A. J. Kuypers, A. P. R. Theuvenet & G. W. F. H. Borst-Pauwels. 1979. Biochim. Biophys. Acta **551:**197–206.

DISCUSSION

G. KIDDER: I noticed in your model for the amino acid leak and its effects, you expressed your results in terms of the cytoplasm to medium ratio. Am I right in inferring that this mechanism would tend to limit cytoplasmic concentrations as the medium concentration of the amino acid was increased?

A. EDDY: Yes.

KIDDER: And, therefore, might be of some potential significance to the operation of the cell?

EDDY: Yes. In the paper we have a computation of that kind showing that the presence of a leak of a certain size will lead to a certain maximum pool content as you raise the external amino acid concentration or substrate concentration. The key issue is this: Is the size of the leak that is needed to give that limitation one that still allows the maximum accumulation observed at very much lower concentrations, or is the leak regulated in relation to the load? We've so far taken the very simplest possibility in which the leak is fixed.

H. R. KABACK: Do you want to suggest where you think these leak pathways are? Are they through carriers? How do they work?

EDDY: That is an extremely important question. If you have a leak through the carrier (which corresponds to the movement of the substrate with the unprotonated carrier) the accumulation ratio, calculated with the usual assumptions, will be independent of the concentration, provided the membrane potential doesn't fall. That is, you will not get the observed limitation of the pool size. However you can get a marked fall in the accumulation ratio by having a leak outside or by uncoupling the pump. I think the simplest possibility is that you have a leak outside. In the α-thioethylglucoside system we've looked for this by varying the level of induction of the pump.

G. SACHS: In the ascites cell, is the leak absolutely sodium dependent?

EDDY: Absolutely sodium independent.

SACHS: Either way?

EDDY: It's been known for many years in the ascites cells there is a component of uptake that is independent of sodium. The question is whether the magnitude of that process can account for the magnitude of the depolarization we observe.

ANION TRANSPORT MECHANISMS IN NEURONS *

John M. Russell

Department of Physiology and Biophysics
University of Texas Medical Branch
Galveston, Texas 77550

Introduction

There are two main issues to be addressed in this presentation. The first of these is whether active transport need be involved to explain chloride movements across certain neuronal membranes. This requires a determination of whether free chloride inside neurons exists at a level different from that expected for a passive thermodynamic equilibrium between the chloride concentration gradient and the membrane potential gradient. If the calculated Nernst equilibrium potential for Cl^- (E_{Cl}) is maintained at a value different than the neuronal membrane potential (V_m), then it follows that active Cl^- transport must be involved.

The second issue follows from the first. If evidence exists for the nonequilibrium transmembrane distribution of chloride across neuronal cell membranes, what sorts of mechanisms are involved and what purposes might these mechanisms serve?

Given the well-known fact that chloride distribution in red blood cells obeys simple Donnan equilibrium predictions, the idea of a nonequilibrium distribution for Cl^- in nerve was slow to gain acceptance. Nevertheless, considerable evidence was gathered during the 1960s and early 70s to suggest that E_{Cl} does not equal V_m in a variety of neuronal preparations. Since nerve is an excitable tissue, it is not surprising that the bulk of this evidence came from analysis of electrical events. Many neurons possess chloride-dependent electrical events, usually inhibitory post-synaptic potentials. The reversal potential for many of these events, that is, the membrane potential at which the chloride-dependent event could elicit no further change in membrane potential, was shown to be different from the resting potential of the neuron.[7, 9, 10, 14, 15, 18, 21, 23–25, 31] This result implied that E_{Cl} did not equal V_m. With the exception of the squid giant axon, which will be discussed below, more direct evidence for such an inequality had to await the development of Cl^--selective microelectrodes. Either miniature Ag-AgCl wires[17] or liquid ion exchanger in a conventional microelectrode[7] were used to make measurements of neuronal intracellular chloride ion activity (a_{Cl}^i) under *in vitro* conditions. Results from such studies seemed to confirm those obtained by the more indirect means. However, both those approaches were subject to one difficult criticism: that impalement of the neuron with one or two electrodes had somewhat depolarized V_m below the normal, *in vivo* conditions, but that due to the relatively low membrane permeability to chloride, the transmembrane chloride distribution had not yet rearranged to conform to the new V_m. That is, during these experiments the cells were neither in equilibrium nor in a steady-state. Obviously, this argument is only meaningful for those neurons that appeared to contain less chloride than expected.

* Supported by NIH Grant NS-11946.

0077–8923/80/0341–0510 $01.75/0 © 1980, NYAS

In the squid giant axon, a number of workers have analyzed extruded axoplasm and found it to be relatively high in chloride content.[1, 4, 11, 16, 19, 26, 34] Here it was important to know whether the total axoplasmic chloride was free in solution or largely bound to cellular proteins. Binding was ruled out by Mauro[22] and Keynes[19] using Ag-AgCl chloride-selective electrode systems. Thus, E_{Cl} for the squid axon is about −30 mV compared to the resting potential of −60 to −70 mV.

Obviously in those neurons with apparently low a_{Cl}^i values, an alternative approach was necessary to confirm or deny the presence of an outwardly-

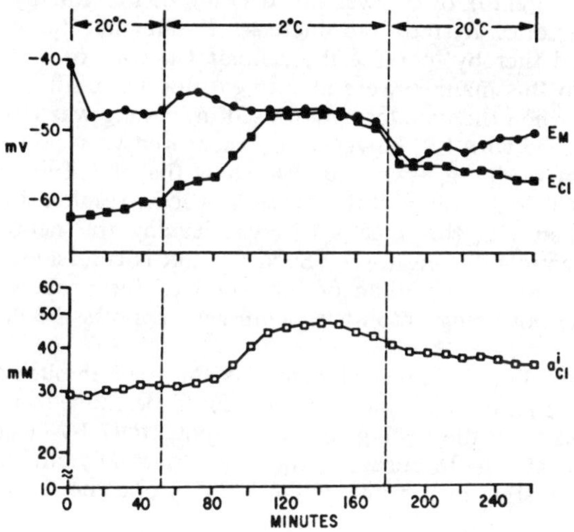

FIGURE 1. Effect of cooling and rewarming on a_{Cl}^i, E_{Cl} and E_m (membrane resting potential) of an *Aplysia* giant neuron. The Cl^--selective microelectrode was calibrated at both temperatures before and after the data were obtained in this experiment and the one in the following figure. (From Russell and Brown.[29] By permission of *Journal of General Physiology*.)

directed chloride pump mechanism. One approach was to raise a_{Cl}^i and see if E_{Cl} would return to a value more negative than V_m, i.e., could a net Cl^- efflux occur against its electrochemical gradient?

Proof of Nonequilibrium Chloride Distribution

In the giant neuron of the *Aplysia* abdominal ganglion, an extensive study revealed that E_{Cl} averaged about −55 mV whereas V_m in the same cell was about −48 mV. These measurements were made 3–6 hours following impalement with the electrodes.[29] However, since these are large cells with a relatively low Cl^- permeability,[13] it was still necessary to see whether such a neuron could eject a load of added Cl^-. Three means were used to raise a_{Cl}^i. Cooling the neuron to below 4° C caused a_{Cl}^i to increase until $E_{Cl} = V_m$. Upon rewarming to 20° C, a_{Cl}^i declined causing E_{Cl} to become more negative than V_m (FIGURE 1). These results imply that cooling inhibited an active extrusion process and a_{Cl}^i increased to its equilibrium level. Warming the cell then reactivated the extrusion process, forcing Cl^- out against its electrochemical gradient. A second method was to bathe the cell for a relatively short time in an artificial seawater solution that contained 100 mM K^+ rather than the normal 10 mM. Such treatment always caused a large net influx of chloride, with the result that a_{Cl}^i was significantly increased. When the normal seawater was returned at 20° C, a_{Cl}^i declined, eventually reaching control values, even though this caused E_{Cl} to become more negative than V_m. However, if the neuron was cooled below 4° C, no such efflux of Cl^- was noted (FIGURE 2). Finally, we have used the polyene antibiotic, nystatin, to increase the membrane permeability of univalent ions and thereby increase the cellular Cl^- content.[30] Neurons whose a_{Cl}^i was trebled in this manner were able to extrude Cl^- until a_{Cl}^i had returned to control values when the counter-cation used in loading was either Na^+ or K^+. When Cl^- was loaded with Li^+, however, a_{Cl}^i recovered very slowly. Thus, three different means of raising a_{Cl}^i were used to show that the *Aplysia* giant neuron could extrude a Cl^- load. Since this extrusion process resulted in E_{Cl} becoming more negative than V_m, this means Cl^- was leaving the neuron against the passive electrochemical driving force. Such a result is consistent with an active transport mechanism. The finding of inhibition of the process by cooling to below 4° C, while not being absolutely definitive, supports the active Cl^- transport hypothesis.

Lux [21] showed that the reversal potential for the inhibitory post-synaptic potential (E_{IPSP}) in cat spinal motoneurons could be moved in a depolarizing direction (i.e., closer to the resting membrane potential) by electrophoretically injecting Cl^- into the motoseurons. E_{IPSP} was normally about 5 mV more negative than the resting membrane potential (V_m) and the Cl^- injection would

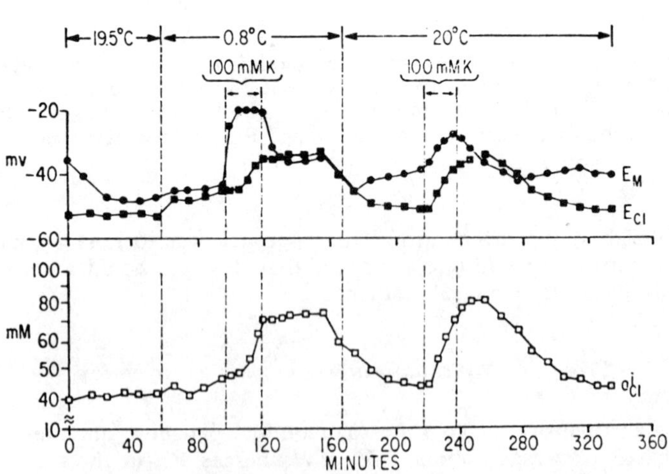

FIGURE 2. Effects of 100 mM potassium-containing artificial seawater on a_{Cl}^i, E_{Cl} and E_m at two temperatures. Notice that at the cooler temperature E_{Cl} remained equal to E_m, but on rewarming Cl^- left the cell against its electrochemical gradient. (From Russell and Brown.[29] By permission of *Journal of General Physiology*.)

render it slightly less negative than V_m. However, the E_{IPSP} would rather quickly recover following such an injection. Such recovery suggests a net Cl^- efflux process in operation. Interestingly enough, this recovery process was significantly inhibited when the animals were intravenously injected with ammonium salts. Similar results have been noted for cat trochlear motoneurons.[20] A possible explanation for this inhibition will be presented later.

The conclusion that comes from these studies is that there is good evidence for the nonequilibrium distribution of chloride across several neuronal cell membranes. Most neurons that have been carefully studied have a lower cellular chloride level than expected; however, the squid giant axon is a conspicuous example of a cell that accumulates chloride against its electrochemical gradient. The next question is, what kinds of mechanisms are responsible for the steady-state nonequilibrium distribution of chloride in neuronal tissue?

Active Cl^- Uptake by Squid Giant Axon

The squid giant axon provides the clearest example of a neuron with a higher cellular chloride content than expected from passive thermodynamic considerations. As mentioned previously, numerous investigators have measured axoplasmic Cl^- concentration and found it to be 3–5 times higher than it should be if $E_{Cl} = V_m$. Its large size imparts technical advantages, which permit a detailed study of the process(es) involved in the transmembrane distribution of Cl^-.

In 1963, Keynes[19] showed clearly that axoplasmic chloride existed free in solution and that the ^{36}Cl uptake rate could be reduced by about 60% if the axon was treated with DNP. Since the effect of DNP was presumably the result of a block of ATP synthesis, thereby depleting cellular ATP stores, it was somewhat disconcerting that treatment with cyanide, an agent which should also result in cellular ATP loss, had very little effect on ^{36}Cl uptake.

Recently I have reexamined the question of an ATP requirement for Cl^- influx using the internal dialysis technique of Brinley and Mullins.[5] By means of this method, the effect of dialyzing out the cellular ATP (its continual synthesis being prevented by treatment with cyanide) on unidirectional Cl^- influx was measured. At 15° C, Cl^- influx into ATP-containing axons is between 10 and 18 picomoles/cm^2 second (pmol/cm^2s), depending on the intracellular chloride concentration (see below). However, after washing out the ATP, the flux falls to 1–2 pmol/cm^2s.[26, 28]

Effect of ATP Analogs on Cl^- Influx

In order to test whether ATP was hydrolyzed during Cl^- uptake, we examined the effects of certain ATP analogs in which one of the oxygens linking the phosphate groups of ATP has been replaced by a methylene group ($-CH_2-$). FIGURE 3 shows the effect of washing out ATP and then presenting the axon with 4 mM alpha, beta methylene ATP (AMP-C-PP). This analog is hydrolyzable by ATPases and has been shown by Dipolo[12] to support calcium efflux from the squid giant axon. As seen in the figure, it was completely without effect on Cl^- influx. Similar results were obtained with beta, gamma methylene ATP (AMPP-C-P). A concentration of 20 mM was used once with

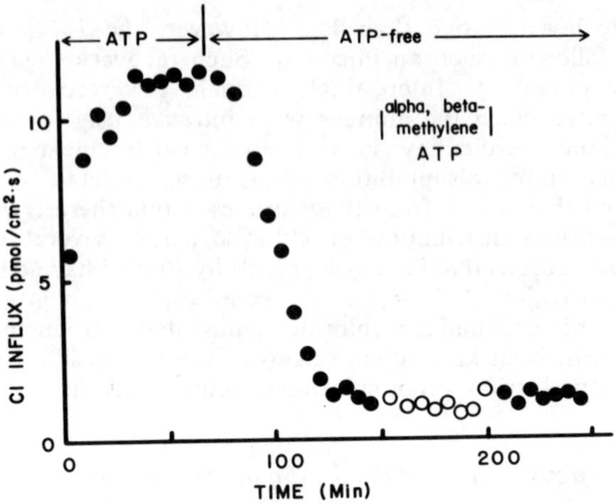

FIGURE 3. Lack of effect of the α,β-methylene-substituted ATP on Cl^- influx into an internally-dialyzed squid giant axon. The experiment was begun with 4 mM ATP and 2 mM cyanide present. When the ATP was removed from the dialysis fluid, Cl^- influx decreased. When the methylene-substituted ATP was presented in a 4 mM concentration, no change in Cl^- influx was noted.

a similar lack of effect; therefore, it seems unlikely that a relatively low potency of the analog can explain its inability to stimulate Cl^- influx. Both analogs caused an inhibition of chloride influx when added to a dialysis fluid containing ATP. This result suggests a certain affinity of the analog for the ATP site(s) involved in chloride influx.

One obvious explanation for these results is that the ATP site for Cl^- influx simply has a very low affinity for the analogs (but see above), and that therefore we have learned nothing about whether ATP is hydrolyzed during chloride uptake. An alternative explanation might be that ATP is required at two sites, but that neither of the ATP analogs used could satisfy the requirements of both sites.

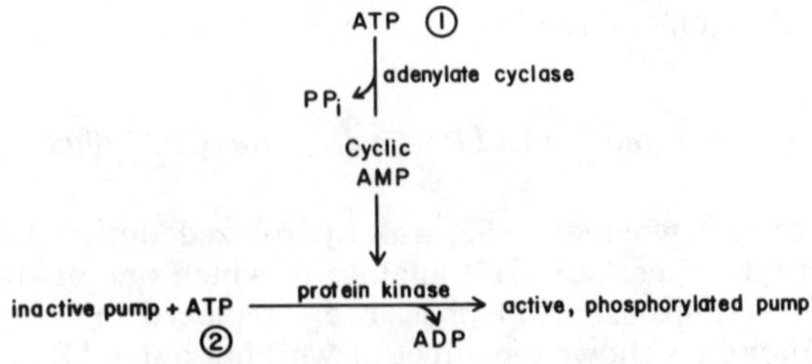

FIGURE 4. Proposed scheme for interaction of cyclic AMP with Cl^- uptake mechanism showing two sites at which ATP would be required. At site one, ATP is cleaved between the α and the β phosphorus molecules. At site two, the point of hydrolysis is between the β and the γ phosphorus molecules.

Role for Cyclic AMP in Chloride Influx?

Recently,[3] we have shown in the barnacle giant muscle fiber that 3′,5′-cyclic AMP (cAMP) greatly stimulates chloride transport. In the squid axon, cAMP was presented to the axon under three different conditions. In five axons dialyzed free of ATP, 10^{-5} M cAMP was presented without any effect on chloride influx. But since cAMP-mediated systems require ATP for phosphorylation by the protein kinase (see below), it is perhaps not too surprising that cAMP alone would be without effect. Therefore, in three axons 10^{-5} M cAMP was added to dialysis fluid containing 4 mM ATP. The results were the same: no effect on chloride influx. In this case, one might argue that 4 mM ATP saturated the influx mechanism so that no further effect would be elicited by addition of cAMP. The third method was based on the premise that the classic cAMP-mediated response requires ATP at two points (FIGURE 4). At the first site, ATP is cleaved between the alpha and beta phosphorus groups, thus α,β-methylene ATP would not be acceptable at this site. However, site two causes cleavage between the β and the γ phosphorus, in which case β,γ-methylene ATP would not suffice. Thus, if the scheme of FIGURE 4 adequately describes our Cl^- uptake system we would expect that the combination of cAMP and α,β ATP would be able to support an extra Cl^- influx relative to ATP-free conditions. This hypothesis was tested in the experiment illustrated in FIGURE 5, as well as in two other axons. As may be seen, no sign of stimulation of Cl^- influx could be detected in the combined presence of 4 mM α, β-methylene ATP plus 10^{-5} M cAMP. Dialysis with 4 mM ATP raised chloride influx to a new steady level within 40 minutes. The agent bumetanide (1 μM) inhibited virtually all the ATP-stimulated influx. In view of the lack of stimulatory effect of the analogs in the presence or absence of cAMP and the sodium requirement for chloride influx (see below), it seems possible that the role of ATP does not require that it be hydrolyzed.

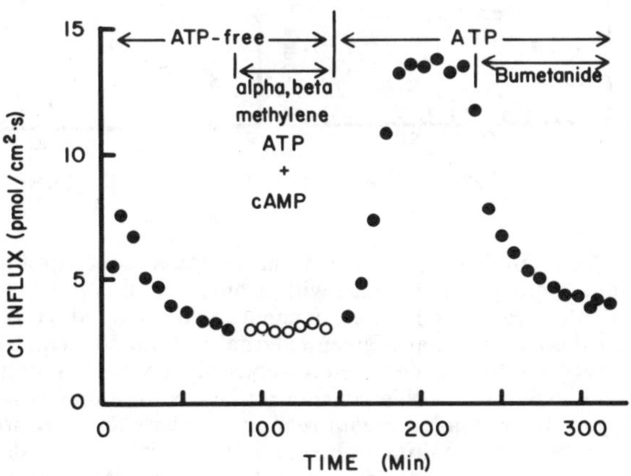

FIGURE 5. Lack of effect of 10^{-5} M cAMP plus 4 mM α,β-methylene ATP on chloride influx into a dialyzed squid giant axon. Dialysis with 4 mM ATP caused by a rapid stimulation of chloride influx that was inhibited by 10^{-6} M bumetanide. Dialysis with ATP-free fluid was begun 30 min before isotope was added to the external fluid.

Relation Between Sodium and Chloride Influxes

It has been known for some time that a portion of the sodium influx into the squid giant axon required ATP but did not seem to be the result of Na/Na exchange.[6, 8] In view of the fact that in epithelial tissue, the fluxes of Na and Cl are often linked, I was interested in examining the possibility of a similar relationship in the squid giant axon.[28] As shown in FIGURE 6A, replacement of external Na with Mg (qualitatively similar results were obtained with Li^+ and choline as replacements) caused chloride influx to decrease. Similarly, when external Cl was replaced by methanesulfonate, sodium influx declined (FIGURE 6B).

TABLE 1 compares the effects of various treatments on the influxes of sodium and chloride. At 15° C, both have a substantial flux that requires ATP. Both chloride and sodium influx are inversely related to the concentration of cellular chloride. Both influxes are reversibly inhibited by furosemide (0.2 mM), but neither is affected by ouabain. The data in TABLE 1 and FIGURE 6 indicate that the sodium fluxes are always slightly less than the comparable chloride

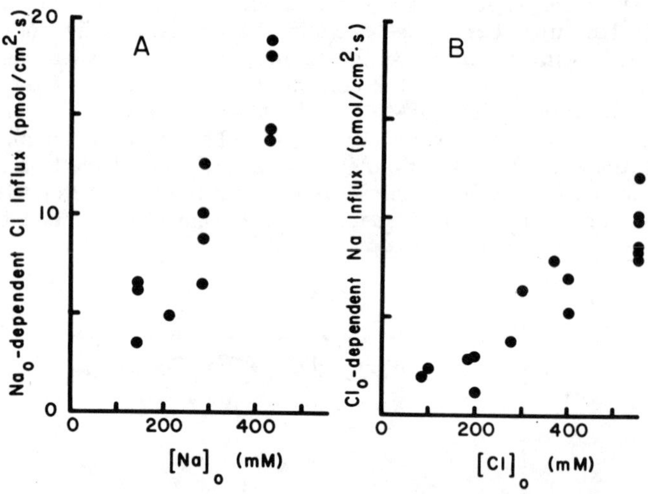

FIGURE 6. **A.** Relation between the external sodium concentration and the Cl^- influx into the squid giant axons dialyzed with a fluid containing 4 mM ATP, 50 mM sodium and nominally free of Cl^-. Na_o-dependent chloride influx is defined as the difference between the Cl^- influx at a given external sodium concentration and the Cl^- influx when external sodium was completely replaced. External sodium was replaced by magnesium. The external chloride concentration was the same as normal seawater, the osmotic difference being made up with mannitol. These data are from 12 different axons. **B.** Relation between sodium influx and the external chloride concentration. Axons were dialyzed with fluid containing 4 mM ATP, 50 mM sodium and nominally free of chloride. Cl_o-dependent sodium influx is defined as the difference between the sodium influx at a given external chloride concentration and the sodium influx when all the external chloride is absent. Methanesulfonate (CH_3-SO_3^-) was used as the substitute anion for chloride. This figure represents data from ten different axons. (From Russell.[28] By permission of *Journal of General Physiology*.)

TABLE 1

A COMPARISON OF CL AND NA INFLUXES INTO THE SQUID GIANT AXON

	ATP Dependent	Cl_i Dependent	Na_o Dependent	Cl_o Dependent	Furosemide Inhibited	Ouabain Inhibited
Cl Influx (pmol/cm²s)	16	14	16.5	—	11	0
Na Influx (pmol/cm²s)	12	11	—	9	8	0 *

* All of the experiments studying Na influx were conducted in the presence of 10^{-5} ouabain as well as 10^{-6} M TTX. Thus, the ouabain-sensitive component is already blocked.

flux. At this time, it is not possible to be sure whether this is meaningful or only reflects the somewhat greater difficulty in doing sodium influx as compared to chloride influx. This difficulty is due to the need to correct for the greater leak of a cation into the negatively charged axon interior, and the fact that small changes in the leak could obscure the sodium influx via the Na-Cl coupled process. Thus, there may be a systematic tendency to underestimate Cl-dependent sodium influx.

Role of Chloride in the Regulation of Intracellular pH

Many neurons have a slightly lower intracellular chloride concentration than expected from passive considerations. Although this fact makes it easier for electrophysiologists to see inhibitory post-synaptic potentials, it is by no means necessary for the function of inhibitory synapses. We, therefore, reasoned that the low cellular chloride levels might reflect the activity of a process that has no direct relation to the electrophysiological processes. In view of the facts that cellular pH (pH_i) is more alkaline than predicted from passive considerations,[2, 32] and that Cl^-/HCO_3^- exchange, albeit passive, has been well-described in the erythrocyte membrane, we sought evidence for an involvement of Cl^- transport in pH_i regulation.

In its simplest form, the model we envisioned involved the exchange of an internal Cl^- for an external HCO_3^-, which would then neutralize a cellular H^+, producing water and CO_2. Thus, if no other systems operated on cellular chloride, one would expect the free chloride levels inside cells to be somewhat less than expected for a passive equilibrium. Any difference between V_m and E_{Cl} would be inversely related to the rate of back-leak of Cl^-.

The squid axon, which was chosen as our experimental subject, has an inwardly directed pump so that the steady-state intracellular chloride concentration does not reflect the activity of our postulated Cl^-/HCO_3^- exchange process. Nevertheless, because of the technical advantages of this preparation, we used it to study relationships between pH_i and Cl^- movements.

Two series of experiments were performed, one to study pH_i and one to study chloride efflux. It had been shown in earlier work [2] that pH_i of the squid axon is normally about 7.3. After it was made acidic, there followed a period during which pH_i returned towards a normal value. This return of pH_i to

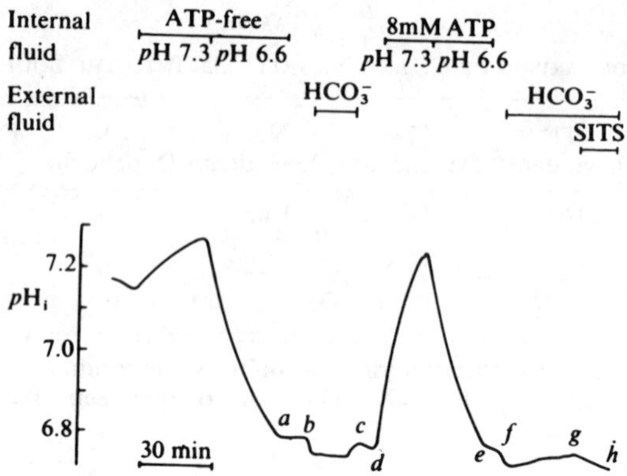

FIGURE 7. Effect of ATP on acid extrusion. The axon was first dialyzed with an ATP-free solution containing 2 mM CN at two different pH values and then with a solution containing 8 mM ATP (150 mM Cl^-) at the same two pH values. pH_i was measured with a glass pH microelectrode. The flow of dialysis fluid was stopped between points a and d and then again between points e and h. Notice that in the absence of ATP, 10 mM HCO_3^-/0.4% CO_2 (pH=8.0) did not cause acid extrusion (segment bc) where after ATP was returned to the axon the HCO_3^-/CO_2 treatment did result in acid extrusion (segment fg) which was blocked by 0.5 mM SITS. (From Russell and Boron.[35] By permission of *Nature*.)

control values (i.e., acid extrusion) could not be predicted from the electrochemical gradients of H^+, OH^- or HCO_3^-. It was also known that HCO_3^- in the external fluid greatly enhanced the rate of acid extrusion. We showed that when the axon was dialyzed free of ATP, acid extrusion could no longer take place, even with HCO_3^- present (FIGURE 7). Thus, an ATP requirement was established for acid extrusion. Next, we showed that even in the presence of ATP and HCO_3^-, if cellular Cl^- was removed by predialysis, no acid extrusion could occur (FIGURE 8). Finally, we showed that acid extrusion could be blocked by 4-acetamido-4-isothicyanostilbene-2,2-disulfonic acid (SITS). Thus, minimal requirements for acid extrusion were: (a) pH_i below pH 7.3; (b) ATP; (c) cellular Cl^-; and (d) external HCO_3^-. In the presence of all these requirements, acid extrusion could be blocked by SITS.

Parallel experiments were conducted on chloride efflux since the model predicts a chloride efflux coupled to acid extrusion, or, more correctly, bicarbonate uptake. In FIGURE 9, *a* shows an experiment in which the pH_i was always 6.8, thus meeting the first criterion for acid extrusion listed above. However, in the absence of ATP, no extra Cl^- efflux is noted when HCO_3^- was applied externally. However, following the return of ATP to the axoplasm, 10 mM HCO_3^- caused almost a doubling of the chloride efflux. This effect was readily reversible when the bicarbonate was removed. Finally, the bottom two panels of FIGURE 9, *b* and *c*, show that the acidotic axons pretreated with SITS were no longer capable of being stimulated by external HCO_3^- even though ATP was present. Thus, the chloride efflux shared the same properties as acid extrusion, thereby suggesting that this extrusion might indeed be by means of a Cl^-/HCO_3^- exchange process.

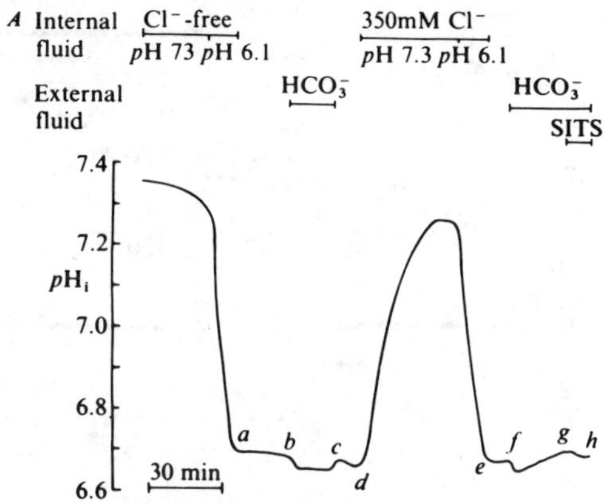

FIGURE 8. Experiment showing the requirement of intracellular chloride for acid extrusion. ATP was present in all dialysis fluids. Experiment was begun by dialyzing with a chloride-free solution, glutamate replacing chloride. When axon was dialyzed to pH 6.7, the dialysis flow was turned off. Exposure to 10 mM HCO_3^-/0.4% CO_2 had no stimulating effect on acid extrusion (segment bc). The axon was then dialyzed with 350 mM chloride and the sequence repeated. Now, in the presence of cellular chloride acid extrusion was stimulated by HCO_3^-/CO_2 applied externally. Again, it was inhibited by 0.5 mM SITS. (From Russell and Boron.[35] By permission of *Nature*.)

Effects of Changing pH_i on Cellular Chloride Levels in Single Neurons

The Cl^-/HCO_3^- exchange model is attractive because of certain testable corollaries. First, it suggests that in cells in which such a mechanism is the dominant one determining cellular chloride content, cellular acidosis ought to promote a net loss of cellular Cl^- (in exchange for HCO_3^-). Second, it suggests that cellular alkalosis might result in the slowing of a neuron's ability to "pump" out a chloride load. Both of these ideas have been tested.

Earlier in this presentation, it was mentioned that Lux[21] and Llinas *et al.*,[20] reported that NH_4^+ treatment blocked an apparent chloride extrusion process from mammalian motoneurons. Both Thomas[32] and Boron and DeWeer[2] have shown that exposure of cells to solutions of NH_4^+ result in a significant alkalinization of pH_i. This is due to the rapid influx of NH_3, which then reacts with cellular protons to yield NH_4^+ and increase the pH_i. Thus, the effect of NH_4^+ treatment on the mammalian motoneurons to raise the intracellular chloride levels may have been secondary to an increased pH_i.

Evidence for such an interpretation comes from some experiments performed on *Aplysia* giant neurons while measuring their a_{Cl}^i with liquid ion-exchange microelectrodes.[27] When these cells were exposed to 50 mM NH_4^+ (pH 7.8) for 30 min., a_{Cl}^i increased by an average of about 3 mM. Thus, under conditions comparable to those of Lux[21] and Llinas *et al.*,[20] direct evidence is obtained for a net gain of chloride. Such a result is compatible with the idea of inhibition of the chloride extrusion process caused by intracellular alkalosis. We believe that the chloride extrusion process is the same as the acid extrusion process.

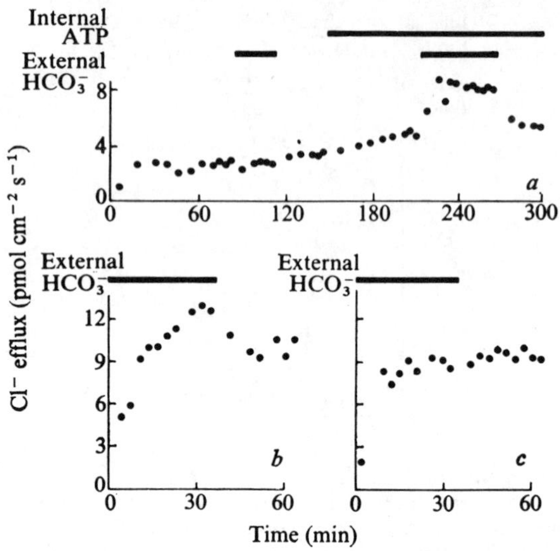

FIGURE 9. Chloride efflux from internally dialyzed squid giant axons measured using ^{36}Cl. *a* The axon was dialyzed with a solution containing 150 mM Cl^- at pH 6.8. At zero time, ^{36}Cl was added to the internal dialysis fluid. In the absence of ATP, 10 mM HCO_3^-/0.4% CO_2 had no effect on Cl^- efflux. However, after dialyzing with 4 mM ATP, the HCO_3^-/CO_2 treatment caused a reversible stimulation of Cl^- efflux. *b c*. Effect of 0.5 mM SITS on HCO_3^-/CO_2-stimulated Cl^- efflux from acidotic axons. In the experiment shown in *b*, Cl^- efflux decreased when HCO_3^-/CO_2 were removed from the bathing media. In *c*, however, the flux in the presence of HCO_3^-/CO_2 was lower and no decrease noted when the HCO_3^-/CO_2 was removed. The fiber in *c* had been pretreated with 0.5 mM SITS prior to the beginning of the flux experiment. (From Russell and Boron.[35] By permission of *Nature*.)

This idea is supported by further experiments on *Aplysia* giant neurons. Thomas[32] and Boron and DeWeer[2] showed that pH_i became relatively acidic during the period following exposure to NH_4^+-containing solutions. The explanation for this is that NH_4^+ has a finite permeability and by its entry it serves as a proton carrier. When the cell is washed free of external NH_4^+, the NH_4^+ in the cell dissociates into NH_3 and H^+, the NH_3 rapidly crosses the neuronal membrane and the result is an intracellular acidosis.[2] In the experiments being discussed here, a 30 min. exposure to 50 mM NH_4^+ (pH 7.8) was followed by an exposure of NH_4^+-free solutions containing 10 mM HCO_3^-/0.4% CO_2 (pH 7.9). In the neuron shown in FIGURE 10, the V_m was voltage clamped to —45 mV to assure that net changes of a_{Cl}^i were not due to the passive electrochemical driving force. It may be seen that a_{Cl}^i fell by about 10 mM during the period when pH_i would bt expected to be acidic, thereby stimulating acid extrusion. The same solution of HCO_3^-/CO_2 was without significant effect if the cells were not first exposed to the NH_4^+ solution. Similar falls of a_{Cl}^i could be obtained by simply exposing neurons to high CO_2 levels, an alternative way of causing cellular acidosis. These results suggest that the lower-than-equilibrium level of a_{Cl}^i may be secondary to the neuron's acid extruding mechanism and that this mechanism causes the net extrusion of a Cl^-. Experiments by

Thomas [33] on snail neurons in which a_{Cl}^i and pH_i were simultaneously measured also support the foregoing concept. He showed that a_{Cl}^i declined when pH_i was made acidic and that reducing a_{Cl}^i to very low levels inhibited the ability of the neuron to readjust its pH_i to normal values.

SUMMARY

Evidence has been presented and reviewed to show that chloride is often not in electrochemical equilibrium across neuronal cell membranes. An ATP- and sodium-dependent uptake mechanism has been described for the squid giant axon. Finally, the role of chloride in the maintenance of pH_i has been discussed. Present evidence favors a net chloride extrusion being involved in the acid extrusion process.

Measurements of free, ionized chloride levels in several neuronal cells have suggested that the transmembrane distribution of chloride does not conform to thermodynamic equilibrium conditions. Net extrusion of Cl^- against its electrochemical gradient has been measured in the giant neuron of the *Aplysia* abdominal ganglion. In the squid giant axon, cellular Cl^- levels are higher than predicted from passive thermodynamic considerations. An ATP requirement as well as a dependence upon extracellular sodium has been demonstrated for

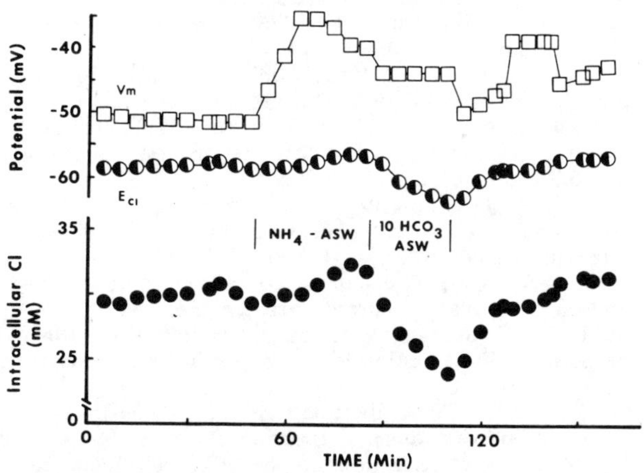

FIGURE 10. Effects of changes in pH_i and external HCO_3^- on intracellular chloride activity in an *Aplysia* giant neuron. E_{Cl} was somewhat more negative than V_m in the control period, a usual finding. When 50 mM NH_4^+ was substituted for Na^+ in the artificial seawater for 30 min., the result was a slight increase of a_{Cl}^i, which averaged 2.3 mM in 23 neurons. During this period, pH_i would be relatively alkaline. The NH_4^+ solution was washed away using a seawater containing 10 mM $HCO_3^-/0.4\%$ CO_2 and a large fall of a_{Cl}^i occurs even though V_m has been voltage clamped to -45 mV. This causes the net loss of Cl^- to be against its electrochemical gradient. It has been shown that pH_i becomes acidic following an exposure to NH_4^+. If the NH_4^+ was washed by a Tris-buffered seawater only a very small change of a_{Cl}^i occurred. Treatment with 0.5 mM SITS completely blocked the net loss of Cl^- seen in this figure. (From Russell.[27] By permission of *Biophysical Journal.*)

chloride influx. Similarly, there is an ATP-, external Cl^--dependent, Na^+ influx. Thus, an ATP-requiring Na-Cl cotransport mechanism appears to account for the high cellular chloride content of the squid giant axon.

Chloride efflux from several neurons appears to be involved in the regulation of intracellular pH (pH_i). When pH_i is made acidic, chloride efflux from the squid giant axon increases and this stimulation requires cellular ATP and external HCO_3^-. When pH_i is measured following an acid load, it is found that pH_i recovery toward normal values requires cellular Cl^-, ATP and external HCO_3^-. Thus, an exchange process between cellular Cl^- and extracellular HCO_3^- appears to play an important role in pH_i regulation. Such a process appears to be responsible for the lower-than-equilibrium levels of chloride found in certain neuronal cells.

References

1. Bear, R. S. & F. O. Schmitt. 1939. Electrolytes in the axoplasm of the giant nerve fibers of the squid. J. Cell. Comp. Physiol. **14:**205–215.
2. Boron, W. F. & P. DeWeer. 1976. Intracellular pH transients in squid giant axons caused by CO_2, NH_3 and metabolic inhibitors. J. Gen. Physiol. **67:**91–112.
3. Boron, W. F., J. M. Russell, M. S. Brodwick, D. W. Keiffer & A. Roos. 1978. Influence of cyclic AMP on intracellular pH regulation and chloride fluxes in barnacle muscle fibers. Nature **276:**511–513.
4. Brinley, F. J. & L. J. Mullins. 1965. Ion fluxes and transference numbers in squid axons. J. Neurochem. **28:**526–544.
5. Brinley, F. J. & L. J. Mullins. 1967. Sodium extrusion by internally dialyzed squid axons. J. Gen. Physiol. **50:**2303–2332.
6. Brown, A. M., J. L. Walker & R. B. Sutton. 1970. Increased chloride conductance as the proximate cause of hydrogen ion concentration effects in *Aplysia* neurons. J. Gen. Physiol. **56:**559–582.
8. Caldwell, P. C., A. L. Hodgkin, R. D. Keynes & T. I. Shaw. 1960. The effects of injecting "energy-rich" phosphate compounds on the active transport of ions in the giant axon of *Loligo*. J. Physiol. **152:**561–590.
9. Chiarandini, D. J. & H. M. Gerschenfeld. 1967. Ionic mechanism of cholinergic inhibition in molluscan neurons. Science **156:**1595.
10. Chiarandini, D. J., E. Stefani & H. M. Gerschenfeld. 1967. Inhibition of membrane permeability to chloride by copper in molluscan neurones. Nature **213:**97.
11. Deffner, G. G. J. 1961. The dialyzable free organic constituents of squid blood, a comparison with nerve axoplasm. Biochim. Biophys. Acta **47:**378–388.
12. DiPolo, R. 1977. Characterization of the ATP-dependent calcium efflux in dialyzed squid giant axons. J. Gen. Physiol. **69:**795–813.
13. Eaton, D. C., J. M. Russell & A. M. Brown. 1975. Ionic permeabilities of an *Aplysia* giant neuron. J. Membr. Biol. **21:**353–374.
14. Eccles, J., R. M. Eccles & M. Ito. 1964. Effects produced on inhibitory postsynaptic potentials by the coupled injection of cations and anions into motoneurons. Proc. R. Soc. B. **160:**197–210.
15. Kehoe, J. 1967. Pharmacological characteristics and ionic bases of a two component post-synaptic inhibition. Nature **225:**820.
16. Koechlin, B. A. 1955. On the chemical composition of the axoplasm of squid giant nerve fibers with particular reference to its ion pattern. J. Biophys. Biochem. Cytol. **1:**511–529.
17. Kerkut, G. A. & R. W. Meech. 1966. The internal chloride concentration of H and D cells in the snail brain. Comp. Biochem. Physiol. **19:**819.

18. KERKUT, G. A. & R. THOMAS. 1964. The effect of anion injection and changes in the external potassium and chloride concentration on the reversal potentials of the IPSP and acetylcholine. Comp. Biochem. Physiol. **11**:199–213.
19. KEYNES, R. D. 1963. Chloride in the squid giant axon. J. Physiol. **169**:690–705.
20. LLINAS, R., R. BAKER & W. PRECHT. 1974. Blockage of inhibition by ammonium acetate action on chloride pump in cat trochlear motoneurons. J. Neurophysiol. **37**:522–532.
21. LUX, H. D. 1971. Ammonium and chloride extrusion: Hyperpolarizing synaptic inhibition in spinal motoneurons. Science **173**:555–557.
22. MAURO, A. 1954. Electrochemical potential difference of chloride ion in the squid axon-seawater systems. Fed. Proc. **13**:96.
23. OZAWA, S. & K. TSUDA. 1973. Membrane permeability changes during inhibitory transmitter action in crayfish stretch receptor cell. J. Neurophysiol. **36**:805–817.
24. PINSKER, H. & E. R. KANDEL. 1969. Synaptic activation of an electrogenic sodium pump. Science **163**:931.
25. PRITCHARD, J. W. 1971. Pentylentetrazol-induced increase in chloride permeability of leech neurons. Brain Res. **27**:414–417.
26. RUSSELL, J. M. 1976. ATP-dependent chloride influx into internally dialyzed squid giant axons. J. Memb. Biol. **28**:335–349.
27. RUSSELL, J. M. 1978. Effects of ammonium and bicarbonate-CO_2 on intracellular chloride levels in *Aplysia* neurons. Biophys. J. **22**:131–137.
28. RUSSELL, J. M. 1979. Cl and Na influx: A coupled uptake mechanism in the squid giant axon. J. Gen. Physiol. In press.
29. RUSSELL, J. M. & A. M. BROWN. 1972. Active transport of chloride by the giant neuron of the *Aplysia* abdominal ganglion. J. Gen. Physiol. **60**:499–518.
30. RUSSELL, J. M., D. C. EATON & M. S. BRODWICK. 1977. Effects of nystatin on membrane conductance and internal ion activities in *Aplysia* neurons. J. Membrane Biol. **37**:137–156.
31. STRUMWASSER, F. 1965. The demonstration and manipulation of a circadian rhythm in a single neuron. *In* Circadian Clocks. J. Aschoff, Ed. :442. North Holland Publishing. Amsterdam.
32. THOMAS, R. C. 1974. Intracellular pH of snail neurones measured with a new pH-sensitive glass micro-electrode. J. Physiol. **238**:159–189.
33. THOMAS, R. C. 1977. The role of bicarbonate, chloride and sodium ions in the regulation of intracellular pH in snail neurones. J. Physiol. **273**:317–338.
34. WEBB, D. A. & J. Z. YOUNG. 1940. Electrolyte contact and action potential of the giant nerve fibre of *Loligo forbesi*. J. Physiol. **98**:299–313.
35. RUSSELL, J. M. & W. F. BORON. 1976. Role of chloride transport in regulation of intracellular pH regulation. Nature **264**:73–74.

CHLORIDE DISTRIBUTION AND EXCHANGE IN VERTEBRATE HEART AND SKELETAL MUSCLE *

Ernest Page, Philip I. Polimeni,† and Donald D. Macchia ‡

Departments of Medicine and of Pharmacological and Physiological Sciences
The University of Chicago
Chicago, Illinois 60637

INTRODUCTION

For the last three years, our laboratory has investigated the distribution and exchange of chloride (Cl) in amphibian and mammalian heart and skeletal muscles. This investigation was stimulated by recent developments in plasma membrane anion transport in red blood cells,[1-3] as well as by significant gaps in the information available about Cl exchange in striated muscle.

For heart muscle, the rate of exchange of cellular Cl in the physiological steady state has not previously been reliably estimated. Although extensive and sophisticated studies of Cl conductance and exchange have been made on skeletal muscle,[4-19] the experimental conditions under which these measurements were made have often differed from the physiological in significant respects. The degree to which the results are applicable to the physiological steady state is therefore uncertain.

In this paper we review experiments designed to measure cellular Cl exchange in heart and skeletal muscle both *in vivo* and *in vitro*. In the course of this work, we have also determined the *in vivo* distribution of Cl between cells and interstitium and reexamined the question of whether Cl is passively distributed.

Distribution of Cl in Muscle

Our experiments have been directed at several questions about Cl distribution in muscle: Since most of the Cl in muscles is outside the cells, how can cellular Cl content and concentration be precisely estimated? Is extracellular Cl distributed uniformly in all muscles or can it be anomalously or inhomogeneously distributed? Is part of cellular Cl sequestered in intracellular membrane-limited compartments? Are the magnitudes of the cellular (or better, cytoplasmic) Cl concentration and activity consistent with a passive distribution of Cl across plasma membrane?

* This work was supported by Grants Nos. HL 10503 and HL 20592, awarded by The National Heart, Lung, and Blood Institute, and the U.S. Public Health Service.

† Current address: Department of Physiology, University of Manitoba, Winnipeg, Manitoba Canada.

‡ Current address: Department of Physiology, Indiana University School of Medicine, Northwest Center for Medical Education, Gary, Indiana 46408.

0077–8923/80/0341–0524 $01.75/0 © 1980, NYAS

Estimation of Extracellular Volume in Muscle

Separation of extracellular from cellular Cl content has long been recognized as a critical problem limiting the precision of estimates of cellular Cl content with chemical or isotopic methods.[20–25] The problem has three components: first, because most of the muscle Cl is extracellular, cellular Cl must be derived from the small difference between two nearly equal quantities (total and extracellular Cl contents), a situation that is statistically unfavorable; second, extracellular volume is unstable and tends to expand progressively in perfused or incubated muscles; and third, molecules used for estimation of extracellular volume may either be imperfectly excluded from muscle cells or equilibrate incompletely with extracellular volume.

TABLE 1

COMPARISON OF SUCROSE SPACE AND MORPHOMETRICALLY DETERMINED EXTRACELLULAR (EC) SPACE IN SKELETAL MUSCLE *

Animal	Muscle	Sucrose Space g EC water / g wet weight	Morphometric Space EC volume / muscle volume
nephrectomized rat	gastrocnemius	0.150±0.007 (10)	0.156±0.007 (7)
toad	semitendinosus	0.192±0.007 (44)	0.203±0.010 (6)

* Sucrose spaces and morphometric data are from a separate series of animals. Figures in parentheses give number of animals and means and S.E.M. of morphometric data averaged over the number of animals. The number of prints processed, number of points counted, and total cross-sectional area analyzed were, respectively, for rat gastrocnemius: 30, 31, 276 and 556×10^4 μm^2; for toad semitendinosus: 5–20, 70, 560, and 125×10^5 μm^2. Muscles fixed in Na cacodylate buffer (pH 7.3), approximately isosmolal with rat or toad plasma, to which glutaraldehyde (final concentration 2%) was added. Muscles were embedded in paraffin and stained with hematoxylin and eosin.

These difficulties have been largely overcome by two innovations in muscle physiology: the application of extracellular tracer measurements to muscle *in vivo*,[25–28] and the use of morphometric techniques[29] to estimate extracellular volume.[25] Although the interstitial space measured with [^{35}S]-SO_4^{-2} progressively increases in incubated or perfused muscles,[23, 24] the [^{35}S]-SO_4^{-2} space measured *in vivo* is constant for rat left ventricular heart muscle,[24] rat gastrocnemius muscle,[27] toad ventricle, and toad and frog skeletal muscle.[26, 27] Morphometry, a second, potentially powerful technique for estimating extracellular volume, is illustrated by the data in TABLE 1 for gastrocnemius muscles of nephrectomized rats and semitendinosus muscles of intact toads. In these muscles, the *in vivo* [^{3}H]-sucrose space agreed well with the volume determined by morphometry performed on fixed and stained sections (obtained after fixation of muscles in 2% glutaraldehyde, added to a Na cacodylate buffer (pH 7.4) whose respective osmolalities were approximately those of rat and toad plasma). Investigators who use morphometry to estimate extracellular volume must bear in mind the possibility that the methods involved in preparing tissue sections may alter cellular and extracellular volumes;[30] however, the data in TABLE 1, as well as those of Polimeni[25] for rat ventricle, suggest that with proper precautions such volume changes do not seriously perturb the *in vivo* volumes. The morphometric method is particularly useful in pathophysiological states when muscles can become permeable to extracellular tracers.

Anomalous Interstitial Anion Distribution in Muscle

In a study designed to compare the interstitial distribution of uncharged extracellular tracers with that of negatively charged extracellular tracers, we made the unexpected observation that divalent anions may be accumulated by or excluded from the interstitial space.[27] The procedure was to measure extracellular volume *in vivo* simultaneously with [^{3}H]-sucrose, an uncharged extracellular tracer, and [^{35}S]-SO_4^{-2}, a tracer carrying a charge of -2. In the rat's left ventricle, the sucrose and sulfate spaces were identical and equal to the volume estimated by morphometry. In rat gastrocnemius muscle, the sucrose space and morphometric space were equal and much larger ($p < 0.0001$) than the [^{35}S]-SO_4^{-2} space. Since the [^{3}H]-sucrose space and the extracellular space identified by morphometry were coextensive, penetration of sucrose into the cells could be ruled out. These observations were interpreted as indicating that the divalent anion was excluded from part of the extracellular volume accessible to sucrose. By contrast, the SO_4^{-2} space in toad skeletal muscle was slightly greater than the sucrose space, but neither space differed significantly from the morphometrically derived extracellular volume.

Cellular Cl Content, Concentration and Activity

Within the limits of the precision of each method, the Cl content of muscles can be accurately estimated by measurement of radioactivity after exchanging all of muscle Cl with radioactive Cl or Br ions,[24, 26, 31, 32] by neutron activation analysis,[32] by coulometry,[26, 32–34] and no doubt by other methods. It is noteworthy that the commonly used coulometric method yields spuriously high values unless precautions are taken to prevent reduction of the electrode by constituents of muscle other than Cl.[26, 32–34] Cellular Cl activity, a_{Cl}^{cell} or, more exactly, cytoplasmic Cl activity, is estimated with Cl ion-selective microelectrodes introduced into cells by impalement;[35] the uncertainties in this method are such that overestimates of the order of 1 mM in a_{Cl}^{cell} due to the presence of HCO_3 ion or other, as yet undefined, cellular constituents cannot be rigorously excluded.

TABLE 2 contains recent data from our laboratory on cellular Cl content, concentration and activity of *in vivo* heart and skeletal muscle, respectively. TABLE 3 is a compendium of selected values from the literature, mostly based on tissues incubated or perfused *in vitro*. Comparison of the data for *in vivo* rat ventricle in TABLE 2 with the *in vitro* measurements in TABLE 3 suggests that the *in vivo* values for $[Cl]_{cell}$ are substantially smaller than those measured *in vitro* on either quiescent[31] or contracting[24] heart muscle. The difference may reflect the more exact measurement of extracellular volume *in vivo*, a lower cell Cl concentration *in vivo*, or both of these factors. For skeletal muscles, the *in vivo* values of a_{Cl}^{cell} (TABLE 2), like those obtained *in vitro*

(TABLE 3), are systematically somewhat larger than the corresponding figure for $[Cl]_{cell}$. Since the activity might be expected to be smaller than the concentration, this discrepancy raises the possibility that the "Cl-selective" electrode may be sensing a small signal from cytoplasmic constituents other than Cl. This possibility is supported by the finding that $[Cl]_{cell}$, unlike a_{Cl}^{cell}, agrees well with the value predicted from the membrane potential by assuming Cl is passively distributed (see below).

TABLE 2

CHLORIDE CONCENTRATION OR ACTIVITY OF MUSCLE CELLS AND PLASMA *In Vivo*

Animal	Muscle	Type of Measurement	Concentration or Activity: Cell (μmol/g cell water)	Concentration or Activity: Plasma (μmol/cell plasma)
Rat ($n=43$)	left ventricle	concentration †	8.2 ±0.5	102.6 ±.5
Toad ($n=16$)	ventricle	concentration	2.3 ±.8	83 ±2
Frog ($n=14$)	ventricle	concentration *	4.8 2.4	80 ±2
Toad ($n=16$)	semitendinosus	concentration *	1.4 ±.3	78 ±3
		concentration ‡	2.8 ±.8	80 ±3
		activity ($n=5$) ‡	3.3 ±.5	81 ±3
		calculated from ($n=5$) $E_m=-88.8\pm2.1$ mV ‡	2.4 ±.2	81 ±3
Frog ($n=28$)	semitendinosus	concentration *	1.7 ±.7	80 ±2

* From Macchia *et al.*[26]
† From Polimeni and Page.[32]
‡ From Macchia and Baumgarten.[51]

Evidence Against a High Cl Content and Concentration in the Sarcoplasmic Reticulum (SR)

The determination of *in vivo* cellular Cl contents given in TABLE 2 was sufficiently precise to serve as evidence against the hypothesis that the Cl concentration in the SR, or in a subcompartment of the SR (the terminal cisterns), is of the same order of magnitude as the interstitial, or plasma Cl, concentration. The implications of this hypothesis and of the associated concept that all or part of SR volume is in continuity with the interstitial space[36-39] are at variance with current ideas about excitation-contraction coupling in skeletal and heart muscle.[40-41] The volumes of the SR and terminal cisterns have been estimated by morphometric techniques for heart muscle[42, 43] and amphibian skeletal

TABLE 3

A SELECTION FROM PUBLISHED VALUES OF CELLULAR CHLORIDE CONCENTRATION OR ACTIVITY FOR VERTEBRATE STRIATED MUSCLE

Animal	Muscle Type	Experimental Conditions	Analytical Method	μmoles Cl / g or ml cell water	Source
Cat (n=31)	ventricle	incubation at 26.5–28° $[K]_o$=5.3 mM, quiescent	^{36}Cl distribution, mannitol space	17 (concentration)	Page [31]
Cat (n=21)	ventricle	same	calculated from resting potential by Nernst equation	7.4 (concentration)	Page [21]
Rat * (n=34)	ventricle	perfusion at 22–24° $[K]_o$=5.93 mM 80–120 contractions/min.	coulometry, SO_4-space	23.8 (concentration)	recalculated from Page & Page [24]
Toad (n=23)	ventricle	*in vivo*	^{32}Br distribution, SO_4-space	5.1 (concentration)	Danielson [23]
Frog (n=72)	ventricle	?	inulin-space	4.1 (concentration)	Verdonck *et al.*[72]
Frog (n=69)	ventricle	superfusion at 20–22° C $[K]_o$=2.5 mM	calculated from measurements of K activity	5.0 (activity)	Walker & Ladle [69]
Frog (n=70)	ventricle	superfusion at 22–24° C $[K]_o$=2.5 mM	intracellular Cl-selective microelectrode	17.6 (activity)	Ladle & Walker [70]
Frog (n=71)	ventricle	incubated at 22–24° C $[K]_o$=2.5 mM	calculated from resting potential by Nernst equation	15.8 (concentration)	calculated by us from Brady & Woodbury [71]
Frog (n=5)	semitendinosus	superfused at 20° C $[K]_o$=2.5 mM	calculated from resting potential by Nernst equation	3.96 (concentration)	Hodgkin and Horowicz [5]
Frog (n=7)	sartorius	incubated at 15–20° C $[K]_o$=2.5 mM	calculated from kinetics of 36-Cl washout	3.1–3.8 (concentration)	Adrian [7]
Toad (n=23)	sartorius	*in vivo*	^{35}Br distribution, SO_4-space	15 (concentration)	Danielson [23]
Frog (n=48)	sartorius	fibers incubated at 25° C $[K]_o$=2.5 mM	intracellular Cl-selective microelectrodes	7.0 (activity)	Armstrong *et al.*[48]
Frog (n=49)	sartorius	fibers incubated at 25° C	intracellular Cl-selective microelectrodes	2.8 (activity)	Bolton & Vaughan-Jones [49]

* Recalculated using correction for nonchloride reduction of electrode as described by Polimeni and Page.[32]

muscle.[44] These volumes may be multiplied by the plasma Cl concentrations to obtain the amount of Cl that the SR and terminal cisterns would contain if the Cl concentration within them were equal to the interstitial or plasma concentrations. By comparing the Cl content of the SR thus calculated with our measured cellular Cl contents, we were able to show that the Cl concentration in the SR and terminal cisterns must be much lower than that of plasma, both for amphibian skeletal muscle [26] and for rat ventricular muscle.[73] This conclusion is consistent with the electron microprobe analysis of terminal cisternal Cl content in normal frog skeletal muscle by the laboratory of Somlyo.[45] Our data and conclusions about the Cl content of the SR in amphibian skeletal muscle [26] have recently been confirmed by Neville.[19]

Is Cl Passively Distributed in Skeletal Muscle in vivo?

The classical studies of Boyle and Conway,[46] and Hodgkin and Horowicz,[5] subsequently confirmed by measurements of a_{Cl}^{cell},[47] showed that, at higher than physiological external K concentrations, Cl ion is passively distributed in *in vitro* preparations of amphibian skeletal muscle. Until recently, *in vivo* measurements of membrane potential, a_{Cl}^{cell}, and $[Cl]_{cell}$ were not available for this material, whereas the measurements made *in vitro* at approximately physiological external K concentration suggested an active accumulation of Cl by muscle cells.[48-50] Macchia and Baumgarten [51] have made *in vivo* insertions of voltage-measuring and Cl-selective microelectrodes into semitendinosus muscles of pithed toads, in which $[Cl]_{cell}$ was also measured at the end of the experiment. Under these conditions the membrane potential, E_m, and the chloride equilibrium potential, E_{Cl}, calculated by the Nernst equation from $[Cl]_{cell}$ and the external Cl concentration, were identical within the dispersion of the method. According to Macchia and Baumgarten,[51] a_{Cl}^{cell} (3.3 ± 0.5 μmol/g cell water) did not significantly exceed $[Cl]_{cell}$ (2.8 ± 0.8 μmol/g cell water). The results of this experiment therefore suggested that Cl is passively distributed across the plasma membrane of the semitendinosus muscle *in vivo;* because of the uncertainty in the interpretation of a_{Cl}^{cell}, a small active accumulation of Cl could not, however, be rigorously ruled out. For skeletal muscle *in vivo,* the settlement of the remaining uncertainty about an unequivocally passive Cl distribution thus requires improvement in the selectivity and interpretation of the Cl activity measurement.

While the question of whether Cl is passively distributed is not unequivocally answerable for the very low cellular Cl concentrations and activities prevailing *in vivo,* nonequilibrium distributions of Cl can be clearly demonstrated under experimental conditions in which skeletal muscle Cl content, $[Cl]_{cell}$ and a_{Cl}^{cell} increase.[52]

Transplasmalemmal Chloride Exchange

Cl Exchange in Mammalian Heart Muscle

Conductance measurements on mammalian ventricular heart muscle [53, 54] and on cardiac Purkinje fibers [55] agree in showing that the plasmalemma is much more permeable to K than to Cl, the permeability ratio (P_{Cl}/P_K) being

0.11 for rabbit papillary muscle.[54] The simplest prediction would therefore be that cellular Cl exchange should be much slower than K exchange, but this prediction turns out to be incorrect. We have recently succeeded in measuring the unidirectional flux of ^{36}Cl out of myocardial cells in isolated, contracting rat ventricles perfused on a Langendorff cannula, under conditions approximating a steady state with respect to cellular Cl content and concentration.[32] Cellular Cl was found to exchange at a single rate. At 38° C, this rate was 31.2 μmoles Cl/(g dry ventricle·min), or 42 pmoles/(cm^2 plasmalemma·sec), assuming Cl exchange per unit plasmalemmal area is the same across the external plasmalemmal envelope and T-tubular plasmalemma. The apparent activation energy of the exchange between 22° and 38° was 10.4 kcal/mole. To our knowledge, the Cl exchange flux at 38° is the most rapid ion flux so far described across cardiac plasma membrane. Moreover, *in vivo* ^{36}Cl influx measurements in rats injected intravenously with Na^{36}Cl show cellular Cl exchange approached completion within five minutes or less,[73] suggesting that the *in vivo* unidirectional influx is of the same order of magnitude as the unidirectional efflux measured in the isolated perfused heart. At 30° C, the value of 20.8 mmoles/(kg dry weight·min) for cellular Cl exchange was significantly larger than the exchange rate of cellular K (5.4–11.8 mmoles/(kg dry weight·min) reported at that temperature for rat ventricles contracting at comparable rates.[56] At 22° C, the rate of Cl exchange was independent of the heart rate over a frequency range of from 80 to 160 contractions per minute. It is, therefore, improbable that the rapid cellular Cl exchange at 22° can be attributed to a Cl current related to the cardiac action potential.[55, 57–64]

The slow rate of net Cl transport as determined by conductance measurements is at variance with the rapid rate of steady state Cl exchange measured with ^{36}Cl. This kind of discrepancy raises the possibility that there may be more than a single transport mechanism for Cl in the cardiac plasma membrane. All mechanisms that are operating to transfer Cl out of the cell would be expected to contribute to the unidirectional Cl efflux measured by Polimeni and Page,[32] but they may not all contribute to permeation of Cl through the membrane channels implicated in conductance measurements.

Whether Cl ion exchanges via a carrier mechanism across the cardiac plasmalemma, whether net movements of Cl ion across the plasmalemma are associated with the cardiac action potential, and whether the distribution and surface density of Cl transport mechanisms in the external plasmalemmal envelope and T-system are similar or differ are specific questions that need to be investigated in more detail. Until these questions are answered, it is premature to speculate about the properties of anion-selective channels or of anion carriers in ventricular muscle.

Cl Fluxes in Amphibian Skeletal Muscle

Our laboratory has introduced a new preparation, the *in vivo* saline-perfused hindlimb of the pithed toad, *Bufo marinus,* in which we have measured $[Cl]_{cell}$, a_{Cl}^{cell}, membrane potential, and exchange of cellular with extracellular Cl.[51, 52, 65] The preparation is also suitable for measuring ionic conductances and net ion fluxes across the plasma membrane of the hindlimb muscles. We initially made the unexpected observation that the rate of efflux of cellular Cl for skeletal muscles perfused with toad Ringer's solution *in vivo* was only about 20% of that for the isolated muscles *in vitro.* We, therefore, had to consider the possibility that the discrepancy reflected a significant difference between Cl transport in the perfused hindlimb and the usual *in vitro* preparation.[4, 5, 7, 13, 53, 66, 67]

In our flux measurements, the hindlimb muscles were loaded *in vivo* by injecting Ringer's solution containing ^{36}Cl intravenously and waiting until the specific activity of muscle Cl was equal to that of plasma Cl.[26] Experiments were done with the hindlimb maximally vasodilated by nitroglycerine. Under these conditions, the rate-limiting process for cellular Cl exchange was translocation of the anion across the plasmalemma, a process which could be shown to be independent of perfusion rate. The mean ± S.E.M. of the unidirectional ^{36}Cl efflux from perfused toad skeletal muscle cells was 9.4 ± 0.5 pmoles/cm^2 of plasma membrane per sec ($n = 13$).[52]

Several aspects of this result merit comment: The dispersion of the measurement was significantly smaller than that of the usual *in vitro* measurements in the muscle literature.[7, 67] The *in vivo* cellular Cl exchange in the perfused hindlimb was only about 25% of that measured by others for amphibian skeletal muscle *in vitro;*[7, 12, 14, 67] the *in vivo* value of efflux obtained by us was about one third the value which we obtained *in vitro* for toad semitendinosus muscles isolated from the hindlimb, loaded with ^{36}Cl, and washed out at 22° C in an incubation solution with the same composition as that used to perfuse the hindlimb; and the efflux rate *in vivo* was approximately 20% of the influx rate measured both *in vivo* and *in vitro*. Moreover, the marked decrease in cellular Cl efflux was associated with a net cellular Cl accumulation. The membrane resting potential remained constant at a value of —89.9 ± 1.9 mV during the net Cl influx. Assuming the activity coefficient for intracellular Cl remained unchanged, Cl ion was thus being accumulated against the electrochemical gradient. The cellular Cl uptake was accompanied by a net uptake of Na and water, the amount of Na accumulated being equal to or somewhat greater than that of Cl. The uptake of Na could not be accounted for by an exchange for cellular K, since cellular K content did not fall. The net influx of Na was therefore not the result of inhibition of the Na-K pump. In isolated toad semitendinosus muscles incubated in toad Ringer's solution *in vitro,* cellular Cl efflux increased significantly with the duration of *in vitro* incubation over a range of incubation times of from one to four hours. Moreover, the Cl efflux could be slowed, with no significant effect on influx, by preloading the muscles with ^{36}Cl *in vivo* and carrying out the *in vitro* efflux in toad plasma instead of in Ringer's solution.

A definitive interpretation of these observations must await additional experiments. The results to date raise two questions about Cl movements in the muscles of the perfused hindlimb: First, what is the driving force moving Cl ion into the cells against the electrochemical gradient? Second, by what mechanism is the marked depression of Cl-36 outflux accomplished? We have considered the possibilities of a primary active transport of Cl with a secondary inward Na movement; of a transport of Na and Cl into the cell via a carrier capable of transporting both Na and Cl; and of a primary movement of Na down its electrochemical gradient with a secondary (passive) cellular uptake of Cl through a Cl "Channel."[52] The depression of Cl-36 efflux might be ascribed to an inhibition or change in affinity for Cl binding sites on a Cl carrier at its intracellular face or to a "single file diffusion" effect on Cl efflux through a Cl channel during a rapid net Cl influx. A single-file diffusion effect is, however, inconsistent with the absence of a membrane potential change and the constancy of E_m during the net uptake of Na and Cl by the cells.

REFERENCES

1. GUNN, R. B., M. DALMARK, D. C. TOSTESON & J. O. WIETH. 1973. J. Gen. Physiol. **61:**185–206.
2. ROTHSTEIN, A., Z. I. CABANTCHIK & P. KNAUF. 1976. Fed. Proc. **35:**3–10.
3. GUNN, R. B. 1979. *In* Transport Across Biological Membranes. G. Giebisch, D. C. Tosteson & H. H. Ussing, Eds. **II:**59–80. Springer-Verlag. New York, New York.
4. LEVI, H. & H. H. USSING. 1948. Acta Physiol. Scand. **16:**232–249.
5. HODGKIN, A. L. & P. HOROWICZ. 1959. J. Physiol. London. **148:**127–160.
6. ADRIAN, R. H. 1960. J. Physiol. London. **151:**154–185.
7. ADRIAN, R. H. 1961. J. Physiol. London. **156:**623–632.
8. ADRIAN, R. H. & W. H. FREYGANG. 1962. J. Physiol. London. **163:**61–103.
9. ADRIAN, R. H. & W. H. FREYGANG. 1962. J. Physiol. London. **163:**104–114.
10. HUTTER, O. F. & A. E. WARNER. 1967. J. Physiol. London. **189:**403–425.
11. HUTTER, O. F. & A. E. WARNER. 1967. J. Physiol. London. **189:**427–443.
12. HUTTER, O. F. & A. E. WARNER. 1967. J. Physiol. London. **189:**445–460.
13. HUTTER, O. F. & A. E. WARNER. 1976. J. Physiol. London. **227:**275–290.
14. MOORE, L. E. 1969. J. Gen. Physiol. **54:**33–52.
15. WOODBURY, J. W. & P. R. MILES. 1973. J. Gen. Physiol. **62:**324–353.
16. VENOSA, R. A., A. C. RUARTE & P. HOROWICZ. 1972. J. Membrane Biol. **9:**37–46.
17. WARNER, A. E. 1972. J. Physiol. London. **227:**291–312.
18. NEVILLE, M. C. 1979. J. Physiol. London. **288:**45–70.
19. NEVILLE, M. C. 1979. J. Physiol. London. **288:**71–84.
20. TASKER, P., S. E. SIMON, B. M. JOHNSTON, K. H. SHANKLEY & F. H. SHAW. 1959. J. Gen. Physiol. **43:**39–53.
21. PAGE, E. 1962. J. Gen. Physiol. **46:**189–200.
22. PAGE, E. 1964. Ann. N.Y. Acad. Sci. **127:**34–48.
23. DANIELSON, B. G. 1964. Acta Physiol. Scand. Suppl. **236:**1–115.
24. PAGE, E. & E. G. PAGE. 1968. Circ. Res. **22:**435–446.
25. POLIMENI, P. I. 1974. Am. J. Physiol. **227:**676–683.
26. MACCHIA, D. D., P. I. POLIMENI & E. PAGE. 1978. Am. J. Physiol. **235:**C122–C127.
27. MACCHIA, D. D., E. PAGE & P. I. POLIMENI. 1979. Am. J. Physiol. **237:**C125–C130.
28. POLIMENI, P. I. & E. PAGE. 1978. Fed. Proc. **37:**300.
29. PAGE, E. 1978. Am. J. Physiol. **4:**C147–C158.
30. EISENBERG, B. R. & B. A. MOBLEY. 1975. Tissue Cell. **7:**383–387.
31. PAGE, E. 1962. J. Gen. Physiol. **46:**201–213.
32. POLIMENI, P. I. & E. PAGE. 1980. Am. J. Physiol. In press.
33. COTLOVE, E. 1963. Anal. Chem. **35:**95–100.
34. COTLOVE, E. 1963. Anal. Chem. **35:**101–105.
35. WALKER, J. L. 1971. Anal. Chem. **43:**89A–93A.
36. VINOGRADOVA, N. A. 1976. Tsitol. **9:**781–790.
37. BIRKS, R. I. & D. F. DAVEY. 1969. J. Physiol. London. **202:**171–188.
38. SPERELAKIS, N., K. SHIGENOBU & R. RUBIO. 1978. Am. J. Physiol. **234:**C181–C190.
39. ROGUS, E. & K. L. ZIERLER. 1973. J. Physiol. London. **233:**227–270.
40. ENDO, M. 1977. Physiol. Rev. **57:**71–108.
41. FABIATO, A. & A. FABIATO. 1977. Circ. Res. **40:**119–129.
42. PAGE, E., J. EARLEY & B. POWER. 1974. Circ. Res. Suppl. 2. **35:**12–16.
43. PAGE, E. & SURDYK-DROSKE, M. 1979. Circ. Res. **45:**260–266.
44. MOBLEY, B. A. & B. R. EISENBERG. 1975. J. Gen. Physiol. **66:**31–46.
45. GONZALEZ-SERRATOS, H., A. SOMLYO, G. MCCLELLAN, H. SHUMAN, L. M. BORRERO & A. P. SOMLYO. 1978. Proc. Nat'l. Acad. Sci. USA. **75:**1329–1333.

46. Boyle, P. J. & E. J. Conway. 1941. J. Physiol. London. **100:**1–63.
47. Baumgarten, C. M. & H. A. Fozzard. 1978. Biophys. J. **21:**184a.
48. Armstrong, W. M., W. Woztkowski & W. R. Bixen. 1977. Biochim. Biophys. Acta. **465:**165–170.
49. Bolton, T. B. & R. D. Vaughan-Jones. 1977. J. Physiol. London. **270:**801–833.
50. Kernan, R. P., M. MacDermott & W. Westphal. 1974. J. Physiol. London. **241:**60P–61P.
51. Macchia, D. D. & C. M. Baumgarten. 1979. Pflügers Arch. **382:**193–195.
52. Macchia, D. D. & E. Page. 1980. Fed. Proc. In press.
53. Hutter, O. F. & D. Noble. 1960. J. Physiol. London. **151:**89–102.
54. Fozzard, H. A. & C. O. Lee. 1976. J. Physiol. London. **256:**663–689.
55. Carmeliet, E. E. 1961. J. Physiol. London. **156:**375–388.
56. Blesa, E. S., G. A. Langer, A. J. Brady & S. D. Serena. 1970. Am. J. Physiol. **219:**747–754.
57. Hutter, O. F. & D. Noble. 1961. J. Physiol. London. **157:**335–350.
58. Dudel, J., K. Peper, R. Rüdel & W. Trautwein. 1967. Pflügers Arch. Gesamte Physiol. **295:**197–212.
59. Reuter, H. 1968. J. Physiol. London. **197:**233–253.
60. Fozzard, H. A. & M. Hiraoka. 1973. J. Physiol. London. **234:**569–586.
61. Hiraoka, M. & M. Hiraoka. 1973. J. Physiol. London. **234:**587–595.
62. McCallister, R. E., D. Noble & R. W. Tsien. 1975. J. Physiol. London. **251:** 1–59.
63. Kenyon, J. L. & W. R. Gibbons. 1979. J. Gen. Physiol. **73:**117–138.
64. Kenyon, J. L. & W. R. Gibbons. 1979. J. Gen. Physiol. **73:**139–158.
65. Macchia, D. D. & E. Page. 1979. Fed. Proc. **38:**1442.
66. Harris, E. J. 1958. J. Physiol. London. **141:**351–365.
67. Harris, E. J. 1963. J. Physiol. London. **166:**87–109.
68. Solomon, A. K. 1960. Compartmental Methods of Kinetic Analysis. *In* Mineral Metabolism, An Advanced Treatise. C. L. Comar & F. Bronner, Eds. **1** (Part A): 119–168. Academic Press. New York, New York.
69. Walker, J. L. & R. O. Ladle. 1973. Am. J. Physiol. **225:**263–267.
70. Ladle, R. O. & J. L. Walker. 1975. J. Physiol. London. **251:**549–559.
71. Brady, A. J. & J. W. Woodbury. 1956. Science. **123:**100–101.
72. Verdonck, F., D. DeClercq & E. Carmeliet. 1965. Arch. Int. Physiol. Biochim. **73:**381–382.
73. Polimeni, P. I. & E. Page. Unpublished observations.

A VOLTAGE-DEPENDENT CHLORIDE CONDUCTANCE CHANNEL FROM *TORPEDO* ELECTROPLAX MEMBRANE *

Christopher Miller and Michael M. White

Graduate Department of Biochemistry
Brandeis University
Waltham, Massachusetts 02254

During the past few years, we have been studying the possibility of incorporating ion-conductance channels from animal cell membranes into artificial bilayers, as a first step towards characterizing these channels in a biochemically defined system. The strategy for this work is to bring about the fusion of native membrane vesicles with a planar phospholipid bilayer, a system upon which electrical conductance may be measured at a controllable applied voltage. Using this approach, we have documented the existence of a voltage-gated K^+-selective channel from rabbit sarcoplasmic reticulum (SR).[1–4] In attempting to extend the technique to other and better-understood membrane systems, we recently showed that "microsac" vesicles of the electric organ of *Torpedo californica,* an electric ray whose electroplax membrane carries a high concentration of acetylcholine receptor, can be made to induce a large increase in conductance of planar bilayers.[5] Quite surprisingly, this microsac-induced conductance had no properties to be expected of the acetylcholine-gated channel, but rather displayed anion selectivity and voltage-dependence, along with clear single-channel fluctuation behavior.

In this paper, we extend the initial results presented on this channel. The major new finding here is that the channel is exquisitely selective to chloride ion and is actually blocked by certain other anions. In addition, we demonstrate the asymmetric inhibition of the channel by those anion-transport workhorses, SITS and DIDS. While we can only speculate upon the physiological role of the system, we consider that the results demonstrate that the Cl^- channel is present in the electroplax of *Torpedo* and that it must be taken into account in any model of the operation of this cell.

Fusion of Microsacs with Planar Bilayers

The experimental setup, shown in FIGURE 1, consists of a phospholipid bilayer separating two aqueous chambers (labelled *cis* and *trans*) containing an appropriate electrolyte (usually 0.1 M KCl). The voltage across the bilayer may be adjusted at will, and the resulting current flow is measured by a simple feedback circuit.[1] Voltage is always defined as zero on the *trans* side of the bilayer, and microsac vesicles are always added to the *cis* side, to final protein concentrations in the range 1–50 μgm/ml. Before addition of vesicles, the bilayers have very low conductance, in the range of $0.5–2 \times 10^{-8}$ mho/cm^2.

* This research was supported by NSF Grant #BNS76–23212 and NIH predoctoral Training Grant #00212–20 (to M.M.W.). C.M. is the recipient of NIH R.C.D.A. No. KO4 AM 00354-01. M.M.W. is the recipient of a Brandeis University Klein Fellowship.

0077–8923/80/0341–0534 $01.75/0 © 1980, NYAS

In FIGURE 2, the responses of bilayers to the addition of microsacs are shown under several different conditions. In general, if Ca^{++} is present on the *cis* side of the membrane, and if the bilayer contains a substantial fraction of acidic phospholipid, *Torpedo* microsacs cause the conductance to increase in a sequence of spike-like jumps. These spikes relax slowly to a nonzero level, and so the average membrane conductance gradually increases, up to four orders of magnitude above its initial level.

We can ask about the factors that control the appearance of these jumps by measuring the number of jumps occurring per unit time under various conditions. With Ca^{++} (0.9 mM) present on the *cis* side of the membrane, the number of jumps in a two-minute interval is reasonably reproducible (within a factor or two), and for the conditions of FIGURE 2A and TABLE 1, falls in the range of 15–20 spikes per two minutes. FIGURE 2A and TABLE 1 (lines 10 and 11) show that addition of Ca^{++} on the *trans* side of the bilayer actually suppresses the appearance of spikes. Ca^{++} on the *cis* side is not an absolute requirement for spikes; however, as is shown in FIGURE 2B and TABLE 1 (lines 8 and 9), occasional spikes are seen even in the presence of EDTA, but the frequency of their occurrence is greatly enhanced by adding *cis* Ca^{++}. We also tested whether osmotic conditions might affect the spike frequency. That this is apparently the case is shown in FIGURES 2C and 2D, and in TABLE 1 (lines 12–15). Addition of 100 mM sucrose to the aqueous medium totally eliminates spike formation, but an equal concentration of glycerol does not. The rationale for this effect is that an osmotic gradient across the vesicle membrane may be a requirement for interaction with the bilayer, such that the inside of the vesicle has higher osmotic strength than the outside. Addition of sucrose to the outside

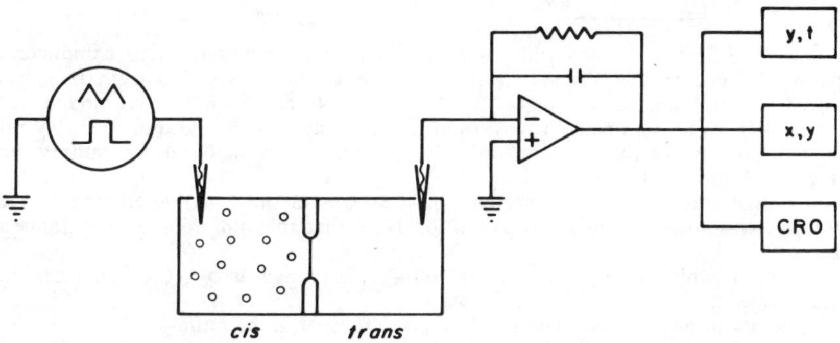

FIGURE 1. A schematic diagram of the planar bilayer assay system which is of the Rudin-Mueller type.[18] Two aqueous chambers, labelled *cis* and *trans,* are separated by a partition containing a 0.8-mm diameter hole. Phospholipid membranes are formed on this hole (70% bovine heart PE—30% egg yolk PG), in the presence of a buffer solution containing 0.1 M KCl, 10 mM HEPES, 0.1 mM EDTA, pH 7.35 or pH 7.0, with addition of Tris base. Planar bilayer conductance was measured as described,[1] and was always less than 2×10^{-8} mho/cm^2. The *trans* chamber is always connected to the amplifier virtual ground, and this chamber is defined as zero-voltage. Voltage command is supplied by a function generator or by a variable battery. *Torpedo* microsac vesicles, prepared as described,[5] were always added to the *cis* chamber.

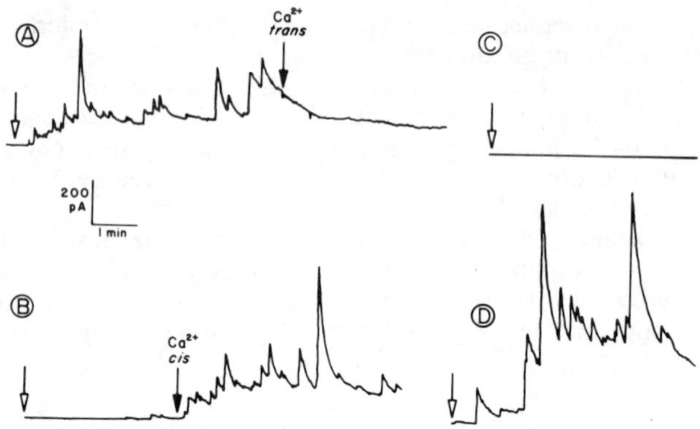

FIGURE 2. Effect of *Torpedo* microsacs on planar bilayer conductance. Conductance responses of planar membranes to the addition of microsacs (55 μgm/ml) were followed at a holding voltage of +30 mV. For all traces, vesicles were added, with constant stirring, at the open arrows. In all traces, unless otherwise indicated, 0.9 mM Ca^{++} was present on the *cis* side only. Recordings were made using buffers with different additions, as follows:

A. Control trace, with Ca^{++} present on the *cis* side only. At filled arrow, Ca^{++} (0.9 mM) was added to the *trans* side also. Note that this inhibits the appearance of spikes.

B. Trace in which no Ca^{++} is present initially. At filled arrow, Ca^{++} (0.9 mM) is added to *cis* side.

C. Conditions as in *A*, but with 0.1 M sucrose present in the buffer.

D. Conditions as in *A*, but with 0.1 M glycerol present in the buffer.

solution would eliminate this gradient, but glycerol would not, since glycerol is permeant to phospholipid vesicles.[6]

These results are important because they are reminiscent of the conditions controlling the fusion of phospholipid vesicles with each other[7, 8] and of the incorporation of sarcoplasmic reticulum vesicles into planar bilayers.[9] In the SR system, we have argued that each unitary conductance jump is the result of the fusion of a *single vesicle* with the planar bilayer, in a process controlled by anionic lipid, divalent cation, and osmotic gradient.[1, 9] We consider that the same conclusion applies to the *Torpedo* microsacs: that each conductance spike represents the incorporation of a single vesicle into the membrane. Of course, we realize that this conclusion must remain tentative in the absence of other assays of fusion in this system. Whatever the particular mechanism of interaction, however, the results show that conductance pathways are becoming inserted into the bilayer *via* a process which is reproducibly controlled by a well-defined set of variables.

Voltage-dependence

After the membrane conductance has increased to a steady level (typically in the order of 10^{-6} to 10^{-4} mho/cm^2), we may then study the properties of the microsac-induced conductance pathway. An immediately eyecatching observation is that the steady-state conductance is voltage-dependent. In FIGURE 3, we present the conductance-voltage (g-V) curves at two different pH values, pH 7.0 and 7.35. For both conditions, we find that at positive voltages, the conductance is low, equal to an ohmic background "leak" conductance,[5] which has been subtracted out in the figure. As the voltage is varied in the negative direction, the conductance increases and then reaches a saturation level. At pH 7.0, the curve is half-saturated at —49 mV, about 14 mV more negative than the half-saturation voltage at pH 7.35. The steepness with which the conductance rises is the same for both pH values.

This conductance-voltage behavior may be understood in terms of a simple two-state "conformational" model of the conductance pathway, which has been shown to operate by a channel mechanism.[5] We assume that the channel can exist only in two states, "open" and "closed" (the former being the conducting state); we also assume that there exists an equilibrium between the two states, and that the equilibrium constant is voltage-dependent solely because of charge movement occurring when the channel undergoes a transition between the open and closed states. Given these assumptions, it follows[3, 10] that the channel-mediated steady-state conductance should follow an expression of the form:

$$g_{\mathrm{rel}} = g(V)/g_{\mathrm{max}} = [1 + \exp[(\Delta G^{\circ}_{i} + zFV)/RT]]^{-1}. \quad \mathbf{(1)}$$

Here, g_{max} is the maximum conductance corresponding to the opening of all

TABLE 1

CONDITIONS CONTROLLING APPEARANCE OF CONDUCTANCE SPIKES †

	1–3 min		3–5 min	
No. Membrane	Addition	No. Spikes	Addition	No. Spikes
1	*	12		
2	*	18		
3	*	16		
4	*	23		
5	*	15		
6	*	11	*	7
7	*	22	*	20
8	*	24	EDTA, *cis*	3
9	EDTA, *cis*	2	Ca,$^{++}$ *cis*	16
10	*	11	Ca,$^{++}$ *trans*	0
11	Ca,$^{++}$ *trans*	4	EDTA, *trans*	23
12	Sucrose	0	Sucrose	0
13	Glycerol	16	Glycerol	13
14	*	16	Sucrose, *cis*	1
15	*	13	Glycerol	6

† Microsac vesicles were added under conditions of FIGURE 2, and the number of conductance spikes were counted in two two-minute time intervals: 1–3 minutes and 3–5 minutes after addition of microsacs. The aqueous medium with no additions was composed of: 0.1 M KCl, 10 mM HEPES, 0.1 mM EDTA, pH 7.35, to which 1 mM $CaCl_2$ was added on the *cis* side only. When no additions were made to this medium, an asterisk (*) is indicated. Marked additions of Ca^{++} were always 0.9 mM in excess of any EDTA present, and additions of EDTA were always 0.2 mM in excess of any Ca^{++} present. Sucrose and glycerol were added to concentrations of 0.1 M; unless otherwise noted, additions were *cis* and *trans*.

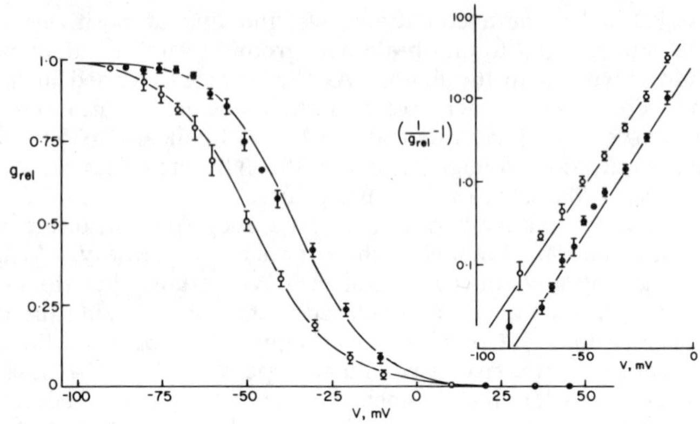

FIGURE 3. Conductance-voltage curves. Planar membranes were modified by microsacs as in FIGURE 2, and conductance was allowed to increase for 10–20 minutes, after which time spikes had spontaneously stopped appearing. Conductances measured at −30 mV were in the range 2–100 μmho/cm,[2] and the relative properties plotted were independent of absolute conductance level. The g-V curves were determined by measuring the steady-state conductance at various applied voltages, each voltage being applied for at least 30 seconds to allow for relaxation to steady-state conditions. Background leak conductance was measured between +30 and +50 mV[5] and was subtracted out. Relative conductance, g_{rel}, is defined with respect to the maximum conductance of the curves measured at highly negative voltages. This maximum conductance was determined by extrapolation as described.[3] Background conductance was typically in the range 4%–10% of the maximum conductance. Solid curves were drawn according to equation 1, with parameters determined by the linearized plot; equation 2 is shown in the inset. Data for each curve were taken from 6–8 different membranes; error bars show standard errors of 3–5 determinations, and points without bars are single determinations. Open points: buffer adjusted to pH 7.0. Parameters determined from linearized plot were: $z=2.2\pm0.1$; $\Delta G_i°=+2.4\pm0.1$ kcal/mol. Closed points: buffer adjusted to pH 7.35. Parameters were: $z=2.3\pm0.1$; $\Delta_{Gi}°_i=+1.9\pm0.1$ kcal/mol.

channels in the membrane, z is the effective charge movement of the opening transition, and $\Delta G°_i$, the "internal free energy of opening," is the chemical part of the free energy difference between open and closed states. The above equation may be linearized so that the parameters z and $\Delta G°_i$ can be determined easily:

$$1n\left[\frac{1}{g_{\mathrm{rel}}}-1\right]=\Delta G°_i/RT+zFV/RT. \qquad \textbf{(2)}$$

Examination of FIGURE 3 shows that the macroscopic g-V curves follow the prediction of this simple model very well. At both pH values, the effective charge movement is 2.3 ± 0.1, indicating a moderately sharp voltage-dependence. The internal free energy of opening is dependent on pH; at pH 7.0, $\Delta G°_i=+2.4$ kcal/mol, while at pH 7.35, $\Delta G°_i=+1.9$ kcal/mol. This means that lowering the pH from 7.35 to 7.0 makes it about 0.5 kcal less favorable for the channel to open, at all voltages.

The voltage-dependence of the steady-state conductance may also be demonstrated via kinetic experiments, an example of which is shown in FIGURE 4.

Here a voltage of +40 mV has been applied for a long time, in order to close all channels. (The nonzero current at this voltage is a reflection of the background "leak" conductance in the system.) The applied voltage is then suddenly jumped to −40 mV, as indicated, and the conductance response is followed in time. Immediately after the voltage-jump, the conductance is equal to the low background level, but a time-dependent increase in conductance begins immediately. After several seconds, the new, higher steady-state conductance is reached. The time course is exponential in all cases we have examined; in this example, the relaxation rate constant is 0.2 sec^{-1}. For the remainder of this report, we confine ourselves to steady-state conditions, i.e., at times long after any voltage change was imposed upon the system.

Single-Channel Behavior

In the experiments described above, many vesicles were allowed to fuse with planar bilayer and increase its conductance by several orders of magnitude. Alternatively, we can perform experiments in which only a single vesicle incorporates into the bilayer, raising the average conductance by a small factor. In this case, when examined at high gain, the conductance fluctuates between discrete levels, indicating the random opening and closing of individual channels;[5] we have previously shown that in each vesicle-fusion event, a "package" of channels (300, for the average-sized spike) is inserted into the membrane,[5] but we have not presented single-channel fluctuation data under steady-state conditions. In FIGURE 5, we now show several traces under such conditions. Channels are seen to open and close randomly between discrete levels, separated by a single-channel conductance of 16 pmho, under these conditions. The

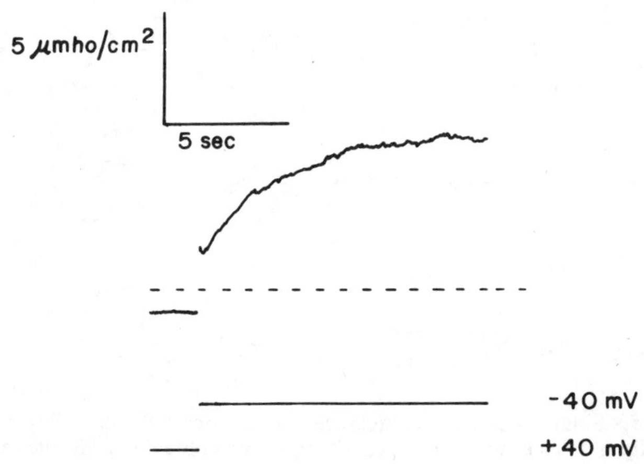

FIGURE 4. Voltage-jump conductance relaxation. Treatment of planar membranes with microsacs was as in FIGURE 3, using buffer at pH 7.35. Voltage was held for 60 seconds at +40 mV, and then jumped to −40 mV where shown on voltage-command trace. Top trace shows the current response; dashed line marks the level of zero current.

single-channel height is the same at both pH 7.0 and pH 7.35 (data not shown).

We can analyze records such as those of FIGURE 5 quantitatively by calculating conductance-level histograms, i.e., by calculating the frequency of appearance of a given conductance. FIGURE 6 shows such a determination for fluctuations recorded from the same membrane at two different voltages. We can draw several firm conclusions from this experiment. First, the histogram is characterized by a series of sharp peaks, equally spaced, with a separation of 16 pmho. The equal spacing of the peaks, which is seen up to as many as seven levels in the figure, indicates strongly that the membrane contains a small number of identical channels, each of which assumes a *single* open state. The second conclusion is that the probability of a channel's opening is dependent on voltage, while the single-channel conductance is not. At —10 mV, most of the time a single channel is in the open state, while occasionally zero or two channels are open; never are more than two channels open simultaneously. At —20 mV, however, we frequently observe up to five channels open. By calculating the time-averaged conductance by integration of histograms such as these, we find that the time-averaged conductance increases e-fold for an approximately 10 mV change in voltage, in the range —10 mV to —25 mV. This is in good agreement with the macroscopic g-V curve (FIGURE 3) and shows that the assumptions of the model leading to equation 1 are reasonable. Finally, we should point out that we have found the single-channel conductance to be constant over a much wider range of voltage than is displayed in FIGURE 6.[5]

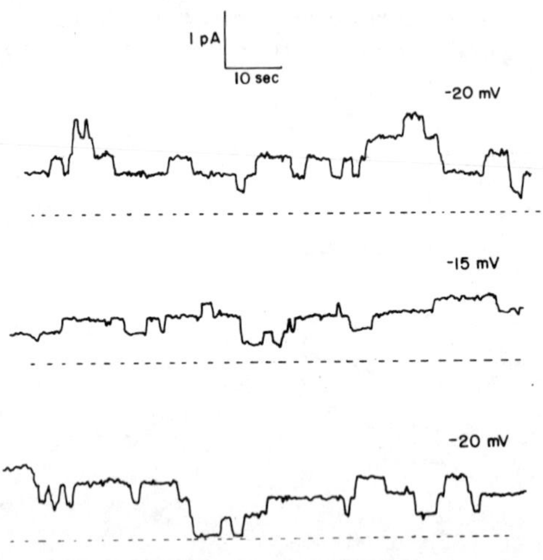

FIGURE 5. Single-channel conductance fluctuations. Planar bilayers composed of PE/cardiolipin (7/3) were observed at high current gain after the occurrence of a single conductance spike, under conditions of FIGURE 4. Traces from different membranes are shown, at holding voltages indicated. Dashed lines mark the level of current corresponding to the background conductance, i.e., with all channels in the closed state. PE/cardiolipin membranes were used for all quantitative single-channel work because they display less background noise than PE/PG membranes. Qualitatively, results in PE/PG are similar. The mean single-channel conductance from the above records is: 16.0±0.3 pmho (S.E. of 25 determinations).

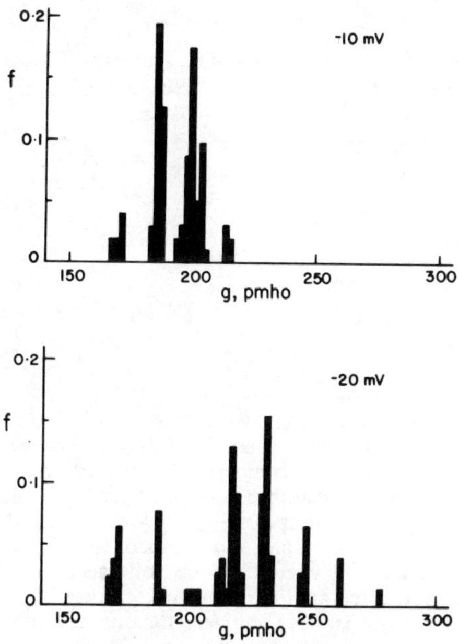

FIGURE 6. Single-channel conductance fluctuations were recorded, as in FIGURE 5, at two different voltages (indicated on figure) from the same membrane. The frequency, *f*, of the appearance of a given conductance, *g*, was calculated by sampling the current record every second, for a total time of 1–2 minutes, the tabulated currents were converted to conductances. The frequency of appearance of conductance in 2-pmho intervals was then calculated and plotted above. The lowest level of conductance, 170 pmho, in this membrane, corresponds to the "zero-level conductance," i.e., the background conductance with no channels open. This level was also determined by measuring current at +40 mV, a voltage at which open channels are almost never observed.

Ion Selectivity

We have already shown [5] that the channel-mediated conductance is ideally selective for Cl^- over K^+; we did not present any data on the interanionic selectivity, however. This may be determined by imposing "bi-ionic" conditions on the system, i.e., with KCl on one side of the bilayer and KX on the other side, where X is the anion to be tested. We then determine the voltage at which no current flows through the channel by the method of "tail currents," as described in our earlier determinations.[5] TABLE 2 lists the permeability ratios of a series of anions, referred to the Cl^- permeability. We find that of all anions tested, only Cl^- and Br^- can permeate the channel, with Br^- being 58% as permeant as Cl^-. This is an extraordinary result, as we know of no other anion conductance channel with such Cl^--specific behavior.

Another striking observation to emerge from these experiments is that SCN^- and I^-, while not permeant themselves, actually block the conductance of the channel to Cl^-. FIGURE 7 shows the result of an experiment in which SCN^- is added to the *cis* chamber, in the presence of a constant concentration

TABLE 2

SELECTIVITY OF VOLTAGE-DEPENDENT CONDUCTANCE †

Ion (X)	P_X/P_{Cl^-}
Cl^-	1.00
F^-	<0.03
Br^-	0.58±0.01
I^-	0* (K_i=30 mM)
SCN^-	0* (K_i=3.3 mM)
NO_3^-	<0.04
CH_3COO^-	<0.02
$SO_4^=$	<0.1
K^+	<0.02

† Ionic permeability ratios, P_X/P_{Cl^-}, were measured by establishing an ionic gradient across a planar bilayer which had been treated with *Torpedo* microsacs, as in FIGURE 2. After the membrane conductance had increased and stabilized in the presence of 0.1 M KCl buffer, the *cis* chamber was perfused with buffer of a different ionic composition, as described.[2] A voltage of −30 mV was then applied for 30–50 seconds, to open the channels. Voltage was then jumped to various voltages, in the range zero-to +50 mV, and the tail currents were followed.[5] Crossover potentials of the tail currents were used to calculate the permeability ratios.[5]

Ions marked with an asterisk (*) indicate that no conductance could be measured after perfusion because the ion blocks the channel, with *cis* inhibition constant, K_i, shown in parentheses.

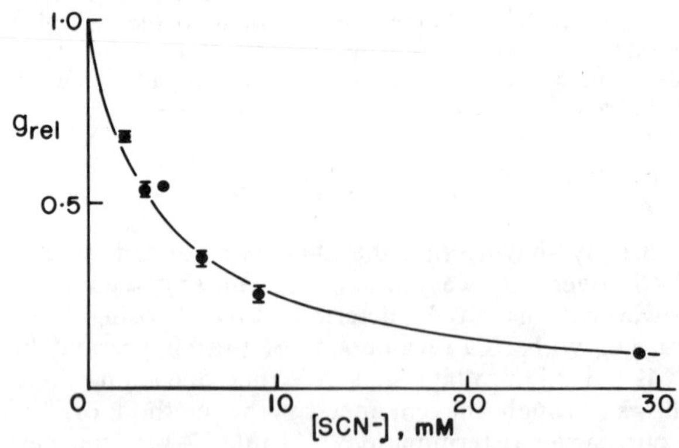

FIGURE 7. Block of macroscopic conductance by SCN^-. Conditions were as in FIGURE 4, and data were taken from five different membranes. After the membrane conductance had stabilized, the control value of conductance at −30 mV was measured, and then KSCN was added to the *cis* chamber, with stirring, to final concentrations shown on graph. Inhibition was immediate and could be reversed by perfusing away the SCN^-. Conductance, g_{rel}, was corrected for background and is plotted relative to the control value. Points without S.E. bars are single measurements. Solid curve is drawn according to: $g_{rel}=[1+(SCN^-)/K_i]^{-1}$, and K_i=3.3 mM.

of Cl^- (0.1 M). It is seen that the Cl^- conductance is progressively inhibited by SCN^-. The inhibition is immediate and reversible (data not shown), and it follows a single-site titration curve, with an inhibition constant of 3.3 mM. SCN^- inhibits from the *trans* side also, but with about 10-fold lower affinity. Similar experiments with I^- show that it also blocks the Cl^- conductance from the *cis* side, with an inhibition constant of 30 mM. We have not observed blocking by *trans* I^-, up to 100 mM. Of all anions listed in TABLE 2, only SCN^- and I^- only show the blocking effect. In an analogy to the blocking of cation-specific channels by impermeant cations, we might postulate that SCN^- and I^- can enter the channel and bind to a "selectivity filter" normally intended for Cl^-, but that the blocking ions cannot pass through the rest of the channel. Much work needs to be done, along the lines of Cs^+ blocking of the SR K^+

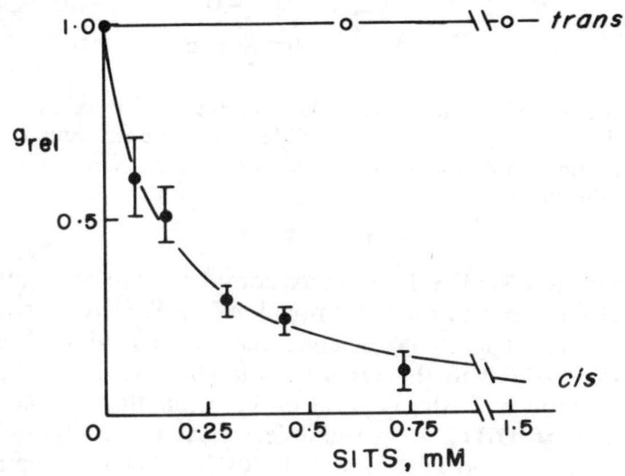

FIGURE 8. Asymmetric inhibition by SITS. Conditions were as in FIGURE 4. Conductance at −30 mV was followed after addition of SITS to either the *cis* or *trans* chambers. Conductance, g_{rel}, was corrected for background leak (which was unaffected by the reagents), and is plotted relative to its value before the addition of the SITS inhibitor. Here, the inhibition was immediate, and the extent of inhibition varied with the concentration of SITS. This extent is plotted on the figure, for *cis* or *trans* additions, as shown. Each titration curve is from three membranes. The solid curve for the *cis* titration is drawn according to a single site inhibition curve, with inhibition constant of 0.12 mM.

channel,[4] before the above suggestion can be solidified into a validated hypothesis.

SITS and DIDS Inhibition

Once the Cl^--selectivity of the channel became apparent, it was obvious that the effects of SITS and DIDS on the channel conductance should be studied. As expected, both of these classical anion transport inhibitors eliminate the Cl^- conductance.[5] Inhibition by SITS added on the *cis* side of the bilayer follows a single-site titration curve, with an inhibition constant of about 100 μM (FIGURE 8). It is noteworthy that SITS has no effect when added to the *trans* chamber. The inhibition by SITS may be reversed by perfusion of the *cis* chamber (data not shown). FIGURE 9 shows that DIDS is a powerful inhibitor.

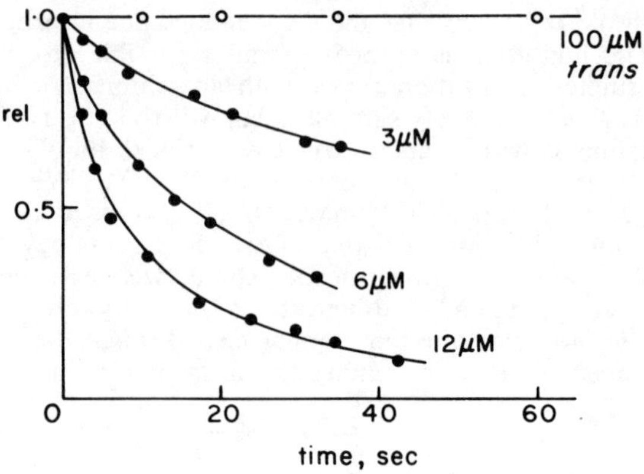

FIGURE 9. Asymmetric inhibition by DIDS. Conditions were as in FIGURE 8. The time course of inhibition was followed for different concentrations of DIDS. For all concentrations, the inhibition was complete at long times. Each time course is taken from a single membrane.

At all concentrations tested, inhibition is complete, but the rate of inhibition varies with DIDS concentration. Removal of DIDS by perfusion does not reverse the inhibition, at least on the time scale of 10 minutes. As with SITS, only when DIDS is added to the *cis* side is it effective as an inhibitor. These results are in harmony with the general observation that the action of SITS is reversible and that of DIDS is irreversible. The possibility that these effects are artifacts due to reaction with amino lipids has been previously ruled out.[5]

The effect of DIDS on the Cl^- conductance is further investigated in FIGURE 10, which shows the g-V curve of a single membrane modified by *Torpedo* microsacs. Before addition of DIDS, the conductance is voltage-dependent, as in FIGURE 3. Note that here we have not subtracted out the background conductance, measured at positive potentials. After reaction with 40 μM DIDS,

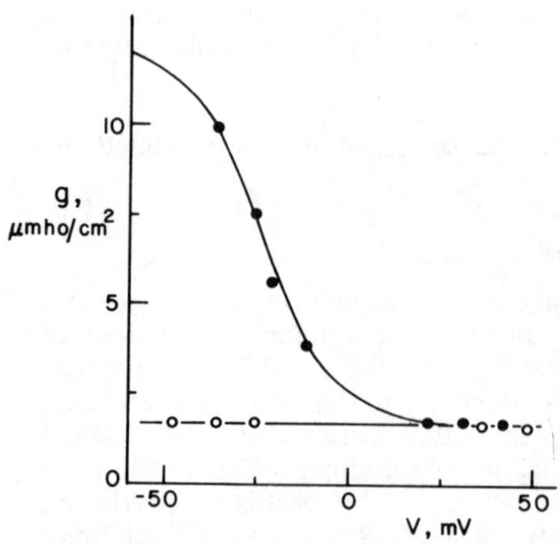

FIGURE 10. Effect of DIDS on conductance-voltage relation. The figure shows the raw data of a partial g-V curve taken on a single membrane, before and after addition of DIDS. Conditions were as in FIGURE 4. After the control g-V curve was determined (filled points), 40 μM DIDS was added to the *cis* chamber and allowed to react for 20 seconds. The g-V curve was again measured (open points). Note that the background leak conductance is unaffected by DIDS. Solid curve through the control points was drawn according to equation 1, with $z=2.6$ and $\Delta G_i^\circ = +1.5$ kcal/mol.

the membrane displays a voltage-independent conductance, equal to the background level. Thus, DIDS specifically eliminates the voltage-dependent conductance, without affecting the background. Similar behavior has been described for SITS.[5]

Effect of pH

As we saw in FIGURE 3, the pH of the aqueous medium controls the position of the g-V curve along the voltage axis. We can obtain further confirmation of this effect by studying the titration curve of conductance measured at a fixed voltage, say —30 mV, at which the channel is partly activated. From FIGURE 3,

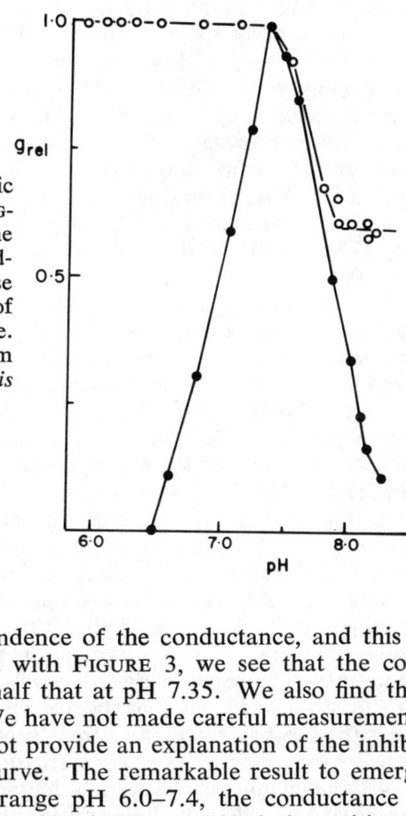

FIGURE 11. Effect of pH on macroscopic conductance. Conditions were as in FIGURE 7, with g_{rel} defined with respect to the conductance at pH 7.35. The pH was adjusted by addition of MES acid or TRIS base and was measured by sampling aliquots of solutions after each addition of acid or base. Each titration, *cis* and *trans,* was taken from two separate membranes. Closed points, *cis* addition; open points, *trans* addition.

we would expect a very sharp pH-dependence of the conductance, and this is confirmed in FIGURE 11. In agreement with FIGURE 3, we see that the conductance measured at pH 7.0 is about half that at pH 7.35. We also find that a sharp maximum exists near pH 7.4. We have not made careful measurements of g-V curves above pH 7.4, so we cannot provide an explanation of the inhibition in the basic range of the titration curve. The remarkable result to emerge from the titration curve is that in the range pH 6.0–7.4, the conductance is strongly sensitive to the pH of the *cis* chamber, but is completely insensitive to that of the *trans* chamber. Above pH 7.4, a partial inhibition by *trans* pH is seen, but we have not investigated its mechanism. All effects between pH 6.0 and pH 8.4 are fully reversible.

It is important to realize that the inhibition in the acid range of the titration curve is an effect of pH not upon the conducting structure of the open channel, but rather upon the probability of channel opening. This conclusion is shown by the shift in half-saturation voltage with pH (FIGURE 3), and by the fact that the single-channel conductance is the same at pH 7.0 as it is at pH 7.35. Finally, we point out that the pH effects cannot be due to possible titration of the ethanolamine lipid in the planar bilayer, as we have checked by measurements of nonactin-mediated K^+ conductance.[11, 12]

Discussion

In this report we have extended the initial observations reported[5] on the conductance conferred upon planar bilayers by *Torpedo* microsac vesicles. There are a number of firm conclusions we wish to state as a result of this work.

1. Channels are incorporated into the planar bilayer in a process resembling fusion of microsac vesicles with the bilayer. A fusion mechanism, while not proven, is likely to be valid on the basis of several lines of study. First, the factors controlling the insertion of channels are identical to those controlling the fusion of phospholipid vesicles with each other,[7, 8] and of SR vesicles with planar bilayers.[1, 9] Second, the incorporation of channels always occurs in discrete packages of many more than one channel; the average spike-height, as in FIGURE 2, represents the insertion of about 300 channels. We have never observed the incorporation of only a single channel. Finally, the complete asymmetry of the resulting conductance properties is most naturally explained by a fusion mechanism. This asymmetry is shown by the g-V curves (FIGURE 3), the SITS and DIDS effects (FIGURE 8) and the pH titration (FIGURE 10).

2. The conductance is mediated by a channel mechanism, in which the channel has only a single open state. This conclusion is immediately apparent from the fluctuation results of FIGURES 5 and 6. Thus, we have added a third channel to the list of channels from animal cells which have been unequivocally shown to display only a single open state: the acetylcholine-gated channel of frog muscle,[13] the K^+-selective channel from SR,[1, 3] and now the Cl^- channel studied here. Of all known channels from higher organisms, only the mitochondrial "VDAC" has been shown to display several open states.[14]

3. The probability of a channel entering the open state is controlled by the voltage across the membrane; the open-state conductance is independent of voltage. The explicit voltage-dependence of the probability of forming an open channel can be described by a simple two-state conformational model, as given by equation 1. We have found that neither the steady-state probability of channel opening nor the voltage-jump kinetics are dependent upon the absolute magnitude of channel-mediated conductance. We take this as strong, though not foolproof, evidence that the aggregation of independently diffusing subunits is *not* the mechanism for channel formation. This is the reason that we refer to the two-state model as a "conformational" model.

4. The channel shows a high degree of selectivity for Cl^- ion, as well as the phenomenon of block by impermeant anions.

5. SITS and DIDS inhibit the channel asymmetrically, in ways and at concentrations consistent with their well-known effects on physiological anion transporters.

We would also like to discuss a few broader conclusions from our results. The first and most basic is that the Cl^- conductance under study here is due to the action of a specific biological transport system; it is not an artifactual conductance induced by vesicles interacting with, and possibly damaging, the planar bilayer. It is necessary to make such a statement because of the justifiable reputation of the planar bilayer system for being subject to conductance artifacts induced by biological materials (the "boiled B.S.A. effect"). We have certainly seen our share of such artifacts while surveying many vesicle systems. With *Torpedo* microsacs, as with SR vesicles,[1] however, we are struck by the reproducibility of the results from membrane to membrane and preparation to preparation. In the case of *Torpedo* microsacs, we should also point out the variety of specific behavior shown by the conductance: Cl^--specificity, voltage-dependence, SITS and DIDS inhibition, blocking by impermeant anions, asymmetric orientation, single-channel behavior, and high sensitivity to pH. From our experience with the planar bilayer, we can say that artifacts do not do these things.

The second broad conclusion is that a Cl^- conductance is involved in some manner in the electrophysiology of the *Torpedo* electroplax. This is unexpected but not surprising. The organ is known to operate as a resting battery with an acetylcholine-gated switch.[15, 16] While the innervated face of the electroplax cell has been known for some time to carry a high density of acetylcholine receptor,[16, 17] the properties of the noninnervated face are largely unknown, due to the impossibility of voltage-clamping the relevant membrane. We are unwilling to postulate a particular function for this channel, but we think it likely that the channel resides in the noninnervated face of the cell. If the innervated face maintained a high Cl^- conductance, the electroplax organ would not be able to operate as a source of high-power electrical pulses. A high Cl^- conductance on the noninnervated membrane, however, would allow the delivery of high power during a pulse by maintaining a low internal resistance of the cell's "battery," which would have as its source the Cl^- gradient across the noninnervated membrane.

Furthermore, if we assume that the Cl^- channel resides in the noninnervated membrane, we can, without too much contortion, wiggle out of a serious and disturbing question: why, if we imagine that the microsac vesicles fuse with the planar bilayer, do we fail to see any acetylcholine-dependent conductance? We need only to postulate that the fusion process, for unknown reasons, *selects* the vesicles derived from the noninnervated membrane.

Given our simple picture of the fusion of microsac vesicles with the planar bilayer, we may ask if the conductance of the average spike can be related to the specific conductance of the electroplax cell membrane. The average spike-height conductance is in the range $2\text{–}4 \times 10^{-9}$ mho, with a wide variation around the mean value. If we make the reasonable assumption that the average vesicle diameter is about 0.4 μm, as determined by negative-stain electron microscopy, we then calculate that the average specific conductance per membrane area is in the range 0.4–0.8 mho/cm^2; the uncertainty in this calculation is great, since we do not know the sizes of those vesicles actually causing the conductance spikes. However, it is worthwhile noting that a calculated specific conductance of 0.1–1.0 mho/cm^2 is not at all unreasonable, since this is the measured conductance of the noninnervated membrane of the *Torpedo* electroplax.[16]

Finally, we wish to mention the results here in connection with similar, but more extensive, studies on K^+-selective channels derived from mammalian SR membrane vesicles. The K^+ channel shows a weaker voltage-dependence ($z = 1.2$) than the Cl^- channel, and its polarity is reversed (i.e., it opens at positive voltages). The K^+ channel is blocked by Cs^+, which is itself nonconductive,[4] just as the Cl^- channel is blocked by the "analogues" SCN^- and I^-. The K^+ channel is unaffected by SITS and DIDS (unpublished observations), but is strongly inhibited by certain sulfhydryl ligands, including organomercurials.[2] Finally, the K^+ channel has a single-channel conductance of about 130 pmho in 0.1 M K^+ (1–4), much larger than the 16 pmho value of the Cl^- channel in 0.1 M Cl^-.

The conclusion to be drawn from this brief comparison is that biological membranes can be made to confer specific conductance properties upon artificial planar bilayers. We have now successfully applied this technique in two separate cases. In each case, we discovered a channel of a type very different from the initial object of our search. This teaches us one of two possible lessons: either that planar bilayers so radically modify the structures of biological ion channels as to give them a completely new set of specific properties, or that there are a great many channels lurking in biological membranes whose existences have not been inferred on the basis of physiological observations. At this point in our work, we prefer the second possibility.

References

1. Miller, C. 1978. J. Membr Biol. **40:**1–23.
2. Miller, C. & R. Rosenberg. 1979. Biochemistry **18:**1138–1145.
3. Miller, C. & R. Rosenberg. 1979. J. Gen. Physiol. **74:**457–478.
4. Coronado, R. & C. Miller. 1979. Nature **280:**807–810.
5. White, M. M. & C. Miller. 1979. J. Biol. Chem. **254:**10161–10166.
6. DeGier, J., J. G. Mandersloot & L. L. M. VanDeenen. 1968. Biochim. Biophys. Acta **150:**666–675.
7. Miller, C. & E. Racker. 1976. J. Membr Biol. **26:**319–333.
8. Miller, C., P. Arvan, J. N. Telford & E. Racker. 1976. J. Membr. Biol. **30:** 271–282.
9. Miller, C. & E. Racker. 1976. J. Membr. Biol. **30:**283–300.
10. Latorre, R., G. Ehrenstein & H. Lecar. 1972. J. Gen Physiol. **60:**72–85.
11. McLaughlin, S. G. A., G. Szabo, G. Eisenman & S. M. Ciani. 1970. Proc. Nat. Acad. Sci. U.S.A. **67:**1268–1275.
12. Latorre, R. & J. E. Hall. 1976. Nature **264:**361–363.
13. Neher, E. & B. Sackmann. 1976. Nature **260:**799–801.
14. Colombini, M. & A. Finkelstein. 1979. This volume.
15. Bennett, M. V. L., M. Wurzel & H. Grundfest. 1961. J. Gen. Physiol. **44:** 757–804.
16. Bennett, M. V. L. 1970. Ann. Revs. Physiol. **32:**471–528.
17. Moreau, M. & J.-P. Changeux. 1976. J. Mol. Biol. **106:**457–467.
18. Mueller, P. & D. O. Rudin. 1969. Curr. Top. Bioenerg. **3:**157–249.

Discussion

P. R. Steinmetz: Is the pH dependence really very different from the voltage dependence?

C. Miller: The drop in macroscopic conductance measured at a fixed voltage caused by lowering the pH is a lot sharper than can be accounted for by a simple titration curve. In the acid range of the inhibition, the primary effect of lowering the pH is to shift the g-V curve to the left, i.e., to make the half-saturation voltage more negative. In the range 6.0–7.5, the conductance of the open channel is independent of pH. So in this range at least, it is the probability of forming an open channel that is controlled by pH. I haven't yet done the necessary measurements in the basic range of the curve, so I just don't know what's going on there.

J. A. ZADUNAISKY: In all the kinetics done for gastric secretion, certain stimulations result, types of interaction, such as inhibition of the chloride channel. Could you comment on the SCN^- effect?

MILLER: It appears that SCN^- blocks the channel with quite high affinity. In 100 mM KCl, you need add only 3 mM SCN^- to reduce the conductance by 50%. To be able to claim that this is really a block, we would have to do single-channel measurements and to see if the SCN^- effect is competitive with Cl^-. We haven't done these measurements yet, but the effect of SCN^- here seems very similar to the blocking of cation channels by impermeant cations. The best example of this is the blocking of the potassium channel in squid axon by cesium, which apparently binds very strongly to what is usually called "a selectivity filter," but then can't squeeze through the rest of the channel. I think it is quite likely that's what's going on here with this pseudohalide. We would predict a voltage-dependence of the block, and we have a lot of work to do on it.

L. J. MANDEL: Did you find any competition between bromide and chloride?

MILLER: I haven't looked. I have only measured the permeability ratio under bi-ionic conditions.

MANDEL: You showed voltage sensitivity in the negative direction. Was your convention negative from the *cis* or the *trans?*

MILLER: I didn't describe the convention because I figured that nobody would remember it anyway. The voltage is defined as zero on the *trans* side always.

MANDEL: Okay. So, the voltage is in the direction to promote chloride movement from the *cis* to the *trans.*

MILLER: That's correct, from the *cis* to the *trans* at negative potentials.

D. G. WARNOCK: Did the DIDS or SITS affect the probability of channel assembly or the single channel conductance?

MILLER: Since DIDS acts irreversibly, it simply kills whatever channel it attacks. It effectively reduces the total number of channels in the membrane. SITS appears to work by binding to the open channel and effectively blocking it reversibly, as SCN^- does; but to really prove this we would have to study the effect in more depth.

WARNOCK: Have you studied single vesicles with an eye to channel noise?

MILLER: Yes, I have done that. If you first set up single channel fluctuations and then add DIDS, the channels just disappear, leaving you with the background conductance. After DIDS treatment, you cannot reactivate the channels by putting on a high voltage.

WARNOCK: There's a chloride conductance in mitochondria described by Dawes *et al.* that is inhibited by local anesthetics.

MILLER: I haven't tried local anesthetics.

WARNOCK: Furosemide?

MILLER: We haven't tried furosemide.

M. LAMPERT (*Princeton University, Princeton, New Jersey*): What do you make of the failure to find cation channels?

MILLER: The electroplax cell is coin shaped. Its membranes are discs about a centimeter in diameter, and the cell is about five or ten microns thick.

The two faces of this cell are functionally different. One is innervated and carries the acetylcholine receptor. The other, the noninnervated face, apparently has no receptors and not much is known about it. It's a very difficult membrane to study electrophysiologically. A few groups have shown that it's a very high conductance membrane, but as far as I know, the ionic basis for that conductance isn't known. Some of the work from Grundfest's group on a related species makes it look as though it may be a chloride conductance.

I think that in the homogenization, the innervated and noninnervated faces form vesicles separately; and although remote, it is possible that the fusion process could select the vesicles of the noninnervated face, which would give us only the Cl^- channels, and no acetylcholine receptor because of dumb luck. It is also possible that we destroy the acetylcholine receptor when it gets into the artificial membrane.

LAMPERT: Isn't it possible to test further?

MILLER: Unfortunately, one cannot do chemical studies of the planar membrane because of its small mass. But, we can do a similar experiment in which it would be easy to purify the noninnervated face vesicles from the preparation, and we are now working on this.

Z. I. CABANCHIK: I know that you didn't want to commit yourself to actual fusion, but is it possible that what you have are vesicles absorbed on the bilayer?

MILLER: It is possible that the vesicles stick onto the bilayer and then somehow transfer their entire content of channels, or a very large proportion of their content of channels, into it. The reason I say this is that if you look carefully at those spikes at a very high sensitivity, you will see that each one of the spikes represents a package of channels. The "average" insertion event dumps about a hundred channels into the bilayer; with the occasional little spikes, you see only about ten channels going in. It is clear that those channels get inserted not singly but in packages. It is also possible, of course, that the process of package insertion occurs by sticking and then letting it all go. I don't have any argument against that because I don't have any assay of exocytosis onto the other side.

CABANCHIK: You seemed to obtain full inhibition in your dose relationship of DIDS. Since it is known to penetrate in this condition, I suppose you used PS there in the membranes.

MILLER: Yes, actually, there's some cardiolipin or phosphatylglycerol. . . .

CABANCHIK: Negatively charged?

MILLER: Yes.

CABANCHIK: Which means all the sites are facing the *cis* side?

MILLER: Yes.

CABANCHIK: Did you do any studies on the vesicles themselves, pretreatment of DIDS?

MILLER: Yes, and so far we haven't achieved good results. However, we've pretreated vesicles with DIDS for only a couple of minutes; and these vesicles still have live channels in them, insofar as they seem to fuse normally and give channel-like conductance properties. I don't understand why.

CABANCHIK: It's DIDS sensitive only after insertion in the bilayer.

MILLER: Although it looks like that, I'm not willing to make the claim before more data are made available.

CABANCHIK: Is it correct you seem to have a marked difference between SITS and DIDS?

MILLER: A marked difference is that the SITS inhibition is reversible, at least on the short time scale of these experiments. We perfuse out the SITS and the conductance comes right back.

CABANCHIK: DIDS is not?

MILLER: DIDS is irreversible.

G. SACHS: About the problem of how the vesicles sit on the membrane, I'm not sure if the value of convention was positive or negative voltage dependence, but if you put vesicles on both sides you get a symmetric curve.

MILLER: Well, not quite. What you're assuming is that I can fuse exactly the same number of vesicles from both sides. You get sort of a skewed U-shaped curve.

SACHS: If you had vesicles stuck on the surface. . . .

MILLER: I think the question is that you might have a voltage dependence because there's a layer of vesicles stuck on the surface of the membrane, and that this makes diffusion difficult from that side.

First of all, the single channel conductance measured at both positive and negative voltages is the same. I think that this is a strong answer to the question because the single channel current is a direct measurement of the transmembrane diffusion process, uncomplicated by gating phenomena.

There's another argument. After inhibition by SITS or DIDS, you get a perfectly ohmic conductance. And finally, you can perfuse out the vesicles in the presence of EDTA after fusion, and there's no change in the conductance properties; the voltage-dependence remains. So I really don't worry anymore about the sticking-vesicle artifact.

Part VII. The Mechanisms and Roles of Anion and Proton Transport in Mitochondrial Function

STRUCTURE AND MODE OF ACTION OF A VOLTAGE DEPENDENT ANION-SELECTIVE CHANNEL (VDAC) LOCATED IN THE OUTER MITOCHONDRIAL MEMBRANE *

Marco Colombini †

Departments of Physiology and Neuroscience
Albert Einstein College of Medicine
Bronx, New York 10461

Introduction

In 1956 Werkheiser and Bartley reported [1] that part of the volume within a mitochondrion, now known as the intramembrane space, is readily accessible to small molecules such as sucrose but inaccessible to a polysaccharide of 50,000 molecular weight. Subsequent work by a number of laboratories supports the conclusion there is a permeability pathway in the outer mitochondrial membrane that selects molecules on the basis of size. Perhaps the simplest entity that could perform this task is a water-filled pore that spans the membrane.

Evidence, for the existence of large pores in the outer membrane of plant mitochondria, was reported in 1965 by Parsons *et al.*[2] Electron micrographs of negatively-stained mitochondrial outer membranes showed the presence of stain-filled channels that measured 25–30 Å in diameter. More recently X-ray diffraction has been used [3] to further substantiate the existence of these pores. However, it was not until 1976 that channel-forming molecules were isolated from mitochondria.[4] At that time the channels (called Voltage Dependent Anion-selective Channels: VDACs for reasons that will become apparent) were not thought of as being associated with the outer membrane. Now there is good evidence [5] that VDAC may indeed be the channel responsible for the permeability characteristics of the outer mitochondrial membrane.

Formation of Channels in Planar Lipid Bilayers

Planar lipid bilayers made in a hole in a plastic partition separating two aqueous phases (either by the method of Mueller *et al.*[6] or by that of Montal and Mueller [7]) is an excellent surrogate membrane on which to insert channel-forming molecules and observe their permeability properties. This basic medium, in which membrane proteins normally find themselves,[8] is highly impermeable to ordinary ions. So impermeable is this bilayer to ions, that the insertion of a single channel-forming entity, into the bilayer, increases its permeability by a measurable amount. The size of the permeability increase is a characteristic property of the particular channel-forming substance.

* This work was supported by National Institutes of Health Grant GM 23578.

† Currently affiliated with Department of Zoology, University of Maryland, College Park, Maryland 20742.

0077–8923/80/0341–0552 $01.75/0 © 1980, NYAS

VDAC channels can be inserted into planar bilayers simply by the addition of an aliquot, of a Triton X-100 extract of mitochondrial membranes, to the aqueous phase bathing the bilayer. The insertion, of single channels into the bilayer, is monitored directly by observing increased permeability to ions in the form of current flowing through the bilayer. FIGURE 1 shows the discreet increases in the bilayer's permeability to ions as a function of time. The permeability, produced by a single channel, is quite constant at the voltage used

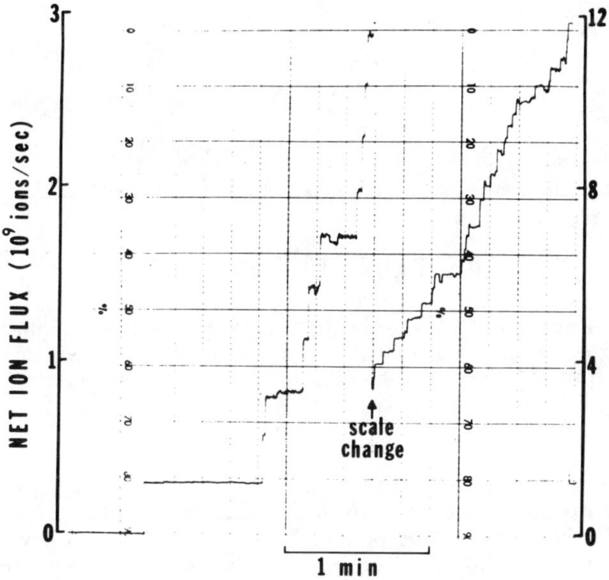

FIGURE 1. The insertion of VDAC channels into a planar lipid bilayer. To an unmodified planar lipid bilayer, made as described previously,[5] was added 5 μl of a 2% Triton X-100 extract containing VDAC from rat liver mitochondria (a purified fraction whose method of purification will be described elsewhere). The aqueous phase contained 1 M KCl 5 mM $CaCl_2$. A 10 mV transmembrane potential was applied continuously and current was measured as described previously.[4] From the current flux, the net ion flux was calculated. This represents the net flow of cations moving in one direction plus the net flow of anions moving in the opposite direction through the membrane. At the point indicated by the arrow the scale was reduced by a factor of 4.

in this record (10 mV) to drive ions across the membrane. Therefore, the permeability of the bilayer increases stepwise as VDAC molecules insert into it. A histogram of the increments shows a uniform population of sizes around (FIGURE 2) 2.8×10^8 ions/sec, the single channel size. This single channel size is the same whether VDAC was extracted from *N. crassa* or rat liver mitochondria.

A characteristic property of VDAC is the voltage dependence of the permeability pathway that it forms in lipid bilayer membranes. FIGURE 3 shows the net ion flux through a lipid bilayer into which a single channel had inserted

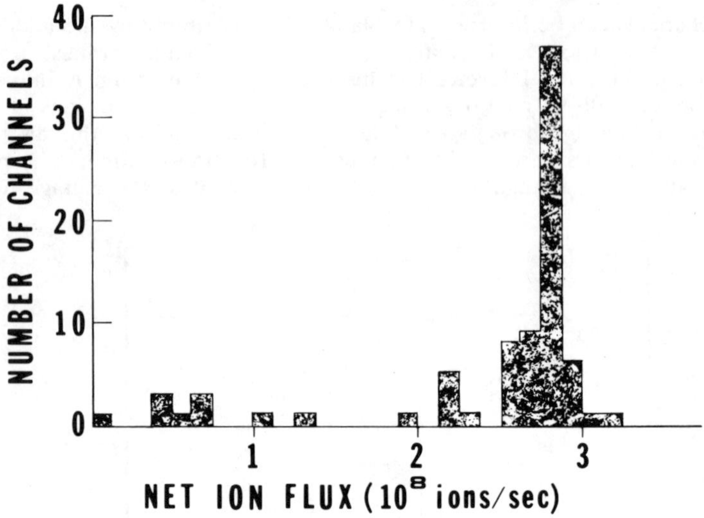

FIGURE 2. Histogram of the sizes of the abrupt changes in membrane permeability observed in experiments similar to that shown in FIGURE 1. All the data was collected with VDAC from rat liver mitochondria and the conditions were as described in FIGURE 1.

at the point indicated by the arrow. The net ion flux, at 0 mV transmembrane potential but in the presence of a KCl gradient, was in the direction expected for anion selectivity. When the potential was increased to 40 mV, the ion flux increased in response to the increased driving force without change in the membrane's permeability to ions. After a short while, VDAC underwent transitions to states of lower permeability, thus reducing the ion flow. This slow "closure" of the channel is in contrast with its rapid opening when the transmembrane potential was returned to zero. In the time scale of the figure, the permeability of the channel returned to its original value immediately. This process of opening and closing can be repeated indefinitely. Note how the same channel can close to a variety of different states. VDAC closes as the voltage is increased in both the positive and negative direction.[4, 5] The rate of closure increases with voltage.

Although VDAC allows both cations and anions to cross the membrane, it does have a distinct preference for anions. In the case of KCl, this preference is two to one for Cl^- over K^+. This value is for VDAC from rat liver inserted into a neutral lipid. In a negatively charged lipid milieu, the selectivity is a little lower while VDAC from *P. aurelia* has a higher selectivity. The selectivity also increases as the hydrated ionic radius of the cation increases. It is interesting to note, however, when VDAC closes, there is a preferential reduction in anion permeability. The permeability to Cl^- is reduced by 70–90% while that for K^+ is only reduced by 25–50% (the value depends on which "closed state" is measured and depends on the source of the channels; the selectivities were determined using the constant field approximation).

The basic properties of the channel are summarized in its name, voltage dependent anion-selective channel, or VDAC. These properties are remarkably constant in VDAC isolated from protozoans, fungi or mammals.

Molecular Structure of VDAC

When VDAC is inserted into a planar bilayer from a Triton extract, a rate of incorporation is observed that is relatively constant for some time. If this rate is taken as a measure of VDAC activity, then one finds that the activity is proportional to the amount of extract added to the aqueous phase (keeping other variables, such as the amount of detergent added, constant). This indicates that the channel exists as a prepackaged entity in the Triton extract. This is in contrast to channel-formers, such as gramicidin, alamecithin, and amphotericin B, which require that a number of subunits come together to form a channel. The amount of permeability produced by these channels is a power function of the quantity of material added.

Attempts were made to measure the size of the channel-forming unit by the behavior of the Triton-solubilized entity in a gel filtration column. A Triton extract of rat liver mitochondrial membranes was run in a CL Sepharose 6B column and the result is shown in FIGURE 4. VDAC activity was assayed by adding aliquots of the effluent to planar bilayers and measuring the rate of channel incorporation. In this way, VDAC activity was localized to the area shown in the figure. When this position is compared to that of standard pro-

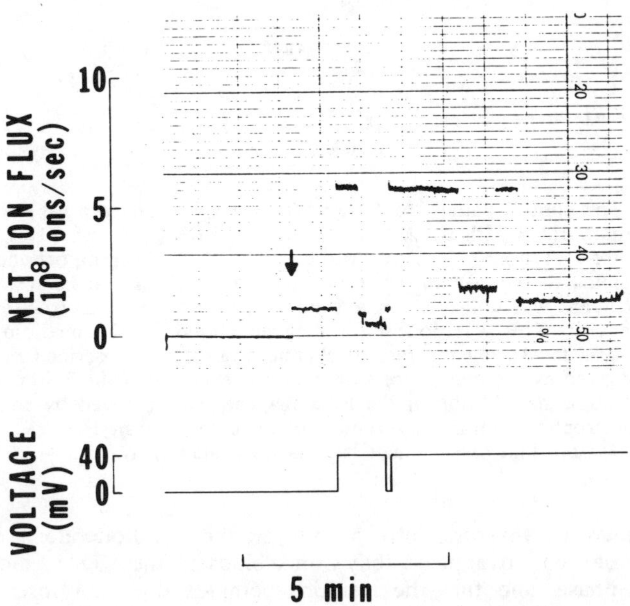

FIGURE 3. The voltage-dependent behavior of a single VDAC channel. The experiment was performed in a similar manner to that described for FIGURE 1, with the following exceptions. A smaller quantity of VDAC extract was added to the medium and the salt concentration was different on the two sides of the membrane. One side contained 1 M KCl and 5 mM $CaCl_2$ while the other had 0.1 M KCl and 5 mM $CaCl_2$. The voltage indicated in the figure was the transmembrane voltage with the sign of the voltage referring to the high salt side. The arrow indicates the point at which the channel entered the membrane. Ion flux changes that occurred in the absence of voltage changes are due to the channel changing its state.

teins run under the same conditions (FIGURE 5), a molecular weight of 110,000 is obtained. This value is only an estimate because the method has the following drawbacks: 1. The size being measured is that of the VDAC-Triton complex and the amount of Triton bound to a membrane protein can be substantial and varies from protein to protein;[9] 2. The amount of detergent bound to the protein standards is equally unknown.

There is compelling evidence that VDAC is a glycoprotein. If concanavalin A (ConA) is added to the aqueous phase bathing a lipid bilayer while

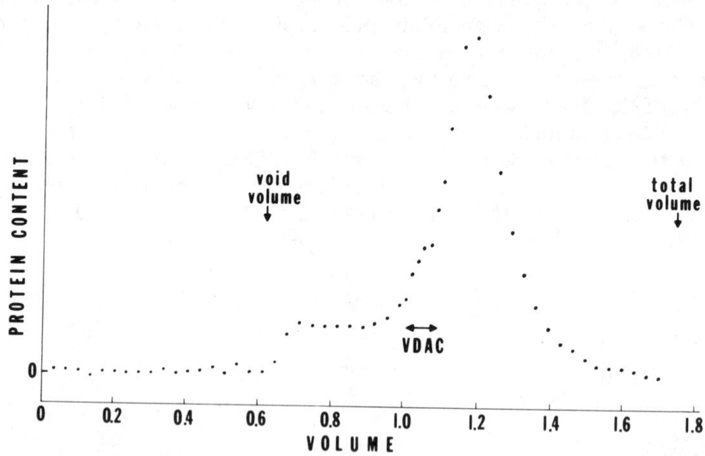

FIGURE 4. The elution pattern of VDAC from a CL Sepharose 6B column. The column was equilibrated with 0.5 M KCl, 1 mM HEPES (Na^+ salt, pH 7.0), 0.02% Na azide and 0.5% Triton X-100. A Triton extract of rat liver mitochondrial membranes was applied to the column after pretreatments, which will be described elsewhere. The effluent was collected on a fraction collector and the fractions assayed for VDAC activity. This was done by adding an aliquot to the medium bathing a planar bilayer and measuring the rate of channel insertion as described in FIGURE 1. Protein was assayed by the method of Lowry *et al.*[21] except that the Triton precipitate, which formed after the addition of the folin reagent, was removed by centrifugation prior to the spectrophotometric measurement of the color intensity. The CL Sepharose 6B was obtained from Pharmacia Fine Chemicals, Piscataway, N.J.

VDAC channels are inserting into the bilayer, the insertion rate is drastically reduced (FIGURE 6). It appears that ConA binds to the VDAC molecules in the aqueous phase, and that the resulting complex does not insert into the bilayer. However, the channels that have already inserted do not appear to be affected. Both the conductance and the voltage dependence are unchanged. In the figure, when the voltage was increased to 40 mV, the channels turned off with the normal rate. (A complete I-V plot was also done and this was not significantly different from controls.) Evidence that ConA binds to VDAC comes from the fact that triton-solubilized VDAC is retained by a ConA-Sepharose column and can be eluted from the column with α-methyl mannoside.

The size of the pore produced by VDAC was obtained by examining the permeability induced to nonelectrolytes of varying size. Bangham *et al.*[10]

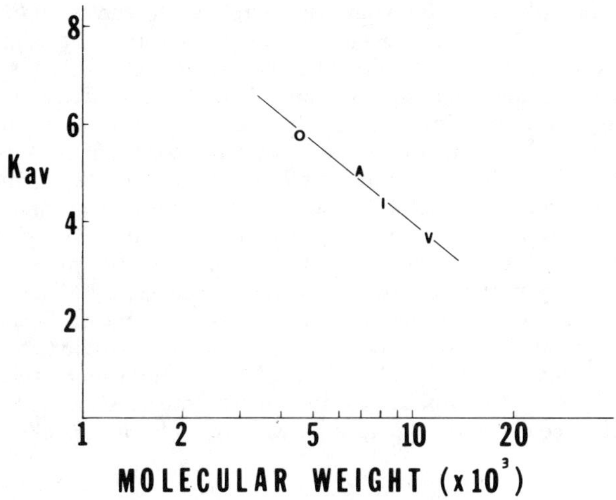

FIGURE 5. The determination of the molecular weight of VDAC. Protein standards were run through the column described in FIGURE 4 under the same conditions as those used for the VDAC run. Ovalbumin (O) and bovine serum albumin (A) were obtained from Polyscience Inc., Warrington, Pa. Human IgG (I) was obtained from Miles Biochemicals, Elkhart, Ind. The symbol (V) indicates the location of VDAC activity. K_{av} is as defined by Pharmacia Fine Chemicals, with the void volume being measured by means of blue dextran.

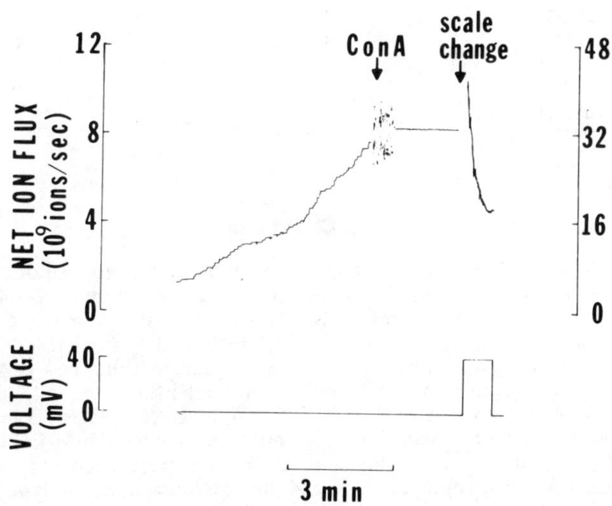

FIGURE 6. Effect of concanavalin A on VDAC insertion into a planar lipid bilayer. The experimental conditions were like those described in FIGURE 3 except that a larger dose of VDAC had been added to the aqueous phase. At the point indicated, 50 μl of 10 mg/ml Con A was added to each of the two aqueous compartments on each side of the membrane. At the point indicated by the arrow, the scale was reduced fourfold, and, at the indicated times, the voltage was changed as shown. Concanavalin A was obtained from Sigma Chemical, St. Louis, Mo.

showed that the volume of liposomes is correlated to the turbidity of the liposome suspension. When the osmotic pressure of the medium bathing the liposomes is changed, the turbidity increases or decreases in response to liposome shrinkage or swelling, respectively. When the tonicity of the medium bathing liposomes containing VDAC channels is increased by the addition of solute, the liposomes shrink to an extent that is dependent on the permeability of the solute. If the solute is completely impermeable, the liposomes will shrink to a smaller volume, as water is drawn out, and remain at that final volume. However, if the solute leaks into the liposomes, then they will reswell as water accompanies the solute back into the liposomes. Thus the physical dimensions, of the permeability pathway induced in the liposomes, by VDAC, can be estimated by the size of the largest permeable solute. FIGURE 7 shows this type of experiment done with three different solutes: α-cyclodextrin (973), inulin (5000) and polyethylene glycol (PEG of 6800 molecular weight). The α-cyclodextrin caused the liposomes to shrink transiently (light transmission decrease) followed by a fairly rapid reswelling. The inulin experiment shows

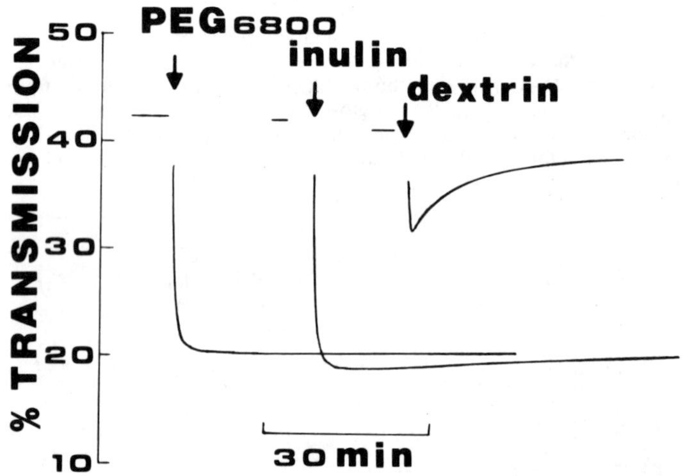

FIGURE 7. The response of VDAC-containing liposomes to osmotic pressure induced with nonelectrolytes. The liposomes were made as follows: To 46 mg of egg phosphatidyl choline, Sigma Chemical, St. Louis, Mo., and 4 mg egg phosphatidic acid, Avanti Biochemicals, Birmingham, Al., which had been dried down together, was added 2.5 ml of mitochondrial membrane suspension (obtained from *N. crassa*) consisting of 2.5 mg protein in 1 mM KCl 1 mM HEPES (Na^+ salt, pH 7.0). The material was sonicated to clarity (at 0 to 5° C) and lyophilized. Liposomes were produced by suspending this material in 2.5 ml 20 mM KCl 1 mM EDTA (Na^+ salt, pH 7.0). Fifty microliters of this liposome suspension were added to 2.0 ml 20 mM KCl 1 mM EDTA (Na^+ salt, pH 7.0) and the percentage transmission at 400 nm was monitored. When the liposomes had stabilized (no change in percentage transmission with time), 500 μl of 20 mM α-cyclodextrin or inulin or PEG6800 were added (as indicated in the figure). The recording just before the arrows indicate the addition has been corrected for the dilution due to the addition of the nonelectrolyte. The inulin from dalia tubers was obtained from Sigma Chemical, St. Louis, Mo., while α-cyclodextrin and the polyethylene glycols were obtained from Aldrich Chemical, Milwaukee, Wis.

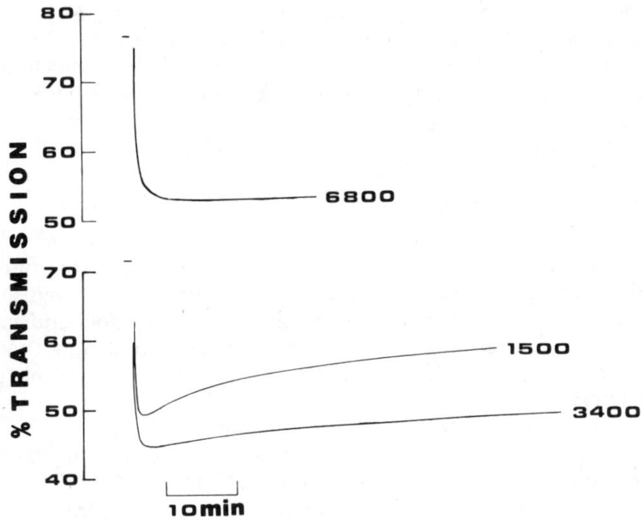

FIGURE 8. The response of VDAC-containing liposomes to osmotic pressures induced with polyethylene glycols of different molecular weights. The liposome suspension was produced as described in FIGURE 7. Fifty microliters of the suspension were added to 2.5 ml of 20 mM KCl 1 mM EDTA (Na^+ salt, pH 7.0). The experiments were performed as described in FIGURE 7 except that 200 μl of 100 mM PEG1500 and PEG3400 and 50 mM PEG6800 were added to the liposomes as indicated in the figure.

a slow reswelling while the PEG6800 experiment shows no detectable reswelling. In FIGURE 8, a series of polyethylene glycols were used. Here PEG-1500 and PEG3400 were permeable while PEG6800 was not.

From the Stokes-Einstein radii [11] of inulin (14 Å), PEG3400 (19 Å) and PEG6800 (29 Å), an estimate of 20 Å for the radius of the VDAC pore is not unreasonable. The smaller end of the size range is chosen because the heterogeneity of the probe molecules tends to overestimate the pore size.

These results are consistent with the known permeability properties of the outer mitochondrial membrane. Werkheiser and Bartley [1] showed that inulin could cross the outer membrane while a 50,000 molecular weight polymer could not. Wojtczak and Zaluska [12] demonstrated that cytochrome *c* (12,500) cannot enter the intramembrane space in intact mitochondria. The molecular dimensions of cytochrome *c* have been reported [13] to be $30 \times 34 \times 34$ Å. Since a square 30×34 cannot be inscribed into a circle of radius 20, steric hindrance may be sufficient to keep cytochrome *c* from entering VDAC. In addition, cytochrome *c* is very basic and VDAC's anion selectivity tends to exclude cations, especially multivalent cations. Finally, preferential exclusion of proteins (probably due to their charge) was also observed by Wallenius [13] when measuring the glomerular permeability. The possibility that the permeability observed in the liposomes is not the same as that studied in the planar bilayer has not been ruled out. However, when material prepared in a very similar manner to that used to make the liposomes (i.e., mitochondrial membranes sonicated with an amount of phospholipid equal to 25 times their protein content and then lyophilized) is used to make planar bilayers (as previously described[4]), not only is most of the ion permeability observed due to VDAC, but also VDAC is the only reproducible permeability observed. Therefore, it is likely to be the pathway that makes the liposomes permeable to the nonelectrolytes.

Similarities between VDAC and Porin

The similarities between bacteria and mitochondria are many and well known. However they are usually limited to the inner mitochondrial membrane and contents of the matrix space. Although the outer mitochondrial membrane is usually thought to resemble the endoplasmic reticulum, now there is at least one aspect of the outer membrane that resembles the outer membrane of gram negative bacteria.

As a result of the very fine work done by a number of laboratories,[14-17] a great deal is now known about a channel-forming protein, called porin or matrix protein, located in the outer membrane of *E. coli* and *S. typhimurium*. Like VDAC, porin forms large pores in vesicles, although these are quite a bit smaller than those produced by VDAC (TABLE 1). These results concur with the observation that the porin channels observed in planar lipid bilayers[16, 17] are half as permeable to ions as those due to VDAC. Porin channels are selective for cations as compared to VDAC's anion selectivity. Porin has a voltage-dependent permeability similar to that of VDAC despite the fact it occurs at a much higher voltage range.[17] VDAC channels begin to "close" at 10 mV and are almost always closed at 30 mV transmembrane potential. Porin, on the other hand, begins to close at 140 mV and is almost always closed at 230 mV.

Porin is a 103 kdal protein made up of three 33 kdal peptides.[18] Although, as mentioned earlier, VDAC exists as prepackaged channel, in Triton X-100, of 110 kdal, preliminary evidence suggests it is made up of 34 kdal peptides (rat liver VDAC dissociated in SDS). Studies done by Mannella and Bonner[19]

TABLE 1

COMPARISON OF VDAC AND PORIN

	VDAC	Porin	Porin Reference
Channel molecular weight	110,000	103,000	18
SDS peptide molecular weight	30,000?	32,600	18
Ion conductance in 1 M KCl (nS)	4.0	2.1	16, 17
$P(K^+)/P(Cl^-)$	0.5	3.5	16
Voltage range of "turn-off" (mV)	10–30	140–230	17
Pore Radius from nonelectrolyte permeability (Å)	20	7	14, 18

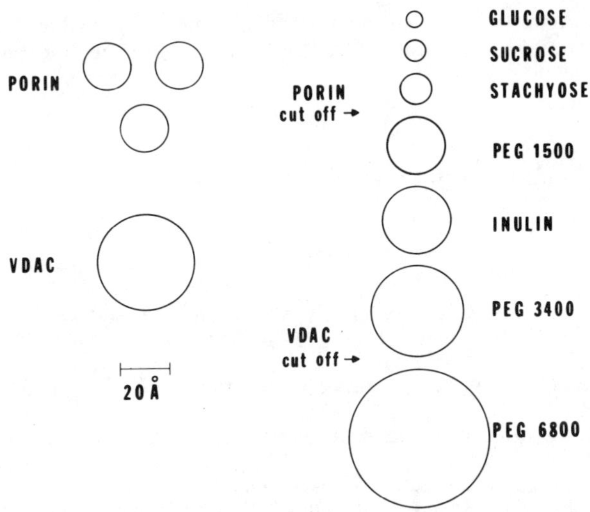

FIGURE 9. Comparison between the molecular dimensions of the marker nonelectrolytes and those of VDAC and porin. The triplet structure of porin is that reported by Stevens *et al.*[15] The spheres shown for the nonelectrolytes correspond to the Stokes-Einstein radii [11] of these molecules. The circle for VDAC represents the proposed size of its pore (20 Å in radius).

indicate that the outer membrane channels in plant mitochondria may be due to an SDS peptide of 30 kdal. The parallelism is remarkable.

The permeability pathway in the *Pseudomonas aeruginosa* [20] appears to have an exclusion limit around 6000 molecular weight. It will be interesting to see how this channel compares to VDAC.

At first it appears somewhat surprising that VDAC and porin should be so similar structurally and yet have quite different pore sizes. High resolution electron micrographs [15] of negatively stained *E. coli* outer membranes, show that one porin molecule may form a triplet of pores located at the apices of an equilateral triangle (drawn to scale in FIGURE 9). When VDAC's pore size is compared to this structure, the overall sizes are quite similar.

SUMMARY

VDAC is a channel-forming protein, located in the outer mitochondrial membrane, whose properties are consistent with the known permeability behavior of that membrane. When extracted with Triton X-100, VDAC exists as a 110,000 molecular weight glycoprotein complex which is a prepackaged channel. When inserted into liposomes, the permeability increase to nonelectrolytes is consistent with a pore radius of 20. In a planar lipid bilayer, VDAC is anion selective and voltage gatable. This suggests that the permeability pathway in the outer mitochondrial membrane might be under physiological control.

Many of VDAC's properties are qualitatively very similar, although quantitatively different, to those of porin, the channel responsible for the permeability of the outer membrane of at least some gram negative bacteria.

ACKNOWLEDGMENT

I would like to thank Dr. Alan Finkelstein for helpful advice and discussions.

REFERENCES

1. WERKHEISER, W. C. & W. BARTLEY. 1957. Biochem. J. **66:**79–91.
2. PARSONS, D. F., W. D. BONNER JR. & J. G. VERBOON. 1965. Can. J. Bot. **43:** 647–655.
3. MANNELLA, C. A. & W. D. BONNER JR. 1975. Biochim. Biophys. Acta **413:** 226–233.
4. SCHEIN, J. S., M. COLOMBINI & A. FINKELSTEIN. 1976. J. Membrane Biol. **30:** 99–120.
5. COLOMBINI, M. 1979. Nature **279:**643–645.
6. MUELLER, P., D. O. RUDIN, H. TI TIEN & W. C. WESCOTT. 1963. J. Phys. Chem. **67:**534–535.
7. MONTAL, M. & P. MUELLER. 1972. Proc. Nat. Acad. Sci. U.S.A. **69:**3561–3566.
8. SINGER, S. J. & G. NICOLSON. 1972. Science **175:**720–731.
9. GUIDOTTI, G. 1976. Trends Biochem. Sci. **1:**11–13.
10. BANGHAM, A. D., J. DEGIER & G. D. GREVILLE. 1967. Lipids **1:**225–246.
11. SCHERRER, R. & P. GERHARDT. 1971. J. Bacteriol. **107:**718–735.
12. WOJTCZAK, L. & H. ZALUSKA. 1969. Biochim. Biophys. Acta **193:**64–72.
13. WALLENIUS, G. 1954. Acta Soc. Med. Ups. **59**(83)**:**5–91.
14. NAKAE, T. 1976. Biochem. Biophys. Res. Commun. **71:**877–884.
15. STEVEN, A. C., B. TEN HEGGELER, R. MULLER, J. KISTLER & J. P. ROSENBUSCH. 1977. J. Cell Biol. **72:**292–301.
16. BENZ, R., K. JANKO, W. BOOS & P. LAUGER. 1978. Biochim. Biophys. Acta **511:** 305–319.
17. SCHINDLER, H. & J. P. ROSENBUSCH. 1978. Pro. Nat. Acad. Sci. U.S.A. **75:** 3751–3755.
18. NAKAE, T., J. ISHII & M. TOKUNAGA. 1979. J. Biol. Chem. **254:**1457–1461.
19. MANNELLA, C. A. & W. D. BONNER JR. 1975. Biochim. Biophys. Acta **413:** 213–225.
20. HANCOCK, R. E. W. & H. NIKAIDO. 1978. J. Bacteriol. **136:**381–390.
21. LOWRY, O. H., N. J. ROSENBROUGH, A. L. FARR & R. J. RANDALL. 1951. J. Biol. Chem. **193:**265–275.

DISCUSSION

A. TURTUNNO (*Downstate Medical Center, Brooklyn, N.Y.*): What is a voltage dependent channel doing in the outer membrane of the mitochondria?

M. COLOMBINI: Clearly, one cannot maintain a potential across the outer membrane by means of an ion gradient. However, it's certainly possible to maintain a Donnan potential across the outer mitochondrial membrane even in the presence of physiological salt concentrations since only about 15 millivolts are required to turn off the channel. The interesting point is the fact that the channel can close and open which implies that the outer mitochondrial membrane may control the function of the mitochondria by controlling the access of metabolites to the inner membrane and efflux of ATP from the mitochondrion.

A. CROFTS: Do you have any idea what the relative sizes of the channel are in the open and closed state from the conductance?

COLOMBINI: The conductance of the closed state is between 20 and 25 percent of the conductance of the open state so one might estimate that the channel area decreases by 75%. However, it's theoretically unsound to estimate sizes from conductance measurements. The physical size of the channel may correlate poorly with the permeability of the ions. For example, changing the electrostatic environment within the channel could change the ion permeability without changing the pore size. In fact, it is possible that the channel doesn't constrict at all.

T. H. WILSON: I was interested in your technique for measuring swelling and shrinkage of the liposomes. What concentration of liposomes do you use and at what wave length do you measure?

COLOMBINI: The technique that I followed was similar to that originally described for liposomes by Bangam and coworkers. I measured turbidity at 400 nanometers and the concentration of the liposomes was approximately a milligram of lipid per ml of solution.

G. SACHS: Can you give an estimate of the water permeability of the liposomes, and how does the channel contribute to the water permeability?

COLOMBINI: I haven't tried to obtain any quantitative measurements from these studies because of the multi-wall nature of the liposomes. A comparison of the water permeability through the bilayer, with and without channels, would require liposomes of uniform size and preferably single wall.

SACHS: Do you see any variation in data with different detergents?

COLOMBINI: Originally the channel was described without the use of detergents. The material was put directly on a monolayer and this was used to make bilayers. No detergent was present and, as far as I can tell, the properties are essentially the same. So far VDAC's properties have been independent of which detergent was used to insert it into the bilayer.

PROTONMOTIVE CYTOCHROME SYSTEM OF MITOCHONDRIA

Peter Mitchell

Glynn Research Institute
Bodmin, Cornwall, PL30 4AU, England

My main aim in this paper is to consider four simple questions about the mitochondrial cytochrome system, originally named and broadly defined by David Keilin:[1] 1. What is it? 2. Where is it? 3. What does it do? 4. How does it do it? The attempted answers mostly depend on a review of published experimental knowledge and theory, but I also make use of the results of a few previously unpublished experiments that have recently been done by Jennifer Moyle in my laboratory. This paper does not follow a historical narrative. Therefore, I would like to remark at the outset that the painstaking work of a multitude of biochemists has contributed to the synthesis attempted here. Many of the threads have been ably drawn together in reviews by Tsoo King.[2,3]

Composition and Topology

The mitochondrial cytochrome system is remarkably compact. It consists of only three proteins: cytochrome *c*, cytochrome *c* reductase (usually called the cytochrome b-c_1 complex or Complex III) and cytochrome *c* oxidase. The polypeptide complexes of cytochrome *c* oxidase and cytochrome *c* reductase each contain about seven types of subunit. Cytochrome *c* oxidase contains four redox components: cytochromes *a* and a_3, and two redox-functional copper ions. Cytochrome *c* reductase contains five redox components: the Rieske ironsulphur center, cytochromes c_1, b_{566}, and b_{562} and ubiquinone (Q). These complexes have been more difficult to recognize as physically discrete protein molecule species than cytochrome *c* because their insolubility in water has presented special technical problems of separation and purification. However, it now appears to be legitimate to regard both cytochrome *c* reductase[4,5] and cytochrome *c* oxidase[6,7] as being well-defined protein molecule species that are soluble, not in (three dimensional) aqueous media, but in the (two dimensional) fluid lipid bilayer medium of the mitochondrial cristae membrane,[8,9] more generally known as the coupling membrane.[10-12] For brevity, I shall sometimes refer to the three proteins of the mitochondrial cytochrome system as *c*, *c*-oxidase and *c*-reductase.

FIGURES 1 and 2 attempt to summarize some of the available information about the composition and topology of the components of the cytochrome system, and suggest possible spatial relationships between the *c*-reductase and *c*-oxidase molecules. In FIGURE 1, the crosshatched central regions of the *c*-reductase and *c*-oxidase molecules, translaid through the bimolecular lipid membrane, represent hydrophobic polypeptides that form a nonaqueous barrier or *B* domain that is physically continuous with the hydrocarbon *B* domain of the lipid membrane. In A, the *c*-reductase and *c*-oxidase molecules, of which

0077-8923/80/0341-0564 $01.75/0 © 1980, NYAS

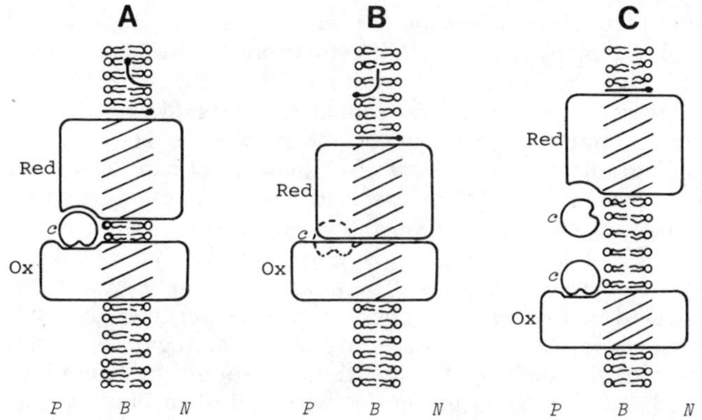

FIGURE 1. Possible relationships between the molecules of cytochrome *c* (*c*), *c*-oxidase (Ox), and *c*-reductase (Red) in the cytochrome system of the mitochondrial cristae membrane. **A** shows indirect complexation of Ox and Red through *c*, which must be free to rotate. **B** shows direct complexation of Ox and Red, with *c* above or below plane of paper. **C** shows surface conduction of *c* between separate Ox and Red molecules. *P* and *N* stand for aqueous domains, and *B* for the nonaqueous osmotic barrier domain of the lipid membrane and of the central (crosshatched) region of the translaid molecules. Ubiquinone is represented as ——•. Further explanations are in the text.

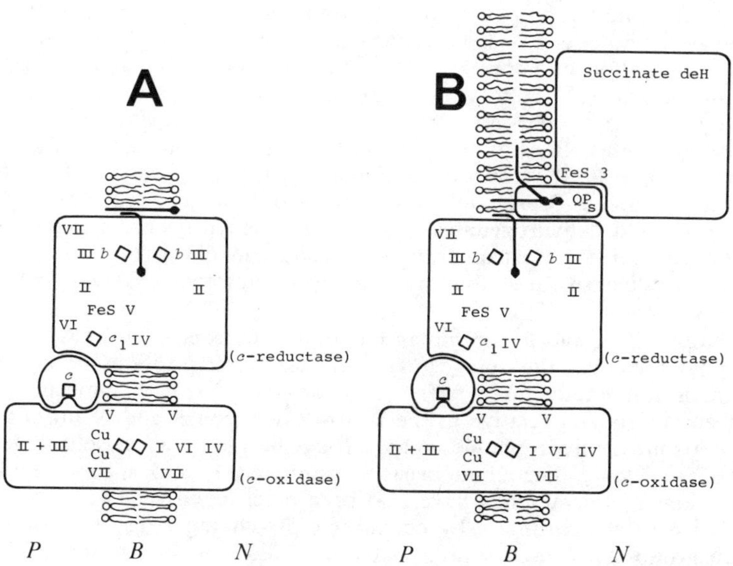

FIGURE 2. Diagrams of cytochrome system topology. **A** represents QH_2 oxidase. **B** represents succinate oxidase. Main conventions are as in FIGURE 1. Other details are in the text.

there are about equal numbers in the membrane, are represented as being complexed only by interaction through cytochrome *c*, which is also supposed to be complexed with the polar head-groups of the lipid molecules in the membrane;[13, 14] but in B, they are represented as being directly complexed by a protein-protein interaction, with cytochrome *c* above or below the plane of the paper in the diagram. I do not know of evidence that can decide between such alternatives at present. In either case, such *c*-reductase-*c*-oxidase complexes may be in dynamic association/dissociation equilibrium in the two-dimensional fluid domain of the membrane. It is also possible that cytochrome *c* may undergo rapid surface conduction[15] between *c*-reductase and *c*-oxidase molecules, as indicated in FIGURE 1C. There is good experimental evidence for the binding of cytochrome *c* to *c*-oxidase and to the c_1 region of *c*-reductase; but a common area of the surface of cytochrome *c* around the haem crevice, which exposes one edge of the haem group,[16, 17] is involved in binding with *c*-oxidase and with *c*-reductase.[18-25] It therefore seems unlikely that cytochrome *c* can form a tight complex with both *c*-oxidase and *c*-reductase simultaneously (but see Chiang and King[26]). A ternary complex of the type represented by FIGURE 1B would presumably be readily dissociable.

In FIGURE 1, A-C, and FIGURE 2A, the *c*-reductase is shown in contact with a Q molecule (——•) of the so-called Q pool[27, 28] of the hydrophobic *B* domain of the lipid membrane, which contains about 10 Q molecules per cytochrome system. In FIGURE 2B, the *c*-reductase is shown complexed with a succinate dehydrogenase molecule via the specific Q-binding protein[3, 5] (QP_s), which, according to computer simulations of electron paramagnetic resonance (EPR) spectra, may bind a pair of interacting Q molecules near to the ironsulphur center 3 (FeS 3) of succinate dehydrogenase. Analysis of the EPR spectra of the coupled semiquinone pair in oriented multilayers made from mitochondria have suggested an edge-to-edge arrangement of the benzenoid rings at a center-to-center separation of about 8 Å, with the intramolecular axis perpendicular to the membrane surface.[29] The positions of the long and bulky isoprenoid chains of the Q molecules associated with QP_s are unknown, and their representation in FIGURE 2B is purely arbitrary, except, of course, that they can be expected to lie in the hydrophobic *B* domain of the membrane system. Similar complexes, involving Q proteins, may be formed between *c*-reductase and other membrane-bound dehydrogenases (e.g., choline dehydrogenase, α-glycerophosphate dehydrogenase, electron transfer flavoprotein dehydrogenase), and also between *c*-reductase and the translaid protonmotive NADH dehydrogenase.[2, 3, 12, 58-60]

In FIGURE 2, A and B, the roman numerals (in order of decreasing molecular weight) represent the approximate, and tentative, positions of polypeptide subunits in the *c*-reductase[4, 5, 30-36] and *c*-oxidase[6, 7, 37-44] molecules. They have been determined relative to the aqueous domains *P* and *N* on either side of the nonaqueous hydrophobic domain *B* by the use of water-soluble labelling reagents (such as *p*-diazonium benzene sulphonate), and specific antibodies. The positions of the subunits have also been determined relative to each other and relative to cytochrome *c* by chemical cross-linking reagents. As different research groups have used source material of different origin and purity, and have employed different media and conditions for dispersing the subunits and separating them by gel electrophoresis, there are uncertainties that include different numbering of some of the subunits. Therefore, the dispositions shown in FIGURE 2, A and B, represent an attempt to sort out partially incompatible data. Duplicate numerals indicate that certain subunits are detectable on both sides of the *B* domain. This may be due either to their static spatial extension across the *B* domain, or to their dynamic mobility to and fro through the *B* domain. In *c*-oxidase, subunit I is the most hydrophobic.[45-47] It is buried in the central *B* domain of the molecule. Subunit II is also relatively hydrophobic,[43, 44] but in *c*-oxidase from yeast mitochondria it contains a reactive -SH group that is accessible to sulphhydryl reagents from the aqueous *P* domain.[48] Cytochrome *c* appears to bind to both subunits II and III of *c*-oxidase from yeast and ox heart mitochondria.[37] In *c*-reductase, subunit II is the most hydrophobic,[46-48] and it is possible that there may be two polypeptides in the *c*-reductase molecule that are described as subunit III or Band III on the gel electrophoresis systems.[28, 48] Subunits IV and V have been positively identified as cytochrome c_1 and the Rieske ironsulphur protein (FeS), respectively,[28, 49-53] as indicated in FIGURE 2. Attempts to identify the subunits normally bound to the haem groups of cytochrome *a* and a_3, and to copper ions of the redox-functional copper in *c*-oxidase, and the subunits normally bound to the haem groups of cytochrome b_{566} and b_{562} in *c*-reductase, must be regarded as speculative at present. The techniques used for dispersal of the subunits are usually so drastic that the noncovalently bound haem molecules and copper atoms are liable to be dissociated from their normal binding sites, and may reassociate with other subunits.[47] Amino acid sequencing studies are showing promise for identifying the haem-binding and Cu-binding regions by sequence homology with known haem and copper proteins, but the work has not quite reached a definitive stage.[45-47] I have not, therefore, explicitly related the *a* cytochrome or redox-functional copper of *c*-oxidase to particular subunits in FIGURE 2. However, I have tentatively attributed subunit III or Band III of *c*-reductase to the *b* cytochromes, because it has been shown that there is a close correspondence between the relatively hydrophobic amino acid composition of the peptide or peptides of Band III and cytochrome *b* isolated by comparatively mild techniques.[3, 51]

The small rectangles in FIGURE 2 represent the haem tetrapyrrole groups of the cytochromes. Using oriented multilayers of *c*-oxidase and *c*-reductase in artificial lipid membranes or in mitochondrial membranes, it has been shown by studies of EPR spectra and polarized optical spectra that the planes of the haem nuclei of cytochromes *a*, a_3, b_{562}, b_{566} and c_1 are nearly normal to the membrane surface.[55-57] This corresponds to an alignment of the EPR *g* tensor giving the largest *g* value (g_z) parallel to the membrane surface. The g_z tensor of the Rieske ironsulphur protein was also found to be parallel to the membrane surface.[57]

In FIGURE 2, the centers of the rectangles representing the haem groups of cytochromes b_{566} and b_{562} of *c*-reductase are arbitrarily placed relative to each other and to the membrane surface, because we have no hard information on this point. However, in *c*-oxidase from yeast, fluorescence energy transfer has been used[48] to measure the distances between the haem *a* groups and the cytochrome *c* bound to subunit III (25 Å), and between a fluorophore covalently linked to subunit II and either the haem *a* (52 Å) or bound cytochrome *c* (35 Å). The haem groups of *c*-oxidase in FIGURE 2 are accordingly shown near the center of the *B* domain, about 25 Å from the nearest approach of cytochrome *c*.

A Q molecule is represented in the *c*-reductase, because preparations of *c*-reductase normally contain bound Q and phospholipid[4, 5] but the Q remains tightly bound after lipid substitution.[58-60] There is also evidence from EPR spectra that *c*-reductase contains at least one Q-reactive site, which is distinct from the QP_s site, at which the Q radical is stabilized by tight binding.[61-64] But there is not, as yet, any direct information about the location of the Q-binding site or sites in or on the *c*-reductase molecule.

The general dimensions and position of the *c*-oxidase molecules relative to the *B* domain of the membrane represented in FIGURES 1 and 2 have been determined from electron micrographs of two-dimensional crystals in vesicular *c*-oxidase preparations obtained from ox heart mitochondria by Triton or deoxycholate extraction.[38, 65] The dimensions of *c*-reductase are known only from negatively-stained electronmicrographs,[4] and the representation in FIGURES 1 and 2 is correspondingly diagrammatic.

To conclude my attempted brief answers to the questions: What is it? and Where is it?, let me emphasize the diagrammatic character of the summary of the cytochrome system represented in FIGURES 1 and 2. These diagrams cannot adequately show the possible dynamic properties and conformational mobility of the components of the system, about which we are still very much in the dark; and I have not attempted to represent the dimension normal to the plane of the paper. Let me, therefore, emphasize that my main aim is to illustrate that the *c*-oxidase and *c*-reductase polypeptide complexes are relatively compact, highly asymmetric, molecules that have a central hydrophobic *B* domain that is continuous with the *B* domain of the bimolecular lipid membrane, and hydrophilic poles that are in contact with the aqueous domains *P* and *N* on either side of the *B* domain. The physical asymmetry of the *c*-oxidase and *c*-reductase molecules, and the fact that all the *c*-oxidase and all the *c*-reductase molecules are translaid the same way round through the mitochondrial cristae membrane, enables us to identify a *P-N* structural polarity of these translaid molecules. I would also emphasize that the *c*-oxidase and *c*-reductase are dissolved in the two-dimensional fluid lipid membrane in a very real sense. This includes the relatively tight binding of a collar of lipid molecules with their hydrophobic chains covering the *B* domain of the protein molecules, and with their polar head-groups linked to polar groups in the hydrophilic extremities of the protein molecules,[58-60, 66-68] much as in the solvation of solutes in three dimensional solutions. The cytochrome *c* may also be regarded as though it were "dissolved" in the surface of the *P* domain, since it may be retained at the surface by polar interaction,[13] but may, nevertheless, have a high lateral mobility in the surface domain.[15] The ubiquinone molecules of the Q pool may also be very mobile in the fluid hydrocarbon *B* domain of the bimolecular lipid membrane,[69, 70] aptly described by Racker as the primary membrane.[71]

Protonmotive Function and Stoicheiometry of the Cytochrome System

The main function of the mitochondrial cytochrome system is illustrated in FIGURE 3. It catalyzes a protonmotive redox reaction in which the transfer of electrons, from the terminal ironsulphur (or other) centers of the respiratory chain-linked dehydrogenases to molecular oxygen, is coupled to the net translocation of protons through the nonaqueous barrier domain *B* from the protonically neutral or negative aqueous domain *N* on one side to the protonically positive aqueous domain *P* on the other side.[12, 72] The dehydrogenase (deH), represented near the top of the diagrams, transfers reducing equivalents to the *c*-reductase via ubiquinone (Q). But it is not known whether the reducing equivalents accepted by the *c*-reductase are electrons (FIGURE 3A) or hydrogen atoms (FIGURE 3B). However, as the diagrams show, this makes no difference to the net protonmotive function of the system, which corresponds to a stoicheiometry (described by the $\leftarrow H^+/e^-$ ratio, or sometimes, more conveniently, as the $\leftarrow H^+/O$ or $\leftarrow H^+/2e^-$ ratio) of $2H^+$ translocated per e^- transferred from the terminal FeS (or other) center of the dehydrogenase to molecular oxygen. That is, $\leftarrow H^+/e^-$ ratio of 2 or a $\leftarrow H^+/O$ ratio of 4.

The protonmotive function of the cytochrome system is generally accepted;[73] but the net stoicheiometry shown in FIGURE 3 is controversial.[72-94] However, the available experimental evidence is surprisingly unanimous about the electron-

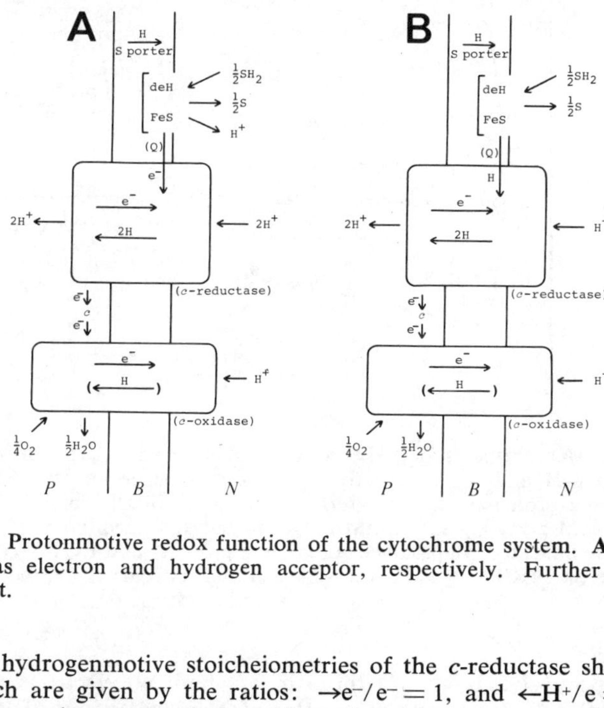

FIGURE 3. Protonmotive redox function of the cytochrome system. **A** and **B** show *c*-reductase as electron and hydrogen acceptor, respectively. Further explanations are in the text.

motive and hydrogenmotive stoicheiometries of the *c*-reductase shown in FIGURE 3, which are given by the ratios: $\rightarrow e^-/e^- = 1$, and $\leftarrow H^+/e = 2$, respectively. Before going on to discuss this, and the more controversial *c*-oxidase function, I must digress briefly to outline the main experimental method on which my thesis depends.

The Respiratory Pulse Method for Measuring Stoicheiometries in Protonmotive Systems

The respiratory pulse method,[95] used to measure $\leftarrow H^+/e^-$, $\rightarrow e^-/e^-$ and $\leftarrow H/e^-$ ratios, is illustrated in FIGURE 4. A lightly buffered suspension of mitochondria or respiratory membrane vesicles is equilibrated anaerobically in a closed thermostatically-controlled glass vessel equipped with a mechanism for vigorous stirring, and with H^+-sensitive and K^+-sensitive electrodes connected to electrometers and recorders that produce a continuous record of the pH and pK of the suspension medium pH_o and pK_o. A pulse of respiration is induced by injecting a small known quantity of air-saturated saline or other standard oxidant solution. In the experiments done in my laboratory with rat liver mitochondria, the amount of oxidant is normally limited to the equivalent of about 1 μg atom O per g of mitochondrial protein. This is less than the quantities of endogenous respiratory chain reductants QH_2, NADH and NADPH, which are about 1.5, 2.5 and 3.5 μmoles per g of mitochondrial

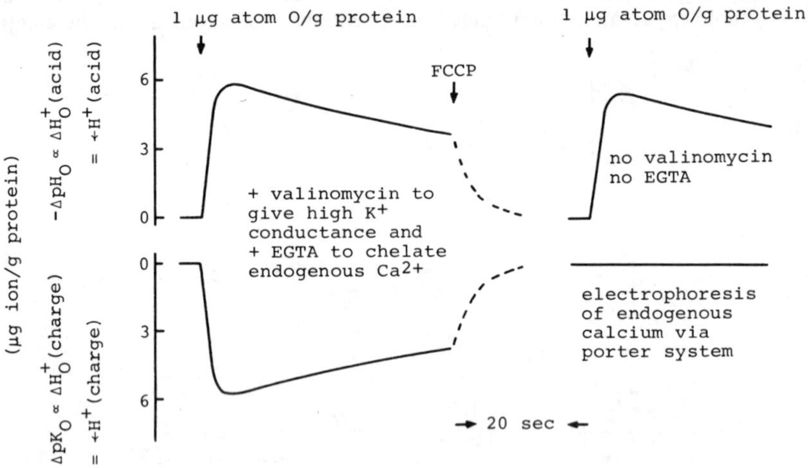

FIGURE 4. Measurements of ΔH_o^+ or $\leftarrow H^+$(acid), and $-\Delta K_o^+$ or $\leftarrow H^+$(charge), equivalent to $\leftarrow H$ and $\rightarrow e^-$, respectively, in O_2-pulse experiments. Rat liver mitochondria (6 mg protein/ml) were preincubated anaerobically in a 150 mM choline chloride, 3.3 mM glycylglycine, 2 mM β-hydroxybutyrate medium at pH_o 7.0–7.1 and at 25°. Where indicated, valinomycin (100 μg/g protein) and EGTA (0.2 mM) were also present; or FCCP (1 μM) was injected. (From Mitchell and Moyle [97] and Mitchell.[98])

protein, respectively. It gives a turnover of about 40 electrons through each *c*-oxidase molecule (assuming 50 nmoles of *c*-oxidase per g of mitochondrial protein) in a pulse lasting about 1 sec, corresponding to an electronic turnover number of about 40 sec^{-1} through the cytochrome system.[96] The reductant oxidized during the pulse may be controlled by inhibitors (antimycin or rotenone) and by adding appropriate oxidants or reductants that react at known sites in the system in appropriate media. In FIGURE 4, the reductant was 2 mM β-hydroxybutyrate in a 150 mM choline chloride medium.

The H^+ ion has two different properties, acidity and electric charge, that can be used to measure its translocation during the respiratory pulse. As illustrated in FIGURE 4, $\leftarrow H^+$ can be measured acidically by the fall of pH_o ($-\Delta pH_o$). It can also be measured electrically (as $\rightarrow e^-$) by the rise of pK_o (ΔpK_o), resulting from the electrophoretic import of K^+ when the membrane is made specifically permeable to K^+ by valinomycin, and the K^+ ion is virtually the only permeant ion species present. To meet the latter requirement with animal mitochondria, the endogenous Ca^{2+} must be immobilized with ethylenediamine tetraacetic acid (EDTA) or ethyleneglycolbis(aminoethyl)-tetraacetic acid (EGTA). By appropriate calibrations, the values of $-\Delta pH_o$ and ΔpK_o, extrapolated to the center of the short respiratory pulse, give values for $\leftarrow H^+$ (acid) and $\rightarrow K^+$, or $\rightarrow e^-$ or $\leftarrow H^+$ (charge), respectively (FIGURE 4). The decays of $-\Delta pH_o$ and ΔpK_o pulses generally have half times of 50–100 sec. It is, therefore, a comparatively simple matter to obtain accurate measurements of $\leftarrow H^+$(acid)/e^- and $\leftarrow H^+$(charge)/e^- ratios. A special advantage of measuring $\rightarrow K^+$, or $\rightarrow e^-$ or $\leftarrow H^+$(charge), as well as $\leftarrow H^+$(acid), is that the $\leftarrow H^+$(charge) value is not depressed by very rapid electroneutral proton-coupled anion-porter activity, such as that of the phosphate-proton symporter at phosphate concentrations around 1 mM. Obviously, measurements of $\rightarrow K^+$ are also valuable in the absence of proton translocation (as in the *c*-oxidase reaction shown in FIGURE 3), when they yield values only for electron translocation $\rightarrow e^-$. The proton translocation values $\leftarrow H^+$ may properly be regarded as the sum of equal and opposite hydrogen and electron translocation values $\leftarrow H$ and $\rightarrow e^-$, respectively.

Function of Cytochrome c *Reductase*

FIGURE 5 shows the results of respiratory pulse experiments done by Jennifer Moyle and me more than a dozen years ago,[97, 98] with rat liver mitochondria in a 150 mM KCl medium, using succinate, in the presence of rotenone, as reductant and either (A) anaerobic saline containing potassium ferricyanide, or (B) air-saturated saline, as the pulsed oxidant. It is well established that ferricyanide does not permeate the cristae membrane,[99] and that it oxidizes cytochrome *c* at the membrane surface, bypassing cytochrome oxidase (see Mitchell,[10] pp. 78–79). Thus, FIGURE 5A shows that, when *c*-reductase is actuated by ferricyanide as oxidant and succinate as reductant, the number of protons (ΔH_o^+) appearing in the *P* domain (see FIGURE 3), outside the membrane, per electron transferred to ferricyanide from succinate, is normally two. But, when the membrane is made permeable to protons by carbonylcyanide *p*-trifluoromethoxyphenylhydrazone (FCCP), only one proton appears outside the membrane (FIGURE 5A). It follows that $2H^+$ normally appear outside and $1H^+$ disappears inside per e^- transferred, as illustrated in FIGURE 3. This is equivalent to 2H travelling outward and one e^- travelling inward through the *B* domain of the *c*-reductase molecule, while one e^- (FIGURE 3A) or one H (FIGURE 3B) is accepted via Q, and one e^- is donated to *c*. This observation that $\rightarrow e^-/e^- = 1$ and $\leftarrow H/e^- = 2$ for *c*-reductase has been confirmed by more recent work in my laboratory.[94] FIGURE 6 illustrates previously unpublished experiments by Jennifer Moyle, using rat liver mitochondria in a medium of low K^+ ion content containing valinomycin and EGTA, with succinate or choline (in the presence of rotenone) as reductant and either O_2 (A), or ferricyanide (B) as oxidant. Under these conditions, the $-\Delta K_o^+/e^- = \rightarrow e^-/e^-$ and $\Delta H_o^+/e^- = \leftarrow H/e^-$ ratios can be measured simultaneously, and the results confirm that with O_2 as oxidant (A) both ratios are close to 2, but with ferricyanide as oxidant (B) the ratios are close to 1 and 2, respectively. In a set of experiments with ferricyanide as oxidant, the values found were: $\rightarrow e^-/e^- = 0.91 \pm 0.04$

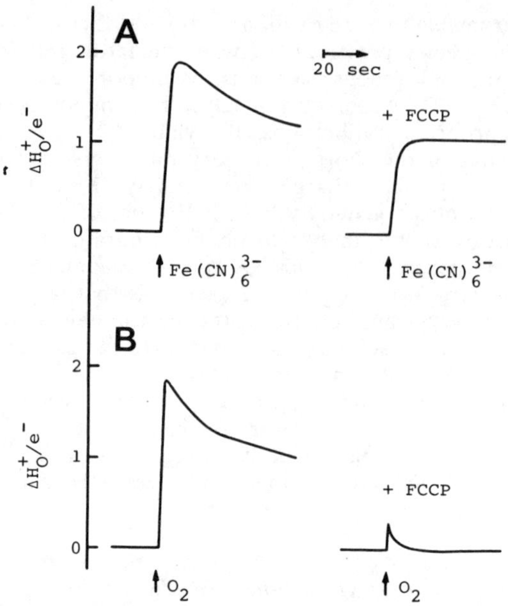

FIGURE 5. Measurements of $\Delta H_o^+/e^-$ ratios in respiratory pulse experiments, using **(A)** ferricyanide or **(B)** oxygen, as oxidant. Rat liver mitochondria (6 mg protein/ml) were preincubated anaerobically in a 150 mM KCl, 3.3 mM glycylglycine, 2 mM succinate medium at pH_o 7.2–7.3 and 25° in the presence of 25 μg rotenone per g of mitochondrial protein. Where indicated, FCCP (1 μM) was present. (From Mitchell and Moyle.[97])

(6 values), and $\leftarrow H/e^- = 1.93 \pm 0.03$ (6 values). In a set of experiments with O_2 as oxidant,[94] the values found were: $\rightarrow e^-/e^- = 1.98 \pm 0.05$ (7 values), and $\leftarrow H/e^- = 2.05 \pm 0.06$ (7 values).

Leung and Hinkle[100] obtained a $\rightarrow e^-/e^-$ value of 0.75 in rotenone-treated ox heart mitochondria, using reduced coenzyme Q_2 as reductant and ferricyanide as oxidant. Leung and Hinkle[100] and Guerrieri and Nelson[101] measured $\rightarrow e^-/e^-$ and $\leftarrow H/e^-$ ratios in phospholipid vesicles translaid with preparations of *c*-reductase from ox heart mitochondria, using reduced coenzyme Q_2 or duroquinol as reductant and ferricytochrome *c* as oxidant. Their results were consistent with $\rightarrow e^-/e^-$ and $\leftarrow H/e^-$ ratios of 1 and 2, respectively. Similar stoicheiometries in rotenone-treated ox heart mitochondria, using duroquinol as reductant and ferricytochrome *c* as oxidant, were also obtained by Papa *et al.*[87] Lehninger *et al.*[77] estimated $\rightarrow e^-/e^-$ ratios, and $\leftarrow H/e^-$ ratios, in suspensions of rat liver mitochondria, in the presence of valinomycin, EGTA, rotenone and *N*-ethylmaleimide (NEM), and ferricyanide, by activating "steady-state" respiration by the addition of succinate.[74] In these experiments, the $\rightarrow e^-/e^-$ and $\leftarrow H_o^+/e^-$ ratios were estimated by the rates of change of pK_o and pH_o (measured with ion-sensitive electrodes) divided by the rate of reduction of ferricyanide (measured spectrophotometrically) during about the first ten seconds after the addition of the succinate. They found values of 0.91 and 1.86 respectively.[77]

There seems to be general agreement that the $\rightarrow e^-/e^-$ and $\leftarrow H/e^-$ ratios for *c*-reductase are close to 1 and 2, respectively, as shown in FIGURE 3.

Function of Cytochrome c *Oxidase*

As illustrated in FIGURE 5B, Jennifer Moyle and I found that when the complete cytochrome system was actuated by oxygen as oxidant and succinate as reductant, the number of protons (ΔH_o^+) appearing in the *P* domain, outside the membrane, per electron transferred from succinate to oxygen was normally two. But when the membrane was made permeable to protons by FCCP, no protons persisted in appearing outside the membrane. It seemed evident that $2H^+$ normally appear outside and $2H^+$ disappear inside per e^- transferred, as illustrated in FIGURE 3. We drew the conclusion from this and other supporting observations a decade ago,[97, 98] (see Mitchell and Moyle [93]) that the reduction and protonation of O_2 to give $2H_2O$, catalyzed by *c*-oxidase, involves the conduction of $4e^-$ from cytochrome *c* at the outer or *P* pole of the molecule to $4H^+$ ions that enter the catalytic site from the inner or *N* pole. Apart from its catalytic role in the reduction of O_2 to H_2O, we therefore attributed a purely electron-conducting function to *c*-oxidase.

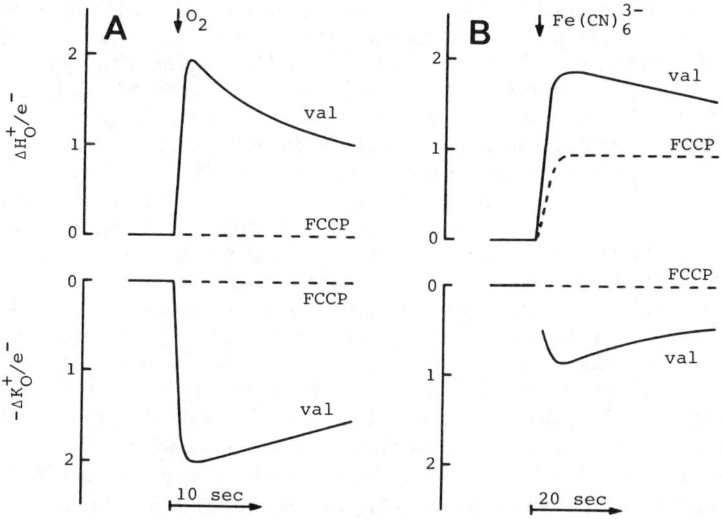

FIGURE 6. Measurements of $\Delta H_o^+/e^-$ and $-\Delta K_o^+/e^-$ ratios in respiratory pulse experiments, using **(A)** air-saturated saline **(B)** anaerobic ferricyanide solution as oxidant. Rat liver mitochondria (6 mg protein/ml) were preincubated anaerobically in media containing 3.3 mM glycylglycine and 0.4 μM rotenone at pH_o 7.0–7.1 and 25° with: **A**, 240 mM sucrose, 10 mM $MgSO_4$, 1 mM EGTA and 0.5 mM succinate; or **B**, 50 mM choline chloride and 1 mM EDTA. Valinomycin (100 μg/g protein) or FCCP (1 μM) were present where indicated. The respiratory pulse was induced by injecting 1 μg atom O in **A**, or 2 μmol $K_3Fe(CN)_6$ in **B**, per g of mitochondrial protein. In **B**, the pK_o traces are corrected for a step-change corresponding to the K^+ injected as $K_3Fe(CN)_6$.

It is, incidentally, noteworthy that, as indicated at the top of the diagrams in FIGURE 3, the reductant substrate (SH_2) carries H into the mitochondria via a solute porter system (S porter).[72] There is a corresponding exit of H, due to exit of $2H_2O$ per O_2 entering the active site of *c*-oxidase. This H is represented in brackets in *c*-oxidase in FIGURE 3, because it is not known whether O_2 and H_2O enter and leave the active site directly through the *P* pole of *c*-oxidase, or whether they may travel via the neighboring region of the lipid membrane. Possible evidence for an oxygen "pocket" in *c*-oxidase may be relevant in this context.[102] Chance *et al.* represented the "pocket" as a channel entering the *c*-oxidase molecule from the *N* pole.[102] Perhaps entry from the (external) *P* pole would be slightly more kinetically favorable, and might be more likely from the evolutionary standpoint. The *c*-oxidase might thus be formally represented as a protonmotive redox loop. But, at all events, these considerations do not affect the observed electron-conducting function and effective stoicheiometry of the *c*-oxidase reaction.

Observations by Wikström *et al.*[103, 104] on acid-base changes accompanying *c*-oxidase activity in whole mitochondria, in submitochondrial vesicles, and in liposomes translaid with *c*-oxidase, have encouraged the view that the *c*-oxidase molecule is equipped with a proton pump that translocates one proton from *N* to *P* per electron transferred from cytochrome *c* to oxygen.[81–84, 86] Lehninger *et al.* went further.[105] They concluded that not one but two protons are pumped by *c*-oxidase per electron transferred,[77] bringing the overall $\leftarrow H^+/O$ ratio of the cytochrome system up to 8,[74–80] and suggesting that our measurements,[72, 91–94] giving a normal $\leftarrow H^+/e^-$ value of 2, or a $\leftarrow H^+/O$ value of 4, for the complete cytochrome system (FIGURES 5A and 6B), are wrong by a quite embarrassing factor of 2. This naturally led us to look hard for the proton-pumping function of *c*-oxidase in mitochondria, since it seemed that we had not merely under-estimated it, but had missed it altogether.

The most direct way of measuring the electronmotive and proton-pumping function of mitochondrial *c*-oxidase is to do respiratory pulse experiments with antimycin-inhibited mitochondria, using O_2 as the pulsed oxidant, and using added ferrocytochrome *c*, or substrates that reduce the mitochondrial cytochrome *c*, such as ferrocyanide, N,N,N′,N′-tetramethyl-*p*-phenylenediamine (TMPD) or 2,3,5,6-tetramethyl-*p*-phenylenediamine (diaminodurene or $DADH_2$), as reductant.[106, 107] However, as illustrated in FIGURE 7, one has to guard against an experimental hazard, recognized many years ago in work on the bypassing of the antimycin-inhibited site by TMPD.[108] Wurster's blue, the stable free radical oxidant form of TMPD, reacts very readily with some component on the reducing side of the antimycin-inhibited site. In control experiments, Jennifer Moyle and I found that, during oxidation of TMPD or ferrocyanide by rat liver mitochondria in the presence of antimycin and FCCP, the rates of consumption of $\frac{1}{2}O_2$ and $2H^+$ were not equal (see FIGURE 7), as they should have been if ferrocyanide or TMPD were the only reductant.[106] There was a substantial excess O consumption attributable to the oxidation of an unidentified mitochondrial component (possibly the Q radical QH) that is in redox contact with endogenous mitochondrial reductants (SH_2 in FIGURE 7) via a protonmotive system. Papa *et al.* reached a similar conclusion, using somewhat different methods.[109] With TMPD, the factor *n* (FIGURE 7), representing the excess O or $2H^+$ shortfall, can amount to as much as 40% of the O reduced. Fortunately, however, *n* is negligibly small when added cytochrome *c* or the two-electron-transfer agent $DADH_2$ is used as reductant in appropriate media.[106, 107]

FIGURE 8 illustrates results of O_2-pulse experiments on rat liver mitochondria in the presence of rotenone and antimycin, using $DADH_2$ as reductant in a 240 mM sucrose, 10 mM $MgSO_4$, 1 mM EGTA medium at pH_O 7.0–7.1. Both pH_o and pK_o changes were measured to enable us to estimate $\rightarrow e^-/O$ and $\leftarrow H^+/O$ values by appropriate use of valinomycin (A,C) and FCCP (B,D). The effect of NEM was also observed (C,D), because, according to Lehninger *et al.*, an underestimation of $\Delta H_o^+/O$ by as much as a factor of 2 is to be expected unless the proton-coupled phosphate porter is inhibited by NEM. After a small correction for the normally protonmotive antimycin-insensitive

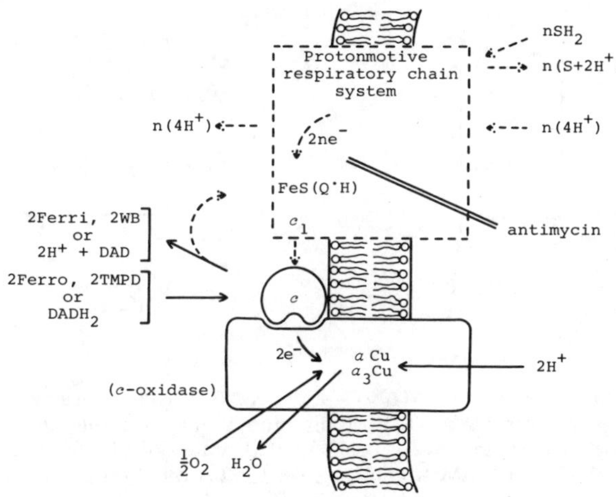

FIGURE 7. Diagram of mitochondrial *c*-oxidase with natural and artificial redox connections. Ferri and ferro stand for ferricyanide or ferricytochrome *c* and ferrocyanide or ferrocytochrome *c*, respectively. WB means Wurster's blue, the free-radical oxidized form of TMPD, and $DADH_2$ means diaminodurene. An unidentified respiratory reductant is represented by SH_2, and FeS stands for the Rieske ironsulphur protein. Other details are in the text.

respiration that proceeds at about 5% of the rate of $DADH_2$ oxidation, the $\Delta H_o^+/O$ ratio was found to be 1.98 ± 0.08 (8 values) and 1.98 ± 0.05 (8 values) in the absence (A) and in the presence (B) of NEM, respectively. The observed pH_O change for oxidation of $DADH_2$ to DAD in the presence of FCCP was negligibly small, confirming that $DADH_2$ oxidation to DAD at the outer surface of the mitochondria under the conditions of these experiments liberates 2.00 H^+ ions. Hence, we observed a proton-pumping function of *c*-oxidase corresponding to —0.02 protons pumped per 0 reduced in the presence or absence of NEM, with a standard deviation not exceeding 0.08 for 16 measurements.[93, 106, 107]

The electrophoretic $\rightarrow K^+/O$ ratio, corresponding to the $\rightarrow e^-/O$ ratio of *c*-oxidase, was found to be 2.03 ± 0.10 (8 values) and 2.01 ± 0.07 (8 values) in the absence and presence of NEM, respectively.[93, 106, 107]

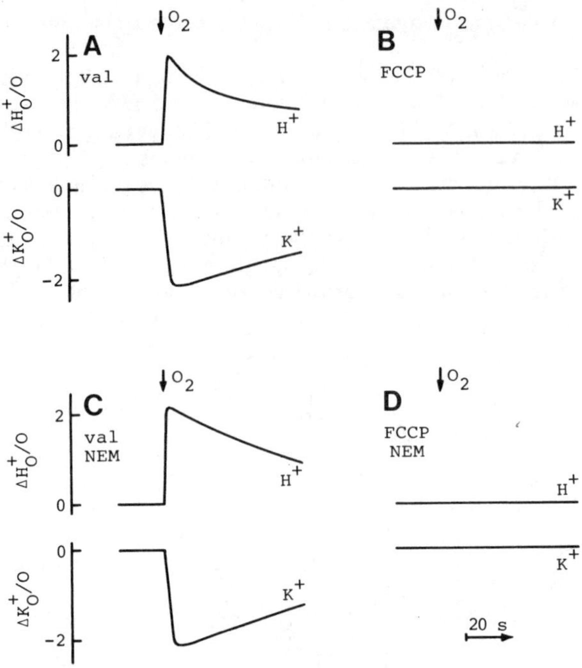

FIGURE 8. Measurement of $\Delta H_o^+/O$ and $\Delta K_o^+/O$ ratios characteristic of *c*-oxidase in O_2-pulse experiments with 0.6 mM diaminodurene as reductant. Rat liver mitochondria (6 mg protein/ml) were preincubated anaerobically in a 240 mM sucrose, 3.3 mM glycylglycine, 10 mM $MgSO_4$, 1 mM EGTA medium at pH_o 7.0–7.1 and 25°. Antimycin (36 μg/g protein) and rotenone (0.4 μM) were present in all experiments. Valinomycin (200 μg/g protein), FCCP (1 μM) and NEM (0.2 mM) were present where indicated. The respiratory pulse was caused by injecting 2.5 μg atom O per g of mitochondrial protein. (From Moyle and Mitchell.[107])

Preliminary O_2-pulse experiments with ferrocytochrome *c* as reductant, using digitonin to facilitate the entry of the ferrocytochrome *c* through the outer mitochondrial membrane,[110] revealed a remarkable, and potentially misleading, net-acidification step-change in pH_o. This net-acidification artefact occurred synchronously with oxidation of part of the ferrocytochrome *c* to ferricytochrome *c* at the surface of the mitochondria.[91, 93, 106, 107] A similar net-acidification artefact associated with oxidation of exogenous ferrocytochrome *c* was observed by Papa *et al.*[88, 89] using a stop-flow method, with suspensions of ox heart mitochondria. It may be explained by the fact that ferrocytochrome *c* binds cations, whereas ferricytochrome *c* binds anions.[3, 111] It is well known that cytochrome *c* binds tightly to the mitochondrial cristae membrane,[13] and we noted that addition of more than 25 μM ferrocytochrome *c* to suspensions of rat liver mitochondria (6 mg of protein/ml) caused a significant increase in the proton conductance of the membrane in the presence of digitonin. Presumably the oxidation of cytochrome *c* induces an effective increase of the acidic dissociation constant of local acid-base groups with which the cytochrome *c* is associated in the mitochondrial membrane system. Accordingly, the net-acidification artefact was sensitive to the ionic composition of the medium. It could be completely eliminated by omitting the digitonin and working with 25 μM ferrocytochrome *c* in a 240 mM sucrose, 10 mM $MgSO_4$, 1 mM EGTA medium.[106, 107]

FIGURE 9 illustrates O_2-pulse experiments like those of FIGURE 8, but using 25 μM horse heart ferrocytochrome *c* as reductant. In this type of experiment, antimycin-insensitive respiration proceeds during the respiratory pulse at about 10% of the rate of ferrocytochrome *c* oxidation. It accounts quantitatively for the ΔH_o^+ pulses observed in A and C, thus confirming that we do not observe a significant proton-pumping function of *c*-oxidase. The secondary electrophoretic $\rightarrow K^+/O$ ratio (A) and $\rightarrow H^+/O$ ratio (B), which represent the $\rightarrow e^-/O$ ratio of *c*-oxidase, were found to be 1.98 ± 0.06 (8 values) and 1.96 ± 0.07 (10 values), respectively. In the presence of NEM, these values were 1.95 ± 0.01 (2 values) and 2.01 ± 0.07 (4 values), respectively.[93, 106, 107]

There are two different ways in which we should have been able to see the proton-pump of *c*-oxidase in the experiments of FIGURES 7 and 8, because the proton has the properties of both electric charge and acidity. Therefore, we should have seen additional charge translocation, over and above the $\rightarrow e^-/O$ value of 2, characteristic of the purely electron-conducting function of *c*-oxidase; and we should have seen a real value of $\leftarrow H^+/O$. In fact, we have succeeded in seeing neither, even when, adopting the precaution advised by

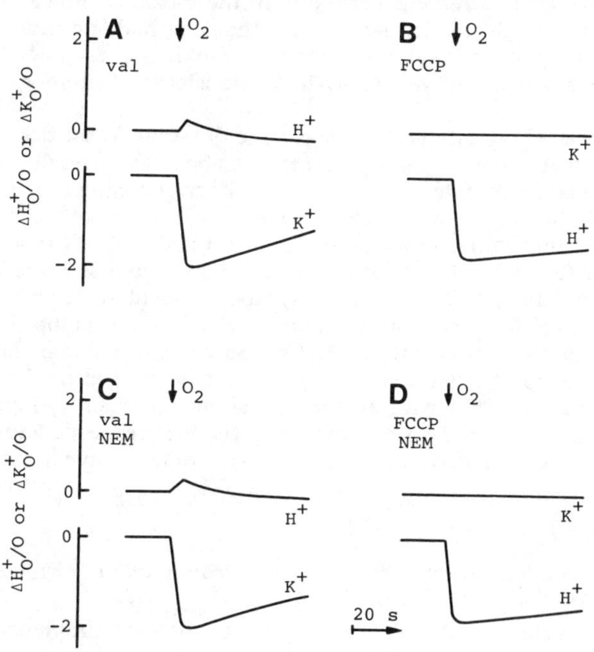

FIGURE 9. Measurements of $\Delta H_o^+/O$ and $\Delta K_o^+/O$ ratios characteristic of *c*-oxidase in O_2-pulse experiments with 25 μM ferrocytochrome *c* as reductant. Other details are as in FIGURE 8.

Lehninger *et al.*,[74] we blocked the proton-linked phosphate porter with NEM. We are not alone in this difficulty. Papa and colleagues have likewise tried hard to observe the proton-pumping function of *c*-oxidase, but have failed to detect it.[87-90] Sorgato and Ferguson [112] and Brand and colleagues [113] thought that they had seen proton pumping by *c*-oxidase in steadily respiring suspensions of antimycin-treated submitochondrial vesicles and intact mitochondria, using TMPD or isoascorbate + ferrocyanide as reductant. But I am afraid that they might well have been deceived by the excess O or bypass type of artefact illustrated in FIGURE 7. The same ambiguity may also apply to the observations by Wikström and colleagues [82, 83, 103, 104] on ferrocyanide oxidation by intact mitochondria, and TMPD oxidation by submitochondrial particles, in the presence of antimycin. Also, in their experiments using ferrocytochrome *c* as reductant, with antimycin-treated rat liver mitochondria, Wikström and Krab [83] added such a high concentration of ferrocytochrome *c* (118 μM) that the external acidification that they observed may well have been due to the net-acidification artefact that we identified, or to some other phenomenon, consequent upon partial disorganization of the mitochondrial cristae membrane.

The acid pulses seen during ferrocytochrome *c* oxidation by suspensions of liposomes translaid with *c*-oxidase [81-84] (and see Casey *et al.*[86]), which appear only at higher cytochrome *c* concentrations under certain conditions,[114] are interesting in themselves, whether they are produced by a net-acidification artefact, or by a real proton-translocation phenomenon. However, in view of the known strong interaction of cytochrome *c* with lipid films,[13, 115] and the presence of residual detergent (cholate) in the *c*-oxidase preparations used for these studies, it might be unwise to infer that one had succeeded in observing the proton-pump of normal mitochondrial *c*-oxidase, even if there were no doubt that the artificial vesicle systems translocated protons under certain conditions.

To summarize my attempt to answer the question, What does it do?, let me remind you that there is general agreement about the overall function of the *c*-reductase, as represented in FIGURE 3. Even Lehninger *et al.*, who have found much higher chemiosmotic stoicheiometries in other regions of the respiratory chain,[74-80] have observed $-\Delta K_o^+/O$ and $\Delta H_o^+/O$ ratios that confirm that the $\rightarrow e^-/e^-$ and $\leftarrow H/e^-$ ratios are close to 1 and 2 respectively for the *c*-reductase reaction.[77] The controversy about *c*-oxidase relates to the proton pump. There are three schools of thought. Some find a pump that pumps two protons per electron transferred. [74-80, 85] Some find a pump that pumps one proton per electron transferred.[81-84, 86] Some have looked hard but have failed to find any pump.[87-95] Owing to the prevailing uncertainty, I propose, in the final section of my paper, to consider only the electron-conducting function of *c*-oxidase represented in FIGURE 3, which is common to all three schools of thought.

Ligand Conduction Mechanisms in the Protonmotive Cytochrome System

For the biochemist, the big question about the cytochrome system has long been, in the recent words of Chiang and King: [26] How does it work?

There are two closely linked functions that we have to explain: one chemical, the other osmotic. They have often been regarded as mechanistically separate: one involving catalytic chemical transformation, the other involving a proton pump (see Boyer *et al.*[73]). In my opinion, we may most easily come to understand what is going on biochemically by regarding the chemical and the osmotic functions as different aspects of one and the same vectorial (or higher tensorial order) chemical process that is catalyzed through a nonaqueous osmotic barrier or *B* domain, so that it has osmotic as well as chemical consequences.[116] Moreover, such a direct chemiosmotic process can be relatively simply conceived in terms of the principle of specific ligand conduction,[72] which is a kinetic extension of the well-known biochemical principle of specific ligand binding in enzymic catalysis and metabolism. Ligand conduction dia-

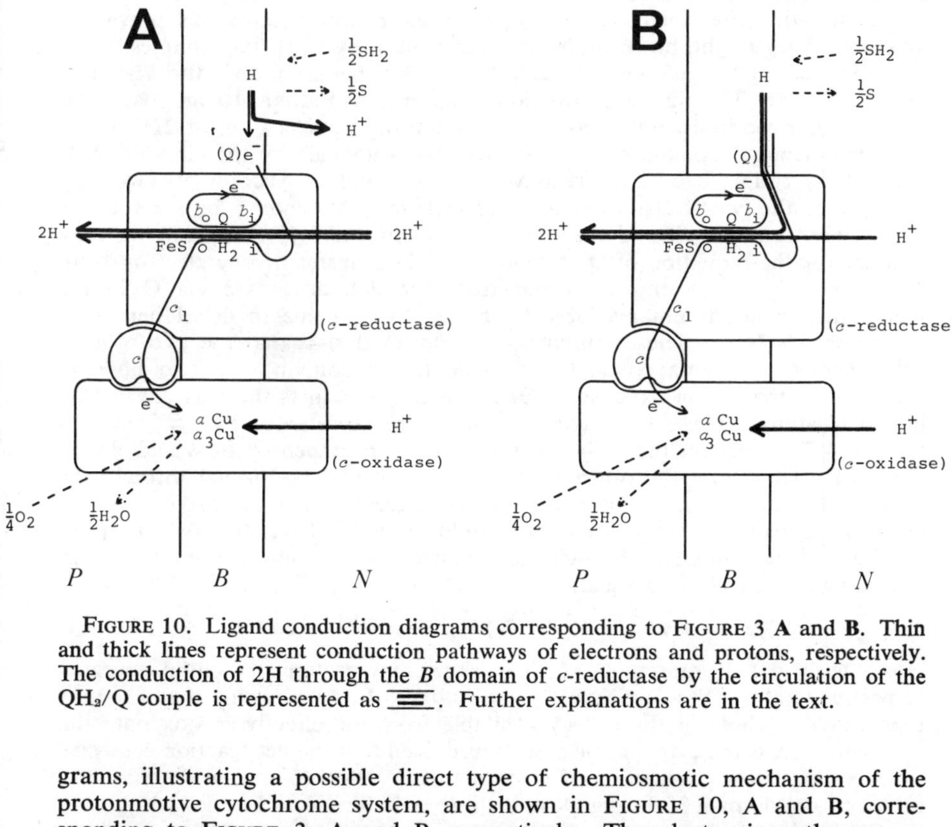

FIGURE 10. Ligand conduction diagrams corresponding to FIGURE 3 **A** and **B**. Thin and thick lines represent conduction pathways of electrons and protons, respectively. The conduction of 2H through the *B* domain of *c*-reductase by the circulation of the QH_2/Q couple is represented as ≡. Further explanations are in the text.

grams, illustrating a possible direct type of chemiosmotic mechanism of the protonmotive cytochrome system, are shown in FIGURE 10, A and B, corresponding to FIGURE 3, A and B, respectively. The electronic pathways are represented by thin lines, and the protonic pathways by thick lines.

In *c*-oxidase, we have to account for the orderly catalytic reduction of O_2 to $2H_2O$, using 4 electrons that enter the molecule from cytochrome *c* at the *P* pole, and $4H^+$ that enter from the *N* pole. The mechanistic details are beginning to emerge from sophisticated spectroscopic and other studies in many laboratories.[6, 7, 116–125] These studies indicate, at least in principle, how the four-electron/four-proton transfer process is catalyzed by the O_2-binding center involving haem a_3, by the four electron-binding centers (haem *a*, haem a_3 and 2Cu centers), and by H^+-binding groups,[126–128] which, judging from pH-dependent midpoint potentials,[126, 127] may include amino acid side-groups at the haem a and a_3 centers. However, an elementary question, that continues to puzzle many workers, is how the electrons are conducted, with a normal turnover number around 50 sec^{-1}, over an estimated distance of 25 Å, from cytochrome c to the haem a groups.[3, 129, 131] Vibronically-coupled electron tunneling is an attractive possibility.[132–134] Another possibility is that specifically articulated mobility of certain subunits may allow a closer approach of cytochrome c to the haem a groups than is generally supposed. It is not always remembered in discussions of this question that the electric membrane potential of respiring mitochondria corresponds to a potential gradient of about half a million volts per cm across the 4 nm thickness of the B domain of the membrane, and that the electrons from cytochrome c flow against this prevailing gradient. They might, however, be shielded from it by local fixed dipoles.[135, 136]

Relatively little is known about the biochemistry of the c-reductase.[4, 5, 53, 61–64] The ligand-conduction diagrams of FIGURE 10 are based on the Q cycle hypothesis,[11, 12] according to which the translocation of $2H^+$ across the nonaqueous B domain of the c-reductase is catalyzed by the circulation of the QH_2/Q couple between Q-reactive centers o and i. Accordingly, we may account for the two-electron oxidation of QH_2 to Q at center o by the transfer of one electron to FeS and the other to cytochrome b_{566} (b_o); and we may account for the reduction of Q at center i by the transfer of one electron from cytochrome b_{562} (b_i) and the other from the dehydrogenase via Q in the membrane. Space is not available to discuss this scheme in detail here (but see [5, 12, 53, 61, 63, 64]). The essential point (FIGURE 10) is that the two protons are conceived as being conducted through the B domain with two bonding electrons on the quinol oxygens of QH_2 that thus conducts them as covalently-bound H atoms. These two hydrogen atoms are represented by $\equiv$ in FIGURE 10. The electron recirculating through the b cytochromes would do no net work. Therefore, the other electron, which would be pushed through the c-reductase via centers i and o by a force corresponding to a steep electrochemical potential energy gradient, would directly drive the two protons, covalently bound in QH_2, through the B domain by a symport type of process, according to the following equation:

$$\Delta\tilde{\mu}e^- \equiv \tilde{\mu}e_i^- - \tilde{\mu}e_o^- \geqq 2(\tilde{\mu}H_P^+ - \tilde{\mu}H_N^+) \equiv 2\Delta\tilde{\mu}H^+ \qquad (1)$$

Here, $\tilde{\mu}e^-$ and $\tilde{\mu}H^+$ represent total electronic and protonic potential energies, respectively, at the sites indicated by the suffixes. It is interesting that $\Delta\tilde{\mu}e$ may represent the whole of the energy available from the effective redox potential difference across the cytochrome system, provided that the net reaction catalyzed by c-oxidase is the conduction of electrons from cytochrome c in domain P to protons from domain N during O_2 reduction, as indicated in FIGURE 10.

Obviously, much more experimental knowledge will be needed to enable us to give a really satisfactory answer to the question: How does it work? But perhaps we are beginning to see glimpses of the general principles of the chemiosmotic molecular mechanisms. Only time can tell.

ACKNOWLEDGMENTS

I thank my research colleague Dr. Jennifer Moyle for constant advice and help, and also Dr. Efraim Racker, Dr. Ian Ragan and Dr. Bernie Trumpower for discussion about the organization of the components of the cytochrome system, and for guidance in the extensive literature. I am indebted to Mr. Robert Harper, Mrs. Stephanie Key and Miss Rosemary Sammons for their help in preparing the manuscript. The general financial support of Glynn Research Ltd. is gratefully acknowledged.

References

1. Keilin, D. 1966. The History of Cell Respiration and Cytochrome. Cambridge University Press. Cambridge.
2. King, T. E. 1966. Adv. Enzymol. **28:**155–236.
3. King, T. E. 1978. FEBS Symp. **45:**17–31.
4. Rieske, J. S. 1976. Biochim. Biophys. Acta **456:**195–247.
5. Trumpower, B. L. & A. G. Katki. 1979. *In* Membrane Proteins in Energy Transduction. R. A. Capaldi, Ed.: 89–200. Marcel Dekker. New York.
6. Erecinska, M. & D. F. Wilson. 1978. Arch. Biochem. Biophys. **188:**1–14.
7. King, T. E., Y. Orii, B. Chance & K. Okunuki, Eds. 1979. Cytochrome Oxidase. Elsevier/North-Holland. Amsterdam.
8. Singer, S. J. & G. L. Nicholson. 1972. Science **175:**720–731.
9. Hackenbrock, C. R. & M. Höchli. 1977. FEBS Symp. **42:**10–36.
10. Mitchell, P. 1966. Chemiosmotic Coupling in Oxidative and Photosynthetic Phosphorylation. Glynn Research. Bodmin.
11. Mitchell, P. 1975. *In* Electron Transfer Chains and Oxidative Phosphorylation. J. M. Tager *et al.,* Eds.: 305–316. Elsevier/North-Holland. Amsterdam.
12. Mitchell, P. 1976. J. Theoret. Biol. **62:**327–367.
13. Nicholls, P. 1974. Biochim. Biophys. Acta **346:**261–310.
14. Cannon, J. B. & J. E. Erman. 1978. Biochem. Biophys. Res. Commun. **84:** 254–260.
15. Roberts, H. & B. Hess. 1977. Biochim. Biophys. Acta **462:**215–234.
16. Swanson, R., B. L. Trus, N. Mandel, O. B. Kallai & R. E. Dickerson. 1977. J. Biol. Chem. **252:**759–775.
17. Takano, T., B. L. Trus, N. Mandel, G. Mandel, O. B. Kallai, R. Swanson & R. E. Dickerson. 1977. J. Biol. Chem. **252:**776–785.
18. Ferguson-Miller, S., H. Weiss, S. H. Speck, D. L. Brautigan, N. Osheroff & E. Margoliash. 1979. *In* Cytochrome Oxidase. T. E. King *et al.,* Eds.: 281–292. Elsevier/North-Holland. Amsterdam.
19. Bisson, R., H. Gutweniger & A. Azzi. 1978. FEBS Lett. **92:**219–222.
20. Rieder, R. & H. R. Bosshard. 1978. J. Biol. Chem. **253:**6045–6053.
21. Rieder, R. & H. R. Bosshard. 1978. FEBS Lett. **92:**223–226.
22. Chiang, Y.-L., L. S. Kaminsky & T. E. King. 1976. J. Biol. Chem. **251:**29–36.
23. Ahmed, A. J., H. T. Smith, M. B. Smith & F. S. Millett. 1978. Biochemistry **17:**2479–2483.
24. Weiss, H. & B. Juchs. 1978. Eur. J. Biochem. **88:**17–28.
25. Speck, S. H., S. Ferguson-Miller, N. Osheroff & E. Margoliash. 1979. Proc. Nat. Acad. Sci. **76:**155–159.
26. Chiang, Y.-L. & T. E. King. 1979. J. Biol. Chem. **254:**1845–1853.
27. Kröger, A. & M. Klingenberg. 1973. Eur. J. Biochem. **34:**358–368.
28. Kröger, A. 1974. BBA Library **13:**103–110.
29. Salerno, J. C., H. J. Harmon, H. Blum, J. S. Leigh & T. Ohnishi. 1977. FEBS Lett. **82:**179–182.
30. Augen, J., S. Power & G. Palmer. 1979. Biochem. Biophys. Res. Commun. **86:**271–277.
31. Bell, R. L., J. Sweetland, B. Ludwig & R. A. Capaldi. 1978. *In* Frontiers of Biological Energetics. P. L. Dutton *et al.,* Eds. Vol. **1:**155–164. Academic Press. New York.

32. Gellerfors, P. & B. D. Nelson. 1977. Eur. J. Biochem. **80:**275–282.
33. Grigolava, I. V. & A. Konstantinov. 1977. FEBS Lett. **78:**36–40.
34. Mendel-Hartvig, I. & B. D. Nelson. 1978. FEBS Lett. **92:**36–40.
35. Nelson, B. D. & I. Mendel-Hartvig. 1977. Eur. J. Biochem. **80:**267–274.
36. Smith, R. J., R. A. Capaldi, D. Muchmore & F. Dahlquist. 1978. Biochemistry **17:**3719–3723.
37. Bisson, R., C. Montecucco, H. Gutweniger & A. Azzi. 1978. *In* Frontiers of Biological Energetics. P. L. Dutton *et al.,* Eds. Vol. **2:**809–816. Academic Press. New York.
38. Frey, T. G., G. Schatz & S. H. P. Chan. 1978. *In* Frontiers of Biological Energetics. P. L. Dutton *et al.,* Eds. Vol. **2:**817–824. Academic Press. New York.
39. Chan, S. H. P. & R. P. Tracy. Eur. J. Biochem. **89:**595–605.
40. Eytan, G. D., R. C. Carroll, G. Schatz & E. Racker. 1975. J. Biol. Chem. **250:**8598–8603.
41. Eytan, G. D. & G. Schatz. 1975. J. Biol. Chem. **250:**767–774.
42. Frey, T. G., S. H. P. Chan & G. Schatz. 1978. J. Biol. Chem. **253:**4389–4395.
43. Hackenbrock, C. R. & K. M. Harmon. 1975. J. Biol. Chem. **250:**9185–9197.
44. Vanderkooi, J., J. S. Leigh, C. S. Owen, P. Glatz & H. Blum. 1978. *In* Frontiers of Biological Energetics. P. L. Dutton *et al.,* Eds. Vol. **1:**54–62. Academic Press. New York.
45. Buse, G., G. J. Steffens, G. C. M. Steffens & R. Sacher. 1978. *In* Frontiers of Biological Energetics. P. L. Dutton *et al.,* Eds. Vol. **2:**799–807. Academic Press. New York.
46. Steffens, G. J. & G. Buse. 1979. *In* Cytochrome Oxidase. T. E. King *et al.,* Eds. **:**79–90. Elsevier/North-Holland. Amsterdam.
47. Yasunobu, K. T., M. Tanaka, M. Haniu, M. Sameshima, N. Reimer, T. Eto, T. E. King, C.-A. Yu, L. Yu & Y.-H. Wei. 1979. *In* Cytochrome Oxidase. T. E. King *et al.,* Eds. **:**91–101. Elsevier/North-Holland. Amsterdam.
48. Dockter, M. E., A. Steinemann & G. Schatz. 1978. J. Biol. Chem. **253:**311–317.
49. Bell, R. L. & R. A. Capaldi. 1976. Biochemistry **15:**996–1001.
50. Katan, M. B., L. Pool & L. Groot. 1976. Eur. J. Biochem. **65:**95–105.
51. Yu, L., C.-A. Yu & T. E. King. 1977. Biochim. Biophys. Acta **495:**232–247.
52. Marres, C. A. M. & E. C. Slater. 1977. Biochim. Biophys. Acta **462:**531–548.
53. Trumpower, B. L. & C. A. Edwards. 1979. FEBS Lett. **100:**13–16.
54. Von Jagow, G., H. Schägger, W. D. Engel, W. Machleidt, I. Machleidt & H. J. Kolb. 1978. FEBS Lett. **91:**121–125.
55. Erecinska, M., D. F. Wilson & J. K. Blasie. 1979. *In* Cytochrome Oxidase. T. E. King *et al.,* Eds. **:**5–8. Elsevier/North-Holland. Amsterdam.
56. Erecinska, M., D. F. Wilson & J. K. Blasie. 1979. Biochim. Biophys. Acta **545:**352–364.
57. Erecinska, M. & D. F. Wilson. 1979. Arch. Biochem. Biophys. **192:**80–85.
58. Ragan, C. I. & C. Heron. 1978. Biochem J. **174:**783–790.
59. Heron, C., C. I. Ragan & B. L. Trumpower. Biochem. J. **174:**791–800.
60. Heron, C., M. Gore & C. I. Ragan. 1979. Biochem. J. **178:**415–426.
61. Konstantinov, A. A. & E. K. Ruuge. 1977. FEBS Lett. **81:**137–141.
62. Siedow, J. N., S. Power, F. F. De La Rosa & G. Palmer. 1978. J. Biol. Chem. **253:**2392–2399.
63. Yu, C. A., S. Nagaoka, L. Yu & T. E. King. 1978. Biochem. Biophys. Res. Commun. **82:**1070–1078.
64. Ohnishi, T. & B. L. Trumpower. 1980. J. Biol. Chem. In press.
65. Goldfarb, W., J. Frank, M. Kessel, J. C. Hsung, C. H. Kim & T. E. King. 1979. *In* Cytochrome Oxidase. T. E. King *et al.,* Eds. **:**161–167. Elsevier/North-Holland. Amsterdam.
66. Eytan, G. D. & R. Broza. 1978. J. Biol. Chem. **253:**3196–3202.
67. Griffith, H. & P. C. Jost. 1979. *In* Cytochrome Oxidase. T. E. King *et al.,* Eds. **:**207–218. Elsevier/North-Holland. Amsterdam.

68. Yu, C. A., L. Yu & T. E. King. 1979. *In* Cytochrome Oxidase. T. E. King *et al.,* Eds. :219–227. Elsevier/North-Holland. Amsterdam.
69. Hauska, G. 1977. *In* Bioenergetics of Membranes. L. Packer *et al.,* Eds. :177–187. Elsevier/North-Holland. Amsterdam.
70. Hauska, G. 1978. *In* Photosynthesis 77. Proc. Fourth Int. Congr. Photosynth. D. O. Hall *et al.,* Eds. :185–196. Biochem. Soc. London.
71. Racker, E. 1970. Essays Biochem. **6:**1–22.
72. Mitchell, P. 1979. Eur. J. Biochem. **95:**1–20.
73. Boyer, P. D., B. Chance, E. Ernster, P. Mitchell, E. Racker & E. C. Slater. 1977. Ann. Rev. Biochem. **46:**955–1026.
74. Lehninger, A. L. 1978. *In* Protons and Ions Involved in Fast Dynamic Phenomena. P. Laszlo, Ed. :435–452. Elsevier/North-Holland. Amsterdam.
75. Lehninger, A. L., B. Reynafarje & A. Alexandre. 1978. *In* Frontiers of Biological Energetics. P. L. Dutton *et al.,* Eds., Vol. **1:**384–393. Academic Press. New York.
76. Reynafarje, B. & A. L. Lehninger. 1978. J. Biol. Chem. **253:**6331–6334.
77. Alexandre, A., B. Reynafarje & A. L. Lehninger. 1978. Proc. Nat. Acad. Sci. **75:**5296–5300.
78. Vercesi, A., B. Reynafarje & A. L. Lehninger. 1978. J. Biol. Chem. **253:** 6379–6385.
79. Reynafarje, B. & A. L. Lehninger. 1978. Proc. Nat. Acad. Sci. **75:**4788–4792.
80. Lehninger, A. L. 1980. Ann. N.Y. Acad. Sci. This volume.
81. Krab, K. & M. Wikström. 1978. Biochim. Biophys. Acta **504:**200–214.
82. Wikström, M. & K. Krab. 1978. *In* Frontiers of Biological Energetics. P. L. Dutton *et al.,* Eds., Vol. **1:**351–358. Academic Press. New York.
83. Wikström, M. & K. Krab. 1979. *In* Cation Flux Across Biomembranes. Y. Mukohata & L. Packer, Eds.: 321–329. Academic Press. New York.
84. Sigel, E. & E. Carafoli. 1978. Eur. J. Biochem. **89:**119–123.
85. Azzone, G. F., T. Pozzan, F. Di Virgilio & V. Miconi. 1978. *In* Frontiers of Biological Energetics. P. L. Dutton *et al.,* Eds., Vol. **1:**375–383. Academic Press. New York.
86. Casey, R. P., M. Thelen & A. Azzi. 1979. Biochem. Biophys. Res. Commun. **87:**1044–1051.
87. Papa, S., F. Guerrieri, M. Lorusso, G. Izzo, D. Boffoli & F. Capuano. 1977. FEBS Symp. **42:**502–519.
88. Papa, S., F. Guerrieri, M. Lorusso, F. Capuano, G. Izzo & D. Boffoli. 1978. *In* Frontiers of Biological Energetics. P. L. Dutton *et al.,* Eds., Vol. **1:**367–374. Academic Press. New York.
89. Lorusso, M., F. Capuano, D. Boffoli, R. Stefanelli & S. Papa. 1979. Biochem. J. **182:**133–147.
90. Papa, S., F. Guerrieri, M. Lorusso, G. Izzo, F. Capuano & D. Boffoli. 1979. *In* Function and Molecular Aspects of Biomembrane Transport. E. Quagliariello *et al.,* Eds.: 197–207. Elsevier/North-Holland. Amsterdam.
91. Mitchell, P. & J. Moyle. 1978. *In* Frontiers of Biological Energetics. P. L. Dutton *et al.,* Eds., Vol. **1:**342–350. Academic Press. New York.
92. Mitchell, P., J. Moyle & R. Mitchell. 1978. *In* Frontiers of Biological Energetics. P. L. Dutton *et al.,* Eds., Vol. **1:**394–402. Academic Press. New York.
93. Mitchell, P. & J. Moyle. 1979. *In* Cytochrome Oxidase. T. E. King *et al.,* Eds. :361–372. Elsevier/North-Holland. Amsterdam.
94. Mitchell, P. & J. Moyle. 1979. Biochem. Soc. Trans. **7:**887–894.
95. Mitchell, P., J. Moyle & R. Mitchell. 1979. Methods in Enzymol. **55:**627–640.
96. Nicholls, P. 1975. *In* Electron Transfer Chains and Oxidative Phosphorylation. E. Quagliariello *et al.,* Eds. :227–232. North-Holland/American Elsevier. Amsterdam/New York.
97. Mitchell, P. & J. Moyle. 1967. *In* Biochemistry of Mitochondria. E. C. Slater *et al.,* Eds. :53–74. Academic Press/PWN. London/Warzawa.

98. MITCHELL, P. 1969. *In* The Molecular Basis of Membrane Transport. D. C. Tosteson, Ed. :483–518. Prentice-Hall. Englewood Cliffs, New Jersey.
99. MITCHELL, P. & J. MOYLE. 1969. Eur. J. Biochem. **9:**149–155.
100. LEUNG, K. H. & P. C. HINKLE. 1975. J. Biol. Chem. **250:**8467–8471.
101. GUERRIERI, F. & B. D. NELSON. 1975. FEBS Lett. **54:**339–342.
102. CHANCE, B., J. S. LEIGH & A. WARING. 1977. BBA Library **14:**1–10.
103. WIKSTRÖM, M. K. 1977. Nature **266:**271–273.
104. WIKSTRÖM, M. K. & H. T. SAARI. 1977. Biochim. Biophys. Acta **462:**347–361.
105. REYNAFARJE, B., M. D. BRAND & A. L. LEHNINGER. J. Biol. Chem. **251:**7442–7451.
106. MOYLE, J. & P. MITCHELL. 1978. FEBS Lett. **88:**268–272.
107. MOYLE, J. & P. MITCHELL. 1978. FEBS Lett. **90:**361–365.
108. PACKER, L. & M. G. MUSTAFA. 1976. Biochim. Biophys. Acta **113:**1–12.
109. PAPA, S., F. GUERRIERI, M. LORUSSO, G. IZZO, D. BOFFOLI & R. STEFANELLI. 1978. FEBS Symp. **45:**37–48.
110. WOJTCZAK, L. & H. ZALUSKA. 1969. Biochim. Biophys. Acta **193:**64–72.
111. OSHEROFF, N., W. H. KOPPENOL & E. MARGOLIASH. 1978. *In* Frontiers of Biological Energetics. P. L. Dutton *et al.,* Eds. Vol. **1:**439–449. Academic Press. New York.
112. SORGATO, M. C. & S. J. FERGUSON. 1978. FEBS Lett. **90:**178–182.
113. BRAND, M. D., W. G. HARPER, D. G. NICHOLLS & W. J. INGLEDEW. 1978. FEBS Lett. **95:**125–129.
114. HINKLE, P. C. 1978. FEBS Symp. **45:**79–83.
115. MORSE, P. D. & D. W. DEAMER. 1973. Biochim. Biophys. Acta **298:**769–782.
116. CLORE, G. M. & B. CHANCE. 1979. Biochem. J. **177:**613–621.
117. LANNE, B., B. G. MALMSTRÖM & T. VANNGARD. 1979. Biochim. Biophys. Acta **545:**205–214.
118. NICHOLLS, P. & V. HILDEBRANDT. 1978. Biochem. J. **173:**65–72.
119. MALMSTRÖM, B. G. 1978. FEBS Symp. **45:**61–68.
120. PETERSEN, L. C. 1978. Eur. J. Biochem. **85:**339–344.
121. POOLE, R. K., A. J. WARING & B. CHANCE. 1979. FEBS Lett. **101:**56–58.
122. SHAW, R. W., R. E. HANSEN & H. BEINERT. 1978. J. Biol. Chem. **253:**6637–6640.
123. SHAW, R. W., R. E. HANSEN & H. BEINERT. 1978. Biochim. Biophys. Acta **504:**187–199.
124. TWEEDLE, M. F., L. J. WILSON, L. GARCIA-INIGUEZ, G. T. BABCOCK & G. PALMER. 1978. J. Biol. Chem. **253:**8066–8071.
125. TYREE, B. & D. A. WEBSTER. 1979. J. Biol. Chem. **254:**176–179.
126. VAN GELDER, B. F., J. L. M. L. VAN RIJN, G. J. A. SCHILDER & J. WILMS. 1977. BBA Library **14:**61–68.
127. ARTZATBANOV, V. Y., A. A. KONSTANTINOV & V. P. SKULACHEV. 1978. FEBS Lett. **87:**180–185.
128. VAN CAMP, H. L., Y. H. WEI, C. P. SCHOLES & T. E. KING. 1978. Biochim. Biophys. Acta **537:**238–246.
129. CUSANOVICH, M. A. 1978. *In* Frontiers of Biological Energetics. P. L. Dutton *et al.,* Eds. Vol. **1:**91–100. Academic Press. New York.
130. KELLER, R. M. & K. WÜTHRICH. 1978. Biochem. Biophys. Res. Commun. **83:** 1132–1139.
131. WHERLAND, S. & I. PECHT. 1978. Biochemistry **17:**2585–2591.
132. HOPFIELD, J. J. 1977. Biophys. J. **18:**311–321.
133. POTASEK, M. J. & J. J. HOPFIELD. 1977. Proc. Nat. Acad. Sci. **74:**3817–3820.
134. VAN LEEUWEN, J. W., A. RAAP, W. H. KOPPENOL & H. NAUTA. 1978. Biochim. Biophys. Acta **503:**1–9.
135. WARSHEL, A. 1978. Proc. Nat. Acad. Sci. **75:**5250–5254.
136. KOPPENOL, W. H., C. A. J. VROONLAND & R. BRAAMS. 1978. Biochim. Biophys. Acta **503:**499–508.

RESPIRATION-COUPLED H^+ EJECTION BY MITOCHONDRIA

Albert L. Lehninger, Baltazar Reynafarje, Adolfo Alexandre,
and Antonio Villalobo

Department of Physiological Chemistry
The Johns Hopkins University
School of Medicine
Baltimore, Maryland 21205

This paper briefly reviews some recent studies in this laboratory on H^+ translocation by the respiratory chain of intact mitochondria, with special emphasis on the coupling of H^+ movements to electron flow from cytochrome c to oxygen. Our observations confirm and extend our earlier conclusion,[1-4] that at least 3 H^+ and probably 4 H^+ are ejected per $2e^-$ passing each of the three energy-conserving sites of the respiratory chain. This value contrasts with the H^+/site ratio of 2, as proposed in the chemiosmotic hypothesis.[5]

Before discussing these data, two important points must be made. In the first place, the new stoichiometric data on H^+ movements coupled to electron flow should not be regarded as a trivial matter, requiring only a minor adjustment of current concepts and hypotheses. The H^+/site ratio is actually a major element in analyzing the mechanisms of both electron transport and oxidative phosphorylation. Biochemical mechanisms are simply not understood until stoichiometry is defined. Secondly, it must be pointed out that estimations of the H^+/site ratio yielding values in excess of 2 are not the product of only one laboratory; over the past dozen years a number of laboratories have reported data that are not accounted for by a H^+/site ratio of 2.0.

Information on the H^+/site ratio of mitochondrial electron transport (defined as the number of vectorial protons translocated from the matrix to the medium per pair of electrons per energy-conserving site) has come from four different types of experimental data: 1. Thermodynamic data on the $Gp/\Delta\tilde{\mu}_{H^+}$ ratio; 2 The stoichiometry and thermodynamics of the inward transport of $P_i^- + ADP^{3-}$ in exchange for ATP^{4-} formed in the matrix; 3. The cation/site uptake ratios during Ca^{2+} or K^+ uptake coupled to electron transport, and 4. Direct stoichiometric measurements of the H^+/site ejection ratio. With regard to the thermodynamic data, measurements have been made in many laboratories of $\Delta\tilde{\mu}_{H^+}$, the electrochemical gradient of protons generated across the membrane by electron transport, which have been compared with the maximum phosphorylation potential (ΔGp) that can be supported by respiring mitochondria in state 4. For synthesis of one molecule of ATP at a ΔGp of 16 kcal (equivalent to 690 emV), a value that is now well-established as a maximum for intact mitochondria *in vitro,* n_1, the number of H^+ ions ejected per pair of electrons per ATP-generating site times $\Delta\tilde{\mu}_{H^+}$, the thermodynamic driving force behind this flow in the steady state, is equal to ΔGp, the gradient that must be overcome for the synthesis of 1 ATP

$$\underset{\text{(electron transport)}}{\text{flow}_1 \times \text{force}_1} = \underset{\text{(ATP synthesis)}}{\text{flow}_2 \times \text{force}_2}$$

$$n_1 \times \Delta\tilde{\mu}_{H^+} = 1 \times \Delta Gp$$

$$n = \Delta Gp/\Delta\tilde{\mu}_{H^+}$$

0077–8923/80/0341–0585 $01.75/0

This relationship tells us that n, the H^+/site ratio, can be theoretically any number, 2, 4, or 10. But if the ratio is low, the thermodynamic driving force must be high, and vice versa. For example, if the H^+/site ratio is 2, a driving force $\Delta\mu_{H^+}$ of 345 emV is required to generate ATP at a ΔGp of 690 emV. However, all reported values of $\Delta\mu_{H^+}$ for intact mitochondria lie in the range between 130–230 mV, indicating that n, the H^+/site ratio, must be greater than 2. Many laboratories have determined the ratio $\Delta Gp/\Delta\mu_{H^+}$ in intact mitochondria. Almost without exception the reported values of $\Delta Gp/\Delta\mu_{H^+}$ (=H^+/site ratio) are in excess of 2 and some are as high as 4.0.[6-10]

The second line of evidence indicating that the average H^+/site ratio is greater than 2 is the stoichiometry of the two transport processes involved in oxidative phosphorylation of external ADP, promoted by the electroneutral $H^+/H_2PO_4^-$ symporter and the electrogenic ADP^{3-}/ATP^{4-} antiporter. To carry out the transport work alone, close to 1 H^+ must enter the mitochondria for each ATP^{4-} (out) generated, a conclusion that emerges from both stoichiometric[11] and thermodynamic measurements of ΔGp (out) and ΔGp (matrix).[12] If the overall H^+/site ratio is 2, this would leave only 1 H^+ per site for carrying out the work of making ATP from ADP and P_i.

The third type of experimental approach comes from stoichiometric experiments on respiration-coupled uptake of the cations K^+ and Ca^{2+}, which enter the matrix in response to the negative-inside membrane potential generated by electron transport. At least 3 laboratories have found that the K^+/site uptake ratio is much greater than 2 and very close to 4.[13-15] Similarly, it has been found[16-19] that the Ca^{2+}/site uptake ratio is close to 2, equivalent to uptake of 4 positive electric charges, on the basis of the bivalent electrophoretic uniport process for Ca^{2+} entry that has now been firmly established,[16, 20-24] contrary to other suggestions.[19]

The fourth type of experimental approach is the direct determination of the H^+/site ejection ratio. We have found that the oxygen pulse method, as originally described[25] for estimating the H^+/site ratio, has a number of disadvantages. 1. It actually measures only one of the two entities required to calculate an H^+/O or H^+/site ratio. 2. It requires arbitrary upward corrections. 3. It is subject to gross underestimation through uptake of H^+ by symport with medium phosphate originating from endogenous sources.[26] 4. Most of the O_2 consumption occurs at less than the assumed state 3 rate, since the O_2 concentration is below K_M for O_2 over much of the reaction period.[47] Although we have been able to reproduce[26] the original oxygen pulse measurements giving values near 2, we have shown that these low values are due to very rapid interfering uptake of H^+ on the H^+/phosphate symporter.[26] This interference can be prevented either by use of *N*-ethylmaleimide or mersalyl as inhibitors of H^+/phosphate symport or by simple depletion and removal of endogenous phosphate.[26] These modifications alone raise the H^+/site ratio to 3 or higher with the oxygen pulse method. However, as indicated above, there are other problems associated with the oxygen pulse procedure.

We have devised a better method[27] for determination of the H^+/O ratio in which *both* H^+ ejection and oxygen consumption are measured, in a medium in equilibrium with atmospheric oxygen, in which O_2 is present at a concentration far above K_M for oxygen. No corrections are required. K^+ (plus valinomycin) or Ca^{2+} serves as the charge-compensating cation to enter the matrix to replace the ejected H^+. Electron flow is initiated in the deenergized rotenone-supplemented mitochondria by addition of an appropriate electron donor, such as succinate. The rates of H^+ ejection and oxygen consumption are then recorded from appropriately calibrated electrodes. The steady-state rate of H^+ ejection (ng-ions per min per mg protein) divided by the corresponding rate of oxygen consumption (ng-atoms per min per mg) gives the H^+/O ratio. Interfering re-uptake of H^+ in symport with medium phosphate is prevented with mersalyl or *N*-ethylmaleimide, which we have shown do not elicit spurious scalar H^+ production under our conditions.[27] Alternatively, H^+ reuptake can be avoided by simply depleting the mitochondria of their inorganic phosphate before testing.[27]

Electrode traces from such experiments on rat liver mitochondria, with K^+ as charge-compensating cation, show a short lag in both oxygen consumption and H^+ ejection after addition of succinate. This lag is not instrumental but is due to the time required for entry of succinate, which is slow under these conditions because the matrix has been depleted of phosphate, with which succinate must exchange in order to enter and gain access to succinate dehydrogenase on the M side of the inner membrane. From the slopes of the linear portions of the H^+ and oxygen traces, the H^+/O rate ratio is calculated. Since 2 energy-conserving sites are traversed, the H^+/site ratio is one-half the H^+/O ratio.

To indicate the reproducibility of the measurements, a recent series of 82 separate sets of unselected data on succinate oxidation in rat liver mitochondria under a variety of optimal and suboptimal conditions gave a mean of 7.4 ± 0.8 (S.D.), equivalent to an average of 3.7 ± 0.4 H^+ per $2e^-$ for sites 2 and 3.

The H^+ ejection accompanying succinate oxidation is completely abolished by FCCP, or nigericin + valinomycin, and is thus the result of a vectorial process; there is no contribution of scalar H^+ to the H^+ ejection. That the H^+ ejection is vectorial has also been verified by simultaneous electrode measurements of either K^+ uptake[15] or Ca^{2+} uptake.[16] The K^+/oxygen uptake ratio (in the presence of valinomycin) is almost exactly equal to the H^+/oxygen ejection ratio,[15] as expected if the ejected H^+ arises from the matrix and is replaced by K^+ to maintain internal electroneutrality. Moreover, the Ca^{2+}/site uptake ratio was found to be close to 2.0,[16] equivalent to 4 + charges entering.

Similar experiments on the H^+/O ratio and cation/O ratios with NAD (nicotinamide-adenine-dinucleotide)-linked substrates such as pyruvate + malate, glutamate, and β-hydroxybutyrate have also been carried out.[15, 16] These are technically more difficult because of the generally lower ratio of oxygen consumption, the strong inhibition of NAD-linked electron transport by *N*-ethylmaleimide, and the high expected H^+/O ratios. Nevertheless H^+/O ratios between 10 and 12 have been repeatedly observed, corresponding to *average* H^+/site ratios of between 3.3 and 4.0 for the 3 sites encompassed.

It may be pointed out here that measurements repeatedly giving exactly 4.0 H^+ per $2e^-$ per site cannot be expected. In the long history of measurements of the P/O ratio of oxidative phosphorylation, which extends over almost 40 years, very few uncorrected measurements yield the expected values of 2.0 for succinate oxidation and 3.0 for NAD-linked substrates, largely because of the occurrence of significant rates of uncontrolled, i.e., uncoupled respiration.[28] For the same reason, i.e., the occurrence of significant uncontrolled or uncoupled electron flow, we cannot expect to realize precisely 4 H^+ per site on measurements of the H^+/site ratio. We are, therefore, very comfortable with our conclusion that the H^+/site ratio is at least 3 and is probably 4.0, on the basis of many measurements and careful analysis of the experimental conditions.

Since nearly all published work on H^+/site ratios has been made on rat liver mitochondria, which promote a number of active metabolic reactions other than electron transport, we have measured average H^+/site ratios in mitochondria from other tissues. Values approaching 4.0 have been observed with mitochondria not only from rat liver,[27] but also from rat heart,[28] Ehrlich ascites tumor cells,[30] and rat brain.[48]

We have also examined the H^+/site ratios of different segments of the chain, with appropriate electron donors and acceptors, and with rotenone, antimycin A, and cyanide used in appropriate ways to limit electron flow to the segment of the chain under question. In such experiments, ferricyanide was used as electron acceptor, together with various site 1 and site 2 substrates as electron donors. TABLE 1 summarizes some of the results. Site 2 with succinate as electron donor and ferricyanide as electron acceptor yielded a $H^+/2e^-$ ejection

TABLE 1

OBSERVED $H^+/2E^-$ RATIOS FOR DIFFERENT SEGMENTS OF THE RESPIRATORY CHAIN

Site(s)	Electron donor	Electron acceptor	H^+/ segment	H^+/ site
1+2+3	Pyruvate and malate	Oxygen	11.9	3.9
1+2	β-hydroxybutyrate	Ferricyanide	7.0	3.5 †
1+2	Pyruvate and malate	Ferricyanide	7.0	3.5 *
1+2	Pyruvate and malate	Ferricytochromes	7.5	3.8 ‡
2+3	Succinate	Oxygen	7.4	3.7
2	Succinate	Ferricyanide	3.9	3.9
3	Ferrocyanide	Oxygen	3.6	3.6
3	Ascorbate and ferrocyanide	Oxygen	3.7	3.7
3	Ascorbate and TMPD	Oxygen	3.7	3.7

* (From A. Villalobo, unpublished data on Ehrlich ascites tumor mitochondria.)
† (From A. Alexandre, unpublished data on rat liver mitochondria.)
‡ Experiment carried out on rat liver mitoplasts.

ratio of very close to 4.0, a $K^+/2e^-$ uptake ratio of 2, and a $Ca^{2+}/2e^-$ ratio of 1.0.[3, 31] In the presence of FCCP, there was formation of 2 H^+ per $2e^-$, corresponding to the 2.0 H atoms removed from succinate. There is now substantial agreement on site 2 measurements with data from the laboratories of Azzone,[14] Wikstrom,[32] and Mitchell,[33] although the latter group places a different interpretation on the H^+/site ratio of 4 for this segment of the chain.[34] Sites 1 + 2, tested with ferricyanide as acceptor and pyruvate + malate, β-hydroxybutyrate as substrates, have given observed average $H^+/2e^-$ ratios close to 7.0.[49]

Although Papa *et al.*[35] and Mitchell and Moyle[34] still claim that no H^+ translocation occurs coupled to electron flow from cytochrome c to oxygen, the experimental conditions used by these groups have been seriously criticized by Wikstrom.[37, 38] The real question is no longer whether H^+ translocation is coupled to electron flow from cytochrome c to oxygen, but rather what the maximum $H^+/2e^-$ ratio is. Our earlier experiments showing that the H^+/O ratio for electron flow from succinate to oxygen is close to 8.0 already indicated that site 3 must yield H^+ ejection with a high $H^+/2e^-$ ratio. Direct experiments in the laboratories of Wikstrom [32, 37-40] and Carafoli [41] have given H^+/O values of 2 or somewhat less, whereas we have reported [3] measurements exceeding 3.0, as have Pozzan *et al.*[14] In addition to these direct stoichiometric measurements, the important experiments of Sorgato and colleagues at Oxford [42] have shown that electron flow through cytochrome oxidase in inverted vesicles generates not only $\Delta\Psi$ but also ΔpH.

We have concentrated much further attention on H^+ ejection coupled to site 3. Our measurements have been made on intact rat liver mitochondria and mitoplasts supplemented with rotenone and antimycin A to limit electron flow to the segment cytochrome *c* to oxygen, EGTA to bind Ca^{2+}, K^+ as charge-compensating cation, valinomycin, oligomycin, and *N*-ethylmaleimide to block H^+ reuptake on the phosphate symporter. We have used ferrocyanide, TMPD, and ferrocytochrome *c* as external reductants of mitochondrial cytochrome *c*. We have used two different methods to measure H^+ ejection (the glass electrode and the spectral changes of phenol red indicator determined by dual wavelength spectrophotometry) and two methods for measuring electron flow (a high-sensitivity oxygen electrode and dual wavelength spectrophotometry of ferricyanide formation), in different combinations. Site 3 experiments are technically difficult and subject to a number of constraints. On the other hand, they have the advantage of providing an excellent situation to test the adequacy of the methods for measuring the rates of H^+ movements and oxygen consumption, since in the presence of excess proton-conducting uncouplers the overall cytochrome oxidase reaction involves H^+ uptake, according to the equation

$$2\ H^+ + 2e^- + \tfrac{1}{2}\ O_2 \rightarrow H_2O$$

The measurements of H^+ uptake and oxygen consumption in the presence of uncouplers must therefore give a H^+/O *uptake* ratio of close to the theoretical 2.0 to be considered stoichiometrically and kinetically valid for measuring the $H^+/2e^-$ translocation ratio in the absence of uncoupler. This criterion has been met in each of the test systems we have optimized. Our measurements with these different procedures have yielded net H^+/O translocation ratios (absence of uncouplers) in the range from a little less than 2.0 to very close to 4.0. In general, most measurements of the H^+/O ratio with TMPD as artificial electron donor tend to give values in the lower range, but under some conditions TMPD yields values significantly over 3.0, as reported earlier.[3] TMPD, as well as its oxidation product, Wurster's blue, are known to inhibit electron transport and phosphorylation. With ferrocyanide as electron donor and appropriate adjustment of reaction conditions, H^+/O translocation ratios close to 4 can be more consistently observed with the optical method. In similar experiments in which uptake of K^+ or Ca^{2+}, the charge-compensating cations, was recorded simultaneously from appropriate electrodes, the K^+/O uptake ratio was significantly greater than 5 and the Ca^{2+}/O ratio greater than 2.5. Overall, these data with ferrocyanide as electron donor, and either electrode or optical measurements of electron flow, show that the H^+/O translocation ratio approaches 4 and the cation$^+/2e^-$ uptake ratio, approaches 6, as expected; since in addition to compensating electrically for the 4 H^+ ejected, the K^+ uptake must also compensate for the 2 H^+ absorbed from the matrix to yield H_2O on reduction of oxygen.

Papa [35] and Mitchell and Moyle [36] have claimed that H^+ ejection coupled to ferrocyanide oxidation in experiments such as these are reflections of an H^+-transporting arm associated with the Q-c_1 region via a postulated "Q cycle" that is assumed to have a step that can be inhibited by 2-heptyl-4-hydroxyquinoline *N*-oxide (HQNO). From titration experiments, we have determined the amount of HQNO required to just inhibit succinate oxidation in the absence of antimycin A, but with rotenone present. This amount of HQNO was found to have *no* effect on the H^+/O translocation ratio in site 3 experiments with ferrocyanide as electron donor.

Perhaps the most direct experiments on site 3 have been made with chemically-reduced ferrocytochrome *c* as direct electron donor in rat liver mitoplasts. Removal of the outer membrane allows medium ferrocytochrome *c* to donate electrons at a very high rate to the mitochondrial cytochrome system. Coupled to this electron flow is H^+ translocation with high H^+/O translocation ratios. In the presence of FCCP only H^+ uptake is observed, corresponding to the formation of H_2O.

Our stoichiometric measurements on the H^+/O ratio of electron flow from ferrocytochrome *c* to oxygen have also yielded new information of significance with regard to energy transduction at site 3. Electron flow through site 3 can occur at much higher rates than from natural substrates, such as succinate—rates considerably higher than expected if uncontrolled oxygen consumption is simply due to back-leakage of H^+. Thus, the rate of back-leakage of H^+ is determined by factors other than a simple ohmic relationship; we suspect a regulatory process. Indeed some of our observations indicate that the cytochrome oxidase reaction is a regulated reaction, in which kinetic parameters and possibly the H^+/O ratio may vary depending on energy state.

Implications

Time does not permit full discussion of the implications of these and other experiments we have recently carried out. However, our data, particularly on site 3, strongly suggest that the 3 energy coupling sites are probably identical in H^+ stoichiometry, contrary to Brand *et al.*[43] Sites 2 and 3 are each capable of translocating up to 4 H^+ per $2e^-$. Site 1 data are less certain, but since sites 1 + 2 together give values between 6 to 7 per $2e^-$, and sites 1 + 2 + 3 give values in the range 10 to 12, it is likely that site 1 has a $H^+/2e^-$ ratio of at least 3. Moreover, since we have determined the intrinsic H^+/ATP ratio during ATP hydrolysis to be 3,[44] we have concluded that of the 4 H^+ equivalents ejected per site, 3 H^+ are used to synthesize one molecule of ATP and the 4th H^+ to furnish energy for the membrane transport steps.[2-4, 44] Any valid hypothesis for the mechanism of energy coupling and oxidative phosphorylation, whether chemical, conformational, or chemiosmotic, will ultimately have to account for the observed H^+ stoichiometry of electron transport and ATP synthesis.

The 3-loop mechanism of H^+ translocation postulated in the chemiosmotic hypothesis can account for translocation of only 2 H^+ per energy-transducing segment of the respiratory chain. How can H^+/site ratios of 4 be rationalized? Three alternative models are possible.[45] 1. The conformational or pump model postulates that all of the translocated H^+ are moved across the membrane by vectorial membrane Bohr effects, in which the energy coupling mode is conformational. 2. A second model supposes that each site contains a loop delivering 2 H^+ per $2e^-$, supplemented with a Bohr pump that provides for the other 2 H^+ per site. 3. The third model postulates that there are actually 6 loops in the electron transport chain, each energy-conserving site containing 2 loops ("miniloops"?), a view that has no supporting evidence but which could provide a rationale for the function of the ~16 known redox centers of the respiratory chain.[45] We have in the past year obtained interesting evidence that site 2 contains a H^+-translocating loop responsible for movement of 2 of the 4 H^+ moved by that site. On the other hand, we are inclined to the view that the cytochrome oxidase segment translocates H^+ by a mechanism other than a ligand-translocating loop mechanism.

Finally, we must add that the H^+ movements occurring during electron transport, or ATP hydrolysis or synthesis, in stoichiometric experiments as described above, in which charge-compensating cations are employed to suppress development of membrane potential, do not necessarily mean that such H^+ movements also occur between the two bulk phases (the matrix and medium) during the actual normal process of oxidative phosphorylation. This should be very clearly understood, since there is increasing evidence from other approaches, particularly in chloroplasts, that charge separation within the membrane coupled to electron flow is orders of magnitude faster than H^+ translocation between the two bulk phases separated by the membrane and that an artificially-impressed membrane potential alone can make ATP.[46] Thus, translocation of H^+ between two bulk phases may not be a necessary event during oxidative phosphorylation; rather, it is possible that charge movements *within* the membrane are the fundamental processes serving as the vehicle of energy transduction. H^+ cycling between two bulk aqueous phases seems to be an unnecessarily circuitous way to bring about energy-coupling in oxidative phosphorylation, although there are certainly some processes, such as H^+-coupled membrane transport systems, in which bulk translocation of H^+ provides the requisite energy. The proton and electron are here to stay as the elementary particles of membrane-associated energy transduction, but much further work is required to establish the *normal* vehicle of energy transduction in oxidative or photosynthetic phosphorylation.

References

1. Reynafarje, B., M. D. Brand & A. L. Lehninger. 1976. J. Biol. Chem. **251:** 7442–7451.
2. Brand, M. D. & A. L. Lehninger. 1977. Proc. Natl. Acad. Sci. **74:**1955–1959.
3. Alexandre, A., B. Reynafarje & A. L. Lehninger. 1978. Proc. Natl. Acad. Sci. **75:**5296–5300.
4. Lehninger, A. L., B. Reynafarje & A. Alexandre. 1977. *In* Structure and Function of Energy-Transducing Membranes. K. van Dam and B. V. van Gelder, Eds.: 95–106. Elsevier. Amsterdam.
5. Mitchell, P. 1976. Biochemical Society Transactions, Vol. 4: 399–430.
6. Lehninger, A. L. & B. Reynafarje. 1978. 12th FEBS Meeting, Dresden, Abstracts, Vol. 1: #1732.
7. Van Dam, K., A. H. C. A. Wiechmann, K. J. Hellingwerf, J. C. Arents & H. V. Westerhoff. 1977. 11th FEBS Meeting, Copenhagen, *In* Membrane Proteins. P. Nicholls *et al.*, Eds., Vol. 45: 121–132.
8. Nicholls, D. G. 1974. Eur. J. Biochem. **50:**305–315.
9. Rottenberg, H. 1975. J. Bioenerg. **7:**61–74.
10. Azzone, G. F., M. Bragadin, T. Pozzan & P. Dell'Antone. 1976. Biochim. Biophys. Acta **459:**96–109.
11. Lanoue, K., S. M. Mizani & M. Klingenberg. 1978. J. Biol. Chem. **253:**191–198.
12. Slater, E. C. 1976. *In* Use of Isolated Liver Cells and Kidney Tubules in

Metabolic Studies. J. M. Tager, H. D. Soling & J. R. Williamson, Eds.: 65–77. North Holland. Amsterdam.
13. Cockrell, R., E. J. Harris & B. C. Pressman. 1966. Biochemistry **5:**2326–2335.
14. Pozzan, T., F. Di Virgilio, M. Bragadin & G. F. Azzone. 1979. Proc. Natl. Acad. Sci. **76:**2123–2127.
15. Reynafarje, B. & A. L. Lehninger. 1978. J. Biol. Chem. **253:**6331–6334.
16. Reynafarje, B. & A. L. Lehninger. 1977. Biochem. Biophys. Res. Communs. **77:**1273–1279.
17. Rossi, C. S. & A. L. Lehninger. 1964. J. Biol. Chem. **239:**3971–3980.
18. Chance, B. 1965. J. Biol. Chem. **240:**2729–2748.
19. Moyle, J. & P. Mitchell. 1977. FEBS Letts. **73:**131–136.
20. Selwyn, M. J., A. P. Dawson & S. J. Dunnett. 1970. FEBS Letts. **10:**1–5.
21. Lehninger, A. L. 1972. *In* The Molecular Basis of Electron Transport. J. Schultz & B. J. Cameron, Eds. Vol. **4:**133–151. Academic Press, Inc. New York, N.Y.
22. Akerman, K. E. O. 1978. FEBS Letts. **93:**293–296.
23. Chappell, J. B., J. D. Arrabaca, R. Deanna & Y. Mathien-Shire. 1978. 12th FEBS Meeting, Dresden, Abstracts #2025.
24. Fiskum, G., B. Reynafarje & A. L. Lehninger. 1979. J. Biol. Chem. **254:** 6288–6295.
25. Mitchell, P. & J. Moyle. 1967. Biochem. J. **105:**1147–1162.
26. Brand, M. D., B. Reynafarje & A. L. Lehninger. 1976. J. Biol. Chem. **251:** 5670–5679.
27. Brand, M. D., B. Reynafarje & A. L. Lehninger. 1976. Proc. Natl. Acad. Sci. **73:**437–441.
28. Vercesi, A., B. Reynafarje & A. L. Lehninger. 1978. J. Biol. Chem. **253:** 6379–6385.
29. Hinkle, P. C. & M. L. Yu. 1979. J. Biol. Chem. **254:**2450–2455.
30. Villalobo, A. & A. L. Lehninger. 1979. J. Biol. Chem. **254:**4352–4359.
31. Alexandre, A. & A. L. Lehninger. 1979. J. Biol. Chem. **254:**11555–11560.
32. Wikstrom, M. & K. Krab. 1978. *In* Energy Conservation in Biological Membranes. G. Schafer & M. Klingenberg, Eds.: 128–139. Springer-Verlag. Berlin.
33. Mitchell, P. & Moyle, J. 1967. *In* Biochemistry of Mitochondria. E. C. Slater *et al.,* Eds.: 53–74. Academic Press and PWN. London and Warsaw.
34. Mitchell, P. 1976. J. Theor. Biol. **62:**327–367.
35. Papa, S., M. Lo Russo, F. Guerrieri & F. Capuano. 1978. *In* The Proton and Calcium Pumps. G. F. Azzone, M. Avron, J. C. Metcalfe, E. Quagliariello & N. Siliprandi, Eds.: 227–238. Elsevier/North-Holland. Amsterdam.
36. Moyle, J. & P. Mitchell. 1978. FEBS Letts. **88:**268–272.
37. Wikstrom, M. K. F. & K. Krab. 1978. FEBS Letts. **91:**8–14.
38. Wikstrom, M. K. F. 1978. *In* The Proton and Calcium Pumps. G. F. Azzone, M. Avron, J. C. Metcalfe, E. Quagliariello & N. Siliprandi, Eds.: 215–226. Elsevier/North-Holland. Amsterdam.
39. Wikstrom, M. K. F. 1977. Nature **266:**271–273.
40. Wikstrom, M. K. F. & H. T. Saari. 1977. Biochim. Biophys. Acta **462:**347–361.
41. Sigel, E. & E. Carafoli. 1978. Eur. J. Biochem. **89:**119–123.
42. Sorgato, M. C. & S. J. Ferguson. 1978. FEBS Letts. **90:**178–182.
43. Brand, M. D., W. G. Harper, D. G. Nicholls & W. J. Ingledew. 1978. FEBS Letts. **95:**125–129.
44. Reynafarje, B. & A. L. Lehninger. 1978. Proc. Natl. Acad. Sci. **75:**4788–4792.
45. Lehninger, A. L. 1978. *In* Protons and Ions Involved in Fast Dynamic Phenomena. P. Laszlo, Ed.: 435–452. Elsevier/North-Holland Biomedical Press. Amsterdam.
46. Witt, H. T. 1979. Biochim. Biophys. Acta **505:**355–427.
47. Reynafarje, B. Unpublished observations.
48. Benavides, J. Unpublished observations.
49. Alexandre, A. & A. Villalobo. Unpublished observations.

KINETICS AND MECHANISMS OF GLUTAMATE TRANSPORT ACROSS THE MITOCHONDRIAL MEMBRANE *

John R. Williamson, Jan B. Hoek, Elizabeth Murphy, and Kathleen E. Coll

Department of Biochemistry and Biophysics
University of Pennsylvania
Philadelphia, Pennsylvania 19104

R. Muturi Njogu

Department of Biochemistry
University of Nairobi
P.O. Box 30197
Nairobi, Kenya

Glutamate is transported across the mitochandrial membrane by two separate systems. Both are carrier mediated and presumably require participation of specific proteins in the membrane. The first is by a glutamate-hydroxyl exchange diffusion (or a glutamate-H^+ symport), and is electroneutral and freely reversible.[1-4] The second is by a glutamate-aspartate antiport, which is electrogenic since it involves an exchange between undissociated glutamic acid (or glutamate plus a proton) and aspartate anion.[5, 6] Influx of glutamate into mitochondria in exchange with aspartate is energy consuming because net movement of negative change from the matrix to the external side of the inner mitochonchondrial membrane dissipates a preexisting membrane potential (negative inside), and requires metabolic energy for its replenishment. Influx of aspartate in exchange with glutamate, as originally described,[1] only occurs at an appreciable rate in deenergized mitochondria when the membrane potential is collapsed.[6] Under normal conditions with respiring mitochondria, efflux of aspartate represents the physiological direction of the carrier, and the obligatory cotransport of glutamate can be balanced by glutamate efflux on the glutamate-hydroxyl carrier.[7]

Current knowledge and ideas concerning these carriers have been reviewed recently.[8-11] They are involved in a number of metabolic functions, as depicted in FIGURE 1. The glutamate-aspartate carrier is intrinsic to the operation of the malate-aspartate cycle for transport of reducing equivalents from NADH into the mitochondria, and accounts for the energy dependence and unidirectionality of this process.[12, 13] Efflux of aspartate from mitochondria is essential for gluconeogenesis from substrates, such as lactate, which are more reduced than glucose.[14] In these metabolic pathways, the glutamate that enters the mitochondria on the glutamate-aspartate carrier transaminates with oxalacetate in the matrix space and both aspartate and α-ketoglutarate exit from the mitochondria. On the other hand, operation of both the glutamate-hydroxyl and glutamate-aspartate carriers are required for nitrogen metabolism and urea synthesis.[7, 15] Ammonia fixation via glutamate dehydrogenase in the mitochondrial matrix produces glutamate that reacts with aspartate aminotransferase to

* This work was supported by NIH Grants AM 22141 to J. B. H. and AM 15120 to J. R. W.

0077–8923/80/0341–0593 $01.75/0 © 1980, NYAS

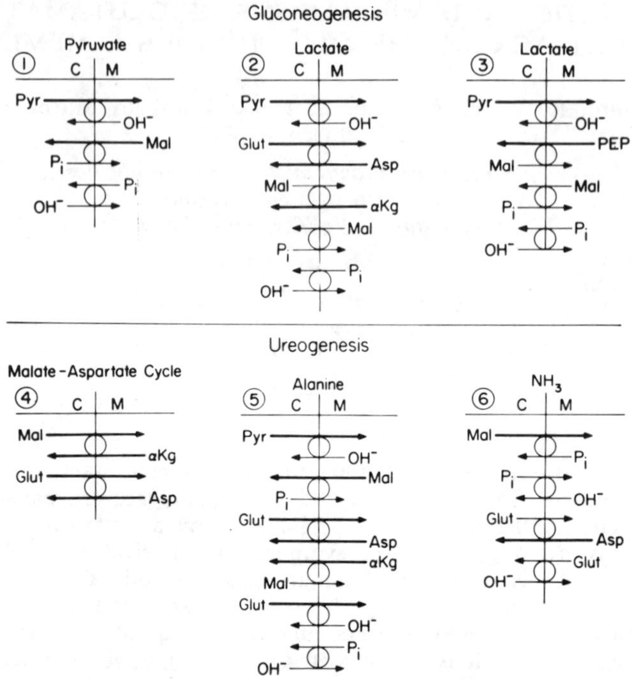

FIGURE 1. Anion transport steps required for different metabolic functions in liver.

regenerate α-ketoglutarate. The aspartate formed is transported out of the mitochondria on the glutamate-aspartate carrier, and the cotransported glutamate is removed from the matrix via the glutamate-hydroxyl carrier. During metabolism of amino acid substrates, such as alanine, ammonia is produced by oxidative deamination of glutamate, and since glutamate is also required for intramitochondrial transamination, the glutamate balance is met by glutamate entry on the glutamate-hydroxyl carrier. These requirements for satisfaction of the stoichiometry of the anion translocators are illustrated in FIGURE 1. The net movement of anions is depicted by large heavy arrows, while ancillary, obligatory transport of other anions are depicted by small arrows. Aspartate efflux from the mitochondria is obligatory for optimum rates of urea synthesis,[15, 16] hence flux of glutamate on either transport system must be sufficient to account for observed rates of ureogenesis.

Of interest, too, is the question of whether transport of metabolites across the mitochondrial membrane can provide a rate-influencing step in the overall metabolic pathway. In this paper, we examine the kinetics of glutamate transport on both systems in isolated rat liver mitochondria with the view of assessing further their potential regulatory role in cellular metabolic processes.

MATERIALS AND METHODS

Rat liver mitochondria from fed, male Sprague-Dawley rats were prepared by conventional procedures.[17, 18] Glutamate influx studies using [U-^{14}C] glutamate were performed using 0.1 mM *N*-ethylmaleimide (NEM) and 1 mM bromcresol purple for inhibitor stop, according to Hoek and Njogu.[19] Incubation of mitochondria, methods used for glutamate and aspartate loading of mitochondria, and methods for the kinetic analysis of metabolite efflux curves are fully described elsewhere.[20, 21] Details of the incubation conditions for individual experiments are given in the legends to the figures and tables.

Results

Kinetics of Glutamate Influx on the Glutamate-Hydroxyl Carrier

Previous work by Hoek and Njogu[19] on ^{14}C-glutamate uptake by rotenone-inhibited rat liver mitochondria demonstrated nonlinear uptake curves and also showed that a 1:1 stoichiometry of glutamate and proton uptake is only observed during the initial rapid phase. These data are illustrated in Figure 2 for mitochondrial incubations at 25°. After the initial linear phase of glutamate uptake, the rate decreased sharply, coincident with a cessation of net proton uptake. The later slow phase of glutamate uptake was sensitive to inhibition by 0.4 mM mersalyl (an inhibitor of the phosphate-hydroxyl translocator)[22] and 2 mM *n*-butyl malonate (an inhibitor of the malate-phosphate translocator).[22] Both phases were inhibited by NEM or bromcresol purple.[2-4] The initial rate of glutamate uptake was unaffected by increasing the proton permeability of the mitochondrial membrane by addition of uncoupling agents.[19] These data suggest that only the initial phase of glutamate uptake is suitable for kinetic analysis of glutamate transport and that glutamate uptake in the absence of normal metabolism in the mitochondria becomes limited by acidification of the matrix space.

The effect of ΔpH on the initial phase of glutamate uptake was further investigated using mitochondria incubated in the presence of nigericin, which

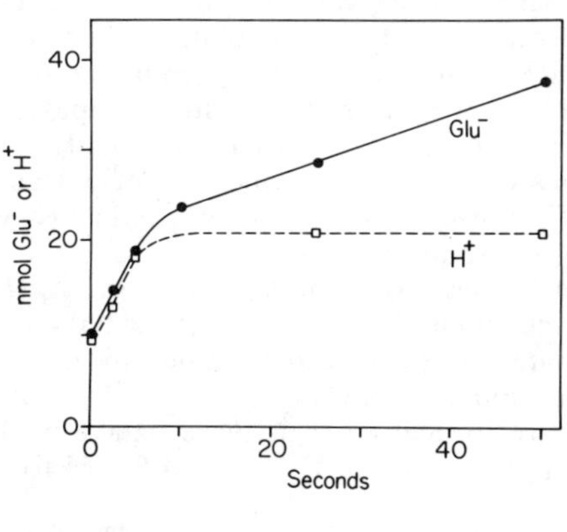

Figure 2. Kinetics of glutamate and proton uptake by rat liver mitochondria. Mitochondria (3.1 mg/ml) were incubated at 25° with 3 mM glutamate in 1 ml of medium containing 100 mM KCl, 2 mM EDTA, 3 mM Tris-HCl pH 7.1, and 5 μg rotenone. At the times indicated, uptake of glutamate was stopped with 0.8 mM NEM. The values for glutamate and H^+ uptake at time zero represent the amounts taken up in an incubation where NEM was added 2 min before the addition of glutamate.

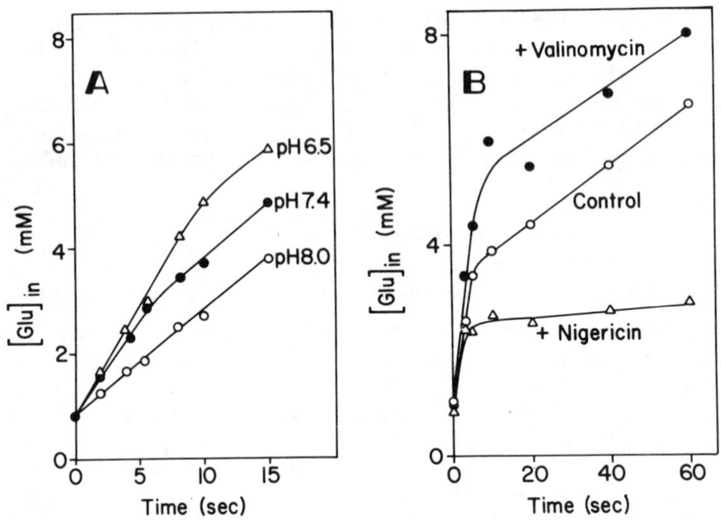

FIGURE 3. Effects of pH, valinomycin and nigericin on glutamate uptake by rat liver mitochondria. Mitochondria (2.5 mg/ml in **A** and 3.8 mg/ml in **B**) were incubated at 25° with 1 mM glutamate in 1 ml of medium containing 45 mM KCl, 2 mM EDTA, 5 mM $MgCl_2$, 50 mM Tris-Cl pH 7.4 (in **B**), and 5 μg rotenone. Valinomycin and nigericin were added at concentrations of 1 μg/ml. Glutamate uptake was stopped by the addition of 0.1 mM NEM and 1 mM bromcresol purple.

collapses the pH gradient by promoting an electroneutral exchange of K^+ for H^+,[33] or valinomycin, which decreases the membrane potential and enhances the ΔpH by promoting the electrophoretic transport of K^+ across the membrane.[24] Thus, FIGURE 3B shows that nigericin largely prevented the accumulation of glutamate in the matrix, while valinomycin extended the initial phase of glutamate uptake and increased the net accumulation. Further studies showed that the log of the glutamate gradient across the membrane at equilibrium was proportional to the log of the proton gradient with a slope of approximately one (data not shown). FIGURE 3A shows that the initial rate of glutamate uptake by mitochondria was decreased by an increase of pH above 7.4. These data suggest that glutamate transport is affected by pH in two ways. The first is a kinetic effect that may relate to protonation of a group on the carrier with a pK above pH 7.4, while the second is influenced by the ΔpH and relates to the capacity for glutamate accumulation in the experimental closed system.

The kinetics of glutamate uptake were determined by measuring the rapid initial rate of ^{14}C-glutamate uptake at a series of external glutamate concentrations (FIGURE 4). Double reciprocal plots of these data were linear and provided values of 125 nmol/mg/min for the V'_{max} and 4.8 mM for the K'_m at 25°.

Kinetics of Glutamate Efflux on the Glutamate-Hydroxyl Carrier

Glutamate efflux from rat liver mitochondria was investigated by measuring the removal of glutamate from the matrix of glutamate loaded mitochondria following dilution of the mitochondria in media at the desired pH and temperature. The mitochondria were then separated from the medium by rapid centrifugation at designated times.[20] FIGURE 5 shows that in contrast to the highly inhibitory effect of NEM on glutamate uptake, 0.2 mM NEM had relatively little effect on efflux. Glutamate efflux was, however, highly inhibited by 0.5 mM bromcresol purple. FIGURE 5 also shows that D-glutamate was transported poorly and was not affected by NEM or bromcresol purple. This experiment illustrates that the glutamate-hydroxyl carrier is stereospecific for L-glutamate.

The effect of varying the external pH of the medium on glutamate efflux is shown in FIGURE 6. Increased external H^+ concentration inhibited efflux in contrast to its stimulatory effect on influx. At time zero, the matrix pH was always more alkaline than that of the medium with the ΔpH being greater with a fall of external pH. Glutamate efflux was associated with an initial further alkalinization of the matrix, which, except for the pH 6.5 condition, was followed by a relative acidification that coincided approximately with depletion of the matrix glutamate. These secondary changes probably represent a redistribution of protons across the mitochondrial membrane via the phosphate-hydroxyl translocator to reestablish the equilibrium disturbed by glutamate efflux.[19, 25]

The effect of varying the matrix pH at constant external pH on glutamate efflux was investigated by addition of nigericin to glutamate-loaded mitochondria. Nigericin was without appreciable effect on glutamate efflux even when the external pH was varied over the range from 6.5 to 7.4 (data not shown, but see TABLE 1). The effect of valinomycin on glutamate efflux using mitochondria incubated in either choline-Cl or KCl media is illustrated in FIGURE 7. Valinomycin in the presence of K^+ partially collapsed the membrane potential (ΔΨ) and increased ΔpH. Again the effects on glutamate efflux were small, indicating the absence of stiking effects induced by the energy state or the matrix pH of the mitochondria.

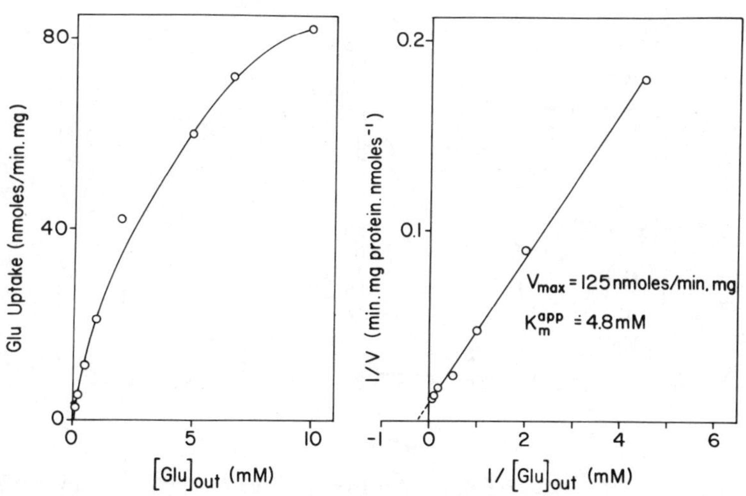

FIGURE 4. Concentration dependence of glutamate uptake by rat liver mitochondria. The incubation conditions (pH 7.4) were the same as in FIGURE 3.

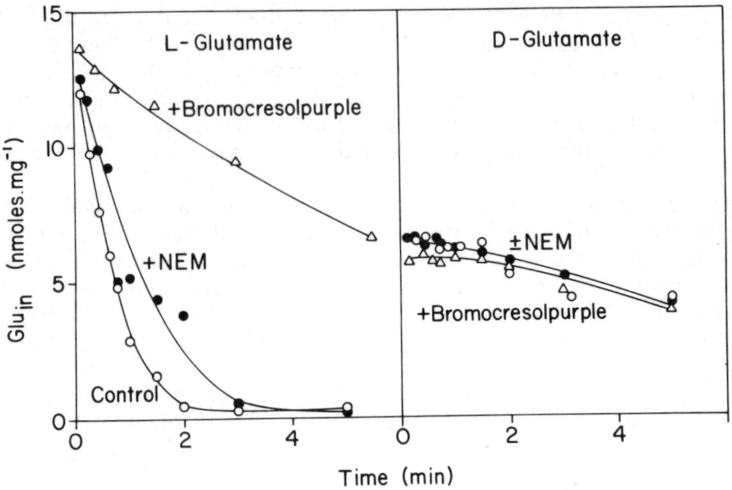

FIGURE 5. Effects of NEM and bromcresol purple on efflux of L and D-glutamate from rat liver mitochondria. Mitochondria (3.6 mg/ml) were incubated at 25°, pH 7.2 in medium containing 50 mM KCl, 50 mM MOPS (2-*N*-morpholinopropanesulfonic acid), 2 mM EDTA, 5 mM $MgCl_2$, 8% (w/v) 40,000 MW dextran, 5 μg/ml rotenone and 1 μg/ml antimycin A. Glutamate efflux was terminated by rapid centrifugation through silicone oil into 14% (w/v) perchloric acid. When present in the incubation medium, the NEM concentration was 0.2 mM and the bromcresol purple concentration was 0.5 mM. L-Glutamate was measured enzymatically and D-glutamate was measured using a Beckman 121 amino acid analyzer.

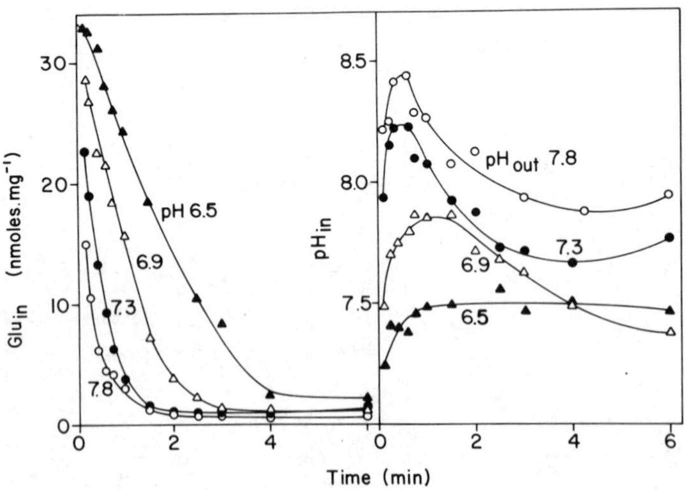

FIGURE 6. Effect of varying external pH on glutamate efflux in rat liver mitochondria. Mitochondria (3.7 μg/ml) were incubated as in FIGURE 5 except the buffer contained 25 mM MOPS and 25 mM MES (2-*N*-morpholinoethanesulfonic acid). The matrix pH was estimated from the distribution of [2-^{14}C] DMO.

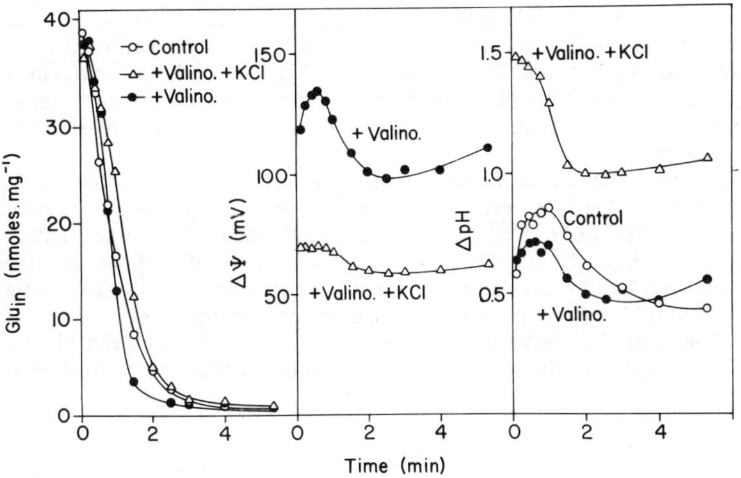

FIGURE 7. Effect of valinomycin on glutamate efflux in rat liver mitochondria. Mitochondria (2.6 μg/ml) were incubated in the same medium as in FIGURE 5, pH 7.3 for the control. Valinomycin (2 μg/ml) was added either to this medium or to K^+-free medium in which the KCl was replaced by choline chloride.

The glutamate efflux curves were resolved kinetically using the computer curve fitting and graphical methods of resolving the generalized translocator integrated rate equation:[20]

$$V'_{max}t - (K'_m + S_f)\ln\frac{S_o - S_f}{S - S_f} - (S_o - S) = 0$$

where V'_{max} and K'_m are apparent kinetic constants, S_o and S_f are the initial and final (equilibrium) concentrations of metabolite in the matrix, and S is the concentration at time t.[20] TABLE 1 illustrates the lack of appreciable effects of nigericin and valinomycin added, either singly or together, on the calculated V'_{max} and K'_m values. In particular, the K'_m/V'_{max} ratio remained remarkably

TABLE 1

EFFECT OF NIGERICIN AND VALINOMYCIN ON THE KINETIC CONSTANTS FOR GLUTAMATE EFFLUX *

Additions	V'_{max} nmol/mg/min	K'_m nmol/mg	K'_m/V'_{max} min
Control	17.1	13.7	0.80
Nig	14.8	13.5	0.91
Val	15.1	10.5	0.70
Nig+Val	19.8	14.3	0.72

* Rat liver mitochondria were incubated in KCl buffer, pH 7.0 at 15° C. See legend to FIGURE 5 for further experimental details. Nigericin and valinomycin were each added at concentrations of 1 μg/ml. V'_{max} and K'_m values were calculated by computer curve fitting of data similar to that of FIGURE 7.

constant when the external pH was unaltered despite a fall of matrix pH induced by nigericin, a rise of matrix pH and fall of $\Delta\Psi$ induced by valinomycin, or a fall of both matrix pH and $\Delta\Psi$ when both nigericin and valinomycin were added together. These data illustrate further the relative insensitivity of glutamate transport to the matrix pH and the electrical potential across the mitochondrial membrane.

In contrast to the above conclusions, a kinetic analysis of the glutamate efflux curves at different external pH shown in FIGURE 6 indicated that the V'_{max} increased with increase of pH_o but that the K'_m did not change (TABLE 2). The variation of K'_m/V'_{max} ratio with pH_o is indicative of intersecting lines in a Lineweaver-Burke plot and suggests a sequential mechanism with formation of a ternary complex between the carrier, glutamate and a proton.

On the basis of the above information, a tentative model of the glutamate-hydroxyl translocator is shown in FIGURE 8. Relevant facts are: A. The sensitivity of both glutamate influx and efflux to external pH; B. The relative insensitivity of glutamate transport to internal pH; C. The lack of effect of the mitochondrial energy state on the V'_{max} of glutamate transport, and D. The effect of an increased external H^+ concentration in causing an increase of the K'_m/V'_{max} ratio for glutamate efflux. Glutamate is assumed to react with the carrier as an anion, after protons are bound first, to produce a ternary complex. Binding and release of glutamate to the carrier on the matrix side of the membrane is assumed not to be affected by the proton concentration. On this model, glutamate is transported as a proton symport rather than a hydroxyl ion antiport.

TABLE 2

EFFECT OF PH ON KINETIC CONSTANTS FOR GLUTAMATE EFFLUX

Rat liver mitochondria were incubated in KCl buffer at the specified pH, at 25° C. See legends to FIGURES 5 and 6 for further experimental details. pH_o and pH_i refer to the pH of the medium and matrix, respectively.

pH_o	pH_i *	V'_{max}	K'_m	K'_m/V'_{max}
		nmol/mg/min	nmol/mg	min
6.5	7.5	16.9	12.6	0.75
6.9	7.9	27.8	11.9	0.43
7.3	8.3	52.1	14.0	0.27
7.8	8.5	51.0	13.5	0.26

* pH_i was taken at its peak value after 30–60 sec of incubation (see FIGURE 6).

Kinetics of Aspartate Efflux on the Glutamate-Aspartate Carrier

Starting with glutamate loaded mitochondria that had been treated with 2 mM NEM, matrix aspartate loading was achieved by addition of 5 mM oxalacetate to allow transamination with glutamate, followed by addition of 5 mM aminooxyacetate to inhibit further transaminase activity.[26] After washing, the mitochondria contained 10–20 nmol/mg of aspartate. Depletion of

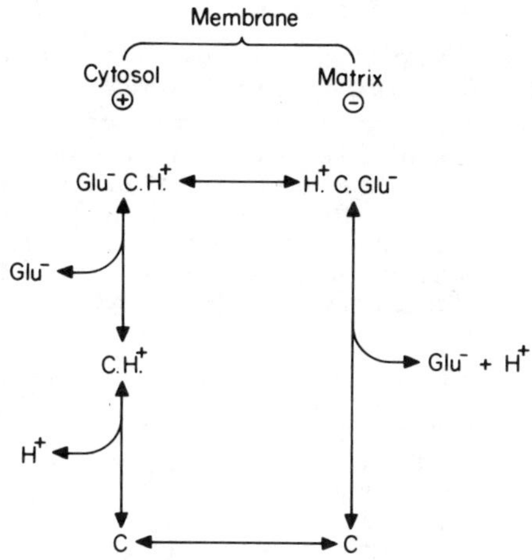

FIGURE 8. Tentative scheme for glutamate transport across the mitochondrial membrane by the glutamate-hydroxyl translocator. The postulated specific mobile carrier is designated by C.

aspartate from the matrix and its appearance in the medium after addition of glutamate at 10° was followed kinetically by rapid centrifugation of the mitochondria through silicone oil.[27] FIGURE 9 illustrates the equivalence of these measurements, thereby indicating that transaminase activity remained inhibited.

In experiments similar to that of FIGURE 9, but with external glutamate varied over the range from 1.5 mM to 15 mM glutamate, a family of curves

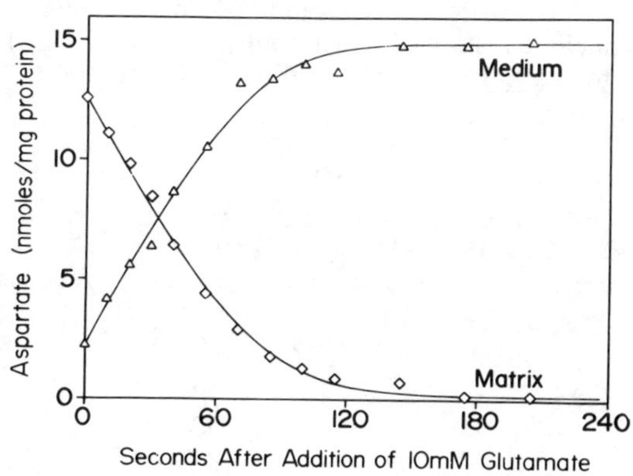

FIGURE 9. Time course of depletion of aspartate from the mitochondrial matrix and appearance in the medium after addition of 10 mM glutamate to aspartate-loaded mitochondria. Rat liver mitochondria (5 mg/ml) were incubated at 10° C in buffer containing 100 mM KCl, 30 mM MOPS, 8% dextran, 10 μM rotenone, 10 mM ascorbate and 0.4 mM TMPD, pH 7.2.

TABLE 3

EFFECT OF GLUTAMATE CONCENTRATION ON GLUTAMATE-ASPARTATE TRANSLOCATOR KINETICS

Aspartate-loaded rat liver mitochondria (5 mg/ml) were incubated at 10° C in buffer containing 100 mM KCl, 30 mM MOPS (2-*N*-morpholinopropanesulfonic acid), 8% (w/v) 40,000 MW dextran, 10 μM rotenone, 10 mM ascorbate and 0.4 mM TMPD (N,N,N′,N′-tetramethyl-p-phenylenediamine) at pH 7.2. Aspartate efflux was initiated by addition of glutamate. The numbers shown are kinetic parameters calculated by the graphical and the computer methods, respectively.[20]

Medium glutamate	V'_{max}	K'_m	K'_m/V'_{max}
mM	nmol/mg/min	nmol/mg	min
1.5	4.7, 4.1	5.8, 5.5	1.28
2.5	6.8, 5.8	7.5, 5.7	1.05
5.0	9.7, 9.4	4.9, 4.4	0.49
10.0	13.9, 12.5	5.0, 3.5	0.32
15.0	17.0, 16.4	5.6, 4.8	0.32

was obtained that were resolved kinetically by the computer curve fitting and by graphical procedures, as described elsewhere.[20, 21] Values for the V'_{max} for glutamate-aspartate exchange at 10° and the K'_m for internal aspartate are shown in TABLE 3. It is seen that as the medium glutamate concentration increases, the K'_m for aspartate remains approximately constant while the V'_{max} increases. The variation of the K'_m/V'_{max} ratio with glutamate concentration indicates that a ternary complex is formed between the carrier, glutamate and aspartate, and that the glutamate aspartate translocator behaves according to a sequential rather than a ping-pong mechanism.

From the apparent kinetic constants given in TABLE 3, it is possible using the equations given in the legend to TABLE 4 to calculate the concentration independent Michaelis constants for external glutamate ($K_m{}^G$), internal aspartate ($K_m{}^A$) and the V_{max} for aspartate efflux at 10° (TABLE 4). The average

TABLE 4

CONCENTRATION INDEPENDENT KINETIC CONSTANTS OF THE GLUTAMATE-ASPARTATE TRANSLOCATOR *

Method	$K_m{}^A$ nmol/mg	$K_m{}^G$ mM	V_{max} nmol/mg/min
Graphical	4.9	5.3	21.2
Computer	3.0	6.3	21.2
Average	3.9	5.8	21.2

* The constants were calculated from replots of the data given in TABLE 3 using the equations: $1/V'_{max} = 1/V_{max} + (K_m{}^G/V_{max})/G_o$

$$K'_m{}^A/V'_{max} = K_m{}^A/V_{max} + K/(V_{max} \cdot G_o)$$

where K is a mechanism-dependent constant indicative of a ternary carrier complex.

glutamate K_m of 5.8 mM compares favorably with the value of 6.4 mM reported by Tischler *et al.*[6] using initial rate proton uptake kinetics. Since the volume of the mitochondrial water space is about 0.8 μl/mg protein, it is seen from TABLE 4 that the glutamate independent K_m for aspartate is about 5 mM. The discrepancy between this value and the value of 38 μM for the external aspartate K_m obtained with submitochondrial particles[28] can probably be ascribed to alterations induced in the membrane during preparation of the submitochondrial particles.

In the above experiments, the proton electrochemical gradient across the mitochondrial membrane during aspartate efflux was maintained high by the presence of ascorbate and TMPD. The data in TABLE 5 show the effect of decreasing $\Delta\Psi$ on aspartate efflux kinetics. The membrane potential was partially collapsed by incubating mitochondria in the presence of valinomycin and

TABLE 5

EFFECT OF DECREASING THE MEMBRANE POTENTIAL ON GLUTAMATE-ASPARTATE TRANSLOCATOR KINETICS *

K^+ mM	V'_{max} nmol/mg/min	K'_m nmol/mg	$\Delta\Psi$ mV	ΔpH	$\Delta\tilde{\mu}H^+$ mV
0	16.9	5.0	147.2	0.99	230
1	11.3	5.1	133.4	1.20	201
5	9.5	5.0	99.9	1.74	197
10	6.0	3.8	73.8	1.99	185
Nig+Val	2.2	4.8	19.1	0.34	44

* Aspartate-loaded mitochondria (5 mg/ml) were incubated at 10° C and pH 7.4 in similar media to that of TABLE 3 except that choline chloride replaced the KCl. Aspartate efflux was initiated by addition of 10 mM glutamate. Valinomycin was present at a concentration of 0.15 μg/ml and nigericin when added was also present at a concentration of 0.15 μg/ml. The membrane potential ($\Delta\Psi$) and the ΔpH were calculated from the distributions of ^{86}Rb and [2-^{14}C] DMO (5,5′-dimethyloxazolidine-2,4-dione), respectively.[21]

varied K^+ concentrations. The partial collapse of $\Delta\Psi$ was associated with an increase of ΔpH, so that the proton electrochemical gradient ($\Delta\tilde{\mu}H^+$) remained approximately constant. The V'_{max} for aspartate efflux decreased with the fall of $\Delta\Psi$, while the K'_m remained approximately constant, even when the $\Delta\Psi$ and $\Delta\tilde{\mu}H^+$ were both decreased by addition of both valinomycin and nigericin (last line of TABLE 5). Calcium loading of mitochondria to 100 nmol/mg protein, which also decreases $\Delta\Psi$ and increases ΔpH, produced similar results (data not shown). FIGURE 10 shows that there was a linear relationship between the V'_{max} for aspartate efflux and $\Delta\Psi$ in the pooled results from different experiments over the range of $\Delta\Psi$ from 70–160 mV. At lower values of $\Delta\Psi$ the relationship became nonlinear, suggestive of a threshold phenomenon. In other experiments, in which aspartate-loaded mitochondria were incubated with 100 mM KCl buffer at pH 6.5 or 7.6 in the absence and presence of nigericin to collapse the ΔpH, no effect of external pH, internal pH, or ΔpH was observed on either the V'_{max} or K'_m for aspartate transport.[21]

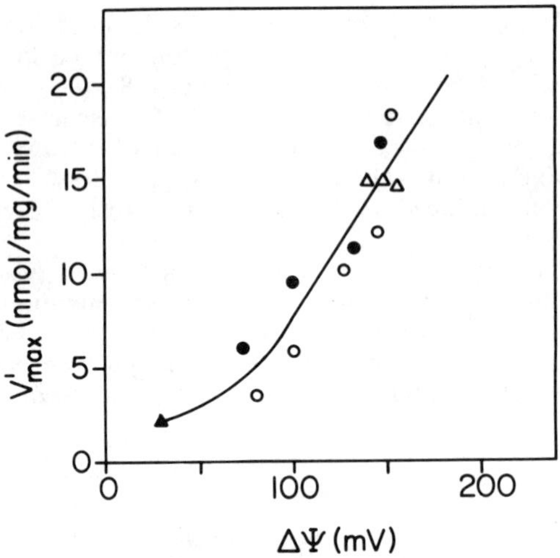

FIGURE 10. Relationship between the apparent V_{max} of the glutamate-aspartate translocator and the mitochondrial membrane potential (ΔΨ). The open circles are taken from data in which ΔΨ was decreased by the electrophoretic uptake of Ca^{2+}, the closed circles and triangle were taken from data in TABLE 3, while the open triangles are further points under control conditions.

On the basis of the data summarized above and presented in detail elsewhere,[21] a minimal model to describe the mechanism of the glutamate-aspartate translocator can be envisioned, as shown in FIGURE 11. The lack of sensitivity of aspartate efflux to pH changes on either side of the membrane suggests that the undissociated form of glutamate binds to the carrier on the external surface

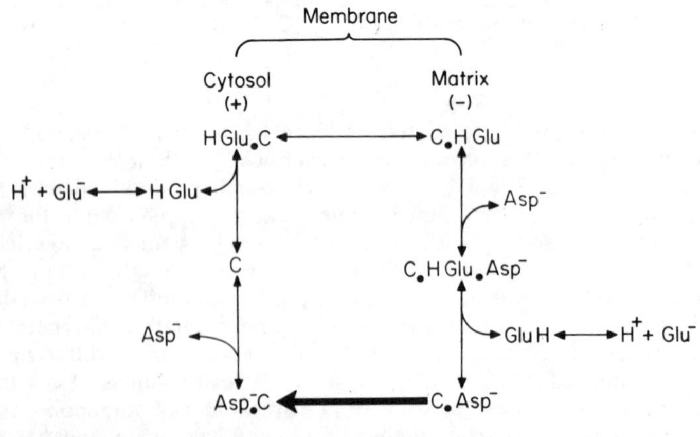

FIGURE 11. Model of the glutamate-aspartate translocator. The symbol *C* represents the mobile carrier in the membrane, while HGlu represents the undissociated form of glutamate, which is assumed to bind directly to the carrier.

of the inner mitochondrial membrane and is also released on the matrix side. Aspartate binds to the H glutamate carrier complex on the matrix side to form a charged ternary complex, with the subsequent release of H glutamate. For energized mitochondria in which there is an electrical potential across the mitochondrial membrane (negative inside), directionality of transport is achieved by the kinetically-limiting step of transport of the charged carrier aspartate complex from the matrix to the cytosol side of the membrane. The available data do not provide information whether the form of the carrier after release of aspartate on the cytosol side of the membrane is the same as that which binds glutamate, as shown in the model, but it is evident that if another step occurs it would not be kinetically rate limiting.

Discussion

The experiments shown in this paper demonstrate that the transport of glutamate across the mitochondrial membrane in either direction shows characteristics of stereospecificity, sensitivity to specific inhibitors, and saturability; properties characteristic of carrier-mediated transport. Glutamate transport was little affected by decreasing the mitochondrial membrane potential [29] and was relatively insensitive to changes of the matrix pH in the region of pH 6.5 to 8.5. However, because glutamate uptake is proton compensated, the limited buffering capacity of the matrix space rapidly limits the rate of glutamate uptake, so that only initial rates prior to acidification of the matrix are a true index of glutamate transport. A lack of appreciation of this fact prior to the studies of Hoek and Njogu [19] explains the earlier discrepancies of glutamate transport kinetics.[3, 4] On the other hand, the clear dependence of glutamate efflux kinetics on the external pH (change of V'_{max} with no change of K'_m) suggests separate binding sites on the carrier for protons and glutamate, with formation of a ternary complex with the carrier on the outside surface of the inner mitochondrial membrane being kinetically rate limiting.

For a nonenergy-linked bidirectional transport process, such as the glutamate-hydroxyl translocator, the kinetics of influx and efflux should be related according to the Haldane relationship:

$$(K'_m/V'_{max})\text{uptake} = (K'_m/V'_{max})\text{efflux} \cdot 10^{-\Delta pH}$$

From FIGURE 4, the (K'_m/V'_{max}) influx ratio at 25° C and pH 7.4 is 4.8/125 = 0.038. From TABLE 2, at 25° C and pH 7.3, the ΔpH = 1 and assuming a matrix space of 0.8 μl/mg protein, the $(K'_m/V'_{max}$ influx$\cdot 10^{-\Delta pH}$ ratio is 17.5·0.1/52.1 = 0.034. These data show, therefore, that the glutamate transport kinetics are in good agreement with the Haldane relationship, thereby providing further confidence in their validity.

The normal ΔpH across the mitochondrial membrane (alkaline inside) allows the glutamate concentration in the matrix to be greater than that in the extramitochondrial space, even in the intact cell,[30] consequently it is of interest that the K'_m for glutamate efflux is about three times greater than that for influx. Conversely, the V'_{max} for glutamate influx at physiological pH values is 2 to 3 times that for efflux. From the values for (V'_{max})influx of 125 nmol/mg/min at 25° C and pH 7.4, and (V'_{max})efflux of 52 nmol/mg/min at 25° C and pH 7.3 quoted above, and an energy of activation of 17.5 kcal/mol[3], V'_{max} for influx and efflux at 37° C are 392 nmol/mg/min and 163 nmol/mg/min, respectively. On the assumption of a mitochondrial protein content of 240 mg/g dry wt in liver, these values correspond to a maximum rate of mitochondrial glutamate influx of 5640 μmol/g dry wt/hr and a maximal rate of glutamate efflux of 2350 μmol/g dry wt/hr at 37° C. Corresponding values for the glutamate-aspartate translocator V'_{max} at 37° C from TABLE 4, using an energy of activation of 15.1 kcal/mol[6] are 3170 μmol/g dry wt/hr. It is apparent that the maximum rates of glutamate transport on the different transport systems are approximately equivalent. Typical values for rates of urea synthesis in isolated hepatocytes at 37° C (and hence estimations of flux through the glutamate carriers) are 300 μmol/g dry wt/hr with alanine as substrate and 800–1000 μmol/g dry wt/hr for maximum rates of urea synthesis from ammonia[15]. It is evident, therefore, that the glutamate carriers have at least a twofold excess capacity for the metabolic requirements of the cell.

It has been a difficult problem to assess whether the mitochondrial translocators provide a rate-limiting step for cellular metabolic processes. For enzymes in a multienzyme sequence that are not affected by allosteric or covalent modification, flux is strongly influenced by the substrate and product concentrations relative to their respective K_m and K_i values.[31] In the steady state, individual metabolite levels adjust to the flux in accordance with the kinetic constants of the enzymes in the sequence. Most of the anion translocators behave like reversible near-equilibrium enzyme systems, and, therefore, may not be expected to provide rate-controlling steps. Thus, when urea synthesis is stimulated by additions of substrate, ammonia and ornithine to liver cells, the concentration of glutamate in the cell increases and that in the mitochondrial matrix even achieves values as high as 50 mM.[15] Under these conditions, glutamate efflux from the mitochondria is obligatory in order to provide a cycle for glutamate cotransport on the glutamate-aspartate translocator (see FIGURE 1). The glutamate gradient remains approximately in equilibrium with the ΔpH of 0.4 units,[15, 30] hence the increase of mitochondrial glutamate concentration may be seen as a metabolic adaptation to allow an increase of net glutamate efflux for a reaction which has a high K_m. The cytosolic glutamate concentration also rises as a secondary consequence of the near equilibrium of the glutamate and proton electrochemical gradients. Since the K_m for glutamate influx is much lower than that for efflux, an overall increase of glutamate concentration in both the cytosol and mitochondria will tend to favor net glutamate efflux. Likewise, the differential effects of pH_{out} and pH_{in} on glutamate efflux and influx could provide a kinetic control of net glutamate transport by alterations of ΔpH,[29] but whether the ΔpH across the mitochondrial membrane changes under physiological conditions in the intact cell has not yet been established. Regulation of glutamate transport by proton permeability may be considered unlikely because of the high capacity of the phosphate-hydroxyl translocator.

Regulation of cellular metabolic processes by the glutamate-aspartate translocator is theoretically more appealing because it represents a disequilibrium step in the metabolic pathway, with asparate efflux being favored over influx as a consequence of the electrogenic nature of the translocator. Regulation of the malate-aspartate cycle by substrate concentrations of the translocators has, in fact, been demonstrated for the special condition of substrate-depleted cells.[10, 16, 32] However, in the presence of substrate to allow build-up of the dicarboxylic amino acids and α-ketoacids, no evidence was obtained that transport of anions across the mitochondrial membrane provided a rate-limiting step in the pathways of either gluconeogenesis or urea synthesis.[15] The special kinetic characteristics of the glutamate-aspartate carrier serve to channel aspartate transport out of the mitochondria, but overall flux is probably regulated at distal steps such as pyruvate carboxylase, which contributes towards determining the oxalacetate availability to mitochondrial aspartate aminotransferase. Feedback control of the glutamate-aspartate translocator is exerted by inhibition due to cytosolic aspartate concentration.[6, 15] We have recently developed a rate equation to describe the kinetics of the glutamate-aspartate translocator, which provided calculated values of aspartate efflux in the hepatocyte during urea synthesis from ammonia that agreed rather closely with the measured rates of urea synthesis.[21] Further studies are in progress to assess in greater detail the validity of the glutamate-aspartate rate equation, and to develop a rate equation to describe glutamate transport on the glutamate-hydroxyl carrier. By this approach, we hope to achieve a greater understanding of the factors affecting the kinetic regulations of glutamate transport in the intact cell.

Summary

The kinetics of glutamate influx and efflux on the glutamate-hydroxyl carrier have been measured and compared in rat liver mitochondria. At pH 7.4 and 25° C, the Michaelis constants and V'_{max} values were in agreement with the Haldane relationship when the ΔpH was accounted for. The K_m values for glutamate influx and aspartate efflux on the glutamate-aspartate translocator are also reported. Extrapolation of the maximum velocities to 37° and the intact liver provide values of 5.6 and 2.4 mmol/g dry wt/hr for glutamate influx and efflux, respectively, on the glutamate-hydroxyl translocator, and 3.2 mmol/g dry wt/hr for the glutamate-aspartate translocator. Both translocators operate by a sequential mechanism with formation of a ternary complex. Their possible regulatory role in urea synthesis by liver is assessed.

Acknowledgment

The authors are indebted to Ms. Gayle Viale for excellent secretarial assistance.

References

1. Azzi, A., J. B. Chappell & B. H. Robinson. 1967. Biochem. Biophys. Res. Commun. **29:**148–152.
2. Meijer, A. J., A. Brouwer, D. J. Reijngoud, J. B. Hoek & J. M. Tager. 1972. Biochim. Biophys. Acta. **283:**421–429.
3. Meyer, J. & P. M. Vignais. 1973. Biochim. Biophys. Acta. **355:**375–384.
4. Bradford, N. M. & J. D. McGivan. 1973. Biochem. J. **134:**1023–1029.
5. Lanoue, K. F. & M. E. Tischler. 1974. J. Biol. Chem. **249:**7522–7528.
6. Tischler, M. E., J. Pachence, J. R. Williamson & K. F. Lanoue. 1976. Arch. Biochem. Biophys. **173:**448–462.
7. Williamson, J. R. 1976. *In* Use of Isolated Liver Cells and Kidney Tubules in Metabolic Studies. J. M. Tager, H. D. Söling and J. R. Williamson, Eds. pp. 79–95. North Holland Publishing Co. Amsterdam.

8. Meijer, A. J. & K. Van Dam. 1974. Biochim. Biophys. Acta. **346:**213–244.
9. Fonyo, A., F. Palmieri & E. Quagliariello. 1976. *In* Horizons in Biochemistry and Biophysics. E. Quagliariello, F. Palmieri and T. P. Singer, Eds. Vol. **2:**60–105. Addisons-Wesley. Reading, Mass.
10. Williamson, J. R. 1976. *In* Gluconeogenesis. M. A. Mehlman & R. W. Hanson, Eds. pp. 165–2201. John Wiley & Sons. New York, New York.
11. Lanoue, K. F. & A. C. Schoolwerth. 1979. Ann. Rev. Biochem. **48:**871–922.
12. Lanoue, K. F. & J. R. Williamson. 1971. Metabolism **20:**119–140.
13. Williamson, J. R., B. Safer, K. F. Lanoue, C. M. Smith & E. Walajtys. 1973. *In* Rate Control of Biological Processes. D. D. Davies, Ed. pp. 241–281. Cambridge University Press. Cambridge.
14. Lardy, H. A., V. Paektau & P. Walter. 1965. Proc. Natl. Acad. Sci. USA **53:**1410–1415.
15. Meijer, A. J., J. A. Gimpel, G. Deleeuw, M. E. Tischler, J. M. Tager & J. R. Williamson. 1978. J. Biol. Chem. **253:**2308–2320.
16. Meijer, A. J., J. A. Gimpel, G. Deleeuw, J. M. Tager & J. R. Williamson. 1975. J. Biol. Chem. **250:**7728–7738.
17. Schneider, W. C. & G. M. Hogeboom. 1950. J. Biol. Chem. **183:**123–128.
18. Myers, D. K. & E. C. Slater. 1957. Biochem. J. **67:**558–572.
19. Hoek, J. B. & R. M. Njogu. 1976. FEBS Lett. **71:**341–346.
20. Williamson, J. R. & R. O. Viale. 1979. Methods Enzymol. **56:**252–278.
21. Murphy, E., K. E. Coll, R. O. Viale, M. E. Tischler & J. R. Williamson. 1979. J. Biol. Chem. **254:**8369–8376.
22. Robinson, B. H. & G. R. Williams. 1969. FEBS Lett. **5:**301–304.
23. Pressman, B. C., E. J. Harris, W. S. Jagger & J. H. Johnson. 1967. Proc. Natl. Acad. Sci. USA. **58:**1949–1956.
24. Padan, E. & H. Rottenberg. 1973. Eur. J. Biochem. **40:**431–437.
25. Debise, R., P. Gachon & R. Durand. 1978. FEBS Lett. **85:**25–29.
26. Lanoue, K. F., A. J. Meijer & A. Brouwer. 1974. Arch. Biochem. Biophys. **161:**544–550.
27. Lanoue, K. F., E. I. Walajtys & J. R. Williamson. 1973. J. Biol. Chem. **248:** 7171–7183.
28. Duszynski, J., G. Mueller & K. F. Lanoue. 1978. J. Biol. Chem. **253:**6149–6157.
29. Hoek, J. B., K. E. Coll & J. R. Williamson. 1979. *In* Function and Molecular Aspects of Biomembrane Transport. E. M. Klingenberg, F. Palmieri & E. Quagliariello, Eds.: 403–406. Elsevier/North Holland Biomedical Press.
30. Tischler, M. E., P. Hecht & J. R. Williamson. 1977. Arch. Biochem. Biophys. **181:**278–292.
31. Higgins, J. 1979. *In* Pharmacology of Ethanol. E. Majchrowicz and E. P. Noble, Eds.: 249–351. Plenum Press. New York, New York.
32. Williamson, J. R., A. J. Meijer & K. Ohkawa. 1974. *In* Regulation of Hepatic Metabolism. F. Lundquist & N. Tygstrup, Eds.: 457–479. Academic Press. New York, New York.

Index of Contributors

(Italicized page numbers refer to comments in discussions)